HANDBOOK

of

ENVIRONMENTAL CONTROL

VOLUME III: WATER SUPPLY and TREATMENT

EDITORS

Richard G. Bond, M.S., M.P.H.
Director, Division of
Environmental Health
University of Minnesota

Conrad P. Straub, Ph. D.
Director, Environmental Health
Research and Training Center
University of Minnesota

COORDINATING EDITOR

Richard Prober, Ph.D.
Associate Professor of Engineering
Case Western Reserve University

CRC PRESS

18901 Cranwood Parkway • Cleveland, Ohio 44128

HANDBOOK OF ENVIRONMENTAL CONTROL

VOLUME III: WATER SUPPLY AND TREATMENT

International Standard Book Number (ISBN)
Complete Set 0-87819-270-0
Volume III 0-87819-273-5

Library of Congress Catalog Card Number 72-92118

PREFACE TO THE ENVIRONMENTAL CONTROL SERIES

Increasing population, developing industry, changing agricultural patterns, and growing community concern, along with a desire for a higher standard of living–physically, materialistically, emotionally, and sociologically–have focused attention on the multiple sources of contamination, pollution, and stress and the need for extensive control of these sources. Furthermore, the recognition that these problems involve many disciplines and that their resolution calls for evaluation of the benefits versus the risks of the specific actions proposed has prompted development of a series of handbooks on environmental control. The aim of the series is to bring together pertinent information in tabular form that will be useful in evaluating the environment, not only from the standpoint of effects on the ecosystem, aquatic and terrestrial, but also on man's relationship to the environment and the environment's relationship to man.

The division of subject matter is along three basic lines: the macroenvironment, which deals primarily with the air, soil, and water milieus; the microenvironment, which is concerned primarily with the community–home, school, work, recreation–and with the factors responsible for stress effects such as noise, temperature, vibration, and radiation; and the personal environment, which relates to the products used and consumed by man. The first segment, dealing with the macroenvironment, will consist of three volumes, one each on the air environment, the soil environment, and the water environment.

The series is designed in a logical sequence to answer the questions *what, why, when,* and *where,* and how to treat them. In raising and answering these questions perhaps a broader, more comprehensive evaluation of the solutions offered and the consequences of particular actions can be reached.

It is our hope that this series on environmental control will prove useful to a wide range of disciplines both technical and scientific, and at the same time be of interest and use to social and political scientists, economists, and attorneys. Hopefully, it will be of value to all decision-makers regardless of their backgrounds, concerns, and environmental affiliations. And in the final analysis, the series should be of use to all students in the environmental and ecological disciplines.

Information in the series of handbooks will be continuously updated, and comments, criticisms, and suggestions regarding scope, format, and selection of subject matter are invited. Any errors or omissions should be brought to the attention of the editors and publishers.

We acknowledge assistance of the Advisory Board and various contributors and wish especially to recognize permission of the many authors and journals for use of published material. Special acknowledgements are made to Thomas Whitson and to the editorial staff of The Chemical Rubber Co., for their intelligent and invaluable assistance in the preparation of the tables.

Minneapolis, Minnesota
August 1972

R.G. BOND
C.P. STRAUB

PREFACE TO VOLUME III: WATER SUPPLY AND TREATMENT

In this, the third volume of the *Handbook of Environmental Control,* the editors have endeavored to select tabular data relating to the overall problem of water supply from source to consumable or usable product. The data presented provide information on water quality characteristics as functions of source, storage, treatment, and distribution.

Water supply and treatment is one aspect of the broader area of water resources management and is of primary concern in producing a product acceptable for a variety of uses – potable, municipal, commercial, industrial, agricultural, and recreational. Attempts have been made to provide information on some of the newer contaminants and approaches being taken regarding their removal.

The information compiled in this volume should be of use to the multidisciplinary worker within the broad scope of environmental management, including ecology, earth sciences, resources, recreation, environmental design and protection. In addition, because of the increased awareness and concern for the environment, students in the above areas and other professions, such as law, business, political science, and sociology as well as industry, should find the volume a source of readily available usable information.

Minneapolis, Minnesota
August 1973

R. G. BOND
C. P. STRAUB

TABLE OF CONTENTS

Preface to the Environmental Control Series vi
Preface to Volume III vii

SECTION 1		**SOURCES AND QUALITY**
	1.	Introduction 3
	1.1	Surface Water 30
	1.1.1	Rivers 37
	1.1.2	Lakes and Impoundments 57
	1.2	Ground Water 82
	1.3	Estuarine and Marine Waters 109
	1.4	Cloud Seeding and Glaciers 115
	1.5	Waste Water 120
SECTION 2		**NEEDS**
	2.	Introduction 131
	2.1	Public Water Supply Needs 148
	2.2	Industrial Water Needs 168
	2.2.1	Manufacturing Industries Water Needs 178
	2.2.2	Mining Industries Water Needs 215
	2.2.3	Water Needs for Electric Power Generation 221
	2.2.4	Industrial Water Reuse and Recycling 228
	2.3	Irrigation and Farm Water Needs 230
SECTION 3		**QUALITY CRITERIA**
	3.	Introduction 247
	3.1	Public Water Supply Quality Criteria 260
	3.1.1	Physical Standards 275
	3.1.2	Bacteriological Standards 277
	3.1.3	Chemical Standards 281
	3.2	Industrial Water Supply Quality Criteria 295
	3.2.1	Food Industries 309
	3.2.2	Manufacturing Industries 312
	3.2.3	Cooling and Steam Generation 324
	3.3	Agricultural Water Supply Quality Criteria 328
	3.3.1	Irrigation 330
	3.3.2	Livestock Watering and General Farmstead Use 339
	3.4	Fish and Wildlife 345
	3.4.1	Aquatic Insects 362
	3.4.2	Shellfish 364
	3.4.3	Fish 366
	3.5	Recreational and Aesthetic 388

SECTION 4 WATER TREATMENT
4.1 Public Water Supply Facilities 391
4.2 Coagulation and Filtration 434
4.3 Iron and Manganese Removal 456
4.4 Removal of Color, Odor, and Taste 463
4.5 Disinfection 474
4.6 Softening and Boiler Water Treatment 493
4.7 Miscellaneous and Nonconventional Processes 500
4.7.1 Reverse Osmosis 511

SECTION 5 WATER DISTRIBUTION
5.1 Water Supply Requirements 521
5.2 Plumbing System Standards 532
5.3 Storage 546

SECTION 6 BIOLOGICAL CONTROL IN WATER SUPPLY SYSTEMS
6. Introduction 555
6.1 Algae 566
6.2 Bacteria 572
6.3 Viruses 580

SECTION 7 RECREATIONAL WATERS
7.1 Swimming and Primary Body Contact 599
7.2 Fishing and Secondary Body Contact 603

SECTION 8 ANALYTICAL METHODS
8. Introduction 617
8.1 Electrometric Methods 652
8.2 Spectrometric Methods 657
8.3 Oxygen Demand 664

SECTION 9 MONITORING
9. Introduction 697
9.1 Public Water Supplies, Wells, and Ground Water 705
9.2 Irrigation Waters, Pesticides, and Salinity Build-up 739
9.3 Heavy Metals 763

Appendix 785

Conversion Factors 817

Acronyms 822

Index 825

ADVISORY BOARD

Section 1

Sources and Quality

1. Introduction
Pages 3–29

1.1 Surface Water
Pages 30–81

1.2 Ground Water
Pages 82–108

1.3 Estuarine and Marine Waters
Pages 109–114

1.4 Cloud Seeding and Glaciers
Pages 115–119

1.5 Waste Water
Pages 120–127

1. SOURCES AND QUALITY

1–1 ESTIMATED WORLD WATER SUPPLY

Water item	Volume, thousands mi³	Volume, thousands km³	Percent of total water
Water in land areas			
Fresh water lakes	30	125	0.009
Saline lakes and inland seas	25	104	0.008
Rivers (average instantaneous volume)	0.3	1.25	0.0001
Soil moisture and vadose water	16	67	0.005
Ground water to depth of 4,000 m	2,000	8,350	0.61
Icecaps and glaciers	7,000	29,200	2.14
Total, in land area (rounded)	9,100	37,800	2.8
Atmosphere	3.1	13	0.001
World ocean	317,000	1,320,000	97.3
Total, all items (rounded)	326,000	1,360,000	100
Annual evaporation[a]			
From world ocean	85	350	0.026
From land areas	17	70	0.005
Total	102	420	0.031
Annual precipitation			
On world ocean	78	320	0.024
On land areas	24	100	0.007
Total	102	420	0.031
Annual runoff to oceans from rivers and icecaps	9	38	0.003
Ground water outflow to oceans[b]	0.4	1.6	0.0001
Total	9.4	39.6	0.0031

[a] Evaporation (420,000 km³) is a measure of total water participating annually in the hydrological cycle.
[b] Arbitrarily set equal to about 5 percent of surface runoff.

Source: U.S. Geological Survey, 1967.

1–2 WATER RESOURCES OF THE CONTINENTAL U.S.

	mi^2	Acre ft
		x 10^6
Gross area of continental U.S.	3,080,809	
Land area, excluding inland water	2,974,726	
Volume of average annual precipitation		4,750
Volume of average annual runoff (discharge to sea)		1,372
Estimated total usable ground water		47,500
Average amount of soil moisture		635
Estimated total lake storage		13,000
Total reservoir storage (capacity of 5,000 acre ft or more)		365

Source: E.A. Ackerman and G.O. Löf, *Technology in American Water Development,* published for Resources for the Future, Inc. by the Johns Hopkins Press, 1959.

1–3 DISTRIBUTION OF WATER IN THE CONTERMINOUS U.S.

	Area, mi^2	Volume, mi^3	Annual circulation, million acre ft/yr	Detention, period, yr
Frozen water				
Glaciers	200	16	1.3	40
Ground ice			(seasonal only)	
Liquid water				
Fresh water lakes[a]	61,000	4,500	150	100
Salt lakes	2,600	14	4.6	10
Average in stream channels	–	12	1,500	0.03
Ground water				
Shallow	3,000,000	15,000	250	200
Deep	3,000,000	15,000	5	10,000
Soil moisture (3 ft root zone)	3,000,000	150	2,500	0.2
Gaseous water				
Atmosphere	3,000,000	45	5,000	0.03

[a] U.S. part of Great Lakes only.

Source: U.S. Geological Survey.

1–4 WORLDWIDE DISTRIBUTION OF FRESH WATER

Source	Quantity, %
Polar ice and glaciers	75
Ground water	
At depths <2,500 ft	11
At depths >2,500 ft, but <12,500 ft	14
Lakes	0.3
Rivers	0.03
Soil (moisture)	0.06
Atmosphere	0.035

Source: Reprinted, with publisher's permission, from *J. Am. Water Works Assn.,* 55:1253, © 1963, by the American Water Works Association, Inc., 2 Park Ave., New York, N.Y. 10016.

1–5 U.S. FRESH WATER RESERVES[1]

Source of water	Amount,[a] 10^6 acre ft	Reserve life,[b] yr
Atmosphere	176	0.6
Soil moisture	635	2.2
Rivers	45	0.1
Reservoirs	365	1.3
Lakes	13,000	46
Ground water	47,500	168
Total		215

[1] Adapted from MacKichan[2,3] and Ackerman and Lof.[4]
[a] In terms of current recoverable reserve discovered.
[b] To support entire national water withdrawal.

REFERENCES

1. L. Koenig, *J. Am. Water Works Assn., 55*(1):59, 1963.
2. K.A. MacKichan, *J. Am. Water Works Assn., 49*(4):369, 1957.
3. K.A. MacKichan, *J. Am. Water Works Assn., 53*(10):1211, 1961.
4. E.A. Ackerman and G.O. Löf, *Technology in American Water Development,* published for Resources for the Future, Inc. by the Johns Hopkins Press, 1959, p. 12.

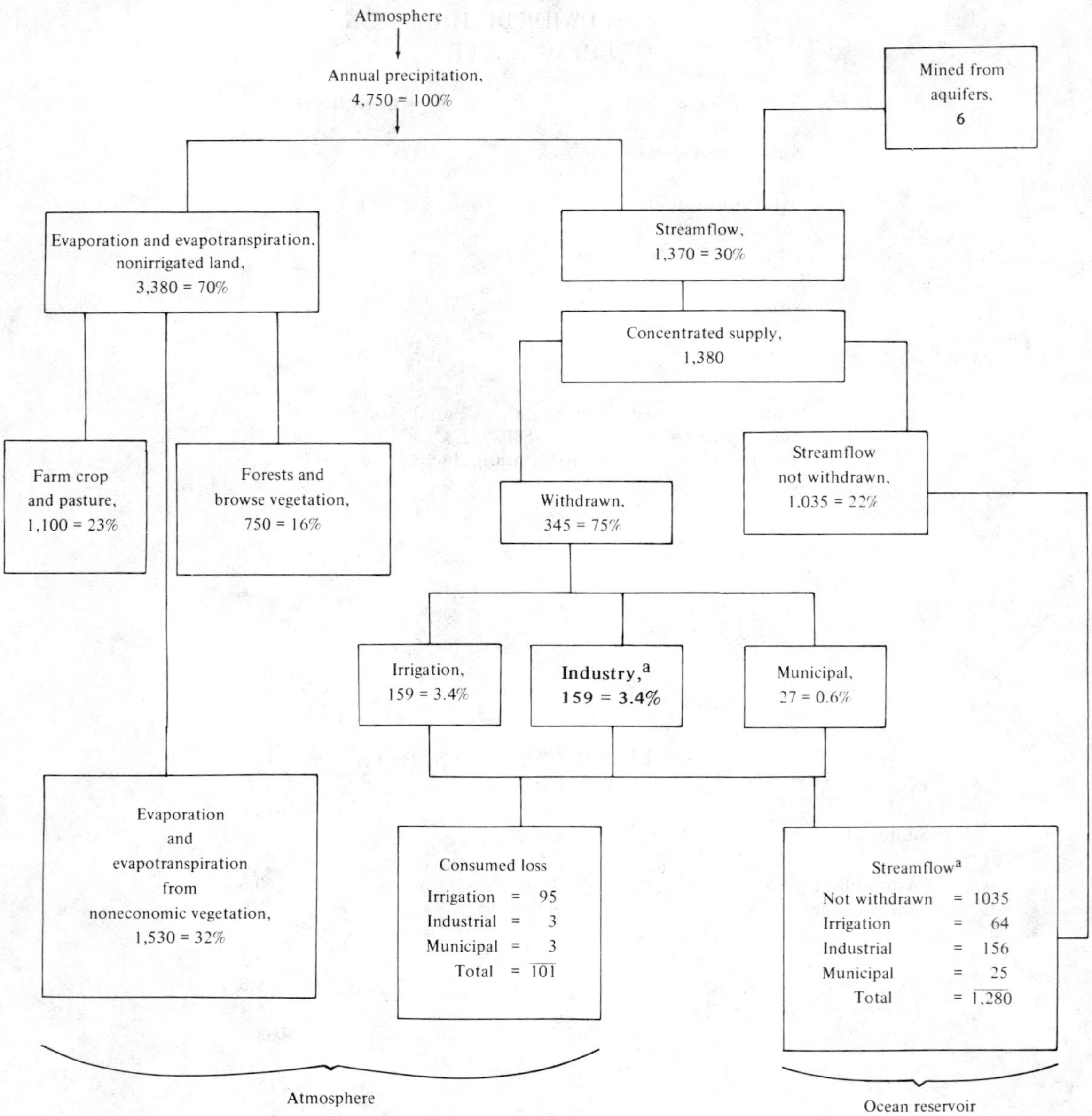

1–6 Average Distribution of Precipitation in the Continental U.S.

Legend:
All figures in units of million acre ft
one acre ft = 43,560 ft^3
one acre ft = 326,700 gal
Total precipitation = $1,552 \times 10^{12}$ gal/year
or $4,200 \times 10^9$ gal/day

[a] The same water may be reused at points spaced along a single stream.

Source: A. Wolman, *Water Resources*, National Academy of Sciences, National Research Council, Committee on Natural Resources, 1962.

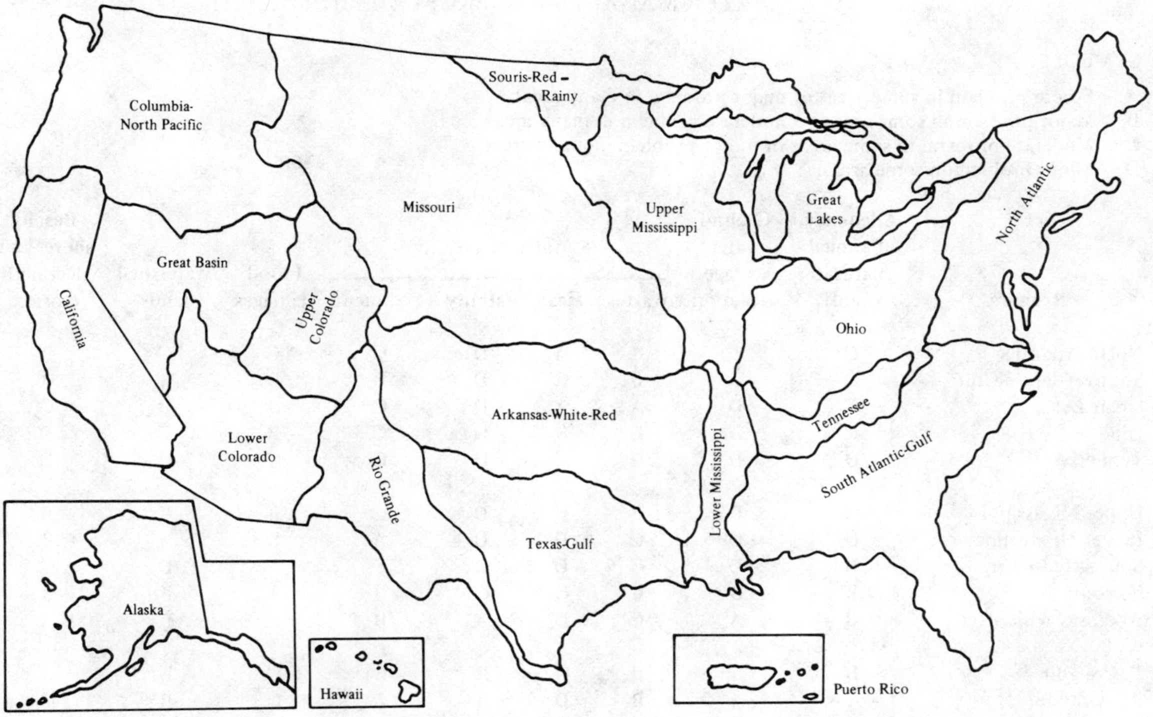

1–7 Water Resources Regions for Water Use in the U.S.

Source: U.S. Water Resources Council, 1968.

1–8 WATER MANAGEMENT PROBLEMS IN THE U.S.

SYMBOLS:

A – Severe problem in some areas or major problem in many areas.
B – Major problem in some areas or moderate problem in many areas.
C – Moderate problem in some areas or minor problem in many areas.
D – Minor problem in some areas.

Region[a]	Adequacy of annual natural runoff	Ground water storage depletion	Water quality				Flood damages	Watershed lands	Beach, shore, and riverbank erosion	Wetlands
			Wastes	Heat	Salinity	Sediment				
North Atlantic	C	C	A	A	D	C	C	C	B	B
South Atlantic-Gulf	D	C	B	C	D	B	B	B	A	A
Great Lakes	C	D	A	A	D	C	D	D	C	C
Ohio	C	D	B	B	D	C	B	C	D	D
Tennessee	D	D	C	C	D	C	C	C	D	D
Upper Mississippi	C	D	B	B	D	C	B	C	B	B
Lower Mississippi	D	D	C	D	D	A	B	B	B	B
Souris-Red-Rainy	B	D	C	D	C	D	B	D	C	B
Missouri	B	B	C	C	C	B	A	B	B	B
Arkansas-White-Red	B	A	C	D	A	B	B	B	A	B
Texas-Gulf	B	A	B	C	B	B	C	B	B	B
Rio Grande	A	A	B	D	A	A	C	B	C	D
Upper Colorado	A	D	C	D	C	B	D	C	D	D
Lower Colorado	A	A	B	D	A	A	C	B	C	D
Great Basin	A	C	B	D	C	C	D	C	D	C
Columbia-North Pacific	C	C	C	B	D	D	B	C	C	B
California	B	B	B	C	B	C	B	B	B	C
Alaska	D	D	C	D	D	D	C	D	B	C
Hawaii	D	C	C	D	D	D	D	C	B	D
Puerto Rico	D	D	C	D	D	B	D	B	C	D

[a] For regions see 1–7.

Source: U.S. Water Resources Council, 1968.

1–9 SUPPLY COMPARED WITH CUMULATIVE, OFF-CHANNEL WATER WITHDRAWALS, 1970

Region	Area,[1] 1,000's mi²	Average runoff[1]		Estimated dependable[2] supply, 1980, bgd	Withdrawals,[a] 1970, bgd	Fresh water consumed, 1970, bgd	Annual flow,[1] in bgd, exceeded in 90 percent of years	Fresh surface water withdrawn,[a] 1970, bgd
		in./yr	bgd					
New England	59	24	67	22	9.7	0.41	49	4.1
Middle Atlantic	102	18	84	36	45	1.4	68	25
South Atlantic-Gulf	270	15	197	75	35	3.3	129	20
Great Lakes	126	12	75	69	39	1.2	54	38
Ohio	163	16	125	48	36	0.92	75	34
Tennessee	41	21	41	14	7.9	0.24	28	7.7
Upper Mississippi	190	7.2	65	31	16	0.76	36	14
Lower Mississippi	96	17	79	25	13	3.6	38	8.5
Souris-Red-Rainy	59	2.2	6.2	3	0.3	0.07	2	0.2
Missouri	515	2.2	54	30	24	12	29	18
Arkansas-White-Red	265	6.0	73	20	12	6.8	36	5.2
Texas-Gulf	175	3.9	32	17	21	6.2	11	7.4
Rio Grande	136	.8	5.0	3	6.3	3.3	2	3.8
Upper Colorado	110	2.5	13	13	8.1	4.1	8	8.0
Lower Colorado	137	0.5	3.2	2	7.2	5.0	1	2.8
Great Basin	185	1.0	7.5	9	6.7	3.2	3	5.5
Columbia-North Pacific	271	16	210	70	30	11	148	26
California-South Pacific	120	9.0	62	28	48	22	30	21
U.S. (conterminous)	3,020	8.3	1,200	515	365	87	747	249
Alaska	590				0.2	0.02		0.2
Hawaii	6.4	44	13		2.7	0.81		0.8
Puerto Rico	3.4				3.0	0.17		0.4
Grand total	3,620				371	88		250

[a] Including some minor interregional diversions.

Source: C.R. Murray and E.B. Reeves, *Estimated Use of Water in the United States in 1970,* Geological Survey Circular 676, U.S. Department of the Interior, 1972.

REFERENCES

1. Modified from Table 31, U.S. Geological Survey Circular 556.
2. D.R. Woodward, *Availability of Water in the United States with Special Reference to Industrial Needs by 1980,* Armed Forces Industrial College, 1957.

1–10 ANNUAL NATURAL RUNOFF BY REGION

Region	Mean annual natural runoff, bgd	Percent of mean[a]
North Atlantic	163	71
Great Lakes	63.2	66
Columbia-North Pacific	210	65
South Atlantic-Gulf	197	59
Tennessee	41.5	58
Upper Colorado	13.7	55
Ohio	125	54
Lower Mississippi[b]	48.4	50
Great Basin	5.89	49
Upper Mississippi	64.6	44
Missouri	54.1	44
Rio Grande	4.9	43
California	65.1	40
Arkansas-White-Red	95.8	36
Souris-Red-Rainy	6.17	32
Texas-Gulf	39.1	30
Lower Colorado[b]	3.19	28
Total	1,200.65	

Note: Water resources:
30″ rainfall over U.S. – 4,200 bgd.
30% runoff (natural runoff) – 1,200 bgd.
Natural runoff for Alaska is estimated at 580 bgd.
Natural runoff for Hawaii is estimated at 13 bgd.

[a] In the one driest year out of 20, the annual natural runoff will be equal or less than the indicated percent of the mean.
[b] Does not include runoff derived from upstream regions.

Source: Blanchard, B., *J. Water Pollut. Control Fed., 41*(8):1407, 1969.

1–11 BASIC HYDROLOGIC DATA

Collected by U.S. Federal Agencies

Table A

Item	Agricultural Research Service	Bureau of the Census	Department of Commerce[a]	Corps of Engineers	Forest Service	Geological Survey	International Boundary and Water Commission
Consumptive use	x				x	x	
Density currents				x		x	
Drought						x	
Evaporation	x			x		x	x
Floods				x	x	x	x
Ground water	x			x		x	
Humidity				x			
Ice				x		x	
Infiltration	x				x		
Interception					x		
Lysimeters	x				x		
Precipitation	x			x	x		x
Chemical quality of water						x	x
River, lake, reservoir stages				x		x	x
Runoff	x			x	x	x	
Sedimentation							
Suspended load	x			x		x	x
Reservoir sedimentation	x			x	x	x	
Snow				x	x	x	
Soil moisture	x						
Solar radiation[e]	x						
Storms				x		x	
Streamflow	x			x	x	x	x
Air temperature					x		x
Soil temperature	x			x	x		
Water temperature[d]			x			x	

[a] Agencies other than the Census Bureau, Coast and Geodetic Survey, and Weather Bureau (very few, special installations).
[d] Also Fishand Wildlife Service.
[e] Also Naval Observatory.

1–11 BASIC HYDROLOGIC DATA (*Continued*)

Table A

Item	Agricultural Research Service	Bureau of the Census	Department of Commerce[a]	Crops of Engineers	Forest Service	Geological Survey	International Boundary and Water Commission
Tides[c]				x		x	
Water use and waste disposal							
Drainage		x					
Industrial use[b]		x	x				
Municipal use			x				
Sewage and waste disposal							
Wind	x			x	x		x
Experimental watersheds	x			x	x	x	

[a] Agencies other than the Census Bureau, Coast and Geodetic Survey, and Weather Bureau (very few, special installations).
[b] Department of the Interior. Agencies other than the Fish and Wildlife Service, Geological Survey, and Bureau of Reclamation (very few, special installations).
[c] Also Coast and Geodetic Survey.

Table B

Item	Mississippi River Commission	Public Health Service	Bureau of Reclamation	Smithsonian Institution	Soil Conservation Service	Tennessee Valley Authority	Weather Bureau
Consumptive use			x		x	x	
Density currents			x		x	x	
Drought						x	x
Evaporation			x			x	x
Floods			x		x	x	x
Ground water			x			x	
Humidity						x	x
Ice							x
Infiltration			x		x	x	
Interception						x	

1–11 BASIC HYDROLOGIC DATA (*Continued*)

Table B

Item	Mississippi River Commission	Public Health Service	Bureau of Reclamation	Smithsonian Institution	Soil Conservation Service	Tennessee Valley Authority	Weather Bureau
Lysimeters					x	x	
Precipitation			x		x	x	x
Chemical quality of water		x	x		x		
River, lake, reservoir stages			x			x	x
Runoff						x	
Sedimentation							
Suspended load			x		x		
Reservoir sedimentation			x		x	x	
Snow			x		x	x	x
Soil moisture					x	x	x
Solar radiation[e]			x	x		x	x
Storms			x		x	x	x
Streamflow	x		x		x	x	
Air temperature	x					x	x
Soil temperature					x	x	x
Water temperature[d]			x			x	
Tides[c]							x
Water use and waste disposal							
Drainage							
Industrial use[b]		x				x	
Municipal use		x					
Sewage and waste disposal		x					
Wind						x	x
Experimental watersheds						x	

[b] Department of the Interior. Agencies other than the Fish and Wildlife Service, Geological Survey, and Bureau of Reclamation (very few, special installations).
[c] Also Coast and Geodetic Survey.
[d] Also Fish and Wildlife Service.
[e] Also Naval Observatory.

Source: W.B. Langbein and W.G. Hoyt, *Water Facts for the Nation's Future – Uses and Benefits of* Hydrologic Data Programs, Ronald Press, New York, 1959.

1–12 EVAPORATION EQUATIONS

The following equations enable pan and lake evaporation to be computed from climatic data at first order weather stations. Daily values of evaporation are obtained using mean daily temperature and vapor pressure data together with data on solar radiation and wind movement as specified.

For pan evaporation, the expression is

$$E_p = \left\{\exp\,[(T_a - 212)\,(0.1024 - 0.01066 \text{ in } R)] - 0.001 + 0.025\,(e_s - e_a)^{0.88}\,(0.37 + 0.0041\,U_p)\right\} \times \left\{0.025 + (T_a + 398.36)^{-2}\;4.7988 \times 10^{10} \exp\,[-7482.6/(T_a + 398.36)]\right\}^{-1}$$

For lake evaporation, the expression is

$$E_L = \left\{\exp\,[(T_a - 212)\,(0.1024 - 0.01066 \text{ in } R)] - 0.001 + 0.0105\,(e_s - e_a)^{0.88}\,(0.37 + 0.0041\,U_p)\right\} \times \left\{0.015 + (T_a + 398.36)^{-2}\;6.8554 \times 10^{10} \exp\,[-7482.6/(T_a + 398.36)]\right\}^{-1}$$

The terms in these expressions are

E_p = pan evaporation, in.
E_L = lake evaporation, in.
T_a = air temperature, °F
e_a = vapor pressure, in. mercury at temperature T_a
e_s = vapor pressure, in. mercury at temperature T_d
T_d = dew point temperature, °F
R = solar radiation, langleys/day
U_p = wind movement, mi/day

Source: U.S. Weather Bureau, 1962.

1–13 MAJOR CHEMICAL CONSTITUENTS IN WATER

Sources, Concentrations, and Effects upon Usability

Constituent	Major sources	Concentration in natural water	Effect upon usability of water	Concentration in public water supplies of 100 cities, mg/l: Untreated water		Treated water		Drinking water should contain less than concentration shown if more suitable supplies are or can be made available[a]
				Max	Min	Max	Min	
Silica, SiO_2	Feldspars, ferromagnesium and clay minerals, amorphous silicachert, opal	Ranges generally from 1.0–30 mg/l, although as much as 100 mg/l is fairly common; as much as 4,000 mg/l is found in brines	In the presence of calcium and magnesium, silica forms a scale in boilers and on steam turbines that retards heat and fluid flow; the scale is difficult to remove. Silica may be added to soft water to inhibit corrosion of iron pipes	72	0.0	72	0.0	
Iron, Fe	**Natural sources** Igneous rocks Amphiboles, ferromagnesian micas, ferrous sulfide (FeS), ferric sulfide or iron pyrite (FeS_2), magnetite (Fe_3O_4) Sandstone rocks Oxides, carbonates, and sulfides of iron clay minerals **Manmade sources** Well casing, piping, pump parts, storage tanks, and other objects of cast iron and steel which may be in contact with the water Industrial wastes	Generally less than 0.50 mg/l in fully aerated water. Ground water having a pH less than 8.0 may contain 10 mg/l, rarely as much as 50 mg/l may occur. Acid water from thermal springs, mine wastes, and industrial wastes may contain more than 6,000 mg/l	More than 0.1 mg/l precipitates after exposure to air; causes turbidity, stains plumbing fixtures, laundry and cooking utensils, and imparts objectionable tastes and colors to foods and drinks. More than 0.2 mg/l is objectionable for most industrial uses	1.90	0.00	1.30	0.00	0.3

[a] U.S. Public Health Service (1962).

1–13 MAJOR CHEMICAL CONSTITUENTS IN WATER (*Continued*)

Constituent	Major sources	Concentration in natural water	Effect upon usability of water	Concentration in public water supplies of 100 cities, mg/l: Untreated water		Treated water		Drinking water should contain less than concentration shown if more suitable supplies are or can be made available[a]
				Max	Min	Max	Min	
Manganese, Mn	Manganese in natural water probably comes most often from soils and sediments. Metamorphic and sedimentary rocks and mica biotite and amphibole hornblende minerals contain large amounts of manganese	Generally 0.20 mg/l or less. Ground water and acid mine water may contain more than 10 mg/l. Water at the bottom of a stratified reservoir may contain more than 150 mg/l	More than 0.2 mg/l precipitates upon oxidation; causes undesirable tastes, deposits on foods during cooking, stains plumbing fixtures and laundry, and fosters growths in reservoirs, filters, and distribution systems. Most industrial users object to water containing more than 0.2 mg/l	0.60	0.00	2.5	0.00	0.05
Calcium, Ca	Amphiboles, feldspars, gypsum, pyroxenes, aragonite, calcite, dolomite, clay minerals	As much as 600 mg/l in some western streams, brines may contain as much as 75,000 mg/l	Calcium and magnesium combine with bicarbonate, carbonate, sulfate, and silica to form heat-retarding, pipe-clogging scale in boilers and in other heat-exchange equipment. Calcium and magnesium combine with ions of fatty acid in soaps to form soapsuds; the more calcium and magnesium, the more soap required to form suds. A high concentration of magnesium has a laxative effect, especially on new users of the supply	145	0.0	145	0.0	
Magnesium, Mg	Amphiboles, olivine, pyroxenes, dolomite, magnesite, clay minerals	As much as several hundred mg/l in some western streams; ocean water contains more than 1,000 mg/l and brines may contain as much as 57,000 mg/l		120	0.0	120	0.0	
Sodium, Na	Feldspars (albite); clay minerals; evaporites, such as halite (NaCl) and mirabilite ($Na_2SO_4 \cdot 10H_2O$); industrial wastes	As much as 1,000 mg/l in some western streams; about 10,000 mg/l in sea water; about 25,000 mg/l in brines	More than 50 mg/l sodium and potassium in the presence of suspended matter causes foaming, which accelerates scale formation and corrosion in boilers. Sodium and potas-	177	1.1	198	1.1	

[a]U.S. Public Health Service, 1962.

1–13 MAJOR CHEMICAL CONSTITUENTS IN WATER (*Continued*)

Constituent	Major sources	Concentration in natural water	Effect upon usability of water	Concentration in public water supplies of 100 cities, mg/l				Drinking water should contain less than concentration shown if more suitable supplies are or can be made available[a]
				Untreated water		Treated water		
				Max	Min	Max	Min	
Potassium, K	Feldspars (orthoclase and microcline), feldspathoids, some micas, clay minerals	Generally less than about 10 mg/l; as much as 100 mg/l in hot springs; as much as 25,000 mg/l in brines	sium carbonate in recirculating cooling water can cause deterioration of wood in cooling towers. More than 65 mg/l of sodium can cause problems in ice manufacture	30	0.2	30	0.0	
Carbonate, CO_3	Limestone, dolomite	Commonly 0 mg/l in surface water; commonly less than 10 mg/l in ground water. Water high in sodium may contain as much as 50 mg/l of carbonate	Upon heating, bicarbonate is changed into steam, carbon dioxide, and carbonate. The carbonate combines with alkaline earths–principally calcium and magnesium–to form a crustlike scale of cal-	21	0	26	0	
Bicarbonate, HCO_3		Commonly less than 500 mg/l; may exceed 1,000 mg/l in water highly charged with carbon dioxide	cium carbonate that retards flow of heat through pipe walls and restricts flow of fluids in pipes. Water containing large amounts of bicarbonate and alkalinity is undesirable in many industries	380	5	380	0	
Sulfate, SO_4	Oxidation of sulfide ores; gypsum; anhydrite; industrial wastes	Commonly less than 1,000 mg/l except in streams and wells influenced by acid mine drainage. As much as 200,000 mg/l in some brines	Sulfate combines with calcium to form an adherent, heat-retarding scale. More than 250 mg/l is objectionable in water in some industries. Water containing about 500 mg/l of sulfate tastes bitter; water containing about 1,000 mg/l may be cathartic	572	0.0	572	0.0	250

[a] U.S. Public Health Service, 1962.

1–13 MAJOR CHEMICAL CONSTITUENTS IN WATER (*Continued*)

Constituent	Major sources	Concentration in natural water	Effect upon usability of water	Concentration in public water supplies of 100 cities, mg/l: Untreated water		Treated water		Drinking water should contain less than concentration shown if more suitable supplies are or can be made available[a]
				Max	Min	Max	Min	
Chloride, Cl	Chief source is sedimentary rock (evaporites); minor sources are igneous rocks. Ocean tides force salty water upstream in tidal estuaries	Commonly less than 10 mg/l in humid regions; tidal streams contain increasing amounts of chloride (as much as 19,000 mg/l) as the bay or ocean is approached. About 19,300 mg/l in sea water; and as much as 200,000 mg/l in brines	Chloride in excess of 100 mg/l imparts a salty taste. Concentrations greatly in excess of 100 mg/l may cause physiological damage. Food processing industries usually require less than 250 mg/l. Some industries–textile processing, paper manufacturing, and synthetic rubber manufacturing–desire less than 100 mg/l	540	0.5	540	0.0	250
Fluoride, F	Amphiboles (hornblende), apatite, fluorite, mica	Concentrations generally do not exceed 10 mg/l in ground water or 1.0 mg/l in surface water. Concentrations may be as much as 1,600 mg/l in brines	Fluoride concentration between 0.6 and 1.7 mg/l in drinking water has a beneficial effect on the structure and resistance to decay of children's teeth. Fluoride in excess of 1.5 mg/l in some areas causes "mottled enamel" in children's teeth. Fluoride in excess of 6.0 mg/l causes pronounced mottling and disfiguration of teeth	7.0	0.0	7.0	0.0	The recommended control limits depend upon annual averages of maximum daily air temperature and range from 0.6–0.8 mg/l at 79.3°–90.5°F and 0.9–1.7 mg/l at 50.0°–53.7°F
Nitrate, NO_3	Atmosphere; legumes, plant debris, animal excrement, nitrogenous fertilizer in soil and sewage	In surface water not subjected to pollution, concentration of nitrate may be as much as 5.0 mg/l but is commonly less than 1.0 mg/l. In ground water the concentration of nitrate may be as much as 1,000 mg/l	Water containing large amounts of nitrate (more than 100 mg/l) is bitter tasting and may cause physiological distress. Water from shallow wells containing more than 45 mg/l has been reported to cause methemoglobinemia in infants. Small	23	0.0	23	0.0	45; In areas in which the nitrate content of water is known to be in excess of the listed concentration, the public should

[a]U.S. Public Health Service, 1962.

1–13 MAJOR CHEMICAL CONSTITUENTS IN WATER (*Continued*)

Constituent	Major sources	Concentration in natural water	Effect upon usability of water	Concentration in public water supplies of 100 cities, mg/l				Drinking water should contain less than concentration shown if more suitable supplies are or can be made available[a]
				Untreated water		Treated water		
				Max	Min	Max	Min	
			amounts of nitrate help reduce cracking of high-pressure boiler steel					be warned of the potential dangers of using the water for infant feeding
Dissolved solids	The mineral constituents dissolved in water constitute the dissolved solids	Surface water commonly contains less than 3,000 mg/l; streams draining salt beds in arid regions may contain in excess of 15,000 mg/l. Ground water commonly contains less than 5,000 mg/l; some brines contain as much as 300,000 mg/l	More than 500 mg/l is undesirable for drinking and many industrial uses. Less than 300 mg/l is desirable for dyeing of textiles and the manufacture of plastics, pulp paper, rayon. Dissolved solids cause foaming in steam boilers; the maximum permissible content decreases with increases in operating pressure	1,580	10	1,580	22	500

[a]U.S. Public Health Service, 1962.

Source: C.N. Durfor and E. Becker, *Geological Survey Water Supply Paper,* U.S. Department of the Interior, 1964. Table prepared by B.P. Robinson.

1–14 MINOR CONSTITUENTS IN WATER

Constituent	Some common sources	Concentration in natural water	Concentration in treated water of public supplies of 100 largest cities, mg/l		Drinking water should contain less than concentration shown if more suitable supplies are or can be made available,[a] mg/l	Concentration in drinking water more than that shown constitutes grounds for rejection of the supply,[a] mg/l
			Max	Min		
Aluminum	Feldspars, micas, and clay minerals; also present as the result of alum treatment in water supplies	Generally less than 0.5 mg/l. Sea water, 7,500 mg/l in acid lakes	1,500	3.3		
Barium	Feldspars, mica, biotite, barite, witherite	Generally less 0.5 mg/l. Sea water, about 0.05 mg/l; brine, as much as 5,000 mg/l	380	1.7		1.0
Boron	Amphiboles, biotite, colemanite, and tourmaline; detergents; industrial wastes	Generally less than a few tenths mg/l. Sea water, about 4.6 mg/l; mineral springs, may exceed 500 mg/l; brine, may exceed 9,000 mg/l	590	2.5		
Chromium	Chromite; industrial pollution	Generally less than 0.1 mg/l. Industrial pollution, as much as 40 mg/l	35	Not detectable		0.05 (hexavalent)
Copper	Native copper, chalcocite, bornite, chalcopyrite, cuprite, many other minerals containing copper. Copper and brass pipes; algae-controlling copper salts; industrial wastes	Generally less than a few tenths mg/l. Sea water, 0.001–0.01 mg/l; industrial pollution (mine water) as much as several thousand mg/l	250	Less than 0.61	1.0	
Lead	Galena, aragonite, feldspars. Dissolution of lead pipes; industrial and mining wastes	Generally less than 0.5 mg/l in streams and in ground water. Sea water, 0.004 mg/l; industrial pollution, several hundred mg/l	62	Not detectable		
Lithium	Mica, amphiboles, pyroxenes	Generally less than 1 mg/l. Sea water, 0.1 mg/l; brine, as much as 80 mg/l	170	Not detectable		
Nickel	Pentlandite, niccolite, chloanthite, garnierite, millerite, olivine, hypersthene	Generally less than 0.1 mg/l	34	Not detectable		
Phosphorus	Apatite; organic wastes; fertilizers; detergents	Generally less than a few mg/l. Sea water, 0.003–0.3 mg/l; industrial pollution, 10 mg/l	370	Not detectable		

[a]U.S. Public Health Service, 1962.

1–14 MINOR CONSTITUENTS IN WATER (*Continued*)

Constituent	Some common sources	Concentration in natural water	Concentration in treated water of public supplies of 100 largest cities, mg/l		Drinking water should contain less than concentration shown if more suitable supplies are or can be made available,[a] mg/l	Concentration in drinking water more than that shown constitutes grounds for rejection of the supply,[a] mg/l
			Max	Min		
Radium	Most igneous rocks, sandstones, shales	Uncontaminated water generally contains less than 10 pCi/l	2.5 pCi/l	Less than 0.1 pCi/l	3 pCi of ^{226}Ra	
Rubidium	Greisen, lepidolite (concealed in many potassium minerals), potassium feldspar, potassium mica	Generally less than 0.1 mg/l	67	Not detectable		
Strontium	Feldspars, apatite, pyroxenes, amphiboles, celestite, strontianite, aragonite, fossils	Generally less than 5.0 mg/l in fresh-water streams; as much as 30 mg/l in ground water. Sea water, about 13 mg/l; brine, as much as 3,500 mg/l	1,200	2.2		
Titanium	Ilmenite, rutile, sphene, phlogopites, amphiboles, biotite, pyroxene	Generally less than 0.1 mg/l	49	Not detectable		
Uranium	Uraninite, carnotite; most igneous rocks, sandstones, and shales	Uncontaminated water generally contains less than 44 mg/l	250	Less than 0.1		
Zinc	Sphalerite; mine waste waters; industrial wastes	Generally in trace amounts. Sea water, 0.005 mg/l; industrial pollution, as much as several thousand mg/l	610	Not detectable		

[a]U.S. Public Health Service, 1962.

REFERENCES

1. C.N. Durfor and E. Becker, *Geological Survey Water Supply Paper*, U.S. Department of the Interior, 1964.
2. K. Rankama and T. G. Sahama, *Geochemistry*, Chicago University Press, 1950.

1–15 SUMMARY RAW WATER CHARACTERISTICS

Public Water Supplies

Constituent	River	Lake	Impoundment	Ground water
SiO_2, mg/l	0.00-21	1.1-5.8	1.0-15	4.3-72
Fe, mg/l	0.00-1.9	0.00-0.16	0.0-0.84	0.00-1.2
Ca, mg/l	2.8-110	2.0-44	0.0-76	3.2-121
Mg, mg/l	0.9-28	0.5-11	0.3-29	0.3-120
Na, mg/l	1.9-131	1.6-24	1.1-99	6.1-129
K, mg/l	0.3-7.8	0.7-4.5	0.2-7.9	0.4-30
HCO_3, mg/l	15-316	9-180	4-256	55-364
CO_3, mg/l	0-1	0	0-21	0-5
SO_4, mg/l	0.0-290	0.8-23	0.6-282	0.8-572
Cl, mg/l	0.5-196	2.6-32	1.0-107	2.0-92
F, mg/l	0.0-0.6	0.0-0.3	0.0-1.0	0.0-7.0
NO_3, mg/l	0.0-8.3	0.0-2.2	0.0-4.7	0.0-17
Dissolved solids (180°C), mg/l	31-657	29-233	10-761	68-1,220
Hardness as $CaCO_3$, mg/l	12-368	7-153	2-336	10-738
Noncarbonate hardness as $CaCO_3$, mg/l	0-206	0-54	0-207	0-446
Specific conductance, μmho 25°C	42-1,270	33-387	8-865	108-1,660
pH	6.2-8.4	6.6-8.3	5.8-9.1	6.7-8.7
Color, units	0-115	1-15	0-50	0-76
Beta activity, pCi/l	16-78	12-66		1.8-130
Radium, pCi/l				<0.1-1.9
Uranium, μg/l				<0.1-250
Ag, μg/l	ND[a]-0.48	<0.18-<0.34	ND-0.74	ND-1.5
Al, μg/l	0-2,400	38-90	17-690	2.9-83
B, μg/l	7.0-130	7.8-80	3.6-250	0-500
Ba, μg/l	3.1-340	31-120	6.2-150	4.6-67
Be, μg/l	ND	ND	ND-0.75	ND
Co, μg/l	ND-2.3	ND	ND-0.99	ND
Cr, μg/l	ND-7.8	0.34-3.6	ND-3.8	ND-1.1
Cu, μg/l	1.1-84	6.0-57	0.78-73	<0.8-15
Fe, μg/l	9.0-3,800	14-100	29-850	1.1-6,600
Li, μg/l	0.08-91	0.43-1.7	0-100	ND-96
Mn, μg/l	2.6-380	0-100	0-450	ND-340
Mo, μg/l	ND-7.8	ND-1.5	ND-5.8	ND-68
Ni, μg/l	ND-15	3.2-21	ND-15	ND-<15
P, μg/l	ND	ND	ND	ND
Pb, μg/l	ND-60	6.4-26	ND-94	ND-38
Rb, μg/l	1.6-16	ND-3.5	ND-5.5	ND-3.7
Sn, μg/l	ND	ND	ND-1.1	ND
Sr, μg/l	2.6-610	80-110	2.4-540	19-6,300
Ti, μg/l	0.5-35	<2-<7.6	0.6-11	ND-<5
V, μg/l	ND-5.1	ND	ND-<5.8	ND-74
Zn, μg/l	ND-210	ND-<340	ND-99	ND-<470
Ga, μg/l	ND-0.0X[b]			<0.X
Zr, μg/l	0.X-X		0.X	ND
Yb, μg/l	0.0X-X	ND	0.X	ND
Y, μg/l	0.X-X	ND	0.X	ND

[a] ND, looked for but not found.
[b] X, semiquantitative determination in digit order shown.

Source: Developed from data in C.N. Durfor and E. Becker, *Geological Survey Water Supply Paper,* U.S. Department of the Interior, 1964.

1–16 TRACE ELEMENT CONCENTRATIONS

383 Water Samples From Miscellaneous Sources

Sample sources	No. samples	Ag	Al	B	Ba	Cd
Surface waters						
Number of positive occurrences	165	37	48	134	153	0
Range[a]		<0.0001–0.003	<0.05–1.2	<0.02–11.0	<0.01–12.0	<0.005
Mean[a]		0.0003	0.6	1.37	0.65	–
Median[a]		0.0001	<0.10	0.11	0.09	–
Well waters from San Joaquin Valley						
Number of positive occurrences	82	51	6	63	69	0
Range		<0.0001–0.0004	<0.02–0.2	<0.02–8.5	<0.01–1.2	<0.005
Mean		0.0002	0.15	0.71	0.31	–
Median		<0.0001	<0.06	0.12	0.08	–
Agricultural drainage waters from Coachella Valley						
Number of positive occurrences	22	22	4	20	1	0
Range		0.0005–0.0010	<0.2–0.8	<0.2–6.0	<0.1–2.2	<0.020
Mean		0.0008	0.45	1.20	2.2	–
Median		0.0008	<0.2	0.70	<0.1	–
Domestic waste waters from California cities						
Number of positive occurrences	22	22	5	22	20	2
Range		0.0006–0.004	<0.1–0.38	0.2–1.3	0.01–0.57	<0.005–0.007
Mean		0.0020	0.15	0.54	0.13	0.0065
Median		0.0020	<0.1	0.52	0.10	<0.005
Spring and well waters surrounding the Salton Sea						
Number of positive occurrences	22	12	2	21	15	0
Range		<0.0001–0.0004	<0.05–0.8	<0.02–11.4	<0.02–3.7	<0.005
Mean		0.0002	0.5	2.60	0.30	–
Median		0.0002	<0.1	0.64	0.02	–
Food processing – Plant 1[b]						
Number of positive occurrences	5	4	5	5	5	0
Range		<0.0001–0.0008	0.9–2.2	0.4–0.7	0.1–0.27	<0.005
Mean		0.0006	1.6	0.54	0.19	–
Median		0.0005	1.7	0.50	0.22	–

[a] Given as mg/l.
[b] Data from live samples taken at monthly intervals.

1–16 TRACE ELEMENT CONCENTRATIONS (*Continued*)

Sample sources	No. samples	Ag	Al	B	Ba	Cd
Food processing – Plant 2[b]						
Number of positive occurrences	5	5	5	5	5	0
Range		0.0005–0.0035	14.0–50.0	0.4–0.8	0.1–0.27	<0.005
Mean		0.0013	25.0	0.58	0.18	–
Median		0.001	16.0	0.60	0.15	–
Metal and chemical processing – Plant 1[b]						
Number of positive occurrences	5	5	3	5	2	0
Range		0.0004–0.0010	<0.1–0.7	0.7–1.9	<0.05–0.04	<0.005
Mean		0.00060	0.50	1.58	0.04	–
Median		0.0005	0.60	1.70	<0.05	–
Plant 2	1	0.008	<2.0	<2.0	<1.0	0.20
Plant 3	1	0.0006	<2.0	<2.0	<1.0	<0.01
Plant 4	1	0.0006	<2.0	4.6	<1.0	<0.01
Plant 5	1	0.0006	<2.0	<2.0	<1.0	<0.01
Plant 6	1	0.0006	<2.0	<2.0	5.8	<0.01
Oil well brines						
Composite of 8 producing wells	1	<0.0001	<0.2	0.4	0.09	<0.001
Brines from desert wells						
Imperial Valley Geothermal Brine	1	1.0	<100.0	570.0	1100.	<1.0
Amboy Salt Well	1	<0.001	<0.05	7.5	<0.2	<0.001
Various water samples from Casa Diablo (includes hot and cold springs, lake waters and streams from area of volcanic activity)[c]						
Number of positive occurrences	27		0	20	15	0
Range		Not determined	<0.2	<0.02–35.0	<0.02–0.9	<0.001
Mean			–	8.41	0.29	–
Median			–	0.63	<0.02	–
Suisun Bay – Lower San Joaquin River water samples						
Number of positive occurrences	10	10	0	2	2	0
Range		0.0002–0.0005	<2.0	<2.0–0.7[d]	<1.0–0.4	<0.01
Mean		0.0003	–	0.65	0.3	–
Median		<0.0003	–	<2.0	<1.0	–

[b] Data from live samples taken at monthly intervals.
[c] Number of positive occurrences, 9; range, <0.04–12; mean, 0.62; median, <0.08.
[d] Boron detected in two samples after concentration fivefold by evaporation.

1–16 TRACE ELEMENT CONCENTRATIONS (*Continued*)

Sample sources	No. samples	Ag	Al	B	Ba	Cd
Suisun Bay – Lower San Joaquin River sediment saturation extracts						
Number of positive occurrences	10	10	0	0	2	0
Range		0.01–0.03	<2.0	<2.0	<1.0–2.4	<0.01
Mean		0.013	–	–	1.3	–
Median		0.013	–	–	<1.0	–

Sample sources	Co	Cr	Cu	Fe	Ga	Mn
Surface waters						
Number of positive occurrences	0	20	165	150	0	154
Range[a]	<0.0005	<0.0005–0.035	0.0009–0.065	0.0013–0.38	<0.0005	0.0005–1.5
Mean[a]	–	0.005	0.006	0.057		0.028
Median[a]	–	<0.001	0.003	0.02		0.004
Well waters from San Joaquin Valley						
Number of positive occurrences	0	55	82	82	0	70
Range	<0.0005	<0.0005–0.009	0.0006–0.047	0.0006–0.5	<0.0005	<0.0005–0.66
Mean	–	0.002	0.006	0.029		0.042
Median	–	<0.001	0.003	0.005		0.003
Agricultural drainage waters from Coachella Valley						
Number of positive occurrences	0	1	22	23	0	13
Range	<0.002	<0.002–0.003	0.003–0.018	0.005–0.048	<0.002	<0.002–1.2
Mean	–	0.003	0.007	0.061		0.159
Median	–	<0.002	0.006	0.008		0.002
Domestic waste waters from California cities						
Number of positive occurrences	18	13	22	22	1	22
Range	<0.001–0.007	<0.001–0.005	0.006–0.065	0.025–1.4	<0.001–0.001	0.002–0.25
Mean	0.0037	0.003	0.024	0.27	–	0.036
Median	0.004	0.002	0.022	0.128	–	0.015
Spring and well waters surrounding the Salton Sea						
Number of positive occurrences	1	8	22	22	6	22
Range	<0.0005–0.005	<0.0005–0.012	0.001–0.28	0.003–0.44	<0.0005–0.003	0.0006–0.24
Mean	–	0.007	0.022	0.063	0.0012	0.025
Median	–	<0 001	0.009	0.032	<0.005	0.004

[a] Given as mg/l.

1–16 TRACE ELEMENT CONCENTRATIONS (*Continued*)

Sample sources	Co	Cr	Cu	Fe	Ga	Mn
Food processing – Plant 1[b]						
Number of positive occurrences	4	4	5	5	5	5
Range	<0.001–0.006	<0.001–0.2	0.005–0.038	1.4–10.5	0.001–0.1	0.003–0.036
Mean	0.004	0.074	0.025	3.54	0.029	0.018
Median	0.003	0.015	0.028	2.10	0.001	0.015
Food processing – Plant 2[b]						
Number of positive occurrences	5	4	5	5	5	5
Range	<0.005–0.008	<0.001–0.10	0.170–0.40	1.2–2.7	<0.001–0.1	0.012–0.047
Mean	0.006	0.026	0.272	1.96	0.041	0.030
Median	0.007	0.002	0.290	2.0	0.003	0.03
Metal and chemical processing – Plant 1[b]						
Number of positive occurrences	5	5	5	5	2	5
Range	<0.005–0.03	0.002–0.2	0.007–0.031	0.075–0.62	<0.001–0.035	0.017–0.055
Mean	0.015	0.050	0.022	0.39	0.035	0.035
Median	0.011	0.005	0.026	0.5	<0.001	0.041
Plant 2	0.013	<0.002	0.38	40.0	0.001	0.040
Plant 3	<0.002	0.002	0.002	0.062	0.001	0.003
Plant 4	0.004	0.002	0.070	0.0175	<0.001	0.065
Plant 5	0.006	0.003	0.013	0.019	<0.001	0.003
Plant 6	0.002	0.004	0.003	0.037	<0.001	0.005
Oil well brines						
Composite of 8 producing wells	<0.001	<0.001	0.002	0.031	<0.001	0.010
Brines from desert wells						
Imperial Valley Geothermal Brine	<1.0	<4.0	6	1200	<1.0	2000
Amboy Salt Well	<0.001	<0.002	0.008	0.35	<0.001	0.095
Various water samples from Casa Diablo (includes hot and cold springs, lake waters and streams from area of volcanic activity)[c]						
Number of positive occurrences	0	4	27	27	0	27
Range	<0.001	<0.0005–0.001	0.001–2.0	0.003–0.9	<0.001	0.001–0.57
Mean	–	0.001	0.079	0.09	–	0.032
Median	–	<0.001	0.004	0.014	–	0.0036

[b] Data from live samples taken at monthly intervals.
[c] Number of positive occurrences, 9; range, <0.04–12; mean, 0.62; median, <0.08.

1–16 TRACE ELEMENT CONCENTRATIONS (*Continued*)

Sample sources	Co	Cr	Cu	Fe	Ga	Mn
Suisun Bay – Lower San Joaquin River water samples						
Number of positive occurrences	0	3	10	10	0	10
Range	<0.02	<0.002–0.005	0.003–0.015	0.005–0.037	<0.01	0.002–0.006
Mean	–	0.003	0.008	0.016	–	0.003
Median	–	<0.002	0.009	0.014	–	0.003
Suisun Bay – Lower San Joaquin River sediment saturation extracts						
Number of positive occurrences	7	1	10	10	5	10
Range	<0.003–0.055	<0.003–0.009	0.010–0.020	0.03–27.0	<0.01–0.01	0.7–30.0
Mean	0.027	0.009	0.013	14.4	0.01	6.9
Median	0.021	<0.003	0.014	4.4	<0.01	2.0

Sample sources	Mo	Ni	Pb	Si	Sr	V
Surface waters						
Number of positive occurrences	149	123	110	164	159	160
Range[a]	<0.0005–0.130	<0.0005–0.025	0.0005–0.018	<0.02–36.4	<0.02–1.8	<0.0005–0.085
Mean[a]	0.009	0.003	0.002	9.6	0.43	0.005
Median[a]	0.003	0.001	0.001	7.5	0.24	0.003
Well waters from San Joaquin Valley						
Number of positive occurrences	78	52	78	82	79	78
Range	<0.0005–0.015	0.0005–0.007	<0.0005–0.01	0.2–28.0	<0.02–1.0	<0.0005–0.66
Mean	0.004	0.002	0.003	8.5	0.37	0.012
Median	0.003	0.0006	0.002	6.9	0.22	0.006
Agricultural drainage waters from Coachella Valley						
Number of positive occurrences	22	22	22	21	22	22
Range	0.007–0.088	0.002–0.008	0.003–0.014	0.7–19.0	0.2–17.2	0.005–0.047
Mean	0.030	0.004	0.005	7.3	3.21	0.011
Median	0.026	0.004	0.004	6.8	2.50	0.01
Domestic waste waters from California cities						
Number of positive occurrences	22	22	21	22	22	22
Range	0.002–0.012	0.003–0.6	<0.001–0.35	2.3–7.5	0.12–1.2	0.003–0.021
Mean	0.006	0.047	0.034	4.4	0.39	0.008
Median	0.005	0.007	0.01	4.0	0.30	0.007

[a] Given as mg/l.

1–16 TRACE ELEMENT CONCENTRATIONS (*Continued*)

Sample sources	Mo	Ni	Pb	Si	Sr	V
Spring and well waters surrounding the Salton Sea						
Number of positive occurrences	21	17	22	22	21	18
Range	<0.002–0.12	<0.0005–0.007	0.008–0.006	1.2–12.0	<0.02–19.0	<0.001–0.085
Mean	0.024	0.003	0.002	5.8	2.2	0.013
Median	0.012	0.001	0.002	4.8	0.60	0.005
Food processing – Plant 1[b]						
Number of positive occurrences	4	5	5	5	5	5
Range	<0.001–0.014	0.018–0.09	0.006–0.015	4.3–9.0	0.35–0.85	0.008–0.016
Mean	0.008	0.046	0.01	5.6	0.64	0.010
Median	0.015	0.044	0.009	5.1	0.70	0.01
Food processing – Plant 2[b]						
Number of positive occurrences	5	5	5	5	5	5
Range	0.01–0.026	0.023–0.05	0.018–0.027	1.6–30.0	0.57–1.1	0.015–0.08
Mean	0.016	0.035	0.020	9.8	0.89	0.032
Median	0.017	0.034	0.018	4.1	0.98	0.024
Metal and chemical processing – Plant 1[b]						
Number of positive occurrences	5	5	5	5	5	5
Range	0.017–0.034	0.004–0.016	0.009–0.040	5.9–25.0	0.30–0.35	0.004–0.008
Mean	0.027	0.009	0.020	12.7	0.32	0.005
Median	0.031	0.01	0.018	12.0	0.32	0.004
Plant 2	0.012	0.053	2.0	<2.0	3.2	0.009
Plant 3	0.002	0.004	0.003	<2.0	<2.0	0.003
Plant 4	<0.001	0.011	0.033	<2.0	<2.0	<0.001
Plant 5	<0.001	0.021	0.003	<2.0	<2.0	<0.001
Plant 6	0.001	0.013	0.006	<2.0	<2.0	<0.001
Oil well brines						
Composite of 8 producing wells	0.006	0.001	0.005	68.0	0.64	<0.001
Brines from desert wells						
Imperial Valley Geothermal Brine	<1.0	<4.0	70.0	<30.0	610.0	6.0
Amboy Salt Well	0.27	0.002	0.001	5.0	15.0	0.0043

[b] Data from live samples taken at monthly intervals.

1–16 TRACE ELEMENT CONCENTRATIONS (*Continued*)

Sample sources	Mo	Ni	Pb	Si	Sr	V
Various water samples from Casa Diablo (inclues hot and cold springs, lake waters and streams from area of volcanic activity)[c]						
Number of positive occurrences	27	21	22	26	12	24
Range	0.005–0.023	<0.0005–0.006	<0.0005–0.014	0.06–68.0	<0.02–0.35	0.0001–0.008
Mean	0.005	0.001	0.003	31.2	0.13	0.004
Median	0.003	0.001	0.001	31.0	<0.04	0.003
Suisun Bay – Lower San Joaquin River water samples						
Number of positive occurrences	9	10	10	4	5	10
Range	<0.001–0.008	0.003–0.014	0.002–0.007	<2.0–3.4	<2.0–3.0	0.006–0.010
Mean	0.003	0.005	0.004	3.0	1.7	0.007
Median	0.002	0.003	0.003	<2.0	<2.0	0.006
Suisun Bay – Lower San Joaquin River sediment saturation extracts						
Number of positive occurrences	10	10	10	10	5	10
Range	0.007–0.026	0.007–0.033	0.010–0.130	2.0–58.0	<2.0–5.2	0.013–0.020
Mean	0.016	0.018	0.056	17.0	3.6	0.021
Median	0.016	0.017	0.038	14.0	<2.0	0.019

[c] Number of positive occurrences, 9; range, <0.04–12; mean, 0.62; median, <0.08.

Source: G.R. Bradford, *Hilgardia, 41*(3):45, 1971.

1.1 SURFACE WATER

1.1–1 CHEMICAL ANALYSES OF TYPICAL CANADIAN SURFACE WATERS

River	pH	Ca	Mg	Dissolved Fe	Na	K	HCO_3	SO_4	Cl	F	NO_3	SiO_2	Hardness as $CaCO_3$, ppm: Non-carbonate	Hardness as $CaCO_3$, ppm: Total	Specific conductance at 25°C, μmho	Alkaline-earth bicarbonates, percent
		Concentration, ppm														
Cordilleran Region																
Yukon at Dawson, Y.T.	7.9	27	6.5		2.1	1.3	87	27	0.5		trace	7.0	22	94	195 (114)[a]	71
Powell at Powell River, B.C.	6.8	1.8	0.3	0.01	1.9	0.2	4.9	4.3	3.6	0.0	0.4	2.5	1.7	5.7	22 (17)	20[b]
Fraser at Mission, B.C.	7.6	15	2.9	0.05	2.0	0.6	57	8.2	0	0.1	0.5	5.7	3.0	50	110 (64)	79
Columbia at Trail, B.C.	8.0	19	3.9	0.03	1.1	0.7	64	10	0.8	0.05	0.5	6.2	7.6	64	135 (76)	83
Peace at Ft. St. John, B.C.	8.1	31	7.0		1.8	0.5	103	17	0.3		0.4	4.6	12	106	214 (119)	84
Oldman at Lethbridge, Alta.	8.2	37	13.5		9.6	1.0	169	29	2.0		3.2	4.0	8.6	148	308 (183)	79
Interior Plains Region																
N. Sask. at Edmonton, Alta.	8.2	65	18		6.2	1.1	224	62	0.8		0.1	10	51	234	429 (273)	77
Battle at Unwin, Sask.	8.4	64	28		120	7.2	459	143	23		2.0	7.5	0	275	973 (621)	50[b]
Milk at Milk R., Alta.	8.2	49	24	0.18	35	2.1	300	54	27		1.2	6.7	0	223	537 (322)	74
Notukeau at Gravelbourg, Sask.	8.1	85	66	0.04	305	13	398	817	16		0.2	4.5	157	483	2,089 (1,503)	28[b]
Mackenzie at Aklavik, N.W.T.	7.6	57	18		6.0	1.3	198	56	4.4		2.4	6.9	55	217	424 (251)	71
Laird at Ft. Simpson, N.W.T.	8.2	38	11		4.0	0.6	134	33	1.4		0.6	10	3.0	139	270 (164)	75
Assiniboine at Headingley, Man.	8.8	84	41		56	8.2	382	182	17		1.2	15	73	379	869 (587)	59
Canadian Shield Region																
Coppermine at mouth, N.W.T.	7.4	9.0	4.0	0.0	0.7	0.4	47	2.7	0.7	0.0	0.0	1.4	0.3	39	83 (42)	91
Great Whate at mouth, Que.	5.9	1.5	0.3	0.04	0.6	0.3	1.6	2.2	2.2	0.0	0.0	1.6	3.7	5.0	20 (9.5)	19[b]
Kosoak at Fort Chimo, Que.	7.1	2.9	1.0	0.02	1.0	0.1	13	2.8	0.9	0.0	0.2	2.0	0.9	11	27 (17)	69
Michipicoten at Wawa, Ont.	7.6	12	2.6		1.1	0.6	39	7.2	1.0		0.5	4.5	8.3	40	84 (48)	78
Great Lakes – St. Lawrence Lowlands Region																
Grand at mouth, Ont.	7.8	87	23		17	2.7	194	158	21	0.3	0.1	0.4	152	311	623 (404)	45[b]
L. Ont. at Port Hope, Ont.	8.0	38	7.9		10	1.9	115	31	19		0.8	2.3	33	127	299 (138)	62

[a]Sum of constituents in parts per million.
[b]Other than alkaline-earth bicarbonate waters.

1.1–1 CHEMICAL ANALYSES OF TYPICAL CANADIAN SURFACE WATERS (*Continued*)

River	Concentration, ppm												Hardness as $CaCO_3$, ppm		Specific conductance at 25°C, μmho	Alkaline-earth bicarbonates, percent
	pH	Ca	Mg	Dissolved Fe	Na	K	HCO_3	SO_4	Cl	F	NO_3	SiO_2	Noncarbonate	Total		
Great Lakes – St. Lawrence Lowlands Region (*continued*)																
Ottawa at mouth, Que.	7.0	8.8	2.1		4.9		29.5	11.5	<0.1		0.8	4.3	6.4	31	74 (47)	68
Rideau at mouth, Ont	7.7	39	11		4.0		159	15	<0.1		0.2	1.6	15	145	(151)	88
St. Lawrence at Levis, Que.	7.3	27	6.7		8.0	1.2	81	21	16		0.1	11	27	99	226 (122)	60
Richelieu at Richelieu, Que.	7.6	15	3.8		2.3	0.9	49	15	2.5		0.4	1.6	12	58	120 (65)	68
Canadian Appalachian Region																
Saint John at Hawkshaw, N.B.	7.4	17	1.6	0.05	1.8	0.5	47	8.7	3.2	0.0	2.8	4.2	10	48	106 (63)	71
Matapedia at Matapedia, Que.	7.8	23	3.3		1.6	0.5	84	4.4	0.6		0.4	5.3	2.6	72	155 (81)	92
Miramichi at Redbank, N.B.	7.2	4.8	0.2		1.3	0.3	16	3.1	0.6		0.6	6.9	0	13	37 (26)	73
Mersey at Milton, N.S.	7.0	1.0	0.9		2.4	0.3	3.7	1.6	4.3		trace	2.0	3.2	6.2	25 (14)	32[b]
Annapolis at Lawrencetown, N.S.	7.0	5.7	1.4	0.22	3.8	1.1	12	11	5.8	0.2	1.2	6.3	10	20	65 (33)	31[b]
Corner Brook at Corner Brook, Nfld.	7.5	9.1	1.4	0.03	2.4	0.4	31	3.3	4.0	0.0	1.2	1.9	3.5	29	75 (39)	71
Piper's Hole Brook at Swift Current, Nfld.	6.4	1.6	0.3	0.09	2.2	0.3	3.9	1.8	3.2	0.0	0.8	3.3	2.0	5.2	23 (16)	32[b]

[a] Sum of constituents in parts per million.
[b] Other than alkaline-earth bicarbonate waters.

Source: Reprinted, with publisher's permission, from *J. Am. Water Works Assn., 56*(9):1173, © 1964, by the American Water Works Association, Inc., 2 Park Ave., New York, N.Y. 10016.

1.1–2 MEAN COMPOSITION OF SURFACE WATERS OF NEW JERSEY

mg/l

	Spring flow				Summer flow			
			Southern				Southern	
Constituent	North	Central	Inner	Outer	North	Central	Inner	Outer
Ca	18.3	11.7	11.0	1.0	31.3	15.8	11.5	0.90
Mg	9.0	4.6	3.5	0.6	15.3	6.5	2.9	0.40
K	1.0	1.3	2.6	0.7	1.3	2.6	3.6	0.60
Na	ND	ND	ND	ND	6.1	9.0	9.6	3.0
Fe	0.07	0.26	0.54	0.51	0.12	0.72	2.0	0.60
Al	0.09	0.21	0.29	0.33	0.09	0.08	0.31	0.26
Si	3.5	8.2	11.2	3.6	3.7	9.3	12.1	4.5
Cl	20.0	13.0	16.0	8.5	12.8	15.5	15.4	6.7
Ni	0.005	0.005	0.009	0.002	0.008	0.016	0.017	0.001
Cu	0.013	0.020	0.018	0.018	0.024	0.030	0.022	0.024
Mn	0.020	0.052	0.055	0.013	0.022	0.056	0.034	0.011
Zn	0.010	0.019	0.014	0.014	0.010	0.085	0.032	0.013
Cr	0.002	0.003	0.003	0.001	0.011	0.013	0.013	0.025
Sr	0.040	0.026	0.027	0.004	0.129	0.065	0.021	0.002

[a] ND – Not determined.

Source: Reprinted, with publisher's permission, from *J. Am. Water Works Assn., 60*(4):455, © 1968, by the American Water Works Association, Inc., 2 Park Ave., New York, N.Y. 10016.

1.1–3 ANALYSES OF TYPICAL U.S. SURFACE WATERS

Surface waters[b]	1			2			3			4		
	Max[a]	Min	Avg	Max	Min	Avg	Max	Min	Avg	Max	Min	Avg
Date of collection	Jan. 1–10, 1962	May 1–10, 1962	Oct. 1961–Sept. 1962	Oct. 1–31, 1961	April 1–30, 1962	Oct. 1961–Sept. 1962	Dec. 1–9 1961	March 1–31, 1962	Oct. 1961–Sept. 1962	Dec. 13–28, 1961	March 29, –April 5, 1962	Oct. 1961–Sept. 1962
Silica (SiO_2)	8.7	2.1	4.4	9.0	7.6	9.5	6.4	5.6	6.2	16	15	15
Iron (Fe)	0.20	0.00	0.01	0.02	0.10	0.07	–	–	–	–	–	–
Manganese (Mn)	0.14	0.00	0.03	–	–	–	–	–	–	–	–	–
Calcium (Ca)	20	8.2	16	8.7	7.5	7.5	46	29	33	83	35	61
Magnesium (Mg)	8.0	2.6	6.0	2.5	1.8	2.4	14	7.1	8.3	24	14	17
Sodium (Na)	10	2.2	7.3	9.0	5.9	7.4	26	7.6	10	64	13	40
Potassium (K)	2.5	1.2	1.9	2.9	1.6	2.1	3.0	2.0	2.0	9.2	4.0	7.6
Bicarbonate (HCO_3)	66	18	51	41	31	36	86	73	80	301	146	195
Carbonate (CO_3)	0	0	–	0	0	–	0	0	–	0	0	–
Sulfate (SO_4)	30	17	26	7.8	6.4	6.9	103	42	49	178	45	124
Chloride (Cl)	9.0	3.2	7.5	7.6	4.1	5.4	36	10	14	21	8.2	14
Fluoride (F)	0.6	0.0	0.2	0.1	0.1	0.1	0.4	0.1	0.2	0.5	0.3	0.4
Nitrate (NO_3)	5.7	2.6	4.1	4.0	1.7	2.3	5.8	4.2	4.4	4.6	3.0	4.0
Dissolved solids	125	46	104	81	59	69	296	154	175	554	220	394
Hardness as $CaCO_3$	83	32	65	32	26	28	173	102	116	307	145	223
Noncarbonate hardness as $CaCO_3$	31	17	24	0	1	0	102	42	50	61	26	66
Specific conductance (μmho at 25°C)	198	86	168	113	84	96	475	249	288	867	346	607
pH	7.8	6.5	–	6.9	7.7	–	8.2	6.9	–	7.6	7.3	–
Color	8	2	4	15	32	25	20	5	11	–	–	–

[a] All concentrations in mg/l.
[b] Surface waters are identified as follows:
1. Delaware River at Trenton, N.J.
2. Roanoke River at Jamesville, N.C.
3. Ohio River at Lock and Dam 53, near Grand Chain, Ill.
4. Missouri River at Nebraska City, Nebr.
5. Mississippi River at St. Francisville, La.
6. Colorado River at Austin, Texas.
7. Colorado River near Grand Canyon, Ariz.
8. Columbia River near The Dalles, Ore.

1.1–3 ANALYSES OF TYPICAL U.S. SURFACE WATERS (*Continued*)

Surface waters[b]	5			6			7			8		
	Max[a]	Min	Avg	Max	Min	Avg	Max	Min	Avg	Max	Min	Avg
Date of collection	**Sept. 1–30, 1962**	**April 1–30, 1962**	**Oct. 1961–Sept. 1962**	**March 1–31, 1962**	**Nov 1–30, 1961**	**Oct. 1961–Sept. 1962**	**Sept. 27–30, 1962**	**May 8–23, 1962**	**Oct. 1961–Sept. 1962**	**March 9–28, 1962**	**June 1–30, 1962**	**Oct. 1961–Sept. 1962**
Silica (SiO_2)	8.7	7.4	9.5	11	9.5	11	19	18	18	18	6.5	13
Iron (Fe)	0.16	0.00	0.03	–	–	–	0.01	0.00	0.00	–	–	–
Manganese (Mn)												
Calcium (Ca)	46	32	40	48	39	43	230	55	77	22	18	20
Magnesium (Mg)	14	8.0	10	18	18	19	51	12	20	6.4	3.9	4.1
Sodium (Na)	26	11	18	37	34	34	212	28	68	10	3.6	6.2
Potassium (K)	3.1	2.1	2.7	3.9	3.8	3.7	4.3	3.2	4.1	1.8	0.8	1.7
Bicarbonate (HCO_3)	160	101	128	193	172	177	239	170	195	101	61	76
Carbonate (CO_3)	0	0	–	0	0	–	0	0	–	0	0	–
Sulfate (SO_4)	47	34	41	37	29	33	730	77	156	20	11	15
Chloride (Cl)	36	11	19	54	51	57	200	20	58	5.0	1.0	3.0
Fluoride (F)	0.2	0.0	0.2	0.3	0.2	0.3	0.5	0.3	0.4	0.3	0.1	0.2
Nitrate (NO_3)	1.8	1.7	2.2	1.8	0.5	1.6	8.9	2.3	3.6	1.5	0.4	1.0
Dissolved solids	279	157	218	330	268	293	1,620	314	531	140	69	99
Hardness as $CaCO_3$	171	113	140	194	172	186	785	185	271	88	54	67
Noncarbonate hardness as $CaCO_3$	40	30	35	36	30	40	585	46	111	5	4	4
Specific conductance (μmho at 25°C)	434	275	352	540	464	516	2,180	477	803	221	125	161
pH	7.6	7.2	–	7.5	7.2	7.3	7.9	7.5	7.7	7.8	7.7	–
Color	10	5	18	–	–	–	–	–	–	–	–	–

[a] All concentrations in mg/l.
[b] Surface waters are identified as follow:

1. Delaware River at Trenton, N.J.
2. Roanoke River at Jamesville, N.C.
3. Ohio River at Lock and Dam 53, near Grand Chain, Ill.
4. Missouri River at Nebraska City, Nebr.
5. Mississippi River at St. Francisville, La.
6. Colorado River at Austin, Texas.
7. Colorado River near Grand Canyon, Ariz.
8. Columbia River near The Dalles, Ore.

Source: L.E. Ciaccio, *Water and Water Pollution Handbook*, Marcel Dekker, New York, 1971.

1.1–4 ANALYSES OF TYPICAL SURFACE WATERS AND GROUNDWATERS IN THE U.S.

Analysis numbers[a]	1	2	3	4	5	6	7	8
Date of collection	8/22/61	9/1/61	10/19/61	1961	8/9/61	9/15/61	8/23/61	7/25/61
Silica (SiO_2)[b]	5.9	2.1	5.5	8.7	25	8.3	9.1	21
Iron (Fe)	0.00	0.17	0.40	–	0.03	0.72	0.10	0.01
Manganese (Mn)	0.00	0.14	0.00	–	–	0.01	0.06	0.00
Calcium (Ca)	8.5	28	37	84	65	12	57	26
Magnesium (Mg)	2.6	7.0	8.9	28	23	6.1	32	6.2
Sodium (Na)	3.6	4.1	17	92	14	7.5	3.4	86
Potassium (K)	1.5	0.9	5.1	4.0	1.8	0.7	1.7	1.9
Bicarbonate (HCO_3)	25	92	128	140	179	78	332	254
Carbonate (CO_3)	0	0	0	1	0	0	0	0
Sulfate (SO_4)	9.0	18	48	285	112	3.8	12	11
Chloride (Cl)	5.5	8.0	10	83	18	3.0	3.5	40
Fluoride (F)	0.1	0.0	0.4	0.4	0.8	0.4	0.1	0.7
Nitrate (NO_3)	4.7	0.5	3.2	1.4	0.2	1.2	0.2	0.0
Dissolved solids	59	129	222	657	410	87	284	318
Hardness as $CaCO_3$	32	99	129	323	256	55	274	90
Noncarbonate hardness as $CaCO_3$	11	24	24	206	110	0	2	0
Specific conductance (μmho at 25°C)	91	213	324	1,040	535	137	498	533
pH	6.4	7.6	7.7	8.4	7.9	6.8	7.6	7.2
Color	5	3	5	–	5	5	1	0

[a] All concentrations in mg/l.
1. Baltimore, Md., North Branch Patapsco River (raw).
2. Detroit, Mich., Detroit River (raw).
3. St. Louis, Mo., Mississippi River (raw).
4. Los Angeles, Calif., Colorado River (raw).
5. Jacksonville, Fla., Composite of several wells (raw).
6. Memphis, Tenn., Allen Well Field (raw)[b].
7. Rockford, Ill., Well No. 15 (raw)[c].
8. Houston, Texas, Southwest Well Field (raw)[d].

[b] Composite of several wells 400–600 ft deep.
[c] 1,355 ft deep.
[d] Composite of 12 wells 490–2,000 ft deep.

Source: C.N. Durfor and E. Becker, U.S. Geological Survey, Water Supply Paper 1812, 1964.

1.1–5 SUMMARY OF DISSOLVED TRACE ELEMENTS

1,577 Surface Waters, October 1, 1962–September 30, 1967

Element	No. positive occurrences	Frequency of detection, %	Observed values, μg/l		
			Min	Max	Mean
Zinc	1,207	76.5	2	1,183	64
Cadmium	40	2.5	1	120	9.5
Arsenic	87	5.5	5	336	64
Boron	1,546	98.0	1	5,000	101
Phosphorus	747	47.4	2	5,040	120
Iron	1,192	75.6	1	4,600	52
Molybdenum	516	32.7	2	1,500	68
Manganese	810	51.4	0.3	3,230	58
Aluminum[a]	456	31.2	1	2,760	74
Beryllium	85	5.4	0.01	1.22	0.19
Copper	1,173	74.4	1	280	15
Silver	104	6.6	0.1	38	2.6
Nickel	256	16.2	1	130	19
Cobalt	44	2.8	1	48	17
Lead	305	19.3	2	140	23
Chromium	386	24.5	1	112	9.7
Vanadium	54	3.4	2	300	40
Barium	1,568	99.4	2	340	43
Strontium	1,571	99.6	3	5,000	217

[a] 1,464 aluminum analyses.

Source: J.F. Kopp in *Trace Substances in Environmental Health,* Vol. 3, D.D. Hemphill, Ed., University of Missouri Press, Columbia, 1969.

1.1.1 RIVERS

1.1–6 MAJOR CONSTITUENTS OF RIVER WATERS USED FOR WATER SUPPLY

Table A

Location	Source of supply	Percent of supply	SiO_2, mg/l	Fe, mg/l	Ca, mg/l	Mg, mg/l	Na, mg/l	K, mg/l	HCO_3, mg/l	CO_3, mg/l	SO_4, mg/l	Cl, mg/l
Birmingham, Ala.	Cahaba River and Lake Purdy	90	5.7	0.03	26	4.9	10	1.7	87	0	31	2.0
Phoenix, Ariz.	Verde River	12	14	0.02	50	14	131	4.8	176	0	51	196
Los Angeles, Cal.	Colorado River[a]	20	8.7	–	84	28	92	4	140	1	285	83
Sacramento, Cal.	Sacramento River	85	21	0.01	10	6.6	10	0.8	69	0	7	8.0
Washington, D.C.	Potomac River	100	5.7	0.00	35	7.9	7.4	1.9	107	0	35	10
Tampa, Fla.	Hillsborough River	97	7.8	0.03	55	5.6	8.0	1.5	150	0	28	16
Atlanta, Ga.	Chattahoochee River	100	8.0	0.07	2.8	1.2	2.3	1.0	15	0	0.0	5.0
Savannah, Ga.	Abercorn Creek	100	10	0.27	4.0	0.9	3.5	1.0	18	0	5.2	4.5
Kansas City, Kans.	Missouri River	100	16	0.01	49	13	26	4.8	169	0	69	14
Topeka, Kans.	Kansas River	100	17	0.03	110	23	72	6.0	316	0	128	95
Louisville, Ky.	Ohio River	100	0.2	0.05	41	9.5	16	2.6	90	0	74	20
New Orleans, La.	Mississippi River	100	5.3	0.01	48	11	17	2.8	148	0	43	26
Detroit, Mich.	Detroit River	100	2.1	0.17	28	7.0	4.1	0.9	92	0	18	8.0
Minneapolis, Minn.	Mississippi River	100	11	0.02	48	16	7.3	1.5	222	0	16	1.6
Kansas City, Mo.	Missouri River	100	7.0	0.10	55	16	35	6.6	196	0	101	12
St. Louis, Mo.	Missouri River	34	6.2	0.20	36	8.8	17	5.1	127	0	47	10
	Mississippi River	66	5.5	0.40	37	8.9	17	5.1	128	0	48	10
Omaha, Neb.	Missouri River	100	16	0.02	74	23	65	6.4	238	0	210	13
Paterson, NJ	Passaic River[b]	47	13	0.47	19	5.8	–	–	63	–	30	15
Philadelphia, Pa.	Delaware River[c]			0.35–1.5					29–50	0–0	21–36	4.0–11
	Schuylkill River[c]			0.34–1.9					50–87	0–0	59–100	8.0–23
Chattanooga, Tenn.	Tennessee River	100	3.8	0.05	23	5.0	8.2	1.1	70	0	13	16

[a] Average analysis, July 1, 1960–June 30, 1961.
[b] Analyzed by the Passaic Valley Water Commission.
[c] Maximum value of constituents in monthly average of analyses by the city of Philadelphia Water Department during 1960.

1.1–6 MAJOR CONSTITUENTS OF RIVER WATERS USED FOR WATER SUPPLY (*Continued*)

Location	Source of supply	Percent of supply	SiO_2, mg/l	Fe, mg/l	Ca, mg/l	Mg, mg/l	Na, mg/l	K, mg/l	HCO_3, mg/l	CO_3, mg/l	SO_4, mg/l	Cl, mg/l
Nashville, Tenn.	Cumberland River	100	2.7	0.24	22	3.1	2.9	0.9	64	0	17	2.0
Austin, Texas	Colorado River	100	9.9[e]		40	20	39 (Na + K)		187	0	34	55
			10[f]		44	19	34	3.8	177	0	34	60
El Paso, Texas	Rio Grande River	14	16	0.46	92	20	165	7.8	236	0	290	127
Houston, Texas	San Jacinto River	24	6.6	0.09	9.5	1.5	11	2.0	29	0	2.4	21
Salt Lake City, Utah	Big Cottonwood Creek	33	5.6	0.00	38	12	4.1	1.0	122	0	41	7.0
Richmond, Va.	James River	100	9.5	0.01	12	2.5	3.1	1.0	41	0	7.6	4.0
Seattle, Wash.	Cedar River	100	9.5	0.05	7.0	0.3	1.9	0.3	26	0	2.4	0.5
Cincinnati, Ohio	Ohio River[d]	100	0.00–0.04	–	25–51	5.5–12	–	–	35–58	0	62–132	15–59
Range			0.00–21	0.00–1.9	2.8–110	0.9–28	1.9–131	0.3–7.8	15–316	0–1	0.0–290	0.5–196

[d] Maximum constituents in monthly analyses by the city of Cincinnati during 1961.
[e] Composite of daily samples, January 1–31, 1962.
[f] Composite of daily samples, June 1–30, 1962.

1.1–6 MAJOR CONSTITUENTS OF RIVER WATERS USED FOR WATER SUPPLY (*Continued*)

Table B

Location	Source of supply	F, mg/l	NO_3, mg/l	Dissolved solids residue at 180°C, mg/l	Hardness as $CaCO_3$, mg/l	Noncarbonate hardness as $CaCO_3$, mg/l	Specific conductance, μmho at 25°C	pH	Color, units	Beta activity, pCi/l
Birmingham, Ala.	Cahaba River and Lake Purdy	0.1	0.3	195.	85.	14.	220.	7.0	0	–
Phoenix, Ariz.	Verde River	0.4	2.7	560	184	40.	1000	7.8	3	–
Los Angeles, Cal.	Colorado River[a]	0.4	1.4	657	323	206	1040	8.4	–	
Sacramento, Cal.	Sacramento River	0.1	0.9	110	52	0	149	7.4	3	
Washington, D.C.	Potomac River	0.1	1.1	172	120	33	279	7.0	5	29[g]
Tampa, Fla.	Hillsborough River	0.2	1.0	214	160	37	339	8.1	25	
Atlanta, Ga.	Chattahoochee River	0.1	0.1	31	12	0	42	6.6	10	13[g]
Savannah, Ga.	Abercorn Creek	0.2	0.7	53	14	0	45	6.6	115	–
Kansas City, Kans.	Missouri River	0.2	8.3	312	174	35	464	7.2	9	70[g]
Topeka, Kans.	Kansas River	0.2	4.9	638	368	109	994	7.5	4	
Louisville, Ky.	Ohio River	0.3	3.2	221	141	67	370	7.3	.5	46[g]
New Orleans, La.	Mississippi River	0.2	0.2	254	163	42	414	6.8	10	38[g]
Detroit, Mich.	Detroit River	0.0	0.5	129	99	24	213	7.6	2	16[g]
Minneapolis, Minn.	Mississippi River	0.2	0.4	228	185	3	266	7.5	17	66[g]
Kansas City, Mo.	Missouri River	0.5	3.8	350	203	42	522	7.8	15	70[g]
St. Louis, Mo.	Missouri River	0.4	3.4	220	126	22	323	7.8	5	48[g]
	Mississippi River	0.4	3.2	222	129	24	324	7.7	5	78[g]
Omaha, Neb.	Missouri River	0.6	0.4	523	279	84	797	7.3	7	78[g]
Paterson, NJ.	Passaic River[b]	–	0.6	152	69	–	–	7.1	40	
Philadelphia, Pa.	Delaware River[c]	–	0.3–0.9	129–345	42–71	–	–	7.2–7.3	8–25	24[g]
	Schuylkill River[c]	–	1.3–1.9	180–320	100–156			7.4–7.6	8–40	30[g]

[a] Average analysis, July 1, 1960–June 30, 1961.
[b] Analyzed by the Passaic Valley Water Commission.
[c] Maximum value of constituents in monthly average of analyses by the city of Philadelphia Water Department during 1960.
[g] Maximum beta activity, raw water, July 1, 1961–June 30, 1962.

1.1–6 MAJOR CONSTITUENTS OF RIVER WATERS USED FOR WATER SUPPLY (*Continued*)

Location	Source of supply	F, mg/l	NO_3, mg/l	Dissolved solids residue at 180°C, mg/l	Hardness as $CaCO_3$, mg/l	Noncarbonate hardness as $CaCO_3$, mg/l	Specific conductance, μmho at 25°C	pH	Color, units	Beta activity, pCi/l
Chattanooga, Tenn.	Tennessee River	0.2	2.9	125	78	20	195	8.0	5	58[g]
Nashville, Tenn.	Cumberland River	0.2	2.8	88	68	15	142	7.6	5	
Austin, Texas	Colorado River	0.3	1.2	294	182	30	526	7.6		
		0.3	0.8	306	188	43	526	7.6		
El Paso, Texas	Rio Grande River	0.6	0.2	831	312	118	1270	7.8	10	34[g]
Houston, Texas	San Jacinto River	0.3	0.0	117	30	6	127	6.2	80	
Salt Lake City, Utah	Big Cottonwood Creek	0.2	0.2	170	144	44	291	7.8	5	
Richmond, Va.	James River	0.0	0.8	70	40	6	96	7.3	5	
Seattle, Wash.	Cedar River	0.0	0.2	40	19	0	47	7.4	5	–
Cincinnati, Ohio	Ohio River[d]	0.1–0.4	–	151–377	91–178	–	–	7.2–8.3	–	20[c]
Range		0.0–0.6	0.0–8.3	31–657	12–368	0–206	42–1270	6.2–8.4	0–115	16–78

[c] Maximum value of constituents in monthly average of analyses by the city of Philadelphia Water Department during 1960.
[d] Maximum constituents in monthly analyses by the city of Cincinnati during 1961.
[g] Maximum beta activity, raw water, July 1, 1961–June 30, 1962.

Source: C.N. Durfor and E. Becker, *Geological Survey Water Supply Paper,* U.S. Department of the Interior, 1962.

1.1–7 MINOR ELEMENTS IN LARGE RIVERS OF NORTH AMERICA

Element	Median, μg/l	Range, μg/l	Atlantic Coast	Gulf Coast	Pacific Coast
Ag	0.09	0–0.94	+	0	0
Al	238	12–2,550	0	+	0
B	10	1.4–58	0	0	0
Ba	45	9–152	–	+	–
Co	0	0–5.8			
Cr	5.8	0.72–84	+	–	–
Cu	5.3	0.83–105	0	+	0
Fe	300	31–1,670	0	+	0
Li	1.1	0.075–37	–	+	0
Mn	20	0–185	+	0	0
Mo	0.35	0–6.9	+	0	+
Ni	10	0–71	+	0	0
Pb	4.0	0–55	0	–	+
Rb	1.5	0–8.0	0	+	–
Sr	60	6.3–802	+	–	0
Ti	8.6	0–107	+	+	–
V	0	0–6.7			
Zn	0	0–215			

Note: + = Significant percentage of determinations above median.
0 = Approximately 50% determinations above and below median.
– = Significant percentage of determinations below median.

Source: L.E. Ciaccio, *Water and Water Pollution Handbook,* Marcel Dekker, New York, 1971.

1.1–8 MINOR CONSTITUENTS OF RIVER WATERS USED FOR WATER SUPPLY

Table A

Location	Source of supply	Ag, μg/l	Al, μg/l	B, μg/l	Ba, μg/l	Be, μg/l	Co, μg/l	Cr, μg/l	Cu, μg/l	Fe, μg/l	Li, μg/l	Mn, μg/l
Birmingham, Ala.	Cahaba River and Lake Purdy	ND[ab]	43	30	50	ND	ND	2.0	3.0	3.7	2.4	15
Phoenix, Ariz.	Verde River	<0.8	130	45	99	ND	ND	ND	2.7	26	91	27
Los Angeles, Cal.	Colorado River[a]											
Sacramento, Cal.	Sacramento River											
Washington, D.C.	Potomac River	<0.25	200	20	85	ND	ND	4.0	4.0	280	1.8	100
Tampa, Fla.	Hillsborough River											
Atlanta, Ga.	Chattahoochee River	<0.03[e]	130	68	11	ND	<0.3	0.59	84	120	0.08	27
Savannah, Ga.	Abercorn Creek	<0.05[e]	110	32	12	ND	ND	0.69	24	140	0.12	21
Kansas City, Kans.	Missouri River			60								
Topeka, Kans.	Kansas River			70								
Louisville, Ky.	Ohio River	<0.3	390	42	140	ND	ND	7.2	17	150	4.8	12
New Orleans, La.	Mississippi River	<0.38	710	74	170	ND	ND	1.6	15	430	4.3	47
Detroit, Mich.	Detroit River	0.21	410	33	100	ND	ND	1.5	43	250	0.79	18
Minneapolis, Minn.	Mississippi River	0.48	88	68	96	ND	ND	1.4	17	29	6.0	24
Kansas City, Mo.	Missouri River	<0.56	2,400[f]	110	340	ND	ND	7.8	6.1	3,800[g]	25	290
St. Louis, Mo.	Missouri River											
	Mississippi River											
Omaha, Neb.	Missouri River		0.	130								0.
Paterson, N.J.	Passaic River[b]											100
Philadelphia, Pa.	Delaware River[c]											0.–190
	Schuylkill River[c]											150–380
Chattanooga, Tenn.	Tennessee River	ND	290	22	67	ND	ND	5.2	4.9	250	1.7	100

[a] Average analysis, July 1, 1960–June 30, 1961.
[b] Analyzed by the Passaic Valley Water Commission.
[c] Maximum value of constituents in monthly average of analysis by the City of Philadelphia Water Department during 1960.
[d] Maximum constituents in monthly analyses by the City of Cincinnati during 1961.
[e] Spectrographic concentrations are based on nonacidified residue on evaporation.
[f] Water on April 11, 1962, had aluminum content of 170 mg/l.
[g] Water on April 11, 1962, had iron content of 8.0 mg/l.

1.1–8 MINOR CONSTITUENTS OF RIVER WATERS USED FOR WATER SUPPLY (*Continued*)

Location	Source of supply	Ag, μg/l	Al, μg/l	B, μg/l	Ba, μg/l	Be, μg/l	Co, μg/l	Cr, μg/l	Cu, μg/l	Fe, μg/l	Li, μg/l	Mn, μg/l
Nashville, Tenn.	Cumberland River	<0.16	310	21	33	ND	2.3	3.7	3.1	260	0.60	100
Austin, Texas	Colorado River											
El Paso, Texas	Rio Grande River											
Houston, Texas	San Jacinto River	<0.15	380	24	57	ND	ND	0.65	2.9	290	0.44	77
Salt Lake City, Utah	Big Cottonwood Creek			130								
Richmond, Va.	James River											10
Seattle, Wash.	Cedar River	0.05	7.0	7.0	3.1	ND	ND	ND	1.1	9.0	–	2.6
Cincinnati, Ohio	Ohio River[d]											
Range		ND–0.48	0.–2,400	7.0–130	3.1–340	ND	ND–2.3	ND–7.8	1.1–84	9.0–3,800	0.08–91	2.6–380

[d] Maximum constituents in monthly analysis by the City of Cincinnati during 1961.

1.1–8 MINOR CONSTITUENTS OF RIVER WATERS USED FOR WATER SUPPLY (*Continued*)

Table B

Location	Source of supply	Mo, μg/l	Ni, μg/l	P, μg/l	Pb, μg/l	Rb, μg/l	Sn, μg/l	Sr, μg/l	Ti, μg/l	V, μg/l	Date of sample collection
Birmingham, Ala.	Cahaba River and Lake Purdy	1.3	<1.9	ND	50	6.0	ND	60	<1.9	ND	8/29/61
Phoenix, Ariz.	Verde River	<2	<8	ND	ND	12	ND	320	<8	ND	7/27/61
Los Angeles, Cal.	Colorado River[a]										7/1/60-6/30/61
Sacramento, Cal.	Sacramento River										7/26/61
Washington, D.C.	Potomac River	1.8	7.8	ND	4.8	<2.5	ND	190	5.2	<7.5	8/25/61
Tampa, Fla.	Hillsborough River										1/12/62
Atlanta, Ga.	Chattahoochee River	ND	1.2	ND	8.7	2.5	ND	2.4	9.6	ND	8/31/61
Savannah, Ga.	Abercorn Creek	ND	0.9	ND	12	1.6	ND	2.0	3.5	ND	8/30/61
Kansas City, Kans.	Missouri River										3/15/62
Topeka, Kans.	Kansas River										3/9/62
Louisville, Ky.	Ohio River	7.8	<3.0	ND	11	5.7	ND	360	4.2	ND	9/3/61
New Orleans, La.	Mississippi River	3.9	9.4	ND	5.9	3.9	ND	200	30	<12	7/21/61
Detroit, Mich.	Detroit River	1.6	11	ND	15	2.1	ND	97	7.5	ND	9/1/61
Minneapolis, Minn.	Mississippi River	1.2	4.4	ND	6.8	16	ND	52	<4.0	ND	7/31/61
Kansas City, Mo.	Missouri River	3.1	15	<560	38	7.8	ND	610	15	<17	10/17/61
St. Louis, Mo.	Missouri River										10/18/61
	Mississippi River										10/19/61
Omaha, Neb.	Missouri River										1/23/62
Paterson, N.J.	Passaic River[b]										1961 avg values
Philadelphia, Pa.	Delaware River[c]										1960 avg values
	Schuylkill River[c]										1960 avg values
Chattanooga, Tenn.	Tennessee River	1.6	<2.2	ND	14	2.5	ND	110	6.5	ND	9/14/61

[a] Average analysis, July 1, 1960 – June 30, 1961.
[b] Analyzed by the Passaic Valley Water Commission.
[c] Maximum value of constituents in monthly average of analysis by the City of Philadelphia Water Department during 1960.

1.1–8 MINOR CONSTITUENTS OF RIVER WATERS USED FOR WATER SUPPLY (*Continued*)

Location	Source of supply	Mo, μg/l	Ni, μg/l	P, μg/l	Pb, μg/l	Rb, μg/l	Sn, μg/l	Sr, μg/l	Ti, μg/l	V, μg/l	Date of sample collection
Nashville, Tenn.	Cumberland River	1.8	1.6	ND	60	2.0	ND	80	4.9	5.1	9/13/61
Austin, Texas	Colorado River										1/1-31/62
											6/1-30/62
El Paso, Texas	Rio Grande River										4/24/62
Houston, Texas	San Jacinto River	ND	3.3	ND	8.2	6.6	ND	11	35	<4.5	7/24/61
Salt Lake City, Utah	Big Cottonwood Creek										7/12/61
Richmond, Va.	James River										1/16/62
Seattle, Wash.	Cedar River		ND	ND	1.1	–	ND	2.6	0.5	ND	10/18/61
Cincinnati, Ohio	Ohio River[d]										1961 avg values
Range			ND–15	ND	ND–60	1.6–16	ND	2.6–610	0.5–35	ND–5.1	

[d] Maximum constituents in monthly analysis by the City of Cincinnati during 1961.

Source: C.N. Durfor, and E. Becker, *Geological Survey Water Supply Paper,* U.S. Department of the Interior, 1962.

1.1–9 SURFACE WATER QUALITY OF RIVERS

Table A

Characteristic	Potomac River[1]	Allen's Fresh Creek[1]	Occoquan Creek[1]	Rappa-hannock River[1]	Des Plaines River, North[2]	Calumet River[2]	Des Plaines River, South[2]	Sanitary Canal[2]
pH	7.3	6.2	6.6	6.6	8.0	7.7	8.1	8.2
Turbidity, units	67	54	300	135	65	42	80	95
Color, units	5	30	50	45				
Conductivity, μmho/cm	180	136	–	60				
Total hardness, mg/l $CaCO_3$	72	24	28	23	508	154	484	134
Alkalinity, mg/l $CaCO_3$	50	4	16	12	330	135	304	130
Chloride, mg/l Cl	7	26	5	4				
Sulfate, mg/l SO_4	14	11	17	9				
Detergent, mg/l ABS	0.10	0.06	0.08	0.03				
Chemical oxygen demand, mg/l	5	13	–	11				
Zeta potential, mV	–24	–22	–23	–21				
Temperature, °C					12	21	16	12
Total solids, mg/l					756	228	798	266
Optical density, mμ					0.111	0.083	0.141	0.143

Table B

Characteristic	Vic Bien Hua Viet Nam Rice paddy[2]	Song Dong Nai River Viet Nam[2]	Meramec River[3]	Little Miami River[4]	Little Miami River[5]	Contra Costa County, Cal. Water District[6]		Little Miami River[7]
						Avg	Range	
pH	6.8	7.1	8.0	7.3–8.5	7.7–8.1	7.9	7.6–8.5	7.5–8.5
Turbidity, units,	130	25	15	5–250	5–1,100	47	15–100	4–27
Color, units								
Conductivity, μmho/cm								
Total hardness, mg/l $CaCO_3$	14	16	172	130–330	125–225	95	65–155	180–270
Alkalinity, mg/l $CaCO_3$	16	16	0–147	85–310	75–300	70	58–94	124–206
Chloride, mg/l Cl			5		10–16	60	25–125	
Sulfate, mg/l SO_4			19		17–50	35	17–68	
Detergent, mg/l ABS								
Chemical oxygen demand, mg/l				5–35	10–50			10–20
Zeta potential, mV								
Temperature, °C	–	–		2–27	3–20	16	8–23	
Total solids, mg/l	212	66						

1.1–9 SURFACE WATER QUALITY OF RIVERS (*Continued*)

Characteristic	Vic Bien Hua Viet Nam Rice paddy[2]	Song Dong Nai River Viet Nam[2]	Meramec River[3]	Little Miami River[4]	Little Miami River[5]	Contra Costa County, Cal. Water District[6]		Little Miami River[7]
						Avg	Range	
Optical density, mμ	0.620	0.086						
Dissolved solids, mg/l			185					
Calcium, mg/l			96			20	15–30	
Magnesium, mg/l			76			11	7–21	
Iron, mg/l			0			0.47	0.1–1.00	
Sodium and potassium, mg/l			2					
Silica, mg/l			7			11	2–21	
Threshold odor at 60°C				3–10				
CCE, μg/l (modified monitor)				185–1,320				
Organic chlorine, μg/l				<0.013–>9.0				
Organic sulfur, μg/l				<0.1–6.3				
Sodium, Na mg/l					10	45	15–100	
Ammonia, NH_3, mg/l					0.45–0.60			
NO_3, mg/l					1.28	1.3	0–5.6	
PO_4, mg/l					0.24–0.40	0.1	0–0.6	
Coliform count, no./ml					15–170			
Manganese, mg/l						0.07	0–0.25	
Potassium, mg/l						3.0	0.5–7.5	
Bicarbonate, mg/l, HCO_3						90	70–110	
CO_2, mg/l						3	0–9	
Fluoride, F^-, mg/l						0.15	0–0.35	
Iodide, I^-, mg/l						0.0222	0.005–0.044	
Total dissolved solids, mg/l						230	140–490	

REFERENCES

1. M. Pressman, *J. Am. Water Works Assn.*, *59*(2):169, 1967.
2. R.B. Grieves, *J. Am. Water Works Assn.*, *59*(7):859, 1967.
3. J.L. Tuepker and H.O. Hartung, *J. Am. Water Works Assn.*, *52*(1):106, 1960.
4. G.G. Rubeck, *et al.*, *J. Am. Water Works Assn.*, *57*(2):181, 1965.
5. G.G.Rubeck, *et al.*, *J. Am. Water Works Assn.*, *56*(2):198, 1964.
6. B.G. Stone, *J. Am. Water Works Assn.*, *64*(6):369, 1971.
7. J. DeMarco *et al.*, *J. Am. Water Works Assn.*, *59*(8):965, 1967.

1.1–10 VARIATIONS IN DISSOLVED SOLIDS, CHEMICAL TYPE, AND SEDIMENT

Rivers in Arid and Semiarid U.S.

Region	Dissolved solids concentrations, mg/l		Prevalent chemical type[a]	Sediment concentrations, mg/l[b]	
	From	To		From	To
Columbia River Basin	<100	300	Ca-Mg, C-b	<200	300
Northern California	<100	700	Ca-Mg, C-b	<200	+500
Southern California	<100	+2,000	Ca-Mg, C-b; Ca-Mg, S-C	<200	+15,000
Colorado River Basin	<100	+2,500	Ca-Mg, S-C; Ca-Mg, C-b	<200	+15,000
Rio Grande Basin	<100	+2,000	Ca-Mg, C-b; Ca-Mg, S-C	+300	+50,000
Pecos River Basin	100	+3,000	Ca-Mg, S-C	+300	+7,000
Western Gulf of Mexico Basins	100	+3,000	Ca-Mg, C-b; Ca-Mg, S-C; Na-P, S-C	<200	+30,000
Red River Basin	<100	+2,500	Ca-Mg, S-C; Na-P, S-C	+300	+25,000
Arkansas River Basin	100	+2,000	Ca-Mg, S-C; Ca-Mg, C-b; Na-P, S-C	+300	+30,000
Platte River	100	+1,500	Ca-Mg, C-b; Ca-Mg, S-C	+300	+7,000
Upper Missouri River Basin	100	+2,000	Ca-Mg, S-C; Na-P, C-b; Na-P, C-b	<200	+15,000

[a] Ca-Mg, C-b = Calcium-magnesium, carbonate-bicarbonate.
Ca-Mg, S-C = Calcium-magnesium, sulfate-chloride.
Na-P, C-b = Sodium-potassium, carbonate-bicarbonate.
Na-P, S-C = Sodium-potassium, sulfate-chloride.

[b] $\text{Sediment concentration} = \dfrac{\text{Annual load}}{\text{Annual streamflow}}$

Source: F.H. Rainwater, *Stream Composition of the Conterminous United States,* U.S. Geological Survey Atlas HA-61, 1962.

1.1–11 HISTORIC WATER QUALITY DATA FROM FOUR WESTERN RIVERS

Date	TDS, ppm	Location
	Colorado River	
1893	833	Yuma, Arizona
	North Platte River	
Sept. 1906	262	North Platte, Nebraska
Oct. 1906	361	Columbus, Nebraska
Oct. 1906	265	Premont, Nebraska
	Arkansas River	
Sept. 1907	1,079	Great Bend, Kansas
Sept. 1907	1,105	Arkansas City, Kansas
Oct. 1907	1,418	Deerfield, Kansas
	Rio Grande River	
1893–1894	399	Mesilla, New Mexico
Sept. 1905	746	Laredo, Texas

Source: F.W. Clark, *The Composition of Rivers and Waters of the United States,* USGS Professional Paper 135, 1924.

1.1–12 INCREASE IN POLLUTANT CONCENTRATIONS

Rhine River

Concentration, mg/l	1965	1972
Chloride ions	130	280
Ammonia (as N_2)	1	3
Total N_2 (Kjeldahl)[a]	na	5
Orthophosphate (as PO_4)	0.6	1.2
Total phosphate	na[c]	2–4
Detergents	0.2	0.4
Phenol[b]	25	40

[a] Organic and inorganic nitrogen except nitrates.
[b] $\mu g/l$.
[c] na – not available.

Note: Samples taken at Lobith, the Netherlands.

Source: Institute for Purification of Waste Water and Industrial Water Treatment, the Netherlands.

1.1–13 ANALYSES OF MISSOURI RIVER WATER

1951–1966

Quantity measured	Omaha, Nebr. Frequency of measurement	Max	Min	Avg	Kansas City, Mo. Frequency of measurement	Max	Min	Avg
Dissolved oxygen, mg/l	Daily	13.9	4.5	9.7	Daily	22.5	0.0	9.4
MBAS, mg/l	Monthly composite	0.1	0.01	0.04	–	–	–	0.1
Orthophosphate, mg/l	Weekly	55[a]	0.02	0.08	–	–	–	–
Phenol, mg/l	Daily during spring runoff	10	0.00	–	–	–	–	–
Threshold odor no.	Daily	50	2	4	Daily	70	2	7
Color, mg/l	Daily	40	4	7	Daily	50	0	9
Suspended solids, mg/l		–	–	–	Daily	13,023	1	983
Alkalinity, mg/l	Daily	254	58	170	Daily	315	74	162
pH	Daily	8.5	7.7	8.2	Daily	8.8	7.7	8.3
Ca, mg/l	Daily	100	29	65	Daily	103.3	24.6	60.3
Mg, mg/l	Daily	29	9	19	Daily	26.8	4.2	17.4
SO_4, mg/l	Weekly	253	46	215	3 x/wk	286	36	152
Cl, mg/l	Weekly	21	6	14	Daily	49	7	21
Dissolved solids, mg/l	Weekly	650	330	500	Daily	2,052	102	443
F, mg/l	Weekly	0.7	0.4	0.5	Daily	0.65	0.03	0.42
NH_3, mg/l	Weekly	13[a]	0.05	0.1	3 x/wk	6.3[b]	0	0.38
NO_3, mg/l	Weekly	8.6	0.2	0.5	Monthly composite	–	–	1.8
Na and K, mg/l	Monthly composite	72	30	60	Monthly composite	–	–	56
Fe, mg/l	–	–	–	–	Daily	1	0.00	0.14
SiO_2, mg/l	2 x/month	16	4.2	8.7	Daily	24	1.5	11.3
Organic – CCE, mg/l	USPHS basic data, 2nd wk each month	268	19	80	–	–	–	–

[a] Fertilizer spill.
[b] 4 hr duration for maximum.

1.1–13 ANALYSES OF MISSOURI RIVER WATER (*Continued*)

1951–1966

	St. Louis County				St. Louis City			
Quantity measured	**Frequency of measurement**	**Max**	**Min**	**Avg**	**Frequency of measurement**	**Max**	**Min**	**Avg**
Dissolved oxygen, mg/l	Weekly	13.8	4.5	9.2	Daily	14.8	3.4	9.1
MBAS, mg/l	Monthly composite	0.14	0.02	0.1	–	–	–	–
Orthophosphate, mg/l	Monthly composite	0.36	0.07	0.24	–	–	–	–
Phenol, mg/l	–	–	–	–	–	–	–	–
Threshold odor no.	–	–	–	–	Daily	16	0	3
Color, mg/l	Daily	40	2	16	Daily	48	4	16
Suspended solids, mg/l	Daily	10,028	1	809	Monthly composite		64	807
Alkalinity, mg/l	Daily	301	62	149	Daily	294	53	148
pH	Daily	8.5	7.2	8.1	Daily	8.9	7.6	8.1
Ca, mg/l	Daily	98.4	22.4	54.9	Daily	101.2	16.0	55.6
Mg, mg/l	Daily	33.6	3.9	16.2	Daily	32.8	4.5	16.2
SO_4, mg/l	Monthly composite	206	35	113	Monthly composite	203	45	118
Cl, mg/l	Monthly composite	40	5	18	Monthly composite	48	8.6	22.6
Dissolved solids, mg/l	Daily	633	111	376	Monthly composite	577	118	376
F, mg/l	Daily since 1961	0.7	0.1	0.40	Monthly composite	0.9	0.2	0.45
NH_3, mg/l	Daily	1.3	0.0	0.12	–	–	–	–
NO_3, mg/l	Monthly composite	9	1	3.4	Monthly composite	11.3	2.1	5.0
Na and K, mg/l	Monthly composite	77	11	43	Monthly composite	86.4	16.8	47.6
Fe, mg/l	Monthly composite	0.5	0.00	0.01	–	–	–	–
SiO_2, mg/l	Monthly composite	20	5	11.7	Monthly composite	13.6	7.5	11.9
Organic – CCE, mg/l								

Source: H.O. Hartung, *J. Water Pollut. Control Fed., 39*(12):1996, 1967.

1.1–14 DOWNSTREAM CHANGES IN A RUNNING STREAM

Weekly Examinations

	Point A	Point B, ½ mi below A	Point C, ¾ mi below A	Point D, 1¼ mi below A	Point E, 2 mi below A
Total solids	340	335	315	315	310
Suspended solids	16	12	8	5	5
Chlorides	26.8	25.8	25	24.8	24.6
Free ammoniacal N	0.18	0.06	0.05	0.04	0.04
Albuminoid N	0.35	0.16	0.17	0.15	0.15
Nitrite N	0.008	0.008	0.006	0.006	0.005
Nitrate N	1.1	1.3	1.3	1.38	1.52
Four hours oxygen absorption	2.15	1.1	1.24	1.2	1.12
Five days BOD	2.80	1.72	1.74	1.30	1.12
Percentage oxygen saturation	101	108	100	103	105
Impurity figure	2.66	1.38	1.46	1.26	1.18

Source: G.V. James, *Water Treatment,* 4th ed., CRC Press, Cleveland, Ohio, 1971.

1.1–15 LABORATORY ANALYSIS OF RIVER WATERS

	Platte River near Venice		Platte River at Ashland		Elkhorn River at O Street	
Quality parameters[a]	**mg/l**	**me/l**	**mg/l**	**me/l**	**mg/l**	**me/l**
Total hardness	163.5	3.27	189.5	3.79	223	4.46
Calcium	46.8	2.34	54	2.7	55.6	2.78
Magnesium	11.3	0.93	13.2	1.09	20.4	1.68
Total alkalinity	157.1	3.14	168.5	3.37	208.5	4.17
Sulfate	67.2	1.4	63.5	1.32	50.2	1.05
Chloride	6.8	0.19	6.8	0.19	10.1	0.28
Sodium	31.6	1.37	33.1	1.44	24.3	1.06
Iron	0.42		0.41		0.35	
Total solids	1,020		1,090		816	
Total dissolved solids	355		350		415	
Orthophosphate	1.07		1.17		1.77	
Organic N	1.7		2.01		2.34	
Specific conductance (6 mho/cm)	409		439		515	
Sodium absorption (ratio)	1.07		1.05		0.711	
Algae count/ml	910		1,610		3,772	
Coliforms/100 ml[b]	2,160		6,610		93,500	

[a] Values in the table are the averages over a 12 month period. For algae counts and phosphate, and organic N values, the averages are for the 2 month period, April–May 1967.
[b] Values for coliforms are geometric means.

Source: Reprinted, with publisher's permission, from *J. Am. Water Works Assn., 61*(4):175, © 1969, by the American Water Works Association, Inc., 2 Park Ave., New York, N.Y. 10016.

1.1–16 RIVERS OF THE WORLD RANKED BY SEDIMENT YIELD

River	Drainage basin, 10^3 km²	Average annual suspended load: Metric tons x 10^6	Average annual suspended load: Metric tons/km²	Average discharge at mouth, 10^3 ft³/sec
Yellow	673	1,887	2,804	53
Ganges	956	1,451	1,518	415
Brahmaputra	666	726	1,090	430
Yangtze	1,942	499	257	770
Indus	969	435	449	196
Ching	57	408	7,158	2
Amazon	5,776	363	63	6,400
Mississippi	3,222	312	97	630
Irrawaddy	430	299	695	479
Missouri	1,370	218	159	69
Lo	26	190	7,308	–
Kosi	62	172	2,774	64
Mekong	795	170	214	390
Colorado	637	135	212	5.5
Red	119	130	1,092	138
Nile	2,978	111	37	100

Source: J. N. Holeman, *Water Resources Res., 4*(4):738, 1968. © by American Geophysical Union.

1.1–17 TRACE ELEMENT COMPOSITION OF SUSPENDED MATERIAL IN RIVERS

River and State	Suspended load, mg/l	ppm: Cr	ppm: Ag	ppm: Mo	ppm: Ni	ppm: Co	ppm: Mn
Brazos, Texas	954	100	0.4	11	30	20	690
Colorado, Texas	150	82	0.6	10	40	17	780
Red, La.	436	37	0.3	5	6	7	320
Mississippi, Ark.	185	150	0.7	18	100	33	2,300
Tombigbee, Ala.	25	220	1.0	22	200	31	5,900
Alabama, Ala.	54	150	4.0	19	100	34	3,700
Chattahoochie, Ga.	71	190	7.0	20	100	35	2,400
Flint, Ga.	12	210	1.0	28	100	39	5,100
Savannah, S.C.	30	460	2.0	35	250	36	4,400
Wateree, S.C.	37	200	1.5	24	100	34	7,000
Pee Dee, S.C.	188	150	0.4	15	100	23	1,300
Cape Fear, N.C.	61	130	0.7	16	70	21	1,700
Neuse, N.C.	36	380	4.0	22	70	30	3,000
Roanoke, N.C.	33	240	4.9	21	100	45	7,900
James, Va.	41	290	7.0	29	300	60	15,000
Rappahannock, Va.	28	140	1.0	31	80	46	2,200
Potomac, Va.	34	170	1.5	23	400	94	7,700
Susquehanna, Pa.	54	290	15.0	32	>1,000	>500	12,000
Rhone, France Avignon, June 1966	296	150	0.7	14	60	29	820
Rio Maipo, Chile Puente Alto, S. of Santiago, September 1966	41	68	1.0	44	40	76	2,400

Source: K.K. Turekian and M.R. Scott, *Environ. Sci. Technol., 1*(11):940, 1967.

1.1–18 TRACE ELEMENT TRANSPORT BY THE SUSQUEHANNA RIVER

	ton/yr transported[a]
Chromium	870
Silver	45
Molybdenum	97
Nickel	3,000
Cobalt	1,500
Manganese	120,000

[a] Assuming 300 x 10^3 ton/yr of suspended sediment – measured for the Juniata River.

Source: K.K. Turekian and M.R. Scott, *Environ. Sci. Technol.*, *1*(11):940, 1967.

1.1–19 GROSS DISSOLVED SOLIDS LOAD

River or area	Drainage area, mi^2	Gross dissolved-solids load, ton/mi^2 yr
Rhine	56,000	695[1]
Congo	1,550,000	39[1]
Mississippi	1,244,000	99[1]
North America	8,245,000	74[1]
Conterminous United States	2,754,000	82[1]

REFERENCES

1. W.H. Durum *et al.*, Int. Assn. Hydrol. Surf. Waters Comm. Publ. 51, General Assembly of Helsinki, Finland, 1960, p. 618.
2. R.B. Dole and H. Stabler, U.S. Geological Survey Water Supply Paper No. 234, 1909, p. 78.

1.1–20 CONCENTRATIONS OF INSECTICIDE FOUND IN MAJOR RIVER BASINS

	Concentration of organo-chlorine insecticide, ng/l							
	Dieldrin		DDT		TDE		γ BHC	
	10	100	10	100	10	100	10	100
U.S.	80	99	82	99	95	100	100	100
Yorkshire and Lancashire	62	87	73	95	71	87	41	96

Source: G.F. Lowden, C.L. Saunders, and R.W. Edwards, *Proc. Soc. Water Treat. Exam.* (in Press).

1.1–21 ORGANICS IN THREE RIVERS

Location	Organics,[a] ppb 1958	1959	1960
Mississippi River			
Minneapolis, Minn.	256	209	195
Dubuque, Iowa	268	195	168
Burlington, Iowa	141	125	127
E. St. Louis, Ill.	173	212	206
Cape Girardeau, Mo.	216	181	170
West Memphis, Ark.	85	131	126
New Orleans, La.	74	86	82
Average	177	163	153
Missouri River			
Williston, N.D.	148	59	80
Bismarck, N.D.	134	156	159
Yankton, S.D.	115	139	147
Omaha, Neb.	123	107	177
St. Joseph, Mo.	156	100	139
Kansas City, Kan.	134	89	89
St. Louis, Mo.	96	117	102
Average	129	110	103
Ohio River			
East Liverpool, Ohio	102	122	241
Huntington, W.V.	160	239	225
Cincinnati, Ohio	404	205	355
Evansville, Ind.	77	76	114
Cairo, Ill.	127	95	182
Average	174	147	223

[a] As alcohol extractable.

Source: Reprinted, with publisher's permission, from *J. Am. Water Works Assn., 55*(3):369, © 1963 by the American Water Works Association, Inc., 2 Park Ave., New York, N.Y. 10016.

1.1–22 SOLUBILITY PARTITION OF ORGANIC REFRACTORIES FROM MISSOURI RIVER WATER

Solvent sequence	Percentage of total extract: Ether insolubles	Water solubles	Amines	Amphoterics	Strong acids	Weak acids	Neutrals	Percentage of neutral fraction: Aliphatic	Aromatic	Oxygenated
Monitor Adsorption Unit Extracts										
Chloroform	7.9	18.9	2.0	6.7	12.7	10.5	37.5	3.8	6.7	86.3
Ethanol	99.0	none	none	none	none	none	none			
Benzene	51.8	29.0	none	none	1.3	7.2	10.0	2.3	69.2	23.8
Basic Adsorption Unit Extracts										
Chloroform	9.5	19.5	1.0	2.8	11.5	12.8	34.3	3.2	7.2	87.8
Ethanol	99.0	none	none	none	none	none	none	–	–	–
Acetone-1[a]	44.2	12.8	none	4.4	12.4	1.5	14.8	3.6	37.8	52.3
Benzene-2[a]	17.6	12.8	none	12.0	none	13.0	34.6	1.8	83.5	9.6
Benzene-1	41.8	16.7	none	3.4	12.8	7.0	10.4	2.6	66.8	27.5
Acetone-2	92.8	0.5	none	1.0	1.1	none	none	–	–	–
Acidic Adsorption Unit Extracts										
Chloroform	26.5	23.2	2.5	7.6	13.2	7.6	23.4	4.6	56.3	33.4
Ethanol	100.0	none	none	none	none	none	none			
Acetone-1	68.0	12.5	none	6.4	5.0	3.3	1.3			
Benzene-2	93.3	none	none	none	none	2.1	3.4			
Benzene-1	65.8	17.5	none	2.6	5.1	2.2	1.9			
Acetone-2	62.7	9.7	none	8.5	3.6	5.8	2.2			
Particulate Matter Extracts – Settleable Materials										
Chloroform	13.5	1.8	none	none	none	15.6	65.6	none	1.5	96.7
Ethanol	27.6	4.5	none	none	10.0	36.3	10.6			
Particulate Matter Extracts – Backwash Materials										
Chloroform	11.5	2.3	none	none	none	20.3	65.0	none	1.3	95.8
Ethanol	25.2	4.8	none	none	9.8	37.5	13.2			

[a] Number 1 or 2 following acetone or benzene refers to the extraction sequence of these solvents following chloroform and ethanol.

Source: Reprinted, with publisher's permission, from *J. Am. Water Works Assn., 55*(6):783, © 1963 by the American Water Works Association, Inc., 2 Park Ave., New York, N.Y. 10016.

1.1–23 FRACTIONATION OF BLACK EARTH CREEK (WISCONSIN) SAMPLE

Gel Permeation Chromatography

Peak	Grade	Molecular weight range	% COD	Molecular size range, radius, cm x 10^{-8}
Ia_1	G-200	>200,000	6	100
Ia_2	G-200	100,000–200,000	4	10–50
Ib	G-100	50,000–100,000	6	5–30
II	G-75	20,000–30,000	11	10–25
IIIa	G-25	3,000–5,000	12	20
IIIb	G-25	–	2	5–15
$IIIc_1$	G-25	100–5,000	13	4
$IIId_1$	G-10		2	
$IIId_2$	G-10		4	0–4
$IIId_3$	G-10	<700	6	
Precipitate not soluble in H_2SO_4			9	
Precipitate soluble in H_2SO_4			9	
Loss on G-75			10	
Distillate			2	
Total			96	

Source: E. Gjessing and G.F. Lee, *Environ. Sci. Technol., 1*(8):631, 1967.

1.1.2 LAKES AND IMPOUNDMENTS

1.1–24 MAJOR CONSTITUENTS OF LAKE WATERS USED FOR WATER SUPPLY

Location	Source of supply	Percent of supply	SiO_2, mg/l	Fe, mg/l	Ca, mg/l	Mg, mg/l	Na, mg/l	K, mg/l	HCO_3, mg/l	CO_3, mg/l	SO_4, mg/l	Cl, mg/l	F, mg/l
Birmingham, Ala.	Inland Lake	10	3.7	0.16	2.0	0.5	2.1	0.9	9.	0	0.8	2.9	0.0
Chicago, Ill.	Lake Michigan	100	1.1	0.03	33	11	3.9	0.7	132	0	20	6.5	0.1
Shreveport, La.	Cross Lake	100	3.7	0.02	10	4.2	24.	2.0	33	0	12	40	0.1
Grand Rapids, Mich.	Lake Michigan	100	1.8	0.01	34	11	4.1	0.9	136	0	21	7.0	0.0
St. Paul, Minn.	Series of lakes	100	2.5	0.02	44	10	6.9	2.8	180	0	14	4.0	0.2
Buffalo, N.Y.	Lake Erie	100	2.0	0.01	38	8.6	9.5	1.4	116	0	23	23	0.1
Rochester, N.Y.	Lake Ontario	21	3.0	0.07[a]	36	9.7	–	–	93	–	17	27	0.3
	Hemlock Lake	79	5.0	0.00	21	5.0	5.2	1.2	62	0	23	7.1	0.0
	Canadice Lake	–	2.9	0.00	12	2.9	3.0	1.0	31	0	19	2.6	0.1
Syracuse, N.Y.	Skaneateles Lake	100	1.7	0.00	34	5.8	1.6	0.9	110	0	17	2.6	0.0
Ft. Worth, Texas	Lake Worth	99	5.8	0.05	44	8.4	20	4.5	153	0	20	32	0.3
Milwaukee, Wisc.	Lake Michigan	100	2.1	0.04	35	10	4.1	1.0	134	0	19	6.5	0.1
Range			1.1–5.8	0.00–0.16	2.0–44	0.5–11	1.6–24	0.7–4.5	9–180	0	0.8–23	2.6–32	0.0–0.3

Location	Source of supply	NO_3, mg/l	Dissolved solids (residue at 180°C) mg/l	Hardness as $CaCO_3$, mg/l	Noncarbonate hardness as $CaCO_3$ mg/l	Specific conductance μmho at 25°C	pH	Color, units	Beta activity pCi/l
Birmingham, Ala.	Inland Lake	0.4	29	7	0	33	6.8	10	
Chicago, Ill.	Lake Michigan	0.4	153	128	20	259	7.8	1	–
Shreveport, La.	Cross Lake	0.3	142	42	15	231	6.6	10	–
Grand Rapids, Mich.	Lake Michigan	0.5	153	130	18	265	7.8	4	–
St. Paul, Minn.	Series of lakes	0.2	199	153	5	312	7.5	15	66[b]
Buffalo, N.Y.	Lake Erie	0.2	177	131	36	306	8.0	1	22[b]
Rochester, N.Y.	Lake Ontario	0.1	233	130	54	–	8.3	–	
	Hemlock Lake	0.7	107	73	22	178	6.9	2	
	Canadice Lake	0.6	62	42	17	104	6.8	2	
Syracuse, N.Y.	Skaneateles Lake	2.2	132	109	19	227	7.1	1	
Ft. Worth, Texas	Lake Worth	0.0	228	144	19	387	7.6		
Milwaukee, Wisc.	Lake Michigan	0.4	159	129	18	258	8.0	1	12[b]
Range		0.0–2.2	29–233	7–153	0–54	33–387	6.6–8.3	1–15	12–66

[a] In solution when collected.

[b] Maximum beta activity, raw water, July 1, 1961 – June 30, 1962.

Source C.N. Durfor and E. Becker, *Geological Survey Water Supply Paper*, U.S. Department of the Interior, 1962.

1.1–25 MINOR CONSTITUENTS OF SURFACE WATERS USED FOR WATER SUPPLY

Lakes

Location	Source of supply	Ag, µg/l	Al, µg/l	B, µg/l	Ba, µg/l	Be, µg/l	Co, µg/l	Cr, µg/l	Cu, µg/l	Fe, µg/l	Li, µg/l	Mn, µg/l	Mo, µg/l
Birmingham, Ala.	Inland Lake												
Chicago, Ill.	Lake Michigan	<0.2	72	7.8	80	ND[a]	ND	1.6	24	14	0.75	<2	<0.8
Shreveport, La.	Cross Lake	<0.18	46	80	78	ND	ND	0.34	6.0	100	1.3	100	ND
Grand Rapids, Mich.	Lake Michigan	<0.25	77	15	74	ND	ND	0.52	52	59	0.71	<2.5	<1.7
St. Paul, Minn.	Series of lakes	<0.34	55	51	120	ND	ND	1.3	44	25	1.7	41	1.5
Buffalo, N.Y.	Lake Erie	<0.26	41	10	31	ND	ND	2.8	22	28	0.92	3.1	0.77
Rochester, N.Y.	Lake Ontario		90									10	
	Hemlock Lake		–									0	
	Canadice Lake		–									0	
Syracuse, N.Y.	Skaneateles Lake											0	
Ft. Worth, Texas	Lake Worth												
Milwaukee, Wisc.	Lake Michigan	<0.24	38	29	36	ND	ND	3.6	57	26	0.43	2.4	1.1
Range		<0.18–<0.34	38–90	7.8–80	31–120	ND	ND	0.34–3.6	6.0–57	14–100	0.43–1.7	0–100	ND–1.5

Location	Source of supply	Ni, µg/l	P, µg/l	Pb, µg/l	Rb, µg/l	Sn, µg/l	Sr, µg/l	Ti, µg/l	V, µg/l	Zn, µg/l	Ga, µg/l	Zr, µg/l	Yb, µg/l	Y, µg/l	Date of sample collection
Birmingham, Ala.	Inland Lake														8/29/61
Chicago, Ill.	Lake Michigan	3.2	ND	14	ND	ND	80	<2	ND	ND					8/25/61
Shreveport, La.	Cross Lake	21	ND	11	3.5	ND	85	2.3	ND	ND			ND	ND	7/27/61
Grand Rapids, Mich.	Lake Michigan	6.2	ND	7.1	ND	ND	84	<2.5	ND	ND					8/30/61
St. Paul, Minn.	Series of lakes	4.1	ND	8.9	<3.4	ND	82	<3.4	ND	<340					7/31/61
Buffalo, N.Y.	Lake Erie	5.1	ND	6.4	ND	ND	110	<2.6	ND	<260					8/22/61
Rochester, N.Y.	Lake Ontario														6/19/61
	Hemlock Lake														6/ 9/61
	Canadice Lake														6/ 9/61
Syracuse, N.Y.	Skaneateles Lake														6/ 7/61
Ft. Worth, Texas	Lake Worth														2/27/62
Milwaukee, Wisc.	Lake Michigan	4.0	ND	26	ND	ND	81	<2.4	ND	ND					8/ 2/61
Range		3.2–21	ND	6.4–26	ND–3.5	ND	80–110	<2–2.3	ND	ND–<340			ND	ND	

[a] ND–looked for but not found.

Source: C.N. Durfor and E. Becker, *Geological Survey Water Supply Paper*, U.S. Department of the Interior, 1962.

1.1–26 SURFACE WATER QUALITY OF PONDS AND LAKES

Characteristics	Miller's Pond[1]	Skokie Lagoons[2]	Lake Austin,[3] Texas (2 samples)	Lake Huron, Saginaw-Midland water supply[4]
pH	6.4	8.5	8.2–8.05	7.7
Turbidity, units	39	63		
Color, units	25			
Total hardness, mg/l $CaCO_3$	14	340	202.2–210	95
Conductivity, μmho/cm	47			
Alkalinity, mg/l $CaCO_3$	6	262	150.2–146.7	0
Chloride, mg/l Cl	4			6
Sulfate, mg/l SO_4	6			10
Detergent, mg/l ABS	0.05		0.12–0.06	
COD, mg/l	11			
Zeta potential, mv	-25			
Total solids, mg/l		600		116
Optical density, mμ		0.109		
Temperature, °C		13		
Total phosphates, mg/l PO_4			1.20–0.11	
Orthophosphates, mg/l PO_4			0.43–0.08	
Silica, mg/l				3
Iron, mg/l				0
Manganese, mg/l				0
Calcium, mg/l				27
Magnesium, mg/l				6.6
Sodium, mg/l				2.0
Potassium, mg/l				0.4
Nitrate, mg/l				0
Bicarbonate alkalinity, mg/l				95
Fluoride, mg/l				0

REFERENCES

1. M. Pressman, *J. Am. Water Works Assn.*, *59*(2):169, 1967.
2. R.B. Grieves, *J. Am. Water Works Assn.*, *59*(7):859, 1967.
3. J.F. Malina, Jr. and S. Tiyaporn, *J. Am. Water Works Assn.*, *56*(6):727, 1964.
4. T.L. Van der Velde and C.C. Crumley, *J. Am. Water Works Assn.*, *54*(12):149, 1962.

1.1–27 SUMMARY OF WATER SAMPLES

Detroit, Michigan

Test	Maximum	Minimum	Average	Number of tests run
Turbidity	10[b]	0.3	1.6	164
Plankton[a]	4,200	64	1,100	164
Chemical oxygen demand	15.3	3.4	8.7	153
Taste	0	0	0	162
Total alkalinity	83	75	80	164
Total hardness	101	90	95	164
Chloride	8	4	6	163
Sulfate	17	12	14	160
pH	8.6	8.0	8.3	164
Coliform index	240	0	6.4	162
Temperature	72	32	56	155
1 hr chlorine demand	0.8	0.2	0.6	13
24 hr chlorine demand	1.23	1.23	1.23	1
Phosphate	0.00	0.00	0.00	48
Organic nitrogen	0.31	0.10	0.20	14

[a] Dominant organisms in order of frequency: *Cyclotella, Rhizosolenia, Synedra Puchella, Tabellaria, Asterionella*.

[b] One sample taken in high winds; next highest sample was Jackson candle turbidity units.

Note: Samples were taken at the site of the Lake Huron raw water intake between May 1964 and December 1966.

Source: Reprinted, with publisher's permission, from *J. Am. Water Works Assn.*, *64*(5):299, © 1972, by the American Water Works Association, Inc., 2 Park Ave., New York, N.Y. 10016.

1.1–28 ANALYSIS OF LAKE WATER

Item	Grand Lake, Ohio		
	Max	Min	Avg
Air temp, max °F	95	5	61.2
Air temp, min °F	69	-20	38.6
Raw water temp, °F	88	34	58
Precipitation, in.	1.94	0	
Raw water pH	9.3	7.2	8.4
ABS (detergents), mg/l	0.5	0	0.14
BOD_5, mg/l	17	4	8.5
Phosphate, mg/l	1	0	–
Raw water threshold odor	140	17	52.8
Tap water threshold odor	24	2	5.5
Raw water turbidity	250	30	114
DO, mg/l	15.8	3.8	9.8
Chlorine demand, mg/l	5.5	0.5	1.7
E. coli, MPN	1,100	0	

Source: Reprinted, with publisher's permission, from *J. Am. Water Works Assn.*, *58*(1):113, © 1966, by the American Water Works Association, Inc., 2 Park Ave., New York, N.Y. 10016.

1.1–29 ANALYSES OF POND WATERS USED FOR $CuSO_4$ EXPERIMENTS

ppm

	Pond A[a]		Pond B[a]		Pond C[a]			
Element	Top	Bottom	Top	Bottom	Top	Bottom	Pond D[b]	Pond E[b]
pH	6.5	7.0	8.2	8.4	7.5	7.5	8.8	7.4
Conductivity[c]	128	142	400	400	170	155	300	220
Cl	37.5	37.5	22.8	15.2	15.2	15.2	52.5	37.5
SO_4	15.0	15.0	90.0	90.0	30.0	30.0	80.0	70.0
Si	12.0	10.0	3.0	3.0	2.0	2.5	6.0	2.0
Al	0.08	0.05	0.05	0.03	0.10	0.05	0.06	0.13
Na	10.4	14.0	4.0	4.2	3.5	3.5	7.8	9.2
K	0.6	1.0	1.6	1.6	0.4	0.4	2.2	5.0
Ca	4.6	6.0	25	25	10.6	10.6	11.2	8.0
Mg	3.0	3.0	28	28	7.4	7.0	14.4	9.0
Cu	0.012	0.023	0.009	0.012	0.009	0.023	0.012	0.015
Mn	0.004	0.720	0.001	0.010	0.050	0.068	0.004	0.007
Zn	0.008	0.078	0.008	1.40	0.013	0.013	0.008	0.032
Fe	0.090	1.250	0.030	0.160	0.160	0.160	0.040	0.013
Ni	0.001	0.003	0.006	0.018	0.001	0.003	0.004	0.005

[a] Ponds A, B, and C located in Warren County, N.J.
[b] Ponds D and E located in Monmouth County, N.J.
[c] μmho/cm

Source: Reprinted, with publisher's permission, from *J. Am. Water Works Assn., 60*(3):367, © 1968, by the American Water Works Association, Inc., 2 Park Ave., New York, N.Y. 10016.

1.1–30 CHEMICAL ANALYSIS OF FORT SMITH LAKE, ARKANSAS

Chemical	Concentration	Chemical	Concentration
Silica (SiO_2), mg/l	4.3	Sulfate (SO_4), mg/l	12
Iron (Fe), mg/l	0.20	Chloride (Cl), mg/l	4.5
Calcium (Ca), mg/l	12	Fluoride (F), mg/l	0.0
Magnesium (Mg), mg/l	2.0	Nitrate (NO_3), mg/l	1.4
Sodium (Na) and Potassium (K), mg/l	2.6	Dissolved solids, mg/l	53
		Total hardness as CaCO, mg/l	38
Carbonate (CO_3), mg/l	0.	Specific conductance[a], μmho	84
Bicarbonate (HCO_3), mg/l	28	pH	7.2

[a] At 25°C.

Source: W.W. Hastings, *Southwest Water Works J., 35*(1):12, 1953.

1.1–31 MAJOR CONSTITUENTS OF WATERS USED FOR WATER SUPPLY

Location	Supply	Percent of supply	SiO_2, mg/l	Fe, mg/l	Ca, mg/l	Mg, mg/l	Na, mg/l	K, mg/l	HCO_3, mg/l	CO_3, mg/l	SO_4, mg/l
Mobile, Ala.	Big Creek (Imp)	100	3.2	0.03	0.0	0.6	2.1	0.3	4	0	0.6
San Diego, Cal.	Lake Hodges Reservoir (Imp)[a]	10	11	0.17	87	29	99	5.6	155	0	282
	Lower Otay Reservoir (Imp)[a]	4	12	0.04	27	21	90	6.3	148	21	43
San Francisco, Cal.	Hetch Hetchy Reservoir[b]	72	2.4	0.02	0.7	0.3	1.1	0.2	4.2	0	0.7
Denver, Colo.	Fraser River and Williams Fork	47	8.2	0.06	6.8	1.9	2.0	0.2	24	0	9.0
	South Platte River and Bear Creek	38	6.8	0.08	27	8.0	24	1.8	86	0	38
	South Platte River	15	7.8	0.21	31	12	34	2.0	104	0	47
Bridgeport, Conn.	Easton Reservoir	21	6.5	0.09	7.0	1.1	3.6	1.2	11	0	13
	Hemlocks Reservoir	51	4.0	0.06[d]	6.4	1.6	3.6	1.0	16	0	13
	Trap Falls Reservoir	26	7.7	0.12	7.3	1.6	3.8	1.0	12	0	14
Ft. Wayne, Ind.	St. Joseph River (Imp)	100	8.0	0.56	76	20	9.7	2.6	256	0	67
Baltimore, Md.	Loch Raven Reservoir	55	5.5	0.01	12	3.2	3.7	1.8	43	0	8.4
	Liberty Reservoir	45	5.9	0.00	8.5	2.6	3.6	1.5	25	0	9.0
Boston, Mass.	Quabbin Reservoir[c]	100	1.0	0.04	2.9	0.6	1.7	0.7	5	0	6.9
Flint, Mich.	Flint River (Imp)	100	4.6	0.24	71	24	15	2.4	254	0	60
Newark, N.J.	Wanaque River (Imp)	45	1.9	0.00	8.8	3.4	3.6	0.8	23	0	17
Albany, N.Y.	Hannacrois Creek (Imp)	92	9.3	0.11	11	2.8	1.8	1.0	34	0	13
Charlotte, N.C.	Catawba River (Imp)	100	10	0.00	3.2	1.2	3.8	1.2	20	0	3.8
Greensboro, N.C.	Lake Brandt (Imp)	100	9.4	0.02	4.4	1.7	2.9	2.4	20	0	7.2
Oklahoma City, Okla.	Lake Hefner (Imp)	100	3.2	0.02	58	26	90	7.6	182	0	152
Tulsa, Okla.	Spavinaw Creek (Imp)	100	–	–	–	–	–	–	–	–	–
Providence, R.I.	Scituate Reservoir (Imp)	92	5.0	0.05	2.6	0.6	2.9	0.6	5	0	6.0
Corpus Christi, Texas	Nueces River (Imp)	100	15	0.10	42	8.3	62	7.9	122	0	45
Dallas, Texas	Garza-Little Elm Reservoir (Imp)	67	2.1	0.06	53	6.2	39	4.4	139	0	44
	Grapevine and Garza-Little Elm	26	2.2	0.04	55	7.1	33	4.4	157	0	52
Salt Lake City, Utah	Deer Creek Reservoir (Imp)	24	6.7	0.00	36	11	4.3	1.0	122	0	32
Norfolk, Va.	Lake Wright (Imp)	62	2.3	0.24	12	4.9	14	3.0	29	0	29
	Lake Prince (Imp)	38	5.8	0.62	11	1.6	4.6	2.0	29	0	10
	Lake Burnt Mills		4.0	0.84	3.6	1.4	4.4	1.7	13	0	5.6
Range			1.0–15	0.0–0.84	0.0–76	0.3–29	1.1–99	0.2–7.9	4–256	0–21	0.6–282

[a] Analyzed by the City of San Diego.
[b] Analyzed by the City of San Francisco.
[c] Analyzed by the Metropolitan District Commission.
[d] In solution when collected.

1.1–31 MAJOR CONSTITUENTS OF IMPOUNDMENT WATERS USED FOR WATER SUPPLY (*Continued*)

Location	Supply	Cl, mg/l	F, mg/l	NO_3, mg/l	Dissolved solids (residue at 180°C), mg/l	Hardness as $CaCO_3$, mg/l	Noncarbonate hardness as $CaCO_3$, mg/l	Specific conductance, μmho at 25°C	pH	Color, units
Mobile, Ala.	Big Creek (Imp)	2.3	0.0	0.2	34	2	0	18	5.8	40
San Diego, Cal.	Lake Hodges Reservoir (Imp)[a]	97	0.4	1.0	761	334	207	–	7.9	–
	Lower Otay Reservoir (Imp)[a]	109	0.5	0.4	461	155	0	–	9.1	–
San Francisco, Cal.	Hetch Hetchy Reservoir[b]	1.0	0.0	0.0	10	3	0	8	7.4	1
Denver, Colo.	Fraser River and Williams Fork	1.0	0.3	0.5	40	25	5	58	7.6	0
	South Platte River and Bear Creek	33	1.0	0.5	172	100	30	332	7.5	0
	South Platte River	44	0.9	0.2	232	127	41	448	7.0	0
Bridgeport, Conn.	Easton Reservoir	5.9	0.1	0.8	51	22	13	73	6.6	9
	Hemlocks Reservoir	5.5	0.1	0.4	47	27	14	78	6.9	6
	Trap Falls Reservoir	6.4	0.1	1.0	57	25	15	79	6.6	12
Ft. Wayne, Ind.	St. Joseph River (Imp)	10	0.4	2.6	334	272	62	535	7.6	22
Baltimore, Md.	Loch Raven Reservoir	5.5	0.0	3.3	68	43	8	111	7.0	3
	Liberty Reservoir	5.5	0.1	4.7	59	32	11	91	6.4	5
Boston, Mass.	Quabbin Reservoir[c]	2.1	0.1	0.0	25	10	–	–	6.3	5
Flint, Mich.	Flint River (Imp)	26	0.1	1.1	348	276	68	578	7.4	17
Newark, N.J.	Wanaque River (Imp)	5.6	0.1	0.6	55	36	17	96	6.5	5
Albany, N.Y.	Hannacrois Creek (Imp)	2.1	0.0	0.0	56	39	11	92	7.1	6
Charlotte, N.C.	Catawba River (Imp)	1.5	0.1	0.3	37	13	0	46	7.2	3
Greensboro, N.C.	Lake Brandt (Imp)	3.5	0.1	1.5	49	18	2	59	7.0	10
Oklahoma City, Okla.	Lake Hefner (Imp)	107	0.6	0.5	518	250	101	865	7.8	2
Tulsa, Okla.	Spavinaw Creek (Imp)	–	–	–	–	–	–	–	–	–
Providence, R.I.	Scituate Reservoir (Imp)	4.5	0.0	0.0	28	9	5	40	6.1	7
Corpus Christi, Texas	Nueces River (Imp)	96	0.4	0.0	343	139	39	597	7.6	0
Dallas, Texas	Garza-Little Elm Reservoir (Imp)	64	0.4	0.5	291	158	44	515	7.5	–
	Grapevine and Garza-Little Elm	42	0.4	1.0	282	166	38	491	7.4	–
Salt Lake City, Utah	Deer Creek Reservoir (Imp)	7.0	0.3	0.3	159	136	36	276	7.8	5
Norfolk, Va.	Lake Wright (Imp)	22	0.1	1.1	123	52	28	174	6.7	35
	Lake Prince (Imp)	9.0	0.1	1.5	73	34	10	102	6.9	32
	Lake Burnt Mills	8.5	0.1	1.0	47	16	6	56	6.7	50
Range		1.0–107	0.0–1.0	0.0–4.7	10–761	2–334	0–207	8–865	5.8–9.1	0–50

[a] Analyzed by the City of San Diego.
[b] Analyzed by the City of San Francisco.
[c] Analyzed by the Metropolitan District Commission.

Note: Imp – Impoundment; Res – Reservoir.

Source: C.N. Durfor and E. Becker, *Geological Survey Water Supply Paper,* U.S. Department of the Interior, 1962.

1.1–32 MINOR CONSTITUENTS OF IMPOUNDMENT WATERS USED FOR WATER SUPPLY

Location	Source of supply	Ag, mg/l	Al, mg/l	B, mg/l	Ba, mg/l	Be, mg/l	Co, mg/l	Cr, mg/l	Cu, mg/l	Fe, mg/l	Li, mg/l
Mobile, Ala.	Big Creek (Imp)										
San Diego, Cal.	Lake Hodges Reservoir (Imp)[a]										
	Lower Otay Reservoir (Imp)[a]										
San Francisco, Cal.	Hetch Hetchy Reservoir[b]										
Denver, Colo.	Fraser River and Williams Fork	ND[c]	23	3.8	6.2	ND	ND	<0.07	16	39	0.24
	South Platte River and Bear Creek	<0.26	190	21	58	ND	<2.6	1.0	31	29	4.2
	South Platte River	ND	86	100	72	ND	ND	ND	1.7	58	5.8
Bridgeport, Conn.	Easton Reservoir		–								
	Hemlocks Reservoir		100								
	Trap Falls Reservoir										
Ft. Wayne, Ind.	St. Joseph River (Imp)	<0.54	180	75	130	ND	ND	2.0	70	110	1.6
Baltimore, Md.	Loch Raven Reservoir	0.74	99	16	39	0.75	0.99	2.6	26	280	0.16
	Liberty Reservoir										
Boston, Mass.	Quabbin Reservoir[e]		150						40		
Flint, Mich.	Flint River (Imp)	<0.53	690	79	150	ND	ND	0.90	17	850	3.0
Newark, N.J.	Wanaque River (Imp)										
Albany, N.Y.	Hannacrois Creek (Imp)	0.12[f]	17	3.6	10.	ND	ND	0.72	73	39	0.07
Charlotte, N.C.	Catawba River (Imp)	<0.04[f]	59	250	8.9	ND	<0.37	0.10	0.78	52	0.06
Greensboro, N.C.	Lake Brandt (Imp)		100								0
Oklahoma City, Okla.	Lake Hefner (Imp)	<0.76	130	170	140	ND	ND	1.9	14	39	20
Tulsa, Okla.	Spavinaw Creek (Imp)	<0.19	330	63	63	ND	<1.9	3.8	8.4	63	0.23
Providence, R.I.	Scituate Reservoir (Imp)										
Corpus Christi, Texas	Nueces River (Imp)										
Dallas, Texas	Garza-Little Elm Reservoir (Imp)										
	Grapevine and Garza-Little Elm										
Salt Lake City, Utah	Dear Creek Reservoir (Imp)	<0.42	80	120	130	ND	ND	0.93	12	80	13
Norfolk, Va.	Lake Wright (Imp)		300								100
	Lake Prince (Imp)		200								0
	Lake Burnt Mills		300								0
Range		ND–0.74	17–690	3.6–250	6.2–150	ND–0.75	ND–0.99	ND–3.8	0.78–73	29–850	0–100

[a] Analyzed by City of San Diego.
[b] Analyzed by City of San Francisco.
[c] ND – looked for but not found.
[e] Analyzed by the Metropolitan District Commission.
[f] Spectrographic concentrations based on nonacidified residue on evaporation.

1.1–32 MINOR CONSTITUENTS OF IMPOUNDMENT WATERS USED FOR WATER SUPPLY (*Continued*)

Location	Source of supply	Mn, μg/l	Mo, μg/l	Ni, μg/l	P, μg/l	Pb, μg/l	Rb, μg/l	Sr, μg/l	Ti, μg/l	V, μg/l	Zn, μg/l	Date of sample collection
Mobile, Ala.	Big Creek (Imp)											4/ 5/62
San Diego, Cal.	Lake Hodges Reservoir (Imp)[a]	230										–
	Lower Otay Reservoir (Imp)[a]	30										–
San Francisco, Cal.	Hetch Hetchy Reservoir[b]											7/ 8/61
Denver, Colo.	Fraser River and Williams Fork	5.5	<0.20	0.7	ND	ND	1.1	4.1	1.0	ND	ND	8/28/61
	South Platte River and Bear Creek	15	3.4	3.9	ND	<2.6	ND	290.	<2.6	ND	ND	8/28/61
	South Platte River	7.2	1.2	ND	ND	<3.6	ND	240	<3.6	ND	ND	8/29/61
Bridgeport, Conn.	Easton Reservoir	10										6/ 1/62
	Hemlocks Reservoir	10										6/ 1/62
	Trap Falls Reservoir	30[d]										6/ 1/62
Ft. Wayne, Ind.	St. Joseph River (Imp)	22	19	9.6	ND	20	ND	540	5.9	ND	ND	8/22/61
Baltimore, Md.	Loch Raven Reservoir	52	0.83	5.8	ND	43	3.3	32	2.7	<3.0	99	8/22/61
	Liberty Reservoir											8/22/61
Boston, Mass.	Quabbin Reservoir[e]											1/15/62
Flint, Mich.	Flint River (Imp)	170	5.8	15	ND	15	ND	260	11	ND	ND	8/31/61
Newark, N.J.	Wanaque River (Imp)	0										7/31/62
Albany, N.Y.	Hannacrois Creek (Imp)	450	ND	2.6	ND	11	1.2	21	0.6	ND	ND	8/17/61
Charlotte, N.C.	Catawba River (Imp)	11	ND	0.7	ND	4.4	1.2	2.4	3.5	<1.1	<37	8/15/61
Greensboro, N.C.	Lake Brandt (Imp)	40										1/16/62
Oklahoma City, Okla.	Lake Hefner (Imp)	7.6	4.5	7.6	ND	13	ND	530	<7.6	ND	ND	7/28/61
Tulsa, Okla.	Spavinaw Creek (Imp)	23	1.7	9.2	ND	94	ND	46	3.1	<5.8	<190	7/27/61
Providence, R.I.	Scituate Reservoir (Imp)											4/18/62
Corpus Christi, Texas	Nueces River (Imp)	10										1/31/62
Dallas, Texas	Garza-Little Elm Reservoir (Imp)											3/22/62
	Grapevine and Garza-Little Elm											3/22/62
Salt Lake City, Utah	Deer Creek Reservoir (Imp)	420	2.4	18	ND	9.7	5.5	220	<4.2	ND	ND	7/12/61
Norfolk, Va.	Lake Wright (Imp)	20										1/17/62
	Lake Prince (Imp)	0										
	Lake Burnt Mills	0										
Range		0–450	ND–5.8	ND–15	ND	ND–94	ND–5.5	2.4–540	0.6–11	ND–<5.8	ND–99	

[a] Analyzed by City of San Diego.
[b] Analyzed by City of San Francisco.
[c] ND – looked for but not found.
[d] In solution when collected.
[e] Analyzed by the Metropolitan District Commission.

Note: Imp – Impoundment; Res – Reservoir.

Source: C.N. Durfor and E. Becker, *Geological Survey Water Supply Paper*, U.S. Department of the Interior, 1962.

1.1–33 CHANGES IN STRIP MINE LAKES AS FUNCTION OF TIME

Lake	Constituent	1940–41[a]	1949–50[b]	1963–65
A_3	pH (median)			
	surface	3.0 (12)[c]	3.3 (10)	3.5 (24)
	bottom	2.8 (12)	3.3 (10)	3.6 (24)
	Bicarbonate (mean), mg/l	0.0 (12)	0.0 (20)	0.0 (24)
	Ca, mg/l	131. (8)	97.7 (8)	64.1 (14)
	Fe, mg/l	62.7 (8)	2.4 (8)	1.2 (6)
	Mn, mg/l	6.0 (8)	1.6 (8)	2.4 (14)
	Na, mg/l	7.5 (8)	2.6 (8)	3.5 (14)
	K, mg/l	10.9 (8)	2.0 (8)	7.5 (14)
	Mg, mg/l	17.3 (8)	9.4 (8)	15.1 (14)
B	pH (median)			
	surface	3.6 (12)	7.0 (10)	7.2 (24)
	bottom	3.6 (12)	6.6 (10)	7.0[d] (24)
	Bicarbonate (mean), mg/l	0.0 (12)	23.0 (20)	73.0 (24)
	Ca, mg/l	87.6 (8)	78.6 (8)	56.6 (14)
	Fe, mg/l		0.3 (3)	0.5 (14)
	Mn, mg/l	4.7 (8)	2.2 (8)	1.4 (13)
	Na, mg/l	9.4 (8)	7.2 (8)	7.0 (14)
	K, mg/l	4.8 (8)	1.9 (8)	7.0 (14)
	Mg, mg/l	24.7 (8)	17.3 (8)	20.7 (14)
D	pH (median)			
	surface	7.3 (12)	7.3 (10)	7.2 (24)
	bottom	7.0 (12)	7.1 (10)	7.1[e] (24)
	Bicarbonate (mean), mg/l		48.0 (20)	57.0 (24)
	Ca, mg/l	29.3 (8)	53.3 (8)	23.0 (14)
	Fe, mg/l		0.4 (6)	0.2 (14)
	Mn, mg/l	0.4 (8)	0.8 (8)	0.8 (13)
	Na, mg/l	8.3 (8)	4.4 (8)	3.5 (14)
	K, mg/l	6.7 (8)	1.6 (8)	4.4 (14)
	Mg, mg/l	16.4 (8)	7.7 (8)	7.2 (14)

[a] Values for 1940–41 are from B.T. Crawford. M.A. Thesis, University of Missouri, Columbia, Mo., 1942.
[b] Values for 1949–50 are from J.R. Heaton. M.A. Thesis, University of Missouri, Columbia, Mo., 1951.
[c] Figures in parentheses are numbers of determinations.
[d] Depth at 1 percent light level, approximated 10 ft (3 m).
[e] Depth at 1 percent light level, approximated 7 ft (2 m).

Source: R.S. Campbell and O.T. Lind, *J. Water Pollut. Control Fed.*, *41*(11):1943, 1969.

1.1–34 FACTORS PERTAINING TO ACID-BASE RELATIONSHIPS

Strip Mine Lakes

		Potential free acidity, mg/l $CaCO_3$		Free CO_2 mg/l		Total alkalinity mg/l $CaCO_3$	
Lake	pH	Range	Mean	Range	Mean	Range	Mean
A_1							
surface	2.3–3.5	227–4,920	2,660	2–70	12	0	0
middle	2.5–3.4	2,090–8,980	5,640	10–24	16	0	0
bottom	2.6–4.1	11,800–16,600	14,200	12–520	116	0	0
A_2							
surface	3.0–3.4	377–745	615	–		0	0
bottom	3.0–3.4	400–787	606	–		0	0
A_3							
surface	3.2–4.0	27–55	35	4–14	9	0	0
bottom	3.3–4.1	28–46	38	15–51	26	0	0
B							
surface	6.1–8.2	0	0	1–100	23	49–92	66
bottom	6.3–7.8	0	0	2–110	29	44–160	80
D							
surface	5.6–7.6	0	0	1–44	8	27–87	47
bottom	6.6–7.4	0	0	1–200	19	35–166	68

Source: R.S. Campbell and O.T. Lind, *J. Water Pollut. Control. Fed., 41*(11):1943, 1969.

1.1–35 IONIC COMPOSITION OF STRIP MINE LAKES

mg/l

	Surface		Middle		Bottom	
Ion	Range	Mean	Range	Mean	Range	Mean
			Lake A_1			
SO_4	1,620–5,030	3,340	1,620–8,170	5,720	3,410–15,200	10,200
Ca	82–253	205	162–403	311	217–420	324
Mg	65–175	115	132–197	175	145–190	166
Na	8.9–23.3	14.0	17.7–29.6	21.5	22–24	28.6
K	0.2–1.6	1.0	0.4–2.8	2.1	<0.1–75.0	31.6
Al	102–360	212	331–465	408	420–490	426
Zn	24.8–86.0	53.8	89–226	139.4	203–318	217.6
Mn	27–95	,56.4	83.5–98.5	91	68.5–85	77.7
Fe^{++}	0–12	5	0–350	120	4,620–7,400	6,010
Fe^{+++}	74–260	177	523–1,050	791	5–206	102
NH_3	0.04–2.0	0.5	0.09–4.74	2.0	6.75–44.2	2.67
SiO_2	24–71.5	45.4	75–114	82.2	105–140	106.8
NO_2	0.001–0.071	0.026	0.001–0.068	0.023	<0.01	
HCO_3	0	0	0	0	0	0
			Lake A_2			
SO_4	216–1,970	1,260			1,190–1,970	1,540
Ca	44–142	98			51–144	107
Mg	10–118	66			17–120	81
Na	6.1–11.9	9.6			6.1–12.0	9.7
K	0.4–1.9	1.2			0.4–1.8	1.1
Al	105–158	129			105–171	135
Zn	0.9–1.8	1.4			1.0–1.8	1.4

1.1–35 IONIC COMPOSITION OF STRIP MINE LAKES (*Continued*)

	Surface		Bottom	
Ion	Range	Mean	Range	Mean
	Lake A_2 (continued)			
Mn	9.4–20.1	13.6	9.5–20.1	13.8
Fe^{++}	0.2–6.0	2.0	0.2–5.5	1.9
Fe^{+++}	7.5–35	15	7.6–33	16
NH_3		<1.0		<1.0
SiO_3	13–26	17	13–28	18
NO_3	<0.01–0.035	0.018	<0.01–0.035	0.019
HCO_3	0	0	0	0
	Lake A_3			
SO_4	108–324	225	191–317	271
Ca	29.4–79.0	55.2	31.6–87.5	66.5
Mg	6.3–17.0	12.3	9.1–19.1	15.4
Na	1.3–4.6	3.1	2.5–4.5	3.4
K	3.35–9.2	6.4	4.47–9.1	7.0
Al	0.54–11.25	2.58	0.52–14.50	3.23
Zn	0.58–1.65	1.08	0.70–2.23	1.36
Mn	0.79–3.18	2.18	1.22–3.00	2.44
Fe^{++}	0.0–0.350	0.168	0.017–0.438	0.226
Fe^{+++}	0.438–1.525	0.978	0.542–1.216	0.939
NH_3		<0.4		<0.4
SiO_2	7.9–29.2	19.2	10.7–28.0	20.6
NO_2	0.001–0.005	0.002	0.001–0.005	0.002
HCO_3	0	0	0	0
	Lake B			
SO_4	134–245	196	161–211	189
Ca	39.5–63.0	55.5	39.2–70.5	59.6
Mg	17.1–25.6	20.0	17.5–25.5	21.2
Na	6.0–7.3	6.6	6.4–8.7	7.2
K	5.7–7.5	6.6	6.4–8.7	7.4
Al	0.001–0.058	0.025	0.001–0.035	0.017
Zn	0.05–0.07		0.05–0.07	
Mn	0.08–2.1	0.56	0.60–8.90	2.30
Fe^{++}	0	0	0	0
Fe^{+++}	0.0–0.024	0.015	0.02–0.132	0.087
NH_3		<0.5		<0.5
SiO_2	0.39–8.1	2.75	0.92–5.68	2.98
NO_2	0.001–0.011	0.003	0.001–0.005	0.015
HCO_3	49–71	56	53–160	80
	Lake D			
SO_4	25–184	83.6	15–295	108.8
Ca	10.6–58.6	28.7	10.6–56.5	30.3
Mg	1.6–20.4	9.5	1.7–19.9	10.0
Na	2.7–6.5	4.1	2.6–6.0	4.2
K	3.4–5.2	3.9	3.4–5.3	3.9
Al	0.001–0.078	0.025	0.001–0.052	0.022
Zn	0.06–0.07		0.06	
Mn	0.01–0.70	0.193	0.01–0.81	0.47
Fe^{++}	0	0	0	0
Fe^{+++}	0.0–1.08	0.40	0.105–0.150	0.127
NH_3		<0.2		<0.3
SiO_2	1.03–6.35	3.89	1.99–7.55	4.65
NO_2	0.001–0.010	0.004	0.001–0.032	0.007
HCO_3	33–82	52	36–97	68

Source: R.S. Campbell and O.T. Lind, *J. Water Pollut. Control Fed., 41*(11):1943, 1969.

1.1–36 VERTICAL EXTINCTION COEFFICIENTS (n'')

Strip Mine Lakes Compared with Selected Non-Acid Lakes

Lake	**Extinction[a] C**
A_1	1.91–4.63
A_2	0.59–2.12
A_3	0.25–0.91
B	0.43–2.34
D	1.09–9.29
Crystal, Wisconsin[2]	0.19
Little Star, Wisconsin[2]	3.90
Ashland Reservoir, Missouri[3]	3.8–24.8
Shallow stock watering ponds, Oklahoma[4]	4.0–164
Michigan[5]	0.23
Douglas, Michigan[5]	0.60

[a] Extinction coefficient–coefficient of surface incident light attenuation in a given depth. Caused by three factors: the water itself, suspended particles, and dissolved matter or color.

Source: R.S. Campbell and O.T. Lind, *J. Water Pollut. Control Fed.*, *41*(11):1943, 1969.

REFERENCES

1. R.S. Campbell and O.T. Lind, *J. Water Pollut. Control Fed.*, *41*(11):1943, 1969.
2. G.E. Hutchinson, *A Treatise on Limnology*, Vol. 1, John Wiley & Sons, New York, 1957.
3. R.S. Campbell, Unpublished data, University of Missouri, Columbia, Mo.
4. G.I. Murphy, *Trans. Amer. Fish Soc.*, *91*(1):69, 1962.
5. G.W. Saunders *et al.*, Great Lakes Res. Div. Publ. No. 8, University of Michigan, 1962.

1.1–37 TYPES OF ALGAE IN GRAND LAKE (OHIO) WATER

Item	Percentage of time present	Algae identification												
		Group[b]							Significance[c]					
		D	G	BG	R	Fl	De	YG	A	S	F	C	P	T
Aphanizomenon	33			x										x
Microcystis (Anacystis)[a]	90			x						x	x		x	x
Gleocapa (Anacystis)	3			x						x	x		x	x
Dinobryon	7					x				x	x	x		x
Cryptomonas	7					x								x
Trachelomonas	43					x					x			
Euglena	23					x				x		x	x	x
Peridinium	23					x					x			x
Pediastrum	50		x							x				x
Scenedesmus[a]	87		x							x			x	x
Ankistrodesmus	40		x							x		x		
Crucigenia	13		x							x				
Closterium	17						x			x	x			
Navicula	3	x								x	x	x	x	
Synedra[a]	57	x								x	x	x		x
Lyngbya	93			x					x	x			x	
Chlamydomonas	7					x							x	x
Cyclotella	53	x								x	x	x		x
Oscillatoria	33			x					x	x	x		x	x
Aphanothece (Coccochloris)	13			x								x		
Celosphaerium[a]	77			x						x				x
Chroococcus (Gomphosphaeria)	57			x						x				x
Aphanocapsa (Anacystis)	3			x						x	x		x	x
Anabena	43			x							x		x	x
Merismopedia (Agmenellum)	67			x								x	x	
Marsoniella	3													
Micractinium	3		x							x				
Tetredon	40		x										x	
Oocystis	47		x							x				
Dictyopherium[a]	73		x								x			x
Golenkinia	7													
Staurastrum	40						x			x		x		x
Fragilaria	3	x								x	x			x
Celastrum	30		x							x				
Schrederia	10		x							x				
Naviculoid diatoms	7	x												
Cosmarium	47						x			x				x
Gleocystis	3		x						x					x

[a] Probably the worst offenders.
[b] Group key: D, diatom; G, green; BG, blue green; R, red; Fl, flagellate; De, Desmid; and YG, yellow-green.
[c] Significance key: A, attached; S, surface; F, filter clogging; C, clean water; P, polluted water; and T, taste and odor.

1.1–37 TYPES OF ALGAE IN GRAND LAKE (OHIO) WATER (*Continued*)

		Algae identification												
		Group[b]							Significance[c]					
Item	**Percentage of time present**	**D**	**G**	**BG**	**R**	**Fl**	**De**	**YG**	**A**	**S**	**F**	**C**	**P**	**T**
Surirella	3	x										x	x	
Spherocystis	3		x							x				
Tabellaria[a]	63	x									x			x
Phacus	13					x				x		x	x	
Ochromonas	33													
Kircherniella	3		x							x				
Selenastrum	33		x							x				
Francia	3													
Melosira	3	x								x	x		x	
Lepocinclis	3					x							x	
Total		8	14	11	0	8	3	0	3	29	16	11	16	23

[a] Probably the worst offenders.
[b] Group key: D, diatom; G, green; BG, blue green; R, red; Fl, flagellate; De, Desmid; and YG, yellow-green.
[c] Significance key: A, attached; S, surface; F, filter clogging; C, clean water; P, polluted water; and T, taste and odor.

Source: Reprinted, with publisher's permission, from *J. Am. Water Works Assn.*, *58*(1):113, © 1966 by the American Water Works Association, Inc., 2 Park Ave., New York, N.Y. 10016.

1.1–38 DISTRIBUTION OF PESTICIDES IN A LAKE

		Average residues				
Attributes	**No. samples**	**DDT and related compounds**	**γ BHC**	**Toxaphene**	**Dieldrin**	**Heptachlor**
Water (ppb)[a]	82	0.62 (0–22.0)	0.01 (0–0.15)	0.02 (0–0.32)	trace	trace
Particles (ppm)[b]	33	14.74 (1.8–78.0)	0	0	0	0
Bottom Sediment (ppm)[b]	39	4.44 (0.01–94.0)	trace	0.03 (0–0.30)	trace	trace

[a] μg/l.
[b] mg/kg.

Source: J.O. Keith and E.G. Hunt, *Trans. 31st North Amer. Wildlife Res. Conf.*, 1966, p. 150.

1.1–39 DISTRIBUTION OF PESTICIDES IN FOUR LAKES

Total Residues

Site	Average ppm DDT and related compounds		
	Suspended material, mg/kg	Filtrate, mg/l	Total water sample, mg/l
Tule lake	6.0	0.0002	0.00045
Lower Klamath	9.0	0.0002	0.00035
Deer Flat	8.3	0.0001	0.0002
McNary	9.3	0.0001	0.0003

Source: J.O. Keith, *J. Appl. Ecol., (Suppl. 3),* 71, 1966.

1.1–40 WATER QUALITY DISTRIBUTION

Coeur d'Alene Lake, Idaho

	Open lake waters		Bays		Adjacent springs	
	Range	Mean	Range	Mean	Range	Mean
pH	7.4–8.2	7.91	7.1–8.1	7.79	6.1–7.9	6.92
Electrical conductivity μmho/cm	42–75	55.88	47–69	53.40	71–222	128.68
Phosphate, ppm	<0.01–0.19	0.049	<0.01–0.16	0.037	<0.01–0.87	0.165
Chloride, ppm	6.0–11.2	8.8	8.8–14.0	9.7	8.8–12.3	10.7

Source: Reprinted, with publisher's permission, from *J. Am. Water Works Assn., 61*(10):525, © 1969 by the American Water Works Association, Inc., 2 Park Ave., New York, N.Y. 10016.

1.1–41 INITIAL EFFECT OF IMPOUNDMENT ON WATER QUALITY

Loch Raven Reservoir

Characteristic	Avg before enlargement for 1918–1922	Avg after enlargement for 1924–1928
Temperature, °C	13.2	10.8
Color, units	24.5 (for 1921–22)	7.4
Turbidity, Jackson units	90.8	19.6
Iron, mg/l	5.0	1.1
Microorganisms,[a] no./cc	150.0 (for 1920–21)	550.0 (for 1925–28)
Bacteria on agar at 20°C, no./cc	12,600.0	390.0

[a] During 1922–1924, the period of filling and initial use, microorganisms averaged 3,000 per cc.

Note: Treatment plant influent water quality data for Montebello filtration plant from *Annual Reports* of the Baltimore Department of Public Works, 1925–1965.

Source: Reprinted, with publisher's permission, from *J. Am. Water Works Assn., 58*(4):508, © 1966, by the American Water Works Association, Inc., 2 Park Ave., New York, N.Y. 10016.

1.1–42 EFFECT OF IMPOUNDMENT ON WATER QUALITY

Howard A. Hanson Reservoir

Characteristic	Inlet	Outlet
Seasonal mean temperature, °C	11.5	12.8
pH	7.5	7.5
DO content, ppm	10.0	10.1
DO saturation, percent	100	100
Total alkalinity as $CaCO_3$, ppm	19	18
Color units	5	7
Turbidity units	1	1
Tannin and lignin content, ppm	0.22	0.25
Orthophosphate content, ppm	0.05	0.04
Total phosphate content, ppm	0.12	0.08
Iron content, ppm	0.06	0.10
Specific conductance at 25°C, μmho/cm	45	46
Ammonia nitrogen content as N, ppm	0.08	0.08
Organic nitrogen content as N, ppm	0.10	0.10
Nitrate nitrogen content as N, ppm	0.06	0.06
Total nitrogen content as N, ppm	0.24	0.24
Potassium content, ppm	0.5	0.4
Chlorophyll a content, ppb	0.32	0.84
Algal organisms, no./ml	2,200	4,100

Note: Comparison of mean incoming and mean outgoing water quality for conservation pool operation 1962 and 1963.

Source: Reprinted, with publisher's permission, from *J. Am. Water Works Assn., 57*(12):1528, © 1965, by the American Water Works Association, Inc., 2 Park Ave., New York, N.Y. 10016.

1.1–43 QUALITY IMPROVEMENT AFTER A THREE MONTH STORAGE

Rhine River, Rotterdam Waterworks

Physical examination	Water quality: During filling	After storage
Color (Pt-Co scale)	28	25
pH	7.5	8.2
Suspended solids, mg/l	30	2
$KMnO_4$ consumption		
Unfiltered samples, mg/l	28	19
Filtered samples, mg/l	22	18
NH_4+-inorganic, mg/l	1.2	0.05
NH_4-organic, mg/l	0.40	0.25
NO_2-, mg/l	0.40	0.06
NO_3-,mg/l	7.0	6.6
$PO_4{}^{3-}$, mg/l	0.25	0.06
Fe, mg/l	1.2	0.06
Mn, mg/l	0.15	0.03
O_2, mg/l	4	11
Threshold taste value	80	3

Note: Storage (standing) period was three months.

Source: Reprinted, with publisher's permission, from *J. Am. Water Works Assn.*, *62*(4):249, © 1970, by the American Water Works Association, Inc., 2 Park Ave., New York, N.Y. 10016.

1.1–44 QUALITY CHANGES IN RESERVOIR DURING OPERATION

Quality characteristic	1966[a] River	1966[a] Reservoir	1967[a] River	1967[a] Reservoir	1968[a] River	1968[a] Reservoir
Threshold taste value	95	14	110	20	70	22
Color (Pt-Co scale)	27	24	26	24	24	24
pH	7.60	8.30	7.55	8.25	7.55	8.10
Suspended solids, mg/l	63	7	118	10	90	8
$KMnO_4$ consumption						
Unfiltered samples, mg/l	31	22	34	23	33	20
Filtered samples, mg/l	22	20	24	21	21	18
$NH_4{}^+$ inorganic, mg/l	1.30	0.35	1.26	0.57	1.46	0.99
NH_4 organic, mg/l	0.40	0.29	0.48	0.29	0.46	0.29
$NO_3{}^-$, mg/l	11.0	9.0	10.8	10.8	10.7	10.9
$PO_4{}^{3-}$, mg/l	0.28	0.12	0.26	0.15	0.36	0.23
SiO_2, mg/l	7	4	8	5	9	6
Fe, mg/l	2.1	0.20	3.6	0.24	3.0	0.31
Mn, mg/l	0.16	0.09	0.22	0.09	0.19	0.08
DO, mg/l	5.7	9.6	5.8	11.0	5.8	11.1
DO, % saturation	54	91	55	102	55	103

[a] Detention period 1966–1967, 30 days; 1968, 20 days.

Source: Reprinted, with publisher's permission, from *J. Am. Water Works Assn.*, *62*(4):249, © 1970, by the American Water Works Association, Inc., 2 Park Ave., New York, N.Y. 10016.

1.1–45 AMMONIA OXIDIZED DURING STORAGE

Rotterdam Waterworks

Temperature, °C	Amount of ammonia oxidized, mg/l
4	0.35 ± 0.05
8	0.5 ± 0.11
12	0.75 ± 0.15
16	1.1 ± 0.2
20	1.6 ± 0.3

Note: Storage period was one month.

Source: Reprinted, with publisher's permission, from *J. Am. Water Works Assn., 62*(4):249, © 1970, by the American Water Works Association, Inc., 2 Park Ave., New York, N.Y. 10016.

1.1–46 NITROGEN LOSSES IN BERENPLAAT RESERVOIR

Total nitrogen, mg/l

Monthly period	River	Reservoir	Nitrogen loss, %
Mar/Apr	4.0	3.1	22
Apr/May	3.7	2.4	35
May/June	3.8	2.3	40
June/July	3.6	2.1	42
July/Aug	3.5	1.8	49
Aug/Sept	3.1	1.2	61
Sept/Oct	3.5	2.0	43
Oct/Nov	4.6	3.6	22
Nov/Dec	4.8	4.4	8

Note: Storage period was one month.

Source: Reprinted, with publisher's permission, from *J. Am. Water Works Assn., 62*(4):249, © 1970, by the American Water Works Association, Inc., 2 Park Ave., New York, N.Y. 10016.

1.1–47 NITROGEN LOSSES IN A THAMES VALLEY RESERVOIR

	N, mg/l		
1969–1970	**River**	**Reservoir**	**Nitrogen loss, %**
Spring	4.89	4.48	8
Summer	4.75	3.13	34
Autumn	5.34	2.59	51
Winter	6.32	5.07	20
1970–1971			
Spring	4.89	4.7	4
Summer	4.97	4.51	9
Autumn	5.09	4.0	21
Winter	6.72	6.43	5

Note: Retention in the reservoir was for about 45 days.

Source: Personal Communication with E. Windle Taylor, 1973.

1.1–48 EFFECT OF VARIOUS SOIL TYPES ON BENTHONIC REGION WATER QUALITY

	Type of soil on reservoir floor							
	Sedimented silt, 2–15% organics		Mineral soil, 8–14% organics		Organic soil (muck), 20–50% organics		Organic soil (peat), 50–90% organics	
	Station no.							
Characteristic	3	8	5	6	1	10	7[a]	9[a]
Mean water depth, ft	88	72	50	30	23	16	17	17
Temperature, °C	13.7	13.5	15.0	16.7	15.3	16.9	16.6	16.8
pH	7.0	7.0	7.1	7.0	7.1	7.1	7.0	7.2
DO, ppm	8.4	8.1	8.5	6.5	8.7	8.1	7.3	8.7
DO saturation, percent	84	80	85	69	90	86	78	93
Total alkalinity as $CaCo_3$, ppm	18	18	18	18	18	20	18	18
Color units	7	7	6	7	8	12	12	8
Turbidity units	1	1	1	1	1	2	1	1
Tannin and lignin content, ppm	0.22	0.26	0.25	0.27	0.24	0.31	0.36	0.30
Orthophosphate content, ppm	0.01	0.02	0.02	0.01	0.01	0.02	0.01	0.01
Total phosphate content, ppm	0.10	0.10	0.12	0.11	0.11	0.12	0.11	0.12
Iron content, ppm	0.05	0.07	0.06	0.11	0.09	0.21	0.20	0.08
Specific conductance at 25°C, μmho/cm	44	44	45	47	44	49	45	44
Ammonia nitrogen content as N, ppm	0.10	0.04	0.11	0.02	0.06	0.05	0.08	0.09
Organic nitrogen content as N, ppm	0.11	0.10	0.12	0.16	0.13	0.12	0.13	0.16
Nitrate nitrogen content as N, ppm	0.05	0.05	0.05	0.05	0.05	0.04	0.04	0.05
Total nitrogen content as N, ppm	0.26	0.19	0.28	0.23	0.24	0.21	0.25	0.30
Potassium content, ppm	0.2	0.3	0.3	0.4	0.3	0.3	0.4	0.3
Chlorophyll A content, ppb	0.32	0.55	0.93	1.05	0.77	0.98	0.83	0.55

[a] Stations 7 and 9 are in the swamp.

Source: Reprinted, with publisher's permission, from *J. Am. Water Works Assn.*, 57(12):1528, © 1965, by the American Water Works Association, Inc., 2 Park Ave., New York, N.Y. 10016.

1.1–49 SUMMARY OF CHEMICAL AND BIOLOGICAL FINDINGS

Wilson Reservoir, Kansas

		1964	1965	1966	1967	1968
Soluble phosphorus,	inlet[a]	20	20	10	–	30
μg/l	dam[b]	30	40	5	–	10
Total phosphorus,	inlet	30	30	25	30	100
μg/l	dam	50	50	10	30	40
Inorganic nitrogen,	inlet	0.26	0.19	0.34	–	0.14
mg/l	dam	0.1	0.26	0.16	–	0.13
Phytoplankton, algae/ml	dam	10,000	86,000	200	780	1,800
Bottom organisms	dam					
No./ft²		40	65	38	2	9
No./m²		430	699	409	21	97
Productivity,	dam					
mg C/m³ day		–	840	55	–	155
Depth of Euphotic zone	dam					
ft	–	–	5.2	5.2	5.2	11
m			1.6	1.6	1.6	3.3
Sulfates	inlet	990	560	540	100	260
	dam	610	148	226	240	291
Chlorides	inlet	3,840	1,180	1,320	226	598
	dam	1,540	237	410	500	580

[a] Farthest upstream impounded station in 1964, station 12; 1965–1968, station 9.
[b] Reservoir station closest to the dam (station 15). Nitrogen and phosphorus concentrations are reported as the element.

Source: Reprinted, with publisher's permission, from *J. Am. Water Works Assn., 62*(7):439, © 1970, by the American Water Works Association, Inc., 2 Park Ave., New York, N.Y. 10016.

1.1–50 WATER QUALITY OF LAKE BLOOMINGTON AND TRIBUTARIES

Bloomington, Illinois

Flow direction station[a]	**M**	**4.5M**	**2.5M →**	**1**	**← 6H**	**7H**	**H**
pH	8.4	8.4	8.4	8.4	8.4	8.4	8.3
Alkalinity[b]	216	181	171	170	171	192	215
Hardness[b]	296	265	251	245	251	289	314
Manganese	0.37	0.11	0.66	0.04	0.05	0.11	0.26
Total iron	2.48	1.06	0.59	0.34	0.47	1.30	2.47
Fe^{++}	–	0.17	0.10	0.04	0.07	0.20	–
Fe^{+++}	–	0.89	0.49	0.30	0.40	1.10	–
Total nitrogen	10.31	8.89	8.81	7.72	7.74	11.71	15.70
NO_3-N	8.29	7.01	6.45	5.68	5.69	9.88	13.35
NH_4-N	Trace	Trace	Trace	Trace	Trace	Trace	Trace
Kjeldahl-N	2.02	1.88	2.36	2.04	2.05	1.83	2.35
Orthophosphate-P	0.02	0.01	0.01	Trace	0.01	0.01	0.01
Polyphosphate-P	0.03	0.02	0.02	0.02	0.02	0.02	0.01
Silica	5.00	3.18	2.20	1.85	1.88	3.86	7.10

[a] Mean values, all expressed in mg/l except pH.
[b] As $CaCO_3$.

Source: Reprinted, with publisher's permission, from *J. Am. Water Works Assn., 62*(8):510, © 1970, by the American Water Works Association, Inc., 2 Park Ave., New York, N.Y. 10016.

1.1–51 CHEMICAL ANALYSIS OF RESERVOIR WATER

Waco Reservoir, Texas

pH	8.2
Total alkalinity	140 mg/l
Total hardness	150 mg/l
Chlorides	22 mg/l
Phosphates	<0.1 mg/l
Sulfates	23.5 mg/l
Dissolved oxygen	7.5 mg/l
Biochemical oxygen demand	1.1 mg/l
Chemical oxygen demand	10.0 mg/l
Ammonia–N	0.53 mg/l
Total dissolved solids[a]	207 mg/l

[a] Estimated from electrical conductance.

Source: Reprinted, with publisher's permission, from *J. Am Water Works Assn., 63*(7):462, © 1971, by the American Water Works Association, Inc., 2 Park Ave., New York, N.Y. 10016.

1.1–52 TYPICAL ALGAE TOTAL COUNT

Waco Reservoir, Texas

Month	Organisms/ml	Month	Organisms/ml	Month	Organisms/ml
Jan	294	May	195	Sept	182
Feb	270	June	159	Oct	268
Mar	294	July	234	Nov	182
Apr	212	Aug	202	Dec	213

Source: Reprinted, with publisher's permission, from *J. Am. Water Works Assn., 63*(7):462, © 1971, by the American Water Works Association, Inc., 2 Park Ave., New York, N.Y. 10016.

1.1–53 HYDROLOGICAL AND LIMNOLOGICAL DATA

Reference	Size
Volume at conservation pool level	104,100 acre ft
Surface elevation at conservation pool level	455 ft
Volume of flood control	553,200 acre ft
Surface elevation at flood control level	500 ft
Length of reservoir	5.8 mi
Width of reservoir at dam site	1.1 mi
Area	7,270 acres
Maximum depth	85 ft
Average depth	35 ft
Water intakes: four each, 60 in. x 72 in.	408 ft
	424 ft
Sluice gate–bottom	440.8 ft
Elevation above sea level	445 ft

Source: Reprinted, with publisher's permission, from *J. Am. Water Works Assn., 63*(7):462, © 1971, by the American Water Works Association, Inc., 2 Park Ave., New York, N.Y. 10016.

1.1–54 COD/TOC RATIOS OF RAW WATER RESERVOIRS[a]

Source	Depth beneath reservoir surface, ft	No. of samples	High	Low	Mean
Briones	5	4	1.3	1.0	1.2
Briones	40	4	1.2	1.0	1.1
Chabot	5	6	0.7	0.4	0.6
Chabot	40	6	0.7	0.5	0.6
Lafayette	5	7	0.9	0.6	0.7
Lafayette	40	7	0.9	0.6	0.7
San Pablo	5	4	1.3	0.9	1.2
San Pablo	40	4	1.3	0.9	1.1
USL[b]	5	3	1.5	0.9	1.2
USL	40	3	1.4	1.2	1.3

[a] COD/TOC–chemical oxygen demand/total organic carbon.
[b] Upper San Leandro Filter Plant.

Note: Data are from the East Bay Municipal Utility District, Oakland, California.

Source: Reprinted, with publisher's permission, from *J. Am. Water Works Assn., 61*(3):107, © 1969, by the American Water Works Association, Inc., 2 Park Ave., New York, N.Y. 10016.

1.1–55 CHEMICAL ANALYSES OF SOURCES OF SUPPLY

Constituent	Date	Orinda (Mokelumne), mg/l[a]	Treatment plants San Pablo,[b] mg/l	Sobrante,[b] mg/l	USL,[c] mg/l
Alkalinity, $CaCO_3$	12/67	20	78	74	84
Hardness, $CaCO_3$	12/67	19.8	93.6	92.6	99.5
Noncarbonate, $CaCO_3$	12/67	2.2	15.9	18.8	15.6
Carbonate, $CaCO_3$	12/67	17.6	77.7	73.8	83.9
Total dissolved solids	12/67	32	148	138	158
Bicarbonate, HCO_3	12/67	12.2	74.4	70.6	82.8
Boron, B	12/67	0.01	0.02	0.02	0.03
Carbonate, CO_3	12/67	5.4	3.3	3.2	1.1
Chloride, Cl	12/67	2	10	10	8
Fluoride, F	12/67	0.07	0.14	0.13	0.15
Nitrate, NO_3	1/68	0.25	0.70	0.54	1.75
Nitrite, NO_2	1/68	0.002	0.002	0.002	0.03
Phosphate, PO_4	1/68	0.000	0.005	0.005	0.005
Silica, SiO_2	12/67	8.0	6.0	7.0	7.1
Sulfate, SO_4	12/67	1.0	34.2	32.4	29.0
Aluminum, Al	12/67	0.00	0.68	0.09	0.08
Calcium, Ca	12/67	6.6	26.6	26.2	26.2
Cadmium, Cd	8/67	0.001	0.001	0.001	0.003
Iron, Fe	12/67	0.03	0.00	0.00	0.01
Magnesium, Mg	12/67	0.8	6.6	6.6	8.3
Manganese, Mn	12/67	0.00	0.00	0.00	0.00
Potassium, K	12/67	0.6	1.6	1.6	1.7
Silver, Ag	8/67	<0.005	<0.005	<0.005	<0.005
Sodium, Na	12/67	1.4	13.6	11.6	11.2
Zinc, Zn	8/67	<0.001	0.003	0.009	0.01
Conductance, μmho	12/67	55	253	247	249
pH	12/67	9.65	8.75	8.6	8.05

[a] Mountain supply.
[b] San Pablo Reservoir supply.
[c] Upper San Leandro Filter Plant.

Note: Data are from the East Bay Municipal Utility District, Oakland, California.

Source: Reprinted, with publisher's permission, from *J. Am. Water Works Assn., 61*(3):107, © 1969, by the American Water Works Association, Inc., 2 Park Ave., New York, N.Y. 10016.

1.1–56 SUSPENDED SOLIDS IN NATURAL WATERS

	General debris		Organic matter	
Order of size, μ	Inorganic	Organic	Plant	Animal
10^6–10^5 Size 5	–	Twigs, leaves, uprooted plants, etc	Large detached clumps of filamentous algae or fungi	Fish, amphibians, etc
10^5–10^4 Size 4	–	Twigs, leaves, uprooted plants, etc plus animal fragments	Large detached clumps of filamentous algae or fungi	Fish, amphibians, worms, etc
10^4–10^3 Size 3	–	Twigs, leaves, uprooted plants, etc	Larger filamentous algae (*Characeae, Rhodophyceae*) fragments	Insects, mollusks, also ova, and larvae of larger animals (mollusks, crustaceans, fish, etc) some worms, bryozoa
10^3–10^2 Size 2	Sand particles	Animal and vegetable fragments	Large diatoms and desmids, etc	Crustaceans, hydroids, many sarcodina and infusoria
10^2–10^1 Size 1	Fine sand and silt	Animal and vegetable fragments, e.g., sponge spicules	Most algae	Smaller crustaceans, and worms, ova, motile reproductive units, etc, many protozoans
10^1–10^0 Size 0	Fine silt or clay particles	Animal and vegetable fragments, e.g., sponge spicules	Small algal forms (e.g., *Chlorella*), swarmers, spores, most bacteria, etc	Motile reproductive units or cells

Source: L.D. Bowen, J.W. Lovell, Jr., and D.K.B. Thistlewayte, 3rd Int. Conf. Water Pollut. Res., Water Pollut. Contr. Fed., 1966.

1.2 GROUND WATER

1.2–1 GROUND WATER COMPARED WITH OTHER RESOURCES

Resource	Estimate adjusted to	Unit	Recoverable reserve discovered: Current[a]	Recoverable reserve discovered: Original[a]	Annual withdrawal[a]	Cumulative withdrawal[a]	Annual depletion[a]	Life of current reserve at annual withdrawal rate, yr	Life of current reserve at annual depletion rate, yr	Cumulative depletion of original reserve, %	Current annual depletion of original reserve, %
Petroleum[4]	1958	billion bbl	32	95	2.6	63	2.6	12	12	66	2.7
Natural gas[4]	1958	trillion ft^3	255	433	12	174	12	22	22	40	2.7
Bituminous coal[4,5,6]	1962	billion long tons	950	990	0.40	40	0.40	2,400	2,400	4	0.04
Potash[6]	1962	million tons K_2O	38	56	0.80	18	0.80	48	48	3.2	1.4
Iron ore[6,b]	1962	million long tons	3,500	8,100	100	4,600	100	35	35	57	1.2
Sulfur and pyrite[6]	1962	million long tons	86	138	3	52	3	29	29	38	2.2
Petroleum[4,c]	1958	billion bbl	183	250	2.6	63	2.6	72	72	25	1.0
Natural gas[4,c]	1958	trillion ft^3	1,330	1,500	12	174	12	128	128	12	0.8
						Cumulative deficit					
Saw timber[2,3,8]	1952	billion fbm	2,000	3,500	48	1,500	0	50	∞	43	0.0
Ground water[7,1]	1955	million acre ft	47,500	47,575	52	75	6.1	880	7,800	0.2	0.01

[a] See unit column.
[b] Measured, indicated, and inferred.
[c] Including reserves to be discovered.

Source: Reprinted, with publisher's permission, from *J. Am. Water Works Assn., 55*(1):59, © 1963, by the American Water Works Association, Inc., 2 Park Ave., New York, N.Y. 10016.

REFERENCES

1. MacKichan, K.A., *J. Am. Water Works Assn., 49*(4):369, 1957.
2. *Our Growing Wood Supply*, American Forest Products Industries, Inc., Washington, D.C., 1957.
3. *People and Timber*, U.S. Department of Agriculture, June 1956.
4. Risser, H.E., *Coal and the Future Energy Market* Illinois State Geolog. Survey, 1960.
5. Netschert, B.C., *The Future Supply of Energy Resources*, Resources for the Future, Inc., December 1959.
6. *Mineral Resources of the United States*, Public Affairs Press, 1947.
7. Piper, A.M., in *The Physical and Economic Foundation of Natural Resources (IV) Subsurface Facilities of Water Management and Patterns of Supply-Type Area Studies,* H.R. Comm. for Interior and Insular Affairs, Washington, D.C., 1953, p. 15.
8. Private Communication, U.S. Forest Service.

1.2–2 ANALYSES OF SUBTERRANEAN WATERS OF DIFFERENT GENETIC TYPES

Comparison with Fresh Lake and Oceanic Water

Radicles	Alpine lake water			I			II			III	
	Grains/gal	Reacting values	Reacting values, %	Grains/gal	Reacting values	Reacting values, %	Grains/gal	React ing values	Reacting values %	Grains/gal	Reacting values
Cl	0.18	0.005	8.9	1.40	0.04	5.7	5.11	0.144	12.5	32.45	0.920
Br	–	–	–	–	–	–	–	–	–	–	–
SO_4	0.23	0.005	8.9	0.98	0.02	2.8	0.14	0.002	0.1	0.11	0.002
HCO_3	–	–	–	17.64	0.29	41.5	26.18	0.429	37.4	14.98	0.250
CO_3	0.57	0.018	32.2	–	–	–	–	–	–	–	–
Sum of negative	0.98	0.028	50.0	20.02	0.35	50.0	31.43	0.575	50.0	47.54	1.172
Na	0.16	0.007	12.5	0.84	0.04 (Na + K)	5.7 (Na + K)	3.92 (Na + K)	0.170 (Na + K)	14.8 (Na + K)	22.68	0.844[a]
K	0.08	0.002	3.6	0.09						0.84	0.021
Ca	0.36	0.017	30.3	4.13	0.20	28.6	5.25	0.261	22.7	4.34	0.217
Mg	0.02	0.002	3.6	1.40	0.11	15.7	1.75	0.144	12.5	0.98	0.080
Sum of positive	0.62	0.028	50.0	6.46	0.35	50.0	10.92	0.575	50.0	28.84	1.172
Total salinity	1.60	–	–	26.48	–	–	42.35	–	–	76.38	–
Source of data			OC[b]		OC				OC		OC

[a] With the approximation.
[b] OC – Source's computations within the accuracy of the slide rule.

1.2–2 ANALYSES OF SUBTERRANEAN WATERS OF DIFFERENT GENETIC TYPES (*Continued*)

Comparison with Fresh Lake and Oceanic Water

		IV			V			Oceanic water		
Radicles	**Reacting values, %**	**Grains/gal**	**Reacting values**	**Reacting values, %**	**Grains/gal**	**Reacting values**	**Reacting values, %**	**Grains/gal**	**Reacting values**	**Reacting values, %**
Cl	39.2	76.65	2.161	32.6	1,050	29.61	47.8	1,354.0	38.110	45.1
Br	–	–	–	–	–	–	–	4.2	0.052	
SO_4	0.1	38.15	0.793	11.9	57.9	1.20	1.9	189.0	3.931	4.7
HCO_3	10.7	21.84	0.358	5.5	9.7	0.16	0.3	–	–	–
CO_3	–	–	–	–	–	–	–	4.9	0.163	0.2
Sum of negative	50.0	136.64	3.312	50.0	1,117.6	30.97	50.0	1,552.1	42.256	50.0
Na	36.0							749.7	32.611	38.5
K	1.4	10.22	0.444	6.7	514.4	22.37	36.1	27.3	0.698	0.8
Ca	9.2	36.96	1.844	27.8	110.6	5.51	8.9	29.4	1.467	1.7
Mg	3.4	12.46	1.024	15.5	37.6	3.09	5.0	91.0	7.480	9.0
Sum of positive	50.0	59.64	3.312	50.0	662.6	30.97	50.0	897.4	42.256	50.0
Total salinity	–	196.28	–	–	1,780.2	–	–	2,449.5	–	–
Source of data			OC			OC			OC	

[b] OC – Source's computations within the accuracy of the slide rule.

Source: I.I. Chebetarev, *Water Water Eng.*, *54*(10):146, 1950.

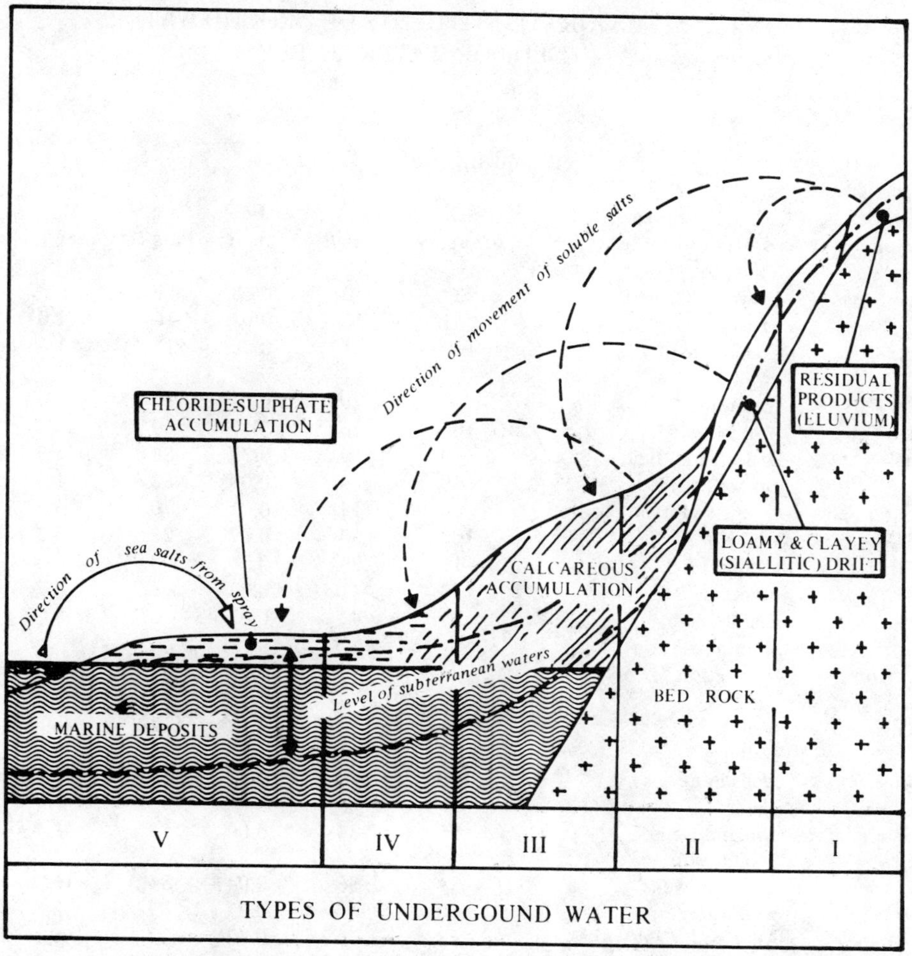

1.2–3 Diagram showing the types of *subterranean waters* in relation to the products of weathering.

Source: I.I. Chebatorev, *Water Water Eng.*, *54*(10):146, 1950.

1.2–4 GENETIC TYPES OF SUBTERRANEAN WATERS IN RELATION TO THEIR SALINITY

Diagram symbols	Types of waters[a]	Classes	Ratio of carbonate to chloride	Practical utility
I	Bicarbonate (alkaline) water	Reversed	>3.0	Excellent in quality
II	Bicarbonate-chloride (alkaline-saline) water	Semireversed	3.0–1.5	Good in quality
III	Chloride-bicarbonate (saline-alkaline) water	Mixed	1.5–0.2	Transitional type; usually of good quality, but often by protracted pumping deteriorates in quality
IV	Chloride-sulfate (saline) water	Brackish	<0.2	Doubtful in quality; sometimes potable
V	Chloride (saline) water	Brine	<0.2	Usually poor in quality

[a] See 1.2–3.

Source: I.I. Chebetarev, *Water Water Eng.*, *54*(10:146, 1950.

1.2–5 MAJOR CONSTITUENTS OF GROUND WATERS USED FOR WATER SUPPLY

Wells and Infiltration Galleries

Location	Source of supply	Percent of supply	SiO_2, mg/l	Fe, mg/l	Ca, mg/l	Mg, mg/l	Na, mg/l	K, mg/l
Montgomery, Ala.	18 wells–Court St.	39	12	0.08	3.5	0.3	51	1.3
	31 wells–Day St.	61	17	0.05	16	1.0	51	1.6
Fresno, Cal.	Composite of wells	100	68	0.00	15	11	17	4.4
Long Beach, Cal.	Well water[a]	–	20	0.04	13	1.0	71	1.0
Jacksonville, Fla.	Well water	100	26	0.15	60	22	14	1.6
Miami, Fla.	Hialeah well field	40	9.5	0.33	88	5.0	22	1.6
	Orr well field	60	4.3	0.00	82	1.8	7.8	0.4
St. Petersburg, Fla.	Cosme well field	100	14.	0.01	70	3.3	6.1	0.7
Savannah, Ga.	Well No. 2	100	53	0.02	28	8.3	9.0	1.7
Honolulu, Haw.	Kaimuki P.S.	13	39	0.74	5.9	6.6	64	2.9
	Beretania P.S.	23	41	0.03	8.7	9.2	35	3.2
	Kalihi underground station	21	42	0.03	3.2	15	34	3.2
Des Moines, Iowa	Infiltration galleries	75	19	0.08	94	33	13	2.8
Topeka, Kans.	Well 2	1	18	0.38	112	18	66	6.4
	Well 3	1	17	0.03	121	21	70	7.0
Wichita, Kans.	Wells in Equus beds	100	22	0.07	66	10	60	3.0
Lincoln, Neb.	Ashland wells–composite	96	34	0.00	58	9.4	25	8.4
	Lincoln wells[b]	4	36	0.04	54	16	25	4.5
Dayton, Ohio	Well water[c]	100	–	0.23	89	33	11	
Memphis, Tenn.	Allen well field	28	8.3	0.72	12	6.1	7.5	0.7
	Sheahan well field	28	8.7	0.75	7.5	3.0	12	0.7
	McCord well field	16	6.8	1.2	7.6	5.1	6.3	0.7
	Parkway well field	28	8.5	0.74	8.4	3.8	17	0.9
Amarillo, Texas	Palo Duro well field		56	0.02	44	39	19	3.8
	McDonald well No. 2		59	0.00	41	37	17	4.9
	Bush well No. 4		72	0.02	39	43	29	6.2
	Westex well No. 3		71	0.01	44	28	27	5.6
	Well 6, section 49		64	0.00	32	34	35	5.6
Lubbock, Texas	City well No. 62, NW well field	7	55	0.00	54	49	129	16
	Well No. 102, shallow water well field	10	43	0.00	98	120	102	30
San Antonio, Texas	Bexar Metropolitan Water District wells		12	0.02	64	17	8.7	1.0
Ranges		.	4.3–72	0.00–1.2	3.2–121	0.3–120	6.1–129	0.4–30

[a] Average analysis by the City of Long Beach, July 1, 1960–June 30, 1961.
[b] Analyzed by the City of Lincoln.
[c] Average analysis by the City of Dayton of monthly composite, 1960.

1.2–5 MAJOR CONSTITUENTS OF GROUND WATERS USED FOR WATER SUPPLY (*Continued*)

Wells and Infiltration Galleries

HCO_3, mg/l	CO_3, mg/l	SO_4, mg/l	Cl, mg/l	F, mg/l	NO_3, mg/l	Dissolved solids (residue at 180°C), mg/l	Hardness as $CaCO_3$, mg/l
118	0	9.2	8.8	0.4	1.9	154	10
149	0	9.6	13	0.4	2.5	203	44
118	0	6.0	5.8	0.1	13	221	82
182	5	13	20	0.4	–	229	37
187	0	87	17	0.7	0.0	373	240
272	0	16	30	0.3	0.1	333	240
248	0	15	12	0.1	0.6	263	212
232	0	0.8	10	0.1	0.7	229	188
133	0	7.2	6.0	0.5	0.0	184	104
81	0	14	76	0.0	0.7	251	42
78	0	8.6	45	0.1	0.0	184	60
62	0	9.0	63	0.0	1.0	215	70
340	0	89	11	0.3	12	464	371
312	0	131	82	0.4	0.5	611	354
338	0	145	87	0.3	0.3	663	390
252	0	68	41	0.4	0.7	404	206
199	0	68	8.5	0.4	0.0	317	183
248	0	59	15	0.3	5.0	333	162
326	0	83	19	–	–	439	356
78	0	3.8	3.0	0.4	1.2	87	55
61	0	3.0	4.2	0.1	0.9	80	31
55	0	5.2	3.0	0.3	0.8	68	40
82	0	3.4	2.0	0.3	0.9	97	36
316	0	37	5.8	4.0	2.8	368	270
305	0	30	3.8	3.3	3.2	352	254
296	0	78	7.8	2.8	4.3	428	274
292	0	31	6.2	3.3	2.0	362	225
313	0	24	6.8	3.5	4.9	364	220
364	0	189	92	3.9	17	794	336
357	0	572	64	7.0	9.4	1,220	738
236	0	25	17	0.4	3.8	278	230
55-364	0–5	0.8–572	2.0–92	0.0–70	0.0–17	68–1,200	10–738

Source: C.N. Durfor and E. Becker, *Geological Survey Water Supply Paper,* U.S. Department of the Interior, 1965.

1.2–6 MINOR CONSTITUENTS OF GROUND WATERS USED FOR WATER SUPPLY

Wells and Infiltration Galleries

Location	Noncarbonate hardness as $CaCO_3$, mg/l	Specific conductance μmho at 25°C	pH	Color, units	Beta activity, pCi/l	Radium (Ra), pCi/l	Uranium (U), μg/l	Silver (Ag), μg/l	Aluminum (Al), μg/l	Boron (B), μg/l	Barium (Ba), μg/l	Chromium (Cr), μg/l	Copper (Cu), μg/l
Montgomery, Ala.	0	243	7.2	0	–	–	–	ND[a]	83	110	67	ND	4.3
	0	305	7.3	0									
Fresno, Cal.	0	249	7.5	3	12	<0.1	0.5	ND	6.2	17	34	0.60	1.0
Long Beach, Cal.	0	–	8.7	76									
Jacksonville, Fla.	87	504	7.9	5	–	–	–	ND	11	34	15	0.71	0.85
Miami, Fla.	17	530	7.8	45	–	–	–	<0.50	12	71	21	ND	2.2
	9	444	7.7	5	–	–	–	ND	6.9	8.5	5.7	ND	1.4
St. Petersburg, Fla.	0	382	7.6	10	–	–	–	ND	13	13	14	<0.36	<0.8
Savannah, Ga.	0	236	8.0	5	–	–	–	<0.26	2.9	10	13	0.31	1.8
Honolulu, Haw.	0	388	6.7	–	1.8	<0.1	<0.1	<0.34	6.7	24	5.4	1.1	1.3
	0	289	7.0	–	2.6	<0.1	<0.1	<0.26	7.9	28	4.6	1.1	3.1
	19	312	6.6	–	4.6	<0.1	<0.1	<0.29	5.5	35	13	1.1	8.4
Des Moines, Iowa	92	706	7.8	3	–	–	–			80			
Topeka, Kans.	98	958	7.3	3						110			
	113	1,030	7.4	4						160			
Wichita, Kans.	0	651	7.9	1						20			
Lincoln, Neb.	20	475	7.4	2						50			
	0	–	–	–						0			
Dayton, Ohio	89	–	7.6	–									
Memphis, Tenn.	0	137	6.8	5									
	0	114	6.9	5									
	0	108	6.7	5									
	0	138	7.4	5									
Amarillo, Texas	12	565	7.1	0									
	4	536	7.1	0									
	32	625	7.1	0									
	0	529	7.1	0									
	0	547	7.2	0									
Lubbock, Texas	38	1,200	7.1	0									
	446	1,660	7.0	0	130	1.9	250	1.5	32	500	30	<1.5	15
San Antonio, Texas	36	461	6.8	0									
Range	0–446	108–1,660	6.7–8.7	0–76	1.8–130	<0.1–1.9	<0.1–250	ND–1.5	2.9–83	0–500	4.6–67	ND–1.1	<0.8–15

[a] ND, looked for but not found.

1.2–6 MINOR CONSTITUENTS OF GROUND WATERS USED FOR WATER SUPPLY

Wells and Infiltration Galleries

Location	Iron (Fe), μg/l	Lithium (Li), μg/l	Magnesium (Mn), μg/l	Molybdenum (Mo), μg/l	Nickel (Ni), μg/l	Lead (Pb), μg/l	Rubidium (Rb), μg/l	Strontium (Sr), μg/l	Titanium (Ti), μg/l	Vanadium (V), μg/l	Zinc (Zn), μg/l	Date of sample collection
Montgomery, Ala.	45	0.12	48	ND	ND	ND	3.3	93	<2.4	ND	ND	8/29/61
Fresno, Cal.	4.8	0.62	ND	<0.9	ND	5.7	2.8	19	ND	12	ND	7/19/61
Long Beach, Cal.												7/ 1/60–6/30/61
Jacksonville, Fla.	1.1	ND	ND	ND	<4.7	ND	ND	400	ND	ND	<470	8/29/61
Miami, Fla.	230	0.50	35	2.8	ND	31	ND	490	<5.0	ND	ND	8/23/61
	450	<0.41	<4.1	ND	<4.1	4.1	ND	280	ND	ND	ND	8/23/61
St. Petersburg, Fla.	79	0.7	<3.6	ND	<3.6	ND	ND	97	ND	ND	ND	8/21/61
Savannah, Ga.	15	1.4	<2.6	ND	ND	<2.6	3.7	39	ND	ND	<262	8/30/61
Honolulu, Haw.	15	0.74	ND	ND	ND	0.6	<3.4	50	<1.0	11	ND	8/29/62
	15	0.56	ND	<0.77	<2.6	4.9	3.3	67	<0.8	7.9	ND	8/29/62
	21	<0.29	3.7	ND	<2.9	5.8	3.7	60	1.0	12	ND	8/29/62
Des Moines, Iowa			60									1/22/62
Topeka, Kans.			180									3/ 9/62
			150									3/ 9/62
Wichita, Kans.												2/13/62
Lincoln, Neb.			340									1/23/62
			0									7/10/61
Dayton, Ohio												1960
Memphis, Tenn.			10									9/15/61
			20									9/15/61
			10									9/15/61
			10									9/15/61
Amarillo, Texas			0									2/12/62
			10									2/12/62
			0									2/12/62
			10									2/12/62
			0									2/12/62
Lubbock, Texas			10									2/14/62
	6,600	96	75	68		38	ND	6,300	ND	74	ND	2/14/62
San Antonio, Texas												2/ 8/62
Range	1.1–6,600	ND–96	ND–340	ND–68	ND–<15	ND–38	ND–3.7	19–6,300	ND–<5	ND–74	ND–<470	

Note: No data was found for beryllium, cobalt, phosphorus, tin, gallium, zirconium, ytterbium, and yttrium.

[a] ND, looked for but not found.

Source: C.N. Durfor and E. Becker, *Geological Water Supply Paper,* U.S. Department of the Interior, 1965.

1.2–7 GROUND WATER QUALITY OF WELLS AND SPRINGS

Characteristic	Arizona well water[1]	Lowell, Fla.[2] Well 1	Lowell, Fla.[2] Well 2	Lowell, Fla.[2] Well 3	Billerica raw water[3] Well 1	Billerica raw water[3] Well 2	Billerica raw water[3] Well 3	Mixture of wells 1, 2 and 3	Lowell, Fla. well No. 1[2]	Hackettstown[4]	Adjusted spring water[5]
Silica, mg/l	33	12	11	11						15	
Iron, mg/l	0.15	0.01	0.03	0.03	10	10.1	7	6		13	<0.01
Calcium, mg/l	445	71	70	64						156	96
Magnesium, mg/l	181	8	7.3	7.5						193	26
Sodium, mg/l	606	2	2	2						33	
Potassium, mg/l	8.5										
Boron, mg/l	1.0								0.0023–0.023		
Bicarbonate, mg/l	360	214	215	205						344	
Sulfate, mg/l	1,040	4.5	5.2	6.5						25	132
Chloride, mg/l	1,290	12	11	5	300	240	170	88		13	14
Fluoride, mg/l	0.6	0.2	0.1	<0.1							
Nitrate, mg/l	45	0.61	0.55	2.2	0	0	0	0.1			
Total solids, mg/l	3,830										460
Hardness, mg/l CaO_3	1,850				92	76	84	74		349	354
Total dissolved solids, mg/l		229	220	230							
Temperature, °C		21.5	21.5	21.5							
pH		7.4	7.4	7.4	5.9	6.0	6.4	6.0		7.2	
Turbidity, units					60	10	40	8			
CO_2, mg/l								55		39	
Nitrites, mg/l					0	0	0.01	0			
Alkalinity, mg/l, methyl orange					20	24	30	29			
Manganese, mg/l					1.3	1.1	1	1.3	0.023–0.23	1.7	<0.05
Chromium, μg/l									23–230		
Copper, μg/l									23–230		<20
Nickel, μg/l									2.3–23		
Strontium, μg/l									23–230		
Titanium, μg/l									23–230		
Silver, μg/l									0.23–2.3		
Hydroxide alkalinity, mg/l										0	
Carbonate alkalinity, mg/l										0	
Alkalinity, mg/l $CaCO_3$											270
Zinc, mg/l											<0.01

REFERENCES

1. E.H. Sieveka, *J. Am. Water Works Assn.*, *52*(5):556, 1960.
2. A.P. Black, et al., *J. Am. Water Works Assn.*, *57*(11):1401, 1965.
3. G.J. Coogan, *J. Am. Water Works Assn.*, *54*(12):1507 1962.
4. S.F. Alling, *J. Am. Water Works Assn.*, *55*(6):749, 1963.
5. J.M. Cohen et al., *J. Am. Water Works Assn.*, *52*(5):660, 1960.

1.2–8 CHEMICAL CHARACTERISTICS OF WELL WATER

Location[a]	Fe, mg/l	pH	Hardness as $CaCO_3$, mg/l	Alkalinity as $CaCO_3$	Total Kjeldahl as N	Free ammonium as N	NO_3, mg/l	Dissolved solids (residue at 180°C), mg/l	Na, mg/l	Cl, mg/l	SiO_2, mg/l	Free CO_2 as CO_2	Ca hardness as $CaCO_3$	SO_4, mg/l
Markham, Canada[1]	0.5	7.6–8.0	150	200	1.4	0.99		220	24	15	19	14	98	
Markham Twp., Canada[1]	0.2–0.7	7.7–7.8	300		0.54	0.02	0.02	460	22	55	20		220	50
Gravenhurst, Canada[1]	1.45	7.1	96	56			0.014	180	12		19		55	28
Kansas City, Mo.[2]	16	7.2	640	460				702		37.5		107.8		103.2
Richmond, Ind.[3]	0.02		308	243						13				
			322	245						12				
Buckeye, Ariz.[4]	0.4	7.6						2,140	598	1,100	5.0			169

[a] All samples are from wells except for Richmond, Indiana which is from filter gallery.

REFERENCES

1. F.J. Dart and P.D. Foley, *J. Am. Water Works Assn.*, *62*(10):663, 1970.
2. J.R. Popalisky and F. Pogge, *J. Am. Water Works Assn.*, *59*(9):1121, 1967.
3. R.S. Wolfe, *J. Am. Water Works Assn.*, *52*(10):1335, 1960.
4. W.C. Hammer, *J. Am. Water Works Assn.*, *56*(12):1537, 1964.

1.2–9 BACTERIOLOGICAL QUALITY OF RAW WATER FROM WELLS

Total coliform		Fecal coliform		Plate count	
Density	Per Cent Occurrence	Density	Per Cent Occurrence	Number	Per Cent Occurrence
0/100 ml	91	0/100 ml	98	0/ml	17
1/100 ml-4/100 ml	6	1/100 ml-4/100 ml	1	1/ml-100/ml	59
5/100 ml-10/100 ml	1	5/100 ml-10/100 ml	<1	101/ml-500/ml	13
11/100 ml-50/100 ml	1	11/100 ml-50/100 ml	<1	501/ml-1,000 ml	4
>50/100 ml	1	>50/100 ml	<1	>1,000/ml	7

Note: For this analysis 621 individual wells were studied.

Source: Reprinted, with publisher's permission, from *J. Am. Water Works Assn., 62*(11):670, © 1970 by the American Water Works Association, Inc., 2 Park Ave., New York, N.Y. 10016.

1.2–10 CONCENTRATION OF TRACE ORGANICS IN MISSOURI WATERS 1965–1966

		Organic extracts, μg/l			
Location	Unit no.	Chloroform	Ethanol	Acetone/Benzene	Benzene/Acetone
Spring Run 1[c]	1	30.4	47.6	24.5/0.0	10.0/10.8
	2	10.8	39.6	28.2/0.0	12.9/10.2
	3	2.5	21.8	18.8/0.0	8.8/ 5.5
	Total	43.7	109.0	71.5/0.0	31.7/26.5
Spring Run 2[c]	1	51.7	76.4	4.0/0.8	2.0/ 1.6
	2	21.8	73.3	9.9/1.0	6.8/17.4
	3	18.6	45.8	15.8/5.6	2.6/14.5
	Total	92.1	195.5	29.7/7.4	11.4/33.5
Well 1[d]	1	4.8	8.0	3.5/1.4	0.0/18.5
	2	0.0	0.0	2.5/1.0	0.0/15.4
	3	0.0	0.0	2.0/1.0	0.0/12.6
	Total	4.8	8.0	8.0/3.4	0.0/46.5
Well 2[e]	1	0.1	2.1	0.2/0.1	1.2/ 1.4
	2	0.0	0.0	0.2/0.2	1.1/ 1.2
	3	0.0	0.0	0.2/0.1	1.1/ 1.1
	Total	0.1	2.1	0.6/0.4	3.4/ 3.7
Missouri River at St. Louis	a	36.2[b]	–	–	–

[a] Only one filter unit was employed.
[b] Average of 24 biweekly samples, individual values ranged from 17–71 μg/l.
[c] Van Buren dolomite.
[d] Potosi limestone, depth 1,150 ft.
[e] Lomotte sandstone, depth 1,745 ft.

Source: S.G. Grigoropoulos *et al.*, in *Trace Substances in Environmental Health*, vol. 3, D.D. Hemphill, Ed., University of Missouri Press, Columbia, 1969.

1.2–11 ELEMENTAL COMPOSITION AND EMPIRICAL FORMULAS OF TRACE ORGANICS RECOVERED FROM MISSOURI WATERS

Extract	Elemental composition, % C	H	O	S	P	N	Empirical formula
Spring							
Run 1 Unit 1							
CCE	59.0[a]	6.96[a]	24.2[a]	<0.1	0.36	0.68	$C_{4.5}H_{6.8}O_{1.5}N_{0.05}P_{0.01}$
CAE	51.2[a]	6.52[a]	36.0[a]	<0.1	0.35	2.53	$C_{4.2}H_{6.4}O_{2.4}N_{0.18}P_{0.01}$
Spring							
Run 2 Unit 2							
CCE	62.5	7.40	24.0	0.20	0.22	0.54	$C_{5.2}H_{7.4}O_{1.5}N_{0.04}P_{0.01}S_{0.01}$
CAE	46.3	7.47	36.6	0.13	2.80	2.04	$C_{3.9}H_{7.5}O_{2.3}N_{0.14}P_{0.09}$
Well 1							
Unit 1							
CCE	52.5[a]	6.74[a]	14.7	27.8[a]	0.22	0.32	$C_{4.2}H_{6.4}O_{0.92}N_{0.02}P_{0.01}S_{1.0}$
CAE	42.9	6.81	36.0	<0.1	0.24	1.36	$C_{3.6}H_{6.8}O_{2.3}N_{0.10}P_{0.01}$
Missouri River							
at St. Louis							
CCE	67.2	7.78	18.5	1.36	0.10	1.68	$C_{5.6}H_{7.8}O_{1.2}N_{0.12}S_{0.04}$

[a] Average of analyses on two separate samples.

Source: S.G. Grigoropoulos, *et al.,* in *Trace Substances in Environmental Health,* vol. 3, D.D. Hemphill, Ed., University of Missouri Press, Columbia, 1969.

1.2–12 OXYGEN DEMAND OF TRACE ORGANICS RECOVERED FROM MISSOURI WATERS

	TOC	COD		BOD, 5 day	
Extract	mg C/mg extract	mg O_2/mg extract	% of TOC	mg O_2/mg extract	% of COD
Spring					
Run 1 Unit 1					
CCE	1.85	1.37	74.0	0.260	19.3
CAE	1.52	1.49	98.0	0.229	15.3
Unit 2					
CCE	–	1.48	–	0.129	8.7
CAE	–	1.30	–	0.104	8.0
Unit 3					
CCE	–	1.35	–	0.150	11.1
CAE	–	1.40	–	0.140	10.0
Spring					
Run 2 Unit 1					
CCE	–	1.32	–	0.176	12.9
CAE	–	1.31	–	0.215	16.5
Unit 2					
CCE	2.01	2.00	99.5	0.164	8.2
CAE	1.48	1.40	94.5	0.191	13.6
Unit 3					
CCE	–	1.72	–	0.144	8.4
CAE	–	1.94	–	0.130	6.7
Well 1					
Unit 1					
CCE	1.72	1.64	95.3	0.04	0.0
CAE	1.55	1.46	94.2	0.04	0.0
Missouri River at St. Louis					
CCE	2.39	2.00	84.0	–	–

Source: S.G. Grigoropoulos *et al.*, in *Trace Substances in Environmental Health*, vol. 3, D.D. Hemphill, Ed., University of Missouri Press, Columbia, 1969.

1.2–13 ODOR POTENTIAL AND CHARACTERISTIC ODOR OF TRACE ORGANICS RECOVERED FROM MISSOURI WATERS

Extract	Characteristic odor	Threshold odor concentration, $\mu g/l$ 20°C	60°C
Spring			
Run 1 Unit 1			
CCE	Musty	2,310	203
CAE	Musty-flowery	3,400	151
Unit 2			
CCE	Musty	110	52
CAE	Earthy-chemical	166	35
Unit 3			
CCE	Musty-chemical	564	110
Spring			
Run 2 Unit 1			
CCE	Musty-medicinal	226	127
CAE	Musty-medicinal	553	140
Unit 2			
CCE	Musty-flowery	110	58
CAE	Musty-flowery	267	140
Well 1			
Unit 1			
CCE	Chemical	15,500	880
CAE	Medicinal-chemical	28,300	3,440

Source: S.G. Grigoropoulos *et al.,* in *Trace Substances in Environmental Health,* vol. 3, D.D. Hemphill, Ed., University of Missouri Press, Columbia, 1969.

1.2–14 CONCENTRATION OF TRACE ORGANICS RECOVERED FROM SUBSURFACE WATERS

Carbon Adsorption Method

Source of water		Carbon filters		Water sampled			Organic extract, μg/l			
Type	Location	Size, ft³	No.	Volume, gal	Rate, gpm	Solvents used in sequence	Total[a]	CCE[b]	CAE[b]	Ref.
Well	Oakwood, Ill	<0.1	3	1,073	1.0	Chloroform,	5,430	2,170	1,140	
	Philo, Ill		2	950		water,	1,500	130	940	
	Clinton, Ill									
	Run 1		2	1,426		ethanol,	3,750	340	2,000	1
	Run 2		3	1,153		ethanol plus	7,290	550	3,310	
	Atwood, Ill		2	1,293		ammonia	4,570	340	1,360	
Deep well	Rolla, Mo	1.5		262,000	6.2	Chloroform	3.2	0.1	2.1	
	Univ. of Mo., Rolla			142,000	6.8	ethanol,	24.2	4.8	8.0	
Spring	Merance Spring, Mo.									
	Run 1		3	129,000	5.5	acetone,	224.2	43.7	109.0	
	Run 2			133,000	4.9	benzene	324.7	92.1	195.5	
Deep well	Rolla Mo			262,000	6.2	Chloroform	9.3	0.1	2.1	2, 3
						ethanol,				
	Univ. of Mo., Rolla			142,000	6.8	benzene,	59.3	4.8	8.0	
Spring	Meramec Spring, Mo.									
	Run 1			129,000	5.5	acetone	210.9	43.7	109.0	
	Run 2			133,000	4.9		332.5	92.1	195.5	
						PHS Drinking Water Standard		200	–	4
						AWWA Potable Water Goals		40	100	5

[a] The sum of CCE, CAE, and other extracts, as appropriate.
[b] CCE–Carbon chloroform extract; CAE–Carbon alcohol extract.

Source: S.G. Grigoropoulos and J.W. Smith, in *Organic Compounds in Aquatic Environments,* S.J. Faust and J.V. Hunter, Eds., Marcel Dekker, Inc., 1971.

REFERENCES

1. L.R. Robinson *et al., J. Am. Water Works Assn., 59*:227, 1967.
2. S.G. Grigoropoulos and J.S. Smith, *J. Am. Water Works Assn., 60*:586, 1968.
3. J.W. Smith, Ph.D. Dissertation, University of Missouri, 1968.
4. *Drinking Water Standards,* U.S. Public Health Service, 1962, p. 31.
5. *J. Am. Water Works Assn., 60*:1317, 1968.

1.2–15 CHEMICAL ANALYSIS OF GROUND WATER

Southern San Joaquin Valley, California

Date sampled[b]	7/16/49	8/2/55	3/6/61	3/6/61	9/12/57	3/7/61	3/7/61	8/29/56	10/11/57
Well depth, ft	468[a]	300	430	510	773	560	704	–	860
Calcium, Ca	366	85	10	169	20	30	42	60	62
Magnesium, Mg	84	12	0.2	18	5.1	1.9	3.2	9.4	5
Sodium, Na	1,610	389	31	81	17	19	27	98	97
Potassium, K	–	1.5	0.9	2.6	1.1	0.9	1.1	1.8	2.8
Bicarbonate, HCO_3	382	104	76	169	107	107	110	116	112
Sulfate, SO_4	1,440	234	9.9	342	10	16	27	108	257
Chloride, Cl	2,100	543	13	112	7	10	40	142	17
Fluoride, F	–	0.4	0.2	0.1	–	0.2	0.1	0.4	2
Nitrate, NO_3	–	1.7	0.6	25	0	5.6	1.3	0.2	0
Boron, B	8.4	0.88	0.1	0.4	–	0.2	0.2	0.2	34
Silica, SiO_2	30	30	19	26	20	24	23	33	26
Total solids	6,040	1,350	122	859	133	161	219	510	524
pH	7.8	8	7.2	7.5	7.6	7.4	7.5	7.7	7.6
Percent sodium	74	76	71	26	39	33	33	53	54
Hardness as $CaCO_3$	504	261	26	496	64	83	118	188	175

[a] This is an example of the brackish water that underlies fresh ground waters in much of the basin.
[b] Measurements are in ppm.

Source: Reprinted, with publisher's permission, from *J. Am. Water Works Assn.*, *59*(12):1513, © 1967, by the American Water Works Association, Inc., 2 Park Ave., New York, N.Y. 10016.

1.2–16 CHEMICAL ANALYSIS OF WATER FROM PRODUCING ZONES

Sara-Sands Area, Siesta Key, Florida

Year drilled	Depth of well, ft	Depth of casing, ft	Date of collection	Concentration, ppm: Calcium Ca	Magnesium Mg	Sodium Na	Sulfate SO_4	Chloride Cl	Fluoride F	Dissolved solids (residue at 180°C)	Hardness as $CaCO_3$	Specific conductance, μmho at 25°C
Zone 1												
1963	230	105	4/25/63	–	–	–	356	62	2.4	–	610	1,080
			5/14/63	–	–	–	320	85	2.5	–	500	1,100
			7/ 2/63	–	–	–	–	70	–	–	–	1,100
			8/27/63	–	–	–	350	57	2.5	–	525	1,070
			9/25/63	100	72	42	350	60	2.4	870	545	1,010
			3/24/64	96	67	–	–	58	–	–	515	1,090
			7/28/64	–	–	–	–	–	–	–	–	1,090
1957	212	168	4/25/63	–	–	–	408	68	2.4	–	630	1,150
			5/14/63	–	–	–	–	–	–	–	–	1,150
			7/ 2/63	–	–	–	–	68	–	–	–	1,150
			8/27/63	–	–	–	424	68	2.4	–	595	1,200
			9/25/63	110	78	51	413	85	2.5	1,060	595	1,210
			7/29/64	–	–	–	412	220	2.6	–	640	1,600
Zone 2												
1963	316	276	4/25/63	–	–	–	564	125	1.6	–	800	1,480
			5/14/63	–	–	–	600	140	1.5	–	860	1,560
			7/ 2/63	190	102	79	656	168	1.6	1,630	895	1,700
			8/27/63	–	–	–	690	180	1.5	–	945	1,830
			9/25/63	202	108	104	672	210	1.4	1,680	950	1,960
			3/24/64	200	110	–	–	180	–	–	953	1,900
			7/28/64	–	–	–	–	–	–	–	–	1,900
1963	364	279	8/27/63	164	94	80	566	146	1.6	1,400	796	1,570
			3/24/64	96	139	–	–	145	–	–	810	1,590
			7/28/64	–	–	–	–	–	–	–	–	1,600
Zones 2 and 3												
1963	452	276	7/ 2/63	308	140	167	1,060	325	1.5	2,440	1,340	2,650
			8/27/63	–	–	–	1,050	320	1.5	–	1,340	2,630
			3/24/64	308	144	–	–	330	–	–	1,360	2,750
			7/30/64	–	–	–	–	–	–	–	–	2,710

Source: Reprinted, with publisher's permission, from *J. Am. Water Works Assn.*, *59*(12):1504, © 1967, by the American Water Works Association, Inc., 2 Park Ave., New York, N.Y. 10016.

1.2–17 CHEMICAL ANALYSIS OF WATER FROM WELLS AFFECTED BY CANALS OR OTHER WELLS

Year drilled	Depth of well, ft	Depth of casing, ft	Date of collection	Concentration, ppm: Calcium Ca	Magnesium Mg	Sodium Na	Sulfate SO_4	Chloride Cl	Fluoride F	Dissolved solids (residue at 180°C)	Hardness as $CaCO_3$	Specific conductance, μmho at 25°C
Zone 1												
1957	250	88	4/25/63	–	–	–	316	628	2.4	–	660	2,610
			5/14/63	–	–	–	312	272	2.4	–	570	1,630
			7/ 2/63	–	–	–	–	108	–	–	–	1,400
			8/27/63	–	–	–	306	81	2.4	–	485	1,070
			9/25/63	112	51	54	306	82	2.4	846	490	1,030
			3/24/64	160	22	–	–	90	–	–	490	1,100
			7/28/64	–	–	–	–	–	–	–	–	1,110
1958	220	134	4/25/63	–	–	–	380	695	2.4	–	805	2,890
			5/14/63	–	–	–	392	845	2.4	–	890	3,240
			7/ 2/63	208	178	856	444	1,710	2.3	4,190	1,250	5,500
			8/27/63	–	–	–	394	1,620	2.3	–	1,220	5,420
			9/25/63	260	222	1,110	450	2,400	2.3	5,400	1,560	7,690
			3/24/64	220	229	–	–	2,400	–	–	1,490	7,600
			5/20/64	–	–	–	403	2,000	2.4	–	–	6,400
			7/28/64	–	–	–	–	–	–	–	–	6,900
Zone 2												
1963	305	134	4/25/63	–	–	–	968	320	1.5	–	1,300	2,460
			5/14/63	–	–	–	–	–	–	–	–	2,440
			7/ 2/63	–	–	–	–	335	–	–	–	2,550
			8/27/63	–	–	–	990	322	1.5	–	1,320	2,590
			9/25/63	298	140	144	1,000	321	1.3	2,390	1,320	2,610
			3/24/64	284	141	–	–	315	–	–	–	2,600
			5/20/64	–	–	–	960	304	1.5	–	–	2,500
			7/28/64	–	–	–	–	–	–	–	–	2,590

Source: Reprinted, with publisher's permission, from *J. Am. Water Works Assn.*, *59*(12):1504, © 1967, by the American Water Works Association, Inc., 2 Park Ave., New York, N.Y. 10016.

1.2–18 ANALYSIS OF NEW WELLS AND TREATED WATER

Lompoc, California

	Value[a]							
					7th St. Well		Plant Flow[b]	
Characteristic	Well No. 1 (2/7/64)	Well No. 2 (2/7/64)	Well No. 3 (2/7/64)	Well No. 4 (2/7/64)	Raw (3/5/64)	Softened[c] (3/5/64)	Influent (2/4/64)	Effluent[c] (2/4/64)
Calcium as Ca	134.4	188.8	163.6	173.2	146.4	46.4	144	32
Magnesium as Mg	70.5	81.3	82.2	90.0	73.0	12.7	67	22
Sodium as Na	19.1	13.6	1.4	41.9	13.8	146.1	83	207
Iron as Fe								
Total	8.5	0.1	9.0	16.7	0.6	0.1	0.90	0.02
Dissolved	0.1	0.1	0.1	0.1	0.2	0.1	–	–
Manganese as Mn	1.9	0.9	3.4	3.1	0.8	0.0	0.35	0.00
Fluoride as F	0.4	0.3	0.6	0.6	0.3	0.2	0.5	0.3
Nitrite as NO_2	3.0	0	1.5	3.5	4.0	Trace	–	–
Nitrate as NO_3	Trace	Trace	Trace	Trace	Trace	Trace	0.18	0.09
Carbonate as CO_3	0.0	0.0	0.0	0.0	0.0	0.0	0.0	0.0
Bicarbonate as HCO_3	327.6	417.9	380.0	397.1	345.3	82.4	325	126
Sulfate as SO_4	248.6	321.6	289.4	405.6	248.2	258.2	398	384
Chloride as Cl	98.6	111.3	96.5	102.8	108.5	105.7	95	95
Total dissolved solids	740.6	926.0	828.7	1,019.3	766.0	621.0	1,085	860
Alkalinity as $CaCO_3$								
Calcium	268	342	312	326	283	62.5	266	80
Magnesium	0	0	0	0	0	0	0	23
Total	268	342	312	326	283	62.5	266	103
Hardness as $CaCO_3$								
Carbonate	268	342	312	326	283	62.5	266	103
Calcium, noncarbonate	68	129	97	107	83	53.5	94	0
Magnesium, noncarbonate	290	334	338	370	300	52	275	67
Total	626	805	747	803	666	168	635	170
Sodium ratio	8.54	4.80	0.54	13.7	5.92	71.3	28.2	79.4
Sodium adsorption ratio	0.33	0.21	0.02	0.68	0.233	4.90	1.43	6.92
Calcium saturation pH	7.11	6.87	6.96	6.91	7.28	8.20	7.10	8.14
Actual pH	7.4	7.6	7.31	7.0	7.4	7.5	8.0	8.1
Saturation index (65° F)	+0.29	+0.73	+0.34	+0.09	+0.12	–0.70	+0.90	–0.06
Conductivity, µmho/cm	1,290	602	1,538	1,492	1,111	961	–	–

[a] Concentrations in parts per million, where applicable.
[b] Represents Well No. 1, different laboratory than later samples.
[c] Normal effluent hardness is 140–150 ppm. Normal effluent saturation index is positive.

Source: Reprinted, with publisher's permission, from *J. Am. Water Works Assn.,* 57(5):607, © 1965, by the American Water Works Association, Inc., 2 Park Ave., New York, N. Y. 10016.

1.2–19 CHEMICAL ANALYSIS OF WATER WELLS

Lompoc, California

Date of test	Calcium as Ca	Magnesium as Mg	Sodium as Na	Total hardness as $CaCO_3$	Total alkalinity as $CaCO_3$	Total solids	Sulfates as SO_4	Chlorides as Cl	Iron as Fe
				Chemical concentration, ppm					
				City Wells[a]					
5/23/60	145.6	68.3	132	644	263	1,114	504	100.7	1.20
10/2/59	165	76	97	725	411	1,345	246	157	trace
10/2/59	179	89	91	811	394	1,400	441	117	trace
5/16/60	145.6	70.3	133.9	652	266	1,129	504	107.8	3.50
10/2/59[b]	158	77	76	708	378	1,321	328	109	trace
				Private Wells					
3/17/60	156	73	78	691	300	1,221	400	102	0.18
3/17/60	136	67	83	623	300	1,166	360	100	0.18
3/17/60	139	67	78	623	295	1,061	366	84	0.24
10/26/60	197	95	92	883	367	1,517	500	140	0.21
10/26/60	212	100	171.6	940	420	1,805	657.6	146.1	2.80

[a] There was no analysis of two of the city wells.
[b] Mixture from distribution system.

Source: Reprinted, with publisher's permission, from *J. Am. Water Works Assn.*, *57*(5):607, © 1965, by the American Water Works Association, Inc., 2 Park Ave., New York, N.Y. 10016.

1.2–20 MINERAL CHARACTERISTICS OF WATER FROM RADIAL COLLECTOR WELLS

West Virginia

	Well									
	Solvay process				Parkersburg		Silicone division			Mobay
Characteristic	No. 1	No. 2	No. 3	No. 4	No. 4	No. 5	South	North	Center	No. 1
Alkalinity ($CaCO_3$), ppm	100	100	100	80	80	80	84	44	22	75
Hardness ($CaCO_3$), ppm	656	424	256	204	200	200	196	190	188	172
Chloride (Cl), ppm	312[a]	580[a]	93	70	28	3	40	36	32	48
Sulfate (SO_4), ppm							96	80	92	70
Calcium (Ca), ppm	262	170	102	70			156	154	156	54
Magnesium (Mg), ppm							40	36	32	9
Silica (SiO_2), ppm							10	7	9	
Iron (Fe), ppm	0.2	0.2	0.2	0.3	0.4	0.3	0.1	0.1	0.3	0.2
Manganese (Mn), ppm	0.3	0.3	0.3	0.3	2.0	0.9				0.9
pH	6.9	6.7	7.0	7.3	7.5	7.3	7.4	6.9	7.1	6.6

[a] Seepage from waste disposal area.

Source: Reprinted, with publisher's permission, from *J. Am. Water Works Assn.*, *52*(9):1206, © 1960, by the American Water Works Association, Inc., 2 Park Ave., New York, N.Y. 10016.

1.2–21 SUMMARY OF DATA ON COLLECTOR WELLS

West Virginia

Well	Year installed	Depth below surface, ft	Static water depth, ft	Collector lines total length, ft	Initial yield, mgd	Yield, 1960, mgd
Solvay Process Division						
Well No. 1	1954	60	15	756	3.7	
Well No. 2	1952	60	15	800	3.7	
Well No. 3	1952	60	15	802	3.7	8.7
Well No. 4	1954	60	15	756	3.7	
Parkersburg						
Well No. 4	1954	52	26	500	4.0	4.0
Well No. 5	1959	48	26	500	4.2	4.2
National Aniline						
Well No. 1	1953	73	15	800	4.3	
Silicones Division						
South Well	1954	49	29	792	2.0	2.0
North Well	1954	50	24	1,046	1.2	1.1
Center Well	1957	46	24	400	1.7	1.4
Mobay Chemical Co.						
Well No. 1	1955	51		850	2.3	2.3
Wells Previously Reported						
W. Va. Ordnance Works						
Well No. 1	1942	73	35	804	2.5	
Well No. 2	1942	66	36	899	2.8	
Well No. 3	1942	61	34	1,081	2.2	
Well No. 4	1942	62	33	1,007	1.8	
Well No. 5	1942	59	31	1,001	2.1	
Well No. 6	1942	59	33	1,219	2.1	
Du Pont						
Well No. 1	1947	92	30	1,259	4.5+	2.25
Parkersburg						
Well No. 1	1946	60	16	800	3.5	
Well No. 2	1949	56	17	800	2.5	1.5
Well No. 3	1949	54	18	800	4.0	3.0
Calco						
Well No. 1	1947	36	24	1,008	4.0	2.3
Well No. 2	1948	51	25	1,052	1.7	1.2
Well No. 3	1952	54	31	787	4.6	2.7

Source: Reprinted, with publisher's permission, from *J. Am. Water Works Assn.*, *52*(9):1206, © 1960, by the American Water Works Association, Inc., 2 Park Ave., New York, N.Y. 10016.

1.2–22 DATA ON WATERS FROM 77 SHALLOW WELLS

Goitrous District North of Hamburg, Germany

		Constituent, ppm					
		Oxygen absorption		NH_3		NO_3	
No. of wells	Urochrome range	Range	Avg	Range	Avg	Range	Avg
16	0.0– 3.5	4.2– 18.4	9.5	0.0– 5.4	0.76	0–204	38.3
24	3.6– 7.0	6.0– 37.6	18.4	0.0–31.0	3.29	0–280	80.9
8	7.1–10.5	20.2– 49.3	32.3	0.3–12.8	4.10	3–244	90.0
14	10.6–14.0	15.5– 96.8	29.0	0.5– 5.0	1.70	0–236	57.7
7	14.1–24.5	35.9– 71.2	51.0	0.1–11.0	1.70	83–170	135.0
8	24.6–35.0	36.6– 91.6	53.0	0.4–10.0	5.00	37–158	94.0
2	37.1–43.7	92.8–100.2	96.5	2.3–61.0	31.60	72–83	77.0

Source: Reprinted, with publisher's permission, from *J. Am. Water Works Assn.*, *55*(7):887, © 1963, by the American Water Works Association, Inc., 2 Park Ave., New York, N.Y. 10016.

1.2–23 ANALYSIS OF WATER FROM SHALLOW WELLS IN AREA OF CONTAMINATION

South Farmingdale, New York

Constituent	Concentration, ppm		
	Well 1[a]	Well 2[b]	Well 3[c]
Total chromium	3.5	1.5	1.0
Hexavalent chromium	3.5	1.5	1.0
Copper	0.12	0.06	0.16
Aluminum	0.1	0.1	0.1
Cadmium	<0.01	<0.01	0.24
Nickel	<0.02	<0.01	<0.01
Cyanide	<0.01	<0.01	<0.01
Iron	1.6	1.6	<0.03

[a] New York City Test Well 3, located 600 ft south of plant.
[b] New York City Test Well 4, located 900 ft south of plant.
[c] Privately owned shallow well at a home at 123 Yoakum Ave.

Source: Reprinted, with publisher's permission, from *J. Am. Water Works Assn.*, *55*(7):887, © 1963, by the American Water Works Association, Inc., 2 Park Ave., New York, N.Y. 10016.

1.2–24 CHARACTERISTICS OF CONTAMINATED AND UNCONTAMINATED WATERS

Characteristic	Concentration, ppm			
	Sample No.[e]			
	1[a]	2[b]	3[c]	4[d]
Turbidity	3.1		<1.0	
Iron	1.44	0.11	0.22	0.18
Manganese	<0.05	<0.05	0.15	0.25
CO_2	24.0	19.0	88.0	57.0
Chromium, hexavalent	<0.01	6.00	<0.01	1.25
Total chromium	<0.01	6.00	<0.01	1.25
Free ammonia nitrogen	1.7	2.0	6.0	4.0
Albuminoid ammonia nitrogen	0.042	0.248	0.136	0.120
Nitrite nitrogen	0.012	0.31	0.094	0.064
Nitrate nitrogen	3.7	2.9	5.1	0.45
Oxygen consumed by permanganate	1.1	1.1	3.3	1.1
Chlorides	20.0	17.4	20.2	13.0
Total hardness as $CaCO_3$	36.0	16.0	86.0	64.0
Total alkalinity as $CaCO_3$	1.0	14.0	65.0	35.0
Total solids	140.0	177.0	211.0	187.0
Copper	<0.05	<0.05	<0.05	<0.05
Cadmium	<0.01	0.40	<0.01	0.03
Nickel	<0.25	0.25	0.25	0.75
ABS	0.39	0.26	1.60	1.10
Color	<5.0	yellow due to hexa-chromium	10.0	yellow due to hexa-chromium
Odor, cold	disagreeable	oily	musty	aromatic
Odor, hot	aromatic	aromatic	aromatic	aromatic
pH	4.88	6.20	6.25	6.10

[a] Uncontaminated well.
[b] Contaminated well.
[c] Uncontaminated stream.
[d] Contaminated stream.
[e] All samples were filtered before analysis.

Source: Reprinted, with publisher's permission, from *J. Am. Water Works Assn.*, *56*(6):739, © 1964, by the American Water Works Association, Inc., 2 Park Ave., New York, N.Y. 10016.

1.2–25 CHEMICAL ANALYSIS OF BRACKISH WATERS

Location	Water depth, ft	Ca, mg/l	Ma, mg/l	Na, mg/l	HCO_3, mg/l	Cl, mg/l	SO_4, mg/l	Total solids, mg/l	pH	Carbon dioxide, mg/l	Ref.
Martin Co., Fla.	1,300	58	61	252	190	495	83	1,225	7.7	8	1
Brevard Co., Fla.	295	88	90	515	171	882	307	2,053	7.48	19	1
Sarasota Co., Fla.	400	248	335	63	200	540	1,150	2,169	7.3	15	1
Coalinga, Calif.		113	71	512		257	1,158	2,392	8.1		2
		129	89	521	162	262	1,260	2,478	7.7	5	3
Miller, SD	1,245	205	54	338	174	83	1,210	2,010	7.3		4
Buckeye, Ariz.		461	197	631	368	1,310	1,060	3,940	7.5		5
Grandfalls, Texas (Pecos River)		744	370		151	4,050	2,940	10,700	8.0		6
San Diego, Calif. Well No. 8		157	59	282		710	95	1,674	7.9	2.8	7

Location	Total hardness, mg/l	Hydrogen sulfide, mg/l	Hardness as $CaCO_3$, mg/l	Alkalinity as $CaCO_3$, mg/l	F, mg/l	NO_3, mg/l	Fe, μg/l	Mn, μg/l	K, mg/l	B, μg/l	Silica, SiO_2, mg/l	Ref.
Martin Co., Fla.	395	Trace										1
Brevard Co., Fla.	590	2.8										1
Sarasota Co., Fla.	1,990	Trace										1
Coalinga, Calif.			574	141	0.4	0.9	0.1	0.02	8.9	3.3		2
	692						0.018			2.75	49.2	3
Miller, S.D.					2.04	0.03	1.3	0.12	18	1.09	14	4
Buckeye, Ariz.					0.5	52	0.06	0.02	8.9	2.2	32	5
Grandfalls, Texas									2,540		14	6
(Pecos River)								0	8		19	
San Diego, Calif. Well No. 8			638	108	0.05	0	0.09					7

REFERENCES

1. K.G. Barnhill, *J. Am. Water Works Assn., 54*(5):526, 1962.
2. E.S. Cary *et al., J. Am. Water Works Assn., 52*(5):585, 1960.
3. J.W. McCutchen and J.S. Johnson, *J. Am. Water Works Assn., 62*(6):346, 1970.
4. West well, city of Miller, South Dakota. Analyses by U.S. Geological Survey, October 31, 1955.
5. Irrigation well, owned by Buckeye Irrigation Company, Buckeye, Arizona. Analyses by U.S. Geological Survey, April 23, 1955.
6. Pecos River below Grandfalls, Texas. Analyses by U.S. Geological Survey, April 1951.
7. E.I. Crossley and F.O. Waters, *J. Am. Water Works Assn., 62*(3):188, 1970.

1.2–26 BRACKISH MUNICIPAL WATER SUPPLIES IN THE U.S.

State	Total no. of supplies	Brackish water supplies: 1,000–3,000 ppm total dissolved solids	Brackish water supplies: 3,000–10,000 ppm total dissolved solids	State	Total no. of supplies	Brackish water supplies: 1,000–3,000 ppm total dissolved solids	Brackish water supplies: 3,000–10,000 ppm total dissolved solids
Alabama	365	8	0	Montana	158	32	2
Alaska	33	0	0	Nebraska	434	0	0
Arizona	400	112	1	Nevada	45	0	0
Arkansas	238	4	0	New Hampshire	138	0	0
California	940	24[a] 44[b]	0	New Jersey	426	1	0
Colorado	257	15	0	New Mexico	198	3	0
Connecticut	145	0	0	New York	1,071[e]	0	0
Delaware	42	0	0	North Carolina	316	2	0
Florida	779	26[c] 28[d]	0	North Dakota	165	83	5
Georgia	600	0	0	Ohio	676	29	0
Hawaii	75	2	0	Oklahoma	421	43	7
Idaho	200	0	0	Oregon	472	1	0
Illinois	923	61	2	Pennsylvania	1,300	0	0
Indiana	375	0	0	Rhode Island	45	0	0
Iowa	712	143	8	South Carolina	153	2	0
Kansas	460	30	0	South Dakota	250	153	3
Kentucky	No data			Tennessee	300	0	0
Louisiana	246	12	0	Texas	1,230	70	0
Maine	165	0	0	Utah	No data		
Maryland	161	1	0	Vermont	190	0	0
Massachusetts	331	0	0	Virginia	171	1	0
Michigan	750	20	0	Washington	1,000	2	0
Minnesota	602	59	0	West Virginia	400	0	0
Mississippi	239	3	0	Wisconsin	570	3	0
Missouri	495	12	2	Wyoming	78	1	0

[a] All water utility sources.
[b] Only one or more of the water utility sources.
[c] Municipal.
[d] Housing subdivision and trailer parks.
[e] Serving 1,737 communities.

Source: Reprinted, with publisher's permission, from *J. Am. Water Works Assn.*, *53*(9):1091, ©1961, by the American Water Works Association, Inc., 2 Park Ave., New York, N.Y. 10016.

1.2–27 BRACKISH MUNICIPAL WATER SUPPLIES

Canada

Province	Total supplies	Brackish water supplies: 1,000–3,000 ppm total dissolved solids	3,000–10,000 ppm total dissolved solids
British Columbia	281	0	0
Manitoba	41	4	0
Newfoundland	29	0	0
Nova Scotia	44	0	0
Saskatchewan	80	32	1

Source: Reprinted, with publisher's permission, from *J. Am. Water Works Assn.*, *53*(9):1091, © 1961, by the American Water Works Association, Inc., 2 Park Ave., New York, N.Y. 10016.

1.2–28 CHARACTERISTICS OF MINERALIZED WATERS

Description of waters	Concentration of dissolved solids, ppm
Slightly brackish	1,000– 3,000
Moderately brackish	3,000–10,000
Highly brackish	10,000–33,000
Sea water	33,000–36,000
Brine[a]	36,000+

[a] As found in the Dead Sea and Salt Lake, for example.

Source: Reprinted, with publisher's permission, from *J. Am. Water Works Assn.*, *54*(5):519, © 1962, by the American Water Works Association, Inc., 2 Park Ave., New York, N.Y. 10016.

1.2–29 DISTRIBUTION OF SODIUM CONCENTRATION IN WATER FROM PRIVATE WELLS INTO SPECIFIC CONCENTRATION GROUPS

Durham County, N.C.

Sodium concentration, mg/l	No. wells
1–25	49
26–50	26
51–75	6
76–100	10
101–125	3
126–150	3

Source: Cantarow and Trumper, *Clinical Biochemistry*, 2nd ed., Saunders, Philadelphia, 1939.

1.2–30 TYPICAL VALUES OF THE STANDARD COEFFICIENT OF PERMEABILITY FOR VARIOUS MATERIALS

Material	Approximate range K_S, gal/day ft^2	K_S, l/day cm^2
Clean gravel	$10^6 - 10^4$	$4.08 \times 10^3 - 4.08 \times 10^1$
Clean sands, mixtures of clean sands and gravels	$10^4 - 10^1$	$4.08 \times 10^1 - 4.08 \times 10^{-3}$
Very fine sands; silts; mixtures of sand, silt, and clay; stratified clays, etc	$10^1 - 10^{-3}$	$4.08 \times 10^{-3} - 4.08 \times 10^{-6}$
Unweathered clays	$10^{-3} - 10^{-4}$	$4.08 \times 10^{-6} - 4.08 \times 10^{-7}$

Source: J.W. Clark and W. Viessman Jr., *Water Supply and Pollution Control,* International Textbook, Scranton, Pa., 1965, p. 85.

1.2–31 APPROXIMATE PERMEABILITIES OF VARIOUS MATERIALS

Typical of material and grain size range, in.	Typical effective size D_{90} mm	in.	Estimated coefficient of permeability, gal/day ft$_2$ Avg	General range	Max
Clay and silt <0.0029		0.001	10	0 to 25	50
		Sand			
Fine (0.029–0.010)	0.10	0.004	175[a]	50 to 350	800
Medium (0.010–0.024)	0.25	0.010	600[a]	100 to 1,200	3,000
Coarse (0.024–0.079)	0.50 to 1.00	0.020 to 0.040	1,500[a]	300 to 2,000	5,000
		Gravel			
Fine (0.079–0.375)	2	0.080	2,500[a]	1,000 to 5,000	10,000
Medium (0.257–0.500)	5	0.200	3,500[a]	3,000 to 10,000	20,000

Medium sand and coarse sand mixture (average) 1,000 (U.C. 2.5)
Coarse sand and fine gravel mixture (average) 1,500 (U.C. 2.5)

[a] Material is clean, well rounded, and generally uniformly graded (U.C. of 1.5 assumed).

Source: Reprinted, with publisher's permission, from *J. Am. Water Works Assn., 59*(10):1292, © 1967 by the American Water Works Association, Inc., 2 Park Ave., New York, N.Y. 10016.

REFERENCE

1. C.F. Tolman, *Ground Water,* McGraw-Hill, New York, 1937, p. 387.

1.3 ESTUARINE AND MARINE WATERS

1.3–1 CHEMICAL COMPOSITION OF SEAWATER

Macrochemical components

Element	**mg/l**	**Element**	**mg/l**
B	4.6–4.8	K	380
Br	65	Mg	1.8×10^{3}
C	28	Na	1.1×10^{4}
Ca	400	O	8.6×10^{5}
Cl	1.9×10^{4}	S	900
F	1.3	Si	0.2–4.0
H	1.1×10^{5}	Sr	8.0–13.0

Microchemical components

Element	**μg/l**	**Element**	**μg/l**
He	5×10^{-2}	Nb	0.01–0.02
Li	100–200	Mo	0.7–12
Be	5×10^{-4}	Ag	0.1–0.3
N	6–700	Cd	0.11
Ne	0.3	In	20
Al	10–70	Sn	3
P	1–100	Sb	0.5
A	400–700	I	50–60
Sc	4×10^{-2}	Cs	0.5–2
Ti	0.02–1	Ba	6–54
V	0.3–2	La	0.3
Cr	0.05–0.25	Ce	0.4
Mn	0.4–10	Y	0.3
Fe	2–50	W	0.1
Co	0.1–0.5	Au	0.004–0.4
Ni	0.1–2	Hg	0.03–0.27
Cu	1–10	Tl	0.01
Zn	5–10	Pb	3–400
Ga	0.007–0.5	Bi	0.2
Ge	0.01–0.1	Rn	9×10^{-12}
As	3–38	Ra	3×10^{-8}
Se	4–6	Th	0.4–0.7
Rb	120–200	U	2–3

Source: L.E. Ciaccio, *Water and Water Pollution Handbook*, vol. 1, Marcel Dekker, New York, 1971.

REFERENCES

1. A.D. Goldberg, *Geol. Soc. Am. Mem.*, *67*(1), 345, 1957.
2. F.W. Clarke, *Data of Geochemistry*, U.S. Geol. Surv. Bull. 770, 1924.
3. H.T. Odum, *Publ. Inst. Marine Sci.*, University of Texas, 4(2):23, 1957.
4. D.W. Hood, *Oceanogr. Marine Biol. Ann. Rev.*, 1:129, 1963.

1.3–2 CHEMICALS IN SEA WATER

Constituent	Concentration,[1] mg/kg	Concentration,[2] mg/l	Residence time, years[2],[a]
Chlorine	18,980	19,000	
Sodium	10,560	10,600	260,000,000
Sulfate	2,560		
Magnesium	1,272	1,350	45,000,000
Sulfur	885		
Calcium	400	400	8,000,000
Potassium	380	380	11,000,000
Bicarbonate	142		
Bromine	65	65	
Carbon		28	
Oxygen		8	
Strontium	13	8	19,000,000
Boron	4.6	4.6	
Silicon	0.04–8.6[b]	3.0	8,000
Fluorine	1.4	1.3	
Nitrogen	0.03–0.9	0.8	
Argon		0.6	
Lithium	0.1	0.17	20,000,000
Rubidium	0.2	0.13	270,000
Phosphorus	0.001–0.10	0.07	
Iodine	0.05	0.06	
Barium	0.05	0.03	84,000
Indium		0.02	
Aluminum	0.16–1.9	0.01	100
Iron	0.002–0.02	0.01	140
Zinc	0.005–0.014	0.01	180,000
Molybdenum		0.01	500,000
Selenium	0.004	0.004	
Copper	0.001–0.09	0.003	50,000
Arsenic	0.003–0.024	0.003	
Tin	0.003	0.003	100,000
Lead	0.004–0.005	0.003	2,000
Uranium	0.00015–0.0016	0.003	500,000
Vanadium		0.002	10,000
Manganese	0.001–0.01	0.002	1,400
Titanium	Trace	0.001	160
Thorium	⩽0.0005	0.0007	350
Cobalt	0.0001	0.0005	18,000
Nickel	0.0001–0.0005	0.0005	18,000
Gallium		0.0005	1,400
Cesium	~0.002	0.0005	40,000
Antimony		0.0005	350,000
Cerium		0.0004	80
Yttrium		0.0003	7,500
Neon		0.0003	
Krypton		0.0003	

[a] Calculated residence time = ratio of amount of element present in the sea to either the gross contribution by rivers or the estimated sedimentation rate.

[b] Silicate.

1.3–2 CHEMICALS IN SEA WATER (*Continued*)

Constituent	Concentration,[1] mg/kg	Concentration,[2] mg/l	Residence time, years[2],[a]
Lanthanum		0.0003	440
Silver		0.0003	2,100,000
Bismuth		0.0002	45,000
Cadmium		0.0001	500,000
Tungsten		0.0001	1,000
Germanium		0.0001	7,000
Xenon		0.0001	
Chromium		0.00005	350
Beryllium		0.00005	150
Scandium		0.00004	5,600
Mercury	0.0003	0.00003	42,000
Niobium		0.00001	300
Thallium		0.00001	
Neodymium		0.000009	270
Helium		0.000005	
Gold		0.000004	560,000
Praeseodymium		0.000002	320
Gadolinium		0.000002	260
Dysprosium		0.000002	460
Erbium		0.000002	690
Ytterbium		0.000002	530
Samarium		0.000002	180
Holmium		0.0000007	530
Europium		0.0000004	300
Radium[3]	8×10^{-11}		

[a] Calculated residence time = ratio of amount of element present in the sea to either the gross contribution by rivers or the estimated sedimentation rate.
[b] Silicate.

REFERENCES

1. From U.S. Geological Survey.
2. B.S. Newell, *Australian Fisheries*, Reprint, June 1969.
3. H.U. Sverdup *et al.*, *The Oceans*, Prentice Hall, Inc., 1942.

1.3–3 PRINCIPAL SOURCES OF MARINE POLLUTION

Symbols: + + important; + significant; (+) slight; ? uncertain; – negligible; 0 potentially harmful

Category of pollutant	Manufacture and use of industrial products – disposal via direct outfalls and rivers	Domestic wastes; disposal via direct outfalls and rivers	Agriculture, forestry, public health via run-off from canal	Deliberate dumping from ships	Operational discharge from ships in course of duties	Accidental release from ships and submarine pipelines	Exploitation of sea bed mineral resources	Military activities	Transfer from the atmosphere
Domestic sewage (including food processing wastes)	+	+ +	–	+	(+)	–	–	–	–
Pesticides	+	+	+ +	?	–	0	–	?	+ +
Inorganic wastes (including heavy metals and other toxic inorganic compounds)	+ +	+	(+)	+	–	0	(+)	?	+
Radioactive materials	+ +	?	–	?	–	0	–	?	+
Oil and oil dispersants	+	(+)	–	+	+	+ +	+	?	–
Petrochemicals and organic chemicals	+ +	(+)	–	+	(+)	0	–	?	–
Organic wastes (including pulp and paper wastes)	+ +	+ +	+	+	–	–	–	–	–

1.3–3 PRINCIPAL SOURCES OF MARINE POLLUTION (*Continued*)

Category of pollutant	Manufacture and use of industrial products – disposal via direct outfalls and rivers	Domestic wastes; disposal via direct outfalls and rivers	Agriculture, forestry, public health via run-off from canal	Deliberate dumping from ships	Operational discharge from ships in course of duties	Accidental release from ships and submarine pipelines	Exploitation of sea bed mineral resources	Military activities	Transfer from the atmosphere
Military wastes	?	–	–	?	?	?	–	?	–
Heat	+ +	–	–	–	–	–	–	–	–
Detergents	(+)	+ +	–	–	–	–	–	–	–
Solid objects	–	–	–	+ +	+ +	(+)	(+)	–	–
Dredging spoil and inert wastes	+	–	–	+	–	–	+ +	–	–

Source: World Health Organization, March 1970.

1.3–4 MAJOR CATEGORIES OF MARINE POLLUTION

Symbols:
- ++ important
- \+ significant
- (+) slight
- ? uncertain
- – negligible

Category	Harm to living resources	Hazards to human health	Hindrance to maritime activities	Reduction of amenities
Domestic sewage (including food processing wastes)	++	++	(+)	++
Pesticides	++	+	–	–
Inorganic wastes (including heavy metals)	++	++	–	(+)
Radioactive materials	?	+	(+)	–
Oil and oil dispersants	+	?	+	++
Petrochemicals and organic chemicals	+	?	(+)	(+)
Organic wastes (including pulp and paper wastes)	++	?	(+)	+
Military wastes	+	?	+	?
Heat	+	–	+	–
Detergents	(+)	–	–	(+)
Solid objects	+	–	+	++
Dredging spoil and inert wastes	+	–	+	+

Source: World Health Organization, March 1970.

1.4 CLOUD SEEDING AND GLACIERS

1.4–1 POTENTIAL PRECIPITATION INCREASES BY CLOUD SEEDING

Western U.S.

State	Total area, acres	Annual precipitation, in.	Estimated part of annual precipitation from orographic clouds, percent	Estimated possible maximum increase in precipitation by artificial seeding: in.	Estimated possible maximum increase in precipitation by artificial seeding: Thousand acre feet	Estimated possible maximum increase in runoff by artificial seeding[a]: in.	Estimated possible maximum increase in runoff by artificial seeding[a]: Thousand acre feet
Arizona	72,901,760	12.5	20	0.3	2,300	0.02	100
California	101,563,520	20.9	50	1.0	8,500	0.4	3,400
Colorado	66,724,480	16.5	40	0.7	3,900	0.2	1,100
Idaho	53,476,480	17.2	60	1.0	4,500	0.5	2,200
Montana	94,168,320	14.8	30	0.4	3,100	0.1	800
Nevada	70,745,600	8.6	20	0.2	1,200	0.02	100
New Mexico	77,866,240	14.3	20	0.3	1,900	0.02	100
Oregon	62,067,840	27.8	40	1.1	5,700	0.6	3,100
Utah	54,346,624	13.2	30	0.4	1,800	0.05	200
Washington	43,642,880	32.0	50	1.6	5,700	1.0	3,600
Wyoming	62,664,960	14.6	30	0.4	2,100	0.1	500
Total	760,168,704	–	–	–	40,700	–	15,200
Total 5 Pacific Coast and Rocky Mountain states[b]							13,400
Percentage in 5 states							88

[a] Computed by use of natural ratio of runoff to precipitation.
[b] Washington, Oregon, California, Idaho, and Colorado.

Source: E.A. Ackerman and G.O. Löf, *Technology in American Water Development*, published for Resources for the Future, Inc. by the Johns Hopkins Press, 1959.

1.4–2 GLACIERS IN THE U.S. EXCLUSIVE OF ALASKA

State	No. of glaciers	Total glacier area		Estimated July-August streamflow derived from glaciers within the state	
		mi^2	km^2	1,000 acre ft	1,000,000 m^3
Washington	797	160	414	1,430	1,790
Wyoming	81	18	47	107	134
Montana	106	10	27	60	75
Oregon	38	8	21	64	80
California	80	7	19	39	49
Colorado	10	1	2	3.4	4
Idaho	4	0.4	1	2.5	3
Nevada	1	0.1	0.3	0.4	0.5
Total	1,117	204	531	1,710	2,140

REFERENCES

1. M.F. Meier, *J. Am. Water Works Assn., 61*(1):8, 1969.
2. M.F. Meier, in *Handbook of Hydrology*, Vente Chow, Ed., McGraw-Hill, 1964.

1.4–3 SNOWPACK AND MELT-WATER RUNOFF

South Cascade Glacier

Year	Spring snowpack[a]			Ice and snow		Melt runoff, % of avg
	in.	mm	% of avg	in.	mm	
1957–1958[b]	91	2,310	77	227	5,760	158
1963–1964[c]	139	3,530	117	91	2,310	64

[a] Water equivalent.
[b] Deficient precipitation and warm summer.
[c] Abnormally high precipitation and a relatively cool summer.

Source: Reprinted, with publisher's permission, from *J. Am. Water Works Assn., 61*(1):8, © 1969, by the American Water Works Association, Inc., 2 Park Ave., New York, N.Y. 10016.

1.4–4 SNOW SURVEY REPORTS IN THE WESTERN U.S.

Report	Frequency of issue	Location	Cooperation
River basins			
Colorado and state of Utah	Monthly (January–May)	Salt Lake City, Utah	Utah State Engineer and other agencies
Columbia	Monthly (January–May)	Boise, Idaho	Idaho State Reclamation Engineer
Upper Missouri and state of Montana	Monthly (February–May)	Bozeman, Mont.	Montana Agricultural Experiment Station
Westwide	October 1, April 1, May 1	Portland, Ore.	All cooperators
States			
Alaska	Monthly (March–May)	Palmer, Alaska	Alaska Conservation District
Arizona	Semimonthly (January 15–April 1)	Phoenix, Ariz.	Salt River Valley Water Users Association, Arizona Agricultural Experiment Station
California	Monthly (February–May)	Sacramento, Cal.	All interested agencies and organizations
Colorado and New Mexico	Monthly (February–May)	Fort Collins, Colo.	Colorado Agricultural Experiment Station, Colorado State Engineer, New Mexico State Engineer
Idaho	Monthly (February–May)	Boise, Idaho	Idaho State Reclamation Engineer
Nevada	Monthly (February–April)	Reno, Nev.	Nevada Department of Conservation and Natural Resources, Division of Water Resources
Oregon	Monthly (January–May)	Portland, Ore.	Oregon Agricultural Experiment Station, Oregon State Engineer
Washington	Monthly (February–May)	Spokane, Wash.	Washington State Department of Conservation
Wyoming	Monthly (February–June)	Casper, Wyo.	Wyoming State Engineer

Note: Snow surveys are conducted cooperatively by Federal, state, and private agencies. Coordination is provided by the U.S. Soil Conservation Service in all states except California where it is provided by the California Department of Water Resources.

Source: U.S. Soil Conservation Service.

1.4–5 HEAT SUPPLIED TO MELTING SNOW BY DIFFERENT PROCESSES

Heat supply	Extreme conditions	Approximate heat supplied[a]
Convection from turbulent air	70° dry bulb, 20 mi wind	600
Condensation of atmospheric moisture	60° dew point, 20 mi wind	600
Absorption of solar radiation	Very moist air, cloudy at night	200
Warm rain	4 in., 50° wet bulb	100
Conduction from soil	New snow	20

[a] cal/cm² day.

Source: W.T. Wilson, *Trans. Am. Geophys. Union, 22*:194, 1964.

1.4–6 RELATION OF SNOW MELT TO SNOW EVAPORATION

Air temperature, °C	Relative humidity, %	Snow evaporated, g cm^{-2} day^{-1}	Heat transfer from air to snow, cal cm^{-2} day^{-1}	Heat required in evaporation process,[a] cal cm^{-2} day^{-1}	Heat available to melt snow, cal cm^{-2} day^{-1}	Melted snow, g cm^{-2} day^{-1}	Melt/ evaporation
5	20	2.02	900	1,370	0	0	
10	20	1.69	1,790	1,150	640	8.0	4.7
15	20	1.25	2,690	850	1,840	23.0	18.4
20	20	.67	3,590	460	3,130	39.1	58.4

[a] Heat required in evaporation process is equal to heat transfer from air to snow plus heat obtained by a lowering of snow-surface temperature.

Source: Corps of Engineers, U.S. Army.

1.4–7 MELTING CONSTANT FOR SNOW

Location	Descriptive notes	Melting period	Melting constant
Albany, N.Y.	Tests of small cylinders	8–12 hours	0.04–0.06
Donner Summit, Cal.	Observations in 1917	April 1–May 6	0.071
Gooseberry Creek, Utah	Field measurements	April 23–May 9	0.091
Gooseberry Creek, Utah	Tests of cores	6-9 hours	0.05–0.07
Finland	All basins, 1934–1937	April	0.108
Soda Springs, Cal.	Average, 1936–1941	April	0.051
New England floods	Studies by Boston Society of Civil Engineers	1–14 days	0.01–0.04
N.Y. and Pa. basins	Flood runoff studies	March or April	0.04–0.07
La Grange Brook, N.Y.	Basin area, 36 acres	March 28–April 6	0.09
New England floods, 1936	Geological Survey, average values	March 9–22	0.03–0.05
Pemigewasset Basin, N.H.	Flood of March 1936	March 17–20	0.16
Crater Lake, Ore.	Small test plots	March 3–June 9	0.153
Crater Lake, Ore.	Small test plots	May 26–June 2	0.658

Note: The melting constant is the depth of water in inches melted per degree day. A degree day is a unit of heat resulting from a day with a mean temperature one degree Fahrenheit above 32° F.

Source: Houk, *Irrigation Engineering*, vol. I, John Wiley, New York, 1951.

1.4–8 PHYSICAL PROPERTIES OF SNOW AND ICE

	Density, g/cm³	Porosity, %	Air permeability, g/cm² sec	Grain size, mm
New snow	0.01–0.3	99–67	>400–40	0.01–5
Old snow	0.2 –0.6	78–35	100–20	0.5 –3
Firn[a]	0.4 –0.84	56– 8	40– 0	0.5 –5
Glacier ice	0.84–0.917	8– 0	0	1 –>100

[a] Firn is snow which has been modified into a dense compact material by deformation, refreezing, recrystallization, and other processes.

Source: Ven-Te Chow, *Handbook of Applied Hydrology,* McGraw-Hill, New York, 1964.

1.5 *WASTE WATER*

1.5–1 CHARACTERISTICS OF WASTE WATERS AND SALINE WATERS

Type	Total dissolved solids, ppm	Total organic content, ppm
Sea water	25–40,000	0– 5
Brackish water	500–10,000	0–150
Secondary sewage effluent	0– 1,500	0– 30
Primary sewage effluent	0– 1,500	0–300

Source: R. Rickles, *Pollution Control*, Noyes Development Corp., 1965, p. 43.

1.5–2 SECONDARY EFFLUENT CHARACTERISTICS USED FOR ADVANCED WASTE TREATMENT

Parameter	Lebanon, Ohio[1]	Batavia, Ohio[1]	Bryan, Texas[2 g]	South Tahoe Public Utility District[3]
Suspended solids, mg/l	20 ± 20	25 ± 20		5–20
Total solids, mg/l	760 ± 140	560 ± 60	485	
Nitrate, mg/l			9.5–43	
Total organic carbon, mg/l	14 ± 4[a]	25 ± 6[a]		
NH_3-N, mg/l	7 ± 5[b]	10 ± 10[b]		
Volatile solids, mg/l			158	
$SO_4^{=}$, mg/l	105 ± 15[c]	90 ± 15[c]		
$PO_4^{\equiv}$-P, total, mg/l	5 ± 3[d]	15 ± 4[d]	18.8–28.8	25–30
Mg^{++}, mg/l	30 ± 4[e]	11 ± 3[e]		
Ca^{++}, mg/l	112 ± 10[e]	50 ± 8[e]		
Hardness as $CaCO_3$, mg/l	405 ± 35[f]	170 ± 30[f]		
Alkalinity as $CaCO_3$, mg/l	340 ± 50	175 ± 50		
Chloride, mg/l			59.5	
Specific conductance, μmho/cm			768	
ABS, mg/l			4.85	1.1–2.9
COD			98.7	80–160
BOD, 5 days, 20°C, mg/l				20–40
Turbidity, units				30–70
Coliform bacteria, MPN[h]/100 ml				150,000
Color				high
Odor				present

[a] TOC-Beckman carbonaceous analyzer.
[b] NH+3-N – alkaline sodium phenate, autoanalyzer.
[c] SO+= – manual turbidimeter.
[d] $PO_+ \equiv$-P – acid molybdenum blue with stannous chloride reduction, autoanalyzer.
[e] Ca^{++} and Mg^{++} – atomic adsorption.
[f] Hardness – Σ (Ca^{++} + Mg^{++}). All other analyses by *Standard Methods for the Examination of Water and Wastewater,* 12th ed., Amer. Publ. Health Assn., 1965.
[g] Average is range of seven samples collected over a period of 4 months during the morning peak hour.
[h] MPN – most probable number.

REFERENCES

1. M.C. Mulbarger *et al., J. Water Pollut. Control Fed., 41*(12):2070, 1969.
2. M.A. Zeitoun *et al., J. Water Pollut. Control Fed., 38*(4):544, 1969.
3. R.L. Culp *et al., J. Water Pollut. Control Fed., 38*(2):147, 1966.

1.5–3 WATER QUALITY OF RECYCLED WATERS

Reverse Osmosis

	Fort Stockton, Texas[a]			Midland, Texas[d]		Kermit, Texas[e]			River Valley Golf Course San Diego, Calif[f]
Component	Feed, ppm[b]	Product, ppm	Brine, ppm	Feed, ppm	Product, ppm	Feed, ppm	Product, ppm	Brine, ppm	Feed, ppm
Total dissolved solids, calculated	1,690[c]	29[c]	6,600[c]	4,270[c]	30[c]	6,565	348	12,557	4,732
Ca	171	1	660	435	1	276	0	695	395
Mg	63		246	268		156	5	328	265
Na	278	9	1,100	530	9	1,410	124	3,199	925
K									25
Fe	0.10	<0.02	<0.02	0.06	<0.02				8.2
Mn	<0.05	<0.05	<0.05	<0.05	<0.05				3.8
CO_3	0	0	0						
HCO_3	243	2	880	227	1	561	23	1,293	450
SO_2	468	4	1,840	1,920	3	1,396	11	3,337	550
Cl	452	11	1,780	840	11	1,760	185	3,785	2,300
F	0.9	0.1	8.0	5.0	<0.1				0.50
NO_3	13.5	2.0	45	40	5				0.84
Total alkalinity as $CaCO_3$	199	2	720	186	1				370
Total hardness as $CaCO_3$	690	2	2,670	2,190	3	1,333	20	3,085	2,070
pH	7.9	5.9	7.5	7.7	5.9	8.9	8.9	8.7	7.2
OH									0
Boron									0.40
Silicon, SiO_2									30

[a] Samples taken February 11, 1966. Feed water source: municipal water supply, Fort Stockton, Texas. Chemical analyses performed by Texas State Department of Health Laboratories.

[b] Taken from feed one–test one analysis.

[c] Includes all of the bicarbonates.

[d] Samples taken January 16, 1966. Feed water source: Cold Park well No. 9, reservoir, Midland, Texas. Chemical analysis performed by Texas State Department of Health Laboratories.

[e] Feed source was Shell Oil Co. deep well at Kermit, Texas. The analysis was performed by Martin Water Laboratories, Monahaus, Texas, February 19, 1966.

[f] Sample source: Unacidified feed. Collected in November 4, 1966. Conductivity, 7750 μmho/cm at 25°C.

Source: Reprinted, with publisher's permission from *J. Am. Water Works Assn.*, *59*(12):1527, © 1967, by the American Water Works Association, Inc., 2 Park Ave., New York, N.Y. 10016.

1.5–4 MINERAL ADDITIONS IN ONE CYCLE OF WATER USE

	Average Range of Increment in Component, mg/l					
Component	**1**	**2**	**Illinois**[3]	**Sunnyfield**[3]	**Wardcliff**[3]	**Amarillo**[4]
Calcium, Ca^{++}	7–109	10–50	17–33	16	6	30[a]
Magnesium, Mg^{++}	1–90	5–20	7–37	7	10	30[a]
Ammonium, NH_4^+	4–45	2–20	10–37	24	13	
Sodium, Na^+	14–742		13–132	104	83	270[a]
Phosphate, PO_4^{-3}	8–65	10–50	24–51	50	28	40[a]
Silica, SiO_2	4–18	10–20	8–21	19	12	0
Chloride, Cl^-	22–1262	25–125	16–34	32	30	280[a]
Nitrate, NO_3^-		1–20	0–1			2[a]
Bicarbonate, HCO_3^-	12–392	18–60	34–118	52	38	70[a]
Sulfate, SO_4^{-2}	10–191	10–40	8–75	74	71	10[a]
Total dissolved solids	120–450	150–500	262–364	346	277	
Ammonia						20

[a] Reported as $CaCO_3$.

REFERENCES

1. R. Stone and J.C. Merrell, *J. Water Pollut. Control Fed., 30*:428, 1918.
2. C.H. Connell and M.C. Forbes, *Water Sewage Works, 111*:397, 1964.
3. R.L. Evans, *J. Am. Water Works Assn., 60*(3):315, 1968.
4. S.L. Terry, *J. Am. Water Works Assn., 63*(3):162, 1971.

1.5–5 INCREMENTS OF COMMON IONS IN DOMESTIC WATER AFTER ONE CYCLE OF WATER USE

mg/l

	California[1][g]				NEIWPCC[3][h]		Sanks and Kaufman[4] Use increment			Wells and Gloyna[5]		Bunch *et al.*[6]	
Component	Min	Max	Normal range	Chanute, Kansas[2]	Concentration in water	Increment in sewage	Min	Max	Avg	Avg	Range	Avg	Range
COD				113	<10	143	22	159	87			143	84–163
COD filtered												95	70–120
COD filtered corrected for Cl^-												82	46–112
BOD				29	<3	100–300	8	27	16	32	11–120		
ABS				6	<0.1	7.4						7.4	4.6–10.1
Total N	12	42	20–40	21						24	22–28	22.0	14.5–28.4
NH_3-N				19	<0.2	16.1	0	36	15			16.1	6.8–22.1
NO_2-N												0.30	0.04–0.58
NO_3-N	0	18	0–18[a]	1			-5	26	10			3.5	1.0[d]–7.7
Ortho PO_4				19								22.8	15.6–33.8
Total PO_4	2[b]	50	20–40	25	<0.1	24.3	7	50	24	24[c]	24[c]	24.3	18.5–34.6
Complex PO_4				7									
TDS	Trace	1,200	100–300	172	10–1,580	290	128	541	320				
Total alkalinity as $CaCO_3$	–	230	100–150	96								122	40–217
$SO_4^{=}$	0	75	15–30	21	0–572	10–191	12	57	28			33	12–52
Cl^-	20	550	20–50	24	0.5–510	22–162	6	200	74	144	96–494	56	40–102
Ca^{++}, as $CaCO_3$	–	250	15–40		0–145	7–109	2.5	125	45			23[e]	1–43[e]
Mg^{++}, as $CaCO_3$	–	118	15–40		0–120	1–90	0	65	26			7[f]	1–11[f]
Na^+	30	290	40–70		0–198	57	8	101	66			57	42–98
K^+	–	22	7–15		0–30	9.3						9.3	7.2–11.7
Bicarbonate, as $CaCO_3$					5–380	122	-44	265	100				
Silica					–	4–18	9	22	15				

[a] Results from biological action in organic nitrogen.
[b] Very low concentrations in natural waters.
[c] Reported as P.
[d] Loss.
[e] As Ca^{++}.
[f] As Mg^{++}.
[g] Summary of mineral constituents between source and disposal for 15 California cities.
[h] New England Interstate Water Pollution Control Commission.

1.5–5 INCREMENTS OF COMMON IONS IN DOMESTIC WATER AFTER ONE CYCLE OF WATER USE (*Continued*)

Component	California[1] [g]				NEIWPCC[3] [h]		Sanks and Kaufman[4] Use increment			Wells and Gloyna[5]		Bunch *et al.*[6]	
	Min	Max	Normal range	Chanute, Kansas[2]	Concentration in water	Increment in sewage	Min	Max	Avg	Avg	Range	Avg	Range
Boron			0.1–0.4		–	0–0.4							
Suspended solids										40	13–74		
Total solids										645	362–1,619	291	128–457
Hydroxylated aromatic (tannic acid)												1.6	0.5–2.9
Carbohydrates, glucose												2.4	0.9–5.0
Organic nitrogen, N												2.2	1.0–4.0
Loss on ignition												69	21–145
Dissolved solids			100–300										

[g] Summary of mineral constituents between source and disposal for 15 California cities.
[h] New England Interstate Water Pollution Control Commission.

REFERENCES

1. Data from *Studies of Waste Water Reclamation and Utilization,* State Water Pollution Control Board, Sacramento, Calif., quoted in *Summary Report – The Advanced Waste Treatment Research Program, June 1960–December 1961,* Public Health Service, U.S. Department of Health, Education, and Welfare, May 1962.
2. G.E. McCallum, *J. Water Pollut. Control Fed., 35*(1):1, 1963.
3. J.W. Masselli, N.W. Masselli, and M.G. Burford, *Controlling the Effects of Industrial Wastes on Sewage Treatment,* Industrial Waste Laboratory, Wesleyan University, June 1970.
4. R.L. Sanks and W.J. Kaufman, Taken from Neal, Robert A. Taft Sanitary Engineering Center, Cincinnati, Ohio.
5. D.M. Wells and E.F. Gloyna, Water Reuse in Texas. Presented at Joint Program of the Rocky Mountain Section AWWA and the WPCA, Albuquerque, October 4, 1965.
6. R.L. Bunch *et al., J. Water Pollut. Control Fed., 36*(11):1411, 1964.

1.5–6 CONCENTRATIONS OF VARIOUS NUTRIENTS IN NATURAL WATERS AND IN WASTE WATER TREATMENT SYSTEMS

System	Nutrient	Concentration, mg/l	Ref.
Sewage	Total hexoses	9.05	1
Soluble fraction	Total amino acids	9.77	
	Total pentose	0.77	
Sewage	Total amino acids	10.0[a]	2
Soluble fraction	Total carbohydrates	90.0[a]	
	Total soluble acids	34.0[a]	
Sewage treatment	Total amino acids	0.12	1
Plant effluent	Total carbohydrates	0.6	
	Total soluble acids	3.3	
Eutrophic lake	Glucose	0.01–0.02	3
Eutrophic pond	Glucose	0.01	
Mildly polluted river	Glucose	0.024	
Polluted river	Glucose	0.059	
Mildly polluted lakes	Glucose	0.009	
Eutrophic coastal ponds	Dissolved carbohydrates	1.57–3.54	4
Coastal Pacific Ocean	Dissolved carbohydrates	0.1–0.4	5
Coastal lagoon		8.0	
Pacific Ocean	Total dissolved amino acids	0.218	6
	Total free amino acids	0.06	

[a] Approximate values calculated from given mg/l C values.

Source: E. Stumm-Zollinger and R.H. Harris, in *Organic Compounds in Aquatic Environments,* S.J. Faust and J.V. Hunter, Eds., Marcel Dekker, Inc., 1971, p. 555.

REFERENCES

1. J.V. Hunter and H. Heukelekian, *J. Water Pollut. Control Fed., 37*:1142, 1965.
2. H.A. Painter *et al., J. Inst. Sewage Pur., 5*:3, 1961.
3. J.E. Hobbie and R.T. Wright, *Limnol. Oceanogr., 10*:471, 1965.
4. G.E. Walsh, *Limnol. Oceanogr., 10*:570, 1965.
5. G.J. Lewis and N.W. Rakestraw, *J. Mar. Res., 14*:253, 1955.
6. E.T. Degens, Symp. on Organic Matter in Natural Waters, University of Alaska, 1968.

1.5–7 COMPARISON OF QUALITY OF RECLAIMED WATER WITH INTERNATIONAL STANDARDS FOR DRINKING WATER AND FOR WATER SOURCES

Chemical substances which serve as criteria	Reclaimed water from double biofiltration	Reclaimed water from primary clarified sewage	World Health Organization Drinking Water Standards	
			Maximum acceptable concentration	Maximum allowable concentration
		Potability		
Total solids	350.0 mg/l	390.0 mg/l	500.0 mg/l	1,500.0 mg/l
Color	Less than 5 units	Less than 5 units	5 units[a]	50 units[a]
Turbidity	Less than 5 units	Less than 5 units	5 units[b]	25 units[b]
Taste	Excellent	Excellent	Unobjectionable	–
Odor	None	None	Unobjectionable	–
Iron (Fe)	0.1 mg/l	0.1 mg/l	0.3 mg/l	1.0 mg/l
Manganese (Mn)	0.0 mg/l	0.0 mg/l	0.1 mg/l	0.5 mg/l
Copper (Cu)	0.0 mg/l	0.0 mg/l	1.0 mg/l	1.5 mg/l
Zinc (Zn)	0.2 mg/l	0.2 mg/l	5.0 mg/l	15.0 mg/l
Calcium (Ca)	20.0 mg/l	30.0 mg/l	75.0 mg/l	200.0 mg/l
Magnesium (Mg)	5.0 mg/l	5.0 mg/l	50.0 mg/l	150.0 mg/l
Sulfate (SO_4)	60.0 mg/l	60.0 mg/l	200.0 mg/l	400.0 mg/l
Chloride (Cl)	100.0 mg/l	100.0 mg/l	200.0 mg/l	600.0 mg/l
pH range	7.0–8.0	7.0–8.0	7.0–8.5	Less than 6.5 or greater than 9.2
Magnesium + Sodium Sulfate	120.0 mg/l	120.0 mg/l	500.0 mg/l	1,000.0 mg/l
Phenolic substances (as phenol)	Nil	Nil	0.001 mg/l	0.002 mg/l
Carbon chloroform extract (CCE: organic pollutants)	0.05 mg/l	<0.5 mg/l	0.2 mg/l	0.5 mg/l[c]
Alkyl benzyl sulfonates (ABS: surfactants)	Nil	Nil	0.5 mg/l	1.0 mg/l
		Hazardous to Health		
Nitrate as NO_3	50.0 to 100.0 mg/l	Nil	–	45.0 mg/l
Fluoride	Less than 0.5 mg/l	Less than 0.5 mg/l	–	1.5 mg/l
		Toxicity		
Phenolic substances	Nil	Nil		–
Arsenic	Nil	Nil		0.05 mg/l
Cadmium	Nil	Nil		0.01 mg/l
Chromium (Cr hexavalent)	Nil	Nil		0.05 mg/l
Cyanide	Nil	Nil		0.2 mg/l
Lead	0.01 mg/l	0.01 mg/l		0.05 mg/l
Selenium	Nil	Nil		0.01 mg/l

[a] Platinum-cobalt scale.
[b] Turbidity units.
[c] Concentrations greater than 0.2 mg/l indicate the necessity for further analysis to determine the causative agent.

1.5–7 COMPARISON OF QUALITY OF RECLAIMED WATER WITH INTERNATIONAL STANDARDS FOR DRINKING WATER AND FOR WATER SOURCES (*Continued*)

Chemical substances which serve as criteria	Reclaimed water from double biofiltration	Reclaimed water from primary clarified sewage	World Health Organization Drinking Water Standards	
			Maximum acceptable concentration	Maximum allowable concentration
		Chemical Indicators of Pollution		
Chemical oxygen demand (COD)	0–10 mg/l	0–30 mg/l	–	–
Biochemical oxygen demand (BOD)	0–5 mg/l	0–20 mg/l	–	–
Total nitrogen exclusive of NO_3	Less than 1 mg/l	Less than 1 mg/l	–	–
NH_3	Nil	Nil	–	–
Carbon chloroform extract (CCE: organic pollutants)	0.05 mg/l	<0.5 mg/l	–	–

Source: E.F. Gloyna and W.W. Eckenfelder, *Water Quality Improvement by Physical and Chemical Processes*, University of Texas Press, Austin, 1970.

Section 2

Needs

2. Introduction
Pages 131–147

2.1 Public Water Supply Needs
Pages 148–167

2.2 Industrial Water Needs
Pages 168–229

2.3 Irrigation and Farm Water Needs
Pages 230–243

2. NEEDS

2–1 ESTIMATED WATER USE IN THE U.S.

Daily Average in Billions of Gallons

	Total water use		Irrigation[a]		Public water utilities		Self supplied uses: Rural domestic[b]		Self supplied uses: Industrial and miscellaneous[c]		Self supplied uses: Steam electric utilities	
Year	Total	Ground	Total	Ground	Total	Ground	Total	Ground	Total	Ground	Total	Ground
1900	40.19	7.28	20.19	2.22	3.00	1.05	2.00	1.60	10.00	2.40	5.00	0.01
1910	66.44	11.68	39.04	5.27	4.70	1.49	2.20	1.76	14.00	3.15	6.50	0.01
1920	91.54	15.78	55.94	8.17	6.00	1.79	2.40	1.94	18.00	3.87	9.20	0.01
1930	110.50	18.18	60.20	9.09	8.00	2.30	2.90	2.40	21.00	4.37	18.40	0.02
1940	136.43	22.56	71.03	11.22	10.10	2.82	3.10	2.64	29.00	5.86	23.20	0.02
1950	202.70	35.19	100.00	19.80	14.10	3.78	4.60	4.09	38.10	7.47	45.90	0.05
1960	322.90	58.17	135.00	35.24	22.00	5.68	6.00	5.58	61.20	11.57	98.70	0.10
1965	269.62	48.57	110.85	30.04	23.74	5.96	4.08	3.86	46.41	8.63	84.54	0.08
1970	327.30	54.27	119.18	33.13	27.03	6.65	4.34	4.13	55.95	10.24	120.80	0.12
1971	338.84	55.27	120.85	33.64	27.69	6.77	4.39	4.18	57.86	10.55	128.05	0.13
1975	384.95	59.29	127.51	35.70	30.31	7.27	4.59	4.37	65.49	11.79	157.05	0.16
1980	442.63	63.98	135.85	38.17	33.60	7.73	4.85	4.61	75.03	13.28	193.30	0.19

[a] Total take, including delivery losses but not including reservoir evaporation.
[b] Rural farm and nonfarm household and garden use, and water for farm stock and dairies.
[c] For 1900–1960, includes manufacturing and mineral industries, rural commercial industries, air conditioning, resorts, hotels, motels, military, and other state and federal agencies, and other miscellaneous uses; thereafter, includes manufacturing, mining and mineral processing, ordnance, and construction.

Source: U.S. Bureau of Domestic Commerce. 1900–1960 based principally on committee prints, *Water Resources Activities in the United States,* for the Senate Committee on National Water Resources, U.S. Senate, 1960; thereafter, based on the first *National Assessment of the Water Resources Council,* 1968.

2–2 PATTERN OF WATER USE

U.S., 1960

Year	Total water usage, bgd	Distribution of water usage, % Agricultural	Industrial	Municipal
1900	40	55.0	37.5	7.5
1960	324	43.6	50.0	6.5
1980 (est.)	547	27.6	65.6	6.8

Source: P.H. McGauhey, *Engineering Management of Water Quality*, McGraw-Hill Book Co., 1968.

2–3 ESTIMATED NATIONAL WATER QUALITY REQUIREMENTS

Use, bgd[a]	1965	1980	2000	2020
Rural	108	135	153	180
Municipal	27	41	56	83
Industrial	55	75	128	206
Steam electric, fresh	53	127	239	408
Steam electric, saline	27	63	216	492
Consumptive use	60	84	100	144
Total withdrawal	330	525	892	1,513[b]

[a] National water requirements were estimated at 270 bgd in 1965.
[b] Available supply based on 30″ runoff = 1,200 bgd.

Source: Extrapolated from B. Blanchard, *J. Water Pollut. Control Fed., 41*(8):1407, 1969.

2–4 WITHDRAWALS AND CONSUMPTIVE USE OF WATER

U.S., 1965

Region	Rural domestic	Municipal	Self supplied industrial	Steam electric power		Agriculture		Total	From ground water sources	From saline sources
				Fresh	Saline	Irrigation	Livestock			
North Atlantic	390	5,446	9,499	10,500	11,400	151	81	37,467	2,562	15,380
South Atlantic-Gulf	504	1,980	3,360	7,600	3,700	3,270	146	20,560	4,221	4,132
Great Lakes	274	3,622	9,069	12,000	–	75	79	25,119	963	25
Ohio	300	1,791	8,606	17,400	–	24	134	28,255	1,760	27
Tennessee	64	253	1,076	4,329	–	8	37	5,767	202	–
Upper Mississippi	203	1,103	1,664	4,800	–	95	314	8,179	1,707	18
Lower Mississippi	58	470	1,884	1,600	200	1,320	39	5,571	1,671	240
Souris-Red-Rainy	14	36	98	200	–	24	19	391	55	–
Missouri	106	969	462	1,400	–	16,039	368	19,344	4,005	–
Arkansas-White-Red	103	687	910	600	–	6,960	150	9,410	5,598	75
Texas-Gulf	40	1,055	5,465	2,100	200	7,450	100	16,410	8,390	3,300
Rio Grande	10	254	215	70	–	6,671	69	7,289	3,130	133
Upper Colorado	6	60	40	20	–	3,880	11	4,017	36	–
Lower Colorado	6	342	140	10	–	6,400	15	6,913	4,147	–
Great Basin	15	274	225	10	–	4,575	16	5,115	935	–
Columbia-North Pacific	148	1,105	1,911	8	–	26,400	59	29,631	4,289	31
California	90	4,010	1,250	70	5,600	26,200	80	37,300	13,610	6,140
Alaska	8	32	102	20	–	–	–	162	25	–
Hawaii	7	115	112	–	300	1,060	3	1,597	706	316
Puerto Rico	5	141	317	1	400	250	6	1,120	157	535
Total	2,351	23,745	46,405	62,738	21,800	110,852	1,726	269,617	58,169	30,352
Consumptive Use										
North Atlantic	186	905	555	75	83	150	69	2,023		
South Atlantic-Gulf	472	363	260	42	19	1,400	139	2,695		
Great Lakes	100	502	362	95	–	68	72	1,199		

2–4 WITHDRAWALS AND CONSUMPTIVE USE OF WATER (*Continued*)

Region	Rural domestic	Municipal	Self supplied industrial	Steam electric power		Agriculture		Total	From ground water sources	From saline sources
				Fresh	Saline	Irrigation	Livestock			
				Consumptive Use						
Ohio	200	230	410	138	–	24	132	1,134		
Tennessee	61	46	174	6	–	8	36	331		
Upper Mississippi	101	162	58	61	–	83	305	770		
Lower Mississippi	52	175	296	16	3	890	38	1,470		
Souris-Red-Rainy	14	11	7	2	–	24	19	77		
Missouri	85	221	71	24	–	9,798	355	10,554		
Arkansas-White-Red	94	241	322	41	–	5,030	146	5,874		
Texas-Gulf	40	400	880	57	2	5,810	100	7,289		
Rio Grande	7	108	46	9	–	4,165	68	4,403		
Upper Colorado	3	14	8	13	–	1,934	10	1,982		
Lower Colorado	5	203	50	8	–	3,170	12	3,448		
Great Basin	9	94	37	2	–	2,100	11	2,253		
Columbia-North Pacific	134	182	100	–	–	10,050	55	10,521		
California	60	1,320	110	70	44	19,290	50	20,944		
Alaska	1	7	4	–	–	–	–	12		
Hawaii	7	39	4	–	3	477	3	533		
Puerto Rico	5	21	10	–	3	225	6	270		
Total	1,636	5,244	3,764	659	157	64,696	1,626	77,782		

Note: All data are in million gal/day. For regions see 1–7.

Source: U.S. Water Resources Council, 1968.

2–5 PROJECTIONS OF WITHDRAWALS AND CONSUMPTIVE USE OF WATER

U.S., 1980

Region	Rural domestic	Municipal	Self supplied industrial	Steam electric power: Fresh	Steam electric power: Saline	Agriculture: Irrigation	Agriculture: Livestock	Total
			Withdrawals					
North Atlantic	400	7,100	14,100	10,900	22,100	230	90	54,920
South Atlantic-Gulf	380	3,300	4,900	28,500	12,000	3,900	200	53,180
Great Lakes	257	5,030	16,700	25,700	–	110	96	47,893
Ohio	350	2,330	11,600	27,300	–	40	129	41,749
Tennessee	128	358	1,600	10,100	–	18	48	12,252
Upper Mississippi	143	1,770	2,800	9,500	–	110	477	14,800
Lower Mississippi	80	647	2,500	5,900	600	3,030	59	12,816
Souris-Red-Rainy	16	49	150	500	–	200	21	936
Missouri	134	1,225	584	1,500	–	19,300	521	23,264
Arkansas-White-Red	213	1,418	1,880	4,100	40	9,400	228	17,279
Texas-Gulf	70	1,890	9,340	5,500	2,700	9,400	180	29,080
Rio Grande	10	430	910	70	–	6,840	70	8,330
Upper Colorado	10	120	200	30	–	5,300	15	5,675
Lower Colorado	7	520	210	40	–	7,700	20	8,497
Great Basin	15	450	340	30	–	6,200	20	7,055
Columbia-North Pacific	148	1,304	4,478	4,000	–	31,400	77	41,407
California	90	5,090	1,660	90	18,300	30,950	110	56,290
Alaska	11	120	200	200	–	4	–	535
Hawaii	5	195	134	–	900	1,420	4	2,658
Puerto Rico	7	250	740	3	2,700	300	10	4,010
Total	2,474	33,596	75,026	133,963	59,340	135,852	2,375	442,626
			Consumptive Use					
North Atlantic	200	1,210	850	120	180	230	80	2,870
South Atlantic-Gulf	355	600	380	190	80	1,600	190	3,395
Great Lakes	85	702	728	184	–	95	87	1,881
Ohio	250	300	550	350	–	40	129	1,619
Tennessee	122	64	258	65	–	16	47	572
Upper Mississippi	94	258	98	166	–	95	392	1,103
Lower Mississippi	72	238	400	60	4	2,180	58	3,012
Souris-Red-Rainy	16	16	8	4	–	150	21	215
Missouri	108	280	90	80	–	12,100	502	13,160
Arkansas-White-Red	194	496	674	95	–	6,800	223	8,482
Texas-Gulf	65	740	1,160	180	20	7,100	170	9,435
Rio Grande	7	220	90	20	–	4,270	69	4,676
Upper Colorado	4	30	35	17	–	2,600	14	2,700
Lower Colorado	5	310	80	35	–	3,630	15	4,075
Great Basin	9	154	56	25	–	3,040	15	3,299
Columbia-North Pacific	134	219	244	13	–	12,900	71	13,581
California	60	4,620	380	80	185	23,800	80	29,205
Alaska	2	24	20	1	–	3	–	50
Hawaii	5	65	5	–	9	640	4	728
Puerto Rico	5	35	20	–	20	270	10	360
Total	1,792	10,581	6,126	1,685	498	81,559	2,177	104,418

Note: All data are in million gal/day. For regions see 1–7.

Source: U.S. Water Resources Council, 1968.

2–6 PROJECTIONS OF WITHDRAWALS AND CONSUMPTIVE USE

U.S., 2000

Region	Rural domestic	Municipal	Self supplied industrial	Steam electric power		Agriculture		Total
				Fresh	Saline	Irrigation	Livestock	
Withdrawals								
North Atlantic	400	10,000	22,800	11,300	68,900	330	130	113,860
South Atlantic-Gulf	340	5,400	8,000	39,800	27,600	6,000	300	87,440
Great Lakes	292	6,900	33,000	56,100	–	170	132	96,594
Ohio	415	3,320	15,900	45,200	–	80	194	65,109
Tennessee	167	560	2,360	10,700	–	23	67	13,877
Upper Mississippi	132	2,760	5,300	21,500	–	200	695	30,587
Lower Mississippi	147	921	3,420	16,700	2,300	4,400	79	27,967
Souris-Red-Rainy	17	82	212	1,100	–	562	29	2,002
Missouri	162	1,481	707	3,200	–	21,600	726	27,876
Arkansas-White-Red	336	2,375	3,147	8,400	40	10,700	318	25,336
Texas-Gulf	120	3,110	15,500	22,000	7,300	9,000	300	57,330
Rio Grande	10	720	1,800	70	–	6,840	70	9,510
Upper Colorado	10	160	1,000	35	–	5,350	20	6,575
Lower Colorado	8	860	280	250	–	7,000	30	8,428
Great Basin	15	745	560	100	–	6,100	30	7,550
Columbia-North Pacific	148	2,100	8,880	22,200	19,200	37,500	107	90,135
California	90	8,280	2,590	300	77,400	31,700	150	120,510
Alaska	12	230	40	250	–	9	–	901
Hawaii	4	320	159	–	2,300	1,910	5	4,698
Puerto Rico	7	400	1,350	3	6,200	350	15	8,325
Total	2,852	50,724	127,365	259,208	211,240	149,824	3,397	804,610
Consumptive Use								
North Atlantic	200	1,750	1,500	320	750	330	110	4,960
South Atlantic-Gulf	320	1,000	620	765	215	2,450	285	5,655
Great Lakes	103	953	1,400	467	–	140	120	3,183
Ohio	290	430	770	775	–	80	194	2,539
Tennessee	158	95	381	115	–	21	64	834
Upper Mississippi	85	403	184	373	–	170	563	1,778
Lower Mississippi	133	343	550	175	7	3,170	75	4,453
Souris-Red-Rainy	17	26	10	10	–	402	29	494
Missouri	130	339	109	200	–	13,500	701	14,979
Arkansas-White-Red	325	832	1,114	205	–	7,800	311	10,587
Texas-Gulf	110	1,200	1,650	495	55	7,100	280	10,890
Rio Grande	7	400	180	65	–	4,270	69	4,991
Upper Colorado	5	35	140	20	–	2,880	20	3,100
Lower Colorado	5	515	130	210	–	3,760	25	4,645
Great Basin	9	255	92	75	–	3,110	21	3,562
Columbia-North Pacific	134	350	481	180	180	15,900	100	17,325
California	60	7,350	620	100	740	23,700	90	32,660
Alaska	2	46	40	2	–	6	–	96
Hawaii	4	106	5	–	20	860	5	1,000
Puerto Rico	5	50	35	–	55	315	15	475
Total	2,102	16,478	10,011	4,552	2,022	89,964	3,077	128,206

Note: All data are in million gal/day. For regions see 1–7.

Source: U.S. Water Resources Council, 1968.

2–7 PROJECTIONS OF WITHDRAWALS AND CONSUMPTIVE USE

U.S., 2020

Region	Rural domestic	Municipal	Self supplied industrial	Steam electric power: Fresh	Steam electric power: Saline	Agriculture: Irrigation	Agriculture: Livestock	Total
Withdrawals								
North Atlantic	400	14,200	37,200	14,700	169,200	420	170	236,290
South Atlantic-Gulf	310	8,300	12,200	56,000	44,800	8,200	380	130,190
Great Lakes	347	9,500	66,000	114,700	–	230	183	190,960
Ohio	490	4,900	23,200	61,200	–	115	258	90,163
Tennessee	212	872	3,100	13,800	–	29	93	18,106
Upper Mississippi	130	4,000	10,000	25,900	–	280	956	41,266
Lower Mississippi	254	1,334	4,740	19,500	7,500	6,000	114	39,442
Souris-Red-Rainy	21	124	297	1,700	–	576	40	2,758
Missouri	190	1,738	829	4,800	–	23,000	1,015	31,572
Arkansas-White-Red	515	3,438	4,553	11,100	40	11,500	443	31,589
Texas-Gulf	160	4,540	22,100	41,000	15,900	8,500	440	92,640
Rio Grande	10	1,090	3,570	100	–	6,840	70	11,680
Upper Colorado	10	240	1,500	50	–	4,900	25	6,725
Lower Colorado	9	1,400	340	600	–	6,500	40	8,889
Great Basin	15	1,005	740	200	–	5,800	40	7,800
Columbia-North Pacific	148	3,240	13,500	44,600	52,600	42,500	147	156,735
California	90	12,690	3,960	300	194,900	32,600	220	244,760
Alaska	13	375	800	300	2,700	18	–	4,206
Hawaii	3	520	188	–	5,300	2,570	6	8,587
Puerto Rico	7	750	1,950	3	10,600	400	20	13,730
Total	3,334	74,256	210,767	410,553	503,540	160,978	4,660	1,368,088
Consumptive Use								
North Atlantic	200	2,550	2,600	580	2,000	420	140	8,490
South Atlantic-Gulf	290	1,500	940	1,440	385	3,350	360	8,265
Great Lakes	125	1,304	2,800	898	–	190	167	5,484
Ohio	340	620	1,100	1,190	–	115	258	3,623
Tennessee	202	140	500	175	–	26	89	1,132
Upper Mississippi	76	580	346	607	–	240	775	2,624
Lower Mississippi	229	497	760	295	40	4,320	110	6,251
Souris-Red-Rainy	21	35	12	20	–	416	40	544
Missouri	153	397	128	320	–	14,400	980	16,378
Arkansas-White-Red	470	1,205	1,611	310	–	8,300	433	12,329
Texas-Gulf	145	1,750	2,160	590	145	7,100	410	12,300
Rio Grande	7	670	350	100	–	4,270	69	5,466
Upper Colorado	5	50	150	35	–	2,880	20	3,140
Lower Colorado	5	840	200	470	–	3,760	35	5,310
Great Basin	9	345	123	160	–	3,110	29	3,776
Columbia-North Pacific	134	537	738	630	740	18,700	137	21,615
California	60	11,300	960	180	1,730	23,800	160	38,190
Alaska	2	75	80	2	13	12	–	184
Hawaii	3	173	6	–	30	1,150	6	1,368
Puerto Rico	5	75	55	–	100	360	20	615
Total	2,481	24,643	15,619	8,002	5,183	96,919	4,238	157,085

Note: All data are in million gal/day. For regions see 1–7.

Source: U.S. Water Resources Council, 1968.

2–8 ESTIMATED WATER WITHDRAWAL AND CONSUMPTION

Composite Regions, 1950–65

Years and regions	Total population, millions	Total water withdrawn, except for hydroelectric power, bgd: Public supplies	Rural domestic and livestock	Irrigation[a]	Thermo-electric power electric utility use	Other self supplied industrial use	Sources of water withdrawn: Ground, fresh	Ground, saline	Surface, fresh	Surface, saline	Reclaimed sewage	Water consumed, total, bgd	Water used for hydroelectric power, bgd
1965													
Fourteen eastern and central water use regions	144.8	15	2.9	5.4	110	36	14	0.25	120	27	0.1	10	1,200
Seven western water use regions	48.0	8.3	1.1	110	18	8.9	45	0.18	83	16.	0.5	66	1,100
Alaska	0.3	0.03	0.01	<0.01	<0.01	0.10	0.03	0	0.12	<0.01	0	0.01	0.75
Hawaii	0.7	0.11	<0.01	1.2	0.59	0.14	0.78	0.04	0.67	0.5	0	0.57	0.36
Total, 50 states	193.8	24	4.0	120	130[b]	45[d]	60	0.47	210	43	0.7	77	2,300
Puerto Rico	2.6	0.14	<0.01	0.25	0.94[c]	0.32	0.15	<0.01	0.43	1.1	0	0.27	0.51
1960													
Fifty states	179.3	21	3.6	110	100	38	50[e]	0.38	190[e]	31	0.1	61	2,000
1955													
Forty-eight states	164	17	3.6[f]	110	72	39	47[f]	0.65	180	18	0.2		1,500
1950													
Forty-eight states	150.7	14	3.6	110[g]	77 (thermoelectric and industrial combined)		34[h] (ground, fresh and saline combined)		170[i] (surface, fresh and saline combined)				1,100

[a] Including conveyance losses.
[b] Includes 34 bgd of saline water, of which 23 bgd was withdrawn in the eastern and central regions.
[c] Entirely saline water.
[d] Includes 8.5 bgd of saline water, of which 4.5 bgd was withdrawn in the eastern and central regions.
[e] Including conveyance losses, estimated to be in the same proportion as in 1955 with respect to sources of water.
[f] Revised.
[g] Including an estimated 30 bgd in irrigation conveyance losses.
[h] Including an estimated 5 bgd in irrigation conveyance losses.
[i] Including an estimated 25 bgd in irrigation conveyance losses.

Source: Table compiled by J.C. Kammerer, U.S. Geological Survey. With permission.

REFERENCES

1. C.R. Murray, *J. Am. Water Works Assn.*, *69*(11):567, 1969.
2. K.A. MacKichan, *Estimated Use of Water in the United States, 1950,* U.S. Geological Survey, U.S. Department of the Interior, 1951.
3. K.A. MacKichan, *J. Am. Water Works Assn.*, *49*(4):369, 1957.
4. K.A. MacKichan and J.C. Kammerer, *Estimated Use of Water in the United States, 1960,* U.S. Geological Survey, U.S. Department of the Interior, 1961.

2–9 CHANGE IN WITHDRAWALS

Use	1950	1955	1960
Rural	3,600	3,600	3,600
Public supplies	14,000	17,000	21,000
Self supplied industrial	77,000	110,000	140,000
Irrigation, excluding conveyance losses	79,000	81,000	84,000
Irrigation, including conveyance losses		110,000	110,000
All uses, excluding conveyance losses	170,000	210,000	250,000
All uses, including conveyance losses		240,000	270,000

Source: Reprinted, with publisher's permission, from *J. Am. Water Works Assn.*, *53*(10):1211, © 1961, by the American Water Works Association, Inc., 2 Park Ave., New York, N.Y. 10016.

2–10 WATER USED IN THE U.S.

1960

Region	Water withdrawn (fresh and saline), mgd[a]				Saline water withdrawn,[b] mgd	Water consumed, mgd
	Ground water	Surface water	Waste water	All water		
New England	350	6,000	0	6,400	3,400	290
Delaware-Hudson	1,200	18,000	0	20,000	9,000	830
Chesapeake	480	6,600	70	7,100	1,900	320
South Atlantic	2,000	11,000	0	13,000	3,800	1,500
Eastern Gulf	700	5,200	0	5,900	880	510
Tennessee-Cumberland	370	7,200	0	7,500	0	400
Ohio	1,300	22,000	0	24,000	0.3	800
Eastern Great Lakes-St. Lawrence	340	12,000	0	13,000	0	420
Western Great Lakes	830	16,000	0	16,000	0	510
Hudson Bay	50	130	0	180	0	54
Upper Mississippi	1,300	10,000	0	11,000	0	570
Upper Missouri	2,800	17,000	0	20,000	5.0	7,400
Lower Missouri	180	1,400	0	1,600	0	200
Lower Mississippi	1,400	3,900	0	5,300	1,600	1,300
Upper Arkansas-Red	2,100	3,400	0	5,500	120	3,200
Lower Arkansas-Red-White	960	4,000	0	4,900	4.0	850
Western Gulf	11,000	11,000	0	22,000	1,400	9,200
Colorado	3,500	11,000	0.2	14,000	0	7,100
Great Basin	910	6,100	0	7,000	5.5	3,300
South Pacific	10,000	23,000	0.5	33,000	9,100	13,000
Pacific Northwest	3,700	25,000	0	29,000	48	8,300
Hawaii	600	970	0	1,600	280	410
Alaska	25	180	0	200	0	0.7
Puerto Rico	210	1,000	0	1,200	650	280
Total	47,000	220,000	71	270,000	32,000	61,000

[a] Values shown are round figures made individually from complete figures and sums.
[b] As used for all purposes except public supplies.

Source: Reprinted, with publisher's permission, from *J. Am. Water Works Assn.*, *53*(10):1211, © 1961, by the American Water Works Association, Inc., 2 Park Ave., New York, N.Y. 10016.

2–11 WATER USE EXCLUDING HYDROELECTRIC POWER

By Regions, 1965

Region[a]	Fresh and saline water withdrawn, mgd				Saline water withdrawn[b] mgd	All water consumed, mgd
	Ground water	Surface water	Waste water	All water		
New England	510	6,700	1.1	7,200	3,500	360
Delaware-Hudson	1,500	23,000	0	24,000	12,000	1,100
Chesapeake	520	8,900	130	9,600	3,300	390
South Atlantic	3,200	19,000	0	22,000	7,700	2,200
Eastern Gulf	1,100	5,600	0	6,700	380	520
Tennessee-Cumberland	240	7,900	0	8,200	0	400
Ohio	1,700	28,000	0.3	30,000	26	940
Eastern Great Lakes-St. Lawrence	340	15,000	0	15,000	13	520
Western Great Lakes	810	17,000	0	18,000	12	600
Hudson Bay	56	210	0	270	3.8	71
Upper Mississippi	1,700	14,000	0	16,000	18	730
Upper Missouri	3,500	15,000	0	19,000	14	10,000
Lower Missouri	240	1,500	0	1,800	0	200
Lower Mississippi	1,300	3,900	0	5,200	380	1,000
Upper Arkansas-Red	4,100	2,900	1.5	7,000	33	4,500
Lower Arkansas-Red-White	1,300	2,200	0	3,400	41	1,200
Western Gulf	14,000	15,000	28	30,000	4,600	14,000
Colorado	4,700	12,000	58	17,000	2.7	8,400
Great Basin	1,600	5,300	51	6,900	13	3,800
South Pacific	13,000	25,000	400	38,000	11,000	15,000
Pacific Northwest	4,200	25,000	2.9	29,000	31	10,000
Hawaii	820	1,200	0	2,000	540	570
Alaska	26	120	0	140	1.0	11
Puerto Rico	150	1,500	0	1,700	1,100	270
Total	61,000[c]	250,000[d]	670	310,000	45,000	78,000

[a] Quantities are shown to two significant figures.
[b] Water containing 1,000 or more mg/l dissolved solids, for industrial cooling.
[c] Includes 470 mgd saline water.
[d] Includes 44,000 mgd saline water.

Source: Reprinted, with publisher's permission, from *J. Am. Water Works Assn., 61*(11):567, © 1969, by the American Water Works Association, Inc., 2 Park Ave., New York, N.Y. 10016.

2–12 SUMMARY OF WATER WITHDRAWN, EXCEPT FOR HYDROELECTRIC POWER

By States, 1970

			Water withdrawn (including irrigation conveyance losses), mgd											
			Ground water			Surface water				All sources				
State	Population, thousands	Per capita use, gpd	Fresh	Saline	Fresh and saline	Fresh	Saline	Fresh and saline	Reclaimed sewage	Fresh	Saline	Fresh and saline	Conveyance losses, mgd	Fresh water consumed,[a] mgd
Alabama	3,444	1,900	280	5.0	280	6,100	290	6,400	0	6,400	300	6,700	0	230
Alaska	302	830	39	0	39	210	1.0	210	0	250	1.0	250	0	16
Arizona	1,772	3,900	4,200	0	4,200	2,600	0	2,600	0	6,800	0	6,800	240	4,800
Arkansas	1,923	1,600	1,500	0	1,500	1,500	0	1,500	0	3,000	0	3,000	90	1,200
California	19,953	2,400	18,000	180	18,000	20,000	8,700	29,000	130	39,000	8,800	48,000	4,900	22,000
Colorado	2,207	6,000	2,100	7.0	2,100	11,000	9.0	11,000	80	13,000	16	13,000	1,400	6,800
Connecticut	3,032	1,200	150	1.0	150	1,400	2,000	3,300	0	1,500	2,000	3,500	0	180
Delaware	548	2,200	68	0	68	110	1,000	1,100	0	180	1,000	1,200	0	26
Florida	6,789	2,300	2,900	87	3,000	3,000	9,400	12,000	0	5,900	9,500	15,000	150	1,900
Georgia	4,590	1,200	630	0	630	4,500	140	4,700	0	5,200	140	5,300	0	420
Hawaii	770	3,500	910	13	920	850	860	1,700	66	1,800	870	2,700	220	810
Idaho	713	22,000	2,500	0	2,500	13,000	0	13,000	2.5	16,000	0	16,000	4,300	4,700
Illinois	11,114	1,400	1,000	40	1,100	15,000	0	15,000	0	16,000	40	16,000	0	360
Indiana	5,194	1,700	470	4.4	470	8,200	0	8,200	0	8,600	4.4	8,600	0	390
Iowa	2,825	750	500	0	500	1,600	0	1,600	0	2,100	0	2,100	0	240
Kansas	2,249	1,700	3,100	0	3,100	640	0	640	0	3,800	0	3,800	43	2,600
Kentucky	3,219	1,400	170	15	180	4,300	0	4,300	0	4,500	15	4,500	0	180
Louisiana	3,643	2,500	1,500	41	1,500	6,700	880	7,500	0	8,100	920	9,100	430	2,200
Maine	994	760	36	0	36	520	200	720	0	560	200	760	0	61
Maryland	3,922	1,300	140	0	140	1,300	3,500	4,900	130	1,600	3,500	5,200	0	140
Massachusetts	5,689	740	350	0	350	1,600	2,300	3,900	0	2,000	2,300	4,200	0	140
Michigan	8,875	1,500	500	400	900	12,000	0	12,000	0	13,000	400	13,000	0	330
Minnesota	3,805	900	780	0	780	2,700	0	2,700	0	3,400	0	3,400	0	310

[a] Excluding irrigation conveyance losses by evapotranspiration.

2–12 SUMMARY OF WATER WITHDRAWN, EXCEPT FOR HYDROELECTRIC POWER (*Continued*)

			Water withdrawn (including irrigation conveyance losses), mgd											
			Ground water			Surface water				All sources				
State	Population, thousands	Per capita use, gpd	Fresh	Saline	Fresh and saline	Fresh	Saline	Fresh and saline	Reclaimed sewage	Fresh	Saline	Fresh and saline	Conveyance losses, mgd	Fresh water consumed,[a] mgd
Mississippi	2,217	950	760	0	760	810	540	1,300	0	1,600	540	2,100	37	410
Missouri	4,677	750	430	0.2	430	3,100	0.9	3,100	0	3,500	1.1	3,500	5.5	310
Montana	694	12,000	150	0	150	7,900	0	7,900	0	8,000	0	8,000	2,200	5,500
Nebraska	1,484	4,100	3,300	0	3,300	2,700	0	2,700	0	6,000	0	6,000	1,300	3,700
Nevada	489	6,700	520	6.2	530	2,700	0	2,700	4.2	3,300	6.2	3,300	1,600	1,500
New Hampshire	738	940	55	0	55	470	160	630	0	530	160	700	0	18
New Jersey	7,168	870	1,000	0	1,000	2,000	3,200	5,200	0	3,100	3,200	6,300	0	380
New Mexico	1,016	3,100	1,600	0	1,600	1,600	0	1,600	22	3,200	0	3,200	160	1,500
New York	18,191	970	890	1.7	890	8,200	8,600	17,000	2.3	9,100	8,600	18,000	0	660
North Carolina	5,082	1,200	430	0	430	5,300	170	5,400	0	5,700	170	5,900	0	480
North Dakota	618	1,000	82	9.6	92	550	0.9	550	0	630	10	650	59	200
Ohio	10,652	1,700	880	0	880	17,000	0	17,000	0	18,000	0	18,000	0	460
Oklahoma	2,559	590	860	37	890	610	13	620	0	1,500	50	1,500	20	830
Oregon	2,091	2,800	970	0	970	5,000	0	5,000	3.1	5,900	0	5,900	1,500	2,600
Pennsylvania	11,794	1,700	770	0	770	19,000	50	19,000	0	20,000	50	20,000	0	450
Rhode Island	950	490	38	0.4	39	110	310	430	0	150	310	460	0	13
South Carolina	2,591	1,300	170	0	170	3,100	100	3,200	0	3,300	100	3,400	0	180
South Dakota	666	910	180	170	350	250	0	250	0	440	170	600	79	250
Tennessee	3,924	1,600	290	0	290	6,100	0	6,100	0	6,400	0	6,400	0	200
Texas	11,197	2,400	9,200	0	9,200	10,000	7,400	18,000	21	19,000	7,400	27,000	480	9,600
Utah	1,059	4,000	680	3.5	680	3,500	5.5	3,500	52	4,200	9.0	4,200	630	2,200
Vermont	445	250	42	0	42	69	0	69	0	110	0	110	0	13
Virginia	4,648	1,200	280	0	280	4,400	850	5,300	0	4,700	850	5,500	0	150
Washington	3,409	2,100	840	0	840	6,300	37	6,300	0	7,100	37	7,200	1,000	2,500

[a] Excluding irrigation conveyance losses by evapotranspiration.

2–12 SUMMARY OF WATER WITHDRAWN, EXCEPT FOR HYDROELECTRIC POWER (*Continued*)

State	Population, thousands	Per capita use, gpd	Water withdrawn (including irrigation conveyance losses), mgd: Ground water			Surface water				All sources			Conveyance losses, mgd	Fresh water consumed,[a] mgd
			Fresh	Saline	Fresh and saline	Fresh	Saline	Fresh and saline	Reclaimed sewage	Fresh	Saline	Fresh and saline		
West Virginia	1,744	3,300	80	0	80	5,700	0	5,700	0	5,800	0	5,800	0	66
Wisconsin	4,418	1,400	490	0	490	5,800	0	5,800	0	6,300	0	6,300	0.7	180
Wyoming	332	17,000	230	21	250	5,500	0	5,500	7.7	5,700	21	5,800	1,500	2,400
District of Columbia	757	1,700	0.8	0	0.8	1,300	0	1,300	0	1,300	0	1,300	0	16
Puerto Rico	2,712	1,100	140	1.5	140	430	2,400	2,900	0	580	2,400	3,000	42	170
U.S.[b]	205,897	1,800	68,000	1,000	69,000	250,000	53,000	300,000	520	320,000	54,000	370,000	22,000	87,000

[a] Excluding irrigation conveyance losses by evapotranspiration.
[b] Including Puerto Rico.

Note: Partial figures may not add to totals because of independent rounding.

Source: C.R. Murray and E.B. Reeves, *Estimated Use of Water in the United States, 1970,* U.S. Geological Survey, U.S. Department of the Interior, 1972.

2–13 SUMMARY OF WATER WITHDRAWN, EXCEPT FOR HYDROELECTRIC POWER

By Regions, 1970

Water Resources Council region	Population, thousands	Per capita use, gpd	Water withdrawn including irrigation conveyance losses, mgd: Ground water: Fresh	Ground water: Saline	Ground water: Fresh and saline	Surface water: Fresh	Surface water: Saline	Surface water: Fresh and saline	Reclaimed sewage	All sources: Fresh	All sources: Saline	All sources: Fresh and saline	Conveyance losses, mgd	Fresh water consumed,[a] mgd
New England	11,520	840	630	1.4	640	4,100	4,900	9,000	0	4,700	4,900	9,700	0	410
Middle Atlantic	38,401	1,200	2,600	1.7	2,600	25,000	17,000	42,000	130	28,000	17,000	45,000	0	1,400
South Atlantic-Gulf	23,445	1,500	4,700	92	4,800	20,000	11,000	30,000	0	24,000	11,000	35,000	150	3,300
Great Lakes	29,104	1,400	1,400	400	1,800	38,000	1.0	38,000	0	39,000	400	39,000	0.1	1,200
Ohio	20,009	1,800	1,700	42	1,800	34,000	16	34,000	0	35,000	58	36,000	0	920
Tennessee	3,234	2,400	170	0	170	7,700	0	7,700	0	7,900	0	7,900	0	240
Upper Mississippi	12,733	1,300	2,200	19	2,300	14,000	0	14,000	0	16,000	19	16,000	0.6	760
Lower Mississippi	6,265	2,100	3,600	41	3,600	8,500	880	9,400	0	12,000	920	13,000	550	3,600
Souris-Red-Rainy	808	390	68	4.8	73	250	0	250	0	310	4.8	320	1.0	71
Missouri	8,489	2,900	5,900	190	6,100	18,000	6.4	18,000	88	24,000	200	24,000	4,800	12,000
Arkansas-White-Red	6,670	1,800	6,600	38	6,700	5,200	18	5,200	11	12,000	55	12,000	310	6,800
Texas-Gulf	9,499	2,200	6,200	0	6,200	7,400	7,200	15,000	15	14,000	7,200	21,000	180	6,200
Rio Grande	1,617	3,900	2,400	0	2,400	3,800	150	3,900	17	6,200	150	6,300	590	3,300
Upper Colorado	404	20,000	110	4.4	110	8,000	0	8,000	0	8,100	4.4	8,100	1,100	4,100
Lower Colorado	2,223	3,300	4,500	0	4,500	2,800	0	2,800	5.1	7,200	0	7,200	300	5,000
Great Basin	1,213	5,500	1,100	9.7	1,100	5,500	5.5	5,500	53	6,700	15	6,700	2,000	3,200
Columbia-North Pacific	6,468	4,700	4,300	0	4,300	26,000	37	26,000	5.6	30,000	37	30,000	7,200	11,000
California-South Pacific	20,009	2,400	18,000	180	18,000	21,000	8,700	30,000	120	39,000	8,800	48,000	5,100	22,000

[a] Excluding irrigation conveyance losses by evapotranspiration.

2–13 SUMMARY OF WATER WITHDRAWN, EXCEPT FOR HYDROELECTRIC POWER (*Continued*)

Water Resources Council region	Population, thousands	Per capita use, gpd	Water withdrawn including irrigation conveyance losses, mgd: Ground water: Fresh	Ground water: Saline	Ground water: Fresh and saline	Surface water: Fresh	Surface water: Saline	Surface water: Fresh and saline	Reclaimed sewage	All sources: Fresh	All sources: Saline	All sources: Fresh and saline	Conveyance losses, mgd	Fresh water consumed,[a] mgd
Alaska	302	830	39	0	39	210	1.0	210	0	250	1.0	250	0	16
Hawaii	770	3,500	910	13	920	850	860	1,700	66	1,800	870	2,700	220	810
Puerto Rico	2,712	1,100	140	1.5	140	430	2,400	2,900	0	580	2,400	3,000	42	170
U.S.[b]	205,897	1,800	68,000	1,000	69,000	250,000	53,000	300,000	520	320,000	54,000	370,000	22,000	87,000

Note: Partial figures may not add to totals because of independent rounding.

[a] Excluding irrigation conveyance losses by evapotranspiration.
[b] Including Puerto Rico.

Source: C.R. Murray and E.B. Reeves, *Estimated Use of Water in the United States, 1970,* U.S. Geological Survey, U.S. Department of the Interior, 1972.

2–14 SUPPLY COMPARED WITH DEMAND[1]

Western U.S.

Region	Estimated dependable supply, bgd, 1980[2]	Total[a] withdrawal, bgd, 1965	Water consumed, bgd, 1965	Fresh surface water withdrawn, bgd, 1965[a]
Missouri	33	21	10.2	16.5
Arkansas	20	10.4	5.7	5.0
Western Gulf	20	30[b]	14	10
Colorado	15	17	8.4	12
Great Basin	9	6.9	3.8	5.3
South Pacific	28	38[c]	15	14
Pacific Northwest	70	29	10	24

[a] Including some minor interregional diversions.
[b] Includes 4.6 bgd saline water.
[c] Includes 11 bgd saline water.

REFERENCES

1. C.R. Murray, *J. Am. Water Works Assn., 61*(11):567, 1969.
2. D.R. Woodward, *Availability of Water in the United States with Special Reference to Industrial Needs by 1980,* Industrial College of Armed Forces, 1957, p. 49.

2–15 PRESENT AND ULTIMATE WATER SUPPLY AND DEMAND DATA

California

Hydrographic area[a]	Percentage of total stream flow[b]	Percentage of present demand[c]	Ultimate annual demand, acre ft (1,000,000's)	Available supply[e] acre ft (1,000,000's)	Ultimate imported need, acre ft (1,000,000's)
North coastal	40.8	2.2	2.1	2.1	0.0
San Francisco Bay	1.7	2.8	3.5	1.4	2.1
Central coastal	3.5	3.0	2.2	1.0	1.2
South coastal	1.7	7.4	5.6	2.7	2.9
Sacramento Basin	31.6	18.9	7.4	7.4	0.0
San Joaquin River and Tulare Lake basins	15.9	46.4	17.2[d]	7.8	9.4
Lahontan	4.5	3.0	6.7	0.6	6.1
Colorado Desert	0.3	16.3	6.4	4.2	2.2
Total	100.0	100.0	51.1	27.2	23.9

[a] See 2–16.
[b] Total annual stream flow, 71,000,000 acre ft.
[c] Present annual demand, 25,000,000 acre ft.
[d] Including delta and operation of salinity control barrier.
[e] Including full rights to Colorado River and potential transfer under existing rights.

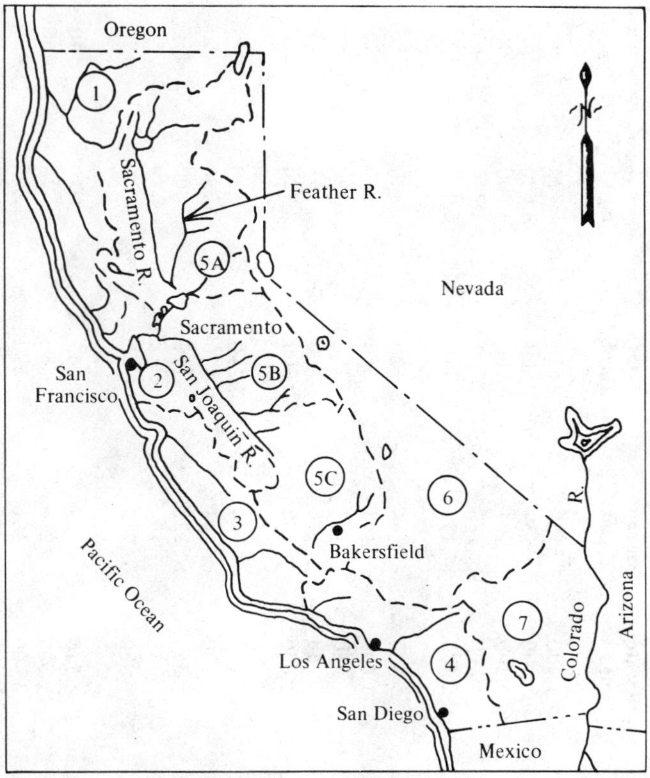

2–16 Hydrographic Boundaries in California.

Note: The dashed lines represent hydrographic-area boundaries.

Source: Reprinted, with publisher's permission, from *J. Am Water Works Assn., 57*(10):1248, © 1965, by the American Water Works Association, Inc., 2 Park Ave., New York, N.Y. 10016.

2.1 *PUBLIC WATER SUPPLY NEEDS*

2.1–1 WATER USED FOR PUBLIC SUPPLIES

By Regions, 1970

Water Resources Council region	Population served: Ground water, thousands	Population served: Surface water, thousands	Population served: All water, thousands	Water withdrawn: Ground water, mgd	Water withdrawn: Surface water, mgd	Water withdrawn: All water, mgd	Water withdrawn: Per capita, gpd	Water delivered: Industrial and commercial uses, mgd	Water delivered: Domestic use and losses,[a] mgd	Water consumed, mgd
New England	2,720	7,360	10,100	330	1,100	1,400	139	610	790	190
Middle Atlantic	8,670	24,900	33,600	1,100	4,100	5,200	154	1,600	3,600	750
South Atlantic-Gulf	7,960	6,560	14,500	1,300	1,400	2,700	187	980	1,700	590
Great Lakes	4,000	19,400	23,400	700	3,700	4,400	188	1,700	2,700	500
Ohio	4,710	10,300	15,000	620	1,500	2,100	139	710	1,400	270
Tennessee	532	1,550	2,080	64	240	300	146	120	180	36
Upper Mississippi	6,350	4,530	10,900	870	690	1,600	143	560	990	190
Lower Mississippi	3,170	1,260	4,430	390	220	610	137	150	460	240
Souris-Red-Rainy	192	209	402	20	25	48	118	6.2	41	19
Missouri	2,850	3,860	6,710	430	590	1,000	152	300	720	250
Arkansas-White-Red	1,780	3,160	4,940	250	490	730	149	210	520	250
Texas-Gulf	3,970	3,950	7,920	590	550	1,100	144	370	780	380
Rio Grande	876	499	1,370	180	130	310	228	110	200	150
Upper Colorado	80	116	196	28	30	58	294	8.1	49	19
Lower Colorado	1,220	576	1,790	250	140	390	219	66	330	190
Great Basin	546	558	1,100	160	160	320	289	45	270	140
Columbia-North Pacific	1,840	2,800	4,630	460	830	1,300	278	580	710	260
California-South Pacific	8,030	10,700	18,700	1,600	1,800	3,400	181	620	2,800	1,400
Alaska	62	64	126	24	35	60	473	7.0	53	11
Hawaii	662	32	694	120	12	140	197	26	110	46
Puerto Rico	396	1,940	2,330	34	170	200	86	77	130	43
U.S.[b]	60,600	104,000	165,000	9,400	18,000	27,000	166	8,800	19,000	5,900

[a] Includes public use.
[b] Including Puerto Rico.

Note: Partial figures may not add to totals because of independent rounding.

Source: C.R. Murray and E.B. Reeves, *Estimated Use of Water in the United States, 1970,* U.S. Geological Survey, U.S. Department of the Interior, 1972.

2.1–2 WATER USED FOR PUBLIC SUPPLIES

By States, 1970

State	Population served			Water withdrawn				Water delivered		
	Ground water, thousands	Surface water, thousands	All water, thousands	Ground water, mgd	Surface water, mgd	All water, mgd	Per capita, gpd	Industrial and commercial uses, mgd	Domestic use and losses,[a] mgd	Water consumed, mgd
Alabama	765	1,420	2,190	100	360	470	213	260	210	36
Alaska	62	64	126	24	35	60	473	7.0	53	11
Arizona	989	509	1,500	190	120	310	208	47	260	160
Arkansas	605	637	1,240	71	95	170	133	63	100	36
California	8,000	10,700	18,700	1,600	1,800	3,400	181	620	2,800	1,400
Colorado	306	1,670	1,980	79	310	390	197	91	300	97
Connecticut	590	1,900	2,490	86	270	360	143	170	190	120
Delaware	217	193	410	30	46	76	185	32	44	20
Florida	4,820	592	5,410	760	120	880	163	170	720	230
Georgia	825	1,350	2,170	190	350	540	250	310	240	130
Hawaii	662	32	694	120	12	140	197	26	110	46
Idaho	407	63	470	96	15	110	237	5.8	110	29
Illinois	3,880	6,790	10,700	720	1,500	2,200	204	570	1,600	210
Indiana	1,500	1,990	3,480	210	280	490	141	140	350	120
Iowa	1,520	505	2,030	180	73	250	123	60	190	37
Kansas	817	823	1,640	130	120	250	155	61	190	74
Kentucky	304	1,860	2,160	24	160	180	83	78	100	26
Louisiana	1,290	1,370	2,670	140	240	380	144	58	330	200
Maine	153	599	752	20	89	110	146	38	72	22
Maryland	368	2,750	3,120	42	380	420	136	85	340	23
Massachusetts	1,460	3,940	5,400	170	590	750	140	320	440	38
Michigan	1,360	5,500	6,870	230	920	1,200	168	640	520	95
Minnesota	1,500	1,250	2,750	160	180	340	125	140	210	34
Mississippi	1,180	213	1,390	160	30	190	134	63	120	70
Missouri	942	3,020	3,970	92	420	510	128	260	250	95
Montana	152	354	507	26	85	110	219	32	79	55
Nebraska	908	211	1,120	150	34	190	168	46	140	38
Nevada	265	176	441	81	54	130	305	39	95	46

[a] Includes public use.

2.1–2 WATER USED FOR PUBLIC SUPPLIES (*Continued*)

State	Population served			Water withdrawn				Water delivered		
	Ground water, thousands	Surface water, thousands	All water, thousands	Ground water, mgd	Surface water, mgd	All water, mgd	Per capita, gpd	Industrial and commercial uses, mgd	Domestic use and losses,[a] mgd	Water consumed, mgd
New Hampshire	261	283	544	32	38	70	128	22	48	3.6
New Jersey	3,030	3,390	6,420	340	560	900	139	110	780	180
New Mexico	645	67	712	130	16	150	204	41	100	66
New York	4,150	12,300	16,500	460	2,200	2,600	161	790	1,900	450
North Carolina	660	2,020	2,680	80	380	460	170	120	340	91
North Dakota	189	206	395	24	26	50	126	1.6	48	28
Ohio	2,480	5,980	8,460	320	1,000	1,300	157	340	990	170
Oklahoma	551	1,450	2,000	72	190	260	130	86	170	100
Oregon	355	842	1,200	67	160	230	188	90	140	48
Pennsylvania	1,350	8,300	9,650	250	1,500	1,800	181	880	870	180
Rhode Island	213	634	847	18	85	100	122	62	41	5.2
South Carolina	221	1,010	1,230	55	240	300	242	120	180	45
South Dakota	284	126	410	42	18	60	145	19	41	16
Tennessee	1,190	1,930	3,120	160	240	400	129	140	260	43
Texas	4,580	4,660	9,250	690	740	1,400	155	470	970	510
Utah	502	444	945	150	130	280	294	25	250	130
Vermont	100	189	289	14	29	42	146	13	30	1.7
Virginia	477	3,040	3,520	74	320	390	111	160	230	42
Washington	1,070	1,780	2,860	290	610	910	317	470	440	160
West Virginia	380	794	1,170	36	140	180	150	71	110	0.2
Wisconsin	1,540	1,570	3,110	220	270	480	155	230	250	48
Wyoming	130	120	250	24	25	49	197	13	37	12
District of Columbia	0	757	757	0	160	160	211	50	110	16
Puerto Rico	396	1,940	2,330	34	170	200	86	77	130	43
U.S.[b]	60,600	104,000	165,000	9,400	18,000	27,000	166	8,800	19,000	5,900

[a] Includes public use.
[b] Including Puerto Rico.

Note: Partial figures may not add to totals because of independent rounding.

Source: C.R. Murray and E.B. Reeves, *Estimated Use of Water in the United States, 1970,* U.S. Geological Survey, U.S. Department of the Interior, 1972.

2.1–3 VARIATIONS IN PER CAPITA WATER DEMAND

Year	Yearly avg, gal/cap day	Ratio, period demand to yearly avg			
		Max month	Max week	Max day	Max hr
1900	60–85	1.15–1.30		1.50–2.00	
1927	50–400	1.25	1.35	1.50	
1940	110			1.50	2.10–2.25
1947	100–120	1.25	1.50	1.75	2.62–3.06
1948	110			1.50–2.00	2.00–3.00
1949	110	1.40		1.50–1.80	2.50–3.50
1953	135	1.28	1.48	1.80	2.70–3.25

Source: Reprinted, with publisher's permission, from *J. Am. Water Works Assn.*, *53*(4):459, © 1961, by the American Water Works Association, Inc., 2 Park Ave., New York, N.Y. 10016.

2.1–4 RATIO OF PEAK DEMANDS TO AVERAGE ANNUAL DEMAND

Area	Average for year	Maximum						Ref.
		Month	Week	Day	6 hrs	3 hrs	1 hr	
Manhasset-Lakeville, L.I.	1.0	1.63	2.05	2.35	4.0	4.6	5.0	1
Port Washington, L.I.	1.0	1.76		3.0		3.4	4.1	1
Westbury, L.I.	1.0	1.55		2.5		3.5	4.5	1
Citizens Water Co., Great Neck, L.I.	1.0	1.77	2.24	3.0		4.9	5.65	1
Albertson, L.I.	1.0	2.5		3.75			8+	1
Levittown, Pa.	1.0	2.0		3.0		8.75	10.0	1
Indianapolis, Ind.	1.0			1.60			2.43	2
Central	1.0			1.50			2.05	2
East	1.0			1.66			2.99	2
North	1.0			2.02			4.00	2

REFERENCES

1. A.D. Henderson, *J. Am. Water Works Assn.*, *48*(4):361, 1956.
2. L.S. Finch, *J. Am. Water Works Assn.*, *48*(4):364, 1956.

2.1–5 WATER REQUIREMENTS DESIGN DATA

Delivery Rates, mil gal/day

City	Design year	Population	10 hr fire rate, mgd 1	Max hr, mgd 2	Max day, mgd 3	Storage, gal x 10^6 4	Total storage, mil gal 5	Used from storage, mil gal 6	Storage ratio total/use 7
Sedalia, Mo.	1985	53,000	8.0	9.0	5.5	3.5	2.0	0.6	3.3
El Dorado, Kans.	1980	16,900	5.4	7.1	4.2	2.9	1.38	0.5	2.8
Colorado Springs, Colo.	1980	273,000	29.0	225.0	127.0	98.0	87.0	16.3	5.3
Madison, Wisc.	1980	204,000	19	116.0	73.0	43.0	37.0	7.2	5.2
Milwaukee, Wisc.	1980	1,127,000	30.0	673.0	403.0	270.0	135.0	45.0	3.0
Bloomington, Ill.	1975	41,000	9.0	14.9	11.2	3.7	17.7	0.6	29.0
Jefferson City, Mo.	1980	35,000	8.1	10.5	6.0	4.5	4.5	0.7	6.4
Newton, Kans.	1985	21,000	6.5	12.0	8.5	3.5	4.4	0.6	7.3
Eden Prairie, Minn.	2000	70,000	11.1	36.4	18.2	18.2	5.0	3.0	1.7
New Brighton, Minn.	2000	39,000	8.6	22.4	12.1	10.3	3.9	1.7	2.3
Bloomington, Minn.	2000	160,000	16.2	84.6	42.3	42.3	36.5	7.0	5.2
Evansville, Ind.	1990	240,000	19.0	104.7	65.5	39.2	28.3	6.5	4.3

Note: Column 2 minus Column 3 = Column 4. Column 7 = $\frac{\text{Column 5}}{\text{Column 6}}$

Source: Reprinted, with publisher's permission, from *J. Am. Water Works Assn., 62*(10):629, © 1970, by the American Water Works Association, Inc., 2 Park Ave., New York, N.Y. 10016.

2.1–6 COMPARISON OF AVERAGE CONSUMPTION RATES OF WATER TO AVERAGE ANNUAL PRODUCTION RATE

Northeast Quadrant of U.S.

Max demand period	Consumption/production, %	Max demand period	Consumption/production, %
1 year	100	5 days	132
9 months	102	1 day	142
6 months	105	18 hrs	150
3 months	109	10 hrs	170
1 month	113	5 hrs	188
2 weeks	120	2 hrs	195
1 week	125	1 hr	200

Source: Reprinted, with publisher's permission, from *J. Am. Water Works Assn., 53*(4):459, © 1961, by the American Water Works Association, Inc., 2 Park Ave., New York, N.Y. 10016.

2.1–7 PEAK WATER DEMANDS IN RESIDENTIAL AREAS

	Type of home				
Lot size, ft²	Group house 2,000–2,400	Detached dwelling 5,000–7,000	Detached dwelling 9,000–12,000	Detached dwelling 15,000–25,000	Detached dwelling 40,000 and over
No. of observations	63	51	67	70	61
Max day, avg rate, gpd/service	327	228	556	1,170	2,320
Peak hour, gpd/service	625	605	1,050	2,630	5,400
Ratio, peak hour to max day	1.91:1	2.67:1	1.89:1	2.25:1	2.33:1
Ratio, peak hour to avg day	3.84:1	3.30:1	4.63:1	7.90:1	10.30:1
Max day, gpcd	93	64	163	278	564
Avg day, gpcd	48	51	67	80	117
Dwelling units/acre[a]	9.0	4.2	3.0	1.6	0.8
Persons/acre	30.6	16.8	10.50	7.46	3.33
Max hr, gpm/acre	3.8	1.8	2.2	2.9	2.7
Max day, gpm/acre	2.9	0.7	1.2	1.3	1.3

[a] Adjusted by taking into account such areas as streets, public buildings, parks, institutions, and drainage courses.

Source: Reprinted, with publisher's permission, from *J. Am. Water Works Assn., 53*(10):1251, © 1961, by the American Water Works Association, Inc., 2 Park Ave., New York, N.Y. 10016.

2.1–8 WATER DEMANDS FOR MEDIUM AND HIGH VALUE HOMES

	Medium value homes[a]		High value homes[b]	
Demand period	Domestic demand, gpd	Sprinkling demand/acre,[c] gpd	Domestic demand, gpd	Sprinkling demand/acre, gpd
Avg day	203	177	195	157
Max day	250	1,900	290	2,540
Peak hours	565	5,750	860	4,430

[a] Pine Valley, suburban area in Baltimore County.
[b] Hampton, residential area of Baltimore.
[c] Average of 0.26 acres/home in medium value area, and 0.8 acres/home in high value area.

Source: Reprinted, with publisher's permission, from *J. Am. Water Works Assn., 56*(4):403, © 1964, by the American Water Works Association, Inc., 2 Park Ave., New York, N.Y. 10016.

2.1–9 WATER USE OF LOW INCOME HOUSING FOR LARGE FAMILIES

Chicago

Project name	No. of apartments	Population	Fixture units	Maximum recorded demand, gpm	Maximum recorded demand, gal/cap day[a]
Clarence Darrow Homes	480	3,312	5,280	282	123
Rockwell Gardens	140	728	1,540	151	299
Robert R. Taylor Homes	474	2,940	5,214	311	152
Midway Gardens	318	670	3,600	307	660

[a] Calculated by the editors.

Source: Reprinted, with publisher's permission, from *J. Am. Water Works Assn., 58*(7):827, © 1966, by the American Water Works Association, Inc., 2 Park Ave., New York, N.Y. 10016.

2.1–10 WATER USE AT CHICAGO HOUSING AUTHORITY PROJECTS FOR THE ELDERLY

Project name	No. of apartments	Population	Demand, gal/min	Demand, gal/cap day[a]
Wicker Park	116	199	125	906
Eckhart Park	181	299	200	964
Blackhawk	252	416	280	970
Midwest Terrace	129	213	160	1,080
Armour Square	198	327	220	971
Garfield Park	151	249	175	1,012
Edgewater Terrace	200	330	185	809
Madison and Pauline	482	795	275	498
Franklin and Drake	157	259	190	1,058
Aldine and Clark	151	249	175	1,012

[a] Calculated by the editors.

Source: Reprinted, with publisher's permission, from *J. Am. Water Works Assn., 58*(7):827, © 1966, by the American Water Works Association, Inc., 2 Park Ave., New York, N.Y. 10016.

2.1–11 DOMESTIC WATER USE DATA

Item	Daily family use, gal[1]	Daily per capita use, gal[1]	Percent[2]	Percent[2]
Drinking and kitchen	8	2.0	2.3	11
Dishwasher	15	3.75	4.3	
Toilet	96	24.0	27.8	41
Bathing	80	20.0	23.2	37
Laundering	34	8.5	9.8	4
Auto washing	10	2.5	2.8	1
Lawn watering	100	25.0	28.9	3
Garbage disposal unit	3	0.75	0.8	–
Cleaning	–	–	–	3
Total	346	86.50	99.9	100
All uses except toilet and lawn watering	150	37.5		

REFERENCES

1. G.W. Reid, *Southwest Water Works J.*, *46*:18, 1965.
2. Akron Bureau of Water Supply, Akron City Water Works, Akron, Ohio.

2.1–12 ESTIMATED POTABLE WATER USE

Type of use	Quantity used, gpcd
Drinking, cooking	1–2
Dishwashing	1–4
Garbage disposal unit	0–4
Laundering, cleaning	3–7
Bathing	10–25
Total	15–42

Source: Reprinted, with publisher's permission, from *J. Am. Water Works Assn.*, *57*(9):1073, © 1965, by the American Water Works Association, Inc., 2 Park Ave., New York, N.Y. 10016.

2.1–13 APPLIANCE WATER USE

Equipment	gpd, max
Automatic washer	58.3
Wringer washer	50.2
Disposer, food	81
Dishwasher	91.3
Air conditioner (water cooling)	2,160–2,880[a]
Wading pools	58.5

[a] Per ton of refrigerator capacity.

Source: Reprinted, with publisher's permission, from *J. Am. Water Works Assn.*, *55*(4):441, © 1963, by the American Water Works Association, Inc., 2 Park Ave., New York, N.Y. 10016.

2.1–14 ESTIMATED RATES OF USE OF HOT WATER

Type of building	gal/day	Type of building	gal/hr
Dwelling, single family		**Industrial plant**	6
No bath	30–60	Lavatory	12
One bath	60–100	Shower	225
Two baths	100–200	Slop sink	20
Dwelling, large family, two baths	200–300	Dishwasher	20–100
Apartment		Demand factor	0.40[b]
One bath	60–125	**Office**	
Two baths	100–200	Lavatory	6
Three baths	150–250	Slop sink	15
Hotel		Demand factor	0.3[b]
Room and bath	20–70	**Hospital**	
Room and lavatory	5–10	Lavatory	6
Restaurant		Shower	75
Hand dishwashing, per meal	1–2	Slop sink	50–150
Machine dishwashing, per meal	1½–4½	Demand factor	0.20[b]
Laundry, per 100 lb dry clothes	80[a]	**Hotel**	
Washing machines[1]		Lavatory	8
100 gal tank	150	Shower	75
150 gal tank	200	Slop sink	30
200 gal tank	300	Dishwasher	50–200
Industrial building, per occupant	5	Demand factor	0.25[b]
Office building, per occupant	5		
Commercial, per worker	2		
Manufacturing, per worker	5		

[a] In addition, 100 gal of cold water, 30 lb of steam to heat the water, and 1 kWh of electrical energy. These minimum figures are exceeded by 50 percent by many average hand laundries and by 100 percent by small mechanically equipped laundries.
[b] Probable number of fixtures in use simultaneously.

Source: H.E. Babbitt, *Plumbing,* McGraw-Hill Book Co., 1960, p. 179.

REFERENCE

1. F.M. Reiter, *Air Conditioning, Heating, and Ventilating,* February 1954, p. 90.
2. F.M. Reiter, Series of articles, *Ind. Gas,* 1954.

2.1–15 WATER REQUIREMENTS FOR DOMESTIC SERVICE, PUBLIC BUILDINGS, SCHOOLS, AND CAMPS

Domestic fixtures	
Fill lavatory	2 gal
Fill bath tub	30 gal
Shower bath	30–60 gal
Flush toilet	6 gal
Dishwashing machine	3 gal per load
Automatic laundry machine	30–50 gal per load
Lawn sprinkler	120 gal/hr
1/2 in. hose and nozzle	240–300 gal/hr
5/8 in. hose and nozzle	270–330 gal/hr
3/4 in. hose and nozzle	300–360 gal/hr
Private homes	
For each member of family including kitchen, laundry, and bath	40 gal/day
Public buildings	gal/hr per fixture
Hotel	50
Apartment houses	20
Hospitals	25
Office buildings	40
Mercantile buildings	35
1½ in. fire hose and nozzle	2,400
Day school	50
Schools and camps	gal/cap day
Schools	15–17
Camps	
with hot and cold running water, kitchen, laundry, shower, bath, and flush toilets	45
with flush toilets	25
without running water or flush toilets	5

Source: *The Water Encyclopedia,* D.K. Todd, Ed., Water Information Center, Port Washington, N.Y., 1970.

2.1–16 INSTITUTIONAL WATER CONSUMPTION DATA

Institution	Suggested 24 hr avg flow, gpd[a]	Flow factors[b]		
		12 hr max	4 hr max	1 hr max
Schools				
Elementary	10	2.0	4.0	5.0
Junior high	12	2.0	4.0	5.0
Senior high	20	1.75	2.5	3.25
Combined	14	1.75	2.5	3.25
Motels	70	1.5	2.5	4.5
Restaurants	9	1.4	1.7	2.0
Shopping centers[c]	250	1.5	1.9	2.1
Hospitals[d]	275	1.25	1.75	1.9

[a] Denominator for schools is student; for motels, guest; for restaurants, meal; for shopping centers, 1000 ft^2; for hospitals, bed.
[b] Factors by which 24 hr average floors are multiplied to obtain various maximum values.
[c] Based on a shopping area with floor area of 133,458 ft^2.
[d] Based on a 120 bed hospital.

Source: Reprinted, with publisher's permission, from *J. Am. Water Works Assn.*, *53*(9):1111, © 1961, by the American Water Works Association, Inc., 2 Park Ave., New York, N.Y. 10016.

2.1–17 HOSPITAL WATER REQUIREMENTS

St. John's Hospital, St. Louis, Missouri

Item[a]	Water use, gpd/bed	Ratio at designated use to avg
Avg daily	300	1
Max daily	375	1.25
Avg 5 hr maximum	530	1.70
5 hr max	675	2.25
2 hr max	800	2.70

[a] Hospital had 525 beds and a staff of 750 (3 shifts). Estimated 92% of waste returned to sewer.

Source: *Design and Construction of Sanitary and Storm Sewers,* Manual of Practice No. 9, Water Pollut. Control Fed., 1969.

2.1–18 COMMERCIAL WATER USE

	Unit	Average annual demand, gpd	Maximum hourly demand rate, gpd	Hour of peak occurrence	Ratio, maximum hourly to average annual	Average annual demand per unit	Ratio, maximum hourly demand to R-40 demand[a]
Miscellaneous residential							
Apartment building	22 units	3,430	11,700	5–6 p.m.	3.41	156 gpd/unit	2.2:1
Motel	166 units	11,400	21,600	7–8 a.m.	1.89	69 gpd/unit	4.0:1
Hotels							
Belvedere	275 rooms	112,000	156,000	9–10 a.m.	1.39	407 gpd/room	29 :1
Emerson	410 rooms	126,000				307 gpd/room	
Office buildings							
Commercial Credit	490,000 ft^2	41,400	206,000	10–11 a.m.	4.89	0.084 gpd/ft^2	38 :1
Internal Revenue	182,000 ft^2	14,900	74,700	11–12 a.m.	5.01	0.082 gpd/ft^2	14 :1
State Office Building	389,000 ft^2	27,000	71,800	10–11 a.m.	2.58	0.070 gpd/ft^2 [b]	13 :1
Shopping centers							
Towson Plaza	240,000 ft^2	35,500	89,900	2–3 p.m.	2.50	0.15 gpd/ft^2	17 :1
Hillendale	145,000 ft^2	26,000				0.18 gpd/ft^2	
Miscellaneous commercial							
Laundries							
Laundromat	10 8-lb washers	1,840	12,600	11–12 a.m.	6.85	184 gpd/washer	2.3:1
Commercial	Equivalent to 10 8-lb washers	2,510	16,200	10–11 a.m.	6.45	251 gpd/washer equivalent	3.0:1
Washmobile	Capacity of 24 cars per hour	7,930	75,000	11–12 a.m.	9.46	330 gpd per car per hr of capacity	14 :1
Service station	1 lift	472	12,500	6–7 p.m.	26.5	472 gpd/lift	2.3:1

[a] Lot type R-40 (one acre) peak hourly demand for single service is 5,400 gpd.
[b] Exclusive of air conditioning.

Source: Residential Water Use Research Project, Johns Hopkins University and Office of Technical Studies of the Architectural Standards Division of the Federal Housing Administration.

2.1–19 COMMERCIAL WATER SUPPLY REQUIREMENTS

Establishment	Average demand, gpd	Rate of max hourly demand gal/hr	Rate of max hourly demand gal/day	Hour of peak occurrence	Ratio of max hourly commercial demand to R-40 demand[a]
Gasoline service station	636	522	12,500	7–9 p.m.[b]	2.3:1
Washmobile[c]	9,370	3,140	75,500	11–12 a.m.	14:1
Laundromat[d]	3,670	522	12,500	12–1 p.m.	2.3:1
Commercial laundry[e]	3,570	825	19,800	10–12 a.m.	3.7:1
Motel (110 units)	5,500	762	18,250	8–9 a.m.	3.4:1
Apartment (22 units)	4,300	486	11,650	5–6 p.m.	2.2:1

[a] Lot type R-40 peak hourly demand for single service is 5,400 gpd.
[b] Erratic, with lower peaks occurring 1–2 p.m. and 5–7 p.m.
[c] Cars washed at rate of 24 per hour. All facilities–steam cleaning of tires, automatic rinsing–available.
[d] Ten washers. Open 8 a.m.–8 p.m.
[e] Facilities equivalent to ten washers. Open 8 a.m.–8 p.m.

Source: Reprinted, with publisher's permission, from *J. Am. Water Works Assn., 53*(10):1251, © 1961, by the American Water Works Association, Inc., 2 Park Ave., New York, N.Y. 10016.

2.1–20 AIRPORT WATER SUPPLY REQUIREMENTS

Airport	Year	Jan	Feb	Mar	Apr	May	June	July	Aug	Sept	Oct	Nov	Dec	Average for year
Dulles International, Va.	1965	35.8	37.2	31.2	29.6	33.1	33.8	50.8	61.2	38.6	29.4	33.3	29.4	36.5
Washington National, D.C.	1966												50.0	
	1967	39.2	43.2											
Kennedy International, N.Y.	1966													39.0
Tulsa International, Okla.	1965	25.0	24.2	27.0	34.8	43.0	56.0	64.0	65.5	48.0	37.0	30.0	22.4	41.2
	1966	24.0	23.0	26.0	28.3	27.0	47.5	67.5	54.3	34.5	30.8	30.8	24.0	34.2
O'Hare International, Chicago, Ill.	1965	21.2	21.2	21.2	25.0	25.0	25.0	24.0	24.0	24.0	21.0	21.0	21.0	23.0
	1966	26.8	26.8	26.8	22.0	22.0	22.0	29.4	29.4	29.4	14.8	14.8	14.8	22.5
Kansas City Municipal, Mo.	1965	24.8	34.1	26.3	26.1	34.1	29.5	41.0	32.8	30.5	29.5	27.0	23.6	30.0
	1966	19.2	21.6	22.8	23.4	24.2	31.2	38.5	34.8	25.4	24.2	27.8	25.6	26.4
Love Field, Dallas, Texas	1964													54.5
	1965													40.2
	1966													32.2
Average for each month		27.0	28.9	24.5	27.1	29.8	35.0	45.0	43.2	33.0	26.7	28.3	26.5	31.6
Max		39.2	43.2	31.2	34.8	43.0	56.0	67.5	65.5	48.0	37.0	33.3	50.0	
Min		19.2	21.2	21.2	22.0	22.0	27.0	24.0	24.0	24.0	14.8	14.8	14.8	

Note: Data are in gal/passenger day. Gal x 3.785 = 1.

Source: E.H. Bryant *et al.*, *J. Water Pollut. Control Fed.*, *41*(5):772, 1969.

2.1–21 SEASONAL PATTERN FOR AIRPORT WATER REQUIREMENTS

Month	Average daily demand, gal/passenger
January–April	28
May	29
June	35
July	45
August	43
September	33
October	31
November, December	28

Note: Gal x 3.785 = l.

Source: E.H. Bryant *et al.*, *J. Water Pollut. Control Fed.*, *41*(5):772, 1969.

2.1–22 REQUIRED FIRE FLOW AND STANDARD HYDRANT SPACING

High Value Districts

Population, 1,000's	Flow required, gpm	Duration, hr	Avg area per hydrant, 1,000 ft²	Population, 1,000's	Flow required, gpm	Duration, hr	Avg area per hydrant, 1,000 ft²
1	1,000	4	120	22	4,500	10	
1.5	1,250	5		27	5,000	10	85
2	1,500	6		33	5,500	10	
3	1,750	7		40	6,000	10	80
4	2,000	8	110	55	7,000	10	70
5	2,250	9		75	8,000	10	60
8	2,500	10		95	9,000	10	55
10	3,000	10	100	120	10,000	10	48
13	3,500	10		150	11,000	10	43
17	4,000	10	90	200[a]	12,000	10	40

[a] Over 200,000 population, 12,000 gpm, with 2,000–8,000 gpm additional for a second fire, for a 10 hr duration.

Source: Reprinted, with publisher's permission, from *J. Am. Water Works Assn.*, *53*(4):458, © 1961, by the American Water Works Association, Inc., 2 Park Ave., New York, N.Y. 10016.

2.1–23 POPULATION NOT SERVED BY PUBLIC WATER FACILITIES

U.S., 1963

State and region	Population: Total,[a] thousands	Not served,[b] thousands	Not served, %
Maine	985	215	22
New Hampshire	646	158	24
Vermont	397	141	36
Massachusetts	5,252	191	4
Rhode Island	877	61	7
Connecticut	2,716	522	19
Northeast	10,872	1,288	12
New York	17,691	2,038	12
New Jersey	6,542	938	14
Pennsylvania	11,408	1,552	14
Middle Atlantic	35,642	4,528	13
Ohio	10,020	2,222	22
Indiana	4,780	1,584	33
Illinois	10,369	1,512	15
Michigan	8,036	1,594	20
Wisconsin	4,059	1,309	32
East North Central	37,265	8,221	22
Minnesota	3,507	1,075	31
Iowa	2,758	861	31
Missouri	4,412	882	20
North Dakota	645	274	42
South Dakota	707	290	41
Nebraska	1,468	427	29
Kansas	2,261	633	28
West North Central	15,758	4,442	28
Delaware	480	70	15
Maryland	3,351	777	23
Virginia	4,288	1,697	40
West Virginia	1,815	644	35
North Carolina	4,786	2,494	52
South Carolina	2,498	1,077	43
Georgia	4,206	1,410	34
Florida	5,532	1,641	30
South Atlantic[c]	26,956	9,810	36
Kentucky	3,121	1,314	42
Tennessee	3,742	1,319	35
Alabama	3,383	1,152	34
Mississippi	2,291	1,309	57
East South Central	12,538	5,094	41

[a] States may not add to regional totals due to independent rounding.
[b] Determined by subtracting number of persons served as shown in *State and Local Public Facility Needs, Vol. 1, Table 5.*
[c] District of Columbia included in U.S. total but omitted from state breakdown because entire population is served by municipal facilities.

2.1–23 POPULATION NOT SERVED BY PUBLIC WATER FACILITIES (*Continued*)

State and region	Population: Total,[a] thousands	Not served,[b] thousands	Not served, %
Arkansas	1,907	842	44
Louisiana	3,410	870	26
Oklahoma	2,450	585	24
Texas	10,257	1,637	16
West South Central	18,024	3,934	22
Montana	701	226	32
Idaho	689	239	35
Wyoming	335	86	26
Colorado	1,913	358	19
New Mexico	990	276	28
Arizona	1,517	134	9
Utah	973	166	2
Nevada	391	104	27
Mountain	7,508	1,589	21
Washington	2,961	533	18
Oregon	1,852	479	26
California	17,556	48	d
Alaska	251	164	65
Hawaii	685	91	13
Pacific	23,304	1,315	6
Total	188,658	40,221	21

[a] States may not add to regional totals due to independent rounding.
[b] Determined by subtracting number of persons served as shown in *State and Local Public Facility Needs, Vol. 1, Table 5.*
[d] Less than 1%.

Source: U.S. Bureau of the Census, *U.S. Census, Population Estimates: 1963. State and Local Public Facility Needs and Financing,* Subcommittee on Economic Progress of the Joint Economic Committee, U.S. Congress, Vol. 1, December 1966.

2.1–24 COMMUNITIES WITH OR WITHOUT PUBLIC WATER FACILITIES

U.S., 1963

			Size of community							
	Total		26–999		1,000–2,499		2,500–5,500		Over 5,500	
State and region	With	Without	With	Without	With	Without	With	Without	With	Without
Maine	337	551	248	546	41	3	23	2	25	–
New Hampshire	171	271	117	270	30	1	9	–	15	–
Vermont	197	266	157	264	23	1	7	1	10	–
Massachusetts	611	198	436	182	56	16	34	–	85	–
Rhode Island	90	66	70	63	3	2	3	1	14	–
Connecticut	274	206	188	200	27	3	20	2	39	1
Northeast	1,680	1,558	1,216	1,525	180	26	96	6	188	1
New York	1,597	1,407	1,058	1,394	224	2	124	10	191	1
New Jersey	563	608	205	603	70	3	92	2	196	–
Pennsylvania	1,107	3,663	346	3,634	322	26	191	2	248	1
Middle Atlantic	3,267	5,678	1,609	5,631	616	31	407	14	635	2
Ohio	954	1,598	454	1,588	201	8	122	2	177	–
Indiana	421	1,392	151	1,390	135	–	52	2	83	–
Illinois	1,018	1,174	448	1,169	246	2	133	3	191	–
Michigan	537	1,102	197	1,090	161	–	72	10	107	2
Wisconsin	517	816	263	802	132	4	56	3	66	7
East North Central	3,447	6,082	1,513	6,039	875	14	435	20	624	9
Minnesota	624	593	363	592	135	1	58	–	68	–
Iowa	725	588	487	588	134	–	47	–	57	–
Missouri	547	1,064	272	1,064	130	–	69	–	76	–
North Dakota	207	298	143	298	49	–	3	–	12	–
South Dakota	254	183	193	182	36	1	14	–	11	–
Nebraska	436	188	324	188	69	–	21	–	22	–
Kansas	496	461	315	461	94	–	44	–	43	–
West North Central	3,289	3,375	2,097	3,373	647	2	256	–	289	–
Delaware	39	173	15	173	14	–	5	–	5	–
Maryland	257	1,159	139	1,157	53	1	22	1	43	–
Virginia	536	1,082	396	1,077	65	4	33	1	42	–
West Virginia	517	1,157	362	1,155	99	2	29	–	27	–

2.1–24 COMMUNITIES WITH OR WITHOUT PUBLIC WATER FACILITIES (*Continued*)

State and region	Total		Size of community: 26–999		1,000–2,499		2,500–5,500		Over 5,500	
	With	Without	With	Without	With	Without	With	Without	With	Without
North Carolina	393	1,175	137	1,151	141	14	57	7	58	3
South Carolina	256	503	87	499	94	3	39	1	36	–
Georgia	451	862	213	855	114	7	68	–	56	–
Florida	391	877	135	848	99	7	71	18	86	4
South Atlantic	2,840	6,988	1,484	6,915	679	38	324	28	353	7
Kentucky	346	1,324	159	1,324	99	–	47	–	41	–
Tennessee	317	1,169	148	1,165	75	4	46	–	48	–
Alabama	386	938	210	935	67	3	59	–	50	–
Mississippi	232	582	97	579	66	2	37	1	32	–
East South Central	1,281	4,013	614	4,003	307	9	189	1	171	–
Arkansas	254	853	131	852	57	1	30	–	36	–
Louisiana	338	652	152	652	85	–	48	–	53	–
Oklahoma	415	464	243	464	83	–	40	–	49	–
Texas	1,117	1,462	539	1,454	260	6	147	2	171	–
West South Central	2,124	3,431	1,065	3,422	485	7	265	2	309	–
Montana	225	168	158	168	37	–	16	–	14	–
Idaho	185	142	117	142	34	–	18	–	16	–
Wyoming	108	64	73	64	16	–	11	–	8	–
Colorado	320	258	227	258	42	–	19	–	32	–
New Mexico	255	183	200	182	21	1	13	–	21	–
Arizona	146	213	81	213	30	–	16	–	19	–
Utah	310	25	228	25	43	–	16	–	23	–
Nevada	56	61	35	61	9	–	6	–	6	–
Mountain	1,605	1,114	1,119	1,113	232	1	115	–	139	–
Washington	493	1,181	335	1,175	79	1	38	3	41	2
Oregon	386	290	279	289	47	–	31	1	29	–
California	1,215	694	684	681	156	–	128	4	247	9
Alaska	33	282	14	278	13	3	2	–	4	1
Hawaii	90	108	51	106	20	2	11	–	8	–
Pacific	2,217	2,555	1,363	2,529	315	6	210	8	329	12
Total	21,750	34,794	12,080	34,550	4,336	134	2,297	79	3,037	31

Source: U.S. Bureau of Census, *U.S. Census, Population Estimates: 1963.*

2.1–25 WATER SUPPLY IN CARIBBEAN, CENTRAL AND SOUTH AMERICAN COUNTRIES

	Urban		Rural	
Country	Population, millions	% served[a]	Population, millions	% served[a]
Caribbean countries				
Barbados	0.1	100	0.1	88
Cuba	5.0	90	2.9	60
Dominican Republic	1.4	58	2.5	8
Haiti	0.8	17	3.9	
Jamaica	0.5	98	1.4	62
Trinidad and Tobago	0.4	100	0.6	95
Central America				
Costa Rica	0.7	100	0.8	48
El Salvador	1.2	80	1.8	28
Guatemala	1.7	82	3.0	10
Honduras	0.7	91	1.8	11
Mexico	24.7	81	20.7	18
Nicaragua	0.8	78	1.0	4
Panama	0.6	91	0.7	18
South America				
Argentina	16.4	72	6.8	5
Bolivia	1.5	55	2.5	1
Brazil	46.2	49	40.2	14
Chile	6.4	83	2.6	2
Colombia	10.4	85	8.7	49
Ecuador	2.0	73	3.4	8
Guyana	0.2	100	0.5	60
Paraguay	0.8	16	1.4	1
Peru	5.8	59	6.4	7
Uruguay	2.3	82	0.5	12
Venezuela	5.9	88	3.4	50

[a] Service by house connection or public hydrant.

Source: *Environ. Sci. Technol., 2*(1):17, 1968.

2.2 *INDUSTRIAL WATER NEEDS*

2.2–1 SELF SUPPLIED INDUSTRIAL WATER USE

By Regions, 1970

Water Resources Council region	Thermoelectric power (electric utility) use, mgd: Water withdrawn: Fresh ground water	Water withdrawn: Surface water: Fresh	Water withdrawn: Surface water: Saline	Water withdrawn: Total fresh water[a]	Water consumed: Fresh	Water consumed: Saline	Other uses, mgd: Water withdrawn: Ground water: Fresh	Ground water: Saline
New England	1.0	1,900	4,600	1,900	3.0	1.0	180	1.4
Middle Atlantic	100	15,000	16,000	15,000	35	17	1,000	1.7
South Atlantic-Gulf	32	15,000	10,000	15,000	120	88	1,500	92
Great Lakes	38	26,000	0	26,000	14	0	300	400
Ohio	54	27,000	0	27,000	50	0	750	42
Tennessee	0	6,100	0	6,100	64	0	45	0
Upper Mississippi	290	12,000	0	12,000	23	0	630	19
Lower Mississippi	66	4,000	140	4,100	190	25	980	41
Souris-Red-Rainy	0.6	140	0	140	0.9	0	5.1	4.8
Missouri	310	3,000	0.9	3,300	34	0.8	360	190
Arkansas-White-Red	46	1,900	0	1,900	82	0	250	38
Texas-Gulf	51	4,700	3,800	4,700	100	24	390	0
Rio Grande	15	6.0	0	21	17	0	110	0
Upper Colorado	0	100	0	100	22	0	12	4.4
Lower Colorado	44	3.4	0	47	36	0	170	0
Great Basin	3.8	130	0	140	6.2	0	85	9.7
Columbia-North Pacific	0	26	0	26	0.1	0	620	0
California-South Pacific	300	1,200	8,200	1,500	24	66	410	180
Alaska	1.4	68	1.0	70	0	0	8.0	0
Hawaii	82	46	860	130	0	0	160	13
Puerto Rico	0.4	0	2,100	0.4	0.4	0	40	1.5
U.S.[b]	1,400	120,000	46,000	120,000	820	220	8,000	1,000

[a] Includes 670 mgd from public supplies.
[b] Including Puerto Rico.

2.2–1 SELF SUPPLIED INDUSTRIAL WATER USE (*Continued*)

By Regions, 1970

Other uses, mgd							All industrial uses, mgd		
Water withdrawn									
Surface water			All water		Water consumed		Water withdrawn		
Fresh	Saline	Re-claimed sewage	Fresh	Saline	Fresh	Saline	Fresh[a]	Saline	Fresh water con-sumed
1,100	320	0	1,200	320	96	18	3,100	4,900	99
5,600	1,300	130	6,700	1,300	330	9.5	22,000	17,000	360
2,100	220	0	3,500	310	540	7.4	19,000	11,000	660
8,300	1.0	0	8,600	400	450	120	35,000	400	460
5,100	16	0	5,800	58	260	0	33,000	58	310
1,300	0	0	1,400	0	72	0	7,500	0	140
1,100	0	0	1,700	19	75	0	14,000	19	98
3,100	740	0	4,000	780	780	78	8,100	920	970
73	0	0	78	4.8	5.6	0.6	220	4.8	6.5
160	5.5	0	520	200	65	22	3,800	200	99
370	18	5.0	630	55	210	50	2,600	55	290
1,000	3,500	1.0	1,400	3,500	580	800	6,100	7,200	690
97	150	0	210	150	97	45	230	150	110
52	0	0	64	4.4	21	0.3	170	4.4	43
42	0	0	210	0	100	0	260	0	140
130	5.5	0	220	15	62	3.4	360	15	68
1,100[c]	37	0	1,800[c]	37	150	4.1	1,800	37	150
48	400	3.5	460	620	170	31	2,000	8,800	190
100	0	0	110	0	4.2	0	180	1.0	4.2
100	3.2	9.4	270	16	4.6	0	390	870	4.6
180	360	0	220	360	18	0	220	2,400	18
31,000	7,100	150	39,000	8,100	4,100	1,200	160,000	54,000	4,900

[a] Includes 670 mgd from public supplies.
[c] Excludes 600 mgd used for log ponds.

Note: Partial figures may not add to totals because of independent rounding. For regions see 1–7.

Source: C.R. Murray and E.B. Reeves, *Estimated Use of Water in the United States, 1970,* U.S. Geological Survey, U.S. Department of the Interior, 1972.

2.2–2 SELF SUPPLIED INDUSTRIAL WATER IN THE U.S.

	Thermoelectric power (electric utility) use, mgd							Other uses, mgd	
	Water withdrawn							Water withdrawn	
	Ground water		Surface water		Ground and surface water			Ground water	
State	Fresh	Saline	Fresh	Saline	Fresh	Saline	Water consumed	Fresh	Saline
Alabama	0	0	5,300	0	5,300	0	0	18	0
Alaska	1	0	1	1	2	1	0	7.1	0
Arizona	18	0	2	0	20	0	15	99	0
Arkansas	6	0	420	0	420	0	5	180	0
California	300	0	660	10,000	960	10,000	18	480	140
Colorado	1	0	150	0	150	0	9	40	6.3
Connecticut	0	0	370	1,200	370	1,200	1	17	0
Delaware	4	0	0	590	4	590	4	27	0
Florida	11	80	1,900	6,100	1,900	6,200	4	730	0.2
Georgia	0	0	1,500	280	1,500	280	0	360	0
Hawaii	31	21	41	500	72	520	0	65	16
Idaho	0	0	0	0	0	0	0	91	0
Illinois	8	0	13,000	0	13,000	0	4	240	34
Indiana	2	0	6,400	0	6,400	0	6	310	5.9
Iowa	2	0	1,500	0	1,500	0	21	130	0
Kansas	25	0	60	0	85	0	33	84	0.4
Kentucky	0	0	2,800	0	2,800	0	5	78	0.6
Louisiana	24	0	1,900	340	1,900	340	13	340	51
Maine	0	0	0	180	0	180	0	16	0
Maryland	0	0	510	1,700	510	1,700	0	40	0
Massachusetts	0	0	440	1,600	440	1,600	2	77	0
Michigan	0	0	5,800	0	5,800	0	4	110	14
Minnesota	0	0	1,300	0	1,300	0	2	270	0
Mississippi	56	0	120	300	180	300	11	180	0
Missouri	5	0	1,600	0	1,600	0	9	160	5.8
Montana	0	0	54	0	54	0	0	24	0
Nebraska	53	0	640	0	690	0	1	35	0
Nevada	1	0	27	0	28	0	0	25	2.7
New Hampshire	0	0	93	140	93	140	0	10	0
New Jersey	1	0	1,100	2,900	1,100	2,900	9	230	9.0
New Mexico	5	0	14	0	19	0	16	73	0
New York	3	0	4,700	5,700	4,700	5,700	10	160	150
North Carolina	0	0	2,900	32	2,900	32	1	70	0
North Dakota	1	0	83	0	84	0	2	3.1	6.7
Ohio	37	0	9,000	0	9,000	0	7	390	0
Oklahoma	4	0	510	6	510	6	17	24	46
Oregon	0	0	5	0	5	0	0	160	0
Pennsylvania	0	0	8,800	0	8,800	0	6	350	0
Puerto Rico	0	0	0	940	0	940	1	38	1.7
Rhode Island	0	0	0	300	0	300	0	24	0.4

2.2–2 SELF SUPPLIED INDUSTRIAL WATER IN THE U.S. (*Continued*)

Other uses, mgd						All industrial uses, mgd				
Water withdrawn						Water withdrawn				
Surface water			All water							
Fresh	Saline	Sewage	Fresh	Saline	Water consumed	Fresh	Saline	Sewage	All water	Water consumed
780	0	0	790	0	82	6,100	0	0	6,100	82
95	0	0	100	0	3.8	100	1.0	0	100	3.8
17	0	0	120	0	37	140	0	0	140	52
130	0	0	310	0	160	730	0	0	730	160
52	540	0.5	530	680	110	1,500	11,000	0.5	13,000	120
120	8.0	0	170	14	29	320	14	0	340	38
60	190	0	78	190	6.1	440	1,400	0	1,900	7.1
23	470	0	51	470	1.2	55	1,100	0	1,100	5.2
170	61	0	900	61	78	2,800	6,200	0	9,000	130
270	340	0	630	340	26	2,100	620	0	2,700	26
51	0	0	120	16	4.1	190	540	0	730	4.1
78	0	0	170	0	12	170	0	0	170	12
1,500	0	0	1,800	34	41	14,000	34	0	14,000	45
2,200	0	0	2,600	6.3	100	9,000	6.3	0	9,000	110
55	0.4	0	180	0.4	19	1,700	0	0	1,700	40
26	0	0	110	0.4	25	200	0.4	0	200	58
180	0.3	0	260	0.9	44	3,000	0.9	0	3,000	48
2,200	0	0	2,600	51	500	4,500	400	0	4,900	520
440	37	0	460	37	29	460	220	0	680	29
450	820	130	620	820	130	1,100	2,500	130	3,800	130
380	140	0	460	140	30	890	1,700	0	2,600	32
1,800	0	0	1,900	14	54	7,700	14	0	7,700	58
1,100	0	0	1,400	0	110	2,700	0	0	2,700	110
71	0	0	250	0	36	430	300	0	730	47
150	0	0	310	5.8	30	1,900	5.8	0	1,900	39
92	0	0	120	0	18	170	0	0	170	18
7.8	0	0	42	0	1.3	730	0	0	730	2.3
28	0	1.2	52	2.7	28	80	2.7	1.2	84	28
160	0	0	170	0	9.0	260	140	0	410	9.0
460	830	0	690	840	170	1,800	3,700	0	5,500	180
9.1	0	0	82	0	55	100	0	0	100	71
1,600	1,400	0	1,800	1,400	130	6,500	7,100	0	14,000	140
270	0	0	340	0	33	3,300	32	0	3,300	34
3.3	0	0	6.4	6.7	3.2	90	6.7	0	97	5.2
4,200	0	0	4,600	0	140	14,000	0	0	14,000	150
27	11	0	51	57	22	560	63	0	620	39
950	0	0	1,100	0	48	1,100	0	0	1,100	48
4,500	50	0	4,900	50	200	14,000	50	0	14,000	10
140	130	0	180	140	9.8	180	1,100	0	1,300	11
21	0	0	45	0.4	4.4	45	300	0	340	4.8

2.2–2 SELF SUPPLIED INDUSTRIAL WATER IN THE U.S. (*Continued*)

	Thermoelectric power (electric utility) use, mgd							Other Uses, mgd	
	Water withdrawn							Water withdrawn	
	Ground water		Surface water		Ground and Surface water			Ground water	
State	Fresh	Saline	Fresh	Saline	Fresh	Saline	Water consumed	Fresh	Saline
South Carolina	0	0	1,000	120	1,000	120	2	38	0
South Dakota	0	0	1	0	1	0	1	12	1.9
Tennessee	0	0	3,300	0	3,300	0	1	130	0
Texas	520	0	2,900	1,400	3,400	1,400	160	400	2.2
Utah	0	0	140	0	140	0	3	52	3.5
Vermont	0	0	53	0	53	0	0	9.1	0
Virginia	0	0	2,900	1,200	2,900	1,200	8	51	0
Washington	0	0	0	0	0	0	0	140	0
West Virginia	0	0	2,700	0	2,700	0	1	100	0.4
Wisconsin	0	0	3,900	0	3,900	0	1	89	0
Wyoming	1	0	170	0	170	0	3	40	0.9
District of Columbia	0	0	200	0	200	0	0	0.8	0
U.S.[a]	1,100	100	91,000	36,000	92,000	36,000	410	6,800	360

[a] Including Puerto Rico.

2.2–2 SELF SUPPLIED INDUSTRIAL WATER IN THE U.S. (*Continued*)

Other uses, mgd						All industrial uses, mgd				
Water withdrawn						Water withdrawn				
Surface water			All water							
Fresh	Saline	Sewage	Fresh	Saline	Water consumed	Fresh	Saline	Sewage	All water	Water consumed
200	33	0	240	33	24	1,300	150	0	1,400	26
12	1.0	0	24	2.9	11	25	2.9	0	28	12
760	0	0	890	0	180	4,200	0	0	4,200	180
620	3,200	7.4	1,000	3,200	390	4,400	4,600	7.4	9,000	540
120	5.1	0	170	8.6	22	320	8.6	0	1,300	25
25	0	0	34	0	1.6	87	0	0	87	1.6
680	72	0	730	72	1.4	3,700	1,300	0	5,000	9.4
300	30	0	450	30	16	450	30	0	480	16
2,000	0	0	2,100	0.4	140	4,800	0.4	0	4,800	140
260	0	0	350	0	8.6	4,200	0	0	4,200	9.6
21	0.1	0	61	1.0	8.4	240	1.0	0	240	11
0.6	0	0	1.4	0	0.3	210	0	0	210	0.3
30,000	8,400	140	37,000	8,800	3,400	130,000	45,000	140	170,000	3,800

Note: Data as of 1965. Sums of detail figures may not agree with totals because of rounding.

Source: U.S. Geological Survey.

2.2–3 SELF SUPPLIED INDUSTRIAL WATER USE

By States, 1970

	Thermoelectric power (electric utility) use, mgd						Other sues, mgd	
	Water withdrawn						Water withdrawn	
		Surface water			Water consumed		Ground water	
State	Fresh ground water	Fresh	Saline	Total fresh water[a]	Fresh	Saline	Fresh	Saline
Alabama	2.6	4,800	200	4,800	25	0.4	93	5.0
Alaska	1.4	68	1.0	70	0	0	8.0	0
Arizona	40	2.0	0	42	32	0	150	0
Arkansas	4.0	950	0	950	3.0	0	330	0
California	300	1,200	8,200	1,500	24	66	410	180
Colorado	30	83	0	110	8.0	0	55	7.0
Connecticut	0	1,100	1,800	1,100	2.0	0	20	1.0
Delaware	0.7	1.4	720	2.1	0	0	22	0
Florida	13	1,700	9,300	1,700	20	86	710	87
Georgia	6.7	3,900	140	3,900	39	1.1	330	0
Hawaii	82	46	860	130	0	0	160	13
Idaho	0	0	0	0	0	0	340	0
Illinois	7.0	11,000	0	11,000	5.0	0	250	40
Indiana	1.0	4,800	0	4,800	5.0	0	140	4.4
Iowa	0	1,400	0	1,400	20	0	150	0
Kansas	38	220	0	260	32	0	120	0
Kentucky	1.9	3,800	0	3,800	21	0	87	15
Louisiana	36	2,700	140	2,800	170	25	460	41
Maine	1.0	20	180	21	0	0	3.0	0
Maryland	0	510	2,700	510	0	0	43	0
Massachusetts	0	580	2,100	580	1.0	1.0	140	0
Michigan	0	9,800	0	9,800	0	0	70	400
Minnesota	280	1,400	0	1,700	0.2	0	160	0
Mississippi	35	410	490	450	35	0	310	0
Missouri	6.0	2,400	0	2,500	13	0	200	0.2
Montana	0	60	0	60	0	0	34	0
Nebraska	270	620	0	890	8.4	0	100	0
Nevada	7.1	49	0	57	7.5	0	39	6.2
New Hampshire	0	250	160	250	0	0	12	0
New Jersey	8.0	1,000	3,200	1,000	26	0	550	0
New Mexico	8.6	20	0	29	27	0	72	0
New York	130	4,700	8,500	4,800	9.6	17	140	1.7
North Carolina	1.0	4,300	170	4,300	30	0	140	0
North Dakota	1.0	350	0.9	350	1.2	0.8	4.7	9.6
Ohio	49	13,000	0	13,000	14	0	390	0
Oklahoma	3.8	180	0	180	28	0	32	37
Oregon	0	22	0	22	0.1	0	110	0

[a] Includes 670 mgd from public supplies.

2.2–3 SELF SUPPLIED INDUSTRIAL WATER USE (*Continued*)

By States, 1970

Other uses, mgd							All industrial uses, mgd		
Water withdrawn									
Surface water		Reclaimed sewage	All water		Water consumed		Water withdrawn		Fresh water consumed
Fresh	Saline		Fresh	Saline	Fresh	Saline	Fresh[a]	Saline	
910	96	0	1,000	100	59	6.1	5,800	300	84
100	0	0	110	0	4.2	0	180	1.0	4.2
24	0	0	180	0	86	0	220	0	120
200	0	0	520	0	210	0	1,500	0	220
28	440	3.5	440	620	170	31	2,000	8,800	190
130	9.0	0	180	16	45	3.3	300	16	53
55	130	0	75	140	6.0	0	1,100	2,000	8.0
64	300	0	87	300	1.4	0	89	1,000	1.4
190	46	0	900	130	140	1.3	2,600	9,500	160
300	0	0	630	0	160	0	4,500	140	200
100	3.2	9.4	270	16	4.6	0	400	870	4.6
110	0	0	450	0	16	0	450	0	16
1,700	0	0	1,900	40	76	0	13,000	40	81
3,100	0	0	3,200	4.4	130	0	8,000	4.4	130
130	0	0	280	0	5.3	0	1,700	0	25
37	0	0	160	0	57	0	420	0	89
280	0	0	370	15	40	0	4,200	15	61
2,900	740	0	3,400	780	600	78	6,100	920	770
400	24	0	400	24	24	1.4	420	200	24
450	820	130	620	820	52	0	1,100	3,500	52
390	160	0	530	160	53	16	1,100	2,300	54
1,600	0	0	1,700	400	120	120	11,000	400	120
1,100	0	0	1,200	0	85	0	2,900	0	85
190	49	0	500	49	53	0	950	540	88
110	0	0	310	0.2	29	0	2,800	0.2	42
120	0	0	150	0	22	0	210	0	22
0.7	0	0	100	0	3.7	0	990	0	12
31	0	0	70	6.2	41	0.4	130	6.2	48
180	0	0	190	0	9.6	0	440	160	9.6
450	0	0	1,000	0	70	0	2,000[a]	3,200	96
14	0	0	86	0	51	0	110	0	78
1,300	64	2.3	1,400	66	120	9.5	6,200	8,600	130
480	0	0	620	0	120	0	5,000	170	150
5.2	0	0	9.9	9.6	1.9	2.8	360	10	3.1
3,300	0	0	3,700	0	110	0	17,000	0	120
95	13	0	130	50	48	50	310	50	76
590[b]	0	0	700[b]	0	26	0	720[b]	0	26

[a] Included 670 mgd from public supplies.
[b] Excludes 600 mgd used for log ponds

2.2–3 SELF SUPPLIED INDUSTRIAL WATER USE (*Continued*)

By States, 1970

	Thermoelectric power (electric utility) use, mgd						Other uses, mgd	
	Water withdrawn				Water consumed		Water withdrawn	
		Surface water					Ground water	
State	Fresh ground water	Fresh	Saline	Total fresh water[a]	Fresh	Saline	Fresh	Saline
Pennsylvania	0	12,000	0	12,000	8.9	0	400	0
Rhode Island	0	0	310	0	0	0	15	0.4
South Carolina	0.7	2,600	68	2,600	14	0	53	0
South Dakota	0.7	2.7	0	3.4	0.6	0	14	170
Tennessee	0	4,900	0	4,900	62	0	88	0
Texas	60	5,800	3,800	5,900	120	24	480	0
Utah	0	87	0	87	3.8	0	56	3.5
Vermont	1.0	4.0	0	5.0	0	0	12	0
Virginia	0	3,100	770	3,100	0.8	0	120	0
Washington	0	4.3	0	4.3	0	0	150	0
West Virginia	0	5,000	0	5,000	1.1	0	25	0
Wisconsin	0	5,300	0	5,300	0	0	110	0
Wyoming	1.0	200	0	200	5.3	0	67	21
District of Columbia	0	1,100	0	1,100	0	0	0.8	0
Puerto Rico	0.4	0	2,100	0.4	0.4	0	40	1.5
U.S.[c]	1,400	120,000	46,000	120,000	820	220	8,000	1,000

[a] Includes 670 mgd from public supplies.
[c] Including Puerto Rico.

2.2–3 SELF SUPPLIED INDUSTRIAL WATER USE (*Continued*)

By States, 1970

Other uses, mgd							All industrial uses, mgd		
Water withdrawn					Water consumed		Water withdrawn		Fresh water con-sumed
Surface water		Re-claimed sewage	All water						
Fresh	Saline		Fresh	Saline	Fresh	Saline	Fresh[a]	Saline	
5,000	50	0	5,400	50	220	0	18,000	50	230
23	0	0	38	0.4	3.8	0	38	310	3.8
300	32	0	350	32	33	0	2,900	100	47
2.6	0	0	17	170	1.9	17	20	170	2.5
960	0	0	1,000	0	47	0	5,900	0	110
1,200	3,600	6.0	1,700	3,600	680	840	7,500	7,400	800
140	5.5	0	190	9.0	45	3.0	280	9.0	48
34	0	0	46	0	2.3	0	51	0	2.3
920	80	0	1,000	80	7.4	0	4,200	850	8.2
410	37	0	560	37	100	4.1	560	37	100
630	0	0	660	0	57	0	5,600	0	58
220	0	0	330	0	10	0	5,600	0	10
20	0	0	87	21	6.0	0.3	290	21	11
0.6	0	0	1.4	0	0.3	0	1,100	0	0.3
180	360	0	220	360	18	0	220	2,400	19
31,000	7,100	150	39,000	8,100	4,100	1,200	160,000	54,000	4,900

[a] Includes 670 mgd from public supplies.

Note: Partial figures may not add to totals because of independent rounding.

Source: C.R. Murray and E.B. Reeves, *Estimated Use of Water in the United States, 1970,* U.S. Geological Survey, U.S. Department of the Interior, 1972.

2.2.1 MANUFACTURING INDUSTRIES WATER NEEDS

2.2–4 WATER USE BY MANUFACTURING INDUSTRIES

U.S., 1963

Industry	Establishments reporting water intake								
	Total	Source of water			Quantity of water intake, mil gal				
		Utility	Other than utility	Source not specified	Under 1	1–9	10–19	20–99	100 and over
U.S., total	130,770	97,922	12,311	20,537	96,738	18,935	4,630	6,035	4,432
Food and kindred products	21,598	15,778	2,600	3,220	11,375	5,754	1,486	2,016	967
Meat products	2,734	1,873	387	474	1,319	695	186	343	191
Meat slaughtering plants	1,398	857	275	266	668	348	102	159	121
Meat processing plants	702	601	37	64	417	187	31	52	15
Poultry dressing plants	634	415	75	144	234	160	53	132	55
Dairies	4,638	3,052	715	871	1,963	1,492	452	577	154
Creamery butter	436	192	91	153	161	130	59	71	15
Natural and process cheese	559	289	111	159	301	153	53	43	9
Condensed and evaporated milk	253	131	63	59	86	49	16	62	40
Ice cream and frozen desserts	668	544	65	59	335	222	45	54	12
Fluid milk	2,722	1,896	385	441	1,080	938	279	347	78
Canned and frozen foods	2,340	1,506	346	488	1,161	502	171	335	171
Canned and cured seafoods	240	149	31	60	153	56	15	13	3
Canned specialties	103	76	17	10	44	11	11	19	18
Canned fruits and vegetables	895	531	162	202	336	203	89	188	79
Dehydrated foods products	120	69	19	32	72	15	6	17	10
Pickles, sauces, salad dressings	288	220	22	46	171	63	17	31	6
Fresh or frozen packaged fish	292	189	24	79	204	59	12	14	3
Frozen fruits and vegatables	402	272	71	59	181	95	21	53	52
Grain mills	2,158	1,452	225	481	1,689	313	61	60	35
Flour mills	339	246	31	62	205	88	18	23	5
Prepared animal feeds	1,608	1,057	163	388	1,364	187	28	21	8
Cereal preparations	37	31	3	3	12	14	–	4	7
Rice milling	60	40	8	12	42	9	4	4	1

2.2–4 WATER USE BY MANUFACTURING INDUSTRIES (*Continued*)

Industry	Establishments reporting water intake								
		Source of water			Quantity of water intake, mil gal				
	Total	Utility	Other than utility	Source not specified	Under 1	1–9	10–19	20–99	100 and over
Blended and prepared flour	78	64	5	9	53	13	9	3	–
Wet corn milling	36	14	15	7	13	2	2	5	14
Bakery products	2,608	2,382	62	164	1,685	653	124	129	17
Bread and related products	2,356	2,165	47	144	1,519	605	113	107	12
Biscuit, crackers, and cookies	252	217	15	20	166	48	11	22	5
Sugar	174	53	99	22	32	11	6	19	106
Raw cane sugar	75	20	39	16	16	9	4	12	34
Cane sugar refining	31	20	7	4	9	2	1	6	13
Beet sugar	68	13	53	2	7	–	1	1	59
Candy and related products	638	543	34	61	407	146	29	45	11
Confectionery products	578	499	27	52	378	133	24	35	8
Chocolate and cocoa products	39	26	5	8	18	9	4	6	2
Chewing gum	21	18	2	1	11	4	1	4	1
Beverages	3,418	2,816	295	307	1,522	1,241	240	278	137
Malt liquors	197	139	42	16	24	31	15	51	76
Malt	43	19	17	7	5	7	1	17	13
Wines and brandy	129	63	45	21	61	30	11	16	11
Distilled liquor except brandy	100	56	29	15	35	17	7	25	16
Bottled and canned soft drinks	2,684	2,307	146	231	1,211	1,107	193	156	17
Flavorings	265	232	16	17	186	49	13	13	4
Miscellaneous foods and kindred products	2,890	2,101	437	352	1,597	701	217	230	145
Cottonseed oil mills	168	127	22	19	66	64	20	12	6
Soybean oil mills	88	52	26	10	17	16	6	18	31
Vegetable oil mills, NC	31	18	7	6	15	5	3	3	5
Animal and marine fats and oils	461	202	162	97	224	112	38	66	21
Roasted coffee	227	212	9	6	183	24	4	8	8
Shortening and cooking oils	100	81	15	4	28	13	9	27	23
Manufactured ice	658	427	134	97	268	255	66	42	27
Macaroni and spaghetti	120	108	7	5	81	27	10	2	–
Food preparations, NC	1,037	874	55	108	715	185	61	52	24

2.2–4 WATER USE BY MANUFACTURING INDUSTRIES (*Continued*)

	Establishments reporting water intake								
		Source of water			Quantity of water intake, mil gal				
Industry	Total	Utility	Other than utility	Source not specified	Under 1	1–9	10–19	20–99	100 and over
Tobacco manufactures	265	237	4	24	146	77	21	14	7
Cigarettes	15	15	–	–	1	1	3	4	6
Cigars	93	83	2	8	50	33	7	3	–
Chewing and smoking tobacco	36	34	1	1	19	12	3	2	–
Tobacco stemming and redrying	121	105	1	15	76	31	8	5	1
Textile mill products	5,259	3,939	501	819	3,253	997	307	430	272
Weaving mills, cotton	367	251	52	64	84	87	51	90	55
Weaving mills, synthetics	310	248	26	36	136	69	29	58	18
Weaving, finishing mills, wool	277	167	63	47	123	59	17	34	44
Narrow fabric mills	299	230	17	52	236	50	3	9	1
Knitting mills	1,933	1,549	92	292	1,439	315	85	76	18
Women's hosiery, except socks	293	203	24	66	185	76	16	13	3
Hosiery, NC	351	266	15	70	209	92	28	19	3
Knit outerwear mills	809	728	25	56	703	75	15	13	3
Knit underwear mills	108	72	4	32	61	28	13	6	–
Knit fabric mills	323	241	21	61	243	34	13	24	9
Knitting mills, NC	49	39	3	7	38	10	–	1	–
Textile finishing, except wool	481	352	76	53	229	86	25	50	91
Finishing plants, cotton	183	119	43	21	82	30	11	14	46
Finishing plants, synthetic	178	135	24	19	81	35	10	22	30
Finishing plants, NC	120	98	9	13	66	21	4	14	15
Floor covering mills	217	158	13	46	139	34	9	20	15
Woven carpets and rugs	44	29	6	9	23	7	3	5	6
Tufted carpets and rugs	118	86	4	28	68	22	5	14	9
Carpets and rugs, NC	55	43	3	9	48	5	1	1	–

2.2–4 WATER USE BY MANUFACTURING INDUSTRIES (*Continued*)

Industry	Total	Source of water: Utility	Source of water: Other than utility	Source of water: Source not specified	Quantity of water intake, mil gal: Under 1	1–9	10–19	20–99	100 and over
Yarn and thread mills	609	415	85	109	318	181	54	42	14
Yarn mills, except wool	293	178	57	58	113	120	34	22	4
Throwing and winding mills	137	106	9	22	99	23	10	5	–
Wool yarn mills	127	91	10	26	80	28	7	7	5
Thread mills	52	40	9	3	26	10	3	8	5
Miscellaneous textile goods	766	569	77	120	549	116	34	51	16
Felt goods, NC	45	35	–	10	19	13	4	5	4
Lace goods	120	84	7	29	107	9	2	2	–
Padding and upholstery filling	117	99	4	14	94	14	3	6	–
Processed textile waste	118	84	12	22	95	18	4	1	–
Coated fabric, not rubberized	117	87	15	15	77	22	3	8	7
Tire cord and fabric	21	14	3	4	2	6	5	7	1
Scouring and combing plants	50	26	20	4	25	5	5	11	4
Cordage and twine	94	71	10	13	71	19	3	1	–
Textile goods, NC	84	69	6	9	59	10	5	10	–
Lumber and wood products	10,605	4,593	1,383	4,629	9,340	854	181	152	78
Logging camps and contractors	1,958	266	175	1,517	1,919	29	7	1	2
Sawmills and planing mills	4,093	1,242	771	2,080	3,472	419	81	72	49
Sawmills and planing mills	3,422	912	661	1,849	2,912	326	69	66	49
Hardwood dimension and flooring	431	235	69	127	327	86	12	6	–
Special product sawmills, NC	240	95	41	104	233	7	–	–	–
Millwork and related products	2,412	1,727	190	495	2,085	221	41	48	17
Millwork plants	1,565	1,226	69	270	1,493	60	5	6	1
Veneer and plywood plants	541	295	98	148	304	145	35	41	16
Prefabricated wood products	306	206	23	77	288	16	1	1	–

2.2–4 WATER USE BY MANUFACTURING INDUSTRIES (*Continued*)

	Establishments reporting water intake								
		Source of water			Quantity of water intake, mil gal				
Industry	Total	Utility	Other than utility	Source not specified	Under 1	1–9	10–19	20–99	100 and over
Wooden containers	578	384	56	138	499	57	14	6	2
Nailed wooden boxes and shook	360	244	38	78	339	15	3	2	1
Wirebound boxes and crates	86	55	10	21	60	18	5	2	1
Veneer and plywood containers	62	33	7	22	43	16	1	2	–
Cooperage	70	52	1	17	57	8	5	–	–
Miscellaneous wood products	1,564	974	191	399	1,365	128	38	25	8
Wood preserving	225	118	62	45	126	62	25	12	–
Wood products, NC	1,339	856	129	354	1,239	66	13	13	8
Furniture and fixtures	5,293	4,411	173	709	4,653	466	101	55	18
Household furniture	3,284	2,704	108	472	2,844	326	72	30	12
Wood furniture not upholstered	1,403	1,077	54	272	1,153	187	41	16	6
Wood furniture, upholstered	987	826	28	133	903	65	12	5	2
Metal household furniture	337	288	16	33	267	45	16	6	3
Mattresses and bedsprings	500	460	8	32	469	26	2	3	–
Household furniture, NC	57	53	2	2	52	3	1	–	1
Office furniture	261	214	10	37	195	41	12	12	1
Wood office furniture	105	83	5	17	89	11	3	2	–
Metal office furniture	156	131	5	20	106	30	9	10	1
Public building furniture	229	180	4	45	190	31	4	3	1
Partitions and fixtures	1,095	938	29	128	1,030	50	8	5	2
Wood partitions and fixtures	791	684	19	88	767	23	1	–	–
Metal partitions and fixtures	304	254	10	40	263	27	7	5	2
Furniture and fixtures, NC	424	375	22	27	394	18	5	5	2
Venetian blinds and shades	229	203	15	11	212	10	2	3	2
Furniture and fixtures, NC	195	172	7	16	182	8	3	2	–

2.2–4 WATER USE BY MANUFACTURING INDUSTRIES (*Continued*)

	Establishments reporting water intake								
		Source of water			Quantity of water intake, mil gal				
Industry	**Total**	**Utility**	**Other than utility**	**Source not specified**	**Under 1**	**1–9**	**10–19**	**20–99**	**100 and over**
Paper and allied products	4,380	3,277	626	477	2,832	703	179	179	487
Pulp mills	47	16	21	10	7	3	2	5	30
Paper mills, except building	342	92	211	39	43	22	8	32	237
Paperboard mills	276	85	145	46	51	28	14	31	152
Paper and paperboard products	1,650	1,357	120	173	1,258	241	57	63	31
Paper coating and glazing	279	208	35	36	174	49	17	28	11
Envelopes	202	183	11	8	158	35	5	3	1
Bags, except textile bags	382	321	21	40	282	71	16	10	3
Wallpaper	51	42	5	4	42	8	1	–	–
Die cut paper and board	276	240	5	31	249	19	3	3	2
Pressed and molded pulp goods	53	37	12	4	19	12	5	9	8
Sanitary paper products	89	54	22	13	63	12	5	5	4
Converted paper products, NC	318	272	9	37	271	35	5	5	2
Paperboard containers and boxes	1,980	1,681	97	202	1,454	392	94	33	7
Folding paperboard boxes	496	425	27	44	399	68	21	6	2
Set-up paperboard boxes	362	315	8	39	338	22	2	–	–
Corrugated shipping containers	768	650	41	77	479	223	53	12	1
Sanitary food containers	160	131	12	17	76	55	11	14	4
Fiber cans, tubes, drums, etc	194	160	9	25	162	24	7	1	–
Building paper and board mills	85	46	32	7	19	17	4	15	30
Chemicals and allied products	7,624	5,560	1,112	952	5,108	1,086	295	451	684
Basic chemicals	1,620	1,012	426	182	687	280	111	168	374
Alkalies and chlorine	49	22	18	9	14	4	1	2	28
Industrial gases	425	292	109	24	200	107	36	33	49
Cyclic (coal tar) crudes	33	28	4	1	15	6	1	7	4
Intermediate coal tar products	130	92	24	14	51	18	7	23	31
Inorganic pigments	80	50	19	11	35	14	3	8	20

2.2–4 WATER USE BY MANUFACTURING INDUSTRIES (*Continued*)

	Establishments reporting water intake								
		Source of water			Quantity of water intake, mil gal				
Industry	Total	Utility	Other than utility	Source not specified	Under 1	1–9	10–19	20–99	100 and over
Organic chemicals, NC	375	229	97	49	179	36	20	36	104
Inorganic chemicals, NC	528	299	155	74	193	95	43	59	138
Fibers, plastics, rubbers	533	325	133	75	251	63	33	61	125
Plastics materials and resins	449	291	93	65	235	59	31	51	73
Synthetic rubber	32	14	13	5	7	–	–	8	17
Cellulosic man-made fibers	26	10	14	2	5	1	2	1	17
Organic fibers, noncellulosic	26	10	13	3	4	3	–	1	18
Drugs	631	489	73	69	436	98	22	45	30
Biological products	77	55	13	9	56	13	3	2	3
Medicinals and botanicals	85	47	23	15	52	18	5	4	6
Pharmaceutical preparations	469	387	37	45	328	67	14	39	21
Cleaning and toilet goods	1,193	1,031	61	101	894	196	39	40	24
Soap and other detergents	353	301	21	31	241	64	17	15	16
Polishes and sanitation goods	430	369	20	41	371	45	6	5	3
Surface active agents	131	109	10	12	81	33	8	6	3
Toilet preparations	279	252	10	17	201	54	8	14	2
Paints and allied products	1,216	1,054	55	107	978	161	27	38	12
Gum and wood chemicals	108	33	34	41	72	14	4	4	14
Agricultural chemicals	1,056	651	188	217	833	115	28	42	38
Fertilizers	287	151	89	47	152	47	24	31	33
Fertilizers, mixing only	515	316	64	135	464	43	2	4	2
Agricultural chemicals, NC	254	184	35	35	217	25	2	7	3
Miscellaneous chemical products	1,267	965	142	160	957	159	31	51	69
Glue and gelatin	247	200	21	26	184	42	4	6	11
Explosives	62	21	26	15	31	5	5	4	17
Printing ink	270	252	4	14	200	51	10	9	–
Carbon black	33	4	21	8	5	5	4	6	13
Chemical preparations, NC	655	488	70	97	537	56	8	26	28

2.2–4 WATER USE BY MANUFACTURING INDUSTRIES (*Continued*)

Industry	Establishments reporting water intake								
		Source of water			Quantity of water intake, mil gal				
	Total	Utility	Other than utility	Source not specified	Under 1	1–9	10–19	20–99	100 and over
Petroleum and coal products	1,387	802	362	223	785	265	59	95	183
Petroleum refining	377	159	175	43	97	43	24	46	167
Paving and roofing materials	772	451	170	151	502	190	28	37	15
Paving mixtures and blocks	575	297	146	132	428	123	15	7	2
Asphalt felts and coatings	197	154	24	19	74	67	13	30	13
Petroleum and coal products, NC	238	192	17	29	186	32	7	12	1
Lubricating oils and greases	194	163	11	20	159	24	4	7	–
Petroleum and coal products, NC	44	29	6	9	27	8	3	5	1
Rubber and plastics products, NC	3,307	2,521	357	429	2,287	558	132	203	127
Tires and inner tubes	121	71	38	12	49	13	8	18	33
Rubber footwear	42	31	6	5	22	11	1	6	2
Reclaimed rubber	20	13	2	5	9	3	1	4	3
Rubber products, NC	771	587	82	102	451	139	50	80	51
Plastics products, NC	2,353	1,819	229	305	1,757	391	72	95	38
Leather and leather products	2,677	2,236	118	323	2,070	392	105	72	38
Leather tanning and finishing	361	278	54	29	181	68	23	53	36
Industrial leather belting	55	49	3	3	43	10	2	–	–
Footwear cut stock	288	251	7	30	255	32	–	1	–
Shoes, except rubber	953	773	19	161	663	208	67	15	–
Footwear, except rubber	850	682	15	153	581	190	64	15	–
House slippers	103	91	4	8	82	18	3	–	–
Leather gloves	104	66	–	38	84	16	3	1	–
Luggage	207	191	8	8	184	17	4	1	1

2.2–4 WATER USE BY MANUFACTURING INDUSTRIES (*Continued*)

	Establishments reporting water intake								
		Source of water			Quantity of water intake, mil gal				
Industry	Total	Utility	Other than utility	Source not specified	Under 1	1–9	10–19	20–99	100 and over
Purses and small leather goods	541	483	19	39	501	35	4	1	–
Handbags and purses	358	327	11	20	331	24	3	–	–
Small leather goods	183	156	8	19	170	11	1	1	–
Leather goods, NC	168	145	8	15	159	6	2	–	1
Stone, clay, and glass products	8,587	5,583	1,551	1,453	5,742	1,894	285	398	268
Flat glass	48	35	10	3	23	1	4	8	12
Pressed and blown glassware	233	175	37	21	67	43	14	77	32
Glass containers	112	80	21	11	13	20	10	54	15
Pressed and blown glass, NC	121	95	16	10	54	23	4	23	17
Products of purchased glass	384	321	23	40	317	33	16	10	8
Cement, hydraulic	198	59	121	18	33	17	10	39	99
Structural clay products	760	416	201	143	414	267	59	16	4
Brick and structural tile	419	202	131	86	232	157	24	6	–
Ceramic wall and floor tile	72	54	9	9	39	20	8	4	1
Clay refractories	133	88	25	20	71	47	11	1	3
Structural clay products, NC	136	72	36	28	72	43	16	5	–
Pottery and related products	323	246	21	56	212	62	16	27	6
Vitreous plumbing fixtures	44	37	4	3	10	11	9	13	1
Vitreous china food utensils	25	17	3	5	12	7	2	3	1
Earthenware food utensils	22	17	2	3	11	6	2	2	1
Porcelain electrical supplies	60	50	4	6	32	15	3	7	3
Pottery products, NC	172	125	8	39	147	23	–	2	–
Concrete and plaster products	5,021	3,259	833	929	3,518	1,223	105	128	47
Concrete block and brick	898	589	119	190	726	154	9	8	1

2.2–4 WATER USE BY MANUFACTURING INDUSTRIES (*Continued*)

	Establishments reporting water intake								
		Source of water			Quantity of water intake, mil gal				
Industry	Total	Utility	Other than utility	Source not specified	Under 1	1–9	10–19	20–99	100 and over
Concrete products	1,474	997	216	261	1,189	248	13	20	4
Ready-mixed concrete	2,453	1,583	423	447	1,522	795	61	49	26
Lime	108	32	51	25	60	18	5	17	8
Gypsum products	88	58	24	6	21	8	17	34	8
Cut stone and stone products	422	242	100	80	311	76	10	20	5
Nonmetallic minerals products	1,198	830	205	163	847	172	51	73	55
Abrasive products	238	184	32	22	176	36	7	10	9
Asbestos products	103	69	20	14	40	15	9	21	18
Gaskets and insulations	182	150	14	18	148	20	2	4	8
Minerals, ground or treated	381	202	105	74	293	53	9	19	7
Mineral wool	90	63	14	13	42	16	10	14	8
Nonclay refractories	86	69	8	9	41	27	11	4	3
Nonmetallic minerals, NC	118	93	12	13	107	5	3	1	2
Primary metal industries	4,951	3,735	587	629	2,948	884	245	432	442
Steel rolling and finishing	694	419	178	97	227	112	41	109	205
Blast furnaces and steel mills	304	130	125	49	67	22	7	46	162
Electrometallurgical products	35	18	13	4	4	8	6	6	11
Steel wire drawing, etc	158	129	13	16	80	31	13	25	9
Cold finishing of steel shapes	78	60	6	12	27	16	7	16	12
Steel pipe and tube	119	82	21	16	49	35	8	16	11
Iron and steel foundries	1,222	944	117	161	716	283	70	102	51
Gray iron foundries	906	710	83	113	593	196	41	43	33
Malleable iron foundries	76	55	9	12	24	26	10	11	5
Steel foundries	240	179	25	36	99	61	19	48	13
Primary nonferrous metal	127	62	52	13	27	16	5	21	58
Primary copper	29	12	16	1	4	2	1	5	17
Primary lead	15	7	6	2	3	3	–	2	7

2.2–4 WATER USE BY MANUFACTURING INDUSTRIES (*Continued*)

Industry	Total	Establishments reporting water intake: Source of water: Utility	Other than utility	Source not specified	Quantity of water intake, mil gal: Under 1	1–9	10–19	20–99	100 and over
Primary zinc	20	12	7	1	4	1	1	3	11
Primary aluminum	28	9	16	3	6	1	–	2	19
Primary nonferrous metals, NC	35	22	7	6	10	9	3	9	4
Secondary nonferrous metals	273	230	21	22	175	61	15	19	3
Nonferrous rolling and drawing	623	441	86	96	288	113	42	88	92
Copper rolling and drawing	111	73	19	19	32	16	10	22	31
Aluminum rolling and drawing	168	111	24	33	87	31	6	18	26
Rolling and drawing, NC	99	78	14	7	59	14	5	11	10
Nonferrous wire drawing, etc	245	179	29	37	110	52	21	37	25
Nonferrous foundries	1,253	1,009	72	172	1,019	157	25	41	11
Aluminum castings	588	464	37	87	484	66	11	21	6
Brass, bronze, copper castings	393	327	20	46	334	43	5	10	1
Nonferrous castings, NC	272	218	15	39	201	48	9	10	4
Primary metal industries, NC	759	630	61	68	496	142	47	52	22
Iron and steel forgings	221	180	22	19	110	50	20	27	14
Nonferrous forgings	33	29	2	2	19	8	1	2	3
Primary metal industries, NC	505	421	37	47	367	84	26	23	5
Fabricated metal products	16,698	13,834	946	1,918	13,893	1,757	414	491	143
Metal cans	250	197	25	28	122	70	20	34	4
Cutlery, hand tools, hardware	1,268	1,060	91	117	982	159	44	53	30
Cutlery	112	91	14	7	85	16	2	5	4
Hand and edge tools	413	331	35	47	335	47	13	15	3
Hand saws and saw blades	65	54	2	9	50	8	2	4	1
Hardware, NC	678	584	40	54	512	88	27	29	22
Plumbing and heating except electric	626	513	48	65	451	102	30	33	10
Plumbing fixtures	64	50	7	7	37	8	6	9	4

2.2–4 WATER USE BY MANUFACTURING INDUSTRIES (*Continued*)

	Establishments reporting water intake								
		Source of water			Quantity of water intake, mil gal				
Industry	Total	Utility	Other than utility	Source not specified	Under 1	1–9	10–19	20–99	100 and over
Plumbing fittings, brass goods	156	135	12	9	108	23	11	11	3
Heating equipment, except electric	406	328	29	49	306	71	13	13	3
Structural metal products	6,048	4,949	317	782	5,444	424	82	78	20
Fabricated structural steel	1,328	1,105	67	156	1,197	95	22	10	4
Metal doors, sash and trim	1,145	916	58	171	1,031	75	18	17	4
Boiler shop products	939	747	88	104	767	120	17	28	7
Sheet metal work	1,856	1,539	69	248	1,741	85	12	13	5
Architectural metal work	420	361	10	49	399	16	5	–	–
Miscellaneous metal work	360	281	25	54	309	33	8	10	–
Screw machine products, bolts, etc	1,624	1,340	90	194	1,424	124	27	36	13
Screw machine products	1,153	950	61	142	1,087	57	4	4	1
Bolts, nuts, rivets, and washers	471	390	29	52	337	67	23	32	12
Metal stampings	1,909	1,593	107	209	1,594	193	45	55	22
Metal services, NC	2,292	1,975	118	199	1,687	370	105	107	23
Plating and polishing	1,656	1,421	97	138	1,131	309	96	98	22
Metal coating, engraving, etc	636	554	21	61	556	61	9	9	1
Fabricated wire products, NC	924	769	52	103	790	87	19	25	3
Fabricated metal products, NC	1,757	1,438	98	221	1,399	228	42	70	18
Metal barrels, drums and pails	122	88	7	27	78	25	8	9	2
Safes and vaults	22	18	–	4	18	2	–	1	1
Steel springs	76	63	7	6	63	6	2	2	3
Valves and pipe fittings	515	436	26	53	350	107	17	32	9
Collapsible tubes	20	16	3	1	12	2	1	4	1
Metal foil and leaf	60	46	5	9	40	10	3	6	1
Fabricated pipe and fittings	231	189	8	34	196	26	2	7	–
Fabricated metal products, NC	711	582	42	87	642	50	9	9	1

2.2–4 WATER USE BY MANUFACTURING INDUSTRIES (*Continued*)

		Establishments reporting water intake							
		Source of water			Quantity of water intake, mil gal				
Industry	Total	Utility	Other than utility	Source not specified	Under 1	1–9	10–19	20–99	100 and over
Machinery, except electrical	18,122	14,734	975	2,413	16,126	1,223	274	335	164
Engines and turbines	135	103	13	19	74	12	9	21	19
Steam engines and turbines	35	27	3	5	18	4	3	4	6
Internal combustion engines	100	76	10	14	56	8	6	17	13
Farm machinery and equipment	891	606	84	201	739	81	28	19	24
Construction and like equipment	1,529	1,232	109	188	1,244	169	47	52	17
Construction machinery	426	335	35	56	316	59	15	23	13
Mining machinery and equipment	142	114	9	19	112	22	2	4	2
Oil field machinery	279	228	17	34	218	31	19	9	2
Elevators and moving stairways	110	101	6	3	95	9	2	4	–
Conveyors	292	233	17	42	258	26	4	4	–
Hoists, cranes, and monorails	88	71	8	9	76	5	4	3	–
Industrial trucks and tractors	192	150	17	25	169	17	1	5	–
Metalworking machinery	5,059	4,113	241	705	4,675	249	50	65	20
Metal cutting machine tools	501	409	21	71	420	46	12	18	5
Metal forming machine tools	243	203	12	28	201	28	5	9	–
Special dies and tools	3,370	2,731	169	470	3,253	86	14	11	6
Machine tool accessories	671	550	26	95	587	53	14	14	3
Metalworking machinery, NC	274	220	13	41	214	36	5	13	6
Special industry machinery	2,114	1,743	106	265	1,835	185	38	44	12
Food products machinery	443	362	18	63	386	39	8	9	1
Textile machinery	361	301	13	47	301	39	8	9	4
Woodworking machinery	168	132	14	22	145	18	2	3	–
Paper industries machinery	156	133	5	18	133	12	6	5	–
Printing trades machinery	292	246	16	30	260	19	7	6	–
Special industry machinery, NC	694	569	40	85	610	58	7	12	7

2.2–4 WATER USE BY MANUFACTURING INDUSTRIES (*Continued*)

Industry	Establishments reporting water intake								
	Total	Source of water			Quantity of water intake, mil gal				
		Utility	Other than utility	Source not specified	Under 1	1–9	10–19	20–99	100 and over
General industrial machinery	2,320	1,962	122	236	1,935	239	42	73	31
Pumps and compressors	415	347	20	48	316	62	9	19	9
Ball and roller bearings	115	81	19	15	48	26	10	18	13
Blowers and fans	156	139	3	14	120	23	5	8	–
Industrial patterns	598	507	32	59	586	9	1	2	–
Power transmission equipment	394	340	20	34	311	50	12	15	6
Industrial furnaces and ovens	159	137	6	16	135	21	–	2	1
General industry machines, NC	483	411	22	50	419	48	5	9	2
Office machines, NC	352	291	23	38	262	45	15	16	14
Computing and related machines	177	147	9	21	126	26	8	9	8
Typewriters	21	18	2	1	9	3	1	4	4
Scales and balances	51	40	4	7	40	8	2	1	–
Office machines, NC	103	86	8	9	87	8	4	2	2
Service industry machines	935	787	47	101	729	127	29	34	16
Automatic vending machines	95	86	3	6	72	16	2	4	1
Commercial laundry equipment	77	67	2	8	59	12	1	4	1
Refrigeration machinery	470	380	27	63	351	66	20	20	13
Measuring and dispensing pumps	36	33	–	3	23	7	2	3	1
Service industry machines, NC	257	221	15	21	224	26	4	3	–
Miscellaneous machinery	4,787	3,897	230	660	4,633	116	16	11	11
Electrical machinery	6,474	5,359	322	793	4,828	856	258	336	196
Electric distribution products	774	624	44	106	596	100	29	35	14
Electric measuring instruments	354	281	20	53	295	38	9	10	2
Transformers	119	94	7	18	82	18	3	12	4
Switchgear and switchboards	301	249	17	35	219	44	17	13	8
Electric industrial apparatus	840	702	40	98	627	102	35	48	28
Motors and generators	282	231	16	35	169	49	18	29	17

2.2–4 WATER USE BY MANUFACTURING INDUSTRIES (*Continued*)

Industry	Total	Establishments reporting water intake: Source of water: Utility	Other than utility	Source not specified	Quantity of water intake, mil gal: Under 1	1–9	10–19	20–99	100 and over
Industrial controls	246	205	12	29	217	19	3	6	1
Welding apparatus	102	88	4	10	77	12	7	5	1
Carbon and graphite products	51	43	2	6	29	7	1	6	8
Electric industrial goods, NC	159	135	6	18	135	15	6	2	1
Household appliances	432	358	25	49	265	71	24	40	32
Household cooking equipment	56	43	2	11	31	11	4	8	2
Household refrigerators	27	22	4	1	7	5	2	2	11
Household laundry equipment	34	23	2	9	13	7	2	6	6
Electric housewares and fans	163	137	9	17	105	25	10	16	7
Household vacuum cleaners	20	18	1	1	11	4	2	–	3
Sewing machines	42	34	4	4	37	3	–	1	1
Household appliances, NC	90	81	3	6	61	16	4	7	2
Lighting and wiring devices	1,229	1,042	58	129	985	146	34	43	21
Electric lamps	69	64	2	3	32	24	3	8	2
Lighting fixtures	753	650	30	73	662	63	13	8	7
Current-carrying devices	266	211	14	41	201	32	12	16	5
Noncurrent-carrying devices	141	117	12	12	90	27	6	11	7
Radio and TV receiving equipment	362	303	16	43	275	38	23	16	10
Radio and TV receiving sets	247	205	9	33	185	27	20	8	7
Phonograph records	115	98	7	10	90	11	3	8	3
Communication equipment	905	752	42	111	679	110	34	42	40
Telephone; telegraph apparatus	67	54	4	9	34	10	5	5	13
Radio, TV communication equipment	838	698	38	102	645	100	29	37	27
Electronic components	1,334	1,061	76	197	979	194	50	75	36
Electron tubes, receiving type	25	21	2	2	10	3	1	8	3
Cathode ray picture tubes	63	56	3	4	49	6	3	1	4
Electron tubes, transmitting	49	41	4	4	24	9	3	7	6
Semiconductors	88	77	4	7	42	15	6	16	9
Electronic components, NC	1,109	866	63	180	854	161	37	43	14

2.2–4 WATER USE BY MANUFACTURING INDUSTRIES (*Continued*)

	Establishments reporting water intake								
		Source of water			Quantity of water intake, mil gal				
Industry	**Total**	**Utility**	**Other than utility**	**Source not specified**	**Under 1**	**1–9**	**10–19**	**20–99**	**100 and over**
Electrical products, NC	598	517	21	60	422	95	29	37	15
Storage batteries	168	151	6	11	92	42	13	17	4
Primary batteries, dry and wet	33	29	–	4	7	15	4	5	2
X-ray apparatus and tubes	67	60	3	4	58	7	1	1	–
Engine electrical equipment	175	143	10	22	133	18	7	9	8
Electrical products, NC	155	134	2	19	132	13	4	5	1
Transportation equipment	4,195	3,253	300	642	3,166	457	115	198	259
Motor vehicles and equipment	1,734	1,422	96	216	1,238	183	50	108	155
Truck and bus bodies	382	293	22	67	338	26	11	5	2
Truck trailers	125	99	8	18	101	17	5	2	–
Motor vehicles and parts	1,227	1,030	66	131	799	140	34	101	153
Aircraft and parts	999	794	83	122	717	118	43	51	70
Aircraft	94	64	13	17	51	9	3	8	23
Aircraft engines and parts	213	170	17	26	132	28	13	15	25
Aircraft propellers and parts	15	11	2	2	10	2	1	2	–
Aircraft equipment, NC	677	549	51	77	524	79	26	26	22
Ship and boat building	873	617	61	195	722	99	12	21	19
Ship building and repairing	283	228	24	31	178	55	11	20	19
Boat building and repairing	590	389	37	164	544	44	1	1	–
Railroad equipment	87	70	6	11	42	17	5	12	11
Locomotives and parts	21	19	–	2	13	2	1	1	4
Railroad and street cars	66	51	6	9	29	15	4	11	7
Motorcycles, bicycles, and parts	47	36	3	8	32	6	1	5	3
Transportation equipment, NC	455	314	51	90	415	34	4	1	1
Trailer coaches	269	172	30	67	238	26	3	1	1
Transportation equipment, NC	186	142	21	23	177	8	1	–	–

2.2–4 WATER USE BY MANUFACTURING INDUSTRIES (*Continued*)

		Establishments reporting water intake							
		Source of water			Quantity of water intake, mil gal				
Industry	Total	Utility	Other than utility	Source not specified	Under 1	1–9	10–19	20–99	100 and over
Instruments and related products	2,321	1,970	128	223	1,918	220	66	77	40
Scientific instruments	416	338	27	51	352	37	9	11	7
Mechanical measuring devices	448	362	28	58	339	59	18	21	11
Mechanical measuring devices	369	300	18	51	289	48	13	14	5
Automatic temperature controls	79	62	10	7	50	11	5	7	6
Optical instruments and lenses	151	130	5	16	129	16	2	3	1
Medical instruments and supplies	683	596	40	47	599	52	14	12	6
Surgical and medical instruments	188	159	17	12	162	19	2	3	2
Surgical appliances and supplies	360	315	16	29	315	24	10	7	4
Dental equipment and supplies	135	122	7	6	122	9	2	2	–
Ophthalmic goods	187	174	3	10	163	12	6	5	1
Photographic equipment	309	254	21	34	235	36	10	17	11
Watches, clocks, and watchcases	127	116	4	7	101	8	7	8	3
Watches and clocks	93	84	3	6	70	6	6	8	3
Watchcases	34	32	1	1	31	2	1	–	–
Miscellaneous manufacturing	6,833	5,968	243	622	6,165	460	99	79	30
Jewelry and silverware	968	893	39	36	902	39	8	13	6
Jewelry, precious metal	668	629	23	16	637	19	5	6	1
Jewelers' findings and materials	107	99	2	6	99	6	1	1	–
Lapidary work	69	64	2	3	68	–	–	1	–
Silverware and plated ware	124	101	12	11	98	14	2	5	5
Musical instruments and parts	178	154	9	15	140	22	11	4	1
Toys and sporting goods	1,239	1,010	65	164	1,055	127	27	23	7

2.2–4 WATER USE BY MANUFACTURING INDUSTRIES (*Continued*)

Industry	Establishments reporting water intake								
	Total	Source of water			Quantity of water intake, mil gal				
		Utility	Other than utility	Source not specified	Under 1	1–9	10–19	20–99	100 and over
Games and toys	447	363	20	64	368	56	9	10	4
Dolls	198	180	11	7	186	9	3	–	–
Children's vehicles	41	35	3	3	25	8	4	4	–
Sporting and athletic goods	553	432	31	90	476	54	11	9	3
Pens, pencils, and office supplies	551	496	15	40	494	34	12	9	2
Pens and mechanical pencils	93	78	4	11	76	9	5	2	1
Lead pencils and art goods	90	81	2	7	73	10	3	3	1
Marking devices	286	267	6	13	282	3	1	–	–
Carbon paper and inked ribbons	82	70	3	9	63	12	3	4	–
Costume jewelry and notions	894	789	29	76	791	73	14	13	3
Costume jewelry	386	360	10	16	346	29	3	8	–
Artificial flowers	158	118	1	39	151	6	1	–	–
Buttons	148	130	8	10	131	14	3	–	–
Needles, pins, and fasteners	202	181	10	11	163	24	7	5	3
Miscellaneous manufactures	3,003	2,626	86	291	2,783	165	27	17	11
Brooms and brushes	302	257	11	34	272	23	3	3	1
Hard surface floor coverings	20	15	3	2	10	4	2	–	4
Matches	24	16	1	7	10	8	3	2	1
Candles	53	36	–	17	45	5	3	–	–
Lamp shades	89	78	–	11	84	4	1	–	–
Morticians' goods	409	353	8	48	381	21	4	1	2
Furs, dressed and dyed	87	74	–	13	81	6	–	–	–
Signs and advertising displays	1,372	1,223	44	105	1,309	58	2	3	–
Umbrellas, parasols and canes	67	64	–	3	64	3	–	–	–
Miscellaneous products, NC	580	510	19	51	527	33	9	8	3
Ordnance and accessories	194	132	23	39	103	32	8	22	29

Note: A dash (–) represents zero; NC is not elsewhere classified.

Source: U.S. Bureau of the Census.

2.2–5 MANUFACTURING WATER WITHDRAWALS BY SOURCE

Year	Total with-drawn, bil gal	Public water systems		Company owned: Surface water		Company owned: Ground water		Brackish water	
		10^9 gal	Percent of total	10^9 gal	Percent of total	10^9 gal	Percent of total	10^9 gal	Percent of total
1954	11,757	1,663	14.1	6,899	58.6	1,573	13.3	1,622	14.1
1959	12,175	1,303	10.7	7,266	59.6	1,164	9.5	2,442	19.8
1964	14,045	1,383	9.8	8,466	60.2	1,360	9.6	2,836	20.1

Source: Reprinted, with publisher's permission, from *J. Am. Water Works Assn., 60*(10):1129, © 1968, by the American Water Works Association, Inc., 2 Park Ave., New York, N.Y. 10016.

2.2–6 BRACKISH WATER USE IN MANUFACTURING

1964

Industrial water use by regions	Total water intake 10^9 gal	Brackish water intake		Water used for cooling and condensing		Water treated before use	
		10^9 gal	Percent of total	10^9 gal	Percent of total	10^9 gal	Percent of total
Delaware and Hudson	1,256	389	30.9	983	54.3	283	22.5
Chemicals	264	101	38.2	217	82.1	121	45.8
Petroleum	359	232	64.6	315	87.7	20	5.5
Primary Metals	300	23	7.6	265	88.3	12	4.0
Chesapeake Bay	757	337	44.5	564	74.5	402	53.1
Chemicals	180	21	11.6	159	88.3	32	17.7
Western Gulf	1,825	1,450	79.4	1,406	77.0	434	23.7
Chemicals	1,218	1,061	87.1	905	74.3	278	22.8
Petroleum	383	271	70.7	339	88.5	73	19.0
California	347	163	46.9	210	60.5	71	20.4
Petroleum	140	110	78.5	127	90.7	16	11.4

Source: Reprinted, with publisher's permission, from *J. Am. Water Works Assn., 60*(10):1129, © 1968, by the American Water Works Association, Inc., 2 Park Ave., New York, N.Y. 10016.

2.2–7 PROJECTED INDUSTRIAL WATER REQUIREMENTS

U.S.

Type of use	1965	1980	2000	2020
Manufacturing				
Gross	90.0	164.3	349	763
Recirculation ratio	2.25	3.0	4.4	6.3
Withdrawal	40.0	54.8	80	121
Discharge	37.4	50.2	70	100
Consumption	2.6	4.6	10	21
Mineral industry				
Gross	11.1	20.0	33.0	88.0
Recirculation ratio	3.43	4.9	7	10
Withdrawal	3.24	4.1	4.7	8.8
Discharge	2.48	2.9	3.3	7.0
Consumption	0.76	1.2	1.4	1.8

Note: Data are in bgd. Values include both publicly supplied and self-supplied water. Conterminous United States only. For regions see 1–7.

Source: U.S. Water Resources Council, 1968.

2.2–8 ESTABLISHMENTS REPORTING VARIOUS ANNUAL WATER INTAKES

1963

Annual water intake, mil gal	No. of establishments
Less than 1	96,181
1–9	18,904
10–19	4,629
20–99	5,996
100 or more	4,584
Total	130,294

Source: Reprinted, with publisher's permission, from *J. Am. Water Works Assn., 60*(10):1129, © 1968, by the American Water Works Association, Inc., 2 Park Ave., New York, N.Y. 10016.

2.2–9 ANNUAL WATER USE BY INDUSTRY

Type of industry	Annual use, gal	Approximate percent of total industrial use
Primary metal manufacture	4.5×10^{12}	32
Chemicals and allied products	3.2×10^{12}	23
Paper industry	2×10^{12}	14
Petroleum and coal products	1.3×10^{2}	9
Food industry	750×10^{9}	5
Stone and vitreous products	248×10^{9}	2
Transportation equipment	247×10^{9}	2
Rubber and plastics	163×10^{9}	1
Nonelectrical machinery	103×10^{9}	
Fabricated metals	57×10^{9}	
Instruments and related products	30×10^{9}	
Leather	16×10^{9}	
Miscellaneous	13×10^{9}	
Tobacco	3×10^{9}	
Furniture	3×10^{9}	

Source: "Water Control News, Commerce Clearing House," *Willing Water*, Am. Water Works Assn., August 1966.

2.2–10 ECONOMIC DATA FOR LARGE WATER USE INDUSTRIES

U.S.

Industry name	No. of establish-ments	Intake water, bil gal	Gross water used, bil gal	Value added, $ millions	Employment
Meat slaughtering plants	237	84	138	1,249	114,444
Raw cane sugar	46	123	252	99	6,661
Beet sugar	58	54	86	196	11,061
Malt liquors	119	59	94	1,189	56,835
Cotton weaving mills	123	27	87	788	127,850
Sawmills and planing mills	93	113	157	279	36,051
Pulp mills	34	279	892	255	12,994
Paper mills, except building	271	1,138	3,095	1,736	120,447
Paperboard mills	183	577	1,845	1,098	56,696
Building paper and board mills	45	31	76	127	9,393
Alkalies and chlorine	31	528	628	381	19,090
Intermediate coal tar products	56	108	277	514	23,105
Organic chemicals[a]	139	1,708	3,319	2,394	75,385
Inorganic chemicals[a]	188	499	1,312	1,541	67,759
Plastics materials and resins	114	218	470	1,120	49,839
Synthetic rubber	24	61	182	330	11,840
Cellulose man-made fibers	18	152	233	356	27,222
Organic fibers, noncellulose	20	98	208	896	40,976
Pharmaceutical preparations	55	28	79	2,089	58,107
Gum and wood chemicals	18	26	117	77	4,115
Fertilizers	59	113	217	176	10,028
Petroleum refining	214	1,388	6,146	2,942	111,644
Tires and inner tubes	51	60	173	1,266	80,320
Rubber products[a]	115	70	105	885	81,023
Cement, hydraulic	145	114	151	660	29,410
Blast furnaces and steel mills	200	3,815	5,510	7,175	481,867
Primary copper	21	32	161	213	9,337
Primary aluminum	20	197	335	470	17,189
Motor vehicles and parts	237	125	395	10,866	541,173
Aircraft engines and parts	36	60	108	1,633	138,944
Total	8,925	14,045	30,599	94,706	6,396,906

[a] Not elsewhere classified.

Source: Reprinted, with publisher's permission, from *J. Am. Water Works Assn.*, *60*(10):1129, © 1968, by the American Water Works Association, Inc., 2 Park Ave., New York, N.Y. 10016.

2.2–11 PROCESS WATER INTAKE BY CHEMICAL AND ALLIED PRODUCT INDUSTRIES

Total Water Intake of 20 bgy or More, 1964

Industry group	Process water intake bgy	Process water intake Percent
Alkalies and chlorine	16	2.8
Intermediate coal tar products	9	1.6
Organic chemicals[a]	314	55.5
Inorganic chemicals[a]	74	13.3
Plastic materials and resins	25	4.4
Synthetic rubber	11	2.0
Pharmaceutical preparations	3	0.5
Soaps and other detergents	2	0.4
Paints and allied products	1	0.2
Gum and wood chemicals	2	0.4
Fertilizers	32	5.6
Explosives	2	0.4
Subtotal	491	87.1
Nonlisted industries[b]	73	12.9
Chemicals and allied products	564	100.0

[a] Not elsewhere classified.
[b] Although the industries selected for study probably determine the range in values of the various quality criteria for process waters for chemical and allied products, it is noted that 3 industries (Cellulosic Man-made fibers, Organic Fibers Noncellulosic, and Glue and Gelatin) use 23, 8, and 6 bgy, which is more than several of the industries under consideration.

Source: *Water Quality Criteria,* U.S. Department of the Interior, Fed. Water Pollut. Control Adm., April 1968, p. 210.

2.2–12 INDUSTRIAL WATER USE

Industry	Usage, %
Chemical	22.0
Metals	21.5
Petroleum and coal	20.3
Pulp and paper	18.5
Food	3.9
Transportation	3.9
Textiles	3.2
Rubber and plastics	3.1
Stone	1.2
Other	2.4

Source: G. Rey, W.J. Lacy, and A. Cywin, *Environ. Sci. Technol.*, *5*:76, 1971.

2.2–13 WATER REQUIREMENTS PER UNIT OF PRODUCT

U.S. industries

Industry and product	Unit of product	Water required per unit, gal
Chemical industries		
Acetic acid	ton of HAc	120,000–290,000
Alcohol	gal	52–138
Ammonia	ton of NH_3	37,500
Ammonium sulfate	–	240,000
Calcium carbide	–	36,500
Calcium metaphosphate	ton of $Ca(PO_3)_2$	2,800
Carbon dioxide	–	24,500
Caustic soda	ton of NaOH (11%)	22,000–25,500
Cellulose nitrate	–	12,000
Charcoal and wood chemicals	ton of $CaAc_2$	79,000
Corn refining	ton of starch	333
Gasoline	gal	7–34
Gunpowder	ton	200,000
Hydrochloric acid	ton of 20 Bé° HCl	3,500
Hydrogen	ton of H_2	800,000
Lactose	–	235,000
Oxygen	100 ft^3	65
Soap	–	300–600
Soda ash	–	18,000–22,000
Sulfuric acid	ton of 100% H_2SO_4	800–6,000
Sulfur	ton	3,000

2.2–13 WATER REQUIREMENTS PER UNIT OF PRODUCT (*Continued*)

Industry and product	Unit of product	Water required per unit, gal
Food and beverage industries		
Beet sugar	–	20,000–25,000
Bread	ton	600–1,200
Beans, green	ton	20,000
Peaches and pears	ton	5,300
Other fruits and vegetables	ton	2,000–10,000
Gelatin	–	15,000–24,000
Meat packing	ton live weight	5,000
Milk products	–	4,000–5,000
Oils, edible	gal	88
Sugar	–	1,200–2,600
Beer	gal	15
Whisky	gal	80
Pulp and paper		
Kraft pulp	ton dry pulp	110,000
Sulfate pulp	–	82,000
Sulfite pulp	–	82,000–230,000
Soda pulp	–	101,000
Paper	–	47,000
Paperboard	–	17,500–103,000
Strawboard	–	31,500
Textile industries		
Cotton	–	20,000–76,000
Cotton bleaching	ton	72,000–96,000
Cotton dyeing	ton	9,500–19,000
Linen	–	200,000
Rayon	ton of yarn	105,000–240,000
Wool scouring	–	40,000–240,000
Metal and metal products		
Rolled steel	–	96,000
Finished steel	–	79,000
Fabricated steel	–	52,000
Steel sheets	–	16,000
Avg, all products	–	20,000–35,000
Aluminum	–	360,000
Miscellaneous		
Electric power	kWh	85–185
Coal washing	–	1,800–4,300
Leather tanning	ton, raw hide	19,500
Synthetic rubber	–	24,000–800,000

REFERENCES

1. R.B. Canlan, "Water Reuse and Conservation," *Betz Indicator, 29*:12, December 1960.
2. B. Symons, "The Disposal of Industrial Wastes," *Water Works and Sewage, 92*:12, December 1945.
3. *Water for Industrial Use*, United Nations.

2.2–14 INDUSTRIAL REQUIREMENTS FOR WATER

Product	Unit produced	Water required, gal/unit	Ref.
Airplane	to test	50,000–125,000	1
Alcohol	gal	100	1
		120[a]	6
Aluminum	lb	3.15[b]	6
		160	1,7
Aviation gas	gal	7–10	1
Brewing			
Beer	1 bbl	470	1
		750	7
		298–2,500	8
Whiskey	gal	80	1
Buildings			
Office	person day	27–45	1
Hospital	bed day	135–350	1,6
Hotels	guest room day	300–525	1
Laundries			
Commercial	lb work	4.3–5.7	1
Institutional	lb work	3.	1
Restaurants	meal	0.5–4.0	1
Butadiene	lb	160	1
		10–300[c]	6
Canning			
Apricots	100 cases No. 2 cans	8,000	1
Asparagus	100 cases No. 2 cans	7,000	1
Beans			
Green	100 cases No. 2 cans	3,500	1
Lima	100 cases No. 2 cans	25,000	1
Pork and beans	100 cases No. 2 cans	3,500	
Beets	100 cases No. 2 cans	2,500	1
Corn	100 cases No. 2 cans	2,500	1
Grapefruit			
Juice	100 cases No. 2 cans	500	1
Sections	100 cases No. 2 cans	5,600	1
Peaches, pears	100 cases No. 2 cans	6,500	1
Peas	100 cases No. 2 cans	2,500	1
Pumpkin (squash)	100 cases No. 2 cans	2,500	1
Sauerkraut	100 cases No. 2 cans	300	1
Spinach	100 cases No. 2 cans	16,000	1
Succotash	100 cases No 2 cans	12,500	1
Tomatoes			
Products	100 cases No. 2 cans	7,000	1
Whole	100 cases No. 2 cans	750	1
Cement	ton	750	1,6
Coke	100 tons	360,000	1
Distilling, grain			
Combined wastes	1,000 bu grain mashed		
Thin slop	1,000 bu grain mashed		
Tailings	1,000 bu grain mashed		
Evaporator condensate	1,000 bu grain mashed	600,000	1
Distilling, molasses	1,000 gal 100 proof	8,400	1
Distilling, cooling water	1,000 gal 100 proof	120,000	1
Electric power	kw	80	1
Electric power, steam		80–170[c]	6
generated	kWh	52–170	8,9
Explosives	lb	100+	1
		100	6

[a] Alcohol industrial (100 proof).
[b] Alumina (Bayer process).
[c] Range from no reuse to maximum recycling.

2.2–14 INDUSTRIAL REQUIREMENTS FOR WATER (*Continued*)

Product	Unit produced	Water required, gal/unit	Ref.
Gasoline	gal	7–10	1
		8.5	2
		20	6
		8.5–25	8
Iron ore (brown ore)	ton	1,000	1,6
Meat			
Packinghouse	100 hogs killed	550	1
		1,100	2
		5.5–11	8
Slaughterhouse	100 hogs killed	550	1
Stockyards	1 acre	160	1
Milk			
Receiving station	1,000 lbs raw milk	180	1
Bottling works	1,000 lbs raw milk	250	1
		4,500	6
Cheese factory	1,000 lbs raw milk	200	1
		2,000	6
Creamery	1,000 lbs raw milk	110	1
		110–250	2
Condensery	1,000 lbs raw milk	150	1
Dry milk factory	100 lbs raw milk	150	1
General dairy	100 lbs raw milk	340	1
Butter	ton	5,000	6
Oil, edible	gal	22	1
Oil field	100 bbl crude	18,000	1
		15.1–1,500	9
Oil refining	100 bbl	77,000	1,2,6
Paper			
Paper mill	1 ton	39,000	1,6
Paperboard	1 ton	100,000	7
		15,000	2
		15,000–90,000[c]	6
		7,700–80,000	9
Pasteboard	1 ton	14,000	1
Strawboard	1 ton	26,000	1,6
Deinking	1 ton	83,000	1
		5,000–85,000	8
Paper pulp			
Groundwood	1 ton dry	5,000	1
		4,000–50,000[c]	6
Kraft	1 ton dry	93,000	6
Soda	1 ton dry	85,000	1,6
Sulfate	1 ton dry	64,000	1
		70,000	6
Sulfite	1 ton dry	60,000	1
		70,000–133,000[c]	6
Poultry	1 bird	1 (per day)	1
Rail freight	ton mi	0.1	1
Records	1 disc	2.4	1
Smokeless powder	ton	50,000	1
Soap factories	ton	500	1,6
Steam power	ton of coal	60,000–120,000	1
Sugar refineries	lb	0.5	1
Tanning			
Vegetable	100 lb raw hide	800	1
Chrome	100 lb raw hide	800	1

[c] Range from no reuse to maximum recycling.

2.2–14 INDUSTRIAL REQUIREMENTS FOR WATER (*Continued*)

Product	Unit produced	Water required, gal/unit	Ref.
Textile			
Cotton	1,000 lb processed	28,000–40,000	8
Sizing	1,000 lb processed	820	1
Desizing	1,000 lb processed	1,750	1
Keiring	1,000 lb processed	1,240	1
Bleaching	1,000 lb processed	300	1
		30,000–80,000	6
Scouring	1,000 lb processed	3,400	1
Mercerizing	1,000 lb processed	30,000	
Dyeing			
Basic	1,000 lb processed	18,000	1
		4,000–8,000	6
Direct	1,000 lb processed	6,400	1
Vat	1,000 lb processed	19,000	1
Sulfur	1,000 lb processed	5,400	1
Developed	1,000 lb processed	14,400	1
Naphthol	1,000 lb processed	4,800	1
Analine black	1,000 lb processed	15,600	1
Print works	1,000 lb processed	4,500	1
Finishing	1,000 lb processed	6	1
Knit goods	lb bleached	8	1
Rayon manufacture	1,000 lb produced	160	1
Rayon hosiery	1,000 lb produced	9,000	1
Woolens	1,000 lb finished	70,000	1,6,8
		40,000–510,000[j]	9
Wool scouring	1,000 lb produced	1,260	2
Steel	ton	65,000	2
		44,000	3
		1,100	4
		15,000–110,000[c,g]	6
		67,000[h]	7
		65,000[i]	8
		6,000–110,000[g]	9
Manufactured ice	ton	244	2
		240–9,000	8
Soft drinks	case	2.5	2
Corn products	bushel	40 (direct)	4
		100–200 (indirect)	4
		140–240[d,e]	6
Ammonium sulfate	ton	200,000	6
Calcium carbide	ton	30,000	6
Carbon dioxide (from flue gas)	ton	20,000	6
Cottonseed oil	gal	20	6
Hydrogen	ton	660,000	6
Oxygen liquid	ft^3	2	6
Soda ash	ton	18,000	6
Sodium chlorate	ton	60,000	6
Sulfuric acid	ton	650–4,875[c]	6
Bread	ton	500–1,000[d]	6
Corn syrup	bu	30–40[d]	6

[c] Range from no reuse to maximum recycling.
[d] Range covers various products or processes involved.
[e] Wet milling.
[f] Live animal.
[g] Steel (rolled).
[h] Steel from ingots.
[i] Steel finished.
[j] Per M yards.

2.2–14 INDUSTRIAL REQUIREMENTS FOR WATER (*Continued*)

Product	Unit produced	Water required, gal/unit	Ref.
Gelatin (edible)	ton	13,200–20,000[d]	6
Meat			
Packing	ton[f]	4,130	6
Packinghouse operation	hog unit	550	6
Sugar			
Beet sugar	ton	2,160	6
		24,000–34,000	8
Cane sugar	ton	1,000	6
		1,000–28,000	8
		4,000–110,000	9
Petroleum refined products	bbl	150–15,000[c]	6
Synthetic fuel			
Coal hydrogenates	bbl	7,296	6
Coal derivatives	bbl	11,150	6
Natural gas derivatives	bbl	3,736	6
Shale derivatives	bbl	873	6
Rayon	ton	220,000	7
		250,000–400,000	8,9
Caper ammonium yarn	ton	90,000–160,000[d]	6
Viscose yarn	ton	200,000	6
Weave, dye and finish	yd	15	6
Coal and coke			
By-product coke	ton	1,500–3,600[a]	6
		3,600	7
Washing	ton	200	6
Rock wool	ton	5,000	6
Sulfur mining	ton	3,000	6
Synthetic rubber	ton	660,000	7
Buna S	ton	631,400	6
GR-S	ton	28,000–670,000[c]	6
Taconite pellets	ton	12,000	7
Ingot iron	ton	18,000	7

[a] Alcohol industrial (100 proof).
[c] Range from no reuse to maximum recycling.
[d] Range covers various products or processes involved.
[f] Live animal.

REFERENCES

1. H.E. Jordan, *J. Am. Water Works Assn., 38*(1):65, 1946.
2. S.T. Powell and H.E. Bacon, *J. Am. Water Works Assn., 42*(8):777, 1950.
3. C.C. Elder, *J. Am. Water Works Assn., 43*(2):124, 1951.
4. W.J. Lauterbach, *J. Am. Water Works Assn., 44*(11):1033, 1952.
5. Task Group Report, *J. Am. Water Works Assn., 45*(12):1249, 1953.
6. Am. Water Works Assn. Distribution Manual, *J. Am. Water Works Assn., 53*(4):459, 1961.
7. E.H. Ruble, *J. Am. Water Works Assn., 57*(7):831, 1965.
8. R.D. Hoak, *Sew. Ind. Wastes, 25*(12):1438, 1953.
9. R. Hodges *et al., J. Water Pollut. Control Fed., 38*(10):1601, 1966.

2.2–15 ESTIMATION OF INDUSTRIAL WATER USAGE

Industry	No. of establishments	Production	Estimated water consumption: Gal per unit	Estimated water consumption: Mill gal, year 1947
Steel (finished)	419	61,857,241 tons	65,000	4,020,721
Oil refining	437	1,887,890,000 bbl	770	1,453,675
Gasoline		791,325,000 bbl	357	791,325
Wood pulp				
Sulfate		5,356,710 tons	64,000	342,829
Sulfite		2,795,960 tons	60,000	167,758
Soda		491,580 tons	85,000	41,784
Groundwood		2,049,814 tons	5,000	10,249
Total wood pulp establishments	226			
Paper		10,646,833 tons	39,000	415,226
Paperboard		9,186,810 tons	15,000	137,802
Total paper establishments	665			
Coke	167	79,146,000 tons	3,600	.284,373
Beer	440	88,027,000 bbl	470	41,373
Whisky	226	246,443,000 gal	80	21,155
Milk, cream, and butter		72,440,000,000 lb	0.11–0.25	14,286
Canning[a] and preserving[b]	2,265	391,546,000 cases	7.5–250	8,520
Manufactured ice[c]	3,423	36,100,000 tons	243.85	8,802
Soft drinks	5,618	927,700,000 cases	2.5	6,250
Woolens and worsted fabrics	495	464,563,000 lb	70	3,252
Wool scouring	74	210,172,000 lb	1.26	2,648
Tanning	561	238,731,000 lb	8	1,910
Soap	249	4,138,001,000 lb	0.25	1,034
Meat packing (hogs)	2,153	51,678,047 hogs	11	568
Cane sugar[d]	25	358,000 tons	1,000	358
Rayon (all types)	38	746,900,000 lb	0.16	119

[a] Estimate of water consumption includes water used in processing 14 fruits and vegetables for which consumption factors are available. This accounts for only 177,321,000 cases. Total water used by this group is probably two or three times the volume given.
[b] Excluding fish.
[c] Includes filling cans and pulling cores.
[d] Includes refining only.

Source: E.B. Besselievre, *Industrial Waste Treatment,* McGraw-Hill Book Co., 1952, p. 113.

2.2–16 WATER CONSUMPTION IN VARIOUS INDUSTRIES

1954

Industry	Water consumed per worker, gal/day
Paper manufacturing	20,600
Petroleum oil production	5,000
Petroleum refining	5,000
Metal smelting and fabrication	2,500
Chemicals production	2,000
Glass production	1,900
Cement production	1,500
Food processing	630
Transportation equipment	400
Machinery (nonelectrical)	400
Printing and publishing	260
Clothing production	60

Source: Reprinted, with publisher's permission, from *J. Am. Water Works Assn.*, *51*(11):1357, © 1959, by the American Water Works Association, Inc., 2 Park Ave., New York, N.Y. 10016.

2.2–17 WATER REQUIRED TO PRODUCE CERTAIN INDUSTRIAL PRODUCTS

Product	Water required[a] tons
Coke	15
Beer	18
Taconite pellets	50
Ingot iron	75
Steel from ingots	270
Paper	420
Rayon	920
Aluminum	1,340
Synthetic rubber	2,750

[a] To produce a ton of the product.

Source: Reprinted, with publisher's permission, from *J. Am. Water Works Assn.*, *57*(7):831, © 1965, by the American Water Works Association, Inc., 2 Park Ave., New York, N.Y. 10016.

2.2–18 RELATIVE USES OF WATER BY SELECTED INDUSTRIES

U.S., 1959

	Cooling, %	Noncooling, % (Process and other uses)
All reporting industries	66	34
Automobile	21	79
Beet sugar	95	5
Chemical	65	35
Distilling	84	16
Meat	25	75
Petroleum	82	18
Soap and detergent	66	34
Steel	66	34
Sugar cane	69	31
Textiles	10	90

Source: National Association of Manufacturers.

2.2–19 COOLING WATER MAKEUP IN INDUSTRY

	Water intake use by function, %		
Industry group	Process	Cooling	Steam
All (U.S. industries)	26	67	7
Chemicals and allied products	15	80	5
Primary metals industries	22	74	4
Petroleum and coal products	6	87	7
Paper and allied products	64	30	6

Source: 1963 Census of Manufacturers.

2.2–20 WATER INTAKE CONSUMED BY SELECTED INDUSTRIES

U.S., 1959

Industry	Water consumption, % of intake
Automobile	6.2
Beet sugar	10.5
Chemicals	5.9
Coal preparation	18.2
Corn and wheat milling	20.6
Distillery	10.4
Food processing	33.6
Machinery	21.4
Meat	3.2
Petroleum	7.2
Poultry processing	5.3
Pulp and paper	4.3
Salt	27.6
Soap and detergents	8.5
Steel	7.3
Sugar, cane	15.9
Textiles	6.7

Source: National Association of Manufacturers.

2.2–21 FOOD INDUSTRY WATER REQUIREMENTS

In-plant use	Water quantity, bgy	Percent of total use
Raw product washing	9.9	15
Product transport[a]	6.6	10
Product preparation[b]	6.6	10
Incorporation in product[c]	4.0	6
Steam and water sterilization of containers	9.9	15
Container cooling	23.7	36
Plant cleanup	5.3	8

[a] Fluming and pumping of raw product.
[b] Blanching, heating, and soaking of product.
[c] Preparation of syrups and brines which enter the container.

Source: *Water Quality Criteria*, U.S. Department of the Interior, Fed. Water Pollut. Control Adm., April 1968, p. 210.

2.2–22 WATER USE IN AN INTEGRATED STEEL MILL

	Volume	
Department	**Finished steel, gal/ton**	**Percent of total**
Blast furnace	10,000	25.0
Open hearths	5,000	12.5
Coke plants	5,000	12.5
Hot mills	10,000	25.0
Finishing mills	8,000	20.0
Sanitary, boiler, and other uses	2,000	5.0
Total	40,000	100.0

Source: Reprinted from the April 1954 issue of *Iron and Steel Engineer, 31*:78, 1954.

2.2–23 PROCESS WATER UTILIZATION IN PRIMARY METALS INDUSTRY

Industry	**Process water used, 1964, bgy**
Iron and steel production	885
Iron and steel foundries	12
Copper industry	36
Aluminum industry	20
All other primary metal industries	43
Total, primary metals	996

Source: *Census of Manufacturers, Water Use in Manufacturing, 1963*, U.S. Department of Commerce, 1966.

2.2–24 RELATIONSHIP BETWEEN GROSS WATER INTAKE AND FINISHED PRODUCT OUTPUT IN PAPER MILLS

Type of mill	No. of mills	Water per ton air dried Finished product, gal		
		Max	Min	Avg
Integrated				
Kraft pulp and paper	3	34,000	13,800	24,100
Kraft pulp, bleachery, paper	14	58,800	9,500	33,000
Kraft pulp, bleachery, fine paper	4	50,000	32,500	39,400
Soda pulp, bleachery, fine paper	2	44,000	28,500	36,300
Groundwood pulp, paper	2	42,300	27,500	34,900
Groundwood pulp, bleachery, fine paper	5	54,000	21,000	36,300
NSSC pulp, bleachery, fine paper	2	90,000	56,000	73,000
Sulfite pulp, bleachery, fine paper	2	83,500	12,400	48,000
Deinking stock, bleachery, paper	2	25,000	19,500	22,250
Rag pulp, bleachery, fine paper	1			55,000
Sulfite and groundwood pulp, paper	2	52,500	42,500	47,500
Kraft and groundwood pulp, bleachery, paper	6	54,200	22,400	32,900
Kraft and groundwood pulp, bleachery, fine paper	2	60,000	55,800	57,900
Kraft and sulfite pulp, bleachery, paper	1			84,500
Kraft, groundwood and semichemical pulp, bleachery paper	1			72,000
Kraft, NSSC pulp, paper	3	22,000	16,000	18,700
Nonintegrated				
Fine paper	8	30,000	7,500	17,700
Paper (wrapping)	8	43,000	4,350	21,000
Paper (roofing felt)	1			7,450
Paper (glassine)	1			6,000
Paper (specialty papers)	3	46,600	10,000	27,200
Paper, bleachery (tissue)	3	62,000	25,000	38,200
Board	3	15,800	2,480	8,860
Soda pulp, bleachery, drying	1			43,800
Kraft pulp, bleachery, drying	4	51,000	38,500	42,900
Kraft pulp, bleachery	1			26,500
NSSC pulp	1			17,700

Source: Reprinted, with publisner's permission, from *J. Am. Water Works Assn., 63*(3):165, ©1971, by the American Water Works Association, Inc., 2 Park Ave., New York, N.Y. 10016.

2.2–25 GROSS WATER INTAKE OF PULP AND PAPER MILLS

Type of mill	No. of mills	Intake, gal/ton[a] Max	Min	Avg
Integrated				
Alkali, sulfite and groundwood pulp; bleachery, paper mill	1			76,500
Alkali and sulfite pulp; bleachery; paper mill	3	116,800	84,000	8,600
Alkali and deinking pulp; bleachery; paper mill	8	102,180	42,000	62,190
Alkali and groundwood pulp; paper mill	1		51,500	51,500
Alkali pulp; paper mill	8	61,200	23,550	36,225
Alkali and sulfite pulp mill	2	95,000	84,000	89,500
Rag pulp; rag paper (book and writing)	2	26,170	23,400	24,985
Rag pulp; rag paper (drafting)	1			125,000
Sulfite and groundwood pulp; bleachery; paper mill (book and fine)	6	77,400	46,250	59,800
Sulfite, groundwood, semichemical pulp; bleachery; paper mill	1			21,000
Sulfite and groundwood pulp; paper mill (news)	2	28,600	26,600	27,600
Sulfite pulp; bleachery; paper mill	4	130,000	65,000	97,600
Groundwood pulp; bleachery; paper mill	4	29,700	13,340	26,740
Groundwood pulp; paper mill	9	108,200	4,210	40,530
Deinking stock; paper mill	7	77,200	22,500	35,750
Semichemical pulp; paper mill	1		21,000	21,000
Nonintegrated				
Fine paper (book writing, litho)	6	84,400	14,600	44,911
Paper (wrapping)	3	42,000	1,500	24,200
Paper (roofing felt)	1			8,675
Paper (facial and sanitary tissue)	3	38,900	925	19,000
Paper (filter absorbent)	2	100,000	100,000	100,000
Paper (glassine)	1			79,000
Board	28	198,000	5,833	16,500
Pulp (sulfite)	1			86,750
Pulp (kraft)	1			50,000
Pulp (soda)	1			34,600
Pulp (groundwood)	1			10,000

[a] **Air dried finished product.**

Source: *Water Consumption in the Pulp and Paper Industry*, Tech. Assn. Pulp Paper Ind. Monograph Series, No. 18, 1957, p. 16.

2.2–26 PULP AND PAPER PROCESS WATER REQUIREMENTS

Gallons per ton of product

	Intake[a]	Recycle	Consumption	Discharge
Mechanical pulping	1,000	47,000	1,000	250
Unbleached chemical pulp and paper	7,000	115,000	1,000	6,000
Bleached chemical pulp and paper	20,000	250,000	1,000	19,000

[a] Does not include about 250 gallons per ton of water present in the woodchips.

Source: *Water Quality Criteria,* U.S. Department of the Interior, Fed. Water Pollut. Control Adm., April 1968.

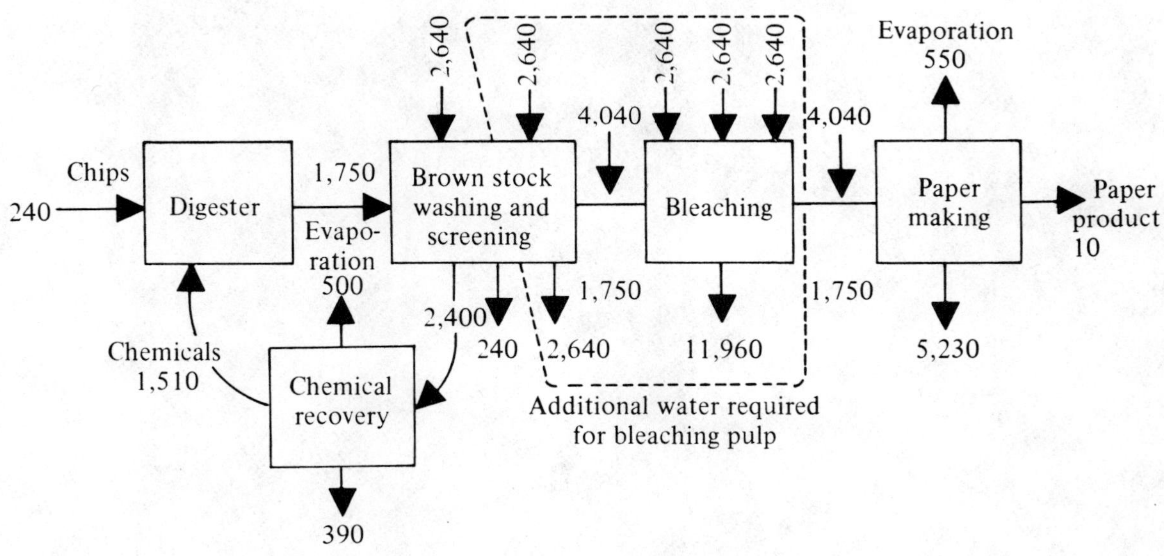

2.2–27 Water Intake, Recycling, and Discharge by a Chemical Pulping Mill.

Note: All data given as gallons of water per ton of product.

Source: *Water Quality Criteria,* U.S. Department of the Interior, Fed. Water Pollut. Control Adm., April 1968.

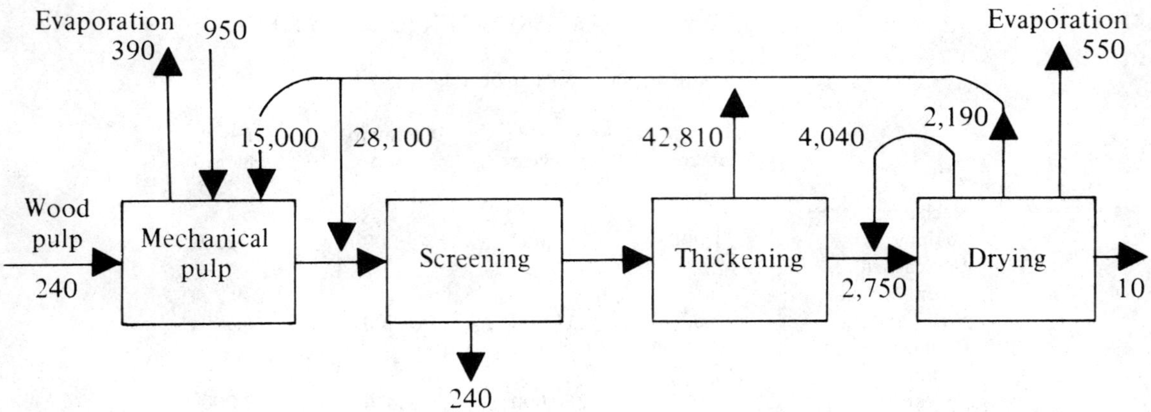

2.2–28 Water Intake, Recycling, and Discharge by a Mechanical Pulping Mill.

Note: All data given as gallons per ton of product.

Source: *Water Quality Criteria*, U.S. Department of the Interior, Fed. Water Pollut. Control Adm., April 1968.

2.2.2 MINING INDUSTRIES WATER NEEDS

2.2–29 WATER USE OF MINERAL INDUSTRIES

U.S., 1963

		Number of mineral establishments reporting water use							
		By principal source of water[a]			By quantity of water intake,[b] mil gal				
Industry	Total	Utility	Other than utility	Source not specified	Under 1	1–9	10–19	20–99	100 and over
All mineral industries	17,107	2,258	11,473	3,376	13,328	1,300	448	1,011	1,020
Metal mining	676	112	386	178	414	44	21	54	143
Iron ores	144	43	87	14	55	12	6	19	52
Copper ores	85	11	48	26	43	7	3	5	27
Lead and zinc ores	93	13	52	28	41	4	5	13	30
Lead ores	33	4	24	5	20	2	–	1	10
Zinc ores	60	9	28	23	21	2	5	12	20
Gold and silver ores	82	10	41	31	62	9	–	2	9
Lode gold	43	6	19	18	38	2	–	1	2
Placer gold	14	1	7	6	6	5	–	1	2
Silver ores	25	3	15	7	18	2	–	–	5
Bauxite	14	1	5	8	12	1	–	1	–
Ferroalloy ores	28	8	9	11	20	2	–	3	3
Manganese ores	11	5	2	4	8	1	–	1	1
Tungsten ores and ferroalloy ores, NC	17	3	7	7	12	1	–	2	2
Metal mining services	54	5	29	20	52	2	–	–	–

[a] Except mine water.
[b] Includes mine water used.

2.2–29 WATER USE OF MINERAL INDUSTRIES (*Continued*)

	Number of mineral establishments reporting water use								
		By principal source of water[a]			By quantity of water intake,[b] mil gal				
Industry	Total	Utility	Other than utility	Source not specified	Under 1	1–9	10–19	20–99	100 and over
Miscellaneous metal ores	176	21	115	40	129	7	7	11	22
Mercury ores	8	1	3	4	6	–	1	–	1
Titanium ores	7	–	7	–	1	–	1	–	5
Uranium-radium-vanadium ores	154	18	100	36	118	5	5	11	15
Metallic ores, NC	7	2	5	–	4	2	–	–	1
Anthracite mining	258	59	127	72	191	18	10	13	26
Anthracite	241	57	119	65	176	16	10	13	26
Anthracite mining services	17	2	8	7	15	2	–	–	–
Bituminous coal and lignite mining	2,457	194	1,446	817	2,055	130	54	110	108
Bituminous coal	2,380	184	1,405	791	1,980	128	54	110	108
Lignite	23	4	14	5	21	2	–	–	–
Bituminous coal and lignite mining services	54	6	27	21	54	–	–	–	–
Oil and gas extraction	9,697	1,273	6,968	1,456	7,879	708	242	531	337
Crude petroleum and natural gas	6,250	634	4,643	973	5,193	389	148	275	245
Crude petroleum	5,265	547	3,907	811	4,272	358	137	263	235
Natural gas	985	87	736	162	921	31	11	12	10
Natural gas liquids	628	77	507	44	158	108	64	218	80
Oil and gas field services	2,819	562	1,818	439	2,528	211	30	38	12
Drilling oil and gas wells	1,442	174	1,091	177	1,203	172	25	36	6
Oil and gas exploration services	173	39	103	31	157	14	1	–	1
Oil and gas field services, NC	1,204	349	624	231	1,168	25	4	2	5
Well surveying, well logging, and cementing wells	115	58	37	20	109	3	–	–	3
Miscellaneous oil and gas field services	1,089	291	587	211	1,059	22	4	2	2

[a] Except mine water.
[b] Includes mine water used.

2.2–29 WATER USE OF MINERAL INDUSTRIES (*Continued*)

Industry	Number of mineral establishments reporting water use								
		By principal source of water[a]			By quantity of water intake,[b] mil gal				
	Total	Utility	Other than utility	Source not specified	Under 1	1–9	10–19	20–99	100 and over
Nonmetallic minerals mining	4,019	620	2,546	853	2,789	400	121	303	406
Dimension stone	146	20	69	57	131	10	3	2	–
Dimension limestone	33	2	24	7	27	4	1	1	–
Dimension granite	45	6	23	16	42	2	1	–	–
Dimension stone, NC	68	12	22	34	62	4	1	1	–
Crushed and broken stone	1,440	222	958	260	1,090	163	44	66	77
Crushed and broken limestone	989	160	651	178	757	107	30	42	53
Crushed and broken granite	115	11	98	6	71	20	3	10	11
Crushed and broken stone, NC	336	51	209	76	262	36	11	14	13
Sand and gravel	1,812	274	1,150	388	1,118	195	59	204	236
Construction sand and gravel	1,679	259	1,048	372	1,042	179	55	193	210
Glass sand	38	4	30	4	8	6	3	8	13
Molding sand	53	6	40	7	44	4	1	1	3
Industrial sand, NC	42	5	32	5	24	6	–	2	10
Clay and related minerals	221	38	135	48	173	14	8	7	19
Bentonite	32	11	19	2	26	4	1	1	–
Fire clay	45	6	25	14	42	1	1	–	1
Fuller's earth	11	3	6	2	6	2	–	1	2
Kaolin and ball clay	37	4	28	5	25	4	–	3	5
Feldspar	19	3	15	1	13	–	–	1	5
Clay and related minerals, NC	77	11	42	24	61	3	6	1	6
Chemical and fertilizer minerals	143	27	102	14	51	6	5	17	64
Barite	30	6	19	5	15	2	–	4	9
Fluorspar	14	5	7	2	7	–	1	2	4

[a] Except mine water.
[b] Includes mine water used.

2.2–29 WATER USE OF MINERAL INDUSTRIES (*Continued*)

Industry	Number of mineral establishments reporting water use								
		By principal source of water[a]			By quantity of water intake,[b] mil gal				
	Total	Utility	Other than utility	Source not specified	Under 1	1–9	10–19	20–99	100 and over
Potash, soda, and borate minerals	17	1	15	1	4	–	–	2	11
Phosphate rock	50	6	39	5	16	1	1	5	27
Rock salt	17	5	12	–	7	2	3	3	2
Sulfur	11	4	7	–	1	–	–	–	10
Chemical and ferilizer mining, NC	4	–	3	1	1	1	–	1	1
Nonmetallic minerals services	81	5	39	37	79	1	–	1	–
Miscellaneous minerals, NC	176	34	93	49	147	11	2	6	10
Gypsum	21	8	8	5	20	–	–	1	–
Mica	9	1	5	3	5	2	–	–	2
Native asphalt and bitumens	12	–	8	4	10	–	1	–	1
Pumice and pumicite	17	4	10	3	17	–	–	–	–
Talc, soapstone, and pyrophyllite	46	4	36	6	38	4	1	2	1
Natural abrasives	8	3	4	1	7	–	–	–	1
Peat	26	6	7	13	25	1	–	–	–
Nonmetallic minerals, NC	37	8	15	14	25	4	–	3	5

[a] Except mine water.
[b] Includes mine water used.

Note: A dash represents zero; NC is not elsewhere classified.

Source: U.S. Bureau of the Census, 1963.

2.2–30 WATER USE IN MINERAL INDUSTRIES

U.S., 1954

Industry group	Water intake including mine water, mil gal	Gross water use including recirculation, mil gal	Indicated % water reuse ($\frac{\text{use} - \text{intake}}{\text{intake}}$ = reuse)
Metal mining, total	137,781	240,669	75
Iron ores	26,855	59,666	122
Copper ores	59,112	82,547	39
Lead and zinc ores			
Valued chiefly for lead	11,015	18,212	65
Valued chiefly for zinc	11,134	19,411	74
Gold and silver ores			
Lode gold	4,277	7,049	65
Placer gold	10,752	16,034	49
Silver ores	2,291	2,299	<1
Bauxite	179	179	0
Ferroalloy ores, except vanadium			
Manganese ores	2,834	3,858	36
Tungsten ores	3,380	4,913	46
Molybdenum ores and ferroalloy ores (except vanadium)[a]	1,447	4,788	231
Metal mining contract services	99	100	1
Miscellaneous metal ores			
Mercury ores	12	12	0
Uranium-radium-vanadium ores	1,601	1,601	0
Titanium ores and metallic ores[a]	2,793	20,000	616

[a] Not elsewhere classified.

Source: *Census of Mineral Industries,* U.S. Bureau of the Census, 1954.

2.2–31 WATER USE IN NONMETAL MINING

Mineral	Water intake, mil gal: Except mine water: Total	Water intake, mil gal: Except mine water: Per production worker	Water intake, mil gal: Including mine water: Total	Water intake, mil gal: Including mine water: Per production worker	Water discharge, mil gal, except mine water	Gross water use including recirculation, mil gal	Per cent of total water intake (except mine water) by principal source: Public water systems	Company surface systems	Company ground systems	Combinations and not specified	Number of mines and pits	Number of preparation plants
Construction sand and gravel	122,617	4.31	179,035	6.29	118,206	247,721	3	54	24	19	3,766	3,156
Crushed limestone	34,821	1.40	53,610	2.15	33,913	57,639	3	53	21	23	1,437	1,356
Kaolin and ball clay	3,788	1.34	3,890	1.38	3,732	4,481	–	35	65	–	61	38
Potash, soda, and borate minerals	10,561	2.23[a]	10,561	2.23	6,455	26,332	–	23	74	3	20	16
Phosphate rock	39,720	8.67	62,618	13.68	40,891[b]	138,399	–	21	69	10	73	54
Rock salt[c]	2,968	1.79[a]	2,968	1.79	2,689	3,086	–	2	75	23	15	13
Sulfur	15,969	5.19	15,969	5.19	15,414	16,735	22	20	30	28	16	5
Total nonmetallic minerals (except fuels)	264,732	2.73	369,092	3.80	253,848	586,865	4	43	34	19	9,273	6,415

[a] Mine water used is included.
[b] Exceeds water intake, in part due to accumulation of rain water in storage facilities.
[c] Includes only operations mining rock salt as such; excludes operations recovering salt by evaporation and from pumped underground brines and solution mining.

Source: *Census of Mineral Industries,* U.S. Bureau of the Census, 1954.

2.2.3 WATER NEEDS FOR ELECTRIC POWER GENERATION

2.2–32 WATER USED FOR HYDROELECTRIC POWER

By Regions, 1970

Water Resources Council region	Mgd	1,000 acre feet per year
New England	130,000	150,000
Middle Atlantic	140,000	160,000
South Atlantic-Gulf	230,000	260,000
Great Lakes	270,000	300,000
Ohio	130,000	150,000
Tennessee	220,000	250,000
Upper Mississippi	110,000	120,000
Lower Mississippi	3,000	3,300
Souris-Red-Rainy	2,000	2,200
Missouri	180,000	200,000
Arkansas-White-Red	41,000	46,000
Texas-Gulf	8,900	9,900
Rio Grande	890	1,000
Upper Colorado	12,000	13,000
Lower Colorado	16,000	18,000
Great Basin	3,800	4,200
Columbia-North Pacific	1,200,000	1,300,000
California-South Pacific	85,000	96,000
Alaska	780	870
Hawaii	330	370
Puerto Rico	610	680
U.S.[a]	2,800,000	3,100,000

[a] Including Puerto Rico.

Source: C.R. Murray and E.B. Reeves, *Estimated Use of Water in the United States, 1970,* U.S. Geological Survey, U.S. Department of the Interior, 1972.

2.2–33 WATER USED FOR HYDROELECTRIC POWER

By States, 1970

State	Mgd	1,000 acre feet per year
Alabama	130,000	150,000
Alaska	780	870
Arizona	20,000	23,000
Arkansas	25,000	28,000
California	84,000	94,000
Colorado	4,000	4,500
Connecticut	6,600	7,400
Delaware	0	0
Florida	11,000	13,000
Georgia	45,000	50,000
Hawaii	330	370
Idaho	84,000	94,000
Illinois	10,000	12,000
Indiana	18,000	21,000
Iowa	35,000	39,000
Kansas	520	590
Kentucky	79,000	88,000
Louisiana	0	0
Maine	80,000	90,000
Maryland	23,000	25,000
Massachusetts	17,000	19,000
Michigan	59,000	66,000
Minnesota	18,000	20,000
Mississippi	0	0
Missouri	9,000	10,000
Montana	72,000	80,000
Nebraska	29,000	32,000
Nevada	4,200	4,800
New Hampshire	22,000	24,000
New Jersey	120	130
New Mexico	430	480
New York	270,000	300,000
North Carolina	88,000	98,000
North Dakota	14,000	15,000
Ohio	370	410
Oklahoma	20,000	22,000
Oregon	350,000	390,000
Pennsylvania	30,000	34,000
Rhode Island	58	65
South Carolina	41,000	46,000
South Dakota	59,000	66,000
Tennessee	130,000	150,000
Texas	9,300	10,000
Utah	3,000	3,300
Vermont	15,000	16,000
Virginia	16,000	18,000
Washington	720,000	810,000
West Virginia	25,000	28,000
Wisconsin	66,000	74,000
Wyoming	6,500	7,200
District of Columbia	6.0	6.7
Puerto Rico	610	680
U.S.[a]	2,800,000	3,100,000

[a] Including Puerto Rico.

Source: C.R. Murray and E.B. Reeves, *Estimated Use of Water in the United States, 1970,* U.S. Geological Survey, U.S. Department of the Interior, 1972.

2.2–34 WATER USED FOR ELECTRIC UTILITY GENERATION OF THERMOELECTRIC POWER

By Regions, 1970

Water Resources Council region	Condenser cooling, mgd					Other uses, mgd					Water consumed, mgd	
	Self supplied					Self supplied						
	Fresh ground water	Surface water		Public supplies	Self supplied and public supplies	Fresh ground water	Surface water		Public supplies	Self supplied and public supplies		
		Fresh	Saline				Fresh	Saline			Fresh	Saline
New England	0	1,900	4,600	0	6,500	1.0	35	0	6.5	42	3.0	1.0
Middle Atlantic	0.6	15,000	16,000	25	31,000	100	460	3.0	27	590	35	17
South Atlantic-Gulf	24	15,000	10,000	2.6	25,000	7.6	200	5.1	12	220	120	88
Great Lakes	0	25,000	0	120	25,000	38	480	0	9.2	530	14	0
Ohio	48	26,000	0	200	26,000	5.7	960	0	13	980	50	0
Tennessee	0	5,700	0	0	5,700	0	400	0	0	400	64	0
Upper Mississippi	280	12,000	0	8.7	12,000	6.0	220	0	2.2	230	23	0
Lower Mississippi	62	3,900	140	2.0	4,100	3.8	160	7.2	2.0	180	190	25
Souris-Red-Rainy	0.6	140	0	0	140	0	0.2	0	0	0.2	0.9	0
Missouri	310	3,000	0.9	85	3,400	2.6	8.3	0	1.1	12	34	0.8
Arkansas-White-Red	45	1,900	0	7.8	1,900	1.0	15	0	0.5	16	82	0
Texas-Gulf	49	4,700	3,800	3.7	8,500	1.7	1.8	0.3	0.1	3.9	100	24
Rio Grande	15	6.0	0	0.1	21	0.2	0	0	0	0.2	17	0
Upper Colorado	0	100	0	0	100	0	2.2	0	0	2.2	22	0
Lower Colorado	44	3.3	0	0	47	0	0.1	0	0	0.1	36	0
Great Basin	3.8	130	0	0.5	140	0	0	0	0.4	0.4	6.2	0
Columbia-North Pacific	0	26	0	0	26	0	0.3	0	0	0.3	0.1	0
California-South Pacific	300	1,200	8,200	130	9,900	0	1.0	0	10	11	24	66
Alaska	0.9	2.5	1.0	0	4.4	0.5	66	0	0	66	0	0
Hawaii	82	46	860	0	980	0	0	0	0	0	0	0
Puerto Rico	0.2	0	2,100	0.5	2,100	0.2	0	0	1.3	1.5	1.5	0
U.S.[a]	1,300	120,000	46,000	590	170,000	170	3,000	16	85	3,300	820	220

[a] Including Puerto Rico.

Note: Partial figures may not add to totals because of independent rounding.

Source: C.R. Murray and E.B. Reeves, *Estimated Use of Water in the United States, 1970,* U.S. Geological Survey, U.S. Department of the Interior, 1972.

2.2–35 WATER USED FOR ELECTRIC UTILITY GENERATION OF THERMOELECTRIC POWER

By States, 1970

State	Condenser cooling, mgd					Other uses, mgd					Water consumed, mgd	
	Self supplied					Self supplied						
	Fresh ground water	Surface water		Public supplies	Self supplied and public supplies	Fresh ground water	Surface water		Public supplies	Self supplied and public supplies		
		Fresh	Saline				Fresh	Saline			Fresh	Saline
Alabama	0	4,600	190	0	4,800	2.6	250	4.0	0	260	25	0.4
Alaska	0.9	2.5	1.0	0	4.4	0.5	66	0	0	66	0	0
Arizona	40	2.0	0	0	42	0	0	0	0	0	32	0
Arkansas	3.0	950	0	0	950	1.0	1.0	0	0	2.0	3.0	0
California	300	1,200	8,200	130	9,900	0	1.0	0	10	11	24	66
Colorado	30	83	0	0	110	0	0.1	0	0.1	0.2	8.0	0
Connecticut	0	1,000	1,800	0	2,900	0	5.0	0	3.0	8.0	2.0	0
Delaware	0.6	1.4	720	2.0	730	0.1	0	0	0.3	0.4	0	0
Florida	12	1,700	9,300	2.6	11,000	1.0	0	0	2.5	3.5	20	86
Georgia	4.4	3,900	140	0	4,000	2.3	28	1.1	9.5	40	39	1.1
Hawaii	82	46	860	0	980	0	0	0	0	0	0	0
Idaho	0	0	0	0	0	0	0	0	0	0	0	0
Illinois	0	11,000	0	1.0	11,000	7.0	320	0	3.0	320	5.0	0
Indiana	1.0	4,700	0	26	4,700	0	110	0	1.0	110	5.0	0
Iowa	0	1,300	0	7.7	1,400	0	42	0	0.2	42	20	0
Kansas	38	220	0	0	260	0	0	0	0	0	32	0
Kentucky	1.0	3,600	0	0	3,600	0.9	230	0	0.7	240	21	0
Louisiana	34	2,600	140	0	2,700	1.8	140	7.2	0	140	170	25
Maine	0	20	180	0	200	1.0	0	0	1.0	2.0	0	0
Maryland	0	500	2,700	0	3,200	0	12	3.0	1.0	16	0	0
Massachusetts	0	550	2,100	0	2,700	0	29	0	2.5	32	1.0	1.0
Michigan	0	9,700	0	0	9,700	0	49	0	0	49	0	0
Minnesota	280	1,400	0	0	1,700	0	0.2	0	0	0.2	0.2	0
Mississippi	34	400	490	0	920	1.0	12	0	0	13	35	0
Missouri	4.0	2,400	0	2.0	2,500	2.0	5.0	0	3.0	10	13	0
Montana	0	60	0	0	60	0	0	0	0	0	0	0
Nebraska	270	620	0	84	970	0	0	0	0	0	8.4	0
Nevada	7.1	49	0	0	56	0	0.1	0	0	0.1	7.5	0

2.2–35 WATER USED FOR ELECTRIC UTILITY GENERATION OF THERMOELECTRIC POWER (*Continued*)

	Condenser cooling, mgd					Other uses, mgd						
	Self supplied					Self supplied					Water consumed	
		Surface water					Surface water					
State	Fresh ground water	Fresh	Saline	Public supplies	Self supplied and public supplies	Fresh ground water	Fresh	Saline	Public supplies	Self supplied and public supplies	Fresh	Saline
New Hampshire	0	250	160	0	410	0	0	0	0	0	0	0
New Jersey	0	1,000	3,200	0	4,200	8.0	0	0	18	26	26	0
New Mexico	8.6	19	0	0.1	28	0	0.9	0	0	0.9	27	0
New York	0	4,400	8,500	22	13,000	130	260	0	3.9	390	9.6	17
North Carolina	0	4,300	170	0	4,500	1.0	41	0	0	42	30	0
North Dakota	1.0	350	0.9	0	350	0	0	0	0	0	1.2	0.8
Ohio	46	13,000	0	300	13,000	3.0	590	0	16	610	14	0
Oklahoma	2.8	180	0	7.8	190	1.0	0.1	0	0.5	1.6	28	0
Oregon	0	22	0	0	22	0	0	0	0	0	0.1	0
Pennsylvania	0	12,000	0	1.9	12,000	0	270	0	7.0	280	8.9	0
Rhode Island	0	0	310	0	310	0	0	0	0	0	0	0
South Carolina	0	2,500	68	0	2,600	0.7	25	0	0	26	14	0
South Dakota	0.7	2.7	0	0.2	3.6	0	0	0	0	0	0.6	0
Tennessee	0	4,500	0	0	4,500	0	340	0	0	340	62	0
Texas	58	5,800	3,800	3.7	9,600	1.9	2.1	0.3	0.1	4.4	120	24
Utah	0	87	0	0.5	87	0	0.2	0	0.4	0.6	3.8	0
Vermont	0	0	0	0	0	1.0	4.0	0	0	5.0	0	0
Virginia	0	3,100	770	0	3,900	0	0	0	0	0	0.8	0
Washington	0	4.0	0	0	4.0	0	0.3	0	0	0.3	0	0
West Virginia	0	4,800	0	0	4,800	0	120	0	0	120	1.1	0
Wisconsin	0	5,300	0	0	5,300	0	0	0	0	0	0	0
Wyoming	0.4	200	0	0	200	0.6	1.0	0	0	1.6	5.3	0
District of Columbia	0	1,000	0	0	1,000	0	60	0	0	60	0	0
Puerto Rico	0.2	0	2,100	0.5	2,100	0.2	0	0	1.3	1.5	0.4	0
U.S.[a]	1,300	120,000	46,000	590	170,000	170	3,000	16	85	3,300	820	220

[a] Including Puerto Rico.
Note: Partial figures may not add to totals because of independent rounding.
Source: C.R. Murray and E.B. Reeves, *Estimated Use of Water in the United States, 1970,* U.S. Geological Survey, U.S. Department of the Interior, 1972.

2.2–36 INDUSTRIAL AND INVESTOR OWNED THERMAL ELECTRIC PLANT WATER INTAKE, REUSE, AND CONSUMPTION

1964

	Water intake, bgy							
	Purpose							
Industrial group	Cooling and condensing	Boiler feed sanitary service, etc	Process	Total	Water recycled, bgy	Gross water use, including recycling, bgy	Water consumed, bgy	Water discharged, bgy
Food and kindred products	392	104	264	760	520	1,280	72	688
Textile mill products	24	17	106	147	163	310	13	134
Lumber and wood products	71	24	56	151	66	217	28	123
Paper and allied products	607	120	1,344	2,071	3,945	6,016	129	1,942
Chemicals and allied products	3,120	202	564	3,886	3,688	7,574	227	3,659
Petroleum and coal products	1,212	99	88	1,399	4,763	6,162	81	1,318
Leather and leather products	1	1	14	16	2	18	1	15
Primary metal industry	3,387	195	996	4,578	2,200	6,778	266	4,312
Subtotal	8,814	762	3,432	13,008	15,347	28,355	817	12,191
Other industries	571	197	271	1,039	1,207	2,246	71	968
Total industry	9,385	959	3,703	14,047	16,554	30,601	888	13,159
Thermal electric plants	34,849	a	–	34,849	5,815	40,664	68	34,781
Total	44,234	959[b]	3,703	48,896	22,369	71,265	956	44,940

[a] Boiler-feed water use by thermal electric plants estimated to be equivalent to industrial sanitary service, etc. water use.
[b] Total boiler feed water.

REFERENCES

1. *Census of Manufacturers, Water Use in Manufacturing, 1963,* U.S. Department of Commerce, 1966.
2. *Water Resources Activities in the United States,* U.S. Senate Committee on National Water Resources, 1960.

2.2–37 PROJECTED FRESHWATER REQUIREMENTS FOR STEAM ELECTRIC POWER GENERATION

1965–2020

	mgd			
	1965	**1980**	**2000**	**2020**
Condenser requirement	84,800	216,400	477,700	806,400
Withdrawn[a]	68,500	141,500	267,200	419,600
Consumptive use	758	1,760	4,645	8,110

[a] Deficit between requirement and withdrawn made up by recirculation.

Source: *The Nation's Water Resources,* U.S. Water Resources Council, 1968.

2.2–38 WATER REUSE–RECYCLE TRENDS

Industry	**Year**	**Gross water use, bil gal/yr**	**Water intake, bil gal/yr**	**Reuse ratio**
Chemicals	1954	4,290	2,690	1.59
and	1959	5,230	3,240	1.61
allied	1964	7,670	3,900	1.96
products	1968	9,460	4,510	2.10
Petroleum	1954	4,150	1,250	3.22
and	1959	5,780	1,320	4.38
coal	1964	6,160	1,400	4.40
products	1968	7,220	1,370	5.25

Source: 1954, 1958, 1963, 1967 Census of Manufacturers.

2.2.4 INDUSTRIAL WATER REUSE AND RECYCLING

2.2–39 RECYCLING OF INDUSTRIAL WATER

Source	Recycle ratios[a]
Auto	2.62
Beet sugar	1.48
Coal preparation	14.91
Corn and wheat milling	1.22
Distilling	1.51
Food processing	1.19
Machinery	5.5
Meat	4.03
Petroleum	7.62
Pulp and paper	3.02
Soaps and detergents	3.08
Steel	1.60
Tanning – leather	1.04
Textiles	1.3
Natural gas transmission	2.32

[a] Recycle ratios

$$= \frac{\text{freshwater intake required without recycle}}{\text{actual freshwater intake using recycle}}$$

or:

= average number of times a gallon of water was recycled.

Source: *Water in Industry,* National Association of Manufacturers and the Chamber of Commerce of the United States, January 1965.

2.2–40 WATER REUSE IN MANUFACTURING

Selected Regions, 1964

Industry description	Total U.S.	Chesapeake Bay Region	Ohio River Region	Eastern Great Lakes Region	Southeast Region	Arkansas White Red Region	Western Gulf Region	California Region
All manufacturers	2.2	2.1	1.6	1.6	2.7	8.6	2.9	3.7
Paper and allied products	2.9	6.0	3.3	1.9	2.8	5.8	–	4.5
Chemicals and allied products	2.0	1.3	1.7	1.3	2.2	11.7	1.9	3.6
Petroleum and coal products	4.4	–	7.4	3.3	12.6	19.6	6.5	4.1
Primary metals	1.5	1.3	1.3	1.4	2.8	–	2.2	11.1

Note: Recirculation rate = gross water used/water intake.

Source: Reprinted, with publisher's permission, from *J. Am. Water Works Assn., 60*(10):1129, © 1968, by the American Water Works Association, Inc., 2 Park Ave., New York, N.Y. 10016.

2.2–41 RESULTS OF WATER CONSERVATION IN INDUSTRIES

	Quantity of water used	
Industry	**Before conservation, gal/ton product**	**After conservation, gal/ton product**
Steel	55,000	1,500–3,000
Paper	7,000–10,000	3,000–4,000
Kraft pulp	45,000	9,000
Newsprint	26,000	6,000
Cannery	2,500	1,200

Source: Reprinted, with publisher's permission, from *J. Am. Water Works Assn.*, *53*(2):119, © 1961, by the American Water Works Association, Inc., 2 Park Ave., New York, N.Y. 10016.

2.2–42 RECIRCULATION OF INDUSTRIAL WATER

Texas, 1964

Standard industrial classification group	**Industries in group reporting recirculation, %**	**Total water requirements served by recirculation, %**
Natural gas	17.0	98.6
Food	10.8	65.8
Printing and publishing	10.8	29.1
Rubber and plastic	10.7	93.3
Lumber and wood	10.7	82.7
Paper	9.6	42.2
Chemicals	8.0	55.2
Petroleum and coal	7.0	92.1

Source: R.G. Fleming and H.D. Jobes, *J. Water Pollut. Control Fed.*, *41*(9):1568, 1969.

2.3 *IRRIGATION AND FARM WATER NEEDS*

2.3–1 RURAL DOMESTIC WATER USE

U.S., 1965

State	With running water, gal/cap day	Without running water, gal/cap day
Arizona	100	10
California	80	20
Connecticut	70	10
Delaware	60	10
Florida	100	10
Hawaii	75	15
Idaho	100	20
Indiana	60	10
Kansas	75	15
Kentucky	55	5
Nebraska	60	10
Nevada	100	10
New Hampshire	60	12
New Jersey	60	10
New York	60	10
North Carolina	100	10
Oregon	75	10
Puerto Rico	25	10
South Dakota	55	25
Washington	75	10
Wisconsin	70	10
All other states	50	10

Source: U.S. Geological Survey, 1965.

2.3–2 WATER FOR RURAL USE

By Regions, 1970

Water Resources Council region	Domestic use, mgd				Livestock use, mgd				Domestic and livestock uses, mgd			
	Withdrawn				Withdrawn				Withdrawn			
	Ground water	Surface water	All water	Water consumed	Ground water	Surface water	All water	Water consumed	Ground water	Surface water	All water	Water consumed
New England	94	2.1	96	47	7.0	4.7	12	12	100	6.8	110	59
Middle Atlantic	340	1.5	340	130	46	33	79	65	380	34	420	190
South Atlantic-Gulf	460	1.9	470	360	110	51	160	150	580	53	630	510
Great Lakes	270	7.1	280	78	62	24	86	82	340	31	370	160
Ohio	240	33	270	180	58	84	140	140	290	120	410	310
Tennessee	51	0.9	52	31	11	20	31	30	62	21	83	62
Upper Mississippi	200	5.7	210	130	200	60	260	250	400	65	460	380
Lower Mississippi	110	0.4	110	100	22	33	55	55	130	34	170	160
Souris-Red-Rainy	19	0.1	19	19	12	2.4	15	15	31	2.5	34	34
Missouri	110	11	120	96	270	170	440	410	380	180	560	500
Arkansas-White-Red	88	5.7	94	84	66	120	180	180	150	120	280	260
Texas-Gulf	80	0	80	80	71	42	110	110	150	42	190	190
Rio Grande	20	0.7	20	13	17	20	37	36	37	20	57	49
Upper Colorado	6.2	1.2	7.4	2.9	6.5	12	19	17	13	14	26	20
Lower Colorado	24	0	24	17	18	10	29	28	43	10	53	45
Great Basin	27	0.6	27	13	35	6.2	41	21	61	6.8	68	34
Columbia-North Pacific	220	29	250	200	18	34	52	47	240	63	300	250
California-South Pacific	120	8.6	130	73	38	54	93	50	160	63	220	120
Alaska	4.5	1.6	6.1	0.2	0	0.1	0.1	0.1	4.5	1.7	6.2	0.3
Hawaii	0.1	0.4	0.5	0.4	1.4	5.9	7.3	6.6	1.5	6.3	7.8	7.0
Puerto Rico	0.4	3.2	3.6	3.0	1.3	7.5	8.8	8.0	1.7	11	12	11
U.S.[a]	2,500	110	2,600	1,700	1,100	790	1,900	1,700	3,600	910	4,500	3,400

[a] Including Puerto Rico.

Note: Partial figures may not add to totals because of independent rounding.

Source: C.R. Murray and E.B. Reeves, *Estimated Use of Water in the United States, 1970,* U.S. Geological Survey, U.S. Department of the Interior, 1972.

2.3–3 WATER FOR RURAL USE

By States, 1970

State	Domestic use, mgd				Livestock use, mgd				Domestic and livestock uses, mgd			
	Withdrawn				Withdrawn				Withdrawn			
	Ground water	Surface water	All water	Water consumed	Ground water	Surface water	All water	Water consumed	Ground water	Surface water	All water	Water consumed
Alabama	63	0	63	63	13	15	27	27	76	15	90	90
Alaska	4.5	1.6	6.1	0.2	0	0.1	0.1	0.1	4.5	1.7	6.2	0.4
Arizona	22	0	22	16	20	8.4	28	28	42	8.4	50	44
Arkansas	49	0	49	49	16	19	34	34	64	19	83	83
California	120	8.6	120	71	38	53	91	49	150	62	220	120
Colorado	9.3	1.4	11	2.1	20	15	35	31	29	16	45	33
Connecticut	38	0.8	39	39	2.1	0.4	2.5	2.5	40	1.2	42	42
Delaware	11	0	11	1.2	1.6	0.1	1.7	1.4	13	0.1	13	2.6
Florida	160	0	160	130	18	12	30	30	180	12	200	160
Georgia	72	1.3	73	8.8	31	0.1	31	31	100	1.4	100	40
Hawaii	0.1	0.4	0.5	0.4	1.4	5.9	7.3	6.6	1.5	6.3	7.8	7.0
Idaho	22	2.5	24	6.1	10	12	22	19	32	14	46	25
Illinois	14	3.2	17	12	32	10	42	42	46	13	60	54
Indiana	76	11	87	61	29	17	46	46	100	28	130	110
Iowa	47	0.1	47	19	110	25	130	130	160	25	180	150
Kansas	48	3.7	52	49	31	47	79	77	79	51	130	130
Kentucky	48	6.4	55	44	3.7	36	40	40	52	42	94	84
Louisiana	67	0	67	67	11	11	22	22	78	11	89	89
Maine	11	1.1	12	3.3	1.1	1.7	2.8	2.6	12	2.8	14	5.9
Maryland	46	0	46	46	10	0.5	11	11	57	0.5	57	57
Massachusetts	28	0	28	2.7	1.3	0.8	2.1	2.3	30	0.8	30	5.0
Michigan	160	0	160	26	24	6.8	31	28	180	6.8	190	54
Minnesota	110	0	110	110	59	9	68	68	170	9.0	170	170
Mississippi	25	0	25	22	16	24	39	39	40	24	64	61
Missouri	29	10	39	18	28	86	110	100	58	96	150	120
Montana	8.8	0.7	9.5	9.5	17	16	33	33	26	17	43	43
Nebraska	22	0	22	22	80	20	100	95	100	20	120	120

2.3–3 WATER FOR RURAL USE (*Continued*)

State	Domestic use, mgd				Livestock use, mgd				Domestic and livestock uses, mgd			
	Withdrawn				Withdrawn				Withdrawn			
	Ground water	Surface water	All water	Water consumed	Ground water	Surface water	All water	Water consumed	Ground water	Surface water	All water	Water consumed
Nevada	5.8	0.2	6.0	2.3	1.6	2.8	4.4	2.4	7.4	3.0	10	4.7
New Hampshire	11	0.2	11	1.1	0.5	0.8	1.3	1.3	12	1.0	13	2.4
New Jersey	80	0	80	40	1.5	0.9	2.4	2.1	81	0.9	82	42
New Mexico	16	0.6	17	8.1	12	32	44	43	29	33	61	51
New York	120	0	120	12	24	13	38	34	140	13	150	45
North Carolina	110	1.1	110	110	43	6.3	50	39	160	7.4	160	150
North Dakota	17	0.1	17	17	9.6	6.2	16	16	26	6.3	32	32
Ohio	88	22	110	100	24	16	40	39	110	38	150	140
Oklahoma	24	4.1	28	25	6.7	46	52	52	31	50	80	78
Oregon	160	14	170	170	2.6	19	22	22	160	33	190	190
Pennsylvania	110	0	110	11	14	14	28	18	120	14	140	29
Rhode Island	4.6	0	4.6	0.7	0.1	0.1	0.2	0.3	4.7	0.1	4.8	1.0
South Carolina	46	0	46	46	4.0	4.9	8.9	9.0	50	4.9	55	55
South Dakota	15	1.0	16	11	81	27	109	93	96	28	120	100
Tennessee	39	0	39	9.9	5.4	29	34	34	45	29	73	44
Texas	95	0	95	95	96	52	150	150	190	52	240	240
Utah	23	0.2	23	11	34	3.2	37	19	57	3.4	60	30
Vermont	10	0.4	11	1.1	5.6	2.7	8.3	8.4	16	3.1	19	9.5
Virginia	73	1.6	74	45	12	17	29	23	84	19	100	68
Washington	43	12	55	19	4.2	2.1	6.3	4.4	48	14	62	24
West Virginia	17	0.6	18	0.2	0.9	6.0	6.9	6.0	18	6.6	25	6.2
Wisconsin	74	0	74	7.3	56	15	71	71	130	15	140	78
Wyoming	5.7	0.7	6.4	5.7	3.8	15	19	18	9.5	16	25	24
District of Columbia	0	0	0	0	0	0	0	0	0	0	0	0
Puerto Rico	0.4	3.2	3.6	3.0	1.3	7.5	8.8	8.0	1.7	11	12	11
U.S.[a]	2,500	110	2,600	1,700	1,100	790	1,900	1,700	3,600	910	4,500	3,400

[a] Including Puerto Rico.

Note: Partial figures may not add to totals because of independent rounding.

Source: C.R. Murray and E.B. Reeves, *Estimated Use of Water in the United States, 1970,* U.S. Geological Survey, U.S. Department of the Interior, 1972.

2.3–4 WATER USED FOR IRRIGATION

By Regions, 1970

Water Resources Council region	Acres irrigated, 1,000 acres	Total water withdrawn, 1,000 acre feet per year: Ground water	Surface water	Re-claimed sewage	All water	Water con-sumed,[a] 1,000 acre ft/yr	Convey-ance loss, 1,000 acre ft/yr	Total water withdrawn, mgd: Ground water	Surface water	Re-claimed sewage	All water	Water consumed,[a] mgd	Convey-ance loss, mgd
New England	77	22	69	0	91	72	0	20	60	0	80	64	0
Middle Atlantic	310	87	57	.2	140	140	0	77	50	0	130	120	0
South Atlantic-Gulf	2,400	1,500	1,200	0	2,700	1,700	160	1,300	1,100	0	2,400	1,500	150
Great Lakes	160	42	60	0	100	99	0.2	37	53	0	90	87	0.1
Ohio	76	10	32	0	42	41	0	8.1	27	0	35	35	0
Tennessee	22	2.7	5.6	0	8.3	8.2	0	2.1	4.5	0	6.6	6.6	0
Upper Mississippi	190	79	41	0	120	110	1.2	69	35	0	100	95	0.6
Lower Mississippi	2,000	2,300	1,300	0	3,600	2,500	620	2,000	1,200	0	3,200	2,200	550
Souris-Red-Rainy	13	9.6	5.0	0	15	13	1.2	8.5	4.2	0	13	12	1.0
Missouri	9,700	5,000	16,000	99	21,000	13,000	5,400	4,500	14,000	88	19,000	12,000	4,800
Arkansas-White-Red	5,400	6,600	2,600	7.0	9,200	6,700	350	5,900	2,300	6.2	8,200	6,000	310
Texas-Gulf	5,000	5,600	1,300	16	6,900	5,500	200	5,000	1,100	14	6,200	4,900	180
Rio Grande	2,500	2,300	4,000	20	6,300	3,400	660	2,000	3,500	17	5,600	3,000	590
Upper Colorado	2,000 ,	60	8,700	0	8,800	4,500	1,200	53	7,800	0	7,900	4,000	1,100
Lower Colorado	1,300	4,400	2,900	5.9	7,300	5,200	340	3,900	2,600	5.1	6,500	4,700	300
Great Basin	1,900	850	5,700	59	6,600	3,300	2,200	760	5,100	53	5,900	2,900	2,000
Columbia-North Pacific	7,100	3,400	27,000	6.4	30,000	11,000	8,000	3,000	24,000	5.6	27,000	10,000	7,200
California-South Pacific	9,000	18,000	20,000	140	38,000	23,000	5,700	16,000	18,000	120	34,000	21,000	5,100

[a] Excluding conveyance losses by evapotranspiration.

2.3–4 WATER USED FOR IRRIGATION (*Continued*)

Water Resources Council region	Acres irrigated, 1,000 acres	Total water withdrawn, 1,000 acre feet per year				Water consumed,[a] 1,000 acre ft/yr	Conveyance loss, 1,000 acre ft/yr	Total water withdrawn, mgd				Water consumed,[a] mgd	Conveyance loss, mgd
		Ground water	Surface water	Reclaimed sewage	All water			Ground water	Surface water	Reclaimed sewage	All water		
Alaska	2.2	0.5	0.5	0	1.0	0.9	0	0.4	0.4	0	0.9	0.8	0
Hawaii	160	610	760	64	1,400	840	250	550	680	57	1,300	750	220
Puerto Rico	91	75	82	0	160	110	47	67	73	0	140	98	42
U.S.[b]	50,000	51,000	91,000	410	140,000	82,000	25,000	45,000	81,000	370	130,000	73,000	22,000

[a] Excluding conveyance losses by evapotranspiration.
[b] Including Puerto Rico.

Note: Partial figures may not add to totals because of independent rounding.

Source: C.R. Murray and E.B. Reeves, *Estimated Use of Water in the United States, 1970,* U.S. Geological Survey, U.S. Department of the Interior, 1972.

2.3–5 WATER USED FOR IRRIGATION

By States, 1970

State	Acres irrigated, 1,000 acres	Total water withdrawn, 1,000 acre feet per year				Water consumed,[a] 1,000 acre ft/yr	Conveyance loss, 1,000 acre ft/yr	Total water withdrawn, mgd				Water consumed,[a] mgd	Conveyance loss, mgd
		Ground water	Surface water	Reclaimed sewage	All water			Ground water	Surface water	Reclaimed sewage	All water		
Alabama	27	6.0	14	0	20	20	0	5.4	12	0	18	18	0
Alaska	2.2	0.5	0.5	0	1.0	0.9	0	0.4	0.4	0	0.9	0.8	0
Arizona	1,200	4,300	2,700	0	7,000	5,000	270	3,800	2,400	0	6,300	4,500	240
Arkansas	1,100	1,200	260	0	1,400	1,000	100	1,100	230	0	1,300	890	90
California	8,700	18,000	19,000	140	37,000	23,000	5,500	16,000	17,000	120	33,000	20,000	4,900
Colorado	4,600	2,100	12,000	90	14,000	7,400	1,600	1,900	11,000	80	13,000	6,600	1,400
Connecticut	14	0.7	6.2	0	6.9	6.9	0	0.5	5.4	0	5.9	5.9	0
Delaware	17	2.6	0.7	0	3.3	3.3	0	2.2	0.5	0	2.7	2.7	0
Florida	1,700	1,400	1,100	0	2,500	1,500	160	1,300	970	0	2,200	1,300	150
Georgia	150	7.6	45	0	53	53	0	6.6	40	0	47	47	0
Hawaii	160	610	760	64	1,400	840	250	550	680	57	1,300	750	220
Idaho	3,700	2,300	15,000	2.8	17,000	5,200	4,800	2,100	13,000	2.5	15,000	4,700	4,300
Illinois	36	17	7.2	0	24	24	0	15	6.0	0	21	21	0
Indiana	34	20	8.9	0	29	29	0	18	7.4	0	25	25	0
Iowa	54	26	3.8	0	30	30	0	23	3.1	0	26	26	0
Kansas	1,800	3,100	230	0	3,300	2,600	48	2,800	200	0	3,000	2,300	43
Kentucky	25	0.4	7.8	0	8.2	8.2	0	0.4	6.7	0	7.1	6.8	0
Louisiana	670	870	880	0	1,700	1,300	480	770	780	0	1,600	1,100	430
Maine	22	0.2	9.9	0	10	10	0	0.2	8.7	0	8.9	8.8	0
Maryland	16	2.5	4.9	0.2	7.6	7.5	0	2.1	4.3	0.2	6.6	6.6	0
Massachusetts	34	21	44	0	65	49	0	18	40	0	58	43	0
Michigan	100	25	41	0	66	66	0	22	36	0	58	58	0
Minnesota	50	14	9.5	0	23	23	0	12	8.0	0	20	20	0
Mississippi	200	240	180	0	420	210	42	220	160	0	370	190	37
Missouri	180	79	8.2	0	87	62	6.7	70	6.9	0	77	55	5.5

[a] Excluding conveyance losses by evapotranspiration.

2.3–5 WATER USED FOR IRRIGATION (*Continued*)

State	Acres irrigated, 1,000 acres	Total water withdrawn, 1,000 acre feet per year				Water consumed,[a] 1,000 acre ft/yr	Conveyance loss, 1,000 acre ft/yr	Total water withdrawn, mgd				Water consumed,[a] mgd	Conveyance loss, mgd
		Ground water	Surface water	Reclaimed sewage	All water			Ground water	Surface water	Reclaimed sewage	All water		
Montana	2,200	71	8,500	0.1	8,600	6,000	2,500	63	7,600	0.1	7,600	5,400	2,200
Nebraska	4,100	3,000	2,300	0	5,300	3,900	1,400	2,700	2,100	0	4,700	3,500	1,300
Nevada	830	430	2,900	4.8	3,400	1,600	1,800	380	2,600	4.2	3,000	1,400	1,600
New Hampshire	3.3	0	3.3	0	3.3	2.5	0	0	2.8	0	2.8	2.0	0
New Jersey	170	63	22	0	85	78	0	56	20	0	76	70	0
New Mexico	1,100	1,500	1,700	25	3,200	1,500	170	1,300	1,500	22	2,800	1,300	160
New York	75	16	15	0	31	31	0	14	13	0	27	27	0
North Carolina	470	56	36	0	92	92	0	50	32	0	82	82	0
North Dakota	74	30	190	0	210	150	66	26	170	0	190	130	59
Ohio	32	11	25	0	35	32	0	9.0	22	0	31	28	0
Oklahoma	620	810	110	0	920	640	22	720	99	0	820	570	20
Oregon	1,900	710	4,700	3.6	5,400	2,600	1,700	630	4,200	3.1	4,800	2,300	1,500
Pennsylvania	35	0.9	11	0	12	12	0	0.8	9.4	0	10	10	0
Rhode Island	3.8	0.5	4.7	0	5.2	3.9	0	0.4	4.1	0	4.5	3.4	0
South Carolina	42	10	22	0	32	32	0	8.9	20	0	29	29	0
South Dakota	150	35	230	0	260	147	88	31	200	0	230	130	79
Tennessee	9.3	1.8	3.6	0	5.4	4.6	0	1.3	2.9	0	4.2	3.8	0
Texas	8,300	8,800	2,800	17	12,000	9,100	540	7,800	2,500	15	10,000	8,100	480
Utah	1,300	470	3,500	58	4,100	2,200	710	420	3,200	52	3,600	2,000	630
Vermont	0.3	0	0.1	0	0.1	0.1	0	0	0.1	0	0.1	0.1	0
Virginia	80	6.1	34	0	40	38	0	5.2	30	0	35	34	0
Washington	1,400	390	5,900	0	6,300	2,500	1,200	350	5,300	0	5,600	2,200	1,000
West Virginia	2.6	0	1.6	0	1.6	1.6	0	0	1.3	0	1.3	1.3	0
Wisconsin	100	37	22	0	60	45	1.2	33	20	0	52	40	0.7
Wyoming	1,700	140	5,900	8.7	6,000	2,600	1,700	130	5,200	7.7	5,400	2,300	1,500

[a] Excluding conveyance losses by evapotranspiration.

2.3–5 WATER USED FOR IRRIGATION (*Continued*)

State	Acres irrigated, 1,000 acres	Total water withdrawn, 1,000 acre feet per year				Water consumed,[a] 1,000 acre ft/yr	Conveyance loss, 1,000 acre ft/yr	Total water withdrawn, mgd				Water consumed,[a] mgd	Conveyance loss, mgd
		Ground water	Surface water	Reclaimed sewage	All water			Ground water	Surface water	Reclaimed sewage	All water		
District of Columbia	0	0	0	0	0	0	0	0	0	0	0	0	0
Puerto Rico	91	75	82	0	160	110	47	67	73	0	140	98	42
U.S.[b]	50,000	51,000	91,000	410	140,000	82,000	25,000	45,000	81,000	370	130,000	73,000	22,000

[a] Excluding conveyance losses by evapotranspiration.
[b] Including Puerto Rico.

Note: Partial figures may not add to totals because of independent rounding.

Source: C.R. Murray and E.B. Reeves, *Estimated Use of Water in the United States, 1970,* U.S. Geological Survey, U.S. Department of the Interior, 1972.

2.3–6 IRRIGATION WITHDRAWALS FOR SEVENTEEN WESTERN STATES

1955

State	Withdrawals for irrigation, 1,000 acre ft
Arizona	7,745
California	25,809
Colorado	7,065
Idaho	16,926
Montana	10,936
Nevada	2,149
New Mexico	2,818
Oregon	7,615
Utah	4,674
Washington	5,638
Wyoming	12,366
Kansas	829
Nebraska	2,858
North Dakota	136
Oklahoma	252
South Dakota	31
Texas	11,466

Source: K.A. MacKichan, *Estimated Use of Water in the United States, 1955,* U.S. Geological Survey, U.S. Department of the Interior, 1957.

2.3–7 PROJECTIONS OF IRRIGATED LAND IN THE U.S.

Region	1960	1965	1980	2000	2020
North Atlantic	240	310	380	550	700
South Atlantic-Gulf	850	1,500	1,800	2,750	3,750
Great Lakes	100	140	230	350	470
Ohio	35	55	90	180	260
Tennessee	15	20	30	40	50
Upper Mississippi	80	140	210	390	550
Lower Mississippi	700	900	2,100	3,050	4,150
Souris-Red-Rainy	10	15	90	240	250
Missouri	6,600	7,400	8,050	9,000	9,600
Arkansas-White-Red	3,100	3,800	5,600	6,400	6,850
Texas-Gulf	5,100	5,500	5,500	5,500	5,500
Rio Grande	1,950	2,000	2,050	2,050	2,050
Upper Colorado	1,370	1,440	1,800	2,000	2,000
Lower Colorado	1,520	1,660	1,750	1,800	1,800
Great Basin	1,700	1,860	1,950	2,000	2,000
Columbia-North Pacific	5,450	6,250	7,700	9,500	11,200
California	8,420	8,850	10,150	10,750	11,100
Total	37,240	41,840	49,480	56,550	62,280

Note: Areas in thousands of acres for the conterminous U.S.; for regions see 1.1–7.

Source: U.S. Water Resources Council, 1968.

2.3–8 U.S. IRRIGATED ACREAGE

Area or State	Acreage, 1956	Acreage, 1954	Increase, %	Sprinkler[a]
U.S. total	36,002,627	30,711,453	+17	3,064,463
17 western states	32,661,501	28,092,947	16	2,010,062
Other states	3,341,126	2,618,506	28	1,054,401
California	7,750,000	7,048,792	10	400,000
Texas	6,962,234	5,439,000	28	575,015
Idaho	2,405,089	2,324,571	3	130,000
Colorado	2,382,000	2,263,000	5	33,110
Nebraska	2,012,320	1,393,733	44	201,230
Montana	1,890,000	1,890,000	–	28,000
Utah	1,612,108	1,072,682	50	3,325
Oregon	1,575,000	1,490,397	6	157,500
Wyoming	1,300,000	1,262,632	3	8,000
Arizona	1,150,000	1,250,000	-9	1,000
Washington	947,000	778,135	22	228,000
Arkansas	892,936	857,390	4	54,756
Florida	821,282	428,282	92	180,000
New Mexico	800,000	649,615	23	3,000
Kansas	722,575	420,000	72	100,000
Louisiana	711,000	707,818	1	39,650
Nevada	700,000	567,498	23	12,000
Oklahoma	285,175	107,981	164	100,882
Mississippi	157,000	132,490	19	54,000
South Dakota	120,000	90,371	33	20,000
Georgia	80,000	27,701	189	79,200
Virginia	45,500	21,805	116	45,050
Iowa	20,000	2,396	818	20,000
Delaware	11,000	5,553	100	11,000
Maine	6,900	1,097	529	6,850

[a] 31.5 percent of irrigation in "other states" use sprinkler while only 6.1 percent of area in western states is sprinkled.

Source: *Irrigation Engineering and Maintenance, 1957 Directory and Buyers' Guide,* H.L. Peace Publications, New Orleans, August 31, 1957.

2.3–9 LAND UNDER IRRIGATION, 1949 and 1954

Area	Irrigated land, acres		Increase, acres	Specified cropland,[a] acres	Rate of increase	
	1949	1954			Total %	Avg %
28 humid states	152,586	587,878	435,292	208,039,591	0.2092	0.0418
U.S.	25,789,453	29,552,155	3,764,700	399,808,371	0.9416	0.1882

[a] Harvested land and land used only for pasture.

Source: Reprinted, with publisher's permission, from *J. Am. Water Works Assn., 50*(3):310, © 1958, by the American Water Works Association, Inc., 2 Park Ave., New York, N.Y. 10016.

2.3–10 ESTIMATED LAND UNDER IRRIGATION

28 Humid States, 1960–2000

Year	Acres irrigated
1960	1,059,410
1970	2,442,472
1980	5,762,415
1990	7,175,774
2000	10,962,095

Source: Reprinted, with publisher's permission, from *J. Am. Water Works Assn., 50*(3):310, © 1958, by the American Water Works Association, Inc., 2 Park Ave., New York, N.Y. 10016.

2.3–11 SOURCES AND ESTIMATED AMOUNTS USED FOR IRRIGATION

	Source of irrigation water	
State	Ground, %	Surface, %
Alaska	10	90
Alabama	20	80
Arizona	80	20
California	50	50
Colorado	90	10
Delaware	40	60
Idaho	15	85
Indiana	44	56
Minnesota	55	45
Missouri	75	25
Montana	–	98
New Jersey	70	30
Oregon	5	95
Pennsylvania	10	90
South Dakota	35	65

Source: Reprinted, with publisher's permission, from *J. Am. Water Works Assn., 55*(6):705, © 1963, by the American Water Works Association, Inc., 2 Park Ave., New York, N.Y. 10016.

2.3–12 IRRIGATION WATER FOR 17 WESTERN STATES

Item	Amount
Average application per season, acre ft/acre	3.4
Range of application, acre ft/acre	2.2–6.1
One acre foot equals (gal)	326,000
Average application per season, gal/acre	1,110,000
Total quantity used, 1956, acre ft	128,000,000 (42 trillion gal)
Assumed season, days	128
Rate of use, acre ft/day	1,000,000
Rate of flow, 1956, ft^3/sec	490,000
Increased by 1970, %	43
Total use, 1970, trillion gal	58.5

Source: Reprinted, with publisher's permission, from *J. Am. Water Works Assn., 35*(5):614, © 1963, by the American Water Works Association, Inc., 2 Park Ave., New York, N.Y. 10016.

2.3–13 WATER REQUIREMENTS FOR FARM ANIMALS AND POULTRY

Usage	gpd
Horse, work	12
Mule	12
Cattle	
Holstein calves (liquid milk or dried milk and water supplied)	
4 weeks of age	1.2–1.4
8 weeks of age	1.6
12 weeks of age	2.2–2.4
16 weeks of age	3.0–3.4
20 weeks of age	3.8–4.3
26 weeks of age	4.0–5.8
Dairy heifers, pregnant	7.2–8.4
Steers	
Maintenance ration	4.2
Fattening ration	8.4
Range cattle	4.2–8.4
Jersey cows[a]	
Milk production 5–30 lb/day	7.2–12
Holstein cows[a]	
Milk production 20–50 lb/day	7.8–22
Milk production 80 lb/day	23
Dry	11
Pigs	
Body weight, 30 lb	0.6–1.2
Body weight, 60–80 lb	0.8
Body weight, 75–125 lb	1.9
Body weight, 200–380 lb	1.4–3.6
Pregnant sows	3.6–4.6
Lactating sows	4.8–6.0
Sheep	
On range or dry pasture	0.6–1.6
On range (salty feeds)	2.0
On rations of hay and grain or hay, roots and grain	0–0.7
On good pasture	Little, if any
Chickens (100 birds)	
1–3 weeks of age	0.4–2.0
3–6 weeks of age	1.4–3.0
6–10 weeks of age	3.0–4.0
9–13 weeks of age	4.0–5.0
Pullets	3.0–4.0
Nonlaying hens	5.0
Laying hens (moderate temperatures)	5.0–7.5
Laying hens (temperature 90°F.)	9.0
Turkeys (100 birds)	
1–3 weeks of age	1.1–2.6
4–7 weeks of age	3.7–8.4
9–13 weeks of age	9–14
15–19 weeks of age	17
21–26 weeks of age	14–15

[a] Allow 15–20 additional gallons per day for each cow for flushing stables and washing dairy utensils.

Source: U.S. Department of Agriculture.

2.3–14 DAILY WATER REQUIREMENTS FOR COMMON FARM ANIMALS

Animal	Minimum daily water requirements, gal		
	PHS[1]	WQC[2]	Lamont[a,3]
Horse, mule or steer	12	7–12	7–12
Dairy cow (drinking only)	15	10–16	12–18
Dairy cow (drinking and dairy servicing)	35	–	36
Sheep and goat	2	1–4	1–2
Hog	4	3–5	4
Chickens, per 100 head	4	8–10	6
Turkeys, per 100 head	7	10–15	–

[a] Values in U.S. gallons.

REFERENCES

1. "Report of Public Health Service Committee on Plumbing Standards," September 1962, U.S. Department of Health, Education, and Welfare, Public Health Service, Washington, D.C., 1963.
2. "Report of the Committee on Water Quality Criteria," Federal Water Pollution Control Administration, U.S. Department of the Interior, April 1, 1968.
3. P. Lamont, *Water and Water Engineering, 54*:19, July 1950.

Section 3

Quality Criteria

3. Introduction
Pages 247–259

3.1 Public Water Supply Quality Criteria
Pages 260–294

3.2 Industrial Water Supply Quality Criteria
Pages 295–327

3.3 Agricultural Water Supply Quality Criteria
Pages 328–344

3.4 Fish and Wildlife
Pages 345–387

3.5 Recreational and Aesthetic
Page 388

3. QUALITY CRITERIA

3–1 VALUE OF WATER FOR VARIOUS USES

Use	Value, \$1/acre-ft Mean	Maximum
Domestic	100.19	235.66
Industrial	40.73	163.35
Irrigation	1.67	27.04
Hydropower	0.71	5.90
Waste disposal	0.63	2.56
Inland navigation	0.05	1.17
Commercial fisheries	0.025	1.06

Note: Hoak[2] calculates fishing-sports value of an acre foot of water as approximately \$21.50, \$23.13, and \$24.68 for 1954, 1980, and 2000, respectively. In these calculations Hoak referred to data provided by Fish and Wildlife and Water Resources, Select Comm. on Natl. Water Resources, U.S. Senate, 86th Congress, 2nd Sess., Comm. Print No. 18, 1960.

REFERENCES

1. E.F. Renshaw, *J. Am. Water Works Assn., 50*(3):303, 1958.
2. R.D. Hoak, *J. Water Pollut. Control Fed., 33*(12):1267, 1961.

3–2 CONCEPT OF WATER QUALITY CONTROL FOR BENEFICIAL USE

Beneficial use to be preserved in river water	Water quality criteria prescribed by control board	Possible conditions if pollution not controlled
Irrigation	700 ppm dissolved minerals, maximum 40 percent sodium ratio, maximum 0.5 ppm boron, maximum	High concentrations: crop damage and loss of production
Bathing	10 coliform bacteria per ml, maximum	High coliform counts; an actual hazard to public health (contamination)
Fishing	5 ppm of dissolved oxygen (DO), minimum	Low dissolved oxygen; reduced fisheries resources or fish kills

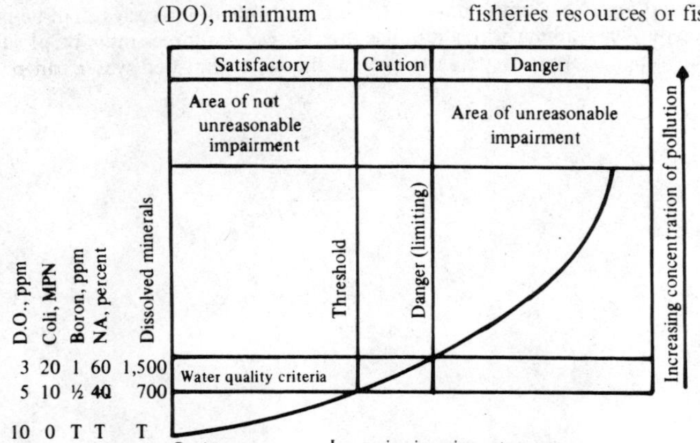

Source: V.W. Bacon and C.A. Sweet, *"Setting Water Quality Criteria in California"*, Am. Soc. Civ. Eng., May 1955.

3–3 GENERAL POLICY, CLASSIFICATION AND STANDARDS OF QUALITY FOR INTERSTATE WATERS

Standards

Water use class and description	Dissolved oxygen	Sludge deposits, solid refuse, floating solids, oil, grease, and scum	Color and turbidity
	Inland Waters		
A – suitable for water supply with treatment by disinfection only, and all other water uses; character uniformly excellent[a,c]	As naturally occurs	None other than of natural origin	None other than of natural origin
B – suitable for bathing and other primary contact recreation. Acceptable for public water supply with appropriate treatment.[c] Suitable for agricultural and certain industrial process and cooling uses. Suitable as an excellent fish and wildlife habitat. Excellent aesthetic value	Minimum 5 mg/l at any time. Normal seasonal and diurnal variations above 5 mg/l will be maintained	None allowable	None in such concentrations that would impair any usages specifically assigned to this class
C – suitable for fish and wildlife habitat, boating, fishing, and certain industrial process and cooling uses. Under some conditions acceptable for public water supply with appropriate treatment. Good aesthetic value[c]	Minimum 5 mg/l any time. Normal seasonal and diurnal variations above 5 mg/l will be maintained. For sluggish, eutrophic waters, not less than 3 mg/l at any time. Normal seasonal and diurnal variations above 3 mg/l will be maintained	None[b]	None in such concentrations that would impair any usages specifically assigned to this class
D – suitable for navigation, power, certain industrial processes and cooling uses, and migration of fish. Aesthetically acceptable	2 mg/l at any time. Normal seasonal and diurnal variations above 2 mg/l will be maintained	None[b]	None in such concentrations that would impair any usages specifically assigned to this class

[a] Class A waters reserved for water supply may be subject to restricted use by state and local regulation.
[b] Sludge deposits, floating solids, oil, grease, and scum shall not be allowed except in that amount that may result from the discharge of appropriately designed and operated sewage and/or industrial waste treatment plants.
[c] Waters shall be free from chemical and radiological constituents in concentrations or combinations which would be harmful to human, animal, or aquatic life for the most sensitive and governing water class use. In areas where fisheries are the governing considerations and approved limits have not been established, bioassays shall be performed as required by the appropriate agencies. For public drinking water supplies the raw water sources must be of such a quality that United States Public Health Service or higher appropriate state agency limits for finished water can be met after conventional water treatment.

3–3 GENERAL POLICY, CLASSIFICATION AND STANDARDS OF QUALITY FOR INTERSTATE WATERS *(Continued)*

Water use class and description	Coliform bacteria	Taste and odor	pH	Allowable temperature increase
		Inland Waters		
A – suitable for water supply with treatment by disinfection only, and all other water uses; character uniformly excellent[a,c]	Total coliforms: Not to exceed a monthly arithmetic mean of 100/100 ml[d]	None other than of natural origin	As naturally occurs	None other than of natural origin
B – suitable for bathing and other primary contact recreation. Acceptable for public water supply with appropriate treatment.[c] Suitable for agricultural and certain industrial process and cooling uses. Suitable as an excellent fish and wildlife habitat. Excellent aesthetic value.	Total coliforms: Not to exceed a monthly median of 1,000/100 ml nor more than 2,400/100 ml in more than 20% of samples collected[d]	None in such concentrations that would impair any usages specifically assigned to this class nor cause taste and odor in edible fish	6.5–8.0	Only such increases that will not impair any usages specifically assigned to this class[e]
C – suitable for fish and wildlife habitat, boating, fishing, and certain industrial process and cooling uses. Under some conditions acceptable for public water supply with appropriate treatment. Good aesthetic value[c]	d	None in such concentrations that would impair any usages specifically assigned to this class nor cause taste and odor in edible fish	6.0–8.5	Only such increases that will not impair any usages specifically assigned to this class[e]
D – suitable for navigation, power, certain industrial processes and cooling uses, and migration of fish. Aesthetically acceptable	None in such concentrations that would impair any usages specifically assigned to this class	None in such concentrations that would impair any usages specifically assigned to this class	6.0–9.0	Only such increases that will not impair any usages specifically assigned to this class[e]

[a] Class A waters reserved for water supply may be subject to restricted use by state and local regulation.

[c] Waters shall be free from chemical and radiological constituents in concentrations or combinations which would be harmful to human, animal, or aquatic life for the most sensitive and governing water class use. In areas where fisheries are the governing considerations and approved limits have not been established, bioassays shall be performed as required by the appropriate agencies. For public drinking water supplies the raw water sources must be of such a quality that United States Public Health Service or higher appropriate state agency limits for finished water can be met after conventional water treatment.

[d] Water quality analyses on interstate waters should include tests for both total and fecal coliform supported by sanitary surveys for the development of background data and the establishment of baselines for these two parameters for the specific waters involved.

[e] The resultant temperature of a receiving water shall not exceed that shown for the most sensitive aspect of the most sensitive specie of aquatic life in that water. Normal seasonal and diurnal temperature variations that existed before the addition of heat of artificial origin shall be maintained. In addition:

1. For streams, the allowable total temperature rise shall not exceed 5°F above ambient unless it can be demonstrated to the satisfaction of the state regulatory agencies that greater rises at various times will not be harmful to fish, other aquatic life or other uses.
2. For lakes, there shall be no discharge or withdrawal of cooling waters from the hypolimnion unless it can be demonstrated to the satisfaction of the state regulatory agencies that such discharges or withdrawals will not be harmful to fish, other aquatic life or other uses. A heated discharge to a lake shall not raise the temperature more than 3°F at the surface immediately outside a designated mixing zone.

More restrictive requirements may be adopted by a state if deemed necessary to meet the class use of the receiving waters.

3–3 GENERAL POLICY, CLASSIFICATION AND STANDARDS OF QUALITY FOR INTERSTATE WATERS *(Continued)*

Water use class and description	Dissolved oxygen	Sludge deposits, solid refuse, floating solids, oil, grease, and scum	Color and turbidity
	Coastal and Marine Waters[f,h,i2]		
SA – suitable for all sea water uses including shellfish harvesting for direct human consumption (approved shellfish areas), bathing, and other water contact sports; excellent aesthetic value	Not less than 5.0 mg/l at any time. Normal seasonal and diurnal variations above 5 mg/l will be maintained	None allowable	None in such concentrations that would impair any usages specifically assigned to this class
SB – suitable for bathing, other recreational purposes, industrial cooling and shellfish harvesting for human consumption after depuration; excellent fish and wildlife habitat, good aesthetic value	Not less than 5.0 mg/l at any time. Normal seasonal and diurnal variations above 5 mg/l will be maintained	None allowable	None in such concentrations that would impair any usages specifically assigned to this class
SC – suitable fish, shellfish and wildlife habitat; suitable for recreational boating, and industrial cooling; good aesthetic value	A minimum of 2 mg/l at any time. Normal seasonal and diurnal variations above 2 mg/l will be maintained	None except for such small amounts that may result from the discharge of appropriately treated sewage and/or industrial waste effluents	None in such concentrations that would impair any usages specifically assigned to this class
SD – suitable for navigation, power, certain industrial processes and cooling uses, and migration of fish; good aesthetic value	Not less than 5 mg/l during daylight hours nor less than 4 mg/l at any time. Normal seasonal and diurnal variations above 4 mg/l will be maintained	None except that amount that may result from the discharge from a waste treatment facility providing appropriate treatment	None in such concentrations that would impair any usages specifically assigned to this class

[f] Costal and marine waters are those generally subject to the rise and fall of the tide.

[g] Surveys to determine coliform concentrations shall include those areas most probably exposed to fecal contamination during the most unfavorable hydrographic and pollution conditions. Water quality analyses on interstate marine and coastal waters should include tests for both total and fecal coliform supported by sanitary surveys for the development of background data and the establishment of a baseline for these two parameters for the specific waters involved.

[h] Waters shall be free from chemical and radiological concentrations or combinations which would be harmful to human, animal, or aquatic life or which would make the waters unsafe or unsuitable for fish or shellfish or their propagation, impair the palatability of same, or impair the water for any other usage. In areas where fisheries are the governing considerations and approved limits have not been established, bioassays shall be performed as required by the appropriate agencies.

3–3 GENERAL POLICY, CLASSIFICATION AND STANDARDS OF QUALITY FOR INTERSTATE WATERS *(Continued)*

Water use class and description	Coliform bacteria	Taste and odor	pH
	Coastal and Marine Waters[f,h,i²]		
SA-suitable for all sea water use including shellfish harvesting for direct human consumption (approved shellfish areas), bathing, and other water contach sports; excellent aesthetic value	Not to exceed a median MPN of 70/100 ml and not more than 10% of the samples shall ordinarily exceed an MPN of 230/100 ml for a 5-tube decimal dilution or 330/100 ml for a 3-tube decimal dilution[g]	None allowable	6.8–8.5
SB-suitable for bathing, other recreational purposes, industrial cooling and shellfish harvesting for human consumption after depuration; excellent fish and wildlife habitat, good aesthetic value	Not to exceed a median of 700/100 ml and not more than 2,300/100 ml in more than 10% of the samples collected in a 30 day period[g]	None in such concentrations that would impair any usages specifically assigned to this class and none that would cause taste and odor in edible fish or shellfish	6.8–8.5
SC-suitable fish, shellfish and wildlife habitat; suitable for recreational boating, and industrial cooling; good aesthetic value	g	None is such concentrations that would impair any usages specifically assigned to this class and none that would cause taste and odor in edible fish or shellfish	6.0–9.0
SD-suitable for navigation, power, certain industrial processes and cooling uses, and migration of fish; good aesthetic value	None in such concentrations that would impair any usages specifically assigned to this class	None in such concentrations that would impair any usages specifically assigned to this class and none that would cause taste and odor in edible fish	6.8–8.5

[i] The temperature of a receiving water shall not exceed that shown below immediately outside a designated mixing zone:

1. *Coastal waters*
The water temperature at the surface of coastal waters shall not be raised more than 4°F over the monthly means of maximum daily temperatures from October through June nor more than 1.5°F from July through September.
2. *Estuaries or portions of estuaries*
The water temperature at the surface of an estuary shall not be raised to more then 90°F at any point provided further, at least 50 percent of the cross sectional area and/or volume of the flow of the estuary including a minimum of 1/3 of the surface as measured from water edge to water edge at any stage of tide, shall not be raised to more than 4°F over the temperature that existed before the addition of heat of artificial origin or a maximum of 83°F, whichever is less. However, during July through September if the water temperature at the surface of an estuary before the addition of heat of artificial origin is more than 83°F, an increase in temperature not to exceed 1.5°F, at any point of the estuarine passageway as delineated above, may be permitted.

More restrictive requirements may be adopted by a state if deemed necessary for the protection of the most sensitive aspect of the most sensitive specie of aquatic life in that water and/or to meet the class use of the receiving water.

Source: New England Interstate Water Pollution Control Commission, *Standards of Water Quality,* 1969.

3–4 CLASSES AND STANDARDS FOR FRESH SURFACE WATERS

New York State

Water standards[b]

Class and best use[a]	Minimum dissolved oxygen, ml/l	Coliform bacteria median, no./100 ml	pH	Toxic wastes, deleterious substances, colored wastes, heated liquids, and taste- and odor-producing substances[c]	Floating solids, settleable solids, oil, and sludge deposits
AA – Source of unfiltered public water supply and any other usage	5.0 (trout) 4.0 (nontrout)	Not to exceed 50	6.5–8.5		None attributable to sewage, industrial wastes, or other wastes
A – Source of filtered public water supply and any other usage	5.0 (trout) 4.0 (nontrout)	Not to exceed 5,000	6.5–8.5	None in sufficient amounts or at such temperatures as to be injurious to fish life or make the waters unsafe or unsuitable	
B – Bathing and any other usages except as a source of public water supply	5.0 (trout) 4.0 (nontrout)	Not to exceed 2,400	6.5–8.5		
C – Fishing and any other usages except public water supply and bathing	5.0 (trout) 4.0 (nontrout)	Not applicable	6.5–8.5	None in sufficient amounts or at such temperatures as to be injurious to fish life or impair the waters for any other best usage	None which are readily visible and attributable to sewage, industrial wastes, or other wastes
D – Natural drainage, agriculture, and industrial water supply	3.0	Not applicable	6.0–9.5	None in sufficient amounts or at such temperatures as to prevent fish survival or impair the waters for agricultural purposes or any other best usage	

[a] Class B and C waters and marine waters shall be substantially free of pollutants that: unduly affect the composition of bottom fauna; unduly affect the physical or chemical nature of the bottom; interfere with the propagation of fish. Class D and SD (marine) will be assigned only where a higher water use class cannot be attained after all appropriate waste treatment methods are utilized. Any water falling below the standards of quality for a given class shall be considered unsatisfactory for the uses indicated for that class. Waters falling below the standards of quality for Class D, or SD (marine), shall be Class E, or SE (marine), respectively, and considered to be in a nuisance condition.

[b] These standards do not apply to conditions brought about by natural causes. Waste effluents discharging into public water supply and recreation waters must be effectively disinfected. All sewage treatment plant effluents shall receive disinfection before discharge to a watercourse and/or coastal and marine waters. The degree of treatment and disinfection shall be as required by the state pollution control agency. The minimum average daily flow for seven consecutive days that can be expected to occur once in ten years shall be the minimum flow to which the standards apply.

[c] Phenolic compounds cannot exceed 0.005 mg/l; no odor-producing substances that cause the threshold-odor number to exceed 8 are permitted; radioactivity limts are to be approved by the appropriate state agency, with consideration of possible adverse effects in downstream waters from discharge of radioactive wastes, and limits in a particular watershed are to be resolved when necessary after consultation between states involved.

Source: I. Grossman, "Experience with Surface Water Quality Standards," *J. Sanit. Eng. Div.*, Am. Soc. Civ. Eng., 94:13, 1968.

3–5 RECEIVING WATER QUALITY CRITERIA

Class	Use	
A	Water supply with chlorination only (fish life, bathing, recreation, process also)	Coliform bacteria – MPN, 50/100 ml maximum Color <10 ppm Turbidity <10 ppm pH – 6.0–8.0 range DO not less than 7.5 ppm; no single observation <6.0 ppm No toxic substances, free acid, debris, odor or taste producers (except natural sources); no sludge deposits of any kind
B	Bathing, fish life, recreation; water supply after complete treatment	Coliform bacteria – MPN (monthly average) between 50 and 500/100 ml Color – 20 ppm maximum desirable Turbidity – 40 ppm maximum desirable pH – 6.0–8.5 5 day BOD – maximum any sample = 3.0 ppm, average <1.5 ppm DO – monthly average >6.5 ppm; no sample <5.0 ppm Other conditions – same as class A
C	Water supply after complete treatment; industrial process, navigation, etc	Coliform bacteria – monthly average, 500–5,000/100 ml Color and turbidity – removable by filtration pH – 6.0 to 8.5 5-day BOD – monthly average <2.0 ppm; single sample <4.0 ppm DO – monthly average >6.5; single sample >5.0 Other conditions – same as class A
D	Navigation; cooling water, etc	Shall not constitute nuisance pH – 6.0–8.5 5-day BOD – monthly average <3.0 ppm; single sample <5.0 ppm Other conditions – no toxic substance, free acid, floating debris

Source: Derived from Potomac River Commission data.

3–6 SURFACE WATER QUALITY LIMIT PROPOSALS

State of Washington

Characteristic	Fresh water[a] Goal	Fresh water[a] Standard	Salt water[a] Goal	Salt water[a] Standard
Alkalinity (phenolphthalein and total)[b]				
Ammonia nitrogen	0.3	0.5	0.0025	0.003
Arsenic	0.003	0.005	0.003	0.004
Bacteria[c]				
Barium	0.01	0.05	0.05	0.06
Bicarbonate[d]				
BOD	1.0	2.0	1.0	2.0
Boron	0.1	0.3	4.7	5.5
Bottom deposits from waste water discharge	None	None	None	None
Cadmium	0.0005	0.001	0.00011	0.00013
Calcium[e]				
Carbonate[f]				
CCE (carbon chloroform extract)	0.00	0.10	0.05	0.10
Chloride	10	20	Natural	120% of natural
Chromium	Trace	0.01	0.00005	0.00006
COD[g]				
Coliforms (domestic sewage origin)	50/100 ml	240/100 ml	50/100 ml	240/100 ml
Color	5 units	5 units over natural	None	5 units
Conductivity	110% of natural	125% of natural	Natural	120% of natural
Copper	0.05	0.02 above background	0.05	0.06
Cyanide	0.005	0.01	None	0.01
Dissolved oxygen	95% saturation	85% saturation	95% saturation	85% saturation
Fecal streptococci[h]				
Floating solids	None	None	None	None
Fluoride	0.5	1.0	1.3	1.5
Hardness (as $CaCO_3$)	20 to 75	20 to 125	–	–
Hydroxide	None	None	None	None
Iron	0.0 above natural	0.1 above natural	0.01 above natural	0.2
Lead	Limit of detection	0.02	Limit of detection	0.004
Magnesium[i]				
Manganese	Trace	0.01	0.002	0.04
Nitrate	0.1 above natural	1.0 above natural	0.5	0.6

[a] All values in mg/l unless otherwise specified.
[b] No specific limits. A waste discharge is not to increase the natural total alkalinity by more than 10 percent or to impart phenolphthalein (CO_3 and OH^-) alkalinity to a receiving water.
[c] No limit specified: refer to coliform organisms.
[d] No limit specified: relates to conductivity and pH.
[e] No limit specified: see hardness.
[f] Although carbonate itself in moderate concentrations is not particularly detrimental, it is associated with high pH values (greater than 8.3). Any carbonate discharge is not to be detectable below the point of discharge.
[g] Since the chemical oxygen demand is related to the BOD, DO, and CCE, it has no limit specified herein.
[h] Fecal bacteria are represented by the coliform group standard.
[i] Controlled by hardness content: no limit specified.

3–6 SURFACE WATER QUALITY LIMIT PROPOSALS *(Continued)*

	Fresh water[a]		Salt water[a]	
Characteristic	Goal	Standard	Goal	Standard
Nitrogen (total)	0.4 above natural	1.0 above natural	0.5	0.6
Threshold-odor number	1.0	3	1.0	3
Oil and tars	None	None	None	None
Pesticides[j]				
pH	7.0–8.0	6.5–8.5	7.5–8.4	7.5–8.4
Phenol	Limit of detectability	0.0005	0.04	0.05
Phosphate (total)	0.03	0.15	0.3	0.4
Potassium	2.5	5.0	380	450
Radioactivity	None	USPHS DWS	None	USPHS DWS
Selenium	Limit of detectability	0.002	0.004	0.005
Silica[k]				
Silver	Limit of detectability	0.003	0.0003	0.0004
Sodium	10 over natural	35 over natural	10,500	12,500
Spent sulfite liquor[l]				
Sulfate	15	30	2700	3200
Surfactants	Trace (LAS)	0.10 (LAS)	Trace (LAS)	0.10 (LAS)
Temperature	Natural + 1°C	Natural + 2°C	Natural + 1°C	Natural + 2°C
Total dissolved solids[m]				
Toxicants, miscellaneous	None detectable	None detectable	None detectable	None detectable
Turbidity	5 units	Natural	3 units	5 units
Viruses[n]				
Zinc	Limit of detectability	Limit of detectability	0.01	0.012

[a] All values in mg/l unless otherwise specified.
[j] Insufficient data.
[k] No standard proposed. Turbidity will include colloidal silica.
[l] Effect covered by other parameters.
[m] No standard proposed. Conductivity standards are related.
[n] None proposed at present.

Source: C.A. Rambow and R.D. Sylvester, "Methodology in Establishing Water Quality Standards," *J. Water Pollut. Control Fed., 39*:1155, 1967.

3–7 RECOMMENDED STREAM QUALITY STANDARDS

Stream quality and uses	Coliforms, MPN/ 100 ml	DO, mg/l	BOD, mg/l	pH	Phenols, 1×10^{-3} mg/l
Excellent recreation (bathing); water supply (chlorination)	50–100	6.5–7.5	0.75–2.0	6.0–8.6	nil
Desirable recreation (bathing and fish life)	100–1,000	5.0–7.0	1.5–3.5	5.8–9.0	1–5
Desirable water supplies (filtration)	1,000–5,000	5.0–6.5	2.0–4.0	5.8–9.0	1–5
Doubtful water supplies (auxiliary treatment); recreation (fish life)	10,000–20,000	3.0–5.5	3.0–7.0	3.8–10.5[a]	10–25
Unsuitable water supplies, recreation, fish life	Over 20,000	Less than 2.0–less than 5.5	Over 4.0–over 10.0	Under 3.8–over 10.5[a]	Over 10–over 25

a Limits recommended by West Virginia Water Commission. Other authorities have more stringent limit requirements.

Source: H.W. Streeter, *Sew. Works J.*, *21*(1):115, 1949.

3–8 RECOMMENDATIONS FOR WATER FOR CERTAIN BENEFICIAL USES TO BE PROTECTED

Prado Water Quality Objectives

	Quality factors				
Water use	**TDS, mg/l**	**Chloride, mg/l**	**Nitrate, mg/l as NO_3**	**Bicarbonate, mg/l HCO_3**	**Hardness, mg/l $CaCO_3$**
Industrial uses					
Boiler feed water					
High pressure	200			48	0
Intermediate pressure	500			120	0
Low pressure	700			170	20
Cooling water	500	500		24	130
Food processing	500	250	10	155	250
Soft drinks		500		53	
Pulp and paper					
Mechanical pulping		1,000			
Unbleached		200			100
Bleached		200			100
Chemical *et al.*					
Organic chem.				128	250
Synth. rubber				155	350
Gums, wood chem.	1,000	500		250	900
Petroleum	1,000	300			350
Agriculture					
Vegetable crops					
Beans	132				
Carrots	132				
Bell peppers	350				
Lettuce	350				

3–8 RECOMMENDATIONS FOR WATER FOR CERTAIN BENEFICIAL USES TO BE PROTECTED *(Continued)*

Water use	Quality factors				
	TDS, mg/l	Chloride, mg/l	Nitrate, mg/l as NO_3	Bicarbonate, mg/l HCO_3	Hardness, mg/l $CaCO_3$
Agriculture (continued)					
Corn	393				
Cabbage	393				
Broccoli	700				
Tomatoes	700				
Spinach	1,000				
Ornamental shrubs					
Roses	362				
Pineapple guava	362				
Viburnum	362				
Pyracantha	700				
Pittosporum	700				
Xylosma	700				
Texas privet	700				
Arbor vitae	1,092				
Spreading juniper	1,092				
Lantana	1,092				
Oleander	1,750				
Bottlebrush	1,750				
Fruit crops					
Rangpur lime, Cleopatra mandarin		437			
Rough lemon tangelo sour orange		262			
Sweet orange, citrange		175			
West Indian avocado rootstock		140			
Mexican avocado rootstock		88			
Thompson seedless grape		437			
Cardinal, Black Rose grape		175			
Boysenberry		175			
Olallie blackberry		175			
Indian summer raspberry		88			
Lassen strawberry		140			
Shasta strawberry		88			

Source: Reprinted, with publisher's permission, from *J. Am. Water Works Assn., 62*(2):106, 1970, by the American Water Works Association, Inc., 2 Park Ave., New York, N.Y. 10016.

3–9 CALIFORNIA REQUIREMENTS FOR WASTE DISCHARGE TO A RECEIVING STREAM

Parameter	Allowable concentration, mg/l
Boron	2.0
Cadmium	0.05
Chromium, Cr^{+6}	0.05
Copper	0.1
Cyanide	0.1
Lead	0.1
Nickel	1.0
Oils	20.0
Phenol	Not to increase more than 0.005 mg/l over that present upstream from discharge
Zinc	0.3
pH	6.5–8.5

Source: R.V. Daigh, *J. Water Pollut. Control Fed.*, *36*(7):895, 1964.

3.1 PUBLIC WATER SUPPLY QUALITY CRITERIA

3.1–1 QUALITY CRITERIA FOR PUBLIC WATER SUPPLIES

Introduction

The National Technical Advisory Subcommittee on Public Water Supplies has found it necessary to make some rather arbitrary decisions in order to proceed with its task of developing raw water quality criteria for public water supplies. Becuase public water supplies commonly involve processing of the raw water to improve its quality before distributing it to consumers, and because treatment processes exist which can, at a price, convert almost any water including sea water and grossly polluted fresh water into a potable product, it is necessary to consider the type of treatment in any discussion of raw water quality criteria for public water supplies.

We have adopted as the considered treatment the most common processes in use in this country in their simplest form for the treatment of surface waters for public use. This may include coagulation (less than about 50 ppm alum, ferric sulfate, or copperas with alkali addition as necessary but without coagulant aids or activated carbon), sedimentation (6 hours or less), rapid sand filtration (3 gal/ft^2 min or less) and disinfection with chlorine (without consideration to concentration or form of chlorine residual). A wide variety of modifications of this basic treatment process are in use for removing various impurities or altering quality characteristics, but we have arbitrarily excluded these modifications in our deliberations because of the difficulty in deciding where to stop in considering the many modifications and elaborations of the basic process.

Definitions

Permissible criteria.

Those characteristics and concentrations of substances in raw surface waters which will allow the production of a safe, clear, potable, aesthetically pleasing, and acceptable public water supply which meets the limits of Drinking Water Standards[10] after treatment. This treatment may include, but will not include more than, the processes described above.

Desirable criteria.

Those characteristics and concentrations of substances in the raw surface waters which represent high quality water in all respects for use as public water supplies. Water meeting these criteria can be treated in the defined plants with greater factors of safety or at less cost than is possible with waters meeting permissible criteria.

Narrative.

The presence of this word in the table indicates that the Subcommittee could not arrive at a single numerical value which would be applicable throughout the country for all conditions. Where this word appears, the reader is directed to the appropriate explanatory text.

Absent.

The most sensitive analytical procedure in Standard Methods[9] (or other approved procedure) does not show the presence of the subject constituent.

Virtually absent.

This terminology implies that the substance is present in very low concentrations and is used where the substance is not objectionable in these barely detectable concentrations.

Discussion

The Subcommittee recognizes that surface waters are used for public water supply without treatment other than disinfection. Such waters at this point of withdrawal should meet Drinking Water Standards[10] in all respects other than bacterial quality.

It should be emphasized that many raw water sources which do not meet these permissible criteria have been and are being used to provide satisfactory public water supplies by suitable additions to and elaboration of the treatment processes defined above. In some instances, however, the water delivered to the customer is of marginal quality. Also the finished water is much more likely to become unsatisfactory if treatment plant irregularities occur. It is recognized that most of the surface water treatment plants providing water for domestic use in the United States are relatively small[7] and without sophisticated technical controls. Marginal quality characteristics, therefore, assume considerable importance to the managers of plants treating such supplies. This is the importance of the factors of safety mentioned in the definition of "desirable criteria." However, managers of all supplies would welcome improved raw water quality.

This Subcommittee believes that the criteria set forth herein can be used in setting standards of raw water quality only with a substantial amount of understanding and discretion. To a considerable extent this is related to the very great regional variations in water quality entirely aside from man-made pollution. In addition, human occupance and activity have inevitable effects on water quality. These facts make it difficult and sometimes impossible to develop uniform numerical criteria suitable for national application.

The criteria selected by the Subcommittee are listed in the table and discussed in the numbered paragraphs cited in the table. The paragraphs also include some rationale of the basis for the criteria. The fact that a substance is not included in these criteria does not imply that its presence is innocuous. It would be quite impracticable to prepare a compendium of all toxic, deleterious, or otherwise unwelcome agents that may enter a surface water supply.

Sampling

Sampling should be of such frequency and of such variety (time of day, season, temperature, river stage or flow, location, depth) as to properly describe the body of water designated for public water supply. Sampling should also be conducted in full cognizance of findings of the sanitary survey. Judgment should be exercised as to the relative desirability of frequent sampling at one point, such as the raw water intake, as compared to less frequent sampling at numerous locations, such as is required for stream profiles or cross sections.

It is clearly not possible to apply these criteria solely as maximum single sample values. The criteria should not be exceeded over substantial portions of time. If they are exceeded, efforts should be made to determine the cause, and corrective measures undertaken.

Analytical Methods

The criteria are based upon those analytical methods described in Standard Methods for the Examination of Water and Wastewater[9] or upon methods acceptable to water pollution control agencies.

Paragraph 1: Color

A limit of 75 color units (platinum-cobalt standard) has been recommended to permit the defined plant to produce water meeting Drinking Water Standards[10] with moderate dosages of coagulant chemicals. At optimum pH the dosage usually required is linearly related to the color of the raw water, and higher color of the type commonly associated with swamp drainage and similar nonindustrial sources can be removed by increasing the coagulant dose. These criteria do not apply to colors resulting from dyes and some other industrial and processing sources which cannot be measured by the platinum-cobalt standard. Such colors should not be present in concentrations which cannot be removed by the defined method of treatment.

Paragraph 2: Odor

The effectiveness of the defined method of treatment in removing odorous materials from water is highly variable depending on the nature of the material causing the odor. For this reason, it has not been feasible to specify any permissible criterion in terms of threshold odor number. The raw water should not have an objectionable odor. Any odors present should be removable by the defined treatment. It is desirable that odor be virtually absent.

Paragraph 3: Temperature

Surface water temperatures vary with geographical location and climatic conditions. Consequently, no fixed criteria are feasible. However, any of the following conditions are considered to detract (sometimes seriously) from raw water quality for public water supply use:

1. Water temperature higher than 85°F;
2. More than 5°F water temperature increase in excess of that caused by ambient conditions;
3. More than 1°F hourly temperature variation over that caused by ambient conditions;
4. Any water temperature change which adversely affects the biota, taste, and odor, or the chemistry of the water;
5. Any water temperature variation or change which adversely affects water treatment plant operation (for example, speed of chemical reactions, sedimentation basin hydraulics, filter wash water requirements, etc);
6. Any water temperature change that decreases the acceptance of the water for cooling and drinking purposes.

Paragraph 4: Turbidity

Turbidity in water must be readily removable by coagulation, sedimentation, and filtration; must not be present in quantities (either by weight or volume) that will overload the water treatment plant facilities; and must not cause unreasonable treatment costs. In addition, turbidity in water must not be frequently changing and varying in characteristics or in quantity to the extent that such changes cause upset in water treatment plant processing.

Customary methods for measuring and reporting turbidity do not adequately measure those characteristics harmful to public water supply and water treatment processing. A water with 30 Jackson turbidity units may coagulate more rapidly than one with 5 or 10 units. Similarly water with 30 Jackson turbidity units sometimes may be more difficult to coagulate than water with 100 units. Sometimes clay added to very low turbidity water will improve coagulation. Therefore, it has not been possible to establish a turbidity criterion in terms of Jackson turbidity units. Neither can a turbidity criterion be expressed in terms of mg/l "undissolved solids" or "nonfilterable solids." The type of plankton, clay, or earth particles, their size and electrical charges are far more determining factors than the Jackson turbidity units. Nevertheless, it must be clearly recognized that too much turbidity or frequently changing turbidity is damaging to public water supply.

The criterion for too much turbidity in water must relate to the capacity of the water treatment plant to remove turbidity adequately and continuously at reasonable cost. Water treatment plants are designed to remove the kind and quantity of turbidity to be expected in each water supply source. Therefore, any increase in turbidity and any fluctuating turbidity load over that normal to a water must be considered in excess of that permissible.

Paragraph 5: Coliform and Fecal Coliform Organisms

Bacteria have been used as indicators of sanitary quality of water since 1880 when *B. coli* and similar organisms were shown to be normal inhabitants of fecal discharges. The coliform group as presently recognized by Drinking Water Standards[10] is defined in Standard Methods for the Examination of Water and Wastewater.[9] This group includes organisms that vary in biochemical and serologic characteristics and in their natural sources and habitats, *i.e.*, feces, soil, water, vegetation, etc.

Because the sanitary significance of the various members of the coliform group derives from their natural sources, differentiation of fecal from nonfecal organisms is important to evaluate raw water quality.[5] Fecal coliforms are characteristically inhabitants of warm-blooded animal intestines. Members of other coliform subgroups may be found in soil, on plants and insects, in old sewage, and in waters polluted some time in the past.

The objective of using the coliform group as an indicator of the sanitary quality of water is to evaluate the disease-producing potential of the water. To estimate the probability of pathogens being contributed from feces, the coliform and fecal coliform content must be quantified.

In relation to raw water sources, the following suggestions are offered to help resolve some of the difficulties of data interpretation.

Fecal coliform organisms may be considered indicators of recent fecal pollution. It is necessary to consider all fecal coliform organisms as indicative of dangerous contamination. Moreover, no satisfactory method is currently available for differentiating between fecal organisms of human and animal origin.

In the absence of fecal coliform organisms, the presence of other coliform group organisms may be the result of less recent fecal pollution, soil runoff water, or, infrequently, fecal pollution containing only those organisms.

In general, the presence of fecal coliform organisms indicates recent and possibly dangerous pollution. The presence of other coliform organisms suggests less recent pollution or contributions from other sources of non-fecal origin.

In the past the coliform test has been the principal criterion of suitability of raw water sources for public water supply. The increase in chlorination of sewage treatment plant effluents distorts this criterion by reducing coliform concentrations without removing many other substances which the defined water treatment plant is not well equipped to remove. It is essential that raw water sources be judged as to suitability by other measures and criteria than coliform organisms concentrations.

The defined water treatment plant is considered capable of producing water meeting Drinking Water Standards[10] bacteriological criteria from these limits. The difference between the suggested concentration of 10,000 coliforms/100 ml and the erstwhile figure of 5,000/100 ml is justified by the difference between the Phelps Index and the MPN. The Subcommittee suggests these numbers and the additional consideration of fecal coliforms in order to provide more realistic parameters in full recognition of modern knowledge and not as a means of sanctioning increased bacterial pollution of waters destined for public water supply use.

Paragraph 6: Alkalinity

Alkalinity in water should be sufficient to enable floc formation during coagulation, must not be high enough to cause physiological distress in humans, and must be proper for a chemically balanced water (neither corrosive nor incrusting). A criterion for minimum and maximum alkalinity in public water supply is related to the relative amounts of bicarbonates, carbonates, and hydroxide ions causing the alkalinity; and also to the pH, filterable (dissolved) solids, and calcium content. Because the permissible criterion for filterable solids is 500 mg/l and the pH range is 6.0–8.5, alkalinity should not be less than about 30 mg/l.

The criterion for maximum alkalinity cannot be expressed in calcium carbonate equivalents as determined from 0.02N H_2SO_4 titration because of the interrelationships stated above. However, alkalinity values higher than about 400 mg/l–500 mg/l would be too high for public water supply use. Within the range of 30 mg/l–500 mg/l, the alkalinity criterion should be that value which is normal to the natural water and which from experience is satisfactory for public water supply use. Frequent variations from normal values are detrimental to public water supply processing control.

Paragraph 7: Ammonia

Ammonia is a significant pollutant in raw water for public water supplies because its reactions with chlorine result in compounds with markedly less disinfecting efficiency than free chlorine. In addition, it is frequently an indicator of recent sewage pollution.

In the early days of waste treatment, the oxidation of ammonia to nitrates was one of the major objectives of waste treatment, but with the development of the BOD test, this objective became neglected. Greater attention to the design and operation of waste treatment plants for the oxidation of ammonia and organic nitrogen is needed to minimize the concentration of pollution forms in these receiving waters.

Paragraph 8: Arsenic, Barium, Cadmium, Chromium (Hexavalent), Copper, Chloride, Cyanide, Iron, Lead, Manganese, Phenols, Selenium, Silver, Sulfate, Zinc, and Radioactive Substances

The significance of these substances as contaminants of drinking water is discussed in Drinking Water Standards.[10] The permissible criteria in this report are those included in Drinking Water Standards. With the possible exception of iron and in some instances copper and zinc, the defined treatment plant does little or nothing to remove these substances.

Paragraph 9: Boron

Boron is found in the natural ground and surface waters in some areas of the U.S., notably in the Western States where as much as 5–15 mg/l are encountered. However, extensive data on boron in both well and surface waters in North America show that the amount of boron normally encountered is less than 1 mg/l. The ingestion of large amounts of boron can affect the central nervous system and protracted ingestion may result in a clinical syndrome known as borism.

Boron is an essential element to plant growth but is toxic to many plants at levels as low as 1 mg/l. The Public Health Service has established a limit of 1 mg/l which provides a good factor of safety physiologically and also considers the domestic use of water for home gardening.

Paragraph 10: Dissolved Oxygen

Criteria for dissolved oxygen are included, not because the substance is of appreciable significance in water treatment or in finished water, but because of its use as an indicator of pollution by organic wastes. It is intended for application to freeflowing streams and not to lakes or reservoirs in which supplies may be taken from below the thermocline.

Annual average of maximum daily air temperature,[a] F	Recommended limit, mg/l
50.0–53.7	1.7
53.8–58.3	1.5
58.4–63.8	1.3
63.9–70.6	1.2
70.7–79.2	1.0
79.3–90.5	0.8

Paragraph 11: Fluoride

The subcommittee recognizes the potential beneficial effects of fluoride ion in domestic water supplies but recommends no "desirable" concentration since any value less than that recommended for the permissible limit would be acceptable from the point of view of a water pollution control program. Rapid fluctuations in raw water fluoride ion levels would create objectionable operating problems for communities supplementing raw water fluoride concentrations. The permissible criterion is the upper limit of the recommended range in Drinking Water Standards.[10]

Paragraph 12: Hardness

A singular criterion for the maximum hardness in public water supply is not possible. Hardness in water is largely the result of geological formations with which the water comes in contact. Public acceptance of hardness varies from community to community. Consumer sensitivity to objectionable hardness is related to the hardness with which he has become accustomed. Consumer acceptance of hardness may also be tempered by economic necessity.

Hardness should not be present in concentrations that will cause excessive soap consumption, or which will cause objectionable scale in heating vessels and pipes. In addition, varying water hardness is objectionable to both domestic and industrial water consumers. With varying hardness, the soap required for laundry, the effect on manufactured products, and the damage to process equipment (such as boilers and cooling coils) cannot be anticipated and compensated without facilities which are not available to most water users. A water hardness criterion must relate to the hardness which is normal to the supply and exclude hardness additions which will cause variations.

A criterion for objectionable hardness must be tailored to fit the requirements of each community. Hardness more than 300–500 mg/l as $CaCO_3$ is excessive for public water supply. Many consumers will object to water harder than 150 mg/l. In other communities, the criterion for maximum water hardness is considerably less than 150 mg/l. A moderately hard water is sometimes defined as having hardness between 60–120 mg/l.

Paragraph 13: Nitrate plus Nitrite

A limit of 10 mg/l (N) of nitrate ion plus nitrite ion will be recommended by Drinking Water Standards.[10] Because the nitrite ion is the substance actually responsible for causing methemoglobinemia, a combined limit on the two ions is more significant than a limit on nitrates only.

Paragraph 14: pH

Most unpolluted waters have a pH values within the range recommended as a permissible criterion. Any pH value within this range is acceptable for public water supply. The further selection of a specific pH value within this range as a desirable criterion cannot be made.

Paragraph 15: Phosphorus

The Subcommittee has considered establishing criteria on phosphorus concentrations but has not been able to establish any generally acceptable limit because of the complexity of the problem. The purpose of such a limit would be twofold:

a. To avoid problems associated with algae and other aquatic plants, and
b. To avoid coagulation problems due particularly to complex phosphates.

Phosphorus is an essential element for aquatic life as well as for all forms of life and has been considered the most readily controllable nutrient in efforts to limit the development of objectionable plant growths. Evidence indicates that high phosphorus concentrations are associated with the eutrophication of waters that is manifest in unpleasant algal or other aquatic plant growths when other growth-promoting factors are favorable; that aquatic plant problems develop in reservoirs or other standing waters at phosphorus values lower than those critical in flowing streams; that reservoirs and other standing waters will collect phosphates from influent streams and store a portion of these within the consolidated sediments; that phosphorus concentrations critical to noxious plant growths will vary with other water quality characteristics, producing such growths in one geographical area but not in another.

Because the ratio of total phosphorus to that form of phosphorus readily available for plant growth is constantly changing and will range from 2–17 times or greater, it is desirable to establish limits on the total phosphorus rather than the portion that may be available for immediate plant use. Most relatively uncontaminated lake districts are known to have surface waters that contain 10–30 μg/l total phosphorus as P; in some waters that are not obviously polluted, higher values may occur.[4] Data collected by the Federal Water Pollution Control Administration, Division of Pollution Surveillance, indicate that total phosphorus concentrations exceeded 50 μg/l (P) at 48 percent of the stations sampled across the nation.[6] Some potable surface water supplies now exceed 200 μg/l (P) without experiencing notable problems due to aquatic growths. Fifty μg/l of total phosphorus (as P) would probably restrict noxious aquatic plant growths in flowing waters and in some standing waters. Some lakes, however, would experience algal nuisances at and below this level.

Critical phosphorus concentrations will vary with other water quality characteristics. Turbidity and other factors in many of the nation's waters negate the algal-producing effects of high phosphorus concentrations. When waters are detained in a lake or reservoir, the resultant phosphorus concentration is reduced by some extent over that in influent streams by precipitation or uptake by organisms and subsequent deposition in fecal pellets or dead organism bodies. See the report of the Subcommittee for Fish, Other Aquatic Life, and Wildlife, and the section on Plant Nutrients and Nuisance Organisms for a more complete discussion of phosphorus associations with the enrichment problem.

At concentrations of complex phosphates of the order of 100 μg/l, difficulties with coagulation are experienced.

Paragraph 16: Total Dissolved Solids (Filterable Residue)

Drinking Water Standards[10] recommended that total dissolved solids not exceed 500 mg/l where other more suitable supplies are available. It is noted, however, that some streams contain total dissolved solids in excess of 500 mg/l. For example, the Colorado River at the point of withdrawal by the Metropolitan Water District of Southern California has a total dissolved solids concentration up to 700 mg/l.

High total dissolved solids are objectionable because of physiological effects, mineral taste, or economic effect. High concentrations of mineral salts, particularly sulfates and chlorides, are associated with corrosion damage in water systems. Regarding taste, on the basis of limited research work underway in California, limits somewhat higher than 500 mg/l are probably acceptable to consumers of domestic water supplies. It is likely that levels set with relation to economic effects are controlling for this parameter.

Increases in total dissolved solids from those normal to the natural stream are undesirable and may be detrimental.

It is recommended that the permissible value for total dissolved solids be set at 500 mg/l in view of the above evaluation. Further, it is recommended that research work be sponsored to obtain more information on total dissolved solids in water relating to physiological effects, consumer attitudes toward taste, and economic considerations.

Paragraph 17: Uranyl Ion

The standard for uranyl ion ($UO_2^{=}$) is established on the basis of its chemical properties rather than on the basis of its being a radioactive material. It is being added to Drinking Water Standards.[10] Uranyl ion is of concern in drinking water because of possible damage to the kidneys. The threshold level of taste and the appearance of color due to uranyl ion occur at about 10 mg/l which is much less than the safe limit of ingestion of this ion insofar as adverse physiological effects are concerned.

The Public Health Service adopted the figure of 5 mg/l which is one half the limit based on taste and color and, therefore, there is a considerable factor of safety in the adoption of 5 mg/l.

Paragraph 18: Carbon Chloroform Extract (CCE)

A limit of 0.2 mg/l carbon chloroform extract in drinking water is recommended in the Drinking Water Standards "as a safeguard against the intrusion of excessive amounts of potentially toxic material into water."[10,3] Although the analytical procedure then in use leaves much to be desired from the standpoint of simplicity, reproducibility, and interpretation, it was the best available at that time. The analytical procedure has been improved since then and the newer technique[1,2] gives substantially higher results than the one originally used. The defined method of treatment generally removes very little of the CCE present in the raw water. In many instances there is an increase during treatment. Whether this is real or apparent is not known.

The permissible criterion of 0.15 mg/l recommended is based on use of the procedure cited in Drinking Water Standards.[10] We do not as yet have sufficient information on which to base a recommended limit using the lower flow rates and sample volumes of the newer procedure. When this information is available, a change in the criterion is advisable. This limit is generally attainable where vigorous efforts at pollution control are carried out.

Paragraph 19: Methylene Blue Active Substances

This is an operationally more precise name for substances discussed in Drinking Water Standards[10] as alkyl benzene sulfonate. The permissible criterion is the same as the limit recommended in those standards. Those standards have been revised to reflect this change in nomenclature.

Paragraph 20: Oil and Grease

It is very important that public water supply be free of oil and grease. The difficulty of obtaining representative samples of these materials from water makes it virtually impossible to express criteria in numerical units. Since even very small quantities of oil and grease may cause troublesome taste and odor problems, the Subcommittee desires that some of this material be present in public water supplies. An additional problem attributable to these agents is the unsightly scum lines on water treatment basin walls, swimming pools, and other containers.

Paragraph 21: Pesticides and Herbicides

Consideration was given by the Subcommittee to three groups of pesticides: the more common chlorinated hydrocarbons, herbicides, and the cholinesterase-inhibiting group which include the organic phosphorus types and the carbamates. The permissible levels are based upon recommendations of the Public Health Service Advisory Committee on Use of the PHS Drinking Water Standards. These values were derived for that Committee by an expert group of toxicologists as those levels which, if ingested over extensive periods could not cause harmful or adverse physiological changes in man. In the case of aldrin, heptachlor, chlordane, and parathion, the Committee adopted even lower than physiologically safe levels; namely, amounts which, if present, can be detected by their taste and odor. It should be noted that this National Technical Advisory Subcommittee on Public Water Supplies is not a group of toxicological experts. Hence, the promulgation of additional criteria by the Public Health Service would also be recommended by this Subcommittee, tempered – as was done above – by its experience and judgment in the area of water treatment, as, for example, in public acceptance or organoleptic properties.

The limit for the cholinergic pesticides is established relative to parathion and is expressed as 0.1 mg/l parathion equivalent. This equivalence is the ratio that a given pesticide of this group has to parathion as unity in its cholinesterase-inhibiting properties. This makes it incumbent upon an administrator of this limit to determine the pesticide involved and to obtain expert toxicological opinion on its parathion equivalence. Nearly all the organophosphorus compounds and the cholinergic carbamates have high acute toxicity to mammals and some have even higher toxicity to fish. Ingestion of small quantities of these compounds over long time periods causes damage to mammalian central nervous systems. Many organophosphorus pesticides hydrolyze rapidly in the environment to harmless or less harmful products. The hazards from the chlorinated hydrocarbon pesticides in water result from both direct effects, because they tend to persist in their original form over long periods, and indirect effects because they may be concentrated biologically in man's food chain. The values which were selected by the Public Health Service as limits for this group of pesticides are, however, set with substantial safety factors insofar as they adversely affect the human body. Generally, fish are more sensitive to this group of pesticides and, therefore, may serve as a rough method for determining when the chlorinated hydrocarbon pesticides content of water is approaching a danger level. See the report of the Fish, Other Aquatic Life, and Wildlife Subcommittee for pesticide limits relative to maintaining healthy and productive aquatic life.

It should be noted that limits for pesticides and herbicides have been set with relation only to human intake directly from a related domestic water supply. The consequence of higher and possibly objectionable concentrations in fish available to be eaten by man due to biological concentration is considered not within the scope of the charge to this Subcommittee.

Source: *Water Quality Criteria,* U.S. Government Printing Office, April 1968.

REFERENCES

1. R.L. Booth, J.N. English, and G.N. McDermott, *J. Am. Water Works Assn., 57,* 215, 1965.
2. A.W. Breidenbach *et al., The Identification and Measurement of Chlorinated Hydrocarbon Pesticides in Surface Waters,* U.S. Department of the Interior, 1966.
3. M.B. Ettinger, *J. Am. Water Works Assn., 52,* 689, 1960.
4. M.F. Gales, Jr., E.C. Julian, and R.C. Kroner, *J. Am. Water Works Assn., 58,* 1363, 1966.
5. E.E. Geldreich, *Sanitary Significance of Fecal Coliforms in the Environment,* U.S. Department of the Interior, 1966.
6. C.B. Gunnerson, *An Atlas of Water Pollution Surveillance in the United States, Oct. 1, 1957, to Sept. 30, 1965,* Federal Water Pollution Control Federation, 1966.
7. L. Koenig, *J. Am. Water Works Assn., 59,* 290, 1967.
8. F.M. Middleton, A.A. Rosen, and R.H. Burttschell, *J. Am. Water Works Assn., 54,* 223, 1962.
9. *Standard Methods for the Examination of Water and Wastewater,* 12th edition, Am. Publ. Health Assn., 1967.
10. *Public Health Service Drinking Water Standards,* U.S. Department of Health, Education and Welfare, 1962.

3.1–2 SURFACE WATER CRITERIA FOR PUBLIC WATER SUPPLIES

Characteristic	Permissible criteria	Desirable criteria	Paragraph
Physical			
Color (color units)	75	<10	1
Odor	Narrative	Virtually absent	2
Temperature[a]	Narrative	Narrative	3
Turbidity	Narrative	Virtually absent	4
Microbiological			
Coliform organisms	10,000/100 ml[b]	<100/100 ml[b]	5
Fecal coliforms	2,000/100 ml[b]	<20/100 ml[b]	5
	(mg/l)	(mg/l)	
Inorganic chemicals			
Alkalinity	Narrative	Narrative	6
Ammonia	0.5 (as N)	<0.01	7
Arsenic[a]	0.05	Absent	8
Barium[a]	1.0	Absent	8
Boron[a]	1.0	Absent	9
Cadmium[a]	0.01	Absent	8
Chloride[a]	250	<25	8
Chromium,[a] hexavalent	0.05	Absent	8
Copper[a]	1.0	Virtually absent	8
Dissolved oxygen	⩾4 (monthly mean) ⩾3 (individual sample)	Near saturation	10
Fluoride[a]	Narrative	Narrative	11
Hardness[a]	Narrative	Narrative	12
Iron (filterable)	0.3	Virtually absent	8
Lead[a]	0.05	Absent	8
Manganese[a] (filterable)	0.05	Absent	8
Nitrates plus nitrites[a]	10 (as N)	Virtually absent	13
pH (range)	6.0–8.5	Narrative	14
Phosphorus[a]	Narrative	Narrative	15
Selenium[a]	0.01	Absent	8
Silver[a]	0.05	Absent	8
Sulfate[a]	250	<50	8
Total dissolved solids[a] (filterable residue)	500	<200	16
Uranyl ion[a]	5	Absent	17
Zinc[a]	5	Virtually absent	8
Organic chemicals			
Carbon chloroform extract[a] (CCE)	0.15	<0.04	18
Cyanide[a]	0.20	Absent	8
Methylene blue active substances[a]	0.5	Virtually absent	19
Oil and grease[a]	Virtually absent	Absent	20
Pesticides			
Aldrin[a]	0.017	Absent	21
Chlordane[a]	0.003	Absent	21
DDT[a]	0.042	Absent	21
Dieldrin[a]	0.017	Absent	21
Endrin[a]	0.001	Absent	21
Heptachlor[a]	0.018	Absent	21

[a] The defined treatment process has little effect on this constituent.

[b] Microbiological limits are monthly arithmetic averages based upon an adequate number of samples. Total coliform limit may be relaxed if fecal coliform concentration does not exceed the specified limit.

3.1–2 SURFACE WATER CRITERIA FOR PUBLIC WATER SUPPLIES (*Continued*)

Characteristic	Permissible criteria	Desirable criteria	Paragraph
Heptachlor epoxide[a]	0.018	Absent	21
Lindane[a]	0.056	Absent	21
Methoxychlor[a]	0.035	Absent	21
Organic phosphates plus carbamates[a]	0.1[c]	Absent	21
Toxaphene[a]	0.005	Absent	8
Herbicides			
2,4-D plus 2,4,5-T, plus 2,4,5-TP[a]	0.1	Absent	21
Phenols[a]	0.001	Absent	8
Radioactivity			
Gross beta[a]	1,000 (pc/l)	<100 (pc/l)	8
Radium-226[a]	3 (pc/l)	<1 (pc/l)	8
Strontium-90[a]	10 (pc/l)	<2 (pc/l)	8

[a] The defined treatment process has little effect on this constituent.

[c] As parathion in cholinesterase inhibition. It may be necessary to resort to even lower concentrations for some compounds or mixture. See paragraph 21.

Source: *Water Quality Criteria,* U.S. Government Printing Office, April 1968.

3.1–3 HEALTH AND AESTHETIC SIGNIFICANCE OF TRACE ELEMENTS AND COMPOUNDS

Occurrence in U.S. Water Supplies

Trace element or compound	Health or aesthetic effect	1962 USPHS drinking water standards limit, mg/l	Occurrence in 2,595 distribution samples, 1969, mg/l	
			Avg	Range
Alkyl benzine sulfonate (ABS)	USPHS limit gives a safety factor of 15,000; not highly toxic	0.5[a]	0.05	0–0.41
Arsenic (As)	Serious systemic poison – 100 mg usually causes severe poisoning, is cumulative, and may cause chronic effects. Late evidence indicates tiny amounts may be beneficial	0.01[a] 0.05[b]	0.0001	<0.03–0.10
Chloride (Cl^-)	Limit set for taste reasons	250.[a]	27.6	<1.0–1,950
Copper (Cu)	Body needs copper at level of about a mg/day for adults; not a health hazard except when large amounts are ingested	1.[a]	0.13	0–8.35
Carbon chloroform extract (CCE)	At limit stated, organics in water are not considered a health hazard	0.2[a]	0.11	0.008–0.56
Cyanide (CN^{-2})	Rapid fatal poison, but limit set provides safety factor of about 100	0.01[a] 0.2[b]	0.00009	<0.1–0.008
Fluoride (F^-)	Beneficial in small amounts; above 2,250 mg dose can cause death	0.7–1.2[a] 1.4–2.4[b]	0.32	<0.2–4.40
Nitrate (NO_3)	Excessive amounts can cause methemoglobinemia (blue baby) in infants	45[a]	6.3	<0.1–127
Sulfate (SO_4)	Above 750 mg/l, usually has laxative effect	250[a]	46	<1.0–770
Zinc (Zn)	Zinc is beneficial in that a child needs 0.3 mg/kg day; 675–2,280 mg/l may be an emetic	5[a]	0.19	0–13.0
Barium (Ba)	Fatal dose is 550–600 mg as the chloride; it is a muscle (including heart) stimulant	1.0[b]	0.034	0–1.55
Cadmium (Cd)	13–15 ppm in food has caused illness	0.01[b]	0.003	<0.2–3.94
Chromium (Cr^{+6})	Limit provides a safety factor. Carcinogenic when inhaled	0.05[b]	0.0023	0–0.079
Lead (Pb)	Serious, cumulative, body poison	0.05[a]	0.013	0–0.64
Selenium (Se)	Toxic to both humans and animals in large amounts. Late research suggests small amounts may be beneficial	0.01[b]	0.003	0–0.07
Silver (Ag)	Can produce irreversible, adverse cosmetic changes	0.05[b]	0.008	0–0.03
Radium (Ra-226)	A bone-seeking, internal alpha emitter that can destroy bone marrow	3 pCi/l[c]	2.2 pCi/l	0–135.9 pCi/l
Strontium (SR-90)	A bone-seeking, internal beta emitter	10 pCi/l[c]	<1.0 pCi/l	0–2.0 pCi/l

[a] Not to be exceeded where other more suitable supplies are, or can be made, available.
[b] Excess constitutes grounds for rejection of the supply.
[c] Consult 1962 *USPHS Drinking Water Standards* for interpretation.

Source: Reprinted, with publisher's permission, from *J. Am. Water Works Assn., 63*(11):728, © 1971 by the American Water Works Association, Inc., 2 Park Ave., New York, N.Y. 10016.

3.1–4 HEALTH AND AESTHETIC SIGNIFICANCE OF TRACE ELEMENTS AND COMPOUNDS IN FINISHED DRINKING WATER

Name of ingredient	Health or aesthetic effect	Unofficial limit, mg/l	Occurrence in U.S. water supplies, mg/l	
			Avg	Range
Substance				
Antimony (Sb)	Similar to arsenic but less acute. Recommended limit not to exceed 0.1 mg/l; routinely below 0.05 mg/l; over longtime periods below 0.01 mg/l	0.05		
Beryllium (Be)	Poisonous in some of its salts in occupational exposure	None		
Bismuth (Bi)	A heavy metal in the arsenic family – avoid in water supplies	None		
Boron (B)	Ingestion of large amounts can affect central nervous system, and protracted ingestion may cause borism	1.0	0.069	0–3.28[a]
Cobalt (Co)	Beneficial in small amounts; about 7 μg/day	None	0.002	0.0–0.019[a]
Molybdenum (Mo)	Necessary for plants and ruminants. Excessive intakes may be toxic to higher animals; acute or chronic effects not well known	None		
Mercury (Hg)	Continued ingestion of large amounts can damage brain and central nervous systems	0.005	0.000	0.000–0.033[b]
Nickel (Ni)	May cause dermatitis in sensitive people; doses of 30–73 mg of $NiSO_4 \cdot 6H_2O$ have produced toxic effects	None	0.005	0.0–0.072[a]
Sodium (Na)	A beneficial and needed body element, but can be harmful to people with certain diseases	20[c]	40 percent above 20 mg/l	0.4–1,900[d]
Tin (Sn)	Long used in food containers without known harmful effects	None		
Uryanyl ion (UO_2)	May cause damage to kidneys	5.0		
Vanadium (V)	Some evidence that vanadium may be beneficial with respect to heart disease	None		
Pesticides				
Aldrin	One or all of these complex organic compounds have severe, acute, adverse health effects when ingested in large amounts. Small amounts accumulate, and long-range effects are generally unknown	0.017[e]	160 samples in NCWSS showed lower values than the limits	
Chlordane		0.003[e]		
DDT		0.042[e]		
Dieldrin		0.017[e]		
Endrin		0.001[e]		
Heptachlor		0.018[e]		
Heptachlor epoxide		0.018[e]		
Lindane		0.056[e]		
Methoxychlor		0.035[e]		
Toxaphene		0.005[e]		
Organic phosphorus plus carbamates, such as Parathion, Malathion, Carbaryl, and others	The organic phosphorus and carbamate pesticides are severe acute poisons affecting the central nervous system; ingestion of small amounts over time can harm the central nervous system	0.1[f]	Same as for pesticides	

[a] National Community Water Supply Study (NCWSS). 2500 samples.
[b] Late data from BWH, USPHS.
[c] The amount at, and above, which people on strict (500 mgd) sodium-restricted diets must include in their daily sodium-intake calculations.
[d] Survey of over 2,000 U.S. municipalities by USPHS in 1963–66.
[e] Report of the Committee on Water Quality Criteria, FWPCA, U.S. Department of the Interior.
[f] As parathion equivalent in cholinesterase inhibition.

3.1–4 HEALTH AND AESTHETIC SIGNIFICANCE OF TRACE ELEMENTS AND COMPOUNDS IN FINISHED DRINKING WATER *(Continued)*

Name of ingredient	Health or aesthetic effect	Unofficial limit, mg/l	Occurrence in U.S. water supplies, mg/l	
			Avg	Range
Herbicides				
Algicide	A group having toxic properties of a generally lower order than the above pesticides. However, they should not be used without great care and should not be found in drinking water	1.0		
Copper key – El		1.0 (Cu)		
Cuprose		1.0 (Cu)		
Cutrine		1.0 (Cu)		
Diquot		0.2		
Dalapon		0.1		
2,4-D		1.0		
2,4,5-T should not be used				
Micro-gard PR		0		
Mogul algicide AG-470		1.0		
Radapon		0.1		
Silvex (2,4,5-TP)		0.2		
Tordon 10K and 101		0		
Twink		0.2		
Ureabor		1.0 (as B)		

Note: Table shows limits not found in 1962 *USPHS Drinking Water Standards.*

Source: Reprinted, with publisher's permission, from *J. Am. Water Works Assn., 63*(11):728, © 1971 by the American Water Works Association, Inc., 2 Park Ave., New York, N.Y. 10016.

3.1–5 POTABLE WATER QUALITY GOALS[a]

American Water Works Association

Characteristic	Goal
Physical Factors	
Turbidity	Less than 0.1 unit
Nonfilterable residue	Less than 0.1 mg/l
Macroscopic and nuisance organisms	No such organisms
Color	Less than 3 units
Odor	No odor
Taste	No taste objectionable
Chemical Factors, mg/l	
Aluminum (Al)	Less than 0.05
Iron (Fe)	Less than 0.05
Manganese (Mn)	Less than 0.01
Copper (Cu)	Less than 0.2
Zinc (Zn)	Less than 1.0
Filterable residue	Less than 200.0
Carbon chloroform extract (CCE)	Less than 0.04
Carbon alcohol extract (CAE)	Less than 0.10
Methylene blue active substances (MBAS)	Less than 0.20
Corrosion and Scaling Factors	
Hardness (as $CaCO_3$)	80 mg/l; a balance between deposition and corrosion characteristics is necessary; a level of 80 mg/l seems best, generally, considering all the quality factors; however, for some supplies, a goal of 90 or 100 mg/l may be deemed desirable
Alkalinity (as $CaCO_3$)	Change of not more than 1 mg/l (decrease or increase in distribution system, or after 12 hr at 130°F in a closed plastic bottle, followed by filtration)
Coupon tests (incrustation and loss by corrosion)	90 day tests (incrustation on stainless steel not to exceed 0.05 mg/cm^2, loss by corrosion of galvanized iron not to exceed 5.00 mg/cm^2)
Bacteriologic Factors	
Coliform organisms (by multiple fermentation techniques)	No coliform organisms
Coliform organisms (by membrane filter techniques)	No coliform organisms

[a] For all health related constituents not stated herein, these goals shall require complete compliance with all recommended and mandatory limits contained in current USPHS Drinking Water Standards. Unless other methods are indicated, analyses shall be made in conformance with the latest edition of *Standard Methods of Examination of Water and Wastewater.*

3.1–5 POTABLE WATER QUALITY GOALS[a] *(Continued)*

Characteristic	Goal
Radiologic Factors	
Gross beta activity	Less than 100 pCi/l

[a] For all health related constituents not stated herein, these goals shall require complete compliance with all recommended and mandatory limits contained in current USPHS Drinking Water Standards. Unless other methods are indicated, analyses shall be made in conformance with the latest edition of *Standard Methods of Examination of Water and Wastewater.*

Note: The data here are not a standard. They implement the areas covered by *Drinking Water Standards.* They cover areas not directly associated with health effects, which are recognized as the prerogative of health agencies.

Source: Reprinted, with publisher's permission, from *J. Am. Water Works Assn., 60*(12):1317, © 1968 by the American Water Works Association, Inc., 2 Park Ave., New York, N.Y. 10016.

3.1–6 RECOMMENDED QUALITY CRITERIA GOALS OF LAKE WATER

At Chicago Intakes[a]

Parameter	Acceptable limits[b]	Permissible limits[d]
No. of coliform organisms/100 ml	<200	2,500
DO, ppm	$\nless$ 8	$>$ 5
BOD, ppm	$<$ 1.5	$\ngtr$ 3
pH	8.2	$\ngtr$ 9 $\nless$ 7.5
Chlorides, ppm	8[c]	$\ngtr$ 15
Color	$<$ 5	$\ngtr$ 15
Total iron, ppm	$<$ 0.10	$\ngtr$ 0.2
Phenols, ppm	$<$ 0.001	$\ngtr$ 0.001
ABS, ppm	$<$ 0.05	$\ngtr$ 0.15
Total dissolved solids, ppm	165	185
Ammonia nitrogen, ppm	$<$ 0.02	0.05
Sulfate, ppm	30	50

[a] Miscellaneous trace contaminants must meet USPHS limits.
[b] Annual average.
[c] Subject to readjustment.
[d] These limits must not be reached more than 12 days per year.

Source: Reprinted, with publisher's permission, from *J. Am. Water Works Assn., 57*(7):841, © 1965 by the American Water Works Association, Inc., 2 Park Ave., New York, N.Y. 10016.

3.1.1 PHYSICAL STANDARDS

3.1–7 PHYSICAL CHARACTERISTICS OF DRINKING WATER

	Area of authority					
		Canada[2]		International[3]		
Characteristic	U.S.[1]	Objective	Acceptable limits	Maximum acceptable	Maximum allowable	Ideal water[4]
Color, TCU[a]	15	<5	15	5	50	3
Odor	3[b]	0[b]	4[b]	Unobjectionable	–	No change on carbon contact
Taste	–	Inoffensive	Inoffensive	Unobjectionable	–	None
Temperature, °C	–	<10	15	–	–	–
Turbidity, JTU[c]	5	<1	5	5	25	0.1
Non-filterable residue	–	–	–	–	–	0.1 unit
pH units[d]	–	–	6.5-8.3	7.0-8.5	>6.5-<9.2	–

[a] True color units, platinum-cobalt scale.
[b] Threshold odor number.
[c] Jackson turbidity units.
[d] The pH units are significant in controlling corrosion and the scaling tendency of water.

Note: No standards are listed for European drinking water, as their standards are concerned with piped supplies.

REFERENCES

1. *Drinking Water Standards,* U.S. Public Health Service, 1962.
2. *Canadian Drinking Water Standards and Objectives – 1968,* Department of National Health and Welfare, Ottawa, October 1969.
3. *International Standards for Drinking Water,* 2nd ed., World Health Organization, Geneva, 1963.
4. E.L. Bean, *J. Am. Water Works Assn., 54*(11):1313, 1962.

3.1–8 RECOMMENDED LEVELS OF RADIOACTIVITY IN DRINKING WATER

Concentration pCi/l

Radionuclide	U.S.[1]	Canada[2,a] Objective	Canada[2,a] Acceptable	Canada[2,a] Max limit	International[3]	European region[4]	Ideal water[5]
^{226}Ra	3				10		3
^{90}Sr	10				30		5
Gross beta activity (^{90}Sr and alpha emitters absent)[b,c]	1,000	<10 pCi/l			1,000	30	100
Gross alpha activity						3	
Specific nuclides		1/10 of ICRP[d] MPC_w for 168 hr week	1/3 of ICRP MPC_w for 168 hr week	The ICRP MPC_w for 168 hr week			

[a] Applicable to all radioactive substances in *raw* as well as treated waters.
[b] Does not include tritium.
[c] The word *absent* means a slightly smaller fraction of the above limits, where the limit for unidentified alpha emitters is taken as the listed value for ^{226}Ra.
[d] ICRP – International Commission on Radiological Protection.
MPC_w – Maximum permissible concentration in water.

REFERENCES

1. *Drinking Water Standards,* U.S. Public Health Service, 1962.
2. *Canadian Drinking Water Standards and Objectives – 1968,* Department of National Health and Welfare, Ottawa, October 1969.
3. *International Standards for Drinking Water,* 2nd ed., World Health Organization, Geneva, 1963.
4. *European Standards for Drinking Water,* 2nd ed., Regional Office for Europe, World Health Organization, Copenhagen, 1970.
5. E.L. Bean, *J. Am. Water Works Assn., 54*(11):1313, 1962.

3.1.2 BACTERIOLOGICAL STANDARDS

3.1–9 LIMITING BACTERIOLOGIC CRITERIA FOR DEVELOPMENT OF SURFACE SUPPLY SOURCES

Authority	Local classification of water	Unit	Limiting concentration of coliform organisms, MPN
Arkansas		avg monthly MPN	5,000
		max monthly MPN	20% samples >5,000
		max monthly MPN	5% samples >20,000
Kansas		max MPN[a]	50[b]
		max MPN	5,000[c,d]
		max MPN	20,000[c,e]
		max MPN	>20,000[f]
Minnesota	I	avg MPN	50
			20% samples >50[a]
	II	avg MPN	5,000
	II	max MPN	20% samples >4,000[d,g]
	II	max MPN	100% samples >2,000
	III	max MPN	20% samples >5,000[c,e]
			not >20,000
New York Water Pollution Control Board	A1 & A2	avg monthly MPN	50[b]
	A4	avg monthly MPN	5,000
			20% samples >5,000[g]
USPHS recommendations	Group II	avg monthly MPN	50[b]
	Group III	avg monthly MPN	5,000[g]
	Group III	max monthly MPN	50% samples >5,000
	Group IV	avg monthly MPN	5,000[c,d]
	Group IV	max monthly MPN	20% samples >5,000
	Group IV	max monthly MPN	5% samples >20,000
	Group V	avg monthly MPN	>20,000[f]
TVA	I	geometric avg MPN	50[b]
	II	geometric avg MPN	5,000[g]
	III	geometric avg MPN	20,000[c]

[a] For ground water.
[b] Minimum treatment is chlorination.
[c] Auxiliary treatment plus postchlorination required.
[d] Prechlorination recommended.
[e] Prechlorination required.
[f] Probable undesirable source.
[g] Minimum treatment is filtration plus postchlorination.

Source: Reprinted with publisher's permission, from *J. Am. Water Works Assn., 52*(11):1416, © 1960 by the American Water Works Association, Inc., 2 Park Ave., New York, N.Y. 10016.

3.1–10 TOTAL COLIFORM STANDARDS FOR DRINKING WATER

Objective	Acceptable limit	Maximum permissible limit
	Most Probable Number Method[a]	
No coliforms	At least 95% of the samples in any consecutive 30 day period should be "negative" for total coliform organisms	At least 90% of the samples in any consecutive 30 day period should be "negative" for total coliform organisms
No coliforms	None of the samples "positive" for total coliform organisms should have an MPN index greater than 4/100 ml	None of the samples "positive" for total coliform organisms should have an MPN index greater than 10/100 ml
	Membrane Filter Method[a]	
No coliforms	At least 95% of the samples in any consecutive 30 day period should be "negative" for total coliform organisms	At least 90% of the samples in any consecutive 30 day period should be "negative" for total coliform organisms
No coliforms	None of the samples "positive" for total coliform organisms should have a MF count greater than 4/200 ml or 10/500 ml portions	None of the samples "positive" for total coliform organisms should have a MF count greater than 6/200 ml or 15/500 ml portions

[a] Where less than 10 samples are analyzed in any consecutive 30 day period, no more than one sample should be "positive" for total coliform organisms. When more than one "positive" sample is encountered, a special series of samples should be collected from the same location and analyzed for total coliform organisms.

Source: *Canadian Drinking Water Standards and Objectives – 1968,* Department of National Health and Welfare, Ottawa, October 1969.

3.1–11 RAW WATER COLIFORM STANDARDS

Objective	Acceptable limit[b]	Maximum permissible limit[b]
	Fecal Coliforms[a]	
At least 95% of the samples in any consecutive 30 day period should have a fecal coliform density of less than 10/100 ml	At least 90% of the samples in any consecutive 30 day period should have a fecal coliform density of less than 100/100 ml	At least 90% of the samples in any consecutive 30 day period should have a fecal coliform density of less than 1,000/100 ml
Treatment by chlorination is required	Complete or partial treatment including chlorination may be required	Complete water treatment is required
	Total Coliforms[c]	
At least 95% of the samples in any consecutive 30 day period should have a total coliform density of less than 100/100 ml	At least 90% of the samples in any consecutive 30 day period should have a total coliform density of less than 1,000/100 ml	At least 90% of the samples in any consecutive 30 day period should have a total coliform density of less than 5,000/100 ml[d]
Treatment by chlorination is required	Complete or partial treatment including chlorination is required	Complete water treatment is required[d]

[a] The Maximum permissible limit of fecal coliforms in a single sample shall be determined by the control agency based upon considerations of the type and degree of enteric pollution and other local conditions within the watershed.

[b] The limits tabulated under the acceptable and maximum permissible columns, apply only where *complete treatment* (flocculation-coagulation, sedimentation, filtration, and disinfection, or any combination of these processes (partial treatment), but always including disinfection) is employed for public supplies; where only disinfection is employed, the limits tabulated under the objective column apply.

[c] The Maximum permissible limit of total coliform organisms in a single sample shall be determined by the control agency based on the type and degree of pollution and other local conditions existing within the watershed.

[d] (1) Applicable to the raw water only where complete treatment of drinking water is employed. (2) Where 5,000 total coliforms per 100 ml are found in more than 10% of the samples examined during the survey period, auxiliary treatment consisting of prechlorination or presedimentation or their equivalents, and postchlorination shall be applied. (3) In all other cases, the limits tabulated under the objective column shall apply.

Source: *Canadian Drinking Water Standards and Objectives – 1968,* Department of National Health and Welfare, Ottawa, October 1969.

3.1–12 WATERBORNE DISEASES IN THE U.S.

1946–1960

Disease	Outbreaks	Cases
Gastroenteritis	126	13,630
Typhoid	39	506
Infectious hepatitis	23	930
Diarrhea	16	5,160
Shigellosis	11	5,653
Salmonellosis	4	24
Amebiasis	2	36
Other	7	45
Total	228	25,984

Source: U.S. Water Resources Council, 1968.

3.1–13 INCIDENCE OF WATERBORNE DISEASES IN THE U.S. BY SOURCE OF POLLUTION

1946–1960

	Private or semipublic supply		Public utility		All systems	
Causes	**Outbreaks**	**Cases**	**Outbreaks**	**Cases**	**Outbreaks**	**Cases**
Untreated surface water						
Contamination of watershed	5	142	4	327	9	469
Overflow of sewage	1	5			1	5
Surface water	10	155	2	197	12	352
Total	16	302	6	524	22	826
Untreated ground water						
Surface pollution near source	15	378			15	378
Overflow of sewage	5	194			5	194
Contamination by flood water	3	30	2	4,200	5	4,230
Faulty construction			1	400	1	400
Creviced limestone	7	483			7	483
Leakage from abandoned well	1	50			1	50
Seepage of sewage	28	1,372	3	26	31	1,398
Undetermined	29	1,378	1	300	30	1,678
Total	88	3,885	7	4,926	95	8,811
Contamination of reservoir or cistern						
New storage tank not disinfected			1	58	1	58
Contamination of tank			1	31	1	31
Dead animal in reservoir	1	100			1	100
Total	1	100	2	89	3	189
Inadequate control of treatment						
Filtration			1	14	1	14
Chlorination alone	14	504	6	2,916	20	3,420
Chlorination and other			5	5,875	5	5,875
Interruption of treatment (chlorination only)	5	169	4	1,292	9	1,461
Total	19	673	16	10,097	35	10,770

Source: Reprinted, with publisher's permission, from *J. Am. Water Works Assn., 56*(8):947, © 1964 by the American Water Works Association, Inc., 2 Park Ave., New York, N.Y. 10016.

3.1.3 CHEMICAL STANDARDS

3.1–14 CHEMICAL STANDARDS OF DRINKING WATER

	Recommended maximum limits,[a] mg/l			Maximum permissible[b] concentrations, mg/l		
Quality factor	**1946**	**1961 revision**	**1962**	**1946**	**1961 revision**	**1962**
Alkylbenzenesulfonate (detergent)	–	0.5	0.5			
Arsenic	–	0.01	0.01	0.05	0.05	0.05
Barium				–	1.0	
Cadmium	–	–		–	0.01	
Carbon chloroform extract	–	0.2	0.2			
Chloride	250	250	250			
Chromium	–	–		0.05	0.05	0.05
Copper	3.0	1.0	1.0			
Cyanide	–	0.01	0.01	–	0.02	
Fluoride	–	1.7[c]	0.7–1.2	1.5	2.2[c]	1.5
Iron plus manganese	0.3					
Iron	–	0.3	0.3			
Lead	–	–		0.1	0.05	0.1
Manganese	–	0.05	0.05			
Nitrate	–	45	45			
Phenols	0.001	0.001	0.001			
Selenium	–	–		0.15	0.01	0.05
Silver	–	–		–	0.05	
Sulfate	250	250	250			
Total dissolved solids	500	500				
Zinc	15	5	5			

[a] Concentrations in water should not be in excess of these limits, when more suitable supplies can be made available.
[b] "Maximum permissible" implies that which constitutes grounds for rejection of supply.
[c] Fluoride temperature concentration relationships are discussed in detail in the text.

REFERENCES

1. *Drinking Water Standards,* U.S. Department of Health, Education, and Welfare, Public Health Service, 1961.
2. *Drinking Water Standards,* U.S. Department of Health, Education, and Welfare, Public Health Service, 1962.

3.1–15 CHEMICAL LIMITS AND RECOMMENDED LEVELS FOR DRINKING WATER

Chemical	U.S.[1,a]	Canada[2,b] Objective	Canada[2,b] Acceptable	International[3,c,d] Max acceptable concentration	International[3,c,d] Max allowable concentration	European region[4,d,p] (Level at which trouble may arise)	Ideal water[5,e]	Manual for evaluating public drinking water supplies[6,f]
Alkyl Benzene Sulfonate (ABS)–surfactants	0.5	<0.2[u]	0.5[u]	0.5	1.0 (0.5)[i]		0.20	0.5[u]
Arsenic (As)	0.01	Not detectable	0.01	0.05	0.05	0.05[s]	0.01	0.01
Carbon Chloroform Extract (CCE)–organic pollutants	0.2	<0.05[w]	0.2[w]	0.2	0.5[c]	0.2–0.5[t]	0.04	0.2
Chloride (Cl)	250.0	<250.	250.	200.	600.	200, 600[r]	–	250.
Copper (Cu)	1.0	<0.01	1.0	1.0	1.5	0.05,[t] 3.0[m,n,o]	0.2	1.0
Cyanide (CN)	0.01	Not detectable	0.01	0.2	0.2	0.05[5]	0.01	0.01
Fluoride (F)	See 1.3–37	See 1.3–37				See 1.3–37	0.7–1.2[f]	See 1.3–37
Iron (Fe)	0.3	<0.05	0.3	0.3	1.0 (50)[g]	0.1[n]	0.05	0.3
Magnesium (Mg)	b	<50	150	50.	150.	<30, 125[q]	–	
Manganese (Mn)	0.05	<0.01	0.05	0.1	0.5 (5)[g]	0.05	0.01	0.05
Nitrate (NO_3)	45.0 (10)[e]	<10[v]	<10[v]		45.0	<50[j] 50–100[k] >100[l,n,p]	5.0[e]	10[v]
Phenols or compounds (as C_6H_5OH)	0.001	Not detectable	0.002	0.001	0.002	<0.001	0.0005	0.001
Sulfate (SO_4)	250.0	<250	500	200.	400.	250	–	250
Total Dissolved Solids (TDS)–filterable residue	500.0	<500	1,000	500.	1,500.		200.	500
Zinc (Zn)	5.0	<1.0	5.0	5.0	15. (1.5)[h]	5.0	1.0	5.0
Calcium (Ca)		<75	200	75.	200.		32 (hardness as Ca)	
pH				7.0–8.5	6.5–9.2		–	
Magnesium and sodium sulfate				500.	1,000.		–	
Total organic phosphorus or chlorinated hydrocarbon insecticides							None	
Aluminum (Al)							0.05	
Alcohol soluble from carbon adsorption								

3.1–15 CHEMICAL LIMITS AND RECOMMENDED LEVELS FOR DRINKING WATER (*Continued*)

Chemical	U.S.[1],[a]	Canada[2],[b]		International[3],[c,d]		European region[4],[d,p] (Level at which trouble may arise)	Ideal water[5],[e]	Manual for evaluating public drinking water supplies[6],[f]
		Objective	Acceptable	Max acceptable concentration	Max allowable concentration			
Hardness (as $CaCO_3$)		<120					0.10	
Alkalinity (as $CaCO_3$)							80	
Coupon tests							<1 mg/l change 3 month tests	
H_2S		Not detectable[p]	0.3			0.05		
Polycyclic Aromatic Hydrocarbons (PAH)						0.0002[r]		
Uranyl ion (UO_2^{++})		<1.0	5.0					
Ammonia as N		0.01	0.5					5.0
Phosphates as PO_4 (inorganic)		<0.2	0.2					

a Not in excess of amounts shown if other supplies are or can be made available.
b Not included.
c Concentrations greater than 0.2 mg/l indicate the necessity for further analyses to determine the causative agent.
d These limiting concentrations are limiting only and can be disregarded in specific instances.
e (10) as N (nitrate and nitrite).
f Temperature range 50–90.5°F (10–32.3°C).
g Maximum allowable limit affecting potability of water.
h These are limiting values for raw water quality and for that reason are lower than the allowable limits for drinking water, where the presence of these metallic substances would probably be the result of the aggressive action of the water on service pipe metals.
i This value has been established on the basis of the maximum sensitivity of the presently acceptable analytical procedures.
j Recommended.
k Acceptable.
l Not recommended.
m After 16 hours contact with new pipes.
n It is advisable that a special sample be collected for examination for nitrate and iron. The sample should be "fixed" at the time of collection by adding one ml of concentrated sulfuric acid for each liter of water.
o In small installations in which removal of iron would be uneconomic, or where the iron is present in a stable form, a level of up to 0.3 mg/l can be permitted.
p If the nitrate content is within the acceptable range and the water is otherwise chemically and bacteriologically satisfactory, it may not give rise to trouble, but physicians in the area should be warned of the possibility of infantile methemoglobinemia occurring. More information is required on the exact circumstances under which infantile methemoglobinemia occurs and the mechanism by which it is produced.
q Not more than 30 mg/l if there are 250 mg/l of sulfate; if there is less sulfate, magnesium up to 125 mg/l may be allowed.
r This limit of 200 mg/l may be exceeded in certain existing conditions but should in no circumstances exceed 600 mg/l.
s Upper limit of concentration.
t Extraction with chloroform, without the use of activated carbon – the "liquid-liquid" method – is preferred. These are suggested values.
u Now classified as detergents (methylene blue active substances).
v As N (nitrate and nitrite).
w Organics as CCE and CAE (total carbon chloroform and carbon alcohol extractibles).

3.1–15 CHEMICAL LIMITS AND RECOMMENDED LEVELS FOR DRINKING WATER (*Continued*)

REFERENCES

1. *Drinking Water Standards,* U.S. Public Health Service, 1962.
2. *Canadian Drinking Water Standards and Objectives – 1968,* Department of National Health and Welfare, Ottawa, October 1969.
3. *International Standards for Drinking Water,* 2nd ed., World Health Organization, Geneva, 1963.
4. *European Standards for Drinking Water,* 2nd ed., Regional Office for Europe, World Health Organization, Copenhagen, 1970.
5. E.L. Bean, *J. Am. Water Works Assn., 54*(11): 1313, 1962.
6. *Manual for Evaluating Public Drinking Water Supplies,* U.S. Public Health Service, 1969.

3.1–16 DRINKING WATER STANDARDS – CHEMICAL LIMITS

mg/l (ppm)

Chemical	U.S. Ref. 1[a]	U.S. Ref. 6	Canada[2,b] Objective	Canada[2,b] Acceptable limit	Canada[2,b] Max permissible limit	International[3,c]	European region[4,d]	Ideal water[5]
Arsenic (As)	0.05	0.01	g	0.01	0.05	0.05	0.05	0.01
Barium (Ba)	1.0	1.0	g	<1.0	1.0	1.0		0.5
Cadmium (Cd)	0.01	0.01	g	<0.01	0.01	0.01	0.01	0.01
Chromium (Cr^{+6})	0.05	0.05	g	<0.05	<0.05	0.05	0.05	0.01
Cyanide (CN)	0.2	0.01	g	0.01	0.20	0.2	0.05	0.01
Fluoride (F)						1.5[e]		0.7–1.2[i]
Lead (Pb)	0.05	0.05	g	<0.05	0.05	0.05	0.1[f]	0.03
Selenium (Se)	0.01	0.01	g	<0.01	0.01	0.01	0.01	0.01
Silver (Ag)	0.05	0.05	–	–	0.05			0.02
Phenolic substances (C_6H_5OH)						0.002		
Nitrates (NO_3)			<10.0[h]	<10.0[h]	10.0[h]	45.0[e]		
Boron (B)		1.0	–	<5.0	5.0			
Mercury (Hg)	0.005[7]							

a Rejection levels.
b Toxic chemicals.
c Toxic substances.
d Toxic substances in piped supplies.
e Components are hazardous to health.
f The upper limit in the supply should be 0.1 mg/l of lead (as Pb), but certain undertakings still use lead piping; in these instances the concentration of lead in the water after prolonged contact with the pipes may be higher. In no instance should the concentration of lead (as Pb) exceed 0.3 mg/l after 16 hours contact with the pipes. If the limit of 0.3 mg/l is regularly exceeded, it will be necessary either to change the piping or to treat the water. Lead is used as a stabilizer in some plastic pipes, and the possibility of its being leached out must be borne in mind.
g Not detectable by the method described in the latest edition of *Standard Methods* (APHA, AWWA, and WPCF) or by any other acceptable method approved by the control agency.
h Nitrate and nitrite as N.
i Over temperature range of 50–90.5°F (10.0–32.3°C).

REFERENCES

1. *Drinking Water Standards,* U.S. Public Health Service, 1962.
2. *Canadian Drinking Water Standards and Objectives – 1968,* Department of National Health and Welfare, Ottawa, October 1969.
3. *International Standards for Drinking Water,* 2nd ed., World Health Organization, Geneva, 1963.
4. *European Standards for Drinking Water,* 2nd ed., Regional Office for Europe, World Health Organization, Copenhagen, 1970.
5. E.L. Bean, *J. Am. Water Works Assn., 54*(11):1313, 1962.
6. *Manual for Evaluating Public Drinking Water Supplies,* U.S. Public Health Service, 1969.
7. *J. Am. Water Works Assn., 62*(5):285, 1970.

3.1–17 EFFECT OF MINIMAL DOSAGE OF FLUORIDE ION

Dosage effect	Total F intake, mg/kg of body weight
Acute (fatal)	14–70
Subacute	3–4
Chronic (long term) (Osteosclerosis–7–20 yr of continuous daily dosage)	0.3–0.6
Mild chronic (Dental mottling if ingested daily first 6–10 yr of life)	0.10–0.16
Caries prevention	0.03[a]–0.06

[a] Assumes a 70 kg adult (approximately 154 lb) consumes 1.6 l of 1 ppm F water per day and includes a food intake of 0.4 mg F.

Source: Lee Powers, *Northwest Medicine, 51*(2):115, February 1952.

3.1–18 RECOMMENDED FLUORIDE LEVELS IN DRINKING WATER

mg/l (ppm)

Annual average of maximum daily air temperatures		U.S.								
		PHS control limits[1]						Europe[4]		
°F	°C	Lower	Optimum	Upper	Ref. 6	Canada[2]	International[3]	Lower	Upper	Ideal waters[5]
50.0–53.7	10.0–12.0	0.9	1.2	1.7			1.0–1.5[f]	0.9[a]	1.7[a]	1.2
					1.8					
53.8–58.3	12.1–14.6	0.8	1.1	1.5				0.8[b]	1.5[b]	1.1
58.4–63.8	14.7–17.6	0.8	1.0	1.3				0.8[b]	1.3[b]	1.0
					1.5	1.4[d]				
63.9–70.6	17.7–21.4	0.7	0.9	1.2				0.7[c]	1.2[c]	0.9
70.7–79.2	21.5–26.2	0.7	0.8	1.0				0.7[c]	1.0[9]	0.8
					1.2	1.2[e]				
79.3–90.5	26.3–32.3	0.6	0.7	0.8						0.7

[a] Generally suitable for northern Europe.
[b] Generally suitable for central Europe.
[c] Generally suitable for southern Europe.
[d] Optimum concentration of fluoride in arctic and subarctic water supplies.
[e] Optimum level of fluoride in treated waters as supplied to the community in all areas of Canada except the arctic and subarctic zones.
[f] Not related to temperature.

REFERENCES

1. *Drinking Water Standards,* U.S. Public Health Service, 1962.
2. *Canadian Drinking Water Standards and Objectives – 1968,* Department of National Health and Welfare, Ottawa, October 1969.
3. *International Standards for Drinking Water,* 2nd ed., World Health Organization, Geneva, 1963.
4. *European Standards for Drinking Water,* 2nd ed., Regional Office for Europe, World Health Organization, Copenhagen, 1970.
5. E.L. Bean, *J. Am. Water Works Assn., 54*(11):1313, 1962
6. *Manual for Evaluating Public Drinking Water Supplies,* U.S. Public Health Service, 1969.

3.1–19 TASTE THRESHOLD FREQUENCIES AT VARYING CONCENTRATIONS OF ZINC ION

Zinc ion concentration, ppm	Cumulative threshold distribution, %				
	Distilled water			Spring water	
	Zinc sulfate	Zinc nitrate	Zinc chloride	Zinc sulfate	Zinc chloride
2	0.5	0.2	0.1	0.07	0.03
4	4.8	2.8	1.9	1	0.06
8	21	14	10	8	4.5
16	47	38	32	29	19
32	78	71	64	59	50
64	96	94	91	64	84

Source: Reprinted, with publisher's permission, from *J. Am. Water Works Assn., 52*(5):660, © 1960 by the American Water Works Association, Inc., 2 Park Ave., New York, N.Y. 10016.

3.1–20 TASTE THRESHOLD FREQUENCIES FOR VARIOUS CONCENTRATIONS OF METALS IN DISTILLED WATER

Ion concentration, ppm	Cumulative threshold distribution, %			
	Zinc sulfate	Cupric chloride	Ferrous sulfate	Manganous sulfate
0.05	ND[a]	ND	6	ND
0.5	ND	ND	26	0.15
1	ND	<0.1	34	0.7
2	0.5	1.4	44	2
4	4.5	19	53	6
8	19	60	63	13
16	47	95	71	26
32	78	>95	81	45
64	96	>95	88	66

[a] ND – Not detected.

Source: Reprinted, with publisher's permission, from *J. Am. Water Works Assn., 52*(5):660, © 1960 by the American Water Works Association, Inc., 2 Park Ave., New York, N.Y. 10016.

3.1–21 TASTE THRESHOLD FREQUENCIES AT VARYING CONCENTRATIONS OF MANGANOUS ION

Manganese concentration, ppm	Cumulative threshold distribution, %	
	Distilled water	Spring water
1	0.7	0
4	6	0
8	14	0
16	26	0.05
32	47	1
64	67	7
128	82	43

Source: Reprinted, with publisher's permission, from *J. Am. Water Works Assn., 52*(5):660, © 1960 by the American Water Works Association, Inc., 2 Park Ave., New York, N.Y. 10016.

3.1–22 RANGE OF DETECTABLE CONCENTRATIONS OF METALS IN DISTILLED WATER

Metal	Threshold distribution	
	5%	95%
	Concentration, ppm	
Zinc	4.3	62
Copper	2.6	15.8
Iron	0.04	256
Manganese	0.9	487

Source: Reprinted, with publisher's permission, from *J. Am. Water Works Assn., 52*(5):660, © 1960 by the American Water Works Association, Inc., 2 Park Ave., New York, N.Y. 10016.

3.1–23 DISTRIBUTION OF REPORTED CASES ON INFANTILE METHEMOGLOBINEMIA BY NITRATE CONCENTRATION IN WATER

Simon Data[1]

	Reported cases	Deaths	Nitrate concentration in water, mg/l				
			Unknown	Known	<50	50–100	>100
Numerical	745	64	496	249	11	29	209
Percent	100	8.6	66.5	33.5	4.4[a]	11.8[a]	83.8[a]

Sattelmacher Data[2]

	Reported cases	Deaths	Unknown	Known	0–40	41–80	81–100	>100
Numerical	1,060	83	593	467	14	16	19	418
Percent	100	7.8	56.0	44.0	3.0[a]	3.4[a]	4.1[a]	89.5[a]

[a] Percent of cases with nitrate concentration known.

Source: Reprinted, with publisher's permission, from *J. Am. Water Works Assn., 63*(2):95, © 1971 by the American Water Works Association, Inc., 2 Park Ave., New York, N.Y. 10016.

REFERENCES

1. C. Simon *et al., Z. Kinderheilk., 91*:124, 1964.
2. P.G. Sattelmacher, "Methemoglobinemia from Nitrates in Drinking Water," *Schriftenreiche des Vereins für Wasser, Boden, und Lufthygiene,* No. 21, 1962.

3.1–24 CRITERIA FOR THE DEVELOPMENT OF DRINKING WATER STANDARDS FOR TRACE ORGANICS[1]

Toxicological

A. Acute and subacute levels to mammals and fish
   1. Effect on this level by
      a. Storage and biological magnification
      b. Environmental stress
      c. Synergism
      d. Chemical changes (biochemical)

B. Chronic levels (as above)

C. Lifetime chronic levels (as above)

Aesthetic

A. Threshold level
   1. Effects on this level by
      a. Synergism
      b. Antagonism
      c. Additivity

Analytical detectability

A. Sensitivity

B. Positive identification

Fate in the environment

A. Persistence – chemical and biological change

B. Documentation of occurrence in the environment

Drinking Water Standards' Philosophy[2]

"In arriving at specific limits, the total environmental exposure of man to a stated specific toxicant has been considered. The Committee has attempted to set limits at the lowest practical level in order to minimize the amount of a toxicant contributed by water, particularly when other sources such as milk, food, or air are known to represent the major exposure of man."

Source: Reprinted, with publisher's permission, from *J. Am. Water Works Assn., 63*(9):605, © 1971 by the American Water Works Association, Inc., 2 Park Ave., New York, N.Y. 10016.

REFERENCES

1. I.H. Suffet, Thesis, Rutgers University.
2. *Drinking Water Standards,* U.S. Department of Health, Education, and Welfare, Public Health Service, 1962.

3.1–25 BIOCIDE LIMITS IN DRINKING WATER

mg/l

Biocide[a]	U.S.[1,b]	Canada[2] Objective and acceptable limits	Canada[2] Max permissible limits[c]	Ideal water[3]
Aldrin	0.017	d	0.017	
Chlordane	0.003	d	0.003	
DDT	0.042	d	0.042	
Dieldrin	0.017	d	0.017	
Endrin	0.001	d	0.001	
Heptachlor	0.018	d	0.018	
Heptachlor epoxide	0.018	d	0.018	
Lindane	0.056	d	0.056	
Methoxychlor	0.035	d	0.035	
Organic phosphates plus carbamates[e]	0.100	d	0.100	
Toxaphene	0.005	d	0.005	
Herbicides (*e.g.*, 2,4-D; 2,4,5-T; (2,4,5-TP)	0.100[f,g]	d	0.100	
Total organic phosphorus or chlorinated hydrocarbon insecticides				None

[a] Conventional water treatment has little effect on these dissolved biocides.
[b] For long-term exposure.
[c] Maximum permissible limits are adopted from the *Report of the National Technical Advisory Committee on Water Quality Criteria to the Secretary of the Interior,* U.S. Department of the Interior, 1968. These limits should be regarded as *tentative standards,* as they are still being evaluated.
[d] Not detectable by an acceptable method of analysis approved by the control agency.
[e] Expressed as parathion equivalents in cholinesterase inhibition.
[f] Individual limits – 0.100 mg/l; sum of any combination of chlorinated phenoxyalkyl pesticide – 0.1 mg/l.
[g] Short period limit only – 2–3 days; no more than once or twice a year.

REFERENCES

1. *Manual for Evaluating Public Drinking Water Supplies,* U.S. Public Health Service, 1969.
2. *Canadian Drinking Water Standards and Objectives – 1968,* Department of National Health and Welfare, Ottawa, October 1969.
3. E.L. Bean, *J. Am. Water Works Assn., 54*(11):1313, 1962.

3.1–26 TASTE LEVEL OF RHODAMINE B DYE IN TAP WATER

Concentration of dye,[a] ppm	Number of persons detecting taste[b]	Description of taste
0.005	0	
0.01	4	Astringent
0.02	6	Astringent
0.05	9	Astringent

[a] Chlorine residual was 0.75 ppm.
[b] Nine persons were on the panel.

Source: Reprinted, with publisher's permission, from *J. Am. Water Works Assn., 56*(6):748, © 1964 by the American Water Works Association, Inc., 2 Park Ave., New York, N.Y. 10016.

3.1–27 GUIDELINES FOR MAXIMUM PERMISSIBLE LEVELS OF PESTICIDES IN FINISHED WATERS

Parts per billion

	1964	1965-6[a]	1967[b]
Lindane		56.0	5.0
Aldrin		32.0	0.25
Heptachlor		78.0	1.0
HCE		18.0	1.0
Dieldrin	<0.1	18.0	0.25
Endrin	<0.1	1.0	0.1
p,p'-DDT plus related compounds	<0.1	42.0	0.5
Methoxychlor		35.0	20.0
Toxaphene		5.0	2.5
Chlordan		52.0	0.25

[a] Based on "maximal acceptable concentrations" of Subcommittee on Toxicology (1965).[1]
[b] Based on the "maximum reasonable stream allowances" of Ettinger and Mount (1967).[2]

Source: M.L. Schafer *et al., Environ. Sci. Technol., 3*:1268, 1969.

REFERENCES

1. H.E. Stockinger, Chrman., Actions Taken by the Subcommittee on Toxicology of the Advisory Committee on Public Health Service Drinking Water Standards, July 1965.
2. M.B. Ettinger and D.I. Mount, *Environ, Sci. Technol.,* 1:203, 1967.

3.1–28 STORAGE OF DDT AND DDE IN BODY FAT AS INFLUENCED BY ENVIRONMENT OR OCCUPATION

Exposure	No. people	DDT storage, ppm	DDE storage, ppm
General population	130	4.0 ± 0.31[a]	7.8 ± 0.50
Environmental	28	4.3 ± 0.66	7.7 ± 0.92
Occupational	14	10.7 ± 2.1	21.6 ± 4.1

[a] Standard error of the mean.

Source: Reprinted, with publisher's permission from *J. Am. Water Works Assn., 57*(10):1311, © 1965, by the American Water Works Association, Inc., 2 Park Ave., New York, N.Y. 10016.

3.1–29 TREND OF STORAGE OF DDT AND DDE IN THE GENERAL POPULATION OF THE U.S.

Year	No. samples	DDT, ppm	DDE, ppm
Before 1942[2]	13	nd[a]	nd
1950[3]	75	5.3	
1953–54[4]	49	7.4	11.2
1954–56[2]	61	4.9	6.1
1961–62[1]	130	4.0	7.8
1962[5]	28	3.7	6.2
1963[6]	282	2.9	7.4

[a] Not detectable.

Source: Reprinted, with publisher's permission, from *J. Am. Water Works Assn., 57*(10):1311, © 1965 by the American Water Works Association, Inc., 2 Park Ave., New York, N.Y. 10016.

REFERENCES

1. G.E. Quinby *et al., J. Am. Med. Assn., 191*:175, 1965.
2. W.J. Hayes, Jr, *et al., Arch. Ind. Health, 18*:398, 1958.
3. E.P. Laug, F.M. Kunze, and C.S. Prickett, *Arch. Ind. Hyg., 3*:245, 1951.
4. W.J. Hayes, Jr., W.F. Durham, and C. Cueto, Jr., *J. Am. Med. Assn., 162*:890, 1956.
5. W.E. Dale and G.E. Quinby, *Science, 142*:593, 1963.
6. W.S. Hoffman, W.I. Fishbein, and M.B. Andelman, *J. Am. Med. Assn., 188:*819, 1964.

3.1–30 CONCENTRATION OF DDT DERIVED MATERIAL IN BODY FAT OF GENERAL POPULATION OF DIFFERENT COUNTRIES

Country	Year	No. samples	DDT, ppm	Total DDT derived material, ppm
Canada	1959–60[1]	62	1.6	5.3
England	1961–62[2]	131		2.2
France	1961[3]	10	1.7	5.2
Hungary	1960[4]	50	5.7	12.4
U.S.	1961–62[5]	130	4.0	12.6
West Germany	1958–59[6]	60	1.0	2.3

Source: Reprinted, with publisher's permission, from *J. Am. Water Works Assn., 57*(10):1311, © 1965 by the American Water Works Association, Inc., 2 Park Ave., New York, N.Y. 10016.

REFERENCES

1. S.T. Read and W.P. McKinley, *Arch. Environ. Health, 3:*209, 1961.
2. C.G. Hunter, J. Robinson, and A. Richardson, *Brit. Med. J., 1:*221, 1963.
3. W.J., Hayes, Jr., W.E. Dale, and R. LeBreton, *Nature, 199*:1189, 1963.
4. A. Denes, *Nahrung* (Ger.), *6*:48, 1962.
5. G.E. Quinby *et al., J. Am. Med. Assn., 191:*175, 1965.
6. H. Maier-Bode, *Med. Exptl.* (Ger.), *1*:146, 1960.

3.1–31 DDT AND DDE CONTENT OF COMPLETE MEALS FOR 1 DAY

Year	Location	Source	No. meals	Total daily content, mg DDT	DDE	Total (as DDT)	DDE, per cent
1953–54[1]	Wenatchee, Wash.	Restaurant	18	0.178	0.092	0.280	37
	Tacoma, Wash.	Prison	7	0.116	0.057	0.179	35
	Mean			0.164	0.083	0.256	36
1954–55[2]	Tallahassee, Fla.	Prison	12	0.202	0.050	0.258	21
1956–57[3]	Walla Walla, Wash.	College, meat abstainers	11	0.041	0.024	0.068	39
1959–60[4]	Anchorage, Alaska	Hospital	3	0.184	0.026	0.213	14
1962–64[5]	Wenatchee, Wash.	Restaurant	12	0.038	0.044	0.087	56
	Wenatchee, Wash.	Household	17	0.314	0.173	0.507	38
	Mean			0.199	0.122	0.335	40

Source: Reprinted, with publisher's permission, from publication and issue, © data by the American Water Works Association, Inc., 2 Park Ave., New York, N.Y. 10016.

REFERENCES

1. K.C. Walker, M.B. Goette, and G.S. Batchelor, *J. Agr. Food Chem., 2*:103, 1954
2. W.J. Hayes, Jr., W.F. Durham, and C. Cueto, Jr., *J. Am. Med. Assn., 162*:890, 1956.
3. W.J. Hayes, Jr. *et al., J. Am. Med. Assn., 191*:175, 1965.
4. W.F. Durham *et al., Science, 134*:1880, 1961.
5. W.F. Durham, J.F. Armstrong, and G.E. Quinby, *Arch. Environ. Health* (in press).

3.1–32 STORAGE OF BHC AND DIELDRIN IN BODY FAT

Country	Year	Dieldrin, ppm	BHC, ppm
U.S.	1961–62[1]	0.15	0.20
U.S.	1963[2]	0.11	0.57
England	1961–62[3]	0.21	
France	1962–63[4]		1.19

Source: Reprinted, with publisher's permission, from *J. Am. Water Works Assn., 57*(10):1311, © 1965 by the American Water Works Association, Inc., 2 Park Ave., New York, N.Y. 10016.

REFERENCES

1. W.E. Dale and G.E. Quinby, *Science, 142*:593, 1963.
2. W.S. Hoffman, W.I. Fishbein, and M.B. Andelman, *J. Am. Med. Assn., 188*:819, 1964.
3. C. G. Hunter, J. Robinson, and A. Richardson, *Brit. Med. J., 1*:221, 1963.
4. W.J. Hayes, Jr., W.E. Dale, and R. LeBreton, *Nature, 199*:1189, 1963.

3.1–33 THRESHOLD ODOR CONCENTRATIONS OF PESTICIDES AND SOLVENTS IN WATER

Pesticides	Threshold odor concentration, ppm
Parathion (technical grade)	0.003
Parathion (pure)	0.036
Endrin	0.009
Lindane	0.330
Formulation components	
Sulfoxide (synergist)	0.091
Aerosol OT (emulsifier)	14.600
Commercial solvents	
Deodorized kerosene	0.082
Solvent 1	0.016
Solvent 2	13.900
Solvent 3	0.090

Source: *The Cost of Clean Water,* U.S. Department of Agriculture report prepared for Federal Water Pollution Control Administration, June 1967.

3.2 INDUSTRIAL WATER SUPPLY QUALITY CRITERIA

CRITERIA FOR INDUSTRIAL WATER SUPPLIES

Introduction

Water quality requirements differ so widely for the hundreds of uses to which water is put industrially that no meaningful criteria for surface water supplies can encompass a majority of such uses. Furthermore, water treatment technology in its present state of development permits the utilization of surface water of literally any available quality to create waters of any desired quality at point of use. Such treatment may be costly, but this cost is usually a small part of the total production and marketing costs. The National Technical Advisory Subcommittee for Water Quality Requirements for Industrial Water Supplies has identified the appropriate quality characteristics of raw waters that have been used and the quality requirements of waters at the point of use for each industry.

Each value given in this report for a quality characteristic of raw water supplies for industrial purposes has occurred in water that has been used somewhere in this country. However, the characteristic values may be altered by treatment to produce the quality of water required at point of use. Hence, a table listing desired water quality criteria at the point of use prior to internal conditioning has been developed for each major industrial requirement.

The Subcommittee has reached the following conclusions regarding the water quality characteristics and requirements for industrial supplies.

1. The quality characteristics of the water supply for an established industry at a given site, if allowed to deteriorate from the range usually experienced for those characteristics of significance to that industry, can cause an undesirable increase in the cost for treatment. On the other hand, an improvement in the quality of the same supply will not significantly decrease the cost of treatment at an existing installation.

2. Marked variations in the quality of an industrial water supply can result in deterioration of product quality for some industries.

3. The water quality requirements at the point of use in each process in each industry as distinguished from the quality characteristics at the point of supply are generally well established for each existing industrial process use. These water quality requirements, however, vary considerably even for the same process depending upon the technological age of the design and other factors.

4. The quality of water employed for process use by different plants in the same industry may vary considerably between plants depending on the cost of treatment, the age of the plant design, operating practices, and the quality and quantity of the available supply.

Minimum Standards

Minimum standards that should be met by all surface waters for all uses include the following. The water should be:

1. free from substances attributable to municipal, industrial, or other discharges, or agricultural practices that will settle to form putrescent or otherwise objectionable sludge deposits;

2. free from floating debris, oil, scum, and other floating materials attributable to municipal, industral, or other discharges, or agricultural practices in amounts sufficient to be unsightly or deleterious;

3. free from materials attributable to municipal, industrial, or other discharges, or agricultural practices producing color, odor, or other conditions in such a degree as to cause a nuisance.

Additional quality characteristics of surface waters that have been used as sources for industrial water supplies are summarized in 1.3–47. The specific water characteristics are maximums and no water will have all of the maximum values shown.

Because of the very extensive use of water for cooling and boiler-feed purposes, the quality characteristics for surface waters for these purposes have been given special emphasis. In general, the surface water quality characteristics for process waters are applicable for the 2-digit SIC group of industries.

Source: *Water Quality Criteria,* U.S. Department of the Interior, Fed. Water Pollut. Control Adm., April 1968.

3.2–2 SUMMARY OF SPECIFIC QUALITY CHARACTERISTICS OF SURFACE WATERS THAT HAVE BEEN USED AS SOURCES FOR INDUSTRIAL WATER SUPPLIES

Table A

	Boiler makeup water		Cooling water			
			Fresh		Brackish[a]	
Characteristic	Industrial 0–1,500 psig	Utility, 700–5,000 psig	Once through	Makeup recycle	Once through	Makeup recycle
Silica (SiO_2)	150	150	50	150	25	25
Aluminum (Al)	3	3	3	3		
Iron (Fe)	80	80	14	80	1.0	1.0
Manganese (Mn)	10	10	2.5	10	0.02	0.02
Copper (Cu)						
Calcium (Ca)			500	500	1,200	1,200
Magnesium (Mg)						
Sodium and Potassium (Na + K)						
Ammonia (NH_3)						
Bicarbonate (HCO_3)	600	600	600	600	180	180
Sulfate (SO_4)	1,400	1,000	680	680	2,700	2,700
Chloride (Cl)	19,000	19,000	600	500	22,000	22,000
Fluoride (F)						
Nitrate (NO_3)			30	30		
Phosphate (PO_4)		50	4	4	5	5
Dissolved solids	35,000	35,000	1,000	1,000	35,000	35,000
Suspended solids	15,000	15,000	5,000	15,000	250	250
Hardness ($CaCO_3$)	5,000	5,000	850	850	7,000	7,000
Alkalinity ($CaCO_3$)	500	500	500	500	150	150
Acidity ($CaCO_3$)	1,000	1,000	0	200	0	0
pH, units			5.0–8.9	3.5–9.1	5.0–8.4	5.0–8.4
Color, units	1,200	1,200		1,200		
Organics						
Methylene blue active substances	2[d]	10	1.3	1.3		1.3
Carbon tetrachloride extract	100	100	e	100	e	100
Chemical oxygen demand (O_2)	100	500		100		200
Hydrogen sulfide (H_2S)					4	4
Temperature, °F	120	120	100	120	100	120

[a] Water containing in excess of 1,000 mg/l dissolved solids.
[d] 1 mg/l for pressures up to 700 psig.
[e] No floating oil.

3.2–2 SUMMARY OF SPECIFIC QUALITY CHARACTERISTICS OF SURFACE WATERS THAT HAVE BEEN USED AS SOURCES FOR INDUSTRIAL WATER SUPPLIES (*Continued*)

Table B

Process water

Characteristic	Textile industry, SIC-22	Lumber industry, SIC-24	Pulp and paper industry, SIC-26	Chemical industry, SIC-28	Petroleum industry, SIC-29	Primary metals industry, SIC-33	Food and kindred products, SIC-20	Leather industry, SIC-31
Silica (SiO_2)			50		50		For the above 2 categories, the quality of raw surface supply should be that prescribed by the NTA Subcommittee on Water Quality Requirements for Public Water Supplies	
Aluminum (Al)								
Iron (Fe)	0.3		2.6	5	15			
Manganese (Mn)	1.0			2				
Copper (Cu)	0.5							
Calcium (Ca)				200	220			
Magnesium (Mg)				100	85			
Sodium and Potassium (Na + K)					230			
Ammonia (NH_3)								
Bicarbonate (HCO_3)				600	480			
Sulfate (SO_4)				850	570			
Chloride (Cl)			200[b]	500	1,600	500		
Fluoride (F)					1.2			
Nitrate (NO_3)					8			
Phosphate (PO_4)								
Dissolved solids	150		1,080	2,500	3,500	1,500		
Suspended solids	1,000	c		10,000	5,000	3,000		
Hardness ($CaCO_3$)	120		475	1,000	900	1,000		
Alkalinity ($CaCO_3$)				500		200		
Acidity ($CaCO_3$)						75		
pH, units	6.0–8.0	5–7	4.6–9.4	5.5–9.0	6.0–9.0	3–9		
Color, units			360	500	25			
Organics								
Methylene blue active substances								
Carbon tetrachloride extract						30		
Chemical oxygen demand (O_2)								
Hydrogen sulfide (H_2S)								
Temperature, °F			95[f]			100		

[b] May be ⩽ 1,000 for mechanical pulping operations.
[c] No large particles ⩽ 3 mm diameter.
[f] Applies to bleached chemical pulp and paper only.

Note: Unless otherwise indicated, units are mg/l and values are maximums. No one water will have all the maximum values shown.

Source: *Water Quality Criteria,* U.S. Department of the Interior, Fed. Water Pollut. Control Adm., April 1968.

REFERENCES

Application of the above values should be based on the following references.

1. *ASTM Book of Standards,* Part 23, Amer. Soc. Test. Mat., Philadelphia, 1966.
2. Am. Publ. Health Assn., Am. Water Works Assn., and Water Pollut. Control Fed., *Standard Methods for the Examination of Water and Wastewater,* Am. Publ. Health Assn., New York, 1965.

3.2–3 WATER QUALITY STANDARDS FOR INDUSTRIAL USE

Table A

Industry and process	Turbidity, JTU	Temperature, °F	Color, ppm	Alkalinity, ppm ($CaCO_3$)	Chloride, ppm	Hardness, ppm ($CaCO_3$)	Iron, ppm
Textiles							
Size			5			25	0.3
Scouring	5		5			25	0.1
Bleaching	5		5			25	0.1
Dyeing	5		5			25	0.1
Paper							
Mechanical		a	30		1,000		0.3
Chemical							
Unbleached		a	30		200	100	1.0
Bleached		95	10		200	100	0.1
Chemicals							
Alkali and chlorine			10	80		140	0.1
Coal tar			5	50	30	180	0.1
Organic			5	125	25	170	0.1
Inorganic			5	70	30	250	0.1
Plastic and resin			2	1.0	0	0	0.005
Synthetic rubber			2	2	0	0	0.005
Pharmaceutical			2	2	0	0	0.005
Soaps and detergents			5	50	40	130	0.1
Paints			5	100	30	150	0.1
Gum and wood			20	200	500	900	0.3
Fertilizer			10	175	50	250	0.2
Explosives			8	100	30	150	0.1
Petroleum					300	350	1.0
Iron and steel		100					
Hot-rolled							
Cold-rolled							
Miscellaneous							
Fruit and vegetable canning			5.0	250	250	250	0.2
Soft drinks			10	85			0.3
Leather tanning			5		250	150	50
Cement				400	250		25

[a] Not considered a problem of concentrations encountered.

3.2–3 WATER QUALITY STANDARDS FOR INDUSTRIAL USE (*Continued*)

Table B

Industry and process	Manganese, ppm	NO_3, ppm	pH	SO_4, ppm	Total dissolved solids, ppm	Suspended solids, ppm	SiO_2, ppm	Ca, ppm	Mg, ppm	HCO_3, ppm
Textiles										
Size	0.05		6.5–10		100	5.0				
Scouring	0.01		3.0–10.5		100	5.0				
Bleaching	0.01		2.0–10.5		100	5.0				
Dyeing	0.01		3.5–10		100	5.0				
Paper										
Mechanical	0.1		6–10							
Chemical										
Unbleached	0.5		6–10			10	50	20	12	
Bleached	0.05		6–10			10	50	20	12	
Chemicals										
Alkali and chlorine	0.1		6–8.5			10		40	8	100
Coal tar	0.1		6.5–8.3	200	400	5		50	14	60
Organic	0.1		6.5–8.7	75	250	5		50	12	128
Inorganic	0.1		6.5–7.5	90	425	5		60	25	210
Plastic and resin	0.005	0	7.5–8.5	0	1.0	2.0	0.02	0	0	0.1
Synthetic rubber	0.005	0	7.5–8.5	0	2.0	2.0	0.05	0	0	0.5
Pharmaceutical	0.005	0	7.5–8.5	0	2.0	2.0	0.02	0	0	0.5
Soaps and detergents	0.1			150	300	10.0		30	12	60
Paints	0.1		6.5	125	270	10		37	15	125
Gum and wood	0.2	5	6.5–8.0	100	1,000	30	50	100	50	250
Fertilizer	0.2	5	6.5–8.0	150	300	10	25	40	20	210
Explosives	0.1	2	6.8	150	200	5	20	20	10	120
Petroleum			6.0–9.0		1,000	10		75	30	
Iron and steel										
Hot-rolled			5–9							
Cold-rolled			5–9			10				
Miscellaneous										
Fruit and vegetable canning	0.2	10	6.5–8.5	250	500	10	50	100		
Soft drinks	0.05									
Leather tanning			6.0–8.0	250				60		
Cement	0.5		6.5–8.5	250	600	500	35			

REFERENCES

1. N.L. Nemerow, *Liquid Wastes of Industry,* Addison-Wesley, Reading, Mass., 1971.
2. *Report of the Committee on Water Quality Criteria,* U.S. Department of the Interior, Fed. Water Pollut. Control Adm., April 1968.

3.2–4 INDUSTRIAL WATER QUALITY TOLERANCES

Table A

Type of industrial use of water	Turbidity, mg/l	Color, mg/l	Oxygen consumed, mg/l	Dissolved oxygen, mg/l	Taste and odor	pH	Hardness as $CaCO_3$, mg/l	Alkalinity as $CaCO_3$, mg/l	Calcium, mg/l	Chlorides, mg/l
Air conditioning										
Baking[a,b]	10	10			None		a		a	
Boiler feedwater										
0–150 psi	20	80	15	1.4		8.0+	80			
150–250 psi	10	40	10	0.1		8.4+	40			
250–400 psi	5	5	4	0		9.0+	10			
over 400 psi	1	2	3	0		9.6+	2			
Brewing										
Light beer[c]	10	10			None	6.5–7.0		75–80	100–200	
Dark beer	10	10			None	6.5–7.0		80–150	200–500	
Carbonated beverages[b]	2	10	1.5	None		250	50–170		250	
Chemical process industries	Requirements vary widely within industry and process used; sometimes distilled water must be used.									
Concrete mixing[b,d]										
Confectionery					Low	7.0+				
Cooling water	50					6.5–7.5	50			
Electroplating							Low			
Fermentation	Same general requirements as brewing. Requirements for gin and spirits equal light beer, and requirements for whiskey equal dark beer.									
Food canning and freezing[b,f]	10				None	7.5+	f	30–250		
Food equipment washing[b]	1	20			None		10			250
Food processing – general[b]	10	10			Low		250	250		
Ice manufacture[b]	5	5	3		None		70	30–50		
Laundering						6.0–6.8	50	60		
Oil well flooding	0					7.0+		Low		
Photographic processing	2	2					100			
Plastics – clear, uncolored	2	2								
Paper and pulp										
Ground woodpulp[g]	50	20–30[c]					182–200[c]	150		25–75
Soda and sulfate pulp[g,h]	25	5					100	75		75

[a] Calcium is necessary for some yeast actions, so some hardness is desirable. Too much hardness, however, retards fermentation and causes soggy bread.
[b] Except as specified, should conform to Public Health Service drinking water standards.
[c] Although the total dissolved solids limit is 500–1,000 ppm, there is a limit of 300 ppm for any one substance.
[d] Water should be free of acids, alkalies, oil, and decayed vegetable matter. Sugar should not exceed 3,400 ppm; SO^3 should not exceed 25 ppm.
[f] No chlorophenols. Hardness varies with product: general canning and freezing 50–85 ppm; peas 200–400 ppm; fruits and vegetables 100–200 ppm; legumes 25–75 ppm.
[g] Turbidity as SiO^2 must not be gritty.
[h] Calcium hardness as $CaCO^3$ should not exceed 50 ppm. Magnesium hardness as $CaCO^3$ should not exceed 50 ppm.

3.2–4 INDUSTRIAL WATER QUALITY TOLERANCES (*Continued*)

Table A

Type of industrial use of water	Turbidity, mg/l	Color, mg/l	Oxygen consumed, mg/l	Dissolved oxygen, mg/l	Taste and odor	pH	Hardness as $CaCO_3$, mg/l	Alkalinity as $CaCO_3$, mg/l	Calcium, mg/l	Chlorides, mg/l
Kraft										
Bleached	40	25					100	75		200
Unbleached	100	100					200	150		200
Fine paper	10	5					100	40–75		
Rayon (viscose)										
Pulp production	5	5					8	50		
Manufacture	0.3					7.8–8.3	55	50–75		
Steel manufacturing						6.8–7.0	50			175
Sugar manufacturing[e]									20	20
Tanning	20	10–100				6.8–8.0	50–513	128–135		
Textile manufacturing										
General[f]	0.3–25	5–70	8				0–50		10	100
Dyeing	5	5–20					20			
Wool scouring		70					20			
Cotton bandage	5	5			Low		20			

[e] Magnesium in any form should not exceed 10 ppm.

[f] No chlorophenols. Hardness varies with product: general canning and freezing 50–85 ppm; peas 200–400 ppm; fruits and vegetables 100–200 ppm; legumes 25–75 ppm.

3.2–4 INDUSTRIAL WATER QUALITY TOLERANCES (*Continued*)

Table B

Type of industrial use of water	Calcium chloride, mg/l	Magnesium chloride, mg/l	Sodium chloride, mg/l	Sulfates, mg/l	Magnesium sulfates, mg/l	Calcium sulfates, mg/l	Sodium sulfate, mg/l	Total dissolved solids, mg/l	Total solids, mg/l	Iron mg/l
Air conditioning										0.5
Baking[a,b]										0.2
Boiler feedwater										
0–150 psi									500–3,000	
150–250 psi									500–2,500	
250–400 psi									100–1,500	
over 400 psi									50	
Brewing							100	500–1,500[j3]		0.1
Light beer[c]	100–200	100–200	275–500		100–200	100–500	100			0.1
Dark beer	100–200	100–200	275–500		100–200	100–500		850		0.1
Carbonated beverages[b]				250						
Chemical process industries										
Concrete mixing[b,d]				Low	Low		Low			
Confectionery								100		0.2
Cooling water				Low						0.5
Electroplating				Low						Low
Fermentation										
Food canning and freezing[b,f]			1,500							0.2
Food equipment washing[b]								850		
Food processing – general[b]								850		0.2
Ice manufacture[b]	300	170–300	300		130–300	300	300	170–1,300		0.2
Laundering										0.2
Oil well flooding										0.1
Photographic processing						<100		Low	200	0.02
Plastics – clear, uncolored									200	0.02

[a] Calcium is necessary for some yeast actions, so some hardness is desirable. Too much hardness, however, retards fermentation and causes soggy bread.
[b] Except as specified, should conform to Public Health Service drinking water standards.
[c] Although the total dissolved solids limit is 500–1,000 ppm, there is a limit of 300 ppm for any one substance.
[d] Water should be free of acids, alkalies, oil, and decayed vegetable matter. Sugar should not exceed 3,400 ppm; SO^3 should not exceed 25 ppm.
[f] No chlorophenols. Hardness varies with product: general canning and freezing 50–85 ppm; peas 200–400 ppm; fruits and vegetables 100–200 ppm; legumes 25–75 ppm.
[j] Not more than 300 mg/l of any one substance.

3.2–4 INDUSTRIAL WATER QUALITY TOLERANCES (*Continued*)
Table B

Type of industrial use of water	Calcium chloride, mg/l	Magnesium chloride, mg/l	Sodium chloride, mg/l	Sulfates, mg/l	Magnesium sulfates, mg/l	Calcium sulfates, mg/l	Sodium sulfate, mg/l	Total dissolved solids, mg/l	Total solids, mg/l	Iron mg/l
Paper and pulp										
Ground woodpulp[g]								500		0.3
Soda and sulfate pulp[g,h]								250		0.1
Kraft										
Bleached								300		0.2
Unbleached								500		1.0
Fine paper								200		0.1
Rayon (viscose)										
Pulp production									100	0.05
Manufacture										0.0
Steel manufacturing										
Sugar manufacturing[e]		e		20	e					0.1
Tanning										0.1
Textile manufacturing										
General[f]		i		100	i					0.1–1.0
Dyeing										0.25
Wool scouring										1.0
Cotton bandage										0.2

[e] Magnesium in any form should not exceed 10 ppm.
[f] No chlorophenols. Hardness varies with product: general canning and freezing 50–85 ppm; peas 200–400 ppm; fruits and vegetables 100–200 ppm; legumes 25–75 ppm.
[g] Turbidity as SiO^2 must not be gritty.
[h] Calcium hardness as $CaCO^3$ should not exceed 50 ppm. Magnesium hardness as $CaCO^3$ should not exceed 50 ppm.
[i] Magnesium should not exceed 5 ppm.

3.2–4 INDUSTRIAL WATER QUALITY TOLERANCES (*Continued*)
Table C

Type of industrial use of water	Manganese, mg/l	Iron and manganese together, mg/l	Aluminum oxide, mg/l	Silica mg/l	Copper, mg/l	Fluoride, mg/l	Carbonate, mg/l	Sodium carbonate, mg/l	Bicarbonate, mg/l	Nitrate, mg/l
Air conditioning	0.5	0.5								
Baking[a,b]	0.2	0.2								
Boiler feedwater										
0–150 psi			5.0	40			200		50	
150–250 psi			0.5	20			100		30	
250–400 psi			0.05	5			40		5	
over 400 psi			0.01	1			20		0	
Brewing	0.1	0.1		50		1.0	50–68	100		30
Light beer[c]	0.1	0.1		50		1.0	50–68	100		30
Dark beer	0.2	0.2				0.2–1.0				
Carbonated beverages[b]										
Chemical process industries										
Concrete mixing[b,d]										
Confectionery	0.2	0.2								
Cooling water	0.5	0.5								
Electroplating							Low			Low
Fermentation										
Food canning and freezing[b,f]	0.2	0.2				1.0				
Food equipment washing[b]		0.2				1.0				
Food processing – general[b]	0.2	0.3				1.0				
Ice manufacture[b]	0.2	0.2		10		1.0			50	
Laundering	0.2	0.2							Low	
Oil well flooding										
Photographic processing	0.02	0.02								
Plastics – clear, uncolored	0.02	0.02								

[a] Calcium is necessary for some yeast actions, so some hardness is desirable. Too much hardness, however, retards fermentation and causes soggy bread.
[b] Except as specified, should conform to Public Health Service drinking water standards.
[c] Although the total dissolved solids limit is 500–1,000 ppm, there is a limit of 300 ppm for any one substance.
[d] Water should be free of acids, alkalies, oil, and decayed vegetable matter. Sugar should not exceed 3,400 ppm; SO^3 should not exceed 25 ppm.
[f] No chlorophenols. Hardness varies with product: general canning and freezing 50–85 ppm; peas 200–400 ppm; fruits and vegetables 100–200 ppm; legumes 25–75 ppm.

3.2–4 INDUSTRIAL WATER QUALITY TOLERANCES (*Continued*)

Table C

Type of industrial use of water	Manganese, mg/l	Iron and manganese together, mg/l	Aluminum oxide, mg/l	Silica mg/l	Copper, mg/l	Fluoride, mg/l	Carbonate, mg/l	Sodium carbonate, mg/l	Bicarbonate, mg/l	Nitrate, mg/l
Paper and pulp										
Ground woodpulp[g]	0.1			50						
Soda and sulfate pulp[g,h]	0.05			20						
Kraft										
Bleached	0.1			50						
Unbleached	0.5			100						
Fine paper	0.05			20						
Rayon (viscose)										
Pulp production	0.03	0.05	8.0	<25	<5					
Manufacture	0.0	0.0								
Steel manufacturing										
Sugar manufacturing[e]									100	
Tanning	0.1	0.2							Low	
Textile and manufacturing										
General[f]	0.05–1.0	0.1–1.0							200	
Dyeing	0.25	0.25								Low
Wool scouring	1.0	1.0								
Cotton bandage	0.2	0.2								

[e] Magnesium in any form should not exceed 10 ppm.
[f] No chlorophenols. Hardness varies with product: general canning and freezing 50–85 ppm; peas 200–400 ppm; fruits and vegetables 100–200 ppm; legumes 25–75 ppm.
[g] Turbidity as SiO^2 must not be gritty.
[h] Calcium hardness as $CaCO^3$ should not exceed 50 ppm. Magnesium hardness as $CaCO^3$ should not exceed 50 ppm.

3.2–4 INDUSTRIAL WATER QUALITY TOLERANCES (*Continued*)

Table D

Type of industrial use of water	Hydroxide, mg/l	Hydrogen sulfide, mg/l	Algae protozoa	Corrosion tendencies	Scale-forming tendencies	Slime-forming tendencies	Free carbon dioxide, mg/l
Air conditioning		None	None	No		No	
Baking[a,b]		0.2					
Boiler feedwater							
0–150 psi	50	5.0		No	No		
250–400 psi	40	3.0		No	No		
over 400 psi	30	0		No	No		
over psi	15	0		No	No		
Brewing							
Light beer[c]		0.2					
Dark beer		0.2					
Carbonated beverages[b]		0.2	None				
Chemical process industries							
Concrete mixing[b,d]				No			20
Confectionery		0.2					
Cooling water			Low	No	**No**	No	
Electroplating							15
Fermentation							
Food canning and freezing[b,f]		1.0					
Food equipment washing[b]							
Food processing – general[b]							
Ice manufacture[b]							
Laundering			None				
Oil well flooding			None				
Photographic processing							
Plastics – clear, uncolored							

[a] Calcium is necessary for some yeast actions, so some hardness is desirable. Too much hardness, however, retards fermentation and causes soggy bread.
[b] Except as specified, should conform to Public Health Service drinking water standards.
[c] Although the total dissolved solids limit is 500–1,000 ppm, there is a limit of 300 ppm for any one substance.
[d] Water should be free of acids, alkalies, oil, and decayed vegetable matter. Sugar should not exceed 3,400 ppm; SO^3 should not exceed 25 ppm.
[f] No chlorophenols. Hardness varies with product: general canning and freezing 50–85 ppm; peas 200–400 ppm; fruits and vegetables 100–200 ppm; legumes 25–75 ppm.

3.2–4 INDUSTRIAL WATER QUALITY TOLERANCES *(Continued)*

Table D

Type of industrial use of water	Hydroxide, mg/l	Hydrogen sulfide, mg/l	Algae protozoa	Corrosion tendencies	Scale-forming tendencies	Slime-forming tendencies	Free carbon dioxide, mg/l
Paper and pulp							
Ground woodpulp[g]							10
Soda and sulfate pulp[g,h]							10
Kraft							
Bleached							10
Unbleached							10
Fine paper							10
Rayon (viscose)							
Pulp production							
Manufacture				No			
Steel manufacturing							
Sugar manufacturing[e]							
Tanning							Low
Textile manufacturing							
General[f]							
Dyeing							
Wool scouring							
Cotton bandage							

[e] Magnesium in any form should not exceed 10 ppm.
[f] No chlorophenols. Hardness varies with product: general canning and freezing 50–85 ppm; peas 200–400 ppm; fruits and vegetables 100–200 ppm; legumes 25–75 ppm.
[g] Turbidity as SiO^2 must not be gritty.
[h] Calcium hardness as $CaCO^3$ should not exceed 50 ppm. Magnesium hardness as $CaCO^3$ should not exceed 50 ppm.

Note: All entries are maximum values unless indicated by plus sign; where range is indicated, both extremes are shown.

Source: *Report of the U.S. Study Commission – Texas,* March 1962.

REFERENCES

1. J. Eller *et al., J. Am. Water Works Assn., 62*(3):149, 1970.
2. T.R. Camp, *Water and Its Impurities,* Reinhold Publishing Corp., 1963, p. 140.
3. Am. Water Works Assn., *Water Quality and Treatment,* 3rd ed., McGraw-Hill, New York, 1971, p. 35.

3.2–5 COLOR LIMITS PERMISSIBLE IN WATER FOR INDUSTRIAL USE

Use	Color as platinum units, ppm
Food products	10
Ice making	5
Bottled beverages	5–10
Pulp and paper	
Alkaline pulps	5
High grade paper	⩽5
Special papers	⩽5
Bleached kraft papers	⩽25
Unbleached kraft papers	⩽100
Groundwood papers	⩽30
Baking	10
Textiles	
Cotton	50
Rayon	5
General	5–70
Boiler feed water	2–80, according to pressure
Brewing	0–10
Food equipment washing	5–20
Food processing	5–10
Tanning	

Source: *Water Quality Criteria,* State Water Pollution Control Board, Sacramento, California, 1963.

3.2.1 FOOD INDUSTRIES

3.2–6 QUALITY REQUIREMENTS OF WATER AT POINT OF USE BY THE SOFT DRINK INDUSTRY

Characteristic	Concentration	Characteristic	Concentration
Alkalinity ($CaCO_3$)	85	Fluoride (F)	c
pH, units	a	Total dissolved solids	a
Hardness ($CaCO_3$)	a	Organics, CCE	0.2[d]
Chlorides (Cl)	500[b]	Coliform bacteria	c
Sulfates (SO_4)	500[b]	Color, units	10[d]
Iron (Fe)	0.3	Taste	d,e
Manganese (Mn)	0.05	Odor	d,e

[a] Controlled by treatment for other constituents.
[b] If present with equivalent quantities of Mg and Ca as sulfates and chlorides, the permissible limit may be somewhat below 500 mg/l.
[c] Not greater than USPHS *Drinking Water Standards.*
[d] In general, public water supplies are coagulated, chlorinated, and filtered through sand and granular activated carbon to insure very low organic content and freedom from taste and odor.
[e] Zero, not detectable by test.

Note: Unless otherwise indicated, units are mg/l and values that normally should not be exceeded. Quality of water prior to the addition of substances used for internal conditioning.

Source: *Water Quality Criteria,* U.S. Department of the Interior, Fed. Water Pollut. Control Adm., April 1968.

Application of the above values should be based upon analytical methods found in the following references.

REFERENCES

1. *ASTM Book of Standards,* Part 23, Amer. Soc. Test. Mat., Philadelphia, 1966.
2. Am. Publ. Health Assn., Am. Water Works Assn., and Water Pollut. Control Fed., *Standard Methods for the Examination of Water and Wastewater,* Am. Publ. Health Assn., New York, 1965.

3.2–7 QUALITY CHARACTERISTICS OF SURFACE WATERS THAT HAVE BEEN USED BY THE FOOD CANNING INDUSTRY

Characteristic	Concentration	Characteristic	Concentration
Alkalinity ($CaCO_3$)	300	Fluoride (F)	a
pH, units	8.5	Organics: Carbon chloroform extract	0.3
Hardness ($CaCO_3$)	310	Chemical oxygen demand (O_2)	b
Calcium (Ca)	120	Odor, threshold number	a
Chlorides (Cl)	300	Taste, threshold number	a
Sulfates (SO_4)	250	Color, units	5
Iron (Fe)	0.4	Dissolved solids	550
Manganese (Mn)	0.2	Suspended solids	12
Silica (SiO_2), dissolved	50	Coliform, count/100 ml	a
Phenols	a		
Nitrate (NO_3)	45		
Nitrite (NO_2)	c		

[a] As specified by National Technical Advisory Subcommittee on Water Quality Criteria for Public Water Supplies.
[b] Accepted as received (if meeting total solids or other limiting values); has never been a problem at concentrations encountered.
[c] Zero, not detectable by test.

Note: Unless otherwise indicated, units are mg/l and values that normally should not be exceeded.

Source: *Water Quality Criteria,* U.S. Department of the Interior, Fed. Water Pollut. Control Adm., April 1968.

Application of the above values should be based upon analytical methods found in the following references.

REFERENCES

1. *ASTM Book of Standards,* Part 23, Amer. Soc. Test. Mat., Philadelphia, 1966.
2. Am. Publ. Health Assn., Am. Water Works Assn., and Water Pollut. Control Fed., *Standard Methods for the Examination of Water and Wastewater,* Am. Publ. Health Assn., New York, 1965.

3.2–8 MAXIMUM CONCENTRATIONS OF VARIOUS CONSTITUENTS DESIRABLE IN WATER USED IN FOOD INDUSTRIES

mg/l

	Washing of butter	Carbonated beverages	Manufac- ture of ice	Canning	Washing equipment in food industries
Turbidity	1.	1.	1.	1.	1.
Color (APHA units)	5.	1.0	5.	–	20.
Taste and odor	0	0	0	0	0
Iron and manganese	0.02	0.3	0.2	0.2	0.2
Chlorides (as Cl)	250	250	–	–	–
Total dissolved solids	850	850	170–350	–	850
Organic matter	0	0	–	–	0
Fluorine	1.	1.	1.	1.	1.

Source: H.V. Miles, Correction of Defects of Water Used for Food Manufacture, *Proc. Inst. Food Technol.*, 5, 59, 1944.

3.2.2 MANUFACTURING INDUSTRIES

3.2–9 WATER QUALITY FOR HYDRAULIC CEMENT INDUSTRY

Characteristic	Concentration, mg/l At point of use	In surface water supply
Acidity ($CaCO_3$)	a	a
Alkalinity ($CaCO_3$)	400	240
Chemical oxygen demand (O_2)	b	b
Coliform bacteria (count/100 ml)	b	b
Color, units	b	b
Hardness ($CaCO_3$)	b	500
Calcium hardness ($CaCO_3$)	b	150
Organics		
Carbon tetrachloride extract	1	1
Iron (Fe)	25	1.8
Manganese (Mn)	0.5	5
pH, units	6.5–8.5	6.9–8.8
Silica (SiO_2)	35	16
Dissolved solids	600	1,120
Suspended solids	500	200
Sulfate (SO_4)	250	235
Chloride (Cl)	250	100

[a] Zero, not detectable by test.
[b] Accepted as received (if meeting total solids or other limiting values), has never been a problem at concentrations encountered.

Source: *Water Quality Criteria,* U.S. Department of the Interior, Fed. Water Pollut. Control Adm., April 1968.

REFERENCES

Applications of the above values should be based upon analytical methods found in the following references.

1. *ASTM Book of Standards,* Part 23, Amer. Soc. Test. Mat., Philadelphia, 1966.
2. Am. Publ. Health Assn., AWWA, and Water Pollut. Contr. Fed., *Standard Methods for the Examination of Water and Wastewater,* Am. Publ. Health Assn., New York, 1965.

3.2–10 QUALITY REQUIREMENTS AND CHARACTERISTICS OF IRON AND STEEL INDUSTRY WATERS

	At point of use[a]				After use[g]
			Selected rinse waters		
Characteristics	**Quenching, hot rolling, gas cleaning**	**Cold rolling**	**Softened**	**Demineralized**	
Settleable solids	100	c	c	c	350
Suspended solids	b	10	c	c	3,000
Dissolved solids	b	b	b	c	1,500
Acidity ($CaCO_3$)					75
Alkalinity ($CaCO_3$)	d	d	d	c	200
Hardness ($CaCO_3$)	d	d	100	c	1,000
pH, units	5–9	5–9	6–9	f	3–9
Chloride (Cl)	b	b	b	c	500
Dissolved oxygen (O_2)	e	e	e	e	
Temperature, °F	100	100	100	100	100
Oil	b	c	c	c	
Floating material	b	c	c	c	30
Organics, carbon tetrachloride extractables					

[a] Unless otherwise indicated, units are mg/l and values that normally should not be exceeded. Quality of water prior to the addition of substances used for internal conditioning.
[b] Accepted as received (if meeting total solids or other limiting values); has never been a problem at concentrations encountered.
[c] Zero, not detectable by test.
[d] Controlled by treatment for other constituents.
[e] Minimum to maintain aerobic conditions.
[f] Concentration not known.
[g] Unless otherwise indicated, units are mg/l and values are maximums. No one water will have all the maximum values shown.

Source: *Water Quality Criteria,* U.S. Department of the Interior, Fed. Water Pollut. Control Adm., April 1968.

REFERENCES

Application of the above values should be based upon analytical methods found in the following references.

1. *ASTM Book of Standards,* Part 23, Amer. Soc. Test. Mat., Philadelphia, 1966.
2. Am. Publ. Health Assn., Am. Water Works Assn., and Water Pollut. Control Fed., *Standard Methods for the Examination of Water and Wastewater,* Am. Publ. Health Assn., New York, 1965.

3.2–11 WATER QUALITY FOR CHEMICAL AND ALLIED PRODUCTS INDUSTRY

At point of use

Characteristic	Alkalies and chlorine	Intermediate coal tar products	Organic chemicals	Inorganic chemicals	Plastic materials and polymers	Synthetic rubber	Drugs and pharmaceutical preparations[a]	Soap and other detergents	Paints and allied products	Gums and wood chemicals	Fertilizers	Explosives	In surface water supply[1,2]
Silica (SiO_2)	b	c	b	b	b	b	b	b	b	50	b	b	b
Iron (Fe)	0.1	b	0.1	b	b	0.1	b	b	b	0.3	b	b	5
Manganese (Mn)	0.1	b	0.1	b	b	0.1	b	b	b	0.2	b	b	2
Calcium (Ca)	b	b	68	b	b	80	b	b	b	100	b	b	200
Magnesium (Mg)	b	b	19	b	b	36	b	b	b	50	b	b	100
Bicarbonate (HCO_3)	b	b	128	b	b		b	b	b	250	b	b	600
Sulfate (SO_4)	b	b	b	b	b	b	b	b	b	100	b	b	850
Chloride (Cl)	b	b	b	b	b	b	b	b	b	500	b	b	500
Nitrate (NO_3)	b	b	b	b	b	b	b	b	b	5	b	b	–
Total solids	b	b	b	b	b	b	b	b	b	1,000	b	b	–
Hardness ($CaCO_3$)	b	b	250	b	b	350	b	b	b	900	b	b	1,000
pH	b	b	6.5–8.7	b	b	6.2–8.3	b	b	b	6.5–8.0	b	b	5.5–9.0
Color	b	b	b	b	b	20	b	b	b	20	b	b	500
Suspended solids	b	b	b	b	b	5	b	b	b	30	b	b	10,000
Odor	b	b	b	b	b	b	b	b	b	c	b	b	b
5 day BOD (O_2)	b	b	b	b	b	b	b	b	b	d	b	b	b
COD (O_2)	b	b	b	b	b	b	b	b	b	d	b	b	b
Dissolved oxygen (O_2)	b	b	b	b	b	b	b	b	b	a	b	b	–
Alkalinity ($CaCO_3$)	b	b	125	b	b	150	b	b	b	200	b	b	500

[a] Potable water.
[b] Accepted as received (if meeting total solids or other limiting values); has never been a problem at concentrations encountered.
[c] No practical limit–any concentration can be handled.
[d] Controlled by treatment for other constituents.

Note: Unless otherwise indicated, units are mg/l and values that normally should not be exceeded. Quality of water prior to the addition of substances used for internal conditioning.

Source: *Water Quality Crit ria.* U.S. Department of the Interior, Fed. Water Pollut. Control Adm., April 1968.

REFERENCES

Application of the above values should be based upon analytical methods found in the following references.

1. *ASTM Book of Standards,* Part 23, Amer. Soc. Test. Mat., Philadelphia, 1966.
2. Am. Publ. Health Assn., Am Water Works Assn., and Water Pollut. Control Fed., *Standard Methods for the Examination of Water and Wastewater,* Am. Publ. Health Assn., New York, 1965.

3.2–12 QUALITY REQUIREMENTS AND CHARACTERISTICS OF PULP AND PAPER INDUSTRY WATERS

	At point of use[a]			After use[b]		
		Chemical pulp and paper			Chemical pulp and paper	
Characteristics	**Mechanical pulping**	**Unbleached**	**Bleached**	**Mechanical pulping**	**Unbleached**	**Bleached**
Silica (SiO_2)	c	50	50	c	50	50
Aluminum (Al)	c	c	c	c	c	c
Iron (Fe)	0.3	1.0	0.1	2.6	2.6	2.6
Manganese (Mn)	0.1	0.5	0.05	c	c	c
Zinc (Zn)	c	c	c	c	c	c
Calcium (Ca)	c	20	20	c	c	c
Magnesium (Mg)	c	12	12	c	c	c
Sulfate (SO_4)	c	c	c	c	c	c
Chloride (Cl)	1,000	200	200	1,000	200	200
Dissolved solids	c	c	c	1,080	1,080	1,080
Suspended solids	c	10[d]	10[d]	c	c	c
Hardness ($CaCO_3$)	c	100	100	475	475	475
pH, units	6-10	6-10	6-10	4.6-9.4	4.6-9.4	4.6-9.4
Color, units	30	30	10	360	360	360
Temperature, °F	c	c	95	c	c	95

[a] Unless otherwise indicated, units are mg/l and values that normally should not be exceeded. Quality of water prior to the addition of substances used for internal conditioning.

[b] Unless otherwise indicated, units are mg/l and values are maximums. No one water will have all the maximum values shown.

[c] Accepted as received (if meeting total solids or other limiting values); has never been a problem at concentrations encountered.

[d] No gritty or color-producing solids.

Source: *Water Quality Criteria*, U.S. Department of the Interior, Fed. Water Pollut. Control Adm., April 1968.

REFERENCES

Application of the above values should be based upon analytical methods found in the following references.

1. *ASTM Book of Standards,* Part 23, Amer. Soc. Test. Mat., Philadelphia, 1966.
2. Am. Publ. Health Assn., Am. Water Works Assn., and Water Pollut. Control Fed., *Standard Methods for the Examination of Water and Wastewater,* Am. Publ. Health Assn., New York, 1965.

3.2–13 COMPARISON OF RAW WATER AND NORMAL MILL WATER ANALYSES

Type of mill	No. of mills	Total hardness as $CaCO_3$, ppm		Total alkalinity as $CaCO_3$, ppm		Turbidity, ppm		Color as color units, ppm		Iron as Fe, ppm	
		Raw[a]	Mill[b]	Raw	Mill	Raw	Mill	Raw	Mill	Raw	Mill
Integrated											
Kraft pulp, paper	3										
Avg		106	71	75	30	95	20	17	7.3	1.2	0.8
Max		270	148	160	36	900	40	40	15	1.5	1.5
Min		10	10	20	15	15	<1	10	5	0.1	0.08
Kraft pulp, bleachery, paper	14										
Avg		58	47	64	57	20	13	31	8	0.55	0.28
Max		494	194	304	304	200	200	710	50	25	1.7
Min		1	0	3	0	0	0	0	0	0	0
Kraft, pulp, bleachery, fine paper	4										
Avg		77	77	60	71	11	3.8	15	3.3	0.29	0.25
Max		174	174	128	128	100	5	60	5	0.9	0.4
Min		10	10	15	15	4.3	1	5	0	0.05	0.1
Soda pulp, bleachery, fine paper	2										
Avg		78	77	51	50	16	2	24	23	0.7	0.7
Max		150	136	96	92	100	4	60	60	1.0	1.0
Min		18	18	4	4	3	2	2	1	0.4	0.4
Groundwood pulp, paper	2										
Avg		36	36	22	7	8.7	1.5	140	5		
Max		44	55	28	10	16	1.7	400	10		
Min		24	35	14	3	7	1.3	40	5		
Groundwood pulp, bleachery, fine paper	5										
Avg		85	57	66	42	14	3.4	41	8	0.1	0.1
Max		188	150	143	90	500	10	433	15	0.1	0.3
Min		10	10	10	10	3	2	15	0	0	0.005
NSSC pulp, bleachery, fine paper	2										
Avg		25	22			15	10	59	22	0.7	0.4
Max		40	23			100	30	100	28	2.0	
Min		16	20			5	5	15	15	0.5	
Sulfite pulp, bleachery, fine paper	2										
Avg		24	24	30	10	7	1.4	22	5	0.44	0.03
Max		40	40	40	15	15		125	10	0.8	
Min		18	18	20	5	0		0	1	0.02	
Deinking stock, bleachery, paper	2										
Avg		425	242	315	315	0	0	0	0	0.27	0.08
Max		450	350			0	0	0	0	0.37	0.25
Min		220	175			0	0	0	0	0.16	0.05

[a] Analysis of water from source.
[b] Normal mill water.

3.2–13 COMPARISON OF RAW WATER AND NORMAL MILL WATER ANALYSES (*Continued*)

Type of mill	No. of mills	Total hardness as $CaCO_3$, ppm		Total alkalinity as $CaCO_3$, ppm		Turbidity, ppm		Color as color units, ppm		Iron as Fe, ppm	
		Raw[a]	Mill[b]	Raw	Mill	Raw	Mill	Raw	Mill	Raw	Mill
Rag pulp, bleachery, fine paper	1										
Avg		170	160	140	140	0.25	0	40	7	0.05	0.05
Max		190	200	180	160	60	2	70	10		
Min		150	120	90	140	10	0	20	3		
Sulfite, groundwood pulp, paper	2										
Avg		55	55	65	65	82	80	115	115	0.2	0.2
Max		94	94	85	85	300	300	300	300	0.5	0.5
Min		38	38	40	40	4	0	20	20	0.02	0.02
Kraft, groundwood pulp, bleachery, paper	6										
Avg		50	44	35	85	28	2	19		0.23	0.1
Max		250	60	200	82	400	4.5	192	27	1.5	0.23
Min		4	1.3	5	17	0	1	0	0	0	0.01
Kraft, groundwood pulp, bleachery, fine paper	2										
Avg		17	17			8	5.4	29	15	0.27	0.2
Max		20	20			160	15	150	35	1.8	0.5
Min		13	13			1	0.2	5	5	0.05	0.05
Kraft, sulfite pulp, bleachery, paper	1										
Avg		10	16	11	18	2	9.25			0.06	
Max		11		18		25				0.1	
Min		10		8		0					
Kraft, groundwood, semichemical, pulp, bleachery, paper	1										
Avg		13	13	7	15	2	1	40	0.5	0.2	0.075
Max		18	18	8	16	10		100		0.3	0.1
Min		10	10	6	14	1	0	15		0.1	0.05
Kraft, NSSC pulp, paper	3										
Avg		37	28	61	80	17	5	10	10	0.14	
Max		100	76	130	93	55	10	40	40	0.9	0.7
Min		7	2	22	50	1	1	0	0	0.007	0
Nonintegrated											
Fine paper	8										
Avg		257	277	190	101	15	3	17	8	0.6	0.04
Max		480	420	249	200	35	6	80	10	2.2	0.1
Min		30	92	3	2.7	0	0	0	5	0.05	0.01
Paper (wrapping)	8										
Avg		204	147	127	103	16	5.6	9	3	0.35	1.2
Max		340	332	320	220	300	25	40	15	1.2	6.6
Min		60	14	60	11	0	0	0	0	0.1	0
Paper (roofing felt)	1										
Avg		132	132	76	76					0.03	0.03
Max											
Min											

[a] Analysis of water from source.
[b] Normal mill water.

3.2–13 COMPARISON OF RAW WATER AND NORMAL MILL WATER ANALYSES (*Continued*)

Type of mill	No. of mills	Total hardness as $CaCO_3$, ppm		Total alkalinity as $CaCO_3$, ppm		Turbidity, ppm		Color as color units, ppm		Iron as Fe, ppm	
		Raw[a]	Mill[b]	Raw	Mill	Raw	Mill	Raw	Mill	Raw	Mill
Paper (glassine)	1										
Avg		95	80	64	20	40				0.2	0.1
Max		150		100		90				0.6	
Min		45		25		20				0.1	
Paper (specialty)	3										
Avg		120	120	103	97	2.5	3.4			0.1	Trace
Max		150	150	130	130	8.0	8.0				
Min		90	90	90	60	0	2.2				
Paper, bleachery (tissue)											
Avg		125	125	114				15		0.9	0.15
Max		190	190	180		75		40		1.5	0.2
Min		78	78	84		10	2	0	0.3	0.3	0.1
Board	3										
Avg		121	121								
Max		250	250								
Min		51	51								
Soda pulp, bleachery, drying	1										
Avg		30	30	72	72	Nil	Nil	15	2.5	0.6	0.05
Max		35	35					20	5	1.0	0.09
Min		25	25					10	0	0.4	0
Kraft pulp, bleachery, drying	14										
Avg		129	65	100	68	40	0.75	15	7.5	2.0	0.08
Max		256	86	250	100	2,000	1.5	60	15	7.0	0.1
Min		30	27.5	45	30	2	0.5	0	0	0.1	0.05
Kraft pulp, bleachery, drying	1										
Avg		4	4	150	150	0	0	3	3	0.1	
Max											
Min											
NSSC pulp	1										
Avg		4.6	18	352	268	15		7.5	5	0.14	
Max								10			
Min								5			

[a] Analysis of water from source.
[b] Normal mill water.

Source: Reprinted, with publisher's permission, from *63*(3):165, © 1971 by the American Water Works Association, Inc., 2 Park Ave., New York, N.Y. 10016.

3.2–14 SUMMARY OF SPECIFICATIONS FOR CHEMICAL COMPOSITION OF PROCESS WATER FOR PAPER MANUFACTURE

Substance, max mg/l	Fine paper	Kraft paper: Bleached	Kraft paper: Unbleached	Groundwood papers	Soda and sulfite pulp
Turbidity	10	40	100	50	25
Color in platinum units	5	25	100	30	5
Total hardness as $CaCO_3$	100	100	200	200	100
Calcium hardness as $CaCO_3$	50				50
Alkalinity to M.O. as $CaCO_3$	75	75	150	150	75
Iron as Fe	0.1	0.2	1.0	0.3	0.1
Manganese as Mn	0.05	0.1	0.5	0.1	0.05
Residual chlorine as Cl_2	2.0				
Silica (soluble) as SiO_2	20	50	100	50	20
Total dissolved solids	200	300	500	500	250
Free carbon dioxide as CO_2	10	10	10	10	10
Chlorides as Cl		200	200	75	75
Magnesium hardness as $CaCO_3$					50

Source: *Water Technology in the Pulp and Paper Industry,* Tech. Assn. Pulp Paper Ind., Monograph Series No. 18, 1957, p. 162. See also Tech. Assn. Pulp Paper Ind. Standards E600 s-48, E601 s-48, and E603 s-49.

3.2–15 RANGES OF DESIRED PROPERTIES OF PROCESS WATER FOR KRAFT PAPER MANUFACTURE

Parameter	Waste: Unbleached	Waste: Bleached
Turbidity, mg/l	5–25	0–5
Color units	10–80	0–5
pH	6.5–8.0	6.8–7.3
Total alkalinity, mg/l	20–150	20–75
Hardness as $CaCO_3$, mg/l	5–200	5–100
Dissolved solids, mg/l	50–500	50–250
Chloride, mg/l	10–150	10–150
Fe, mg/l	< 0.5	< 0.2
Mn, mg/l	< 0.3	< 0.1
COD, mg/l	0–12	0–8
BOD, mg/l	0–5	0–2

Source: R. Smith, *J. Water Pollut. Control Fed., 40*(9):1546, 1968.

3.2–16 WATER QUALITY PROBLEMS IN THE PULP AND PAPER INDUSTRY

Process	Problem	Cause
Mill supply	Piping system deposits and deterioration	Iron content and corrosion
	Turbidity	Floc carryover from treatment plant
	Deposits	Iron content
	Excessive color and organism carryover into product	High algae content and excessive color
	Low brightness	High organic color due to spring and fall runoff
	Scale	High hardness
Bleaching	Cost, low brightness	Color
	High color has adverse effect on bleaching	Occasional high iron, manganese, and color
	Low brightness	High color and iron
	Pulp brightness	High hardness and turbidity
Bleach plant–second extraction stage	Scale	High hardness
Paper machines	Paper quality	Surface runoff causes short periods of high turbidity
	Reduction in brightness	Color
	Low brightness	Nonsettleable fiber
	Machine operation	High alkalinity
	Low brightness	High iron
	Low brightness	Algae
	Black spots on paper	Manganese
	Paper quality	Radioactivity
	Bacteria slime	High algae and bacteria
Stock washing, screening	Foaming, poor washing, screening	High chlorides
Washing	Acetylation properties	High hardness
Wash and drying	Pulp brightness	High hardness
Felt showers for paper machines	Deposits	High iron content
Paper machine water filters	Plugged filters	High iron
Paper machine felts and wires	Pluggage and deposits	Turbidity carryover

3.2–16 WATER QUALITY PROBLEMS IN THE PULP AND PAPER INDUSTRY (*Continued*)

Process	Problem	Cause
Pulp mill	Low paper brightness	High turbidity and color
Pulp and papermaking	Low brightness	Color and turbidity
Paper mill	Low paper brightness	High turbidity and color
Coating mill	Low paper brightness	High turbidity and color
Heat exchangers	Scale	High hardness and alkalinity
Rosin sizing	Fouling	High hardness
Digesting	Scale on heating surfaces	High hardness

Source: Reprinted, with publisher's permission, from *J. Am. Water Works Assn.*, *63*(3):165, © 1971, by the American Water Works Association, Inc., 2 Park Ave., New York, N.Y. 10016.

3.2–17 QUALITY REQUIREMENTS AND CHARACTERISTICS OF PETROLEUM INDUSTRY WATERS

Characteristic	Concentration at point of use[a]	Concentration after use[b]
Silica (SiO_2)	50	c
Iron (Fe)	15	1
Calcium (Ca)	220	75
Magnesium (Mg)	85	30
Total sodium and potassium (Na, K)	230	c
Bicarbonate (HCO_3)	480	c
Sulfate (SO_4)	570	c
Chloride (Cl)	1,600	300
Fluoride (F)	1.2	c
Nitrate (NO_3)	8	c
Dissolved solids	3,500	1,000
Suspended solids	5,000	10
Hardness ($CaCO_3$)	900	350
Color, units	25	c
pH, units	6.0–9.0	6.0–9.0
Noncarbonate hardness ($CaCO_3$)		70

[a] Unless otherwise indicated, units are mg/l and values that normally should not be exceeded. Quality of water prior to the addition of substances used for internal conditioning.

[b] Unless otherwise indicated, units are mg/l and values are maximum. No one water will have all the maximum values shown.

[c] Accepted as received (if meeting total solids or other limiting values); has never been a problem at concentrations encountered.

Source: *Water Quality Criteria,* U.S. Department of the Interior, Fed. Water Pollut. Control Adm., April 1968.

REFERENCES

Application of the above values should be based upon analytical methods found in the following references.

1. *ASTM Book of Standards,* Part 23, Amer. Soc. Test Mat., Philadelphia, 1966.
2. Am. Publ. Health Assn., Am. Water Works Assn., and Water Pollut. Control Fed., *Standard Methods for the Examination of Water and Wastewater,* Am. Publ. Health Assn., New York, 1965.

3.2–18 QUALITY REQUIREMENTS OF WATER AT POINT OF USE BY THE LEATHER TANNING AND FINISHING INDUSTRY

Characteristic	Tanning processes	General finishing processes	Coloring
Alkalinity ($CaCO_3$)	a	a	a
pH, units	6.0–8.0	6.0–8.0	6.0–8.0
Hardness ($CaCO_3$)	150	b	c, d
Calcium (Ca)	60	b	c, d
Chloride (Cl)	250	250	e
Sulfate (SO_4)	250	250	e
Iron (Fe)	50	0.3	0.1
Manganese (Mn)	e	0.2	0.01
Organics: carbon chloroform extract	e	0.2	c
Color, units	5	5	5
Coliform bacteria	f	f	e
Turbidity	c	c	c

[a] Accepted as received (if meeting total solids or other limiting values); has never been a problem at concentrations encountered.
[b] Lime softened.
[c] Zero, not detectable by test.
[d] Demineralized or distilled water.
[e] Concentration not known.
[f] *Drinking Water Standards,* U.S. Public Health Service, 1962.

Note: Unless otherwise indicated, units are mg/l and values that normally should not be exceeded. Quality of water prior to the addition of substances used for internal conditioning.

Source: *Water Quality Criteria,* U.S. Department of the Interior, Fed. Water Pollut. Control Adm., April 1968.

REFERENCES

Application of the above values should be based upon analytical methods found in the following references.

1. *ASTM Book of Standards,* Part 23, Amer. Soc. Test. Mat., Philadelphia, 1966.
2. Am. Publ. Health Assn., Am. Water Works Assn., and Water Pollut. Control Fed., *Standard Methods for the Examination of Water and Wastewater,* Am. Publ. Health Assn., New York, 1965.

3.2–19 QUALITY REQUIREMENTS AND CHARACTERISTICS OF TEXTILE INDUSTRY WATERS

At point of use[a]

Characteristic	Sizing suspension	Scouring	Bleaching	Dyeing	Concentration after use[c]
Iron (Fe)	0.3	0.1	0.1	0.1	0.3
Manganese (Mn)	0.05	0.01	0.01	0.01	1.0
Copper (Cu)	0.05	0.01	0.01	0.01	0.5
Dissolved solids	100	100	100	100	150
Suspended solids	5	5	5	5	1,000
Hardness ($CaCO_3$)	25	25	25	25	120
pH, units					6.0–8.0
Cotton	6.5–10	9.0–10.5	2.5–10.5	7.5–10.0	
Synthetics	6.5–10	3.0–10.5	b	6.5–7.5	
Wool	6.5–10	3.0–5.0	2.5–5.0	3.5–6.0	
Color, units	5	5	5	5	d

[a] Water quality prior to addition of substances used for internal conditioning. Unless otherwise indicated, units are mg/l and values that normally should not be exceeded.
[b] Not applicable.
[c] Unless otherwise indicated, units are mg/l and values are maximum. No one water will have all the maximum values shown.
[d] Accepted as received (if meeting total solids or other limiting values); has never been a problem at concentrations encountered.

Source: *Water Quality Criteria,* U.S. Department of the Interior, Fed. Water Pollut. Control Adm., April 1968.

REFERENCES

Application of the above values should be based upon analytical methods found in the following references.

1. *ASTM Book of Standards,* Part 23, Amer. Soc. Test. Mat., Philadelphia, 1966.
2. Am. Publ. Health Assn., Am. Water Works Assn., and Water Pollut. Control Fed., *Standard Methods for the Examination of Water and Wastewater,* Am. Publ. Health Assn., New York, 1965.

3.2.3 COOLING AND STEAM GENERATION

3.2–20 QUALITY REQUIREMENTS OF WATER AT POINTS OF USE FOR STEAM GENERATION AND COOLING IN HEAT EXCHANGERS

	Boiler feed water[b]			Cooling water makeup for recirculation
Characteristic[a]	Low pressure, 0–150 psig	Intermediate pressure, 150–700 psig	High pressure, 700–1,500 psig	Fresh
Silica (SiO_2), mg/l	30	10	0.7	50
Aluminum (Al), mg/l	5	0.1	0.01	0.1
Iron (Fe), mg/l	1	0.3	0.05	0.5
Manganese (Mn), mg/l	0.3	0.1	0.01	0.5
Calcium (Ca), mg/l	c	d	d	50
Magnesium (Mg), mg/l	c	d	d	c
Ammonia (NH_4), mg/l	0.1	0.1	0.1	c
Bicarbonate (HCO_3), mg/l	170	120	48	24
Sulfate (SO_4), mg/l	c	c	c	200
Chloride (Cl), mg/l	c	c	c	500
Dissolved solids, mg/l	700	500	200	500
Copper (Cu), mg/l	0.5	0.05	0.05	c
Zinc (Zn), mg/l	c	d	d	c
Hardness ($CaCO_3$), mg/l	20	d	d	130
Free mineral acidity ($CaCO_3$), mg/l	d	d	d	d
Alkalinity ($CaCO_3$), mg/l	140	100	40	20
pH, units	8.0–10.0	8.2–10.0	8.2–9.0	c
Color, units	c	c	c	c
Organics				
Methylene blue active substances	1	1	0.5	1
Carbon tetrachloride extract	1	1	0.5	1
Chemical oxygen demand (O_2)	5	5	0.5	75
Dissolved oxygen (O_2)	2.5	0.007	0.007	c
Temperature, °F	c	c	c	c
Suspended solids	10	5	d	100

[a] Unless otherwise indicated, these values normally should not be exceeded. No one water will have all the maximum values shown.
[b] Quality is of water prior to addition of substances used for internal conditioning.
c Accepted as received (if meeting total solids or other limiting values); has never been a problem at concentrations encountered.
[d] Zero, not detectable by test.

Source: *Orange County Progress Report,* Orange County Planning Department, Orange County, Cal., March 1968.

3.2–21 QUALITY CHARACTERISTICS OF SURFACE WATERS THAT HAVE BEEN USED FOR STEAM GENERATION AND COOLING IN HEAT EXCHANGERS

	Boiler makeup water			Cooling water			
	Industrial		Electric utilities	Once through		Makeup for recycling	
Characteristic	Low and medium pressure, 0–700 psig	High pressure, 700–1,500 psig	High pressure, > 1,500 psig	Fresh	Brackish[a]	Fresh	Brackish[a]
Silica (SiO_2)	150	150	150	50	25	150	25
Aluminum (Al)	3	3	3	3	b	3	b
Iron (Fe)	80	80	80	14	1.0	80	1.0
Manganese (Mn)	10	10	10	2.5	0.02	10	0.02
Copper (Cu)	b	b	b	b	b	b	b
Zinc (Zn)	b	b	b	b	b	b	b
Calcium (Ca)	b	b	b	500	1,200	500	1,200
Magnesium (Mg)	b	b	b	b	b	b	b
Ammonia (NH_3)	b	b	b	b	b	b	b
Bicarbonate (HCO_3)	600	600	600	600	180	600	180
Sulfate (SO_4)	1,400	1,400	1,400	680	2,700	680	2,700
Chloride (Cl)	19,000	19,000	19,000	600	22,000	500	22,000
Nitrate (NO_3)	b	b	b	30	b	30	b
Phosphate (PO_4)	b	b	50	4	5	4	5
Dissolved solids	35,000	35,000	35,000	1,000	35,000	1,000	35,000
Hardness ($CaCO_3$)	5,000	5,000	5,000	850	7,000	850	7,000
Acidity ($CaCO_3$)	1,000	1,000	1,000	c	c	200	c
Alkalinity ($CaCO_3$)	500	500	500	500	150	500	150
pH, units	b	b	b	5.0–8.9	5.0–8.4	3.5–9.1	5.0–8.4
Color, units	1,200	1,200	1,200	b	b	1,200	b
Organics							
Methylene blue active substances	1	2	10	1.3	b	1.3	1.3
Carbon tetrachloride extract	100	100	100	d	d	100	100
Chemical oxygen demand (O_2)	100	100	500	b	b	100	200
Odor	b	b	b	b	b	b	b
Hydrogen sulfide (H_2S)	b	b	b	b	4	b	4
Dissolved oxygen (O_2)	b	b	b	b	b	b	b
Temperature, °F	120	120	120	100	100	120	120
Suspended solids	15,000	15,000	15,000	5,000	250	15,000	250

a Brackish water–dissolved solids more than 1,000 mg/l by definition, *1963 Census of Manufacturers,* U.S. Department of Commerce.
b Accepted as received (if meeting total solids or other limiting values); has never been a problem at concentrations encountered.
c Zero, not detectable by test.
d No floating oil.

Note: Unless otherwise indicated, units are mg/l and values are maximums. No one water will have all the maximum values shown.

Source: *Water Quality Criteria,* U.S. Department of the Interior, Fed. Water Pollut. Control Adm., April 1968.

REFERENCES

Application of the above values should be based upon analytical methods found in the following references.

1. *ASTM Book of Standards,* Part 23, Amer. Soc. Test. Mat., Philadelphia, 1966.
2. Am. Publ. Health Assn., Am. Water Works Assn., and Water Pollut. Control Fed., *Standard Methods for the Examination of Water and Wastewater,* Am. Publ. Health Assn., New York, 1965.

3.2–22 BOILER WATER QUALITY RECOMMENDATIONS AT INCREASING PRESSURES

Table A

Constituent	Low	Intermediate			High			
	to 250 psi	250–300 psi	300–450 psi	450–500 psi	500–600 psi	600–750 psi	750–900 psi	900–1,000 psi
Total hardness, as $CaCO_3$	0	0	0	0	0	0	0	0
Total alkalinity, as $CaCO_3$		700[b]		500[b]	500[b]	400[b]	300	250
Hydroxide, as $CaCO_3$	150–300		130	100	100	100	100	50–75
Silica[c] as SiO_2	100–133	50–67	50–67	50–67	19–25	19–25	7.5–10	7.5–10
Dissolved oxygen, as O	0.015			0.007				
Solids, dissolved	3,000–5,000	3,500[b]	3,000	2,500[b]	2,500[b]	2,000[b]	1,500[b]	1,250[b]
Solids, suspended	300–600	300[b]		150[b]	150[b]	100[b]	60[b]	40[b]
pH[a]	9.5			9.7				
Oil	1.0			0.1				
Iron, as Fe								
Copper, as Cu								
Sodium sulfite,[a] as Na_2SO_3	30–60							
Phosphate,[a] as PO_4	40–80	30–50	30–50	20–40	20–40	20–40	20–40	20–40
Organic dispersant[a]	50–75							

Table B

Constituent	High		Very high			Supercritical, over 3,000 psi
	1,000–1,250 psi	1,250–1,500 psi	1,500–2,000 psi	2,000–2,500 psi	2,500–3,000 psi	
Total hardness, as $CaCO_3$	0	0	0	0	0	0
Total alkalinity, as $CaCO_3$	200[b]	200[b]	150[b]	100[b]	100[b]	
Hydroxide, as $CaCO_3$	50–75	50–75				
Silica,[c] as SiO_2	3.8–5	2.1–2.9	0.8–1.1	0.3–0.5	0.09–0.13	0.02
Dissolved oxygen, as O						0.005
Solids, dissolved	1,000[b]	1,000[b]	750[b]	500[b]	500[b]	0.05–0.5[c]
Solids, suspended	20[b]	20[b]				
pH[a]	10.3–10.8	10.3–10.8	10.3–10.8			9.0–9.6
Oil						
Iron, as Fe						0.01[c]
Copper, as Cu						0.01[c]
Sodium sulfite,[a] as Na_2SO_3						
Phosphate,[a] as PO_4	20–40	20–40	20–40			
Organic dispersant[a]						

[a] These values minima. All others maxima.
[b] American Boiler Manufacturers Association recommendation.
[c] Values are for feedwater, not boiler water. For silica, this applies to pressures over 500 psi.

Note: All values except pH are in ppm.

Source: *Water Quality and Treatment,* 3rd ed., McGraw-Hill, New York, 1971, p. 510.

3.2–23 REQUIRED WATER QUALITIES FOR BOILER FEED

Impurity	Desired quality, for stated boiler pressure, lb/in.^{2} 150	400	2,000 sub-critical	3,200 critical
pH	8.0	9.00	9.00	9.00
TDS, mg/l	500.	100.	0.50	0.05
HCO_3, mg/l	50.	5.0	0.	0.
SiO_2, mg/l	40.	5.0	0.02	0.01
Ca + Mg (Hardness, mg/l as $CaCO_3$)	80.	10.	0.	0.
Fe + Al, mg/l	5.0	0.05	–	–
Cu, mg/l	–	–	0.01	0.005
Conductance, μmho/cm	–	–	1.0	0.10
Suspended matter, mg/l	20.	5.	–	–

Source: E.F. Gloyna and W.W. Eckenfelder, Jr., *Water Quality Improvement by Physical and Chemical Processes,* University of Texas Press, Austin, 1970, p. 259.

3.3 AGRICULTURAL WATER SUPPLY QUALITY CRITERIA

3.3–1 MAXIMUM ALLOWABLE SALINITIES AS ELECTRICAL CONDUCTIVITY (EC) OF THE SOIL SOLUTIONS

Crops Representing the Full Range of Salt Tolerance Among Crop Plants

Crops	**Maximum permissible EC in soil solution**	**Maximum permissible EC at indicated sensitive stages**[a]
Bermuda grass, tall wheat grass	18	–
Barley	18	10(S)
Sugar beets	16	8(G)
Cotton	16	–
Tomato, broccoli, spinach, alfalfa	8	–
Rice	8	4(S,F)
Potato, corn, flax	6	–
Bean, clovers (*Trifolium* spp.)	4	–
Most pome and stone fruits, citrus	4	–
Strawberries, blackberries, boysenberries	2–4	–

[a] G – germination; S – seedling; F – flowering.

Note: One mmho/cm is equivalent, on the average, to approximately 640 mg salt/l in the soil solution.

Source: L. Bernstein, *Spec. Tech. Publ. 418,* Am. Soc. Test. Mat., 1967, p. 51.

3.3–2 PERMISSIBLE LIMITS FOR TOTAL CONCENTRATION, PERCENT SODIUM AND BORON IN SEVERAL CLASSES OF IRRIGATION WATER

Classes of water	Total concentration as: Conductance, mho/cm x 10^5 at 25°C	Total concentration as: Dissolved solids, ppm	Percent sodium	Boron, ppm Crops group: Sensitive	Boron, ppm Crops group: Semitolerant	Boron, ppm Crops group: Tolerant
Excellent, less than	25	175	20	0.33	0.67	1.0
Good	20– 75	175– 525	20–40	0.33–0.67	0.67–1.33	1.0–2.0
Permissible	75–200	525–1,400	40–60	0.67–1.00	1.33–2.00	2.0–3.0
Doubtful	200–300	1,400–2,100	60–80	1.00–1.25	2.00–2.50	3.0–3.75
Unsuitable, more than	300	2,100	80	1.25	2.50	3.75

Source: L.V. Wilcox, *Sewage Works J.*, *20*(1):24, 1948. Based on data suggested by C.S. Schofield, *Detailed Analyses Showing Quality of Irrigation Waters,* South Coastal Basin Investigation, California Department of Public Works, 1933.

3.3–3 SUGGESTED GUIDELINES FOR SALINITY IN IRRIGATION WATER

Crop response	TDS, mg/l	EC,[a] mmho/cm
Water for which no detrimental effects will usually be noticed	<500	<0.75
Water which can have detrimental effects on sensitive crops	500–1,000	0.75–1.50
Water that may have adverse effects on many crops and requiring careful management practices	1,000–2,000	1.50–3.00
Water that can be used for salt tolerant plants on permeable soils with careful management practices	2,000–5,000	3.00–7.50

[a] Electrical conductivity.

Source: *Water Quality Criteria,* U.S. Department of the Interior, Fed. Water Pollut. Control Adm., April 1968.

3.3.1 IRRIGATION

3.3–4 PERMISSIBLE NUMBER OF IRRIGATIONS IN HUMID AREAS WITH SALINE WATER BETWEEN LEACHING RAINS

Crops of Different Salt Tolerance[a]

Irrigation water		Number of irrigations for crops having		
Total salts, mg/l	Electrical conductivity, mmho/cm at 25°C	Low salt tolerance	Moderate salt tolerance	High salt tolerance
640	1	7	15	–
1,280	2	4	7	11
1,920	3	2	4–5	7
2,560	4	2	3	5
3,200	5	1	2–3	4
3,840	6	1	2	3
4,480	7	–	1–2	2–3
5,120	8	–	1	2

[a] Based on a 50% yield decrement.

Source: J. Lunin *et al.*, *Use of Brackish Water for Irrigation in Humid Regions,* U.S. Department of Agriculture, 1960.

3.3–5 MAXIMUM PERMISSIBLE CHLORIDE CONTENTS IN SOIL SOLUTION

Various Fruit Crop Varieties and Rootstocks

Crop	Limit of tolerance to chloride in soil solution, me/l	Limit of concentration in irrigation water, assuming 25% leaching requirement me/l	ppm
Rootstocks			
Citrus			
Rangpur lime, Cleopatra mandarin	50	12.5	437
Rough lemon, tangelo, sour orange	30	7.5	262
Sweet orange, citrange	20	5.0	262
Avocado			
West Indian	16	4.0	140
Mexican	10	2.5	88
Stone fruit			
Marianna	50		
Lovell, Shalil	20		
Yunnan	14		
Varieties			
Grape			
Thompson seedless, Perlette	50	12.5	437
Cardinal, Black Rose	20	5.0	175
Berries			
Boysenberry	20	5.0	175
Olallie blackberry	20	5.0	175
Indian summer raspberry	10	2.5	88
Strawberry			
Lassen	16	4.0	140
Shasta	10	2.5	88

REFERENCES

1. *Water Quality Criteria,* U.S. Department of the Interior, Fed. Water Pollut. Control Adm., April 1968.
2. D.C. Baier, *J. Am. Water Works Assn., 62*(2):106, 1970.
3. L. Bernstein, *Quantitative Assessment of Irrigation Water Quality with Reference to Soil Properties, Climate, Irrigation, Management and the Salt Tolerance of Plants,* Am. Soc. Test. Mat., 1967.

3.3–6 SALT TOLERANCE OF VEGETABLE CROPS

Crop	EC_e[b]	Limit of concentration in irrigation water, assuming 25% leaching requirement[a]	
		EC_w	ppm[c]
Low salt tolerance			
Beans	750	188	132
Medium salt tolerance			
Carrots	750	188	132
Bell peppers	2,000	500	350
Lettuce	2,000	500	350
Corn	2,250	562	393
Cabbage	2,250	563	393
Broccoli	4,000	1,000	700
Tomatoes	4,000	1,000	700
High salt tolerance			
Spinach	5,750	1,442	1,001

[a] The indicated salt tolerances are the level at which the yield is reduced by 10 percent.
[b] EC electrical conductivity of soil solution, μmho/cm
[c] Salinity of irrigation water, using a conversion of 1 mho = 700 ppm.

Source: *Orange County Progress Report,* Orange County Planning Department, Orange County, Cal., 1968.

REFERENCE

1. *Water Quality Criteria,* U.S. Department of the Interior, 1968.

3.3–7 SALT TOLERANCE OF ORNAMENTAL SHRUBS

Shrub[a]	EC_e[b]	Limit of concentration in irrigation water	
		EC_w[c]	ppm[d]
Sensitive			
Roses	3,000	517	362
Pineapple guava	3,000	517	362
Viburnum	3,000	517	362
Moderately sensitive			
Pyracantha	5,000	1,000	700
Pittosporum	5,000	1,000	700
Xylosma	5,000	1,000	700
Texas privet	5,000	1,000	700
Moderately salt tolerant			
Arbor vitae	8,000	1,560	1,092
Spreading juniper	8,000	1,560	1,092
Lantana	8,000	1,560	1,092
Salt tolerant			
Oleander	10,000	2,000	1,400
Bottlebrush	10,000	2,000	1,400

[a] The indicated salt levels are the maximum tolerance by the U.S. Department of Agriculture.
[b] EC electrical conductivity of soil solution, μmho/cm.
[c] Salinity of irrigation water required, as derived from author's table, assuming a 25% leaching requirement.
[d] Required irrigation water in ppm, using a conversion of 1 mho = 770 ppm.

Source: *Water Quality Criteria,* U.S. Department of the Interior, Fed. Water Pollut. Control Adm., April 1968.

3.3–8 RELATIVE TOLERANCE OF PLANTS TO BORON

Tolerant	**Semitolerant**	**Sensitive**
Athel (Tamarix asphylla)	Sunflower (native)	Pecan
Asparagus	Potato	Black walnut
Palm (Phoenix canariensis)	Acala cotton	Persian (English) walnut
Date palm (P. dactylifera)	Pima cotton	Jerusalem artichoke
Sugar beet	Tomato	Navy bean
Mangel	Sweetpea	American elm
Garden beet	Radish	Plum
Alfalfa	Field pea	Pear
Gladiolus	Ragged Robin rose	Apple
Broadbean	Olive	Grape (Sultanina and Malaga)
Onion	Barley	Kadota fig
Turnip	Wheat	Persimmon
Cabbage	Corn	Cherry
Lettuce	Milo	Peach
Carrot	Oat	Apricot
	Zinnia	Thornless blackberry
	Pumpkin	Orange
	Bell pepper	Avocado
	Sweet potato	Grapefruit
	Lima bean	Lemon

Source: U.S. Salinity Laboratory Staff, *Diagnosis and Improvement of Saline and Alkali Soils,* U.S. Department of Agriculture, 1954.

3.3–9 TRACE ELEMENT TOLERANCES FOR IRRIGATION WATERS

Element	For water used continuously on all soils, mg/l	For short term use[a] on fine textured soils only, mg/l
Aluminum	1.0	20.0
Arsenic	1.0	10.0
Beryllium	0.5	1.0
Boron	0.75	2.0
Cadmium	0.005	0.05
Chromium	5.0	20.0
Cobalt	0.2	10.0
Copper	0.2	5.0
Fluorine	b	b
Iron	b	b
Lead	5.0	20.0
Lithium	5.0	5.0
Manganese	2.0	20.0
Molybdenum	0.005	0.05
Nickel	0.5	2.0
Selenium	0.05	0.05
Tin	b	b
Tungsten	b	b
Vanadium	10.0	10.0
Zinc	5.0	10.0

[a] The term "short term" means a period of time as long as two decades.
[b] Effects vary with chemical form. Some, such as tin and tungsten, may be excluded by plants.

Source: Basic information on trace elements was supplied by J.F. Hodgson of the U.S. Soil, Plant, and Nutrition Laboratory, Ithaca, N.Y.

3.3–10 PESTICIDES IN IRRIGATION WATERS

Insecticides, fungicides, rodenticides, and herbicides, as a group, include both organic and inorganic compounds, all of which can directly or indirectly have a bearing upon the irrigation water in which they are found. The effects of some of these can be detrimental to crops, livestock, wildlife, and man. Some are easily broken down and disappear quickly while others are persistent. Some are only sparingly soluble in water, but all cause problems if accidental spillage produces high concentrations in water or if they become adsorbed on colloidal particles subsequently dispersed in water.

Compounds derived from petroleum are used directly for pest control or are involved in formulation and synthesizing other pesticides. Many of these substances produce no serious pollution hazards because they break down rapidly. Synthetic materials developed within the last 20 years produce most of the hazard potential. There are several types such as halogenated hydrocarbons, organophosphates, carbamates, phenoxys, thiocyanates, substituted ureas, and triazines. Many of the halogenated hydrocarbons appear to be quite persistent in the environment. Aldrin and dieldrin, chlorinated hydrocarbon insecticides, have been found to be absorbed by vegetable crops from contaminated soil. DDT, a widely used insecticide for many years, has been found to be very persistent and can be transported in runoff from agricultural areas as well as being transported by air currents.[1]

Herbicides are used widely in agriculture directly on the crop and on the soil, on cropped and noncrop areas in the vicinity of agricultural areas, and for aquatic weed control. Petroleum solvents are effective aquatic weed killers which are rapidly dissipated and degraded. These aromatic solvents are widely used for keeping irrigation canals clear of weeds and are not harmful to crops.[2] Copper sulfate is also widely used in irrigation for algae and other aquatic weed control. The copper concentration is maintained at a low level and has little or no history of producing harmful effects on crops. The fate of copper applied for weed control in irrigation canals is being studied in cooperative aquatic weed research programs by the U.S. Bureau of Reclamation and Agricultural Research Service.

The phenoxy acid herbicides, 2,4-D and 2,4,5-T, have become suspect as contaminants of irrigation water. Some research has been initiated by various groups. These materials are known to be subject to rapid biological degradation in soils, but their fate in irrigation water and runoff water is still not well understood. Recent findings indicate that water temperature and dissolved oxygen content may influence the rate of biological decomposition of these herbicides in impounded water.[3]

Millions of acres of farm, range, and forest lands are treated annually with millions of pounds of pesticides. It has been predicted that present use of pesticides will increase tenfold in the next 20 years. There is too little known about the ultimate fate of the many compounds and their influence on irrigated agriculture as well as the total environment. This stresses the need for further work to determine the potential effect of the many pesticides in irrigation water. Some work is currently underway by industry, universities, and federal agencies to study the fate of these pesticides in irrigated agriculture; but as yet the state of the science is very incomplete.

There is little evidence to indicate that under normal use insecticide contamination of irrigation water would be detrimental to plant growth or accumulate in or on plants in toxic concentrations. Herbicides, on the other hand, could be harmful to crop growth if misused. Since many herbicides break down in water, permissible limits should be established for the point of application to crops. Suggested permissible levels are shown in 3.3–11 along with information on treatment rates of application and estimated concentrations in water reaching the field or crop. These levels are tentative and subject to change as indicated by future research.

Source: *Water Quality Criteria,* U.S. Government Printing Office, April 1968.

REFERENCES

1. S.R. Weibel *et al.*, "Pesticides and Other Contaminants in Rainfall and Runoff," *J. Am. Water Works Assn.*, 1966.
2. V.F. Burns *et al.*, *The Use of Aromatic Solvents for Control of Submersed Aquatic Weeds in Irrigation Channels,* U.S. Department of Agriculture, 1955.
3. J. De Marco, J. Symons, and G. Robeck, *Behavior of Synthetic Organics in Stratified Impoundments,* U.S. Department of Interior, Fed. Water Pollut. Control Fed., 1967.

3.3–11 TOLERANCE OF CROPS TO HERBICIDES USED IN AND AROUND WATER

Herbicide	Site of use	Formulation	Treatment rate	Concentration that may occur in irrigation water[a]	Crop injury threshold in irrigation water, mg/l[b]
Acrolein	Irrigation canals	Liquid	15 mg/l for 4 hr	10–0.1 mg/l	Flood or furrow: beans-60, corn-60, cotton-80, soybeans-20, sugar beets-60
			0.6 mg/l for 8 hr	0.4–0.02 mg/l	Sprinkler: corn-60, soybeans-15, sugar beets-15
			0.1 mg/l for 48 hr	0.05–0.1 mg/l	
Aromatic solvents (xylene)	Flowing water in canals or drains	Emulsifiable liquid	5–10 gal/ft³ sec (350–750 mg/l) applied in 30–60 min	700 mg/l or less	Alfalfa >1,600, beans-1,200, carrots-1,600, corn-3,000, cotton-1,600, grain sorghum >800, oats-2,400, potatoes-1,300, wheat >1,200
Copper sulfate	Canals or reservoirs	Pentahydrate crystals	Continuous treatment 0.5–3.0 mg/l, slug treatment: 1/3–1 lb (0.15–0.45 kg)/ft³ sec water flow	0.04–0.8 mg/l during first 10 mi, 0.08–9.0 mg/l during first 10–20 mi	Threshold is above these levels
Dalapon	Banks of canals and ditches	Water soluble salt	15–30 lb/A or 17–34 kg/ha	Less than 0.2 mg/l	Beets >7.0, corn >0.35
Diquat	Injected into water or sprayed over surface	Liquid	3–5 mg/l, 1–1.5 lb/A, or 1.2–1.7 kg/ha	Usually less than 0.1 mg/l	Beans-5.0, corn-125
Diuron	Banks and bottoms of small dry powder ditches	Wettable powder	Up to 64 lb/A or 72 kg/ha	No data	No data

[a] Herbicide concentrations given in this column are the highest concentrations that have been found in irrigation water, but these levels seldom remain in the water when it reaches crop land.

[b] Unless indicated otherwise, all crop tolerance data were obtained by flood or furrow irrigation. Threshold of injury is the lowest concentration causing temporary or permanent injury to crop plants even though, in many instances, neither crop yield nor quality was affected.

3.3–11 TOLERANCE OF CROPS TO HERBICIDES USED IN AND AROUND WATER (*Continued*)

Herbicide	Site of use	Formulation	Treatment rate	Concentration that may occur in irrigation water[a]	Crop injury threshold in irrigation water, mg/l[b]
Dichlobenil	Bottoms of dry canals	Granules or wettable powder	7–10 lb/A or 7.9–12.6 kg/ha	No data	Alfalfa-10, corn >10, soybeans-1.0, sugar beets-1.0–10
Endothall	Ponds and reservoirs	Water soluble Na or K salts	1–4 mg/l	Absent or only traces	Corn-25, field beans-1.0, alfalfa >10.0
Entothall amine salts	Reservoirs and static water canals	Liquid or granules	0.5–2.5 mg/l	Absent or only traces	Corn >25, soybeans >25, sugar beets-25
Fenac	Bottoms of dry canals	Liquid or granules	10–20 lb/A or 12.6–25.2 kg/ha	Absent or only traces	Alfalfa-1.0, corn-10, soybeans-0.1, sugar beets-0.1–10
Monuron	Banks and bottoms of small dry powder ditches	Wettable powder	Up to 64 lb/A or 72 kg/ha	No data	No data
Silvex	Woody plants and brambles on floodways, along canal, stream, or reservoir banks	Esters in liquid form	2–4 lb/A or 2.2–4.4 kg/ha	No data. Probably well under 0.1 mg/l	Corn >5.0, sugar beets and soybeans >0.02
	Floating and emersed weeds in southern waterways	Esters in liquid form	2–8 lb/A or 2.2–8.8 kg/ha	0.01–1.6 mg/l 1 day after application	

[a] Herbicide concentrations given in this column are the highest concentrations that have been found in irrigation water, but these levels seldom remain in the water when it reaches crop land.

[b] Unless indicated otherwise, all crop tolerance data were obtained by flood or furrow irrigation. Threshold of injury is the lowest concentration causing temporary or permanent injury to crop plants even though, in many instances, neither crop yield nor quality was affected.

3.3–11 TOLERANCE OF CROPS TO HERBICIDES USED IN AND AROUND WATER (*Continued*)

Herbicide	Site of use	Formulation	Treatment rate	Concentration that may occur in irrigation water[a]	Crop injury threshold in irrigation water, mg/l[b]
TCA	Banks of canals and ditches	Water soluble salt	Up to 64 lb/A or 72 kg/ha	Usually less than 0.1 mg/l	No injury observed at levels used
2,4-D amine	On banks of canals and ditches	Liquid	1–4 lb/A or 1.1–4.4 kg/ha	0.01–0.10 mg/l	Field beans >1.0, grapes-0.7, sugar beets >0.2, soybeans >0.02, corn-10, cucumbers, potatoes, sorghum, alfalfa, peppers >1.0
	Floating and emersed weeds in southern canals and ditches	Liquid	2–4 lb/A or 2.2–4.4 kg/ha	No data. Probably less than 0.1 mg/l	
Picloram	For control of brush on watersheds	Liquid or granules	1–3 lb/A or 1.1–3.3 kg/ha	No data	Corn >10, field beans-0.1, sugar beets >1.0

[a] Herbicide concentrations given in this column are the highest concentrations that have been found in irrigation water, but these levels seldom remain in the water when it reaches crop land.

[b] Unless indicated otherwise, all crop tolerance data were obtained by flood or furrow irrigation. Threshold of injury is the lowest concentration causing temporary or permanent injury to crop plants even though, in many instances, neither crop yield nor quality was affected.

Source: *Water Quality Criteria, 1972,* Manuscript, National Academy of Sciences, U.S. Government Printing Office, to be published in 1973.

REFERENCES

1. H.F. Arle, *Proc. West. Weed Control Conf., 12*:58, 1950.
2. H.F. Arle and G.N. McRae, *West. Weed Control Conf. Res. Progr. Report,* 1959, p. 72.
3. H.F. Arle and G.N. McRae, *West. Weed Control Conf. Res. Progr. Report,* 1960, p. 61.
4. V.F. Bruns, *Weeds, 3*:359, 1954.
5. V.F. Bruns, *Weeds, 5*:250, 1957.
6. V.F. Bruns, *Calif. Weed Conf. Proc., 16*:40, 1964.
7. V.F. Bruns, *Proc. Wash. State Weed Conf.,* Agr. Exp.Serv., Washington State University, Pullman, 1969, p. 33.
8. V.F. Bruns and W.J. Clore, *Weeds, 6*:187, 1958.
9. V.F. Bruns and J.H. Dawson, *Weeds, 7*:333, 1959.
10. V.F. Bruns *et al., The Use of Aromatic Solvents for Control of Submersed Aquatic Weeds in Irrigation Channels,* U.S. Department of Agriculture, 1955.
11. V.F. Bruns, R.R. Yeo, and H.F. Arle, *Tolerance of Certain Crops to Several Aquatic Herbicides in Irrigation Water,* U.S. Department of Agriculture, 1964.
12. P.A. Frank, R.J. Demint, and R.D. Comes, *Weed Sci., 18*(6):687, 1970.
13. U.S. Department of Agriculture, Agriculture Research Service, Suggested Guide for Weed Control, Agricultural Handbook No. 332, 1969.
14. R.R. Yeo, *Response of Field Crops to Acrolein,* Research Committee, West. Weed Control Conf., Salt Lake City, Utah, Research Progress Report, 1959, p. 71.
15. V.F. Bruns, J.M. Hodgson, and N.F. Arle, Response of Several Crops to Six Herbicides in Irrigation Water, 1971, unpublished data.

3.3.2 LIVESTOCK WATERING AND GENERAL FARMSTEAD USE

3.3–12 CRITERIA FOR FARMSTEAD WATER SUPPLIES

In view of the wide variety of sources used for farmstead water supplies and uncontrolled influences of geographic location and climatic conditions, no single set of values can realistically be established as criteria for farmstead supplies and, accordingly, most of the values are given in the form of acceptable ranges at point of use. In developing the criteria summarized in Table 3.3–13 considerable reliance has been placed on the U.S. Public Health Service Drinking Water Standards. Water which meets these standards is generally safe and acceptable to the user. Farmstead water supply includes water to be used for all household purposes, washing of raw agricultural commodities, and for milk sanitation. Specific requirements above those for general farmstead use are indicated also.

Quality Considerations

TASTE AND ODOR

The water supply should be substantially free of substances offensive to sight, taste, or smell. Taste and odor in water may result from the presence of a wide variety of substances including organic compounds, inorganic compounds, and algae. Knowledge concerning the source of taste and odor components is useful in determining what treatment, if any, is necessary to make the water acceptable.

The odor of water is usually due to the presence of dissolved gases such as hydrogen sulfide and volatile organic compounds. Threshold odor values in excess of three units[1] are generally considered objectionable. Dissolved inorganic salts of iron, zinc, manganese, copper, sodium, potassium, and others may be detected by taste. Limits for many of these ions are listed later in this section.

COLOR

The water supply should be substantially free of color. Dissolved organic material from decaying vegetation and certain inorganic matter cause color in water. Occasionally, excessive blooms of algae or the growth of aquatic microorganisms may also impart color. While color itself is not usually objectionable from the standpoint of health, its presence in excess of 15 color units[1] is aesthetically objectionable and suggests that the water needs appropriate treatment.

TEMPERATURE

The temperature of the water supply is not an important quality consideration for most farmstead uses. Where large volumes of water are to be used for hydrocooling farm products, however, the natural temperature can be a factor influencing its acceptability for such use.

pH

The pH of waters can range from 5.5–9.0[2] but most surface waters fall between pH 7.0 and 8.5[3] usually due to the presence of bicarbonate and carbonate ions. Waters with a pH below 6.0 can cause excessive corrosion in plumbing systems while waters with a pH above 8.5 suggest excessive sodium. Knowledge of the water pH is useful in determining necessary measures for corrosion control, sanitation, and adequate disinfection. It is recommended that the pH of farmstead water for milkhouse use fall between 6.8 and 8.5.

TOTAL DISSOLVED INORGANIC COMPOUNDS

Firm standards for total dissolved inorganic solids are not realistic in view of the naturally occurring difference in waters from various sources and geographical locations. The importance of total dissolved inorganic solids in farmstead waters for domestic use relates largely to taste, hardness, and laxative properties. It is desirable that the total dissolved inorganic solids not exceed 500 mg/l. Chlorides and sulfates should not exceed 250 mg/l. No general recommendations are appropriate for sodium, magnesium, potassium, phosphorus, sulfur, or calcium.

Although in excess of the above recommendations, waters containing up to 5,000 mg/l total dissolved inorganic solids can be used if alternate sources are not available. Under these conditions, however, the acceptability of the water depends upon the ionic composition of the dissolved solids and the feasibility of treatment to remove objectionable ions.

DISSOLVED ORGANIC COMPOUNDS

Determination of total dissolved organic compounds in water by measurement of carbon chloroform extractable substances (CCE) is too involved and expensive to be considered for application to farmstead water supplies. If surface waters are being used, an estimate of the CCE may be obtained from nearby municipalities using the same or similar sources. If deep ground waters are being used on the farmstead, the problem of total dissolved organic compounds may be ignored unless there is reason to expect contamination.

The organic compounds of possible concern in connection with farmstead supplies are persistent chlorinated organic pesticides. Although the data currently available[4-6] indicate that contamination of both ground and surface water with any of these materials rarely exceeds 0.1 μg/l and, accordingly, is negligible in terms of human health, it is highly desirable that water supplies remain in such a condition. Some criteria, such as those suggested by Ettinger and Mount,[7] are too low to be broadly applied at the present time in water quality evaluation. It is considered appropriate, therefore, that the permissible levels of specific pesticides in a farmstead water supply should not exceed the limits suggested by the PHS advisory committee, on use of the PHS drinking water standards.

TURBIDITY

The presence of suspended material such as clay, silt, finely divided organic material, and plankton contributes to the cloudiness or turbidity of water. Turbidity in excess of 5 units[1] is easily detectable in a glass of water and is usually objectionable for aesthetic reasons. Clay or other suspended particles may not adversely affect health, but water containing such particles may require treatment to make it suitable for certain uses. Following a rainfall, variations in the ground water turbidity may be considered an indication of surface pollution.

TRACE ELEMENTS

Attention has been given only to those trace elements which commonly occur in water supplies or those which are regarded as being particularly hazardous. The development of standards for every potentially hazardous substance which may occur in a water supply is an impossible task. The user of farmstead water supplies should ascertain, however, that there are no geological or environmental conditions which render the supply unsafe due to the presence of substances not covered herein.

Water containing more than 2.5 mg/l fluoride is detrimental during tooth formation and should not be used without suitable treatment.[8,9]

RADIONUCLIDES

All radiation exposure is regarded as harmful and any unnecessary exposure to ionizing radiation should be avoided. The acceptability of a farmstead water supply containing radioactive materials should be based upon the determination that the intake of radioactive substances from such water when added to that from all other sources is not likely to result in exposure greater than that recommended by the Federal Radiation Council.[10] Supplies containing radium-226 and strontium-90 are acceptable without consideration of other sources of radioactivity if the concentrations of these radionuclides do not exceed 3 and 10 pCi/l, respectively. In the known absence of strontium-90 and alpha-emitting radionuclides, the water supply is considered acceptable if the gross beta activity does not exceed 1,000 pCi/l. If the gross beta activity is in excess of this amount, a more complete radiochemical analysis is required to determine that the sources of radiation exposure are within the limits of the radiation protection guides.

PATHOGENIC AND NONPATHOGENIC MICROORGANISMS

The presence of any coliform organisms in a water supply suggests fecal contamination. The common technique used to measure coliform organisms, however, is based on a probability formula and compliance with the criterion is met if the sample is found to contain no more than one coliform organism per 100 ml of water. If routine chlorination or other effective means of disinfection are used, levels up to 100 coliform organisms per 100 ml in the raw water supply can be tolerated. Coliform densities above this level may require special treatments.

It should be kept in mind that negative test results cannot be considered as an assurance of a continuously safe supply unless the results of sanitary surveys and subsequent negative tests support such a position.

In addition to limitations for coliform organisms, water used for dairy sanitation should contain no more than 20 organisms of all other types per milliliter and no more than 5 proteolytic and/or lipolytic organisms per milliliter.[11a]

[a] *Standard Methods for the Examination of Dairy Products,* 11th ed., Am. Publ. Health Assn., New York, 1960 and B.W. Hammer and F.J. Babel, *Dairy Bacteriology,* 4th ed., John Wiley & Sons, 1957, 15.

Source: *Water Quality Criteria,* U.S. Department of the Interior, Fed. Water Pollut. Control Adm., April 1968.

REFERENCES

1. *Standard Methods for the Examination of Water and Wastewater Including Bottom Sediment and Sludges,* Am. Publ. Health Assn., New York, 1965.
2. *PHS Manual of Individual Water Supply Systems,* U.S. Public Health Service, U.S. Department of Agriculture, 1962.
3. *Water Pollution Surveillance System,* Division of Water Supply and Pollution Control, U.S. Public Health Service, U.S. Department of Agriculture, vol. 1–2, 1962–63.
4. S.D. Faust and O.M. Aly, *J. Am. Water Works Assn., 56*:267, 1964.
5. H.P. Nicholson *et al., Limnol. Oceanogr., 9*:310, 1964.
6. *Monitoring Agricultural Pesticide Residues – A Preliminary Report of Studies on Soil, Sediment and Water in Mississippi River Delta,* U.S. Department of Agriculture, 1966.
7. M.D. Ettinger and D.I. Mount, *Environ. Sci. Technol., 1*:203, 1967.
8. T.W. Bendixen, *Agr. Eng., 44*:366, 1963.
9. *Water System and Treatment Handbook,* 3rd ed., Water System Council, Glenville, Ill.
10. *Staff Report No. 1,* Federal Radiation Council, 1960.
11. E.P. Merrill and B.L. Arnold, *Farm Water Supplies,* Coop. Ext. Ser., University of Vermont, 1963.

3.3–13 KEY WATER QUALITY CRITERIA FOR FARMSTEAD USES

Characteristic	Recommendations (at point of use): General farmstead uses		Additional special use requirements
Taste and odor	Substantially free		
Color	Substantially free		
pH	6.0–8.5		6.8–8.5 dairy sanitation
Total dissolved inorganic solids	500 mg/l (under certain circumstances, higher levels are acceptable)		
Dissolved organic compounds	No recommendations for total organics		
	The concentration of persistent chlorinated organic pesticides should not exceed the following:		
	Compound, μg/l		
	Endrin	1	
	Aldrin	17	
	Dieldrin	17	
	Lindane	56	
	Toxaphene	5	
	Heptachlor	18	
	Heptachlor epoxide	18	
	DDT	42	
	Chlordane	3	
	Methoxychlor	35	
Turbidity	Substantially free		
Hazardous trace elements	Levels in excess of those shown are grounds for rejection of a supply:		
	Substances, mg/l		
	Arsenic	0.05	
	Barium	1.00	
	Cadmium	0.01	
	Chromium	0.05	
	Cyanides	0.2	
	Lead	0.05	
	Selenium	0.01	
	Silver	0.05	
Other trace elements	Levels shown below should not be exceeded if alternate sources are available:		
	Substances, mg/l		
	Manganese	0.05	In dairy sanitation, water should contain <20 mg/l potassium and <0.1 mg/l iron and copper
	Iron	0.3	
	Copper	1.0	
	Zinc	5.0	In tooth formation, water should contain <2.5 mg/l fluorides
	Fluoride	0.7–1.2	
	Nitrate	45.0	

3.3–13 KEY WATER QUALITY CRITERIA FOR FARMSTEAD USES (*Continued*)

Characteristic	Recommendations (at point of use): General farmstead uses		Additional special use requirements
	pCi/l		
Radionuclides	Strontium-90	10	
	Radium-226	3	
	In absence of above radionuclides, 1,000 pCi/l gross β activity		
Nonpathogenic microorganisms	To conform to USPHS drinking water standards		For dairy sanitation, water should not contain more than 20 organisms/ml and contain not more than 5 lypolytic and/or proteolytic organisms/ml

Source: *Water Quality Criteria,* U.S. Department of the Interior, Fed. Water Pollut. Control Adm., April 1968.

3.3–14 KEY WATER QUALITY CRITERIA FOR LIVESTOCK USE

Characteristic	Recommendations
Total dissolved solids (TDS)	< 10,000 mg/l, depending upon animal species and ionic composition of the water
Hazardous trace elements, maximum allowable concentrations	
Arsenic	< 0.05 mg/l
Cadmium	< 0.01 mg/l
Chromium	< 0.05 mg/l
Fluorine	< 2.40 mg/l
Lead	< 0.05 mg/l
Selenium	< 0.01 mg/l
Organic substances	
Algae (water bloom)	Avoid abnormally heavy growth of blue-green algae
Parasites and pathogens	Conform to epidemiological evidence
Dissolved organic compounds	Biological accumulation from environmental sources, including water, shall not exceed established, legal limits in livestock products
Radionuclides	Conform to recommendations for farmstead water supplies

Source: *Water Quality Criteria,* U.S. Department of the Interior, Fed. Water Pollut. Control Adm., April 1968.

3.3–15 PROPOSED SAFE LIMITS OF SALINITY FOR LIVESTOCK

Animal	Threshold salinity concentration,[a] TDS, mg/l
Poultry	2,860
Swine	4,290
Horses	6,435
Dairy cattle	7,150
Beef cattle	10,000
Sheep (adult, dry)	12,000

[a] Total salts, mainly NaCl.

Source: *Water Quality Criteria,* U.S. Department of the Interior, Fed. Water Pollut. Control Adm., April 1968.

REFERENCES

1. *J. Agr. W. Australia, 27*(2):156, 1950.
2. *J. Agr. W. Australia, 3*(2), 1959.

3.3–16 PROPOSED TOXIC DOSE RANGES FOR ARSENIC

Animal	Toxic dose of As, g/animal
Poultry	0.05– 0.10
Dogs	0.01– 0.20
Swine	0.50– 1.00
Sheep, goats, horses	10.00–15.00
Cattle	15.00–30.00

Source: J.A. Wadsworth, *Vet. Med.,* 7(10):412, 1952.

3.4 FISH AND WILDLIFE

3.4–1 AQUATIC LIFE WATER QUALITY CRITERIA

Recommendations

Dissolved oxygen

Habitats for well-rounded warm water fish populations: DO concentration shall not be less than 4.0 mg/l at any time or any place.

Habitats for commercial, nongame, and forage fish species: DO concentration shall not be less than 3.0 mg/l at any time or any place.

Habitats for cold water fish: DO concentration shall not be less than 5.0 mg/l at any time or at any place.

Oxygen demand of the stream shall be maintained at a level which permits normal diurnal and seasonal variation above the minimum levels specified.

Hydrogen – ion concentration

pH values should be maintained between 6.0 and 8.5.

Daily fluctuations which exceed pH 8.5 and are correlated with photosynthetic activity may be tolerated. However, any sudden drop below pH 6.0 or sudden rise above 8.5, not related to photosynthesis, indicates abnormal conditions which should be investigated immediately.

Mixing zones

No separate criteria be established for those sections of streams into which pollutants are being discharged.

Zones of admixture shall not extend entirely across a stream, but shall leave at all times and throughout the entire length of the zone a passageway adequate for migration, but consisting of not less than ¼ of the width and ¼ of the cross section of the stream.

Toxic wastes shall be mixed thoroughly and rapidly with the receiving water in the mixing zone.

Temperature

To maintain well-rounded warm water and commercial nongame and forage fish habitats

a. Stream temperatures shall not exceed 93° F at any time or any place, and a daily mean of 90° F should not be exceeded.

a. Stream temperatures shall not exceed 93°F at any time or any place, and a daily mean of 90°F should not be exceeded.

b. The temperature shall be below 55° F during December, January, and February.

c. The months of March, April, October, and November shall be transition periods during which the temperature can be changed gradually by not more than 7°F per day.

To maintain trout habitats stream temperature shall not exceed 55°F during the months of October through May, nor exceed 68°F during the months of June through September, and insofar as possible, the temperature should not be raised in streams used for natural propagation of trout.

Toxicity

The final concentration of any waste in the receiving water should not exceed 1/10 of the 96 hr median tolerance limit (TL_m), except that other limiting concentrations (for example, application factor) may be used in specific cases when justified on the basis of available evidence and approved by the appropriate regulatory agency.

3.4–1 AQUATIC LIFE WATER QUALITY CRITERIA *(Continued)*

Recommendations

For water containing wastes composed chiefly of stable components of moderate toxicity without measurable oxygen demand, the static bioassay may be used to determine TL_m value. For water containing unstable or volatile components with a measurable oxygen demand, the flow through bioassay shall be used.

Radioactivity

The concentration of radionuclides in an effluent shall have a gross activity (alpha, beta, and gamma radiation) not exceeding 100 pCi/1.

Source: *Environ. Sci. Technol., 1*(11):888, 1967.

3.4–2 EFFECT ON ALGAE AND ON INVERTEBRATES IN RIVERS CHURNET AND DOVE

Discharge from Two Outfalls of Waste Waters Containing Copper

Position	No. algae/mm^2	Animals	
		No. species	No. animals per unit area
River Churnet			
Just above 1st outfall	1,600	15	> 5,000
½ mile below 1st outfall	150	0	0
Just below 2nd outfall; 3½ miles below 1st	100	0	0
2 miles below 2nd outfall; 5½ miles below 1st	16,700	0	0
8 miles below 2nd outfall; 11½ miles below 1st	58,200	2	2
River Dove			
Before confluence with R. Churnet	2,800	30	1,193
Below confluence; 11¾ miles below 1st outfall	27,600	4	42
19½ miles below 1st outfall	55,100	8	59
30½ miles below 1st outfall	53,300	12	371

Source: Ministry of Agriculture and Fisheries, *Observations on the Conditions of Rivers Churnet and Dove in 1938,* Trent Fishery Board, Annual Report, 1938, Appendix 1.

3.4–3 COPPER TOXICITY IN ENRICHED FILTER STERILIZED SEAWATER IN THE ABSENCE AND PRESENCE OF NTA[a]

Growth in the Absence of NTA

	OD[c] control	Cu, μg/l								
		50	100	150	200	250	300	350	400	450
D. tertiolecta[b]	0.195	95	95	90	80	80	72	82	51	51
C. nana	0.190	18	8	–	–	–	–	–	–	–
A. carteri	0.273	–	–	–	–	–	–	–	–	–
O. luteus	0.110	13	–	–	–	–	–	–	–	–
S. costatum	0.184	91	64	–	–	–	–	–	–	–
I. galbana	0.520	45	9	5	–	–	–	–	–	–

Growth in the Presence of 10 mg/l NTA

	OD[c] control	Cu, μg/l								
		50	100	150	200	250	300	350	400	450
D. tertiolecta[b]	0.325	78	120	130	144	121	126	127	125	132
C. nana	0.259	97	90	82	92	89	81	85	75	77
A. carteri	0.475	83	77	83	85	87	83	71	63	52
O. luteus	0.620	94	100	98	101	84	90	91	97	89
S. costatum	0.370	113	106	109	99	100	108	106	106	102
I. galbana	1.030	93	102	93	94	83	83	90	87	85

[a] As percent of control, OD at 420 mμ.
[b] 7 day growth, all others 14 days.
[c] OD–Optical density.

Source: S.J. Erickson *et al.*, *J. Water Pollut. Control Fed.*, *42*(8) (Part 2):R270, 1970.

3.4–4 COPPER TOXICITY IN STEAM STERILIZED MEDIA IN THE ABSENCE OF NTA[a]

E-1A Synthetic Seawater

	OD[c] control	Cu, μg/l								
		50	100	150	200	250	300	350	400	450
D. tertiolecta[b]	0.580	102	100	98	97	93	93	91	83	84
C. nana	0.455	100	98	89	73	77	69	62	46	30
A. carteri	0.570	90	7	3	2	–	–	–	–	–
O. luteus	0.035	57	54	43	29	6	–	–	–	–
S. costatum	0.318	102	106	104	112	114	118	106	96	97
I. galbana	0.780	89	81	74	58	67	57	51	45	42

Steam Sterilized Enriched Seawater

	OD[c] control	Cu, μg/l								
		50	100	150	200	250	300	350	400	450
D. tertiolecta[b]	0.390	105	112	108	101	106	101	104	98	89
C. nana	0.445	106	104	102	120	108	109	111	87	18
A. carteri	0.485	102	97	93	96	93	83	39	3	3
O. luteus	0.055	73	60	36	24	15	–	–	–	–
S. costatum	0.390	82	80	76	79	–	83	77	82	89
I. galbana	0.850	98	96	90	75	59	14	5	2	3

[a] As percent of control, OD at 420 mμ.
[b] 7 day growth, all others 14 days.
[c] OD – optical density.

Source: S.J. Erickson *et al., J. Water Pollut. Control Fed., 42*(8) (Part 2):R270, 1970.

3.4–5 MEDIAN TOLERANCE LIMITS OF CERTAIN CHEMICALS FOR VARIOUS ANIMALS IN DIFFERENT DILUTION MEDIA

		TL_m, mg/l			
Compound and animal tested	**Medium**	**24 hr**	**48 hr**	**72 hr**	**96 hr**
Acetic acid					
Culex sp. larvae	RDW	1,500	1,500	–	–
D. magna	SRW	47	–	–	–
L. macrochirus	RDW	>100<1,000	–	–	–
Acetone					
D. magna	RDW	10	10	–	–
Adipic acid					
L. macrochirus	RDW	<330	–	–	–
Ammonium chloride					
C. carassius	SRW	640	–	–	–
D. magna	ULW	202	161	67	50
D. magna	SRW	–	–	–	139[a]
L. macrochirus	SRW	725	725	725	725
Lymnaeae sp. egg	ULW	241	173	73	70
Ammonium hydroxide					
D. magna	SRW	60[a]	32[a]	–	20[a]
Ammonium sulfate					
D. magna	SRW	423[a]	433[a]	–	292[a]
Ammonium sulfite					
D. magna	SRW	299[a]	273[a]	–	203[a]
Barium chloride					
Rana sp. eggs	ULW	244.3 mg/l–91.67% development 2,443.0 mg/l–89.55% development 24,430.0 mg/l– 0.00% development			
Butyric acid					
D. magna	SRW	–	61	–	–
L. macrochirus	RDW	200	–	–	–
Calcium chloride					
D. magna	SRW	3,526[a]	3,005[a]	–	–
D. magna	ULW	1,838	759	759	649[a]
L. macrochirus	SRW	8,350	–	–	–
Lymnaea sp. eggs	ULW	4,485	3,094	3,308	2,573
Capric acid					
L. macrochirus	RDW	I			
Caproic acid					
L. macrochirus	RDW	>150<200	–	–	–
Caprylic acid					
L. macrochirus	RDW	I	–	–	–
Chromic sulfate					
D. magna	ULW	0.1	0.03	–	–
Chromic sulfate plus sodium dichromate					
Lymnaea sp. eggs	ULW	0.17–0.22	0.17–0.22	–	–
Ferric chloride					
D. magna	SRW	36[a]	21[a]	–	15[a]
Formaldehyde					
D. magna	RDW	>100<1,000	–	–	–
Formic acid					
L. macrochirus	RDW	175	–	–	–
Glutaric acid					
L. macrochirus	RDW	330	–	–	–

[a] Denotes 25 hr when placed after a value in the 24 hr column, 50 hr when in the 48 hr column, and 100 hr when in the 96 hr column.

3.4–5 MEDIAN TOLERANCE LIMITS OF CERTAIN CHEMICALS FOR VARIOUS ANIMALS IN DIFFERENT DILUTION MEDIA (*Continued*)

		TL_m, mg/l			
Compound and animal tested	**Medium**	**24 hr**	**48 hr**	**72 hr**	**96 hr**
Magnesium chloride					
D. magna	SRW	3,391[a]	3,699[a]	–	3,484[a]
Magnesium sulfate					
D. magna	ULW	963	929	861	788
D. magna	SRW	–	–	–	3,803
L. macrochirus	SRW	19,000	–	–	–
Lymnaea sp. eggs	ULW	10,530	6,525	6,300	6,250
Malonic acid					
L. macrochirus	RDW	150	–	–	–
Phenol					
D. magna	RDW	100	100	–	–
D. magna young	SRW	17[a]	7[a]	–	–
D. magna adult	SRW	61[a]	21[a]	–	–
L. macrochirus	SRW	>10<15[a]	–	–	–
M. latipinna	SRW	63[a]	22[a]	–	–
Potassium chloride					
D. magna	SRW	–	–	–	679[a]
D. magna	ULW	343	337	117	29
L. macrochirus	SRW	5,500	–	–	–
Lymnaea sp. eggs	ULW	1,941	1,492	1,018	1,100
Potassium cyanide					
D. magna	ULW	2	2	0.7	0.4
Lymnaea sp. eggs	ULW	796	796	147	130
Potassium dichromate					
C. carassius	SRW	705	–	–	–
D. magna	SRW	–	–	–	0.4[a]
L. macrochirus	SRW	739	–	–	–
Potassium ferricyanide					
D. magna	ULW	905	549	0.6	0.1
Potassium nitrate					
D. magna	ULW	490	490	226	39
D. magna	SRW	–	–	–	900[a]
L. macrochirus	SRW	5,500	–	–	–
Lymnaea sp. eggs	ULW	1,941	1,492	1,018	1,100
Propionic acid					
Culex sp. larvae	RDW	>1,000	>1,000	–	–
D. magna	SRW	–	50	–	–
L. macrochirus	RDW	188	–	–	–
Pyridine					
D. magna	RDW	2,114	944	–	–
Sodium acetate					
L. macrochirus	RDW	5,000	–	–	–
Culex sp. larvae	RDW	7,500	7,425	–	–
Sodium anthraquinone alpha-sulfonate					
D. magna	ULW	186	186	186	50
D. magna	SRW	–	–	–	12[a]
Lymnaea sp. eggs	ULW	186	186	186	186
Sodium bicarbonate					
Culex sp. larvae	RDW	2,000	2,000	–	–

[a] Denotes 25 hr when placed after a value in the 24 hr column, 50 hr when in the 48 hr column, and 100 hr when in the **96 hr** column.

3.4–5 MEDIAN TOLERANCE LIMITS OF CERTAIN CHEMICALS FOR VARIOUS ANIMALS IN DIFFERENT DILUTION MEDIA *(Continued)*

		TL_m, mg/l			
Compound and animal tested	**Medium**	**24 hr**	**48 hr**	**72 hr**	**96 hr**
Sodium bisulfate					
Culex sp. larvae	RDW	300	300	–	–
Sodium bisulfite					
D. magna young	SRW	116[a]	81[a]	–	–
D. magna adult	SRW	–	–	–	102
D. magna adult	ULW	171	119	97	82
Dugesia sp.	ULW	179	179	179	179
Lymnaea sp. eggs	ULW	179	59	59	59
M. latipinna	SRW	241[a]	220[a]	–	–
Sodium bisulfite plus sodium carbonate					
D. magna	ULW	1,484–665	1,484–665	1,060–475	99–45
D. magna	SRW	–	–	–	436–85
Sodium bisulfite plus sodium chromate					
D. magna	SRW	–	–	–	67–0.278[a]
Sodium bisulfite plus sodium silicate					
D. magna	SRW	14,210–950	11,723–784	222–15	222–15
Sodium bisulfite plus sodium sulfate					
D. magna	SRW	–	–	–	82–3,654
Sodium bisulfite plus sodium carbonate and sodium chromate					
D. magna	SRW	–	–	–	86–441–0.354[a]
Sodium bisulfite plus sodium carbonate and sodium silicate					
D. magna	SRW	–	–	–	39–198–93[a]
Sodium bisulfite plus sodium carbonate and sodium sulfate					
D. magna	SRW	–	–	–	57–296–2,569[a]
Sodium bisulfite plus sodium chromate and sodium silicate					
D. magna	SRW	–	–	–	224–0.86–506[a]
Sodium bisulfite plus sodium chromate and sodium sulfate					
D. magna	SRW	–	–	–	78–0.32–3,443[a]
Sodium bisulfite plus sodium silicate and sodium sulfate					
D. magna	SRW	–	–	–	52–126–2.326[a]
Sodium *p*-bromobenzene sulfonate					
D. magna	ULW	2,347	1,943	971	809
D. magna	SRW	–	–	–	523[a]
L. macrochirus	SRW	<1,560	–	–	–
Lymnaea sp. eggs	ULW	2,590	2,590	2,590	2,590

[a] Denotes 25 hr when placed after a value in the 24 hr column, 50 hr when in the 48 hr column, and 100 hr when in the 96 hr column.

3.4–5 MEDIAN TOLERANCE LIMITS OF CERTAIN CHEMICALS FOR VARIOUS ANIMALS IN DIFFERENT DILUTION MEDIA *(Continued)*

		TL_m, mg/l			
Compound and animal tested	**Medium**	**24 hr**	**48 hr**	**72 hr**	**96 hr**
Sodium butyl sulfonate					
D. magna	ULW	8,000	8,000	5,400	2,700
Sodium butyrate					
L. macrochirus	RDW	5,000	–	–	–
Sodium carbonate					
Amphipoda	ULW	360	176	67	67
Culex sp. larvae	RDW	1,820	600	–	–
D. magna	ULW	347	265	–	–
D. magna	SRW	607[a]	565	–	524
Dugesia sp.	ULW	384	360	360	341
L. macrochirus	ULW	385	–	–	–
Lymnaea sp. eggs	ULW	403	403	395	411
M. latipinna	SRW	405[a]	297[a]	–	–
Sodium carbonate plus sodium chromate					
D. magna	SRW	–	–	–	420–0.34[a]
Sodium carbonate plus sodium chromate and sodium silicate					
D. magna	SRW	–	–	–	187–0.15–88
Sodium carbonate plus sodium chromate and sodium sulfate					
D. magna	SRW	–	–	–	240–0.19–2,078[a]
Sodium carbonate plus sodium silicate					
D. magna	ULW	265–1,776	265–1,776	265–1,776	265–1,776
Sodium carbonate plus sodium silicate and sodium sulfate					
D. magna	SRW	–	–	–	161–76–1,396[a]
Sodium carbonate plus sodium sulfate					
D. magna	ULW	198–666	172–577	172–577	66–222
Sodium chloride					
C. carassius	SRW	13,750	–	–	–
Culex sp. larvae	RDW	10,500	10,200	–	–
D. magna	SRW	6,447[a]	5,874[a]	–	3,114[a]
D. magna	ULW	3,412	3,310	–	–
L. macrochirus	SRW	14,125	–	–	–
Lymnaea sp. eggs	ULW	3,412	3,388	–	–
M. latipinna	SRW	18,735[a]	16,595	–	–
Sodium *p*-chlorobenzene sulfonate					
D. magna	ULW	8,600	7,659	3,964	2,150
D. magna	SRW	–	–	–	2,394[a]
L. macrochirus	SRW	<3,219	–	–	–
Lymnaea sp. eggs	ULW	8,600	7,633	6,343	5,053
Sodium 2-chlorotoluene-4-sulfonate					
L. macrochirus	SRW	<1,374	–	–	–

[a] Denotes 25 hr when placed after a value in the 24 hr column, 50 hr when in the 48 hr column, and 100 hr when in the 96 hr column.

3.4–5 MEDIAN TOLERANCE LIMITS OF CERTAIN CHEMICALS FOR VARIOUS ANIMALS IN DIFFERENT DILUTION MEDIA *(Continued)*

		TL_m, mg/l			
Compound and animal tested	**Medium**	**24 hr**	**48 hr**	**72 hr**	**96 hr**
Sodium 2-chlorotoluene-5-sulfonate					
D. magna young	SRW	0.8[a]	0.6[a]	–	0.4[a]
D. magna adult	SRW	3.3[a]	1.3[a]	–	–
Lymnaea sp. eggs	SRW	30[a]	–	–	–
M. latipinna	SRW	115.2[a]	66.1[a]	–	–
Sodium chromate plus sodium silicate					
D. magna	SRW	–	–	–	0.21–130[a]
Sodium chromate plus sodium silicate and sodium sulfate					
D. magna	SRW	–	–	–	0.28–122–2,255[a]
Sodium chromate plus sodium sulfate					
D. magna	SRW	–	–	–	0.28–3,044[a]
Sodium 2,5-dichloro-benzene sulfonate					
D. magna	ULW	4,931	4,931	2,490	938
D. magna	SRW	–	–	–	1,468[a]
L. macrochirus	SRW	<3,750	–	–	–
Lymnaea sp. eggs	ULW	4,981	4,513	3,984	3,144
Sodium dichromate					
D. magna	ULW	22	10	–	–
Sodium formate					
L. macrochirus	RDW	5,000	–	–	–
Sodium nitrate					
C. carassius	SRW	12,150	–	–	–
D. magna	ULW	5,980	3,581	2,125	665
D. magna	SRW	–	–	–	4,206
L. macrochirus	SRW	12,800	–	–	–
Lymnaea sp. eggs	ULW	6,375	6,460	5,950	3,251
Sodium *m*-nitrobenzene sulfonate					
D. magna	ULW	8,665	8,665	6,017	5,067
D. magna	SRW	–	–	–	2,235[a]
L. macrochirus	SRW	<1,350	–	–	–
Sodium 4-nitrochloro-benzene-2-sulfonate					
D. magna	ULW	4,698	3,483	948	948
D. magna	SRW	–	–	–	1,474
L. macrochirus	SRW	6,375	–	–	6,375[a]
Lymnaea sp. eggs	ULW	3,532	4,439	3,736	3,208
Sodium 4-nitrotoluene-2-sulfonate					
L. macrochirus	SRW	<1,440	–	–	–
Sodium oxalate					
L. macrochirus	RDW (without $CaCl_2$)	4,000	–	–	–
Sodium *p*-phenolsulfonate					
D. magna	ULW	13,510	13,510	3,494	1,471
L. macrochirus	SRW	–	–	–	19,616[a]
Lymnaea sp. eggs	ULW	10,700	9,122	8,828	8,828

[a] Denotes 25 hr when placed after a value in the 24 hr column, 50 hr when in the 48 hr column, and 100 hr when in the 96 hr column.

3.4–5 MEDIAN TOLERANCE LIMITS OF CERTAIN CHEMICALS FOR VARIOUS ANIMALS IN DIFFERENT DILUTION MEDIA *(Continued)*

		TL_m, mg/l			
Compound and animal tested	**Medium**	**24 hr**	**48 hr**	**72 hr**	**96 hr**
Sodium mono-hydrogen phosphate					
D. magna	SRW	1,154[a]	1,089[a]	–	426[a]
Sodium mono-hydrogen phosphate plus sodium pyrophosphate					
D. magna	ULW	3,580–433	3,580–391	–	–
Lymnaea sp. eggs	ULW	2,685–63	2,954–223	2,828–60	2,685–248
Sodium phosphate					
D. magna	SRW	237[a]	177[a]	–	126[a]
Sodium propionate					
Culex sp. larvae	RDW	–	2,320	–	–
L. macrochirus	RDW	5,000	–	–	–
Sodium pyrophosphate					
D. magna	ULW	433	391	–	–
Sodium silicate					
Amphipoda	ULW	895	263	261	160
D. magna	ULW	575	494	413	216
D. magna	SRW	–	–	–	247
Lymnaea sp. eggs	ULW	632	630	630	632
Sodium sulfate					
Amphipoda	ULW	2,380	1,110	880	880
Culex sp. larvae	RDW	11,430	13,350	–	–
D. magna	ULW	8,384	2,564	725	630
D. magna adult	SRW	–	–	–	4,547
D. magna young	SRW	6,800	6,100	–	–
L. macrochirus	SRW	17,500	–	–	–
Lymnaea sp. eggs	ULW	5,401	5,400	5,400	3,553
M. latipinna	SRW	20,040	15,996	–	–
Sodium sulfide					
D. magna	SRW	16[a]	13[a]	–	9[a]
Sodium sulfite					
D. magna	SRW	299[a]	273[a]	–	203[a]
Sodium thiosulfate					
D. magna	SRW	2,245[a]	1,334[a]	–	805[a]
Sodium valerate					
L. macrochirus	RDW	5,000	–	–	–
Valeric acid					
D. magna	SRW	–	45	–	–
Xylene					
D. magna	RDW	>100<1,000	–	–	–
Zinc stearate					
L. macrochirus	RDW	I			

[a] Denotes 25 hr when placed after a value in the 24 hr column, 50 hr when in the 48 hr column, and 100 hr when in the 96 hr column.

Note: TL_m mg/l – median tolerance limit.
SRW – Standard Reference Water.[1]
RDW – Reference Dilution Water.[2]
ULW – glass-wool filtered University Lake water.
I – chemical to insoluble in water to be toxic.

Source: B.F. Dowden and H.J. Bennett, *J. Water Pollut. Control Fed.*, *37*(9):1310, 1965.

REFERENCES

1. L. Freeman, *Sew. Ind. Wastes*, *25*(7):845, 1953.
2. B.F. Dowden, *Proc. Nat. Acad. Sci.*, *23*:77, 1960.

3.4–6 48 HOUR TL_m VALUES OF PESTICIDES FROM STATIC BIOASSAY FOR VARIOUS TYPES OF FRESH WATER ORGANISMS

To provide reasonably safe concentrations of these materials in receiving waters, application factors ranging from 1/10–1/100 should be used with these values depending on the characteristic of the pesticide in question and used as specified in *Water Quality Criteria,* U.S. Government Printing Office, 1968, p. 58. Concentrations thus derived tentatively may be considered safe under the environmental conditions recommended.

mg/l

Insecticides

	Stream invertebrate[a]		**Cladocerans[b]**		**Fish[c]**		**Gammar lacustris,[d]**
Pesticide	**Species**	**TL_m**	**Species**	**TL_m**	**Species**	**TL_m**	**TL_m**
Abate	Pteronarcys californica	100			Brook trout	1,500	640
Aldrin[e]	P. californica	8	Daphnia pulex	28	Rainbow trout	,3	12,000
Allethrin	P. californica	28	D. pulex	21	Rainbow trout	19	20
Azodrin					Rainbow trout	7,000	
Aramite			D. magna	345	Bluegill	35	100
Baygon[e]	P. californica	100			Fathead	25	50
Baytex[e]	P. californica	130	Simocephalus serrulatus	3.1	Brown trout	80	70
Benzene hexachloride (lindane)	P. californica	8	D. pulex	460	Rainbow trout	18	88
Bidrin	P. californica	1,900	D. pulex	600	Rainbow trout	8,000	790
Carbaryl (sevin)	P. californica	1.3	D. pulex	6.4	Brown trout	1,500	22
Carbophenothion (trithion)			D. magna	0.09	Bluegill	225	28
Chlordane[e]	P. californica	,55	S. serrulatus	20	Rainbow trout	10	80
Chlorobenzilate			S. serrulatus	550	Rainbow trout	710	
Chlorthion	4.5		D. magna	4.5			
Coumaphos			D. magna	1			0.14
Cryolite			D. pulex	5,000	Rainbow trout	47,000	
Cyclethrin			D. magna	55			
DDD (TDE)[e]	P. californica	1,100	D. pulex	3.2	Rainbow trout	9	1.8
DDT[e]	P. californica	19	D. pulex	0.36	Bass	2.1	2.1
Delnav (dioxathion)					Bluegill	14	690
Delmeton (systex)				14	Bluegill	81	
Diazinon[e]	P. californica	60	D. pulex	0.9	Bluegill	30	500
Dibrom (naled)	P. californica	16	D. pulex	3.5	Brook trout	78	160
Dieldrin[e]	P. californica	1.3	D. pulex	240	Bluegill	3.4	1,000
Dilan			D. magna	21	Bluegill	16	600
Dimethoate (cygon)	P. californica	140	D. magna	2,500	Bluegill	9,600	400
Dimethrin					Rainbow trout	700	
Dichlorvos (DDVP)[e]	P. californica	10	D. pulex	0.07	Bluegill	700	1
Disulfoton (di-syston)	P. californica	18			Bluegill	40	70
Dursban	Peteronareella badia	1.8			Rainbow trout	20	0.4
Endosulfan (thiodan)	P. californica	5.6	D. magna	240	Rainbow trout	1.2	64
Endrin[e]	P. californica	0.8	D. pulex	20	Bluegill	0.2	4.7
EPH			D. magna	0.1	Bluegill	17	36
Ethion ion	P. californica	14	D. magna	0.01	Bluegill	230	3.2
Ethyl guthion[e]			D. pulex		Rainbow trout		

[a] Stonefly bioassay was done at Denver, Colorado, and at Salt Lake City, Utah. Denver tests were in soft water (35 mg/l TDS), nonaerated, 60°F. Salt Lake City tests were in hard water (150 mg/l TDS), aerated, 48–50°F. Response was death.

[b] Daphnia pulex and Simocephalus serrulatus bioassay were done at Denver, Colorado, in soft water (35 mg/l TDS), nonaerated, 60°F. Daphnia magna bioassay was done at Pennsylvania State University in hard water (146 mg/l TDS), nonaerated, 68°F. Response was immobilization.

[c] Fish bioassay was done at Denver, Colorado, and at Rome, N.Y. Denver tests were with 2 in. fish in soft water (35 mg/l TDS), nonaerated; trout at 55°F; other species at 65°F. Rome tests were with 2–2½ in. fish in soft water (6 mg/l TA:pH 5.85–6.4), 60°F. Response was death.

[d] Gammarus bioassay was done at Denver, Colorado, in soft water (35 mg/l TDS), nonaerated, 60°F. Response was death.

[e] **Becomes bound to soil when used according to directions, but highly toxic (reflected in numbers) when added directly to water.**

3.4–6 48 HOUR TL_m VALUES OF PESTICIDES FROM STATIC BIOASSAY FOR VARIOUS TYPES OF FRESH WATER ORGANISMS *(Continued)*

Insecticides

Pesticide	Stream invertebrate[a]		Cladocerans[b]		Fish[c]		Gammar lacustris,[d]
	Species	TL_m	Species	TL_m	Species	TL_m	TL_m
Fenthion	P. californica	39	D. pulex	4			
Guthion[e]	P. californica	8	D. magna	0.2	Rainbow trout	10	0.3
Heptachlor[e]	P. badia	4	D. pulex	42	Rainbow trout	9	100
Kelthane (dicofol)	P. californica	3,000	D. magna	390	Rainbow trout	100	
Kepone					Rainbow trout	37.5	
Malathion[e]	P. badia	6	D. pulex	1.8	Brook trout	19.5	1.8
Methoxychlor[e]	P. californica	8	D. pulex	0.8	Rainbow trout	7.2	1.3
Methyl parathion[e]			D. magna	4.8	Bluegill	8,000	
Morestan	P. californica	40			Bluegill	96	
Ovex	P. californica	1,500			Bluegill	700	
Paradichlorobenzene					Rainbow trout	880	
Parathion[e]	P. californica	11	D. pulex	0.4	Bluegill	47	6
Perthane			D. magna	9.4	Rainbow trout	7	
Phosdrin[e]	P. californica	9	D. pulex	0.16	Rainbow trout	17	310
Phosphamidon	P. californica	460	D. magna	4	Rainbow trout	8,000	3.8
Pyrethrins	P. californica	64	D. pulex	25	Rainbow trout	54	18
Rotenone	P. californica	900	D. pulex	10	Bluegill	22	350
Strobane[e]	P. californica	7			Rainbow trout	2.5	
Tetradifon (tedion)					Bluegill	1,100	140
TEPP[e]					Fathead	390	52
Thanite			D. magna	450			
Thimet					Bluegill	5.5	70
Toxaphene	P. californica	7	D. pulex	15	Rainbow trout	2.8	70
Trichlorfon (dipterex)[e]	P. badia	22	D. magna	8.1	Rainbow trout	160	60
Zectran	P. californica	16	D. pulex	10	Rainbow trout	8,000	76
Herbicides, Fungicides, Defoliants, Algicides							
Ametryne					Rainbow trout	3,400	
Aminotriazole							
Aquathol					Bluegill	257	
Atrazine			Daphnia magna	3,600	Rainbow trout	12,600	
Azide, potassium					Bluegill	1,400	10,000
Azide, sodium					Bluegill	980	9,000
Copper chloride					Bluegill	1,100	
Copper sulfate					Bluegill	150	
Dichlobenil	Pteronarcys californica	44,000	Daphnia pulex	3,700	Bluegill	20,000	1,500
2,4-D, PGBEE			D. pulex	3,200	Rainbow trout	960	1,800
2,4-D, BEE	P. californica	1,800			Bluegill	2,100	760
2,4-D, isopropyl					Bluegill	800	
2,4-D, butyl ester					Bluegill	1,300	
2,4-D, butyl + ispropyl ester					Bluegill	1,500	
2,4,5-T isooctyl ester					Bluegill	16,700	
2,4,5-T isopropyl ester					Bluegill	1,700	
2,4,5-T PGBE					Bluegill	560	
2(2,4-DP) BEE					Bluegill	1,100	

[a] Stonefly bioassay was done at Denver, Colorado, and at Salt Lake City, Utah. Denver tests were in soft water (35 mg/l TDS), nonaerated, 60°F. Salt Lake City tests were in hard water (150 mg/l TDS), aerated, 48–50°F. Response was death.

[b] Daphnia pulex and Simocephalus serrulatus bioassay were done at Denver, Colorado, in soft water (35 mg/l TDS), nonaerated, 60°F. Daphnia magna bioassay was done at Pennsylvania State University in hard water (146 mg/l TDS), nonaerated, 68°F. Response was immobilization.

[c] Fish bioassay was done at Denver, Colorado, and at Rome, N.Y. Denver tests were with 2 in. fish in soft water (35 mg/l TDS), nonaerated; trout at 55°F; other species at 65°F. Rome tests were with 2–2½ in. fish in soft water (6 mg/l TA:pH 5.85–6.4), 60°F. Response was death.

[d] Gammarus bioassay was done at Denver, Colorado, in soft water (35 mg/l TDS), nonaerated, 60°F. Response was death.

[e] Becomes bound to soil when used according to directions, but highly toxic (reflected in numbers) when added directly to water.

3.4–6 48 HOUR TL_m VALUES OF PESTICIDES FROM STATIC BIOASSAY VARIOUS TYPES OF FRESH WATER ORGANISMS *(Continued)*

Herbicides, Fungicides, Defoliants, Algicides

	Stream invertebrate[a]		Cladocerans[b]		Fish[c]		Gammar lacustris,[d]
Pesticide	**Species**	**TL_m**	**Species**	**TL_m**	**Species**	**TL_m**	**TL_m**
Dalapon	P. californica Very low toxicity		D. magna	6,000		Very low toxicity	
Dead-X	P. californica	5,000	D. pulex	3,700	Rainbow trout	9,400	5,600
DEF	P. californica	2,300			Bluegill	36	230
Dexon	P. californica	42,000			Bluegill	23,000	6,000
Dicamba					non-toxic		5,800
Dichlone			D. magna	26	Rainbow trout	48	11,500
Difolatan	P. californica	150			Channel cat	31	6,500
Dinitrocresol	P. californica	560			Rainbow trout	210	
Diquat					Rainbow trout	12,300	
Diuron	P. californica	2,800	D. pulex	1,400	Rainbow trout	4,300	380
Du-ter					Bluegill	33	
Dyrene			D. magna	490		15	
Endothal, copper					Rainbow trout	290	
Endothal, dimethylamine					Rainbow trout	1,150	
Fenac, acid	P. californica	70,000			Rainbow trout	16,500	
Fenac, sodium	P. californica	80,000	D. pulex	4,500	Rainbow trout	7,500	18,000
Hydram (molinate)	P. californica	3,500			Rainbow trout	290	
Hydrothol 191					Rainbow trout	690	1,000
Lanstan (korax)					Rainbow trout	100	5,500
LFN isopropyl ester					Rainbow trout	79	
Paraquat	P. californica Very low toxicity		D. pulex	3,700	Very low toxicity		18,000
Propazine					Rainbow trout	7,800	
Silvex, PGBEE			D. pulex	2,000	Rainbow trout	650	
Silvex, isooctyl					Bluegill	1,400	
Silvex, BEE					Bluegill	1,200	
Simazine	P. californica	50,000			Rainbow trout	5,000	21,000
Sodium arsenite	P. californica Very low toxicity		Simocephalus serrulatus	1,400	Rainbow trout	36,500	
Tordon (picloram)					Rainbow trout	2,500	48,000
Trifuralin	P. californica	4,200	D. pulex	240	Rainbow trout	11	5,600
Vernam (vernolate)[e]					Rainbow trout	5,900	25,000

[a] Stonefly bioassay was done at Denver, Colorado, and at Salt Lake City, Utah. Denver tests were in soft water (35 mg/l TDS), nonaerated, 60°F. Salt Lake City tests were in hard water (150 mg/l TDS), aerated, 48–50°F. Response was death.

[b] Daphnia pulex and Simocephalus serrulatus bioassay were done at Denver, Colorado, in soft water (35 mg/l TDS), nonaerated, 60°F. Daphnia magna bioassay was done at Pennsylvania State University in hard water (146 mg/l TDS), nonaerated, 68°F. Response was immobilization.

[c] Fish bioassay was done at Denver, Colorado, and at Rome, N.Y. Denver tests were with 2 in. fish in soft water (35 mg/l TDS), nonaerated; trout at 55°F; other species at 65°F. Rome tests were with 2–2½ in. fish in soft water (6 mg/l TA:pH 5.85–6.4), 60°F. Response was death.

[d] Gammarus bioassay was done at Denver, Colorado, in soft water (35 mg/l TDS), nonaerated, 60°F. Response was death.

[e] Becomes bound to soil when used according to directions, but highly toxic (reflected in numbers) when added directly to water.

Source: *Handbook of Analytical Toxicology,* I. Sunshine, Ed., Chemical Rubber Co., Cleveland, 1969, p. 582.

3.4–7 COMPARATIVE TOXICITIES OF INSECTICIDES TO FISH AND MAMMALS AND SOLUBILITY OF INSECTICIDES IN WATER

Compound	96-hr TL_m[a] flathead minnows, mg/l[b,1,5]	LD_{50} for male white rats,[c] mg/kg[1]	LD_{50} for rats/kg body weight, mg[2,4]	Rats LD_{50} mg/kg[f,3]	Bluegills TL_m 96 hr mg/l[g,3]	Solubility in water, mg/l	
						Ref. 1	Ref. 3[h]
Chlorinated hydrocarbons							
Endrin	0.0013	17.8	5–45[d]	7.5	0.0006	0.1	0.19–0.26
Toxaphene	0.0051	90	40	–	–	Nil	
Dieldrin	0.016	46	60	46	0.0079	0.1	
Aldrin	0.028	67	50	–	–	0.1	0.14–0.18
DDT	0.034	113	250[e]	118	0.016	0.1	0.016–0.040[i]
Methoxychlor	0.035	6,000	5,000	–	–	Nil	
Heptachlor	0.056	100	90	–	–	Nil	
Lindane	0.056	125	125	91	0.077	10	0.5–6.6
Chlordane	0.069	335	500	–	–	Nil	
BHC	2.0	1,000	–	–	–	5–10	
Organic phosphorus							
EPN	0.20	12-40	13.6-42[d]	–	–	Slight	
Para-Oxon	0.33	3–3.5	3.5	–	–	20	
Parathion	1.4	13	6–15[d]	3.6	0.700	20	18–31
TEPP	1.7	2	1.2–2.0	–	–	High	
Chlorothion	3.2	880	1,500	–	–	40	
Systox	3.6	6.2	6–12	–	–	20	
Methyl parathion	8.3	14–42	9–25	–	–	50	
Malathion	12.5	1,375	1,156	–	–	145	
OMPA	121	10	–	–	–	High	
Dipterex	180	630	450	–	–	130,000	

[a] Median tolerance dose, or concentration at which 50% of organisms die.
[b] As 100% active compound.
[c] Oral feeding. The lethal dose at which 50% of the rats die.
[d] Males are more tolerant than females.
[e] LD_{50} man.
[f] Acute oral toxicity to female rats, in mg pesticide/kg body weight.
[g] 96 hr median tolerance limits, in mg of pesticide/l water.
[h] Apparent solubility in distilled water, varies with particle size.
[i] Technical grade DDT, includes *o,p′* and *p,p′* isomers + DDE.

REFERENCES

1. R.T. Skrinde *et al.*, *J. Am. Water Works Assn.*, *54*(11):1407, 1962.
2. Compiled by H.J. Webb, *J. Am. Water Works Assn.*, *54*(1):83, 1962.
3. G.G. Robeck *et al.*, *J. Am. Water Works Assn.*, *57*(2):181, 1965.
4. Pesticide Chemicals Official Compendium, Assn. Am. Pest. Control Off., College Park, Md.
5. C. Henderson *et al.*, *The Toxicity of Organic Phosphorous and Chlorinated Hydrocarbon Insecticides to Fish,* a paper presented at the second seminar on Biological Problems in Water Pollution, Cincinnati, Ohio, 1959.

3.4–8 EFFECT OF ALKYL-ARYL SULFONATE, INCLUDING ABS, ON AQUATIC ORGANISMS

Organisms	Concentration, mg/l	Time	Effect	Ref.
Trout	5.0	26–30 hr	Death	1
	3.7	24 hr	TL_m	
	5.0		Gill pathology	2
Bluegills	4.2	24 hr	TL_m	3
	3.7	48 hr	TL_m	
	0.86		Safe	
	16.0	30 days	TL_m	4
	5.6	90 days	Gill damage	5, 19
	17.0	96 hr	TL_m	
Fathead minnows	2.3		Reduced spawning	6
	13.0	96 hr	TL_m	7
	11.3	96 hr	TL_m	8
Fathead minnow fry	3.1	7 days	TL_m	6
Pumpkinseed sunfish	9.8	3 months	Gill damage	9
Salmon	5.6	3 days	Mortality	10
Yellow bullheads	1.0	10 days	Histopathology	11
Emerald shiner	7.4	96 hr	TL_m	8
Bluntnose minnow	7.7	96 hr	TL_m	8
Stoneroller	8.9	96 hr	TL_m	8
Silver jaw	9.2	96 hr	TL_m	8
Rosefin	9.5	96 hr	TL_m	8
Common shiner	17.0	96 hr	TL_m	8
Carp	18.0	96 hr	TL_m	8
Black bullhead	22.0	96 hr	TL_m	8
"Fish"	6.5		Minimal lethality	12
Trout sperm	10.0		Damage	13
Daphnia	5.0	96 hr	TL_m	14
	20.0	24 hr	TL_m	15
	7.5	96 hr	TL_m	15
Lirceus fontinalis	10.0	14 days	6.7 percent survival (hard water)	16
Crangonyx setodactylus[a]	10.0	14 days	0 percent survival (hard water)	16
Stenonema ares	8.0	10 days	20–33 percent survival	16
	16.0	10 days	0 percent survival	16
Stenonema heterotarsale	8.0	10 days	40 percent survival	16
	16.0	10 days	0 percent survival	16
Isonychia bicolor	8.0	9 days	0 percent survival	16
Hydropsychidae (mostly cheumatopsyche)	16.0	12 days	37–43 percent survival	16
	32.0	12 days	20 percent survival	16
Orconectes rusticus	16.0	9 days	100 percent survival	16
	32.0	9 days	0 percent survival	16
Goniobasis livescens	16.0	12 days	40–80 percent survival	16
	32.0	12 days	0 percent survival	16

[a] Misidentified originally as Synurella.

3.4–8 EFFECT OF ALKYL-ARYL SULFONATE, INCLUDING ABS, ON AQUATIC ORGANISMS *(Continued)*

Organisms	Concentration, mg/l	Time	Effect	Ref.
Snail	18.0	96 hr	TL_m	9
	24.0	96 hr	TL_m	9
Chlorella	3.6		Slight growth reduction	17
Nitzchia linearis	5.8		50 percent reduction in growth in soft water	18
Navicula seminulum	23.0		50 percent reduction in growth in soft water	18

Source: *Water Quality Criteria,* U.S. Department of the Interior, Fed. Water Pollut. Control Adm., April 1968.

REFERENCES

1. J. Wurtz-Arlet, *Vortraege Originalfassung Intern. Kongr. Grenz Flaechenaktive Stoffe, 3*(329):10374, 1960.
2. O.J. Schmid and H. Mann, *Nature* (London), *192*:675, 1960.
3. H. Turnbull, J.G. DeMann, and R.F. Weston, Symp. Waste Disposal Petrol. Ind., *Ind. Eng. Chem., 46*:324, 1954.
4. A.E. Lemke and D.I. Mount, *Trans. Am. Fish. Soc., 92*(4):372, 1963.
5. J. Cairns, Jr. and A. Scheier, *Proc. 17th Ind. Waste Conf.,* Purdue Univ., 1963.
6. Q.H. Pickering, *Int. J. Air Water Pollut., 10*:585, 1966.
7. C. Henderson, Q.H. Pickering, and J.M. Cohen, *Sew. Ind. Wastes, 31*(3):295, 1959.
8. T.O. Thatcher, *Int. J. Air Water Pollution, 10*:535, 1966.
9. J. Cairns, Jr. and A. Scheier, *Notulae Naturae, 367,* 1964.
10. G.Z. Holland *et al.,* State of Washington, Department of Fish., Res. Bull. 5, 1960.
11. J.E. Bardach, M. Fujiya, and A. Holl, *Science, 148:*1605, 1965.
12. E. LeClerc and F. Devlaminck, *Bull. Cen. Belge et Document Ezux, 17*:165, 1952.
13. H. Mann and O.J. Schmid, *Int. Rev. Hydrobiol., 46*(419), 1961.
14. F. Sierp and H. Theile, *Vom Wasser, 21*:197, 1954.
15. S. Godzch, *Acta Hydrobiol., 3*(4):281, 1961.
16. E.W. Surber and T.O. Thatcher, *Trans. Am. Fish. Soc., 92*(2):152, 1963.
17. T.E. Maloney, *J. Water Pollut. Control Fed., 38*(1):38, 1966.
18. J. Cairns, Jr., A. Scheier, and N.E. Hess, *Water Wastes, 9*(1):22, 1964.
19. A.E. Berry, *Sew. Ind. Wastes, 23*(4):508, 1951.

3.4–9 TOXICOLOGICAL EFFECTS OF SURFACTANTS ON TEST ALGAE SPECIES

Species	Day	ABS	LAS	AE 7.0	AE 7.4	ABE 9.5
Anabaena cylindrica	7	3[a]	3	3	4	4
	14	1	1	3	3	4
	21	1	3	4	4	4
Anabaena variabilis	7	2	3	1	1	1
	14	4	4	2	2	1
	21	4	4	3	4	1
Anacystis nidulans	7	1	1	1	2	3
	14	3	4	1	2	3
	21	3	4	3	3	4
Calothrix parietina	b	b	b	b	b	b
Cylindrospearmum sp.	7	3	4	3	3	3
	14	1	2	2	2	3
	21	2	4	4	3	3
Gloeocapsa alpicola	7	3	2	3	2	2
	14	4	4	3	2	2
	21	4	4	4	3	3
Nostoc muscorum	7	4	4	4	4	4
	14	3	3	4	4	3
	21	4	4	4	4	4
Oscillatoria borneti	7	3	3	2	2	2
	14	3	4	2	2	2
	21	4	4	2	2	2
Oscillatoria chalybia	7	2	3	3	3	2
	14	2	3	4	3	2
	21	3	4	3	3	2
Oscillatoria formosa	7	2	4	3	3	3
	14	2	3	4	4	3
	21	3	4	4	4	4
Oscillatoria tenuis	7	2	3	2	2	2
	14	1	2	3	3	2
	21	3	3	4	4	4
Phormidium faveolarum	7	2	4	4	2	4
	14	3	4	4	3	4
	21	4	4	4	4	4
Ankistrodesmus braunii	7	4	4	2	2	1
	14	4	4	3	4	4
	21	4	4	4	4	4
Chlorella pyrenoidosa	7	2	2	1	1	1
	14	4	4	3	3	3
	21	4	4	4	4	4
Chlorella vulgaris	7	2	3	1	1	1
	14	4	4	1	1	2
	21	4	4	2	2	2
Scenedesmus obliquus	7	3	4	3	3	2
	14	4	3	4	4	4
	21	4	4	4	4	4
Scenedesmus quadricauda	7	3	3	4	4	3
	14	4	4	4	4	4
	21	4	4	4	4	4
Ulothrix fimbriata	7	1	1	2	1	2
	14	4	4	1	2	2
	21	4	4	3	3	4

[a] Numbers indicate comparison with growth in control cultures: 4 – full potential similar to control; 3 – slight inhibition, turbidity, and chlorophyll decrease; 2 – moderate inhibition, turbidity, and chlorophyll decrease with yellowing of cells; and 1 – marked inhibition, turbidity, and white to yellow coloration, with few living cells.

[b] No change from normal throughout test period.

Source: E.M. Davis and E.F. Gloyna, *J. Water Pollut. Control Fed.*, *41*(8) (Part I):1494, 1969.

3.4.1 AQUATIC INSECTS

3.4–10 THE CONCENTRATIONS OF HEAVY METALS HAVING AN EFFECT ON AQUATIC ORGANISMS

Metal	Metal concentration, mg/l	Type of water	Organism	Effect	Ref.
Arsenic	3-14	–	Mayfly	No harm	1
	10-20	–	Damsel flies	No harm	1
Barium	29	Lake Erie	*Daphnia magna*	Immobilized	2
Cadmium	0.01	River Havel	*Daphnia magna*	Threshold toxic	3
Chromium	0.70	River Havel	*Daphnia magna*	Threshold toxic	3
Cobalt	5.00	River Havel	*Daphnia magna*	Threshold toxic	3
Copper	0.027	Lake Erie	*Daphnia magna*	Immobilized	3
Lead	5.00	River Havel	*Daphnia magna*	Threshold toxic	3
Mercury	0.03	River Havel	*Daphnia magna*	Threshold toxic	3
Nickel	6.00	River Havel	*Daphnia magna*	Threshold toxic	3
Silver	0.03	River Havel	*Daphnia magna*	Threshold toxic	3
Zinc	1.80	River Havel	*Daphnia magna*	Threshold toxic	3
	0.30	–	Mayfly nymphs	Killed	1
	0.33	–	Stoneflies, caddisflies	No effect	1

Source: Compiled by S.L. Warnick and H.L. Bell, *J. Water Pollut. Control Fed.*, *41*(2):280, 1969.

REFERENCES

1. W. Rudolfs *et al.*, *Sew. Ind. Wastes, 22*(9):1157, 1950.
2. B.G. Anderson, *Trans. Am. Fish Soc.*, *78*:96, 1948.
3. G. Bringmann and R. Kuhn, *Gesundheits-Ingenieur, 80*:115, 1959.

3.4–11 METAL TOXICITY–BIOASSAY DATA[a]

Metal	Insect	96 hr TL_m, mg/l	50% survival		Physical and chemical data of the 16 mg/l test water after two weeks								
			days	mg/l	C°	DO mg/l	pH	Alkalinity, mg/l	Acidity, mg/l	Hardness, mg/l	N, mg/l	Metal, mg/l	Mg metal absorbed by insects
Cu++ from $CuSO_4 \cdot 5H_2O$	acroneuria	8.3			18.5	8.0	6.8	54	20	40	4.9	12.3	0.06
	ephemerella	0.32 (48 hr)			18.5	8.0	6.9	42	6	40	1.5	10.4	0.14
	hydropsyche		14	32.0	18.5	8.0	6.8	40	4	46	0.9	14.5	0.22
Zn++ from $ZnSO_4 \cdot 7H_2O$	acroneuria		14	32.0	18.5	8.0	7.6	46	12	50	6.5	5.5	0.01
	ephemerella		10	16.0	18.5	8.0	7.6	30	8	54	1.7	7.9	0.12
	hydropsyche		11	32.0	18.5	8.0	7.6	30	8	52	1.1	4.7	0.18
Cd++ from $CdSO_4 \cdot 8H_2O$	acroneuria		14	32.0	18.5	8.0	7.3	60	12	52	7.2	14.1	0.42
	ephemerella	2.0			18.5	8.0	7.0	56	8	54	1.6	15.2	0.05
	hydropsyche		10	32.0	18.5	8.0	7.0	54	6	56	1.5	14.0	0.13
Pb++ from $PbSO_4$	acroneuria		>14	64.0	18.5	8.0	7.3	60	12	54	7.6	2.2	0.05
	ephemerella		7	16.0	18.5	8.0	7.0	56	6	52	3.1	3.0	0.12
	hydropsyche		7	32.0	18.5	8.0	7.1	42	6	54	3.5	3.2	0.08
Fe++ from $FeSO_4$	acroneuria		9	16.0	18.5	8.0	7.7	72	14	48	14.1	1.2	0.02
	ephemerella	0.32			18.5	8.0	8.2	54	6	48	0.95	1.4	0.01
	hydropsyche		7	16.0	18.5	8.0	8.1	46	6	50	1.28	1.6	0.01
Ni++ from $NiSO_4 \cdot 6H_2O$	acroneuria	33.5			18.5	8.0	7.0	54	10	40	6.0	6.4	0.04
	ephemerella	4.0			18.5	8.0	7.0	46	6	42	1.5	7.2	0.03
	hydropsyche		> 14	64.0	18.5	8.0	7.0	42	6	48	1.1	13.0	0.02
Co++ from $CoSO_4 \cdot 7H_2O$	acroneuria		8	32.0	18.5	8.0	7.2	66	12	50	12.3	9.5	0.01
	ephemerella	16.0			18.5	8.0	6.9	46	6	50	1.2	10.6	0.03
	hydropsyche		7	32.0	18.5	8.0	7.0	46	6	46	1.3	14.7	0.01
Cr+++ from $CrCl_3 \cdot 6H_2O$	acroneuria		7	32.0	18.5	8.0	6.8	50	8	50	9.2	4.7	0.03
	ephemerella	2.0			18.5	8.0	6.0	46	14	50	4.7	7.7	0.01
	hydropsyche	64.0			18.5	8.0	6.4	46	14	42	2.2	5.2	0.01
Hg++ from $HgCl_2$	acroneuria	2.0			18.5	8.0	7.8	52	8	46	6.5	–	–
	ephemerella	2.0			18.5	8.0	7.6	46	6	42	1.6	–	–
	hydropsyche	2.0			18.5	8.0	7.6	46	6	42	1.4	–	–

[a] DO concentration of 8.0 mg/l maintained.

Source: S.L. Warnick and H.L. Bell, *J. Water Pollut. Control Fed.*, *41*(2):280, 1969.

3.4.2 SHELLFISH

3.4–12 TL_m VALUES FOR SHRIMP

Organochloride pesticides	μg/l
Aldrin	0.04
BHC	2.0
Chlordane	2.0
Endrin	0.2
Heptachlor	0.2
Lindane	0.2
DDT	0.6
Dieldrin	0.3
Endosulfan	0.2
Methoxychlor	4.0
Perthane	3.0
TDE	3.0
Toxaphene	3.0
Organophosphorus pesticides	**μg/l**
Coumaphos	2.0
Dursban	3.0
Fenthion	0.03
Naled	3.0
Parathion	1.0
Ronnel	5.0

Note: The 48 hour TL_m is listed for each chemical in μg/l.

Source: *Water Quality Criteria,* U.S. Department of the Interior, Fed. Water Pollut. Control Adm., April 1968.

3.4–13 MEAN CONCENTRATIONS AFFECTING LARVAE GROWTH AND SURVIVAL

Concentration, mg/l active ingredient

	Clams		Oysters		Both Species	
Type	Mean	Range	Mean	Range	Mean	Range
Anionic						
Alkyl aryl sulfonate[a]	1.55	0.55–3.00	0.76	0.14–1.50	1.15	0.14–3.00
Alkyl sulfate	1.22	0.73–1.46	1.07	0.29–1.46	1.15	0.29–1.46
Cationic	0.34	0.01–1.00	0.25	0.05–0.50	0.30	0.01–1.00
Nonionic	2.66	1.00–5.00	2.00	1.00–2.50	2.33	1.00–5.00
For all surfactants	1.44	0.01–5.00	1.02	0.05–2.50	1.23	0.01–5.00

[a] AAS-1 not entered in these figures.

Note: Means were compiled by averaging lowest concentrations of all compounds tested that significantly reduced, in all replicate experiments, survival of developing embryos and growth and survival of developed veliger larvae.

Source: H. Hidu, *J. Water Pollut. Control Fed., 37*(2):262, 1965.

3.4–14 CONCENTRATIONS OF SYNTHETIC DETERGENTS EFFECTING A 50 PERCENT REDUCTION IN LARVAE DEVELOPMENT

	Active ingredient, mg/l			Active ingredient, mg/l	
Test compound[a]	**Clams**	**Oysters**	**Test compound[a]**	**Clams**	**Oysters**
Anionic			Cationic		
Alkyl aryl sulfonate			C-1	1.27	0.49
AAS-1[b]	5.83	1.63	C-2	0.0085	0.09
AAS-2	0.94	0.27			
AAS-3	1.03	0.39	Nonionic		
Alkyl sulfate			N-1	0.77	0.86
AS-1	0.47	0.37	N-2	1.75	1.60

[a] Characteristics of the various materials tested:

Material	Composition
Anionic	
Alkyl aryl sulfonates	
AAS-1	active alkyl naphthalene sulfonate polyphosphate fatty alcohol sulfates inactive not available (percentages not available)
AAS-2	active alkyl benzene sulfonate 54.8% inactive Na_2SO_4 40.3% Free oil 0.5% NaOH 1.3% Na_2CO_3 0.7% H_2O 26%
AAS-3	active triethanolamine dodecyl benzene sulfonate 60% inactive H_2O 40%
Alkyl sulfate	
AS-1	active alkyl sulfate 26.0 to 28.9% free fatty alcohol 1.0 to 2.5% inactive (probably H_2O) 68.6 to 73%
Cationic	
C-1	active para diisobutyl phenoxyethoxy ethyl dimethyl benzyl ammonium chloride monohydrate 98.8% inactive H_2O 1.2%
C-2	active lauryl pyridinium chloride 100%
Nonionic	
N-1	active iso-octyl phenoxy poly-ethoxy ethanol 100%
N-2	active alkyl polyether alcohol 100%

[b] Expressed as mg/l gross product (percentage active ingredient unavailable).

Note: These concentrations produced a 50% reduction in the percentage of fertilized clam and oyster eggs developing to normal 48 hr straight hinge larvae. These values are comparable to 48 hr TL_m values of acute toxicity studies.

Source: H. Hidu, *J. Water Pollut. Control Fed.*, *37*(2):262, 1965.

3.4.3 FISH

3.4–15 PROVISIONAL MAXIMUM TEMPERATURES RECOMMENDED AS COMPATIBLE WITH THE WELL-BEING OF VARIOUS SPECIES OF FISH AND THEIR ASSOCIATED BIOTA

Temperature, °F	Response
93	Growth of catfish, gar, white or yellow bass, spotted bass, buffalo, carpsucker, threadfin shad, and gizzard shad
90	Growth of largemouth bass, drum, bluegill, and crappie
84	Growth of pike, perch, walleye, smallmouth bass, and sauger
80	Spawning and egg development of catfish, buffalo, threadfin shad, and gizzard shad
75	Spawning and egg development of largemouth bass, white and yellow bass, and spotted bass
68	Growth or migration routes of salmonids and for egg development of perch and smallmouth bass
55	Spawning and egg development of salmon and trout (other than lake trout)
48	Spawning and egg development of lake trout, walleye, northern pike, sauger, and Atlantic salmon

Note: Recommended temperatures for other species, not listed above, may be established if and when necessary information becomes available.

Source: *Water Quality Criteria,* U.S. Department of the Interior, Fed. Water Pollut. Control Adm., April 1968.

3.4–16 AVERAGE TURBIDITIES FOUND TO BE FATAL TO FISH

Species	Length of exposure, days	Turbidity, mg/l
Large mouth bass	7.6	101,000
Pumpkin seed sunfish	13	69,000
Channel catfish	9.3	85,000
Black bullhead	17	222,000
Golden shiner	7.1	166,000

Source: I.E. Wallen, *Bull. Okla. Agr. Mech. Coll., Biol. Ser., 48*:27, 1951.

3.4–17 SURVIVAL OF FATHEAD MINNOWS IN FOUR WOOD FIBERS

Various Oxygen Levels[a] and Temperatures

Bioassay number	Temperature, °F	Oxygen,% saturation[b]	Control 0	Wood fiber concentration, mg/l 74	100	272	738	2,000
Aspen groundwood								
63-B	65	65	100 (5)	100 (5)	100 (5)	–	–	–
63-C	65	40	90 (10)	70 (10)	60 (10)	–	–	–
63-F	75	35	70 (10)	80 (10)	90 (10)	90 (10)	–	–
62-I	55	100	100	–	100	100	98	96
62-G	65	100	98	–	100	100	100	98
62-S	65	100	100	–	–	–	94	–
62-H	75	100	100	–	100	100	100	96
Conifer groundwood								
63-B	65	65	100 (5)	100 (10)	100 (5)	100 (5)	–	–
63-C	65	40	90 (10)	80 (10)	90 (5)	80 (10)	–	–
63-F	75	35	70 (10)	80 (10)	80 (10)	100 (10)	–	–
62-J	55	100	96	–	100	100	100	88
62-K	65	100	100	–	100	98	90	33
62-S	65	100	100	–	–	–	78	–
62-P	75	100	100	–	98	96	72 (60)	22 (60)
Conifer kraft								
62-M	55	100	100	–	100	100	98	80
62-N	65	100	100	–	100	100	98	86
62-S	65	100	100	–	–	–	96	–
62-O	75	100	92	–	90	90	88 (60)	76 (60)
Conifer sulfite								
63-B	65	65	100 (5)	100 (10)	100 (5)	–	–	–
63-C	65	40	90 (10)	70 (10)	90 (10)	–	–	–
63-F	75	35	70 (10)	70 (10)	90 (10)	80 (10)	–	–
62-Q	65	100	96	–	98	98	98	94
62-S	65	100	100	–	–	–	100	–

[a] 72 hr bioassays at low oxygen levels; 96 hr bioassays at 100% saturation.
[b] Levels designated as 100% varied from 85–110% during course of the run.

Note: Each number represents 50 fish, unless otherwise noted in parentheses.

Source: L.L. Smith *et al.*, *J. Water Pollut. Control Fed.*, *37*(1):130, 1965.

3.4–18 SURVIVAL OF WALLEYE FINGERLINGS IN THREE WOOD FIBERS

Various Oxygen Levels[a]

Bioassay number	Temperature, °F	Oxygen,% saturation	Control 0	Wood fiber concentration, mg/l: 74	100	272	738	2,000
Aspen groundwood								
63-B	65	65	80 (5)	40 (5)	20 (5)	–	–	–
63-D	65	35	100 (10)	70 (10)	40 (10)	–	–	–
63-E	75	40	80 (10)	30 (10)	10 (10)	10 (10)	–	–
62-A	65	100	86 (14)	–	100 (10)	60 (10)	60 (10)	40 (10)
62-E	65	100	78 (18)	–	25 (12)	–	–	–
62-C	75	100	92 (12)	–	90 (10)	88 (8)	100 (8)	–
62-F	75	100	100 (15)	–	87 (15)	–	–	–
Conifer groundwood								
63-B	65	65	80 (5)	–	0 (5)	–	–	–
63-D	65	35	100 (10)	10 (10)	10 (10)	0 (10)	–	–
63-E	75	40	80 (10)	20 (10)	0 (10)	0 (10)	–	–
62-A	65	100	86 (14)	–	90 (10)	0 (10)	0 (10)	0 (10)
62-E	65	100	78 (18)	–	33 (18)	0 (12)	0 (12)	–
62-B	75	100	100 (6)	–	100 (6)	50 (6)	–	–
62-C	75	100	92 (12)	–	67 (12)	0 (8)	0 (8)	–
62-F	75	100	100 (15)	–	67 (15)	–	–	–
Conifer sulfite								
63-B	65	65	80 (5)	–	40 (5)	60 (5)	–	–
63-D	65	35	100 (10)	30 (10)	80 (10)	– –	–	
63-E	75	40	80 (10)	80 (10)	30 (10)	10 (10)	–	–
62-A	65	100	86 (14)	–	93 (15)	60 (10)	44 (9)	30 (10)
62-E	65	100	78 (18)	–	44 (18)	12 (17)	6 (18)	0 (18)
62-C	75	100	92 (12)	–	92 (12)	88 (8)	75 (8)	88 (8)
62-F	75	100	100 (15)	–	100 (15)	–	–	–

[a] 72 hour bioassays.
Note: Number of fish in parentheses.

Source: L.L. Smith *et al., J. Water Pollut. Control Fed., 37*(1):130, 1965.

3.4–19 LETHAL OXYGEN LEVELS FOR FISH AT VARIOUS TEMPERATURES

Species	Lethal oxygen level at given temperature, mg/l: 25.0°C	18.5°C	12.0°C
Northern swellfish[a]	1.36	0.92	0.70
Winter flounder[b]	1.03	0.87	0.66
Common silverside[c]	1.28	0.93	0.77

[a] *Spheroides maculalus.*
[b] *Pseudopleuronectes americanus.*
[c] *Menidia menidia.*

Source: J.G. Hoff, *J. Water Pollut. Control Fed., 39*(2):276, 1967.

3.4–20 OBSERVED LETHAL CONCENTRATION OF SELECTED CHEMICALS IN AQUATIC ENVIRONMENTS

Chemical	Organism tested	Lethal concentration, mg/l	Exposure time, hr
ABS (100 percent)	Fathead minnow	3.5–4.5	96
ABS (100 percent)	Bluegills	4.2–4.4	96
Household syndets	Fathead minnow	39–61	96
Alkyl sulfate	Fathead minnow	5.1–5.9	96
LAS (C12)	Bluegill fingerlings	3	96
LAS (C14)	Bluegill fingerlings	0.6	96
Acetic acid	Goldfish	423	20
Alum	Goldfish	100	12–96
Ammonia	Goldfish	2–2.5 NH_3	24–96
Ammonia	Perch, roach, rainbow trout	3N	2–20
Sodium arsenite	Minnow	17.8 As	36
Sodium arsenate	Minnow	234 As	15
Barium chloride	Goldfish	5000	12–17
Barium chloride	Salmon	158	–
Cadmium chloride	Goldfish	0.017	9–18
Cadmium nitrate	Goldfish	0.3 Cd	190
CO_2	Various species	100–200	–
CO	Various species	1.5	1–10
Chloramine	Brown trout fry	0.06	–
Chlorine	Rainbow trout	0.03–0.08	–
Chromic acid	Goldfish	200	60–84
Copper sulfate	Stickleback	0.03 Cu	160
Copper nitrate	Stickleback	0.02 Cu	192
Cyanogen chloride	Goldfish	1	6–48
H_2S	Goldfish	10	96
HCl	Stickleback	pH 4.8	240
HCl	Goldfish	pH 4.0	4–6
Lead nitrate	Minnow, stickleback, brown trout	0.33 Pb	–
Mercuric chloride	Stickleback	0.01 Hg	204
Nickel nitrate	Stickleback	1 Ni	156
Nitric acid	Minnow	pH 5.0	–
Oxygen	Rainbow trout	3 cc/liter	–
Phenol	Rainbow trout	6	3
Phenol	Perch	9	1
Potassium chromate	Rainbow trout	75	60
Potassium cyanide	Rainbow trout	0.13 Cn	2
Sodium cyanide	Stickleback	1.04 Cn	2
Silver nitrate	Stickleback	70 K	154
Sodium fluoride	Goldfish	1000	60–102
Sodium sulfide	Brown trout	15	–
Zinc sulfate	Stickleback	0.3 Zn	120
Zinc sulfate	Rainbow trout	0.5	64
Pesticides			
Chlorinated hydrocarbons			
Aldrin	Goldfish	0.028	96
DDT	Goldfish	0.027	96
DDT	Rainbow trout	0.5–0.32	24–36
DDT	Salmon	0.08	36

3.4–20 OBSERVED LETHAL CONCENTRATION OF SELECTED CHEMICALS IN AQUATIC ENVIRONMENTS (*Continued*)

Chemical	Organism tested	Letal concentration, mg/l	Exposure time, hr
DDT	Brook trout	0.032	36
DDT	Minnow, guppy	0.75 ppb	29
DDT	Stoneflies (species)	0.32–1.8	96
BHC	Goldfish	2.3	96
BHC	Rainbow trout	3	96
Chlordane	Goldfish	0.082	96
Chlordane	Rainbow trout	0.5	24
Dieldrin	Goldfish	0.037	96
Dieldrin	Bluegill	0.008	96
Dieldrin	Rainbow trout	0.05	24
Endrin	Goldfish	0.0019	96
Endrin	Carp	0.14	48
Endrin	Fathead minnow	0.001	96
Endrin	Various species	0.03–0.05 ppb	–
Endrin	Stoneflies (species)	0.32–2.4 ppb	96
Heptachlor	Rainbow trout	0.25	24
Heptachlor	Goldfish	0.23	96
Heptachlor	Bluegill	0.019	96
Heptachlor	Redear sunfish	0.017	96
Methoxychlor	Rainbow trout	0.05	24
Methoxychlor	Goldfish	0.056	96
Toxaphene	Rainbow trout	0.05	24
Toxaphene	Goldfish	0.0056	96
Toxaphene	Carp	0.1	–
Toxaphene	Goldfish	0.2	24
Toxaphene	Goldfish	0.04	170
Toxaphene	Minnows	0.2	24
Organic phosphates			
Chlorothion	Fathead minnow	3.2	96
Dipterex	Fathead minnow	180	96
EPN	Fathead minnow	0.2	96
Guthion	Fathead minnow	0.093	96
Guthion	Bluegill	0.005	96
Malathion	Fathead minnow	12.5	96
Parathion	Fathead minnow	1.4–2.7	96
TEPP	Fathead minnow	1.7	96
Herbicides			
Weedex	Young roach	40–80	1 month
Weeda Zol	and trench	15–30	1 month
Weeda Zol T.L.		20–40	1 month
Simazine (no plants present)	Minnow	0.5	< 3 days
Atrazine (A361) (plants present)	Minnow	5.0	24
Atrazine in Gesaprime	Minnow	3.75	24
Bactericides			
Algibiol	Minnow	20	24
Soricide tetraminol	Minnow	8	48

Source: P.H. McGauhey, *Engineering Management of Water Quality,* McGraw-Hill, New York, 1968.

3.4–21 TOXICITY OF CHEMICALS TO FISH

Substance	Fish tested	Toxic concentration, ppm	Ref.
Acetic acid	Goldfish	423	1
Alum (potash)	Goldfish	100	1
Aluminum nitrate	Stickleback	0.1 (as Al)	2
Ammonium hydroxide	Goldfish	2–2.5 (as NH_3)[a]	1
Ammonium chloride	Rainbow trout	500 (as NH_3)	3
Ammonium sulfate	Rainbow trout	1,000 (as NH_3)	3
Arsenic (sodium arsenite)	Minnow	17.8 (as As)	3
Arsenic (sodium arsenate)	Minnow	234 (as As)	3
Barium nitrate	Stickleback	500 (as Ba)	2
Bromine	Goldfish	20	1
Cadmium nitrate	Stickleback	0.3 (as Cd)	2
Carbon monoxide	Various	Less than 1.2 ml/l	4
Chlorine	Goldfish	2	1
	Rainbow trout	Less than 0.03	5
Citric acid	Goldfish	894	1
Chromic acid	Goldfish	200	1
Copper nitrate	Stickleback	0.02 (as Cu)	2
Copper sulfate	Stickleback	Less than 0.03 (as Cu)	2
Cresylic acid	Goldfish	1	1
Hydrogen sulfide	Goldfish	Less than 100	1
Lead nitrate	Brown trout	0.33 (as Pb)	6
	Stickleback	0.1	2
Mercuric chloride	Stickleback	0.01 (as Hg)	2
Nickel nitrate	Stickleback	1 (as Ni)	2
Ozone	Various	0.01	7
	Goldfish	100	1
Phenol	Rainbow trout	6	8
p-Cresol	Rainbow trout	5	8
Potassium cyanide	Goldfish	0.1 to 0.3	1
	Rainbow trout	0.13 (as CN)	8
Silver nitrate	Stickleback	0.004 (as Ag)	2
Sodium cyanide	Stickleback	1.04 (as CN)	9
Sodium sulfide	Stickleback	4.5 (as S)	9
	Brown trout	1 (as S)	10
Tannic acid	Goldfish	100	1
Zinc sulfate	Stickleback	0.3 (as Zn)	11
	Rainbow trout	Less than 25 (as Zn)	3
Detergents: Anionic	–	6–7	12
Nonionic	–	2–3	12
Soaps	–	Up to 1,000	12
Pesticides: DDT	Goldfish	0.1	13
BHC	–	0.035	14
Aldrin	–	0.02	15
Parathion	Bluegills	0.2	16
Suspended solids (Which may be gravely adverse to fisheries but less adverse for angling.)	–	20	

[a] The effect depends on the pH.

Source: G.V. James, *Water Treatment,* 4th ed., CRC Press, Cleveland, 1971.

3.4–21 TOXICITY OF CHEMICALS TO FISH (*Continued*)

REFERENCES

1. Grindley, *Ann. Appl. Biol., 33*:103, 1946.
2. Wells, *Bull. Ill. Lab. Nat. History,* II:557, 1918.
3. Taylor and James, U.S. Bureau of Fisheries, Document No. 1045; U.S. Comm. Fish, App. 7, 322, 1928.
4. Carpenter, *Brit. J. Exp. Biol., 4*:378, 1927.
5. Hubbs, *Physiol. Zoöl, 3*:441, 1930.
6. Alexander, Southgate, and Bassindale, *Tech. Paper, No. 5, Water Pollut. Res.* 1935.
7. Jones, *J. Exp. Biol., 23*:298, 1947.
8. Longwell and Pentelow, *J. Exp. Biol., 12*:I, 1935.
9. Jones, *J. Exp. Biol., 15*:394, 1938.
10. Southern Sea Fisheries Committee, Report in *Dorset Daily Echo,* 11th July 1952.
11. Kalabina, *Z. Fischerei, 33*:295, 1935.
12. Hubault, *Compt. Rend. Acad. Agr. France, 21*:228, 1935.
13. Galstoff *et al., U.S. Bureau of Fisheries, Bull. No. 18*, 1935 p. 48.
14. Butcher, *Avon Biol. Res. Ann. Rep.,* 1937-38.
15. Someren, *Avon Biol. Res. Ann. Rep.,* 1937-38.
16. Fitch *et al., Cornell Veterinarian, 24*:30, 1934.

3.4–22 LIMITING CONCENTRATIONS OF SUBSTANCES TOXIC TO FISH

Substances	ppm
Sulfides (as/S) and H_2S	1.0
Cyanide (as/CN)	0.3
Phenol and cresols	5.0
Higher tar acids	10.0
Naphthalene	10.0
Ammonium thiocyanate	200.0
Ammonium thiocyanate (as NH_3)	45.0
Ammonium sulfate (as NH_3)	262.0
Ammonium chloride (as NH_3)	194.0
Ammonium carbonate (as NH_3)	35.0
Free ammonia (i.e., NH_4OH)	10.0
pH value, safe range	5.0–9.0

Source: G.R. Nellist, *Gas World, 135*(3516) Coking Supplement:7, January 5, 1952.

3.4–23 TOXICITY OF ACIDS TO GOLDFISH

Acid	Concentration, parts per 100,000	pH value	Period of survival of fish
Sulfuric	13.4	4.3	6.2–96 hr
	13.4	4.5	> 4 days
Hydrochloric	16.6	4.0	4.5–6.5 hr
	15.9	4.5	> 4 days
Nitric	75.0	3.4	0.5–0.8 hr
	20.0	4.9	> 4 days
Acetic	10.0	6.8	48 hr–> 4 days
	1.0	7.3	> 4 days
Lactic	65.4	4.0	6–43 hr
	43.0	4.6	> 4 days
Oxalic	100.0	2.6	0.4–0.5 hr
	20.0	5.3	> 4 days
Citric	89.4	4.0	4–48 hr
	62.5	4.5	> 4 days

Source: M.M. Ellis, *Detection and Measurement of Stream Pollution,* U.S. Department of Commerce, Bureau of Fisheries, 1937.

3.4–24 OBSERVED LETHAL CONCENTRATIONS OF COMPLEX CYANIDES TO FISH

Chemical	TL_m, mg/l	Exposure time, hrs	Remarks
Sodium cyanide, NaCN	0.25 CN^-	24	Soft water, 20°C
	0.24 CN^-	48	Soft water, 20°C
	0.23 CN^-	96	Soft water, 20°C
Sodium cyanide + zinc sulfate, NaCN + $ZnSO_4$	0.20 CN^-	24	Soft water, 20°C
	0.19 CN^-	48	Soft water, 20°C
	0.18 CN^-	96	Soft water, 20°C
Sodium cyanide + cadmium sulfate, NaCN + $CdSO_4$	0.23 CN^-	24	Soft water, 20°C
	0.21 CN^-	48	Soft water, 20°C
	0.17 CN^-	96	Soft water, 20°C
Sodium cyanide + nickelous sulfate, NaCN + $NiSO_4$	460 CN^- [a]	24	Soft water, 20°C
	310 CN^- [a]	48	Soft water, 20°C
	2.5 CN^- [b]	48	Soft water, 20°C
	0.95 CN^- [b]	96	Soft water, 20°C
	0.65 CN^- [b]	168	Soft water, 20°C
Sodium cyanide + cupric sulfate, NaCN + $CuSO_4$	2.2 CN^-	24	Soft water, 20°C
	2.0 CN^-	48	Soft water, 20°C
	1.5 CN^-	96	Soft water, 20°C
	1.8 Cu	24	Soft water, 20°C
	1.6 Cu	48	Soft water, 20°C
	1.2 Cu	96	Soft water, 20°C
Sodium cyanide + zinc sulfate + nickel sulfate, NaCN + $ZnSO_4$ + $NiSO_4$	1.0 CN^-; 0.6 Zn; 0.13 Ni	16	Soft water, 20°C
Sodium cyanide + zinc sulfate + copper sulfate	1.0 Zn; 0.025 Cu; 0.05 CN^-		Not killed in 96 hrs
	1.0 Zn; 0.25 Cu; 0.33 CN^-		Not killed in 96 hrs
Zinc sulfate and copper sulfate	1.0 Zn; 0.025 Cu	14	

[a] Secondary TL_m–control pH 9.0.
[b] Primary TL_m–dilute nearly neutral solutions.

Source: P. Doudoroff, *Sewage Ind. Wastes, 28*(8):1020, 1956.

3.4–25 THE TOXICITY OF SOME METALS TO FISH

Metal	Metal concentration, mg/l	Ref.
Zn	0.78 – 33.4	1
Zn	0.01 – 10.1	2, 8,[a] 10[a]
Zn	0.3 – 0.7	3
Cu	0.015– 3.0	2, 8,[a] 10[a]
Cu	0.022– 1.76	1
Cu	0.01 – 0.02	3
Co	1.0 – 10.0	4
Co	7.0 – 15.0	3
Ni	0.08 – 1.0	3, 8,[a] 9[a]
Ni	4.45 – 44.5	1
Pb	0.1 – 6.3	6
Pb	5.58 –482.0	1
Pb	0.1 – 0.4	3
Cd	0.01 – 10.0	7[a]
Cd	0.63 – 73.5	1
Cd	0.03	3
Hg	0.01 – 0.02	3
Cr (III)	41	8,[a] 9[a]
Cr (IV)	20	8,[a] 9[a]
CN^-	0.02	8,[a] 9[a]
NH_3, free	2	7,[a] 10[a]

[a] Approximate limit for game fish.

Source: S.L. Warnick and H.L. Bell, *J. Water Pollut. Control Fed., 41*(2):280, 1969.

REFERENCES

1. Q.H. Pickering and C. Henderson, *Proc. 19th Ind. Wastes Conf.,* Purdue University Ext. Serv., *117:* 578, 1964.
2. W. Rudolfs *et al., Sew. Ind. Wastes, 22*(9):1157, 1950.
3. J.R.E. Jones, *J. Exp. Biol.* (Britain), *15*:394, 1938.
4. G. Schweiger, *Arch. Fischwiss.* (Germany), *8*:54, 1957.
5. J.R.E. Jones, *J. Exp. Biol.* (Britain), *16*:425, 1939.
6. Anon., *Report of the Water Pollution Research Board, with the report of the Direction of the Water Pollution Research Laboratory for the year 1959,* Department of Scientific and Industrial Research, Her Majesty's Stationery Office, London, England.
7. J.E. McKee and H.W. Wolf, *Water Quality Criteria,* a report prepared for the State of California Water Pollution Board, 1963.
8. E.F. Barth *et al., J. Water Pollut. Control Fed., 37*(1):87, 1965.
9. P. Doudoroff and M. Katz, *Sew. Ind. Wastes, 25*(7):802, 1953.
10. Estimated by C.M. Tarzwell, Robert A. Taft Sanitary Engineering Center, Cincinnati, Ohio.

3.4–26 METAL CONCENTRATIONS FOR VARIOUS SURVIVAL TIMES OF LARGE STICKLEBACKS[a]

		Concentration of metal at which survival time averaged, mg/l				Lethal concentration limit[b]	
Metal	Compound tested	1 day	2 days	4 days	1 week	mg/l	g atoms/l
Na	$NaNO_3$	3,000	1,500	800	600	500	2.17×10^{-2}
Ca	$Ca(NO_3)_2$	4,500[c]	3,000	1,500	1,000	800	2.00×10^{-2}
Sr	$Sr(NO_3)_2$	10,000	7,000	3,000	1,500	1,200	1.37×10^{-2}
Mg	$Mg(NO_3)_2$	2,000	1,500	500	350[c]	300	1.23×10^{-2}
Ba	$Ba(NO_3)_2$	3,000	2,500	1,000	500	400	2.90×10^{-3}
K	KNO_3	500	200	100	70	50	1.27×10^{-3}
Mn	$Mn(NO_3)_2$	300	150	100	50	40	7.28×10^{-4}
Co	$Co(NO_3)_2$	150	50	20	15	10	1.68×10^{-4}
Cr	$Cr_2(SO_4)_3$	5.0	2.0	1.4	1.3	1.2	2.30×10^{-5}
Ni	$Ni(NO_3)_2$	10.0	3.0[d]	1.5	1.0	0.8	1.34×10^{-5}
Zn	$ZnSO_4$	1.5[c]	0.9[c]	0.7	0.4	0.3	4.60×10^{-6}
Al	$Al(NO_3)_3$	0.3	0.2[c]	0.13[c]	0.1	0.07	2.60×10^{-6}
Au	$HAuCl_4$	1.5[d]	1.4[c]	1.0	0.6	0.4	2.00×10^{-6}
Cd	$Cd(NO_3)_2$	7.0	3.0	0.7	0.3	0.2	1.78×10^{-6}
Pb	$Pb(NO_3)_2$	1.0	0.7	0.3	0.2	0.1	4.80×10^{-7}
Cu	$Cu(NO_3)_2$	0.3	0.1[d]	0.06	0.02[d]	0.015	2.36×10^{-7}
Hg	$HgCl_2$	0.6[c]	0.45[c]	0.2	0.02	0.008	4.00×10^{-8}
Ag	$AgNO_3$	0.1	0.02[d]	0.01	0.004	0.003	2.80×10^{-8}

[a] Concentrations of metals at which the survival time of large sticklebacks averaged about 1 day (19–30 hr), 2 days (38–60 hr), 4 days (76–120 hr), and 1 week (140–200 hr), according to the data of Jones.[1,2]
[b] Jones' "lethal concentration limits," or concentrations of the metals at which the average survival time nearly equalled that of controls (about 10 days).
[c] Interpolated value; suitable empirical value lacking.
[d] Corresponding values derived by interpolation are greater than the nearest empirical values shown by about 30–40%.

Note: Stickleback–*Gasterosteus aculeatus.*

Source: P. Doudoroff and M. Katz, *Sewage Ind. Wastes, 25*(7):802, 1953.

REFERENCES

1. J.R.E. Jones, *J. Exp. Biol., 15*:394, 1938.
2. J.R.E. Jones, *J. Exp. Biol., 16*:425, 1939.

3.4–27 EXAMPLES OF FISH AS INDICATORS OF WATER SAFETY FOR LIVESTOCK

Material	Toxic levels, mg/l for fish	Toxic effects on animals
Aldrin	0.02	3 mg/kg food (poultry)
Chlordane	1.0 (sunfish)	91 mg/kg body weight in food (cattle)
Dieldrin	0.025 (trout)	25 mg/kg food (rats)
Dipterex	50.0	10.0 mg/kg body weight in food (calves)
Endrin	0.003 (bass)	3.5 mg/kg body weight in food (chicks)
Ferban, fermate	1.0–4.0	
Methoxychlor	0.2 (bass)	14 mg/kg alfalfa hay, not toxic (cattle)
Parathion	2.0 (goldfish)	75 mg/kg body weight in food (cattle)
Pentachlorophenol	0.35 (bluegill)	60 mg/l drinking water not toxic (cattle)
Pyrethrum (allethrin)	2.0–10.0	1,400–2,800 mg/kg body weight in food (rats)
Silvex	5.0	500–2,000 mg/kg body weight in food (chicks)
Toxaphene	0.1 (bass)	35–110 mg/kg body weight in food (cattle)

Source: *Water Quality Criteria,* U.S. Department of the Interior, Fed. Water Pollut. Control Adm., April 1968. Compiled from data in *Water Quality Criteria,* Cal. State Water Qual. Control Board, 1963.

3.4–28 ACUTE TOXICITY CONCENTRATIONS OF FISH POISONS AFFECTING FATHEAD MINNOWS

Toxicant	96 hr TL_m, mg/l	½ 96 hr TL_m, mg/l
Toxaphene	0.013	0.007
Rotenone	0.006	0.003
Cube root	0.066	0.03
Sulfoxide	0.74	0.37
Formulation A[a]	0.10	0.05
Formulation B[a]	0.066	0.03
Formulation C[a]	0.066	0.03
Formulation D[a]	0.10	0.05

[a] Formulation A contains active ingredients of 2.0% rotenone and 7.0% toxaphene.
Formulation B contains active ingredients of 5.0% rotenone and 15% other cube extractives.
Formulation C contains active ingredients of 2.5% rotenone, 5% cube extractives, and 2.5% sulfoxide.
Formulation D contains active ingredients of 5.0% rotenone and 10% other cube extractives.

Source: Reprinted, with publisher's permission, from *J. Am. Water Works Assn., 52*(12):1551, © 1960, by the American Water Works Association, Inc., 2 Park Ave., New York, N.Y. 10016.

3.4–29 TOXICITY OF SOME ORGANIC PESTICIDES TO GOLDFISH

	Die in 10 days, ppm	Turn over in 24 hr, ppm
Toxaphene dust (20% toxaphene)	0.025	0.125
Aldrin dust (2.5% aldrin equivalent)	0.8	2
BHC–DDT dust (5% DDT and 8% BHC, including 3% gamma isomer)	3	25

Source: P. Doudoroff *et al., J. Water Pollut. Control Fed., 25*(7):843, July 1953.

3.4–30 MEDIAN TOLERANCE LIMITS FOR PETROCHEMICALS OBTAINED WITH THE MOVING AVERAGE-ANGLE AND GRAPHICAL INTERPOLATION METHODS

Compound	Dilution water	Test fish	24 hr TL_m, mg/l	24 hr Confidence limits, mg/l	48 hr TL_m, mg/l	48 hr Confidence limits, mg/l	96 hr TL_m, mg/l	96 hr Confidence limits, mg/l	Graphical interpolation 96 hr TL_m, mg/l
Benzene (reagent)	Soft	Fatheads[b]	35.56	25.81–45.83	35.08	26.74–43.67	33.47	25.19–41.66	31
	Hard	Fatheads	34.42	26.10–42.83	32.00	22.33–41.16	32.00	22.33–41.16	31
	Soft	Bluegills[c]	22.49	17.50–28.37	22.49	17.50–28.37	22.49	17.50–28.37	22
	Soft	Goldfish[d]	34.42	26.10–42.83	34.42	26.10–42.83	34.42	26.10–42.83	32
	Soft	Guppies[e]	36.60	28.63–54.33	36.60	28.63–54.33	36.60	28.63–54.33	32
Chlorobenzene B.P. 130–132	Soft	Fatheads	29.12	23.24–38.49	29.12	23.24–38.49	29.12	23.24–38.49	29
	Soft	Fatheads	33.93	27.61–43.27	33.93	27.61–43.27	33.93	27.61–43.27	35
	Hard	Fatheads	39.19	34.09–47.64	34.98	29.02–44.23	33.93	27.61–43.27	35
	Soft	Bluegills	24.00	18.88–30.51	24.00	18.88–30.51	24.00	18.88–30.51	20
	Soft	Goldfish	73.03	55.56–146.77	56.00	56.00–56.00	51.62	40.10–71.72	45
	Soft	Guppies	45.53	36.35–58.19	45.53	36.35–58.19	45.53	36.35–58.19	44
O-Chlorophenol (practical)	Soft	Fatheads	21.96	17.82–31.25	19.12	15.47–24.55	11.63	8.86–14.66	12
	Hard	Fatheads	21.52	18.46–26.61	18.00	15.04–21.54	14.48	11.43–18.74	14
	Soft	Bluegills	11.31	9.44–14.34	10.59	9.07–12.60	10.00	8.32–12.01	10
	Soft	Goldfish	14.48	11.80–21.56	12.37	10.69–15.19	12.37	10.69–15.19	14
	Soft	Guppies	22.17	19.21–27.10	20.78	18.37–24.21	20.17	17.68–23.64	23
3-Chloropropene (B.P. 45–45.5°)	Soft	Fatheads	24.00	18.88–30.51	24.00	18.88–30.51	19.78	14.97–24.78	24
	Hard	Fatheads	25.86	20.52–33.29	24.00	18.88–30.51	24.00	18.88–30.51	22
	Soft	Bluegills	59.30	50.87–70.34	42.33	33.52–53.47	42.33	33.52–53.47	42
	Soft	Goldfish	26.56	21.12–34.37	20.87	15.99–26.18	20.87	15.99–26.18	22
	Soft	Guppies	57.68	48.87–68.76	53.54	43.01–71.14	51.08	41.03–67.02	48
O-Cresol (98% practical)	Soft	Fatheads	Not found	Not found	Not found	Not found	12.55	9.72–15.92	13
	Hard	Fatheads	18.00	15.04–21.54	13.42	10.50–17.15	13.42	10.50–17.15	14
	Soft	Bluegills	22.17	19.21–27.10	20.78	18.37–24.21	20.78	18.37–24.21	24
	Soft	Goldfish	Not found	Not found	Not found	Not found	23.25	17.41–30.66	23
	Soft	Guppies	49.13	38.10–99.60	25.31	20.04–32.45	18.85	14.07–23.61	20

[b] Fathead minnows – *Pimephales promelas.*
[c] Bluegills – *Lepomis macrochirus.*
[d] Goldfish – *Carassius auratus.*
[e] Guppies – *Lebistes reticulatus.*

3.4–30 MEDIAN TOLERANCE LIMITS FOR PETROCHEMICALS OBTAINED WITH THE MOVING AVERAGE-ANGLE AND GRAPHICAL INTERPOLATION METHODS (*Continued*)

			24 hr		48 hr		96 hr		
Compound	Dilution water	Test fish	TL_m, mg/l	Confidence limits, mg/l	TL_m, mg/l	Confidence limits, mg/l	TL_m, mg/l	Confidence limits, mg/l	Graphical interpolation 96 hr TL_m, mg/l
Cyclohexane (C.P.)	Soft	Fatheads	35.08	26.74–43.67	35.08	26.74–43.67	32.71	23.03–42.07	30
	Hard	Fatheads	42.33	33.52–53.47	42.33	33.52–53.47	42.33	33.52–53.47	32
	Soft	Bluegills	42.33	33.52–53.47	40.60	30.58–53.04	34.72	24.99–44.69	31
	Soft	Goldfish	42.33	33.52–53.47	42.33	33.52–53.47	42.33	33.52–53.47	33
	Soft	Guppies	57.68	48.87–68.76	57.68	48.87–68.76	57.68	48.87–68.76	48
Ethyl Benzene (B.P. 134–136)	Soft	Fatheads	48.51	38.90–62.83	48.51	38.90–62.83	48.51	38.90–62.83	40
	Hard	Fatheads	42.33	33.52–53.47	42.33	33.52–53.47	42.33	33.52–53.47	36
	Soft	Bluegills	35.08	26.74–43.67	32.00	32.00–32.00	32.00	32.00–32.00	29
	Soft	Goldfish	94.44	79.62–110.08	94.44	79.62–110.08	94.44	79.62–110.08	73
	Soft	Guppies	97.10	81.45–114.58	97.10	81.45–114.58	97.10	81.45–114.58	78
Isoprene (practical)	Soft	Fatheads	86.51	74.19–97.96	86.51	74.19–97.96	86.51	74.19–97.96	75
	Hard	Fatheads	74.83	58.75–95.32	74.83	58.75–95.32	74.83	58.75–95.32	75
	Soft	Bluegills	42.54	32.50–50.15	42.54	32.50–50.15	42.54	32.50–50.15	39
	Soft	Goldfish	180.00	150.41–215.41	180.00	150.41–215.41	180.00	150.41–215.41	180
	Soft	Guppies	240.00	188.77–305.14	240.00	188.77–305.14	240.00	188.77–305.14	240
Methyl Methacrylate (monomer, practical)	Soft	Fatheads	421.2	357.4–551.4	338.2	291.7–398.7	159.1	125.5–190.7	150
	Soft	Fatheads	455.1	374.7–664.9	455.1	374.7–664.9	160.2	136.3–183.4	130
	Hard	Fatheads	498.6	394.7–877.8	338.2	291.6–398.7	311.0	248.3–418.7	350
	Hard	Fatheads	391.9	340.9–476.4	368.1	326.4–426.8	320.0	268.7–381.1	320
	Soft	Bluegills	368.1	326.4–426.9	357.5	314.4–417.1	232.2	153.9–341.8	250
	Soft	Goldfish	423.3	335.2–534.7	423.3	335.2–534.7	277.1	237.9–313.5	240
	Soft	Guppies	368.1	326.4–426.9	368.1	326.4–426.9	368.1	326.4–426.9	420
Phenol (U.S.P. 100%)	Soft	Fatheads	40.60	30.58–53.04	40.60	30.58–53.04	34.27	22.83–45.86	29
	Hard	Fatheads	38.62	30.10–48.29	38.62	30.10–48.29	32.00	23.78–39.87	32
	Soft	Bluegills	25.85	21.17–38.18	23.88	20.16–31.49	23.88	20.16–31.49	26
	Soft	Goldfish	49.86	39.47–87.78	49.13	38.10–99.60	44.49	36.10–68.76	46
	Soft	Guppies	49.86	39.47–87.78	49.86	39.47–87.78	39.19	34.09–47.64	44
O-Phthalic anhydride	Soft	Fatheads	Not found	Not found	Not found	Not found	Not found	Not found	>56[a]
	Soft	Fatheads	Not found	Not found	Not found	Not found	Not found	Not found	>56
	Soft	Fatheads	Not found	Not found	Not found	Not found	Not found	Not found	>56

[a] Not all in solution.

3.4–30 MEDIAN TOLERANCE LIMITS FOR PETROCHEMICALS OBTAINED WITH THE MOVING AVERAGE-ANGLE AND GRAPHICAL INTERPOLATION METHODS (*Continued*)

Compound	Dilution water	Test fish	24 hr TL_m, mg/l	24 hr Confidence limits, mg/l	48 hr TL_m, mg/l	48 hr Confidence limits, mg/l	96 hr TL_m, mg/l	96 hr Confidence limits, mg/l	Graphical interpolation 96 hr TL_m, mg/l
Styrene	Soft	Fatheads	56.73	47.67–67.83	53.58	43.04–71.21	46.41	37.11–59.54	51
(stabilized	Hard	Fatheads	62.81	54.99–73.69	62.81	54.99–73.69	59.30	50.87–70.34	60
B.P. 43–45)	Soft	Bluegills	25.05	19.03–33.53	25.05	19.03–33.53	25.05	19.03–33.53	22
	Soft	Goldfish	64.74	57.17–75.48	64.74	57.17–75.48	64.74	57.17–75.48	68
	Soft	Guppies	74.83	58.75–95.32	74.83	58.75–95.32	74.83	58.75–95.32	68
Toluene	Soft	Fatheads	46.31	37.03–59.40	46.31	37.03–59.40	34.27	22.83–45.86	44
reagent	Hard	Fatheads	56.00	44.73–67.11	56.00	46.73–67.11	42.33	33.52–53.47	45
	Soft	Bluegills	24.00	18.88–30.51	24.00	18.88–30.51	24.00	18.88–30.51	24
	Soft	Goldfish	57.68	48.87–68.76	57.68	48.87–68.76	57.68	48.87–68.76	62
	Soft	Guppies	62.81	54.99–73.69	60.95	52.84–71.93	59.30	50.87–70.34	66
Vinyl acetate	Soft	Fatheads	24.00	18.88–30.51	24.00	18.88–30.51	24.00	18.88–30.51	24
(stabilized,	Soft	Fatheads	22.17	19.21–27.10	20.31	17.01–25.62	19.73	16.28–25.11	20
practical)	Hard	Fatheads	39.19	34.09–47.64	39.19	34.09–47.64	39.19	34.09–47.64	44
	Hard	Fatheads	36.81	32.64–42.69	36.81	32.64–42.69	35.75	31.44–41.71	41
	Soft	Bluegills	18.53	15.73–22.07	18.00	15.04–21.54	18.00	15.04–21.54	18
	Soft	Goldfish	42.33	33.52–53.47	42.33	33.52–53.47	42.33	33.52–53.47	42
	Soft	Guppies	31.08	26.10–36.63	31.08	26.10–36.63	31.08	26.10–36.63	26
Xylene	Soft	Fatheads	28.77	25.62–32.64	27.71	24.58–31.25	26.70	23.53–29.97	21
	Hard	Fatheads	28.77	25.62–32.64	28.77	25.62–32.64	28.77	25.62–32.64	22
	Soft	Bluegills	24.00	18.88–30.51	24.00	18.88–30.51	20.87	15.99–26.18	22
	Soft	Goldfish	36.81	32.64–42.69	36.81	32.64–42.69	36.81	32.64–42.69	24
	Soft	Guppies	34.73	30.26–40.75	34.73	30.26–40.75	34.73	30.26–40.75	39

Source: Q.H. Pickering and C. Anderson, *J. Water Pollut. Control Fed.*, *38*(9):1419, 1966.

3.4–31 COMPARISON OF ACUTE TOXICITY OF PETROCHEMICALS TO DIFFERENT SPECIES OF FISH

Compound	96 hr TL_m and 95 percent confidence limit for organism,[a] mg/l			
	Fatheads[b]	Bluegills[c]	Goldfish[d]	Guppies[e]
O-Chlorophenol	11.63(8.86–14.66)	10.00(8.32–12.01)	12.37(10.69–15.19)	20.17(17.68–23.64)
Vinyl acetate	24.00(18.88–30.51)	18.00(15.04–21.54)	42.33(33.52–53.47)	31.08(26.10–36.63)
	19.73(16.28–25.11)			
Xylene	26.70(23.53–29.97)	20.87(15.99–26.18)	36.81(32.64–42.69)	34.73(30.26–40.75)
O-Cresol	12.55(9.72–15.92)	20.78(18.37–24.21)	23.25(17.41–30.66)	18.85(14.07–23.61)
Benzene	33.47(25.19–41.66)	22.49(17.50–28.37)	34.42(26.10–42.83)	36.60(28.63–54.33)
Toluene	34.27(22.83–45.86)	24.00(18.88–30.51)	57.68(48.87–68.75)	59.30(50.87–70.34)
Chlorobenzene	29.12(23.23–38.49)	24.00(18.88–30.51)	51.62(40.10–71.72)	45.53(36.35–58.19)
	33.93(27.61–43.27)			
Phenol	34.27(22.83–45.86)	23.88(20.16–31.49)	44.49(36.10–68.76)	39.19(34.00–47.64)
Styrene	46.41(37.11–59.54)	25.05(19.03–33.53)	64.74(57.17–75.48)	74.83(58.75–95.32)
Ethyl benzene	48.51(38.90–62.83)	32.00(32.00–32.00)	94.44(79.62–110.1)	97.10(81.45–114.6)
Cyclohexane	32.71(23.03–42.07)	34.72(24.99–44.69)	42.33(33.52–53.47)	57.68(48.87–68.76)
3-Chloropropene	19.78(14.97–24.78)	42.33(33.52–53.47)	20.87(15.99–26.18)	51.08(41.03–67.02)
Isoprene	86.51(74.19–97.96)	42.54(32.50–50.15	180.0(150.4–215.4)	240.0(188.8–305.1)
Methyl methacrylate	159.1(125.5–190.7)	232.2(153.9–341.8)	277.1(237.9–313.5)	368.1(326.4–426.9)
	160.2(136.3–183.4)			
O-Phthalic anhydride	Not found	Not found	Not found	Not found

[a] Under standard conditions–soft water as diluent; temperature 25°C.
[b] Fathead minnow–*Pimephales promelas.*
[c] Bulegills–*Lepomis macrochirus.*
[d] Goldfish–*Carassius auratus.*
[e] Guppies–*Lebistes reticulatus.*

Source: Q.H. Pickering and C. Henderson, *J. Water Pollut. Control Fed.*, *38*(9):1419, 1966.

3.4–32 ACUTE TOXICITY OF THREE PETROCHEMICALS TO FATHEAD MINNOWS[a]

Different Ages

		TL_m mg/l				Dilution water characteristic		
Compound	**Age**	**24 hr**	**48 hr**	**96 hr**	**Date**	**pH**	**Alkalinity, mg/l**	**Hardness, mg/l**
Isoprene	1 day	75	75	75	6/12/61	8.1	148	184
		77	77	77	7/12/61	8.6	146	168
		85	85	85	7/12/61	8.1	130	158
		83	80	80	7/31/61	8.3	126	156
	2 days	75	75	75	7/12/61	8.6	146	168
		85	80	80	7/21/61	8.1	130	158
	10 days	82	78	78	7/21/61	8.1	130	156
		76	75	75	7/31/61	8.3	126	156
	Adult	75	75	75	11/16/61	7.5	20	20
		75	75	75	11/16/61	8.2	320	352
Methyl methacrylate	1 day	410	410	410	6/ 2/61	8.6	136	170
		560	530	400	7/21/61	8.1	130	156
	2 days	420	420	410	7/12/61	8.6	146	168
		560	500	430	7/21/61	8.1	130	158
	4 days	520	490	460	7/12/61	8.6	146	168
	Adult	450	380	150	2/29/60	7.5	18	22
		480	450	130	6/27/60	7.5	20	24
		510	380	350	2/29/60	8.1	294	356
		440	420	320	6/27/60	8.2	286	370
Vinyl acetate	1 day	18	18	14	6/ 2/61	8.6	136	170
		16	15	15	6/21/61	8.2	136	164
		14	14	14	7/10/61	8.2	144	172
	2 days	18	17	15	6/21/61	8.2	136	164
		15	15	15	7/10/61	8.2	144	172
	4 days	25	24	23	6/21/61	8.2	136	164
		28	27	26	7/31/61	8.3	126	156
	Adult	25	22	20	12/ 7/59	7.5	20	24
		24	24	24	7/31/61	7.5	18	20
		42	42	41	12/ 7/59	8.1	302	362
		44	44	44	7/31/61	8.2	260	312

[a] Fathead minnows, *Pimephales promelas.*

Source: Q.H. Pickering and C. Henderson, *J. Water Pollut. Control Fed., 38*(9):1419, 1966.

3.4–33 COMPARISON OF THE ACUTE TOXICITY OF PETROCHEMICALS TO FATHEAD MINNOWS[a]

Different Dilution Waters

Compound	Soft water		Hard water	
	96 hr TL_m,[b] mg/l	95% confidence limits, mg/l	96 hr TL_m[b]	95% confidence limits
O-Chlorophenol	11.63	8.86–14.66	14.48	11.43–18.74
Vinyl acetate	24.00	18.88–30.51	39.19	34.09–47.64
	19.73	16.28–25.11	35.75	31.44–41.72
Xylene	26.70	23.53–29.97	28.77	25.62–32.64
O-Cresol	12.55	9.72–15.92	13.42	10.50–17.15
Benzene	33.47	25.19–41.66	32.00	22.33–41.16
Toluene	34.27	22.83–45.86	42.33	33.52–53.47
Chlorobenzene	29.12	23.24–38.49	33.93	27.61–43.27
	33.93	27.61–43.27		
Phenol	34.27	22.83–45.86	32.00	23.78–39.87
Styrene	46.41	37.11–59.54	59.30	50.87–70.34
Ethyl benzene	48.51	38.90–62.83	42.33	33.52–53.47
Cyclohexane	32.71	23.03–42.07	42.33	33.52–53.47
3-Chloropropene	19.78	14.97–24.78	24.00	18.88–30.51
Isoprene	86.51	74.19–97.96	74.83	58.75–95.32
Methyl methacrylate	159.1	125.5–190.7	311.0	248.3–418.7
	160.2	136.3–183.4	320.0	268.7–381.1
O-Phthalic anhydride	Not found		Not found	

[a] Fathead minnows, *Pimephales promelas.*
[b] Values obtained by the moving average-angle method.

Source: Q.H. Pickering and C. Henderson, *J. Water Pollut. Control Fed., 38*(9):1419, 1966.

3.4–34 ACUTE TOXICITY OF TRACE ORGANICS

Test material	Ratio, CCE/CAE	Test fish Length, cm/ Weight, g	TL_m value, mg/l 24 hr	48 hr	96 hr	120 hr
Spring[a] Run 1						
Unit 1	1/1.56	Trout, 9.4/9.7	138	130	96	92
		Red shiners, 5.0/2.0		no effect up to 240		
Unit 2	1/3.65	Red shiners, 2.9/1.3		no effect up to 200		
Spring,[a] Run 2						
Unit 1	1/1.48	Trout, 10.3/16.3	130	125	95	82
		Red shiners, 2.8/1.2		no effect up to 305		
		Red shiners, 5.6/2.2		no effect up to 305		
		Sunfish, 7.5/9.1	166	141	115	103
Unit 2	1/3.35	Trout, 10.5/16.0	88	75	61	56
		Trout, 5.0/2.6	201	186	155	130
		Red shiners, 5.7/2.3	195	170	148	20
		Golden shiners, 6.4/4.0	180	171	160	52
		Sunfish, 7.0/8.7	137	121	114	00
		Mosquito fish, 2.0/0.2		no effect up to 270		
Unit 3	1/2.46	Trout, 10.1/16.2		no effect up to 180		
		Trout, 5.2/2.0		no effect up to 180		
Missouri River, CCE	N/A	Trout, 5.3/2.8	36	32	28	24
		Golden shiners, 6.6/4.3	59	52	39	33
		Sunfish, 7.7/11.7	56	49	45	39
Sevin[b]	N/A	Trout, 10.0/15.0	2.3	1.5	1.0	0.7
		Red shiners, 4.6/1.4	13	12	9.2	7.4
Malathion[c]	N/1	Trout, 10.0/15.0	0.0050	0.0046	0.0028	0.0023
		Red shiners, 4.8/1.9	0.040	0.036	0.025	0.023

[a] Combined extract.
[b] Based on 50 percent active ingredients.
[c] Based on 57 percent active ingredients.

Source: Reprinted, with publisher's permission, from *J. Am. Water Works Assn., 60*(8):969, © 1968, by the American Water Works Association, Inc., 2 Park Ave., New York, N.Y. 10016.

3.4–35 LONG-TERM TOXICITY OF TRACE ORGANICS

Test material	Test fish length, cm, weight, g	Total accumulated test time for 50 percent kill, days / Test material concentration, mg/l: 10		7.5		4.2		1.0			96 hr TL_m value, mg/l
		0[a]	1.0[a]	0	0.75	0	0.42	0	0.10	Control	
Spring Run 2, Unit 2	Trout, 8.2/11	30	20	37	23	48	29	>54	>54	>54[b]	61–155[c]
Missouri River CCE	Trout, 5.3/2.2	10	5	13	11	19	13	>19	>19	>19[d]	23
		24		13.5		5.6					
		0	2.4	0	1.35	0	0.56			Control	
Spring Run 2, Unit 2	Red shiners, 4.7/1.5	30	20	39	>65	45			>65[d]		148
		0.0010		0.00075		0.00056		0.00024		Control	
		0	0.00010	0	0.000075	0	0.000056	0	0.000024		
Malathion	Trout, 6.0/3.8	7	5	15	8	>17	10	>17	>17	>19	0.0028

[a] Test material concentration in recovery water, mg/l.
[b] Fish died because of loss of temperature control.
[c] Values for trout having 10.5/16 and 5./2.6, length/weight, respectively.
[d] Fish showed signs of disease: test discontinued.

Source: Reprinted, with publisher's permission, from *J. Am. Water Works Assn.*, *60*(8):1969, © 1968 by the American Water Works Association, Inc., 2 Park Ave., New York, N.Y. 10016.

3.4–36 TOXICITY FACTORS FOR TEST FISH EXPOSED TO SUBSURFACE WATER CCE AND CAE

			Test fish characteristics				Toxicity Factor, $y_t \times 10^{-7}$	
Time, days	TL_m, mg/l	Type of fish	Length, cm	Weight, g	GSA, mm^2	K x 10,[2] g/cm^3	Experimental[a]	Computed[b]
1	201	Trout	5.0	2.6	2,400	2.08	2.48	
	88		10.5	16.0	3,100	1.38	2.48	2.50
2	186		5.0	2.6	2,400	2.08	2.24	
	75		10.5	16.0	3,100	1.38	2.24	2.29
4	155		5.0	2.6	2,400	2.08	1.79	
	61		10.5	16.0	3,100	1.38	1.79	1.87
5	130		5.0	2.6	2,400	2.08	1.56	
	56		10.5	16.0	3,100	1.38	1.56	1.70
30	10		8.2	11.0	4,300	1.99	0.26	0.22
37	7.5		8.2	11.0	4,300	1.99	0.20	0.17
48	4.2		8.2	11.0	4,300	1.99	0.11	0.12
60	–		–	–	–	–	–	0.11
70	–		–	–	–	–	–	0.10
1	195	Red Shiners	5.7	2.3	2,320	1.24	78	78
2	170		5.7	2.3	2,320	1.24	68	70
4	148		5.7	2.3	2,320	1.24	59	58
5	120		5.7	2.3	2,320	1.24	48	52
30	24		4.7	1.5	2,780	1.45	9.6	10
39	13		4.7	1.5	2,780	1.45	5.4	5.4
50	–		–	–	–	–	–	4.5
60	–		–	–	–	–	–	4.2
70	–		–	–	–	–	–	4.1
80	–		–	–	–	–	–	4.0

[a] Computed using the following equation:

$$y_t = C_t \cdot \left[\frac{GSA}{K}\right]^{1.25}$$

where, y_t = toxicity factor relating the toxicant TL_m value at time t with test fish characteristics
C_t = TL_m value at time t, mg/l
GSA = gill surface area, mm^2
K= condition factor, g/cm^3.

[b] Computed using the following equations. For trout:

$$y_t = 0.10 \times 10^7 + (2.7 \times 10^7)\, e^{-0.11t}$$

where, y_t = same as above
t = test time in days.

For Red Shiners:

$$y_t = 4.0 \times 10^7 + (82.0 \times 10^7)\, e^{-0.11t}$$

where terms are the same as above.

Note: Test material used was Spring Run 2, Unit 2, CCE and CAE (ratio 1/3.55).

Source: Reprinted, with publisher's permission, from *J. Am. Water Works Assn., 62*(8):499, © 1970, by the American Water Works Association, Inc., 2 Park Ave., New York, N.Y. 10016.

3.4–37 TOXICITY EQUATIONS AND LONG-TERM TL_m VALUES OR ORGANIC MICROPOLLUTANTS AND OTHER TOXICANTS

Test material	Ratio, CCE/CAE	Test fish	y_t defined as	Toxicity equation	Long-term TL_m reached at time
Spring[a] Run 2					
Unit 1	1/1.48	Trout		$y_t = 0.85 \times 10^7 + (2.95 \times 10^7)e^{-0.17t}$	33.1 mg/l–40 days
		Sunfish		$y_t = 1.8 \times 10^7 + (11.6 \times 10^7)e^{-0.15t}$	25.4 mg/l–50 days
Unit 2	1/3.35	Trout		$y_t = 0.10 \times 10^7 + (2.7 \times 10^7)e^{-0.11t}$	3.9 mg/l–70 days
		Sunfish	$C_t \cdot (\frac{GSA}{K})^{1.25}$	$y_t = 4.2 \times 10^7 + (9.0 \times 10^7)e^{-0.13t}$	47.0 mg/l–50 days
		Red shiners		$y_t = 4.0 \times 10^7 + (82.0 \times 10^7)e^{-0.11t}$	10.3 mg/l–50 days
		Golden shiners		$y_t = 47.5 \times 10^7 + (18.0 \times 10^7)e^{-0.32t}$	141.0 mg/l–20 days
Missouri River					
CCE	N/A	Trout		$y_t = 0.30 \times 10^2 + (20.7 \times 10^2)e^{-0.11t}$	0.44 mg/l–80 days
		Sunfish		$y_t = 11.8 \times 10^2 + (13.2 \times 10^2)e^{-0.27t}$	30.1 mg/l–30 days
		Golden shiners		$y_t = 5.0 \times 10^2 + (41.0 \times 10^2)e^{-0.18t}$	7.5 mg/l–40 days
Acrylonitrile[1]	N/A	Flathead minnows		$y_t = 4.0 \times 10^2 + (56.0 \times 10^2)e^{-0.38t}$	3.2 mg/l–20 days
Benzonitrile[1]	N/A	Flathead minnows		$y_t = 5.0 \times 10^3 + (32.0 \times 10^3)e^{-0.25t}$	40.0 mg/l–30 days
Adiponitrile[1]	N/A	Sunfish	C_t/K	$y_t = 1.15 \times 10^4 + (3.85 \times 10^4)e^{-0.30t}$	359.0 mg/l–30 days
3-Chloroprene[2]	N/A	Guppies		$y_t = 2.4 \times 10^3 + (1.88 \times 10^3)e^{-0.10t}$	34.0 mg/l–60 days
Toluene[2]	N/A	Guppies		$y_t = 2.0 \times 10^3 + (2.55 \times 10^3)e^{-0.04t}$	28.2 mg/l–140 days
Malathion[3]	N/A	Trout		$y_t = 0.10 \times 10^{-2} + (46.7 \times 10^{-2})e^{-0.23t}$	0.02 μg/l–60 days
		Red shiners		$y_t = 1.90 \times 10^{-1} + (24.8 \times 10^{-1})e^{-0.15t}$	3.3 μg/l–50 days
Sevin[3]	N/A	Trout		$y_t = 0.20 \times 10^2 + (2.06 \times 10^2)e^{-0.44t}$	0.3 mg/l–20 days
		Red shiners		$y_t = 1.4 \times 10^2 + (9.1 \times 10^2)e^{-0.18t}$	2.0 mg/l–40 days
		Golden shiners[4]		$y_t = 9.5 \times 10^2 + (74.9 \times 10^2)e^{-0.42t}$	8.8 mg/l–20 days

[a] Combined CCE and CAE extract.

Note: NA–Not applicable.

REFERENCES

1. C. Henderson *et al.*, *Proc. 15th Ind. Waste Conf.*, Purdue University, *106*:121, 1961.
2. Q. Pickering and C. Henderson, *J. Water Pollut. Control. Fed.*, *38*:1419, 1966.
3. J.W. Smith and S.G. Grigoropoulos, *J. Am. Water Works Assn.*, *60*(8):969, 1968.
4. D. Evans, *The Toxic Effects and Fate of Selected Pesticides in Natural Waters*, M.S. Thesis, University of Missouri, Rolla, 1954.

3.4–38 TOXICITY FACTORS FOR TEST FISH EXPOSED TO A SURFACE WATER CCE AND ACRYLONITRILE

			Test fish characteristics				Toxicity Factor $y_t \times 10^{-2}$	
Time, days	Test material	TL_m, mg/l	Type of fish	Length, cm	Weight, g	K x 10^2, g/cm³	Experimental[a]	Computed[b]
1	Missouri River CCE	36.0	Trout	5.3	2.8	1.86	19.3	18.9
2		32.0		5.3	2.8	1.86	17.2	16.9
4		27.6		5.3	2.8	1.86	15.0	13.6
5		24.0		5.3	2.8	1.86	13.0	12.3
10		10.0		5.3	2.2	1.48	6.8	7.0
13		7.5		5.3	2.2	1.48	5.4	5.0
19		4.2		5.3	2.2	1.48	3.8	2.9
30		–		–	–	–	–	1.1
40		–		–	–	–	–	0.6
60		–		–	–	–	–	0.3
80		–		–	–	–	–	0.3
1	Acrylonitrile[1]	34.0	Flathead minnows	5.7	1.5	0.81	42.5	42.4
2		22.0		5.7	1.5	0.81	27.5	30.2
3		18.0		5.7	1.5	0.81	22.5	22.0
4		10.0		5.7	1.5	0.81	12.5	16.3
5		8.0		5.7	1.5	0.81	10.0	12.2
10		7.0		5.7	1.5	0.81	8.8	5.2
15		5.0		5.7	1.5	0.81	6.3	4.2
20		4.0		5.7	1.5	0.81	5.0	4.0
25		3.5		5.7	1.5	0.81	4.4	4.0
30		2.6		5.7	1.5	0.81	3.3	4.0
40		–		–	–	–	–	4.0

[a]Computed using the following equation:

$y_t = C_t/K$, where y_t = toxicity factor relating the toxicant TL_m value at time t with test fish characteristics;
C_t = TLM value at time t, mg/l, and
K = condition factor, g/cm³.

[b]Computed using the following equations. For the surface water CCE:

$y_t = 0.30 \times 10^2 + (20.7 \times 10^2)e^{-0.11t}$, where y_t = same as above test, and
t = time in days.

For acrylonitrile:

$y_t = 4.0 \times 10^2 + (56.0 \times 10^2)e^{-0.38t}$, where terms are the same as above.

Source: Reprinted, with publisher's permission, from *J. Am. Water Works Assn.*, *62*(8):499, © 1970, by the American Water Works Association, Inc., 2 Park Ave., New York, N.Y. 10016.

REFERENCE

1. C. Henderson *et al.*, *Proc. 15th Ind. Waste Conf.*, Purdue University, *106*:121, 1961.

3.5 RECREATIONAL AND AESTHETIC

3.5–1 CONCENTRATION OF PHENOLIC COMPOUNDS THAT CAUSE TAINTING OF FISH FLESH

Compound or waste	Concentration affecting taste and odor, mg/l	Fish tested	Ref.
Pure compounds			
Phenol	15–25	Trout, carp, tench, chub, eel, minnow, perch, bluegill, pike, goldfish	2
Cresols	10	Tench, carp, eel, trout, minnow	1
Xylenols	1–5	Roach, perch, carp	1
Pyrocatechol	2–5[a]	Perch, carp, roach	1, 3
Pyrogallol	20–30	Roach, carp	1
p-Quinone	0.5	Carp, tench, roach	1
Pyridine	5	Roach, carp	1
Naphthalene	1	Roach	1
α-Naphthol	0.5	Roach, carp	1
Quinoline	0.5–1.0	Roach, carp	1
Chlorophenol	0.1		1
p-Toluidine	20.0		3
β-Naphthol	1.0		3
α-Naphthylamine	3.0		3
Resorcinol	30.0		3
Phloroglucinol	10.0		3
Mixed phenolic wastes			
Coal coking wastes	0.02–0.1	Freshwater fish	2
Coal tar wastes	0.1	Freshwater fish	2
Phenols in polluted river	0.02–0.15	Minnows	2
Sewage containing phenols	0.1	Freshwater fish	2

[a] 2.5. See reference 2.

REFERENCES

1. H.J. Bandt, *Wasserwirtsch.-Wassertech.*, *9*:1, 1955
2. Ohio River Valley Sanitation Commission, Aquatic Life Advisory Committee, *Aquatic Life Water Quality Criteria*, 3rd Progress Report, *J. Water Pollut. Control Fed.*, *32*(1):65, 1960.
3. H.J. Brandt, *Eine Monographische Studie,* Wiss. Abhandl., Akademie Verlag, Berlin, 1958.

Section 4

Water Treatment

4.1 Public Water Supply Facilities
Pages 391–433

4.2 Coagulation and Filtration
Pages 434–455

4.3 Iron and Manganese Removal
Pages 456–462

4.4 Removal of Color, Odor, and Taste
Pages 463–473

4.5 Disinfection
Pages 474–492

4.6 Softening and Boiler Water Treatment
Pages 493–499

4.7 Miscellaneous and Nonconventional Processes
Pages 500–518

4. WATER TREATMENT

4.1 PUBLIC WATER SUPPLY FACILITIES

See Section 9.1 for more definitive data on supplies, particularly municipal water supplies.

4.1–1 SOURCES AND TREATMENTS OF PUBLIC WATER SUPPLIES

100 Largest Cities in the U.S., 1962

		Population served	
Source and treatment	**Number of cities**	**Thousands**	**Percent of population of 100 largest cities**
Surface water			
Chlorination	66	39,939	65.9
Sedimentation and coagulation	54	27,772	45.8
Rapid sand filtration	51	26,511	43.8
Slow sand filtration	7	2,536	4.2
Pressure filtration	2	356	0.6
Iron removal	4	1,475	2.4
Softening			
With lime[a]	10	4,649	7.7
With lime-soda ash	9	3,359	6.0
Ground water			
No treatment	1	150	0.2
Chlorination	19	5,545	9.2
Sedimentation and coagulation	7	2,127	3.5
Rapid sand filtration	7	2,970	4.9
Iron and manganese removal	5	1,474	2.4
Softening			
With lime	3	1,055	1.7
With lime-soda ash	1	320	0.5
Mixed surface and ground water			
No treatment	1	72	0.1
Chlorination	13	14,015	23.1
Sedimentation and coagulation	7	8,770	14.5
Rapid sand filtration	8	1,921	3.2
Slow sand filtration	1	33	0.1
Iron and manganese removal	1	500	0.8
Softening			
With lime	2	340	0.6
With lime-soda ash	1	259	0.4
By cation exchange	2	698	1.2

[a] At least one city supplements lime softening with soda ash during critical periods.

Source: C.N. Durfor and E. Becker, *Geological Survey Water Supply Paper 1812*, U.S. Department of the Interior, 1962.

4.1–2 PUBLIC WATER SUPPLIES FOR COMMUNITY WATER SYSTEMS IN THE U.S.

			Communities having a population served of				
Item	Unit	Total	500,000 or more	250,000–499,999	100,000–249,999	50,000–99,999	25,000–49,999
Communities and districts	number	926	43	35	137	251	460
Estimated population served	1,000	104,810	50,648	9,977	18,525	12,372	13,288
Ownership of public water supply	number	926	43	35	137	251	460
Municipal or district–public	number	772	38	27	113	215	379
Private	number	154	5	8	24	36	81
Average output of water[a]	mgd	15,694,	7,614	1,481	2,947	1,725	1,927
Water used, max month[a]	mil. gal	621,606	304,701	57,453	113,707	68,010	77,735
Water used, max day[a]	mil. gal	25,287	11,994	2,416	4,771	2,850	3,256
Impounded storage	mil. gal	5,982,917	2,440,578	585,871	1,232,258	506,462	1,217,748
Transmission capacity, max	mgd	32,770	19,414	3,538	4,293	3,000	2,525
Dependable yield, ground water	mgd	8,300	1,337	857	1,972	1,550	2,584
Raw water storage	mil. gal	378,062	311,943	50,751	1,960	10,703	2,705
Raw water pumping capacity	mgd	25,858	12,699	2,174	5,116	2,990	2,879
Treatment plant							
Rated capacity	mgd	25,511	12,125	2,586	4,593	2,870	3,337
Overload capacity[a]	mgd	31,176	13,701	3,016	7,177	3,428	3,854
Service pumping capacity	mgd	31,412	13,766	3,226	6,659	3,942	3,819
Distribution storage							
Pumped to mains	mil. gal	6,558	2,194	443	1,361	1,610	950
Elevated	mil. gal	76,188	58,135	1,706	11,515	2,521	2,311
Number of services[a]	1,000	23,933	8,732	2,384	4,881	3,704	4,232
Number of meters[a]	1,000	21,317	7,086	2,359	4,481	3,495	3,896

[a] Approximately 5% estimated.

Note: Data as of 1963. Includes only systems serving a population of 25,000 or more. Estimated total number of public water systems in the U.S. (conterminous area) is 21,000, serving a population of approximately 144 million. An estimated 16.0 percent of the water used by manufacturing industries is supplied by public water systems.

Source: U.S. Department of Commerce.

4.1–3 SOURCES AND TREATMENT OF WATER SUPPLY

Communities of 25,000 or More

Source	No. facilities	Percentage of total	Estimated population served	Percentage of total
1958[1]				
All facilities				
Surface	393	60.0	68,475,090	75.9
Ground	211	32.2	13,935,144	15.5
Surface and Ground	51	7.8	7,776,685	8.6
Total	655	100.0	90,186,919	100.0
Facilities supplying untreated water				
Surface	0	0	0	0
Ground	20	3.1	1,068,019	1.2
Surface and ground	0	0	0	0
Total	20	3.1	1,068,019	1.2
Facilities supplying treated water				
Surface	393	60.0	68,475,090	75.9
Ground	191	29.1	12,867,125	14.3
Surface and ground	51	7.8	7,776,685	8.6
Total	635	96.9	89,118,900	98
1960[2]				
All facilities				
Surface	419	57.6	67,491,426	70.3
Ground	251	34.5	20,178,118	21.0
Surface and ground	57	7.9	8,306,903	8.7
Total	727	100.0	95,976,447	100.0
Facilities supplying untreated water				
Surface	0	0	0	0
Ground	21	2.9	1,393,903	1.5
Surface and ground	0	0	0	0
Total	21	2.9	1,393,903	1.5
Facilities supplying treated water				
Surface	419	57.6	67,491,426	70.2
Ground	230	31.6	18,784,215	19.6
Surface and ground	57	7.9	8,306,903	8.7
Total	706	97.1	94,582,544	98.5
1962[3]				
All facilities				
Surface	446	49.3	68,888,167	68.2
Ground	382	42.2	22,864,556	22.7
Surface and ground	77	8.5	9,187,297	9.1
Total	905	100.0	100,940,020	100.0
Facilities supplying untreated water				
Surface	0	0	0	0
Ground	90	9.9	2,193,955	2.2
Surface and ground	0	0	0	0
Total	90	9.9	2,193,955	2.2
Facilities supplying treated water				
Surface	445	49.2	68,861,759	68.2
Ground	295	32.6	20,934,806	20.7
Surface and ground	75	8.3	8,949,500	8.9
Total	815	90.1	98,746,065	97.8

4.1–3 SOURCES AND TREATMENT OF WATER SUPPLY *(Continued)*

Source	No. facilities	Percentage of total	Estimated population served	Percentage of total
		1964[4]		
All facilities				
Surface	523	34.6	71,547,530	64.5
Ground	865	57.1	23,158,370	20.9
Surface and ground	126	8.3	16,265,619	14.6
Total	1,514	100.0	110,971,519	100.0
Facilities supplying untreated water				
Surface	1	0.1	43,225	a
Ground	414	27.3	3,382,140	3.0
Surface and ground	3	0.2	340,385	0.3
Total	418	27.6	3,765,750	3.4
Facilities supplying treated water				
Surface	522	34.5	71,504,305	64.5
Ground	438	28.9	19,349,230	17.4
Surface and ground	118	7.8	15,765,434	14.2
Total	1,078	71.2	106,618,969	96.1

[a] Less than .05.

REFERENCES

1. K.H. Jenkins, *J. Am. Water Works Assn., 53*(1):31, 1961.
2. K.H. Jenkins and H.W. Owen, *J. Am. Water Works Assn., 55*(1):35, 1963.
3. K.H. Jenkins, *J. Am. Water Works Assn., 55*(12):1485, 1963.
4. K.H. Jenkins, *J. Am. Water Works Assn., 57*:1360, 1965.

4.1–4 SERVICES AND METERS

Communities of 25,000 or More

Ownership	Estimated population served	No. services	No. meters	Persons per service	Percentage of services metered
			1958[1]		
Public	58,872,344	14,103,513	12,205,183	4.2	86.5
Private	9,122,425	2,540,246	2,266,487	3.6	89.2
Not stated	54,400	18,220	18,198	3.0	99.9
Total	68,049,169	16,661,979	14,489,868	4.1	87.0
			1960[2]		
Public	57,642,483	15,135,771	12,982,069	3.8	85.8
Private	8,760,445	2,529,450	2,276,284	3.5	90.0
Not stated	57,000	18,540	18,518	3.1	99.8
Total	66,459,928	17,683,761	15,276,871	3.8	86.4
			1962[3]		
Public	67,534,385	17,856,340	15,266,297	3.8	85.5
Private	10,860,624	2,986,976	2,742,521	3.6	91.8
Not stated	413,754	149,253	151,702	2.8	101.6
Total	78,808,763	20,992,569	18,160,520		

REFERENCES

1. K.H. Jenkins, *J. Am. Water Works Assn., 53*(1):31, 1961.
2. K.H. Jenkins and H.W. Owen, *J. Am. Water Works Assn., 55*(1):35, 1963.
3. K.H. Jenkins, *J. Am. Water Works Assn., 55*(12):1485, 1963.

4.1–5 FACILITIES WITH CHEMICAL AND/OR BACTERIOLOGICAL LABORATORIES

Population size group served	1960[1]		1962[2]		1964[3]	
	No. with laboratories[a]	Percent of total	No. with laboratories[a]	Percent of total	No. with laboratories[a]	Percent of total
Under 25,000	46	52	20	67	107	58.8
25,000–50,000	194	70	290	68	287	71.2
50,000–100,000	115	76	167	80	156	76.8
Over 100,000	151	72	196	81	247	80.2
Total	506	70	673	74	797	72.7

[a] Verified number.

REFERENCES

1. K.H. Jenkins and H.W. Owen, *J. Am. Water Works Assn., 55*(1):35, 1963.
2. K.H. Jenkins, *J. Am. Water Works Assn., 55*(12):1485, 1963.
3. K.H. Jenkins, *J. Am. Water Works Assn., 57*:1360, 1965.

4.1–6 CONVENTIONAL TREATMENT PROCESSES

Communities of 25,000 or More

Process	1958[1]		1960[2]		1962[3]		1964[4]	
	No. plants	Estimated population served	No. plants	Estimated population served	No. plants	Estimated population served	No. plants	Estimated population served
Filtration								
Rapid sand (gravity)	316	47,835,760	338	49,711,716	358	48,911,243	474	58,845,020
Rapid sand (pressure)	19	2,712,970	22	1,864,390	21	1,902,572	33	2,152,295
Slow sand	23	5,752,875	2	97,265	19	5,386,790	25	4,152,490
Anthracite	45	5,208,330	49	5,517,605	65	6,704,995	87	6,839,710
Diatomaceous earth	–	–	–	–	–	–	2	362,000
Aeration	154	12,242,585	157	12,782,050	185	14,721,204	233	15,250,335
Softening								
Lime soda	43	6,235,350	42	6,020,867	42	6,168,298	67	10,015,730
Lime	36	6,052,150	46	7,294,978	58	7,497,542	66	8,456,320
Cation exchange	12	1,295,260	16	1,529,700	15	1,333,659	22	3,260,620
Taste and odor control	226[a]	34,294,615	249[a]	35,518,613	285[a]	36,514,982[a]	348	43,985,135
Activated carbon	182	28,060,625	198	29,650,550	232	30,876,855		
Chlorine dioxide	37	5,586,130	39	5,643,374	45	5,303,549		
Sulfur dioxide	–	–	–	–	5	672,208		
Others and not stated	38	6,637,760	40	4,858,720	40	4,308,083		
Corrosion control	236[a]	32,924,240	285[a]	34,791,144	333[a]	36,614,370[a]	421	39,464,595
Phosphate compounds	95	9,502,270	112	10,790,574	138	11,508,938		
Alkali for pH adjustment	161	24,781,475	195	25,392,858	216	26,608,720		
Chlorine gas	11	792,960	12	929,680	19	977,586		
Hypochlorites	–	–	–	–	3	16,225		
Sodium silicate	4	622,500	5	687,000	5	850,964		

[a] Figures entered under "Taste and odor control" and "Corrosion control" are not additive, because more than one chemical may be used.

REFERENCES

1. K.H. Jenkins, *J. Am. Water Works Assn.*, *53*(1):31, 1961.
2. K.H. Jenkins and H.W. Owen, *J. Am. Water Works Assn.*, *55*(1):35, 1963.
3. K.H. Jenkins, *J. Am. Water Works Assn.*, *55*(12):1485, 1963.
4. K.H. Jenkins, *J. Am. Water Works Assn.*, *57*:1360, 1965.

4.1–7 DATA ON TYPICAL PUBLIC WATER SUPPLY SYSTEMS

System	A	B	C	D	E	F	G
Population served	1,200	4,000	4,000	7,000	17,000	35,000	113,000
Water consumption, mgd							
Avg	0.12	0.44	0.54	0.90	1.74	12.00	11.00
Max day	0.36	1.08	0.81	2.70	4.17	18.00	18.90
Peak hour	0.72	1.98	1.35	5.40	9.40	32.45	34.60
Peak max	2.00	1.83	1.67	2.00	2.26	1.80	1.83
Area served, mi^2	0.47	1.16	6.27	15.60	13.26	24.25	20.75
Replacement cost of fixed assets							
General service and fire protection	490,000	995,000	1,953,000	4,527,000	8,370,000	15,650,000	31,290,000
General service only	321,000	761,000	1,501,000	3,560,000	7,294,000	14,249,000	28,365,000
Incremental cost of fire protection	169,000	234,000	452,000	967,000	1,076,000	1,401,000	2,925,000
Incremental cost as a percent	34.5	23.5	23.2	21.4	12.9	9.0	9.3

Source: Reprinted, with publisher's permission, from *J. Am. Water Works Assn., 62*(7):407, © 1970, by the American Water Works Association, Inc., 2 Park Ave., New York, N.Y. 10016.

4.1–8 DESIGN CHARACTERISTICS OF WATER TREATMENT PROCESSES

	Design characteristic				
Treatment method	Loading rate, gpm/ft²	Site area per mgd, ft²	Head loss, ft	Proportion of wash water to design capacity, %	Cost[a] per mgd, $1,000
Filtration					
Rapid sand	3	375	6	1.5	15
Anthracite	3	375	6	1	16
Sand-anthracite	6	260	6	1.5	12.5
Diatomaceous earth	1.5	150	20	1	25
Microstraining	15	75	1.5	1	20
Sedimentation					
Conventional rate	0.85	1,050	1	0.5	12.5
High rate	1.25	630	1	0.5	8

[a] Estimated for 40 mgd capacity facility.

Source: Reprinted, with publisher's permission, from *J. Am. Water Works Assn., 56*(5):535, © 1964 by the American Water Works Association, Inc., 2 Park Ave., New York, N.Y. 10016.

4.1–9 CHARACTERISTICS OF RAW WATER AND EFFLUENTS OF VARIOUS WATER TREATMENT UNITS

	Water quality characteristics[a]				
Water sample	Suspended solids,[b] mg/l	Turbidity units	Chlorides mg/l	COD,[c] mg/l	Threshold odor number
Raw	330.0	380	16.7	30.1	13
Settled	17.0	75	16.6	12.0	2
Diatomite filtered	<0.1	2	16.5	8.6	2
Carbon filtered					
1st filter	<0.1	2	16.5	4.0	1
2nd filter	<0.1	2	16.5	2.85	0

[a] Averages of daily composite samples.
[b] Particulate matter as measured by membrane filter.
[c] Corrected for chloride concentration.

Source: Reprinted, with publisher's permission, from *J. Am. Water Works Assn., 55*(6):783, © 1963 by the American Water Works Association, Inc., 2 Park Ave., New York, N.Y. 10016.

4.1–10 RAW WATER QUALITY AND CHEMICAL COST DATA FOR PUBLIC WATER SUPPLY SYSTEMS

Location	Chemical costs, cents/ 1,000 gal	BOD, ppm	COD, ppm	Hardness, ppm	Color scale, units	DO deficit, %	TDS, ppm	Cl demand, ppm	Cl^-, ppm	Turbidity scale, units
Philadelphia, Pa.	1.73	2.9	10.5	97.2	26.1	18.7	218	4.2	9.4	65.4
Washington, D.C.	0.89	2.1	12	100	12.2	10.2	148	2.1	10	53.4
Buffalo, N.Y.	0.22	1.0	6.8	129	0	–4.0	205	0.6	21.1	16.5
Detroit, Mich.	0.31	0.5	3.9	99.5	0	5.8	127	0.6	7.2	18.3
Omaha, Nebr.	2	1.9	27	235	16.4	18.9	445	3.1	9.8	238
Duluth, Minn.	0.21	0.5	4.5	43.4	2.5	1	59	0.9	2	43
Minneapolis, Minn.	2.67	3.5	33	194	28	25.3	262	0.7	10	43
Pasco, Wash.	0.35	1.5		70.6	6	–7		1.1		4.8
Wenatchee, Wash.	0.18			65			89			
St. Louis, Mo.[a]	1.86	2.3	30	201	18.7	15.2	397	2.4	20.1	822
Kansas City, Mo.[a]	1.84	3	58.6	228	16.5	17.7	414	3.3	17.2	850
Cincinnati, Ohio	1.20	2	21	143	8.2	7	263	3.1	34.2	67
Yankton, S.D.[a]	3.46	1.4		222	46.6	0		2.7	16.5	46.1
New Orleans, La.[a]	1.91	0.7	16.6	148	15	11.8	236	1	27.4	239

[a] Treatment includes softening.

REFERENCES

1. G.K. Young, Jr. *et al., J. Am. Water Works Assn., 57*(3):293, 1965.
2. W. Gandenberger, *Gas u. Wasserfach, 97*:504, 1956.
3. *A Survey of Operational Data for Water Works in 1960,* Am. Water Works Assn., 1964.

4.1–11 SUMMARY OF QUALITY OF PUBLIC WATER SUPPLIES

100 Largest Cities in the U.S., 1962

	Raw water supplies[a]		Treated water supplies	
	Population served, (millions)	No. cities	Population served, (millions)	No. cities
Hardness (ppm)				
Less than 61	21	29	23	30
61–120	15	16	22	41
121–180	16	22	11	16
More than 180	8	27	3.7	13
Dissolved solids (ppm)				
Less than 100	21	27	21	27
101-250	23	38	28	48
251-500	11	29	8	22
More than 500	1.5	6	1	3
pH				
Less than 7.0	16	18	14	9
7.0-9.0	42	80	38	74
More than 9.0			7	17

[a] A few cities are not included because data are lacking.

Source: C.N. Durfor and E. Becker, *Geological Survey Water Supply Paper 1812,* 1962.

4.1–12 TYPICAL ANALYSIS FOR CALIFORNIA PUBLIC WATER SUPPLY SYSTEMS SURVEYED

Town	Sources[a]	TDS mg/l	Ca, mg/l	Mg, mg/l	Na, mg/l	HCO_3, mg/l	Cl, mg/l	NO_3, mg/l	SO_4, mg/l
1	1	84	4	1	4	34	1	1	1
2	3	96	14	3	6	73	2	2	1
3	2	98	17	4	7	55	7	2	20
4	1	104	8	4	0	59	2	2	1
5	2	119	5	5	17	32	26	9	3
6	7	164	15	5	36	130	3	1	14
7	1	215	10	9	30	124	10	4	8
8	2	265	52	10	27	202	15	1	46
9	3	332	42	15	48	223	47	1	17
10	2	371	71	13	37	227	31	6	67
11	2	379	59	12	46	220	35	2	69
12	1	400	75	15	40	240	37	1	68
13	3	452	31	35	69	278	49	14	63
14	1	472	110	25	11	267	38	96	48
15	3	640	88	37	43	246	44	41	156
16	3	672	77	38	55	162	79	102	133
17	1	766	120	24	93	293	116	20	181
18	3	814	101	44	75	221	134	73	175
19	3	832	108	40	71	201	35	5	365
20	2	848	119	28	100	210	114	7	299
21	5	896	106	32	121	170	60	12	411
22	3	964	112	48	127	302	160	29	247
23	4	994	136	35	147	405	142	100	150
24	3	998	78	59	134	205	88	18	408
25	2	1,001	140	53	80	266	45	12	438
26	1	1,063	144	54	93	257	53	16	470
27	4	1,150	191	65	60	315	132	13	402
28	4	1,180	81	16	295	184	424	5	147
29	3	2,303	319	129	130	282	145	20	1,110

[a] Number of sources of supply serving specific town.

Source: Reprinted with publisher's permission from *J. Am. Water Works Assn.*, *61*(11):575, © 1969, by the American Water Works Association Inc., 2 Park Ave., New York, N.Y. 10016.

4.1–13 CHEMICAL CHARACTERISTICS OF RAW AND FINISHED WATERS

Surface Waters of Denver Water Supply

Element or chemical	Date of collection	Moffat Treatment Plant[a] mg/l		Domestic taps, mg/l		North Marston Plant,[b] mg/l		South Marston Plant,[b] mg/l		Kassler Plant,[b] mg/l		Kassler Galleries,[c] mg/l		Domestic taps, mg/l	
		Raw	Finished	No. 1 Finished	No. 2 Finished	Raw	Finished	Raw	Finished	Raw	Finished	Raw	Finished	No. 3 Finished	No. 4 Finished
Silica	5/13/66	7.8	7.0	6.8	7.4	6.4	7.4	7.0	7.0	8.0	7.8	9.6	10	6.8	7.4
	6/10/66	7.0	7.0	6.5	6.5	6.0	6.5	7.0	7.0	8.0	6.5	9.5	9.5	6.0	7.5
	7/8/66	8.0	8.4	8.0	8.0	7.2	7.4	8.0	7.8	10.0	7.8	11	11	7.2	7.4
	8/5/66	6.4	6.4	7.0	7.2	5.8	5.6	7.4	6.6	9.2	7.0	10	10	6.2	6.6
	9/16/66	6.0	5.8	6.4	6.4	3.6	3.9	4.4	4.2	8.2	6.9	9.6	10	4.4	6.9
Calcium	5/13/66	8.6	13	13	13	30	29	29	30	26	32	34	34	30	30
	6/10/66	7.9	12	12	12	30	30	29	29	24	31	33	33	30	30
	7/8/66	8.3	11	12	12	31	31	29	29	32	34	32	32	29	28
	8/5/66	8.0	13	13	13	28	32	29	29	29	31	31	31	26	29
	9/16/66	7.0	12	11	11	25	26	26	26	24	25	29	28	26	26
Magnesium	5/13/66	1.8	1.8	1.9	1.9	9.2	9.0	8.8	8.8	7.9	9.6	10	9.8	8.1	9.0
	6/10/66	1.8	1.8	2.2	2.1	9.9	9.6	9.5	9.7	7.4	11	10	10	10	10
	7/8/66	1.6	1.7	1.6	1.6	10	10	10	10	11	11	10	11	11	11
	8/5/66	1.5	1.5	1.6	1.6	9.8	9.8	10	10	8.4	9.3	10	10	7.3	9.4
	9/16/66	1.6	1.6	1.5	1.5	8.0	8.2	8.1	8.5	5.6	5.8	8.6	8.8	8.3	6.3
Sodium	5/13/66	2.5	2.8	2.8	3.4	22	22	22	22	20	26	26	26	22	22
	6/10/66	2.4	2.8	3.1	3.0	26	26	24	24	22	27	28	28	26	27
	7/8/66	2.0	2.4	2.4	2.1	26	26	25	25	27	28	28	28	26	26
	8/5/66	1.5	2.1	2.6	3.3	25	26	27	27	20	23	25	26	18	24
	9/16/66	1.3	1.8	2.0	1.9	19	18	18	18	11	11	19	20	18	13
Potassium	5/13/66	1.2	1.2	1.2	1.2	2.1	2.1	2.1	2.0	2.0	2.1	2.1	2.2	2.0	2.0
	6/10/66	0.6	1.0	0.6	0.8	1.6	1.6	1.8	1.8	1.8	1.8	2.2	2.0	1.8	1.8
	7/8/66	1.1	1.0	1.0	0.9	2.0	2.2	1.8	1.8	2.0	1.8	2.0	1.9	1.9	2.0
	8/5/66	0.6	1.1	0.5	0.6	1.4	1.4	1.7	1.4	1.4	1.4	1.6	1.5	1.0	1.3
	9/16/66	0.6	0.7	0.7	0.5	1.9	1.8	1.8	1.9	1.7	1.8	1.7	1.9	1.9	1.7
Bicarbonate	5/13/66	31	31	30	30	96	93	90	92	83	103	108	107	90	94
	6/10/66	29	21	32	32	102	100	100	98	84	99	100	100	94	97
	7/8/66	28	28	28	29	103	100	102	100	108	110	104	102	94	94
	8/5/66	26	28	26	29	94	100	98	96	100	100	101	101	75	91
	9/16/66	24	25	24	25	80	77	78	77	68	65	90	90	78	65

[a] Frazer and Williams Fork Rivers via Moffat Tunnel.
[b] South Platte River and Dillon Reservoir (Snake River, Blue River, and Tunmile Creek).
[c] Underground seepage from South Platte River and Kassler filter beds.

4.1–13 CHEMICAL CHARACTERISTICS OF RAW AND FINISHED WATERS (*Continued*)

Element or chemical	Date of collection	Moffat Treatment Plant[a], mg/l		Domestic taps, mg/l		North Marston Plant,[b] mg/l		South Marston Plant,[b] mg/l		Kassler Plant,[b] mg/l		Kassler Galleries,[c] mg/l		Domestic taps, mg/l	
		Raw	Finished	No. 1 Finished	No. 2 Finished	Raw	Finished	Raw	Finished	Raw	Finished	Raw	Finished	No. 3 Finished	No. 4 Finished
Sulfate	5/13/66	8.0	18	18	20	37	39	38	40	34	44	44	46	40	40
	6/10/66	8.0	18	17	17	43	46	41	42	39	47	50	50	46	48
	7/8/66	8.0	14	16	15	43	45	40	41	47	49	48	49	47	49
	8/5/66	7.0	16	17	17	43	46	43	45	35	43	46	47	38	44
	9/16/66	6.0	15	13	14	39	40	39	43	35	39	43	48	42	41
Chloride	5/13/66	0.6	1.8	1.5	1.5	30	29	29	28	26	32	32	32	29	29
	6/10/66	0.6	2.6	1.8	1.7	31	31	29	30	21	33	33	33	32	33
	7/8/66	0.7	1.5	1.3	0.9	33	33	30	31	34	34	33	34	33	33
	8/5/66	0.6	1.3	1.3	1.7	29	30	30	31	23	26	28	29	21	28
	9/16/66	0.7	1.1	1.1	1.3	22	22	22	21	11	14	19	20	21	14
Fluoride	5/13/66	0.2	1.0	1.0	1.0	0.8	1.0	1.0	1.0	1.0	0.8	1.0	1.0	1.0	0.9
	6/10/66	0.2	0.9	0.9	0.8	0.8	0.8	0.8	0.8	0.8	0.8	0.9	0.9	0.9	0.8
	7/8/66	0.2	1.0	1.0	1.0	0.9	0.9	1.0	1.0	0.8	0.8	1.0	1.0	0.9	0.9
	8/5/66	0.2	1.0	1.0	1.0	0.9	1.6	1.0	1.0	1.0	0.9	1.0	1.0	1.0	0.9
	9/16/66	0.2	1.0	0.9	0.9	0.8	0.8	0.8	0.8	0.6	0.7	0.9	0.9	0.8	0.7
Nitrate	5/13/66	0.3	0.3	0.4	0.3	0.1	0.2	0.2	0.3	0.3	0.7	1.1	1.1	0.2	0.3
	6/10/66	0.1	0.3	0.4	0.4	0.2	0.2	0.2	0.4	0.7	0.6	0.7	1.1	0.3	1.5
	7/8/66	0.0	0.2	0.2	0.2	0.0	0.2	0.1	0.2	0.2	0.7	0.7	1.0	0.1	0.4
	8/5/66	0.4	0.3	0.4	0.4	0.2	0.3	0.4	0.5	0.4	0.6	0.9	1.0	0.4	1.0
	9/16/66	0.0	0.1	0.4	0.4	0.1	0.2	0.0	0.4	0.3	0.5	0.8	1.2	0.4	1.2
Dissolved solids (residue at 180°C)	5/13/66	54	72	70	72	191	188	193	193	176	212	226	224	197	184
	6/10/66	49	61	65	61	207	208	200	204	174	217	232	232	214	222
	7/8/66	50	64	62	61	192	191	212	180	207	200	212	206	193	197
	8/5/66	42	56	56	59	201	203	207	206	177	193	212	214	160	199
	9/16/66	36	52	50	51	167	166	176	146	123	134	172	176	165	141
Hardness as $CaCO_3$	5/13/66	29	40	41	41	114	110	109	112	98	120	127	126	109	112
	6/10/66	27	38	39	39	116	115	112	113	91	123	124	124	117	117
	7/8/66	27	35	37	37	119	119	114	114	126	131	122	126	118	116
	8/5/66	26	39	39	39	111	121	114	114	108	116	119	119	95	111
	9/16/66	24	37	34	34	96	99	99	101	83	87	109	107	99	91
Noncarbonate hardness as $CaCO_3$	5/13/66	4	15	16	16	35	34	35	36	30	36	38	39	35	35
	6/10/66	3	21	13	13	33	33	30	33	22	42	42	42	40	37
	7/8/66	4	12	14	13	35	37	31	32	37	41	37	42	41	39
	8/5/66	5	16	18	15	34	39	34	36	26	34	36	36	34	37
	9/16/66	4	16	14	13	30	36	35	38	28	34	35	33	35	38

a Frazer and Williams Fork Rivers via Moffat Tunnel.
b South Platte River and Dillon Reservoir (Snake River, Blue River, and Tunmile Creek).
c Underground seepage from South Platte River and Kassler filter beds.

Source: Reprinted, with publisher's permission, from *J. Am. Water Works Assn., 61*(2):60, © 1969, by the American Water Works Association, Inc., 2 Park Ave., New York, N.Y. 10016.

4.1–14 AMMONIA NITROGEN IN INTAKE WATER SUPPLY AND ACTIVATED CARBON AND CHLORINE APPLIED

Abnormal Odor Pollution Periods, 1950–1968

Year	Maximum ammonia nitrogen, ppm	Maximum activated carbon dosage, lb/mil gal	Maximum chlorine dosage, lb mil/gal
1950	0.276	1,158	84.0
1951	0.196	446	45.5
1952	0.248	590	45.1
1953	0.120	266	35.4
1954	0.148	356	20.8
1955	0.125	279	24.3
1956	0.180	415	24.4
1957	0.248	325	26.8
1958	0.680	503	19.4
1959	0.290	712	24.5
1960	0.272	324	16.0
1961	0.358	388	20.3
1962	0.590	370	27.2
1963	0.230	680	42.8
1964	0.496	745	56.1
1965	0.198	320	53.1
1966	0.260	385	53.9
1967	0.260	557	70.0
1968	0.250	411	51.1

Source: Reprinted with publisher's permission from *J. Am. Water Works Assn., 61*(11):603, © 1969, by the American Water Works Association Inc., 2 Park Ave., New York, N.Y. 10016.

4.1–15 ACTIVATED CARBON APPLIED FOR REMOVAL OF ODORS

Lake Water, 1950–1968

Year	Activated carbon applied, lb	Average carbon dosage, lb/mil gal	Maximum hourly carbon dosage during odor period, lb/mil gal
1950	2,874,905	26	1,158
1951	3,545,552	31	446
1952	3,203,426	26	590
1953	1,775,063	14	266
1954	2,011,359	16	356
1955	2,057,781	17	279
1956	1,981,108	16	415
1957	3,032,729	24	325
1958	2,762,540	22	503
1959	3,035,509	23	712
1960	2,727,005	21	324
1961	2,632,923	21	388
1962	2,865,541	22	370
1963	3,194,443	23	680
1964	3,773,665	27	745
1965	3,094,606	23	320
1966	4,678,661	33	385
1967	4,455,273	32	557
1968	4,879,309	33	411

Source: Reprinted with publisher's permission from *J. Am. Water Works Assn., 61*(11):603, © 1969, by the American Water Works Assocation Inc., 2 Park Ave., New York, N.Y. 10016.

4.1–16 DESIGN FACTORS FOR WATER TREATMENT PLANTS

					Sedimentation basin or clarifier								
Plant no.[f]	Design capability (Qd), mgd	Year of last major addition to plant	Mixing basin detention, min	Floccu-lation basin detention, min	Detention, min	Surface loading, gpm/ft²	Average transport rate, ft/min	Clarifier loading, gpm/ft²	Basins housed or covered[e]	Filters rate, gpm/ft²	Filters housed or covered[e]	Design manpower, man hr/mg	Days 32° and below, days/year
1	6.00	1959	1.3	30	391	0.651	0.102		o	1.70	0	3.1	11
2	9.00	1953	5.0	30	240	0.313	0.430		o	1.65	½o	9.8	17
3	8.00	1950	0.6	40	283	0.333	0.588		o	1.85	½o	6.6	48
4	6.00	1952	–	65	288	0.366	0.327		o	2.00	h	10.0	33
5	0.500	1953	MI[c]	51	305	0.329	0.199		o	1.93	h	48.0	81
6	0.500	1951	MI[c]	52	346	0.336	0.221		o	1.93	h	48.0	63
7	0.500	1950	10.0	43	315	0.273	0.144		o	1.72	h	48.0	63
8	0.500	1939	with floc	20	240	0.345	0.184		o	1.93	h	48.0	80
9	6.00	1957	–	57	447	0.225	0.278		o		h	4.0	37
10	0.300	1962	with floc	18	220	0.284	0.251		o	1.64	h	15.0	85
12	12.00	1958	–	17[b]	206[b]	0.290[d]	0.580		o	3.31	h	2.53	94
					81[d]	d							
13	4.00	1946	MI[c]	10	212	0.495	1.16		o	1.96	h	7.5	100
14	0.432	1947	b	b	67[b]	b		1.59	h	1.85	h	23.1	148
15	12.00	1960	b	b	102[b]	b	0.161[b]	1.10	h	1.87	h	5.1	144
17	12.00	1961	with floc[c]	23	202	0.778	0.448		c	1.54	h	9.6	144
18	0.300	1948	b	b	47[b]	b		2.19	h	0.075	c	12.0	144
19	8.00	1946	1.6	33	125	0.702	0.925		o	1.93	½c	7.9	144
							0.127						
21	7.50	1955	0.7	51	335	0.321			o	1.94	o	6.8	11
22	6.81[a]	1955	b	b	102[b]	b	0.800[b]	1.18	o	2.00[a]	c	5.0	65
23	0.576	1938	–	7	327	0.177	0.566		o	1.94	h	42.0	65

[a] Based on an arbitrary design filter rate as shown.
[b] Suspended solids contact clarifier.
[c] MI signifies mixing basins detention could not be computed because mixing is done in the line. "With floc" indicates mixing basin not separated from flocculation basin and flocculation detention includes both:
[d] This plant operates sedimentation basins and a clarifier in parallel. Data applicable to the clarifier are shown in the right hand column. Certain dimensions of the clarifier were not available.
[e] Housed, covered, open.
[f] See 4.1–17 for description of plants.

4.1–16 DESIGN FACTORS FOR WATER TREATMENT PLANTS *(Continued)*

Plant no.[f]	Design capability (Qd), mgd	Year of last major addition to plant	Mixing basin detention, min	Flocculation basin detention, min	Sedimentation basin or clarifier: Detention, min	Surface loading, gpm/ft²	Average transport rate, ft/min	Clarifier loading, gpm/ft²	Basins housed or covered[e]	Filters rate, gpm/ft²	Filters housed or covered[e]	Design manpower, man hr/mg	Days 32° and below, days/year
24	0.504	1930	Ml	22	249	0.326			o	1.82	h	47.0	83
25	0.504	1937	Ml	20	283	0.298	0.340		o	1.82	h	48.0	73
26	0.500	1965	b	b	229[b]	b	0.092[b]	0.491	o	1.62	h	28.0	70
27	0.486	1948	1.1	142	6960	0.011	0.033		o	3.00[a]	h	18.0	70
28	1.00	1964	b	b	280[b]	b	0.161[b]	0.354	o	1.49	o	26.6	66
							0.427						
29	8.00	1958	2.4	39	261	0.457			o	1.95	o	7.9	59
30	6.81[a]	1955	0.18		102		0.80		o	2.00[a]	c	5.0	65

[a] Based on an arbitrary design filter rate as shown.
[b] Suspended solids contact clarifier.
[e] Housed, covered, open.
[f] See 4.1–17 for description of plants.

Source: Reprinted, with publisher's permission, from *J. Am. Water Works Assn., 59*(3):290, © 1967 by the American Water Works Association, Inc., 2 Park Ave., New York, N.Y. 10016.

4.1–17 WATER TREATMENT PLANT DESCRIPTION

Table A

	Plant no.													
	1	2	3	4	5	6	7	8	9	10	12	13	14	15
Source														
Lake or reservoir			x		x								x	
River	x	x		x	x	x	x		x	x	x	x		x
Wells														
Reservoir (>20 day detention at $\bar{Q}$)								x						
Pumps														
Low-lift	x	x		x	x	x	x		x			x	x	x
High-lift	x	x	x	x	x	x	x	x	x	x	x	x	x	x
Chemical dosage for coagulation														
Alum	x	x	x	x	x	x	x	x	x	x	x	x	x	x
NaOH											x	x		
Iron salts		x												
Lime		x	x	x		x	x	x	x			x		
Soda ash					x					x				x
Activated silica														
Other coagulant														x
Disinfection														
Chlorine gas	x	xx	x	x	x	xx	x	xx	x	xx	x	x	x	xx
Chlorine dioxide														
Mixing device or tank														
Baffle mix								x						
Turbulence of influent														
In raw water line					x	x						x	x	
Slow mechanical mix														
Rapid mechanical mix	x	x	x	x			x		x	x	x			x
Filters														
Anthrafilt														
Gravity (slow)														
Covered														
Open	x	½x	½x	x										
Gravity (rapid)	x	x	x	x	x	x	x	x	x	x	x	x	x	x
Sand	x	x	x	x	x	x	x	x	x	x	x	x	x	x
Backwash														
Automatic										x				
From high-line												x		x
With pump (direct or via storage)	x	x	x	x	x	x	x	x	x		x		x	
Fluoride adjustment														
Hydrofluosilicic acid										x				
Sodium silicofluoride	x		x									x		
Sodium fluoride				x					x					
Sedimentation			x	x	x	x	x	x	x	x	x	x		
Basins, baffle (other than inlet or outlet)		x												
Covered basins (other than housed)													x	x
Mechanical sludge removal	x													
Plain		x												
Upflow cylindrical tanks (MTPS SV) – suspended solids contact clarifiers											x		x	x

4.1–17 WATER TREATMENT PLANT DESCRIPTION *(Continued)*

Table A

	Plant no.													
	1	2	3	4	5	6	7	8	9	10	12	13	14	15
Chemical taste and odor control														
Activated carbon	x	x	x	x									x	
Chlorine dioxide									x					
$KMnO_4$	x										x			
Flocculation basin														
Baffled		x		x				x		x	x			
Slow mechanical	x	x	x		x	x	x		x		x	x		
Recarbonation														

Table B

	Plant no.												
	17	18	19	21	22	23	24	25	26	27	28	29	30
Source													
Lake or reservoir		x		x	x	x	x	x				x	x
River	x		x							x	x		
Wells													
Reservoir (>20 day detention at Q)									x				
Pumps													
Low-lift	x	x	x	x	x	x	x	x		x	x	x	
High-Lift	x	x	x	x	x	x	x	x	x	x	x	x	x
Chemical dosage for coagulation													
Alum	x	x	x		x	x	x	x	x	x	x	x	x
NaOH											x		
Iron salts				x									
Lime				x	x	x		x	x		x	x	x
Soda ash							x						
Activated silica					x						x		x
Other coagulant			x										
Disinfection													
Chlorine gas	x	x	xx	x	x	x	x	x	x	x	x	x	x
Chlorine dioxide													
Mixing device or tank													
Baffle mix													
Turbulence of influent			x		x	x			x				x
In raw water line						x	x	x		x	x		
Slow mechanical mix													
Rapid mechanical mix				x								x	
Filters													
Anthrafilt							x						
Gravity (slow)		x											
Covered		x	½x		x								x
Open				x							x	x	
Gravity (rapid)	x		x	x	x	x	x	x	x	x	x	x	x
Sand	x	x	x	x	x	x		x	x	x	x	x	x
Backwash													
Automatic													
From high-line					x	x	x	x		x			x
With pump (direct or via storage)	x		x	x					x		x	x	

4.1–17 WATER TREATMENT PLANT DESCRIPTION *(Continued)*

Table B

	Plant no. 17	18	19	21	22	23	24	25	26	27	28	29	30
Fluoride adjustment													
Hydrofluosilicic acid													
Sodium silicofluoride			x		x								x
Sodium fluoride													
Sedimentation				x	x	x		x		x			x
Basins, baffle (other than inlet or outlet)			x				x						
Covered basins (other than housed)	x												
Mechanical sludge removal												x	
Plain													
Upflow cylindrical tanks (MTPS SV) – suspended solids contact clarifiers		x			x				x		x		
Chemical taste and odor control													
Activated carbon			x			x		x			x	x	
Chlorine dioxide	x		x										
$KMnO_4$													
Flocculation basin													
Baffled			x			x			x	x			
Slow mechanical	x			x			x	x				x	
Recarbonation				x							x		

Note: The above table summarizes the characteristics of the water plants listed in 4.1–16 to 4.1–20.

Source: Reprinted, with publisher's permission, from *J. Am. Water Works Assn.*, *59*(3):290, © 1967 by the American Water Works Association, Inc., 2 Park Ave., New York, N.Y. 10016.

4.1–18 PRODUCTION FACTORS FOR WATER TREATMENT PLANTS

					Average for Period												
					Filters			Average raw water quality									
Plant no.[g]	Average production, (Q) mgd	Utilization factor U, fraction	Daily average operation time, hr	Sedimentation detention, min	Utilization, average, hr/day	Rate, gpm/ft²	Run (backwashed), Kgal/ft²	Turbidity, ppm	Temperature, °F	Total alkalinity, ppm	Phenolphthalein alkalinity, ppm	Free CO_2, ppm	pH	Total hardness (as $CaCO_3$), ppm	Noncarbonate hardness (as $CaCO_3$), ppm	Total dissolved solids, ppm	Color
1	1.87	0.31	24	1,255	9.9	1.29	5.0	68	63	129	0	–	7.8	180	–	–	–
2	7.09	0.79	24	305	24	1.30	3.8	250	62	85	–	2.8	7.7	119	–	160[d]	0
3	5.82	0.73	24	658	24	1.34	7.0	5[d]	62	16	8[b]	12	7.3	12	–	–	–
4	4.71	0.79	24	367	24	1.60	8.1	72	64	20	–	5	6.8	32	–	44[f]	14[f]
5	0.166	0.33	8.1	309	8.1	1.89	6.7	–	–	15	–	–	6.9	–	–		
6	0.163	0.33	8.7	382	8.7	1.75	3.8	–	–	133	–	0	8.1	149[b]	9[b]	152[b]	7[b]
7	0.342	0.68	17	327	17	1.65	7.6	85	60	142	0[b]	1.8	7.6	159	–	–	25[b]
8	0.157	0.31	7.7	240	7.7	1.89	4.4	10[d]	53	9		5	6.8	10	0[b]	134[b]	5
9	1.57	0.26	15	1,074	15			27[f]	64[e]	17[f]	0[f]	–	6.9[f]	12.5[f]	0[f]	43[f]	10[d]
10	0.078	0.26	24	846	6.2	1.64		5.3	63	24[b]	–	–	7.1				38
12	11.3	0.94	24	219	24	3.12	1.5	44	58	46	–	–	7.3	63[b]	–	135[b]	33
13	1.87	0.47	16	302	16	1.38	3.9	25	58	101	–	–	7.9	122	–	–	
14	0.198	0.46	24	146	11	1.85	9.5	10	49	97[e]	–	1[d]	7.8	113[e]	–	130[e]	9.8[a]
15	4.91	0.41	23	239	23	0.81	4.7	5	54	102	–	2.2	7.8	121	–	190[d]	–
17	9.25	0.77	24	214	20	1.46	6.5	13	49[e]	156	–	–	8.2	91	0	190[d]	–
18	0.169	0.56	24	83	24	0.042	1.8	26	49[e]		–	–	–	–	–	190[d]	–
19	5.66	0.71	24	176	15	2.30	3.8	15	49	156[e]	–	–	7.9	91[e]	–	190[d]	–
21	8.87	1.18	24	283	24	2.29	9.0	7	68	133	–	–	7.6	152	–	–	–
22	2.11[a]	0.31	24	331	21	1.36	2.0	10[c]	62	161	3	–	8.0	181	14[b]	216[b]	10[b]
23	0.129	0.22	24	1,460	17	0.61	3.2	5.8	–		0	–	7.8	118			–

[a] Does not include that portion of the supply which comes from well water, except chlorine unit consumption based on total raw water. This plant has been operated in two different manners with the changeover in March 1965. The two methods of operation are described as Plant No. 22 prior to 3/65 and No. 30 after 3/65.
[b] Production estimated from average production in first month during which distribution system was metered.
[c] From two to four spot analyses in separate months during the data period. (Similar cases from a single spot analysis identified by final note to this table.)
[d] Average data from sources other than the interview campaign. Period of average may not cover the data period and sampling location may not be precisely at intake.
[e] Data from a nearby plant using the same source.
[f] Data from a single spot analysis.
[g] See Table 4.1–17 for description of plants.

4.1–18 PRODUCTION FACTORS FOR WATER TREATMENT PLANTS *(Continued)*

Average for Period

					Filters			Average raw water quality									
Plant no.[g]	Average production, (Q) mgd	Utilization factor U, fraction	Daily average operation time, hr	Sedimentation detention, min	Utilization, average, hr/day	Rate, gpm/ft²	Run (backwashed), Kgal/ft²	Turbidity, ppm	Temperature, °F	Total alkalinity, ppm	Phenolphthalein alkalinity, ppm	Free CO_2, ppm	pH	Total hardness (as $CaCO_3$), ppm	Noncarbonate hardness (as $CaCO_3$), ppm	Total dissolved solids, ppm	Color
24	0.328	0.65	15	239	15	1.90	6.0	155[c]	–	78[f]	0[f]	–	7.6[f]	84[f]	6[f]	130[f]	350[f]
25	0.127	0.25	11.5	538	11.5	0.82	0.7	23	–	39	0[f]	–	–	68	20[f]	145[f]	10[f]
26	0.174	0.35	18	494	7.4	1.83	0.8	7	80	82	–	7.9	7.5	111	84	143	15
27	0.440[b]	0.89	15	4,802	15	4.3	3.9	6[f]		104	6	–	7.6	146	42	230	5
28	0.297	0.30	8	314	8	1.36	4.3	50	75	26	0	–	6.1	25	0[f]	40[f]	149
29	5.29	0.66	24	394	24	1.29	5.4	36	67	69	–	9	7.2	84	–	–	–
30	3.51	0.52	24	198	21	1.36	4.0	10[b]	70	166	1	–	7.9	185	14[b]	216[b]	10[b]

[b] Production estimated from average production in first month during which distribution system was metered.
[c] From two to four spot analyses in separate months during the data period. (Similar cases from a single spot analysis identified by final note to this table.)
[f] Data from a single spot analysis.
[g] See Table 4.1–17 for description of plants.

Source: Reprinted, with publisher's permission, from *J. Am. Water Works Assn., 59*(3):290, © 1967, by the American Water Works Association, Inc., 2 Park Ave., New York, N.Y. 10016.

4.1–19 UNIT CONSUMPTIONS FOR WATER TREATMENT PLANTS

Table A

	Manpower		Energy					Water			
Plant no.[i]	Total man-hr/mgal	Cleaning basins, man-hr/mgal	Electrical total, kWh Kgal	Raw water, kWh Kgal	Finished water, kWh Kgal	Oil, gal/mgal	Gas, Kcf/mgal	Back-wash, kWh/mgal[j]	Back-wash, %	Sludge drain, clarification, %	Sedimentation basin, %
1	6.7	–	1.73[a]	–	–	–	–	–	–	–	0
2	12.7	0.159	2.46	–	–	–	–	3.8	3.4	–	0.30
3	9.0	0.049	0.33[c]	–	–	–	5.47[c]	–	1.9	–	0.76
4	12.7	0.020	2.04	1.26	0.78	–	–	9.1	1.1	–	0.16
5	49	0.132	2.33	0.94	1.39	–	–	–	1.5	–	0.41
6	49	0.504	2.5[e]	–	–	–	–	–	–	–	0.70
7	70	–	3.25	1.29	1.96	–	–	2.6	1.9	–	–
8	76	–	2.15	–	–	–	–	2.4	1.2	–	–
9	7.3	0.035	3.27	–	–	–	–	–	–	–	0.38
10	58	0.211	2.39	–	–	–	–	0[f]	–	–	0.60
12	2.7	0.014	1.42	–	–	–	–	–	2.8	–	0.94
13	16	0.211	1.32	–	–	–	–	15	2.3	–	0.18
14	51	–	2.37	–	–	119	–	–	1.3	–	–
15	12.5	–	2.32	0.41	1.91	23.1	0.005	39	3.8	–	–
17	12.5	0.004	1.41	0.70	0.71	11	–	7.0	1.2	–	0.06
18	21	–	1.60	–	–	27	–	–	–	–	–
19	11	0.036	0.82	–	–	18	–	4.6	2.0	–	0.09
21	5.8	0.005	3.03	–	–	–	1.41[g]	4.7	2.1	–	0.14
22	16	–	2.89	–	–	–	–	45	5.2	–	–
23	186	0.936	–	–	–	–	–	19	4.4	–	1.31
24	46	0.600	–	–	–	–	–	4.6	1.3	–	1.01
25	94	0.650	–	–	–	–	–	14	2.3	–	0.75
26	33	–	3.60	–	–	–	–	54	9.2	–	–
27	13	0.300	1.22	–	–	–	–	16	2.7	–	0.86

[a] Plant is operated in parallel with a softening plant. Unit consumption given is for total operation.
[c] Plant uses gas engine regularly and electric energy for peaking
[e] Based on one month only.
[f] Automatic backwash and gravity raw water supply.
[g] This gas used for recarbonation, not heating.
[i] See Table 4.1–17 for plant description.
[j] Refers to raw water.

4.1–19 UNIT CONSUMPTIONS FOR WATER TREATMENT PLANTS *(Continued)*

Plant no.[i]	Manpower: Total man-hr/mg	Manpower: Cleaning basins, man-hr/mg	Energy: Electrical total, kWh/Kgal	Energy: Raw water, kWh/Kgal	Energy: Finished water, kWh/Kgal	Energy: Oil, gal/mg	Energy: Gas, Kcf/mg	Water: Backwash, kWh/mgRW	Water: Backwash, %	Water: Sludge drain, clarification, %	Water: Sedimentation basin, %
28	34	–	2.78	1.11	1.68	–	2.3	–	3.4	1.9	–
29	12	0.100	1.72	0.82	0.45	–	0.4	3.8	1.9	–	0.29
30	9.8	0.032	1.65	–	–	–	–	19	2.2	–	0.17

Table B

Chemical – ib/mg

Plant no.[i]	Coagulant: Alum	Coagulant: Iron sulfate	Coagulant auxiliary: Clay	Coagulant auxiliary: $CaCO_2$	Coagulant auxiliary: Na_2SiO_3	Coagulant auxiliary: $(NH_4)_2SO_4$	Alkali: Hydrated lime equiv.	Alkali: Type[d]	Disinfection Cl	Taste and Odor: C	Taste and Odor: $KMnO_4$	Taste and Odor: $NaClO_2$
1	157							None	39[a,b]	18[a]	2.2[a]	
2	69	270					267	h	43			
3	239						287	q	34			
4	116						109	h	22	2.5		
5	184						204	s	10			
6	514						267	h	30			
7	429						99	h	36			
8	100						19	h	22			
9	80						71	h	29			1.3
10	585						471	s	31			
12	339						15	c	14		1.1	
13	220						68	h	9.4	4.4		

a Plant is operated in parallel with a softening plant. Unit consumption given is for total operation.
b Chlorine consumption for both plants adjusted by amount required to oxidize ferrous sulphate used in softening plant.
c Plant uses gas engine regularly and electric energy for peaking.
d Symbols stand for hydrated lime, cuicklime, soda ash, and caustic soda.
h Plant sometimes is operated as a softening plant. Unit consumption applies only to the months in which softening was not carried on.
i See Table 4.1–17 for plant description.

4.1–19 UNIT CONSUMPTIONS FOR WATER TREATMENT PLANTS *(Continued)*

Table B

Chemical – lb/mg

Plant no.[i]	Coagulant		Coagulant auxiliary				Alkali		Disinfection	Taste and Odor		
	Alum	Iron sulfate	Clay	$CaCO_2$	Na_2SiO_3	$(NH_4)_2SO_4$	Hydrated lime equiv.	Type[d]	Cl	C	$KMnO_4$	$NaClO_2$
14	0.8							None	13	0.4		
15	115			119				None	30			0.07
17	31							None	19			
18	153							None	8			
19	77							None	12	1.4		
21	75						31[h]		36			
22	235				88		14	h	40			
23	357						107	s	41	19		
24	486						335	h	15			
25	761						428	h	40			
26	345						150	h	55			
27	75							None	22			
28	361				13	2.3	241	h	37			
29	200						284	h	69	31	3.1	
30	146				64			h	22			

[d] Symbols stand for hydrated lime, quick lime, soda ash, and caustic soda.
[h] Plant sometimes is operated as a softening plant. Unit consumption applies only to the months in which softening was not carried on.
[i] See Table 4.1–17 for plant description.

Source: Reprinted, with publisher's permission, from *J. Am. Water Works Assn., 59*(3):290, © 1967 by the American Water Works Association, Inc., 2 Park Ave., New York, N.Y. 10016.

4.1–20 AVERAGE UNIT COSTS FOR PERIOD

¢/Kgal

Plant no.	Chemicals, excluding fluoride	Energy			Heating, ¢/Kgal of Qd	Manpower	Manpower, ¢/Kgal of Qd	Maintenance and repair	Miscellaneous	Miscellaneous, ¢/Kgal of Qd	Total operating costs	Capital cost (7.787%), actual 1964 investment	Capital costs, ¢/Kgal of Qd	Total cost at actual U	Total cost at U = 1.0
		Total	Pumping	Heating											
1	0.88	1.59	1.59	–	–	1.41	0.44	0.10	0.10	0.03	4.08	10.4	3.23	14.48	6.23
2	1.27	1.89	1.89	–	–	2.0	1.58	0.17	0.33	0.26	5.66	–	–		
3	1.04	0.74	0.74	–	–	1.44	1.05	0.47	0.10	0.07	3.79	4.70	3.42	8.49	6.72
4	0.65	2.04	2.04	–	–	7.74	2.16	0.08	0.25	0.20	5.76	9.36	7.34	15.12	12.46
5	1.30	3.26	3.26	–	–	8.5	8.5	0.88	0.33	0.11	14.27	–	–		
6	2.30	3.50	3.50	–	–	6.1	6.0	0.88	0.33	0.11	13.11	20.7	6.87	33.81	19.28
7	2.00	3.25	3.25	–	–	11.00	7.80	0.61	2.42	1.65	19.28	15.1	10.4	34.38	25.60
8	0.80	4.30	4.30	–	–	7.60	7.40	1.3	0.18	0.06	14.18	8.31	2.65	22.49	15.94
9	0.53	3.90	3.90	–	–	1.5	0.91	0.57	0.12	0.03	6.62	–	–		
10	4.40	4.9	4.8	0.15	0.04	7.6	1.98	1.14	0.42	0.11	18.46	53.6	13.9	72.06	25.81
12	1.78	1.45	1.42	0.03	0.03	0.95	0.89	0.65	0.12	0.11	4.95	1.30	1.27	6.25	6.12
13	0.75	1.81	1.72	0.09	0.04	4.62	2.20	0.50	0.23	0.11	7.91	4.72	2.94	12.63	8.10
14	0.23	6.41	4.63	1.78	0.82	12.6	5.80	4.31	0.24	0.11	23.79	25.5	11.7	49.29	26.22
15	0.61	2.01	1.79	0.22	0.09	4.30	1.76	0.05	0.27	0.11	7.24	9.40	3.85	16.64	8.24
17	0.24	1.08	0.97	0.11	0.08	3.69	2.84	0.04	0.03	0.02	5.08	–	–		
18	0.81	3.8	3.2	0.58	0.32	5.90	3.30	2.1	0.16	0.09	12.77	11.40	6.42	24.17	15.71
19	0.33	0.93	0.75	0.18	0.13	5.90	2.16	0.70	0.19	0.13	6.05	4.21	2.98	10.26	7.07
21	0.44	2.68	2.68	–	–	1.18	1.39	0.44	0.10	0.12	4.84	4.01	4.75	8.85	9.86
22	1.22	2.02	2.02	–	–	2.85	0.88	0.36	0.35	0.11	6.80	8.60	2.66	15.40	7.09
23	2.20	–	–	–	–	33.0	7.26	–	–	–	–	–	–	–	–
24	3.50	–	–	0.01	0.006	6.7	6.7	0.22	0.03	0.02	–	7.15	4.65	–	–
25	–	–	–	–	–	16.7	7.38	0.67	–	–	–	24.5	6.18	–	–
26	2.70	4.30	4.30	–	–	5.71	4.82	0.28	0.14	0.05	13.13	17.4	6.05	30.53	18.09
27	0.60	1.46	1.46	–	–	2.25	3.08	0	0.18	0.16	4.49	8.87	8.77	13.36	14.07
28	1.77	2.96	2.89	0.07	0.02	5.65	4.80	0	0.37	0.11	10.75	17.8	5.29	28.55	14.80
29	1.23	1.63	1.61	0.02	0.01	2.71	1.79	0.20	0.16	0.11	5.93	8.67	4.41	12.60	9.32
30	0.69	1.15	1.15	–	–	1.72	0.89	0.28	0.21	0.11	4.01	5.17	2.66	9.19	5.67

Source: Reprinted, with publisher's permission, from *J. Am. Water Works Assn.*, *59*(3):290, © 1967, by the American Water Works Association, Inc., 2 Park Ave., New York, N.Y. 10016.

4.1–21 AVERAGE SERVICE LIVES OF MASS PLANT ITEMS

Several Water Utilities in Northeastern U. S.

Property classification	Avg service life, yrs		
	Range	Avg	Bulletin F[1]
Driven pipe wells		35.0	100
Electric power pumping equipment	30–45	36.7	50
Gasoline pumping equipment	35–45	40.0	25
Chlorinators		25.0	
Distribution mains			
4 in. and smaller	50–75	56.0	50
6 in. and larger	135–150	141.3	65–100
All sizes	130–150	137.5	
Services	50–60	57.5	40
Meters	40–65	54.0	30
Hydrants	45–90	63.0	50
Office equipment	15–35	24.0	19
Transportation equipment	5–7	6.0	10
Shop equipment	20–45	32.5	25
Tools and work equipment	15–45	27.0	20
Laboratory equipment	20–30	25.0	25

Source: Reprinted, with publisher's permission, from *J. Am. Water Works Assn.*, *56*(7):881, © 1964, by the American Water Works Association, Inc., 2 Park Ave., New York, N.Y. 10016.

REFERENCE

1. *Income Tax Depreciation and Obsolescence, Estimated Useful Lives and Depreciation Rates,* U.S. Treasury Department, Bureau of Internal Revenue, 1942.

4.1–22 PROCESS ENERGIES IN WATER TREATMENT UPSETTING PULSES

Description	Energy, ft lb/mil gal
Temperature rise, 2°C	2×10^4
Turbidity rise from 10–1,000 JTU	3×10^4
Salinity rise from 0–500 mg/l	1×10^4

Source: Reprinted, with publisher's permission, from *J. Am. Water Works Assn.*, *64*(6):382, © 1972, by the American Water Works Association, Inc., 2 Park Ave., New York, N.Y. 10016.

4.1–23 PROCESS ENERGIES IN WATER TREATMENT

Steady-state Sedimentation

Process parameter	Energy, ft lb/mil gal
Weight of floc	
Formed in clear water	5 x 10^2
Formed with 80 mg/l kaolin	5 x 10^3
Conventional basin inlet	1 x 10^5
Drag in conventional basin	10
Drag in tube settlers	2 x 10^3
Slurry recirculation reactors	
Weight of slurry	5 x 10^4
Mixing power input	5 x 10^6

Source: Reprinted, with publisher's permission, from *J. Am. Water Works Assn.*, *64*(6):382, © 1972, by the American Water Works Association, Inc., 2 Park Ave., New York, N.Y. 10016.

4.1–24 CHEMICALS USED FOR COAGULATION OR SOFTENING IN PLANTS SERVING COMMUNITIES OF 25,000 OR MORE

	1958[1]		1960[2]		1962[3]		1964[4]	
Chemical	**Number of facilities**	**Estimated population served**	**Number of facilities**	**Estimated population served**	**Number of plants**	**Estimated population served**	**Number of plants**	**Estimated population served**
Alum	316	46,630,355	332	48,502,502	364	49,776,177	460	55,757,215
Lime	262	37,472,950	273	37,617,316	315	39,003,857	412	45,578,520
Soda ash	68	9,247,890	69	9,191,961	82	10,508,615	87	11,207,455
Iron salts	42	10,975,900	49	10,607,605	62	11,112,542	70	11,282,820
Activated silica	28	6,316,610	29	6,343,247	41	7,131,463	46	8,635,285
Others	30	3,346,510	39	4,042,049	40	4,089,057	42	3,632,360
Total[a]	366	52,055,265	389	54,376,252	441	56,612,473	558	65,089,070

[a] Columns are not additive because more than one chemical may be used.

REFERENCES

1. K.H. Jenkins, *J. Am. Water Works Assn.*, *53*(1):31, 1961.
2. K.H. Jenkins and H.W. Owen, *J. Am. Water Works Assn.*, *55*(1):35, 1963.
3. K.H. Jenkins, *J. Am. Water Works Assn.*, *55*(12):1485, 1963.
4. K.H. Jenkins, *J. Am. Water Works Assn.*, *57*(11):1360, 1965.

4.1–25 PHASES OF TREATMENT IN WHICH VARIOUS CHEMICALS ARE USED

Chemical	Coagulant or aid	Alkalinity and pH adjustment	Disinfection	Dechlorination	Mineral oxidation	Taste and odor control	Algae control	Corrosion control	Softening	Fluoridation[a]
Activated carbon				x		x	x			
Aluminum sulfate	x									
Ammonia, anhydrous			x	x						
Ammonium hydroxide			x	x						
Ammonium sulfate	x		x							
Bentonite	x									
Calcium carbonate	x	x								
Calcium hydroxide		x						x	x	
Calcium oxide		x						x	x	
Carbon dioxide		x							x	
Chlorine			x		x	x	x			
Chlorine dioxide			x		x	x	x			
Copper sulfate							x			
Ferric chloride	x									
Ferric sulfate	x									
Ferrous sulfate	x									
Fluosilicic acid										x
Hydrochloric acid		x								
Hypochlorite (Ca)			x							
Hypochlorite (Na)			x							
Ozone			x		x	x				
Potassium permanganate						x				
Sodium aluminate	x									
Sodium bisulfite[b]				x						
Sodium carbonate		x						x	x	
Sodium chloride									x	
Sodium chlorite			x			x				
Sodium fluoride										x
Sodium hexametaphosphate[b]								x	x	
Sodium hydroxide[c]		x						x	x	
Sodium silicate	x							x		
Sodium silicofluoride										x
Sodium sulfite[b]				x						
Sodium tripolyphosphate								x		
Sulfuric acid		x							x	
Sulfur dioxide[c]				x						

[a] Chemicals utilized in processes for removal of fluoride include activated alumina, calcium hydroxide, magnesium oxide and sodium bisulfate.
[b] Also used for treatment of boiler water.
[c] Also used for cleaning filter sand.

Source: E.S. Hopkins and E.L. Bean, *Water Purification Control,* 4th ed., Williams & Wilkins Co., Baltimore, 1966, p. 310.

4.1–26 PROPERTIES OF CHEMICALS MOST EXTENSIVELY USED IN WATER TREATMENT AND SEWAGE PLANTS

Chemical	Formula	Trade name	Use	Available forms and appearances	Density, lb/ft^2 and commercial strength	Solubility in water, g/100 cc
Activated carbon	C	–	Adsorbent, taste and odor control, removal of trace organics	Black powder or granules (4–6 mesh), highly porous	15–20, 95% C	0
Aluminum sulfate	$Al_2(SO_4)_3 \cdot 14\ H_2O$	Filter alum	Coagulation, sludge conditioning	Light tan to gray-green lumps, granules, or powder. Liquid alum also available	60 to 75 (Powder is lighter) 17% Al_2O_3 (Minimum) Liquid–80 lb/ft^2, 8.3% Al_2O_3 49% dry aluminum sulfate	60.0 at 0°C, 65.3 at 10°C, 71.0 at 20°C, 78.8 at 30°C
Anhydrous ammonia	NH_3	–	Nitrogen feed for bacteria, pH adjustment	Liquefied gas	38.5 at 60°F, 99% NH_3	44 at 0°F, 32 at 70°F, 26 at 100°F
Ammonium hydroxide	NH_4OH	Aqua ammonia	Nitrogen feed for bacteria, pH adjustment	Liquid, colorless	56, 29.4% NH_3	NH_4OH solutions completely miscible
Bentonite (clay)	$(Al_2O_3 \cdot Fe_2O_3 \cdot 3MgO) 4SiO_2 \cdot x\ H_2O$	–	Coagulant aid	Pellets or powder	60	Forms colloidal sol
Calcium carbonate	$CaCO_3$	Limestone, calcite		White to gray powder, 100 mesh	100–135, 96–99% $CaCO_3$	

4.1–26 PROPERTIES OF CHEMICALS MOST EXTENSIVELY USED IN WATER TREATMENT AND SEWAGE PLANTS (*Continued*)

Chemical	Formula	Trade name	Use	Available forms and appearances	Density, lb/ft^2 and commercial strength	Solubility in water, g/100 cc
Calcium hydroxide	$Ca(OH)_2$	Slaked lime	Coagulation, pH adjustment, sludge conditioning	White powder	25–35 $Ca(OH)_2$ 97–99%[a] CaO 71–74%	0.18 at 0°C, 0.16 at 20°C, 0.15 at 30°C
Calcium hypochlorite	$Ca(OCl)_2$	–	Disinfection, corrosion caused by biological growths, odor control	Lumpy white or yellowish-white powder, granules, or pellets	50–55, 70% available Cl_2	21.88 at 0°C, 22.7 at 20°C, 23.4 at 40°C
Calcium oxide	CaO	Quicklime	Coagulation, pH adjustment, sludge conditioning	White powder or porous white to light gray lumps	55–60, 90–96% CaO, Below 88% is poor quality	Reacts to form $Ca(OH)_2$. See solubility of calcium hydroxide
Carbon dioxide	CO_2	Carbonic acid gas, solid (dry ice)	Recarbonation, pH control	Colorless, odorless gas; subliming white solid	Solid, 97.5; 100% CO_2	3.6 at 0°C, 1.8 at 20°C
Chlorinated ferrous sulfate	$Fe_2(SO_4)_3 \cdot FeCl_3$	Chlorinated copperas	Coagulation sludge, conditioning	Yellow solution		
Chlorine	Cl_2	–	Odor control, disinfection, corrosion control	Liquefied compressed gas	Gas–0.201, Liquid–91.70, 99.8% Cl_2	1.46 at 0°C, 0.98 at 10°C, 0.716 at 20°C, 0.57 at 30°C
Chlorine dioxide	ClO_2	–	Disinfection	Greenish–yellow gas, condenses to red liquid	–	>1.0

4.1–26 PROPERTIES OF CHEMICALS MOST EXTENSIVELY USED IN WATER TREATMENT AND SEWAGE PLANTS (*Continued*)

Chemical	Formula	Trade name	Use	Available forms and appearances	Density, lb/ft^2 and commercial strength	Solubility in water, g/100 cc
Copper sulfate	$CuSO_4$ $5H_2O$	Blue vitriol	Algae control	Blue crystals or powder	75–93, 25.2% Cu	19.2 at 0°C, 31.2 at 30°C
Dolomitic hydrated lime	$Ca(OH)_2 \cdot Mg(OH)_3$	–	pH adjustments, sludge conditioning	White powder	35–40, 46.3% CaO, 33.7% MgO, Typical assay	See $Ca(OH)_2$
Dolomitic lime	$CaO \cdot MgO$	Quicklime	pH adjustment, coagulant, sludge conditioning	White (light gray, tan) pebble, crushed, lump, ground, pulverized	55–60, Average 57, CaO 55–57.5% MgO 37.6–40.5%	See $Ca(OH)_2$, Reacts to form $Ca(OH)_2 \cdot Mg(OH)_2$
Ferric chloride, $FeCl_3$ solution	$FeCl_3 \cdot 6H_2O$–crystal, $FeCl_3$–anhydrous	Liquid ferric chloride, crystals, ferric chloride	Coagulation, sludge conditioning, odor control (H_2S)	Dark brown solution Yellow, brown crystals Green, black granules	Solution–11.6 to 12.7 lb/gal 37–47% $FeCl_3$ Crystal–60–64 lb/ft^2 60% $FeCl_2$ Anhydrous–65 to 70 lb/ft^2 96–97% $FeCl_3$	Solution–completely Crystal–64.4 at 10°C 91.1 at 20°C Anhydrous–74.4 at 0°C
Ferric sulfate	$Fe_2(SO_4)_3 \cdot$ 2-$3H_2O$	Partly hydrated ferric sulfate	Coagulation	Reddish tan-gray granules	70–72 $Fe(SO_4)_3 \cdot 3H_2O$ 68% $Fe_2(SO_4)_3$ 18.5% Fe $Fe(SO_4)_3 \cdot 2H_2O$ 76% $Fe_2(SO_4)_3$ 21% Fe	Very soluble
Ferrous sulfate	$FeSO_4 \cdot 7H_2O$	Copperas, sugar sulfate	Coagulation	Green to brownish yellow granules, crystals, powder, lumps	63–66 55% Fe_2SO_4 20% Fe	28.7 at 0°C, 37.5 at 10°C, 48.5 at 20°C, 60.2 at 30°C

4.1–26 PROPERTIES OF CHEMICALS MOST EXTENSIVELY USED IN WATER TREATMENT AND SEWAGE PLANTS (*Continued*)

Chemical	Formula	Trade name	Use	Available forms and appearances	Density, lb/ft^2 and commercial strength	Solubility in water, g/100 cc
Fluosilicic acid	H_2SiF_6	Liquid fluoride	Fluoridation	Colorless to straw liquid	76–84, 24–36% H_2SiF_6	Completely miscible
Ozone	O_3	–	Disinfection	Colorless gas, sweetish odor	–	<0.5
Potassium permanganate	$KMnO_4$	–	Taste and odor control, removal of trace organics	Purple granules	90, 97–99% $KMnO_4$	3.0 at 0°C, 6.0 at 20°C
Sodium aluminate	$NaAlO_2$	Soda alum	Coagulation	White to gray crystals, colorless liquid,	45–60 (solid) 76 (liquid) 32% Al_2O_3 · Na_2O	3.6 at 20°C
Sodium bisulfate	$Na_2S_2O_3$	Meta-bisulfite	Deaeration	White crystalline powder, sulfur odor	74–85; 93–99%	5.4 at 20°C
Sodium carbonate	Na_2CO_3	Soda ash	Softening	White powder, free flowing	Extra light, 23; light, 35; dense, 65 99% Na_2CO_3	7.2 at 0°C, 10.8 at 10°C, 17.9 at 20°C
Sodium chloride	NaCl	Salt	Ion-exchange softening regenerant	Grayish crystals or rock	62–68 (crystals), 70–75 (rock) 98% NaCl	3.46 at 0°C, 3.6 at 20°C
Sodium hypochlorite	NaOCl	Chlorine bleach	Disinfection	Light yellow liquid	117–120, 10–15% available chlorine	Completely miscible
Sodium fluoride	NaF	–	Fluoridation	White crystals or powder	60–75 (powder), 75–90 (crystals) 97% NaF	<4.2

4.1–26 PROPERTIES OF CHEMICALS MOST EXTENSIVELY USED IN WATER TREATMENT AND SEWAGE PLANTS *(Continued)*

Chemical	Formula	Trade name	Use	Available forms and appearances	Density, lb/ft^2 and commercial strength	Solubility in water, g/100 cc
Sodium hexametaphosphate	$Na_6(PO_3)_6$	Glassy phosphate	Sequestrant	Glassy flakes, pellets, granules or powder	Variable 65–67.5% P_2O_5	>24
Sodium hydroxide	NaOH	Caustic soda, lye	pH control	White crystalline, flakes, liquid solution	Variable 98.9% NaOH; solution 50% NaOH	29 at 0°C, 53 at 20°C
Sodium silicate	$2Na_2O \cdot x\ SiO_2 + H_2O$; x = 1–7.5	Water glass		Syrupy opaque liquid	84–87	20–55
Sodium fluosilicate	Na_2SiF_6	Sodium silicofluoride	Fluoridation	White to yellow powder, free-flowing	75–95; 59.7% F	0.43 at 1.7°C; 0.75 at 23.9°C
Sodium sulfite	Na_2SO_3	Sulfite	Deaeration	White crystals or powder	80–91; 93–98% Na_2SO_3	12–24
Sodium tripolyphosphate	$Na_5P_3O_{10}$		Sequestrant	White granules or powder	60–70, 57.5% P_2O_5	<15 at 20°C
Sulfuric acid	H_2SO_4	Oil of vitriol	Sludge conditioning, pH adjustment	Colorless syrupy liquid	60° Be′ 106.4 lb/ft^2, 66° Be′ 114.5 lb/ft^2, 60° Be′ 77.7% H_2SO_4, 66° Be′ 93.2% H_2SO_4	Completely miscible

REFERENCES

1. *Manual of Practice No. 8*, Water Pollut. Control Fed., 1967.
2. E.S. Hopkins and E.L. Bean, *Water Purification Control,* Williams and Wilkins, Baltimore, 1966.

4.1–27 STORAGE AND FEEDING OF CHEMICALS USED IN WATER TREATMENT AND SEWAGE PLANTS

Chemical	Shipping containers and their dimensions, in.	Properties likely to cause trouble	Storage space required	Storage container materials	Best feeding form	Best feed regulation	Strength of solution or suspension (%) and properties	Suitable handling material for solutions
Activated carbon	Bags	Dusty; flammable		Iron; steel	Slurry			Stainless steel, rubber, silicon iron
Aluminum sulfate	Bags 100 lb 28 x 17 x 7 200 lb Bulk	Dusty	33 ft^3/ton in bags or bulk	Concrete Steel Wood Moisture proof	Ground, granules, or liquid	Dry or solution feed	0.25–6.0 Acid and corrosive	Lead Rubber Acid-resisting tile Duriron Plastics Stainless steel 316 Asphalt Cypress Steel Plastics
Anhydrous ammonia	Steel cylinders 100 lb} 58 *X* 150 lb} 13 *D* Green gas label	Irritating liquefied gas Maximum air concentration (MAC), 100 ppm	13 ft^2/ton in 150 lb cylinder Avoid heat and sunlight Noncompatible with halogens	Black iron Steel	Gas	Ammoniator		
Ammonium hydroxide		Evolution of ammonia gas above 80°F						
Bentonite (clay)	Bags 100 lb	Dusty						Iron, steel
Calcium carbonate		Dusty						
Calcium hydroxide	Bags 50 lb 30 x 16 x 9 Bulk	Dusty	70 ft^3 per ton in bags or bulk based on average bulk density of 30 lb/ft^3	Concrete Steel Wood	Finer particle sizes more efficient but more difficult to handle and feed	Dry feed or slurry	Up to 10 Alkali and very encrusted	Rubber Iron Concrete

4.1–27 STORAGE AND FEEDING OF CHEMICALS USED IN WATER TREATMENT AND SEWAGE PLANTS (*Continued*)

Chemical	Shipping containers and their dimensions, in.	Properties likely to cause trouble	Storage space required	Storage container materials	Best feeding form	Best feed regulation	Strength of solution or suspension (%) and properties	Suitable handling material for solutions
Calcium hypochlorite	Cans 5 lb 15 lb Drums 100–24¼ x 14½ *D* 300 lb 800 lb Barrels 415 lb Cases 45 lb 21 x 13½ x 9½ (9 5-lb cans) Yellow label	Gives off Cl_2 on standing exposed to air, especially when not cool Irritant Explosive above 100°C MAC as gas, 1 ppm	120 ft^2/ton Cl_2 in 100-lb drums Cool Dry Noncompatible with organics	Store in shipping containers	Granules or powder Pellets dissolve more slowly	Solution	Up to 2.0 as available Cl_2 Strong alkaline and corrosive	Rubber Glass Plastics Ceramics
Calcium oxide	Bags 80 lb–34 x 19 x 9 Barrels Bulk	Dusty; slakes upon standing long, causing expansion	38 ft^3/ton in bags or bulk based on average bulk density of 57 lb/ft^3	Concrete Steel	Crushed to pass 1 in. ring, fines included	Thin, milk-like slurry	Up to 25 Dilute to 10 after slaking Alkali and very Encrustant	Rubber Iron Cement
Carbon dioxide		Respiratory hazard, drums (dry ice)						
Chlorinated ferrous sulfate							1 part Cl_2 to 7.8 parts $FeSO_4$ by weight Corrosive	Rubber Ceramics Plastics
Chlorine	Steel cylinders 100 lb 36 x 10 *D* 150 lb 54 x 10 *D* 2,000 lb 81½ x 30 *D* Bulk Green gas label	Moist Cl_2 is extremely corrosive Irritant to eyes Toxic MAC, 1 ppm	Cool Ventilated	Store in shipping containers	Gas	Approved chlorinator	0.0025 (1 lb to 45–50 gal) Acid and corrosive	Dry Cl_2 Black iron, copper, monel, steel Wet Cl_2 Tantalum Silver Glass Hard rubber

4.1–27 STORAGE AND FEEDING OF CHEMICALS USED IN WATER TREATMENT AND SEWAGE PLANTS (*Continued*)

Chemical	Shipping containers and their dimensions, in.	Properties likely to cause trouble	Storage space required	Storage container materials	Best feeding form	Best feed regulation	Strength of solution or suspension (%) and properties	Suitable handling material for solutions
Chlorine dioxide	Produce in plant due to instability	Strong respiratory irritant; toxic						
Copper sulfate		Dusty; toxic						
Dolomitic hydrated lime	Bags 50 lb 5½ x 15¼ x 25½ Bulk	Moisture causes caking Dusty	50 ft^3/ton	Steel Wood Concrete Paper	Powder	Dry or solution feed	Up to 28 Alkali Encrustant	Steel Rubber Concrete Wood Plastics
Dolomitic lime	Bags 80 lb Barrels 200 lb 400 lb Drums 200 lb Bulk	Dusty Alkaline Caustic Affinity for water Expands with hydration	32–54 ft^3/ton Must be dry and free of high humidity	Iron Steel Concrete Wood Waterproofed paper bags	Pebble (small) or crushed to pass ¾ in. mesh, fines included, or solution as a slurry	Dry feed with anti-arching hoppers Solution feed	Up to 25% Dilute after slaking to 10% Caustic	Iron Steel Rubber Concrete Wood Plastic
Ferric, chloride	Carboys 12 gal Tank cars 4,000 gal 8,000 gal Barrels 435 lb Drums 100 lb 23½ x 14-1/8 *D* 150 lb 450 lb–33 x 21 *D*	Solution–acid, corrosive, stains Crystal–melts at 37°C, hygroscopic, corrosive, stains Anhydrous–deliquescent, corrosive, stains	Solution Carboys 20 ft^2/ton Bulk 61 ft^2/ton Barrels 70 ft^2/ton Drums 40 ft^2/ton	Rubber-lined steel, wood, or concrete Store in shipping containers	Solution	Solution feed only	Up to 45% $FeCl_2$ Acid and corrosive	Acid-proof brick Ceramics Stoneware Rubber Glass Plastics
Ferric sulfate	Bags 100 lb 25 x 16 x 6 30 x 16 x 5	Stains	28 ft^3/ton in bags or bulk	Concrete Steel Wood	Solution	Dry feed Solution feed with special dissolving equipment	1–25 Acid and corrosive	Rubber Duriron 18-8 stainless steel

4.1–27 STORAGE AND FEEDING OF CHEMICALS USED IN WATER TREATMENT AND SEWAGE PLANTS (*Continued*)

Chemical	Shipping containers and their dimensions, in.	Properties likely to cause trouble	Storage space required	Storage container materials	Best feeding form	Best feed regulation	Strength of solution or suspension (%) and properties	Suitable handling material for solutions
Ferric sulphate (continued)	Drums 250 lb 29½ x 15½ *D* 24½ x 17 *D* 400 lb Bulk							Lead Ceramics
Ferrous sulfate	Bags 100 lb 12 x 12 x 25 Barrels 400 lb 30 x 36 *D* Bulk	Stains Cakes above 20° C in moist air	31 ft^3/ton in bags or bulk Cool Ventilated	Concrete Steel Wood	Granules	Solution feed Dry feed with hopper agitation	0.25–6.0 Acid and corrosive	Rubber Duriron 18-8 stainless steel Lead
Fluosilicic acid		Corrosive vapors; respiratory hazard, skin irritant						
Ozone	Produce in plant due to instability	Respiratory hazard						
Potassium permanganate		Dusty; toxic; fumes from solution						
Sodium aluminate	100, 150 lb bags 250, 325, 400 lb drums	Alkaline skin burns						Iron, steel, rubber, plastics
Sodium bisulfite		Toxic; throat and lung irritant						
Sodium carbonate		Dusty						
Sodium chloride		Corrosive						

4.1–27 STORAGE AND FEEDING OF CHEMICALS USED IN WATER TREATMENT AND SEWAGE PLANTS (*Continued*)

Chemical	Shipping containers and their dimensions, in.	Properties likely to cause trouble	Storage space required	Storage container materials	Best feeding form	Best feed regulation	Strength of solution or suspension (%) and properties	Suitable handling material for solutions
Sodium fluoride		Toxic; dusty						
Sodium hexametaphosphate		Eye damage						
Sodium hydroxide		Skin burns; fumes respiratory hazard						
Sodium hypochlorite		Irritating fumes						
Sodium silicate	Drums, trucks tank cars	Nonhazardous						Cast iron, steel, rubber
Sodium sulfite		Toxic						
Sulfuric acid	Carboys 5 gal 13 gal 30 x 20½ *D* Chemikegs 13 gal 16 x 23½ *D* Drums 55 gal 35 x 25 *D* 110 gal Bulk Liquid white label	Hazardous strong acid, corrosive Causes severe burns MAC mist, 1 ppm	20 ft^2/ton carboys Avoid sunlight	Store in shipping containers	Solution	Solution feed	Up to 25 Acid and corrosive	Lead Rubber Glass Porcelain Ceramics Plastics Acid brick

REFERENCES

1. *Manual of Practice No. 8,* Water Pollut. Control Fed., 1967.
2. E.S. Hopkins and E.L. Bean, *Water Purification Control,* Williams and Wilkins, Baltimore, 1966.
3. E.B. Bessilieare, *Industrial Waste Treatment,* McGraw-Hill, New York, 1952.

4.1–28 CHARACTERISTICS OF COMMON GASES ENCOUNTERED IN SEWERS, WASTEWATER PUMPING STATIONS, AND TREATMENT PLANTS

Table A

Class	Gas	Chemical formula	Common properties[a]	Specific gravity or vapor density (Air = 1)	Physiological effect[a]
1	Carbon dioxide	CO_2	Asphyxiant. Colorless, odorless. When breathed in large quantities may cause acid taste. Nonflammable. Not generally present in dangerous amounts unless oxygen deficiency exists	1.53	Cannot be endured at 10% more then few min, even if subject is at rest and oxygen content normal. Acts on respiratory nerves
1b	Carbon monoxide	CO	Chemical asphyxiant. Colorless, odorless, tasteless, flammable. Poisonous	0.97	Combines with hemoglobin of blood. Unconsciousness in 30 min at 0.2–0.25%. Fatal in 4 hr at 0.1%. Headache in few hr at 0.02%
2	Chlorine	Cl_2	Irritant. Yellow green color. Choking odor detectable in very low conc. Nonflammable	2.49	Irritates respiratory tract. Kills most animals in very short time at 0.1%
3	Gasoline	C_5H_{12} – C_9H_{20}	Volatile solvent. Colorless. Odor noticeable at 0.03%. Flammable	3.0–4.0	Anesthetic effects when inhaled. Rapidly fatal at 2.4%. Dangerous for short exposure at 1.1–2.2%
1a	Hydrogen	H_2	Simple asphyxiant. Colorless, odorless, tasteless. Flammable	0.07	Acts mechanically to deprive tissues of oxygen. Does not support life
2 and 4	Hydrogen sulfide	H_2S	Irritant and poisonous volatile compound. Rotten egg odor in small concentration. Exposure for 2–15 min at 0.01% impairs sense of smell. Odor not evident at high conc. Colorless. Flammable	1.19	Impairs sense of smell rapidly as concentration increases. Death in few min at 0.2%. Exposure to 0.01–0.1% rapidly causes acute poisoning. Paralyzes respiratory center
1a	Methane	CH_4	Simple asphyxiant. Colorless, odorless, tasteless. Flammable	0.55	Acts mechanically to deprive tissues of oxygen. Does not support life
1a	Nitrogen	N_2	Simple asphyxiant. Colorless, tasteless. Nonflammable. Principal constituent of air (about 79%)	0.97	Physiologically inert
–	Oxygen (in air	O_2	Colorless, odorless, tasteless. Supports combustion	1.11	Normal air contains 20.8% of O_2. Man can tolerate down to 12%. Min safe 8 hr exposure, 14–16%. Below 10% dangerous to life. Below 5–7% probably fatal

[a] Percentages shown represent volume of gas in air.

4.1–28 CHARACTERISTICS OF COMMON GASES ENCOUNTERED IN SEWERS, WASTEWATER PUMPING STATIONS, AND TREATMENT PLANTS *(Continued)*

Table A

Class	Gas	Chemical formula	Common properties[a]	Specific gravity or vapor density (Air = 1)	Physiological effect[a]
Mainly 1a	Sludge gas	c	Mostly a simple asphyxiant. May be practically odorless, colorless. Flammable	Variable	Will not support life

Table B

Class	Max safe 60 min exposure, % by volume in air	Max safe 8 hr exposure, % by volume in air[b]	Explosive range, % by volume in air: Lower limit	Explosive range, % by volume in air: Upper limit	Likely location of highest concentration	Most common sources
1	4.0–6.0	0.5	–	–	At bottom; when heated may stratify at points above bottom	Products of combustion, sewer gas, sludge. Also issues from carbonaceous strata
1b	0.04	0.005	12.5	74.0	Near top, especially if present with illuminating gas	Manufactured gas, flue gas, products of combustion, motor exhausts. Fires of almost any kind
2	0.0004	0.0001	–	–	At bottom	Chlorine cylinder and feed line leaks
3	0.4–0.7	varies	1.3	6.0	At bottom	Service stations, garages, storage tanks, and houses
1a	–	–	4.0	74.0	At top	Manufactured gas, sludge digestion tank gas, electrolysis of water. Rarely from rock strata

[a] Percentages shown represent volume of gas in air.
[b] Conforms to "Threshold Limit Values of Air-Borne Contaminents for 1968" adapted at the 30th Annual Meeting, American Conference of Governmental Industrial Hyigenists, St. Louis, May 13, 1968.
[c] Mostly methane and carbon dioxide, with small amounts of nitrogen sulfide, and oxygen; occasionally traces of carbon monoxide.

4.1–28 CHARACTERISTICS OF COMMON GASES ENCOUNTERED IN SEWERS, WASTEWATER PUMPING STATIONS, AND TREATMENT PLANTS *(Continued)*

Class	Max safe 60 min exposure, % by volume in air	Max safe 8 hr exposure, % by volume in air[b]	Explosive range, % by volume in air		Likely location of highest concentration	Most common sources
			Lower limit	Upper limit		
2 and 4	0.02–0.03	0.001	4.3	46.0	Near bottom, but may be above bottom if air is heated and highly humid	Coal gas, petroleum, sewer gas. Fumes from blasting under some conditions. Sludge gas
1a	Probably no limit, provided O_2% is sufficient for life	–	5.0	15.0	At top, increasing to certain depth	Natural gas, sludge gas, manufactured gas, sewer gas. Strata of sedimentary origin. In swamps or marshes
1a	–	–	–	–	Near top, but may be found near bottom	Sewer gas, sludge gas. Also issues from some rock strata
–	–	–	–	–	Variable at different levels	Oxygen depletion from poor ventilation and absorption, or chemical consumption of oxygen
Mainly 1a	No data. Would vary widely with composition	–	5.3	19.3	Near top of structure	From digestion of sludge

[b] Conforms to "Threshold Limit Values of Air-Borne Contaminents for 1968" adapted at the 30th Annual Meeting, American Conference of Governmental Industrial Hygienists, St. Louis, May 13, 1968.

Source: *Manual of Practice No. 1,* Water Pollut. Control Fed., 1969.

4.1–29 COST COMPARISON AND TRENDS OF ALKALINE AGENTS

				(1951)		(1954)[2]		(1965)[3]	
Agent	Grade	Container	Basicity factor	Cost, $/ton of basicity	Relative cost	Cost, $/ton	Relative cost	Cost, $/ton	Relative cost
Sodium hydroxide	Flake (76%)	Drums	0.687	109.90	7.13	154	13	152	8.5
Sodium carbonate	Dense (58%)	Bags	0.507	63.11	4.10	112	9.3	73	4.1
Ammonia	Aqua (35%)	Drums	1.647	37.64	2.44				
Ammonia	Anhydrous	Tank Car	1.647	48.36	3.14				
Magnesium oxide	Powder	Bags	1.306	57.43	3.73	64	5.3		
High calcium hydrate	Chemical	Bags	0.710	20.45	1.33	20	1.7	24	1.4
Dolomitic hydrate	Chemical	Bags	0.912	20.13	1.31	15	1.25	19	1.1
High calcium quicklime	Pulverized	Bags	0.941	19.13	1.24				
	Pulverized	Bulk	0.941	15.41	1.00	12 1.0	18	1.0	
Dolomitic quicklime	Ground	Bags	1.110	13.94	0.90				
	Ground	Bulk	1.110	10.79	0.70	10	0.9	15	0.85
High calcium limestone	Pulverized	Bags	0.489	10.23	0.66				
	Pulverized	Bulk	0.489	7.15	0.46				
Dolomitic limestone	Pulverized	Bags	0.564	8.86	0.57	8	0.7	21.50	1.2
	Pulverized	Bulk	0.564	6.21	0.40	7	0.6	18.60	1.05

REFERENCES

1. N.L. Nemerow, *Theories and Practices of Industrial Waste Treatment,* Addison-Wesley, Reading, Mass., 1963.
2. R.D. Hoak, Sewage Ind. Wastes, 22, 212, 1950.
3. R.D. Ross, *Industrial Waste Disposal,* Van Nostrand Reinhold, 1968.

4.1–30 ANNUAL CUMULATIVE FINDINGS ON THE INSTITUTION OF CONTROLLED FLUORIDATION

1945–1970

	Fluoridation status at end of each year		
Year	No. communities	Number of water supply systems	Population[a]
1945	6	3	231,920
1946	12	8	332,467
1947	16	11	458,748
1948	26	13	581,683
1949	49	29	1,062,779
1950	100	62	1,578,578
1951	368	171	5,079,321
1952	751	353	13,875,005
1953	1,007	482	17,666,339
1954	1,194	572	22,336,884
1955	1,347	672	26,278,820
1956	1,583	772	33,905,474
1957	1,717	879	36,215,208
1958	1,890	995	38,461,589
1959	1,990	1,081	39,628,377
1960	2,111	1,172	41,169,412
1961	2,197	1,249	42,201,115
1962	2,321	1,350	44,045,392
1963	2,612	1,482	46,678,380
1964	2,758	1,573	48,363,066
1965	3,030	1,692	59,855,024
1966	3,145	1,785	62,437,390
1967	3,827	2,091	71,916,682
1968	4,229	2,372	74,579,666
1969	4,834	2,653	80,096,860
1970(Est)	5,400	2,950	83,000,000

[a] Includes adjustment for population growth during intercensal years 1951–1959 based on results of the 1960 census of population.

Source: F.J. Maier, *Crit. Rev. Environ. Control, 2*(3):387, 1971.

4.1–31 STATUS OF FLUORIDATION THROUGHOUT THE WORLD

Country or territory	Year first project started	No. communities reporting fluoridated water	Population served by fluoridated water
Australia	1956	23	4,159,000
Belgium	1956	1	10,000
Brazil	1953	86	2,400,00
Canada	1945	459	6,063,000
Chile	1953	64	3,300,000
Colombia	1953	6	2,401,000
Czechoslovakia	1958	30	1,000,000
El Salvador	1956	1	1,380,000
Federal Republic of Germany	1952	1	6,000
Finland	1959	1	60,000
Hong Kong	1961	11	3,570,000
Ireland	1964	44	1,200,000
Japan	1952	1	30,000
Kuwait	1968	1	676,000
Malaysia	–	6	3,000,000
Mexico	1960	5	1,750,000
Netherlands	1953	15	3,000,000
New Zealand	1953	29	1,205,000
Panama	1950	8	510,000
Papua and New Guinea	1966	1	38,000
Paraguay	1959	1	135,000
Poland	–	1	500,000
Puerto Rico	1953	59	1,808,000
Romania	–	1	100,000
Ryuku Islands	–	2	740,000
Sarawak	–	–	180,000
Singapore	1958	1	2,000,000
Sweden	1952	2	130,000
Switzerland	1960	3	250,000
United Kingdom	1955	15	2,250,000
U.S.	1945	5,300	83,000,000
USSR	1960	24	13,000,000
Venezuela	1952	22	60,000

Source: F.J. Maier, *Crit. Rev. Environ. Control, 2*(3):387, 1971.

4.2 COAGULATION AND FILTRATION

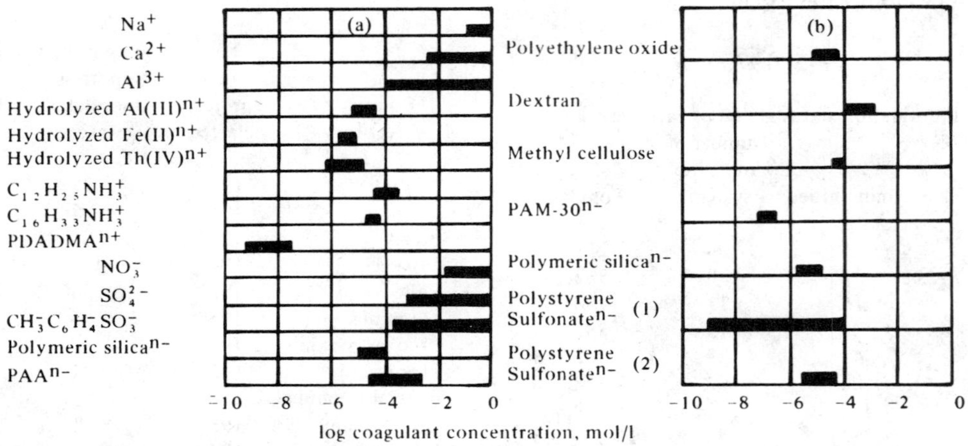

4.2–1 Typical Destabilization Regions for a Few Representative Coagulants.

Note: (a) contains coagulation regions observed when the colloid and the coagulant are of opposite charge. (b) Contains data obtained for the coagulation of colloids by uncharged molecules or by ions of like charge. These simplified figures are based on experimental results by various investigators who have used different experimental procedures and systems; the coagulation regions are therefore not always precisely comparable. The locations of the coagulation regions are exemplifications, that is, they attempt to show essential features and are meaningful in a semiquantitative way only. With sorbable species, the exact location of the coagulation region on the concentration axis depends upon the concentration and surface potential of the colloids, the pH, the presence of other cations and anions, and the temperature. Interactions with polymers are dependent upon molecular weight and steric factors.

Data

Coagulant	Colloid	Charge reversal	Restabilization	Remarks	Ref.
Na^+	$(AgBr)^-$, 10^{-4} M	No	No		1
Ca^{2+}	$(AgBr)^-$, 10^{-4} M	No	No		1
Al^{3+}	$(AgBr)^-$, 10^{-4} M	No	No		1
Hydrolyzed $Al(III)^{n+}$	$(AgI)^-$ 10^{-4} M	Yes	Yes	pH = 5	2
Hydrolyzed Fe $(III)^{n+}$	$(SiO_2)^2$, 0.2 g/l	Yes	Yes	pH = 5	3
Hydrolyzed $Th(IV)^{n+}$	$(AgBr)^-$, 10^{-4} M	Yes	Yes	pH = 4.5	4
$C_{12}H_{25}NH_3{}^+$	$(AgI)^-$, 8.1 x 10^{-3} M	Yes	Yes		5
$C_{16}H_{33}NH_3{}^+$	$(AgI)^-$, 8.1 x 10^{-3} M	Yes	Yes		5
Polydiallyldimethyl-ammonium^{n+} $(PDADMA)^{n+}$	$(Kaolin)^-$, 73.2 mg/l	Yes	Yes	Coagulant dosage as moles $C_8H_{16}NCl$/l	6
$NO_3{}^-$	$(AgBr)^+$, 10^{-4} M	No	No		7
$SO_4{}^{2-}$	$(AgBr)^+$, 10^{-4} M	No	No		7
$CH_3-C_6H_4-SO_3{}^-$	$(FeO(OH))^4$, 2.6 x 10^{-3} M	No	No		8
Polymeric or "activated" silica	$(AgBr)^+$, 5 x 10^{-4} M	Yes	Yes		9
Polyacrylic acid^{a-}, $(PAA)^{n-}$	$(AgBr)^+$, 2 x 10^{-3} M	Yes	Yes		10
Polyethylene oxide	$(Kaolin)^-$, 56.4 mg/l	No	Yes		11
Dextran	A. aerogenes, ca. 100 mg/l	No	Yes	pH = 5	12
Methyl cellulose	$(AgBr)^-$, 2 x 10^{-3} M	No	Yes		10
Polyacrylamide^{n-}, 30% Hydrolyzed $(PAM\text{-}30)^{a-}$	$(Kaolin)^-$, 33.3 mg/l	No	Yes	Ca^{2+} = 2.25 x 10^{-3} M	13
Polymeric or "activated" silica^{n-}	$(AgBr)^-$, 5 x 10^{-4} M	No	Yes		9
Polystyrene sulfonate^{n-}, (1)	$(AgBr)^-$, 5 x 10^{-4} M	No	Yes	Ca^{2+} = 4.45 x 10^{-4} M	14
Polystyrene sulfonate^{n-}, (2)	A. aerogenes, ca. 100 mg/l	No	Yes		12

Source: Reprinted with publisher's permission, from *J. Am. Water Works Assn., 60*(5):514, © 1968, by the American Water Works Association, Inc., 2 Park Ave., New York, N.Y. 10016.

REFERENCES

1. B. Tezak, *J. Phys. Chem.*, *57*:301, 1953.
2. E. Matijevic *et al.*, *J. Colloid Sci.*, *19*:333, 1964.
3. C.R. O'Melia and W. Stumm, *J. Colloid Interface Sci.*, *23*:437, 1967.
4. E. Matijevic *et al.*, *J. Phys. Chem.*, *64*:1157, 1960.
5. B. Tamamushi and K. Tamaki, *Kolloid-Zeitschrift, 163*:122, 1959.
6. A.P. Black *et al.*, *J. Colloid Interface Sci.*, *21*:626, 1966.
7. E. Matijevic and M. Kerker, *J. Phys. Chem.*, *62*:1271, 1958.
8. H. Freundlich and G.V. Slottman, *Zeitschrift für physik Chemie, 129*:305, 1927.
9. W. Stumm *et al.*, *Environ. Sci. Technol.*, *1*:221, 1967.
10. S. Krahtohvil *et al.*, *J. Colloid Sci.*, *19*:373, 1964.
11. J.J. Morgan and F.B. Birkner, *Flocculation Behavior in Dilute Clay Polymer Systems,* Progress Report, U.S. Public Health Service, Research Grant No. WP-00942-02, August 1966.
12. P.L. Busch and W. Stumm, *Environ. Sci. Technol.*, *2*:49, 1968.
13. A.P. Black *et al.*, *J. Am. Water Works Assn.*, *57*(12):1547, 1965.
14. A. Sommeraner *et al.*, *Kolloid-Zeitschrift und Zeitschrift für Polymere* (in press).

4.2–2 MODES OF DESTABILIZATION AND THEIR CHARACTERISTICS

Phenomena	Physical double layer theory (coagulation)	Chemical bridging model (flocculation)	Aggregation by hydrolyzed metal ions
Electrostatic interaction	Predominant	Subordinate	Important
Chemical interactions and adsorption	Absent	Predominant	Important
Zeta potential for optimum aggregation	Near zero	Usually not zero	Not necessarily zero
Addition of an excess of destabilizing species	No effect	Restabilization due to complete surface coverage	Restabilization usually accompanied by charge reversal; may be blurred by hydroxide precipitation
Fraction of surface coverage (θ) by destabilization for optimum aggregation	Negligible	$\theta = 0.5$ (La Mer (9)) in general, $0 < \theta < 1$	$0 < \theta < 1$, θ not necessarily = 0.5
Relationship between optimum dosage of destabilizing species and the concentration of colloid (or concentration of colloidal surface)	CCC virtually independent of colloid concentration	Stoichiometry, a linear relationship between flocculent dose and surface area	Stoichiometry possible, but does not always occur
Physical properties of the aggregates which are produced	Dense, great shear strength, but poor filtrability in cake filtration	Flocs of 3 dimensional structure; low shear strength, but excellent filtrability in cake filtration	Flocs of widely varying shear strength and density

4.2–3 COAGULATION SLUDGE CHARACTERISTICS

Plant	Treatment	BOD (5 day), mg/l	COD, mg/l	pH	Total solids, mg/l	Volatile solids, mg/l	Total suspended solids, mg/l	Volatile suspended solids, mg/l
A	Alum Coagulation and sedimentation	41 72[a] 144[b]	540	7.1	1,159	571	1,110	620
B	Alum Coagulation, clarifier	90	2,100	7.1	10,016	3,656	5,105	2,285
C[c]	Alum Coagulation, clarifier	108	15,500	6.0	16,830	10,166	19,044	10,722
D	Alum Coagulation, clarifier	44	–	6.0	–	–	15,790	4,130
E	Diatomite Filter	105	340	7.6	7,466	275	7,560	260
F[1]	Alum Coagulation, sedimentation	380	1,320 (1,162–1,580)[e]	6.6 (6.5–6.7)	17,350 (4,380–28,580)	4,120 (1,390–6,330)		
G[2]	Alum Coagulation, sedimentation[d]	–	–	7.0	8,050	3,320	7,750	3,200
H[f,3]	Sedimentation basin sludge	57	669	7	4,350	2,310	3,650	
I[g,3]	Sedimentation basin sludge	45	1,048	7	4,260	1,020	3,580	
J[h,3]	Sedimentation basin sludge	–	–	7	14,386	3,450	13,640	
K[i,4]	Sedimentation basin sludge	36–77	500–1,000	7	4,300		3,610	

[a] BOD after 7 days.
[b] BOD after 27 days.
[c] Activated carbon in sample.
[d] Milwaukee, Howard Avenue Plant.
[e] Range of values.
[f] Shoremont Plant, Rochester, N.Y.
[g] Wolcott, N.Y.
[h] Ithaca, N.Y.
[i] Average data from Shoremont and Wolcott.

Source: H.B. Russelmann, *Water Sewage Works, 115*(R.N.):R64, 1968.

REFERENCES

1. H.M. Bugg, *et al., J. Am. Water Works Assn., 62*(12):792, 1970.
2. A.E. Albrecht, *J. Am. Water Works Assn., 64*(1):46, 1972.
3. E.R. Sutherland, *J. Am. Water Works Assn., 61*(4):186, 1969.
4. W.K. Neubauer, *J. Am. Water Works Assn., 60*(7):819, 1968.

4.2–4 GT VALUES

Flocculation

	Time, min	Volume agitation, ft^3	Water, hp	hp dissipated per unit volume $W = \frac{hp \times 550}{volume}$	Velocity gradient $G = \sqrt{\frac{W}{\text{Absolute viscosity}}}$ sec^{-1}	Gt ($\times 10^3$)
Atlanta, Ga. (Chattahoochee)[a]						
9 flocculation tanks	50	3–14,000 each	0.87	0.045	46	125
3 rect clarifiers		(3 hor. reels)	0.70	0.036	41	(Total)
14,000 gpm			0.44	0.023	33	
					(per row)	
Melbourne, Fla.[a]						
1 90 ft dial x 15 ft swd	19	14,000	8.5	0.332	126	144
0.94 gpm/ft^2	(40)	(propeller pumps)	(8.5)	(0.332)	(65)	(144)
5,600 gpm	(@ 2,800 gpm)					
Muskegon, Mich.[a]						
2 70 ft x 70 ft x 15 ft swd	28	15,700	6.0	0.234	93	157
4,200 gpm each, 1 gpm/ft^2		(propeller pumps)				
Detroit S.W.[a]						
(Wayne County)						
20 flocculation tanks	40	163,500	17.5	0.05	46	111
Rect clarifier		(5 hor. plate turbine)			(for 5 rows)	
12,500 gpm						
Omaha MUD[b]						
6 90 ft x 90 ft x 19 ft swd	22	30,500	32.0	0.577	145	190
4 10,500 gpm, 1.5 gpm/ft^2		(propeller pumps)				
2 15,800 gpm, 2.3 gpm/ft^2						
Thomasville, Ga.[b]						
1 54 ft diam x 15 ft swd	25	14,100	4.5	0.176	92	137
4,200 gpm, 2.25 gpm/ft^2		(propeller pumps)	changed			
			to			
			10.0	0.39	136	204
Minot, N.D.[b]						
1 84 ft diam x 19 ft swd	29	32,300	15.0	0.256	97	169
8,350 gpm, 1.3 gpm/ft^2		(propeller pumps)	changed			
			to			
			20.0	0.34	111	193
Ann Arbor, Mich.[b]						
1 130 ft diam	25	51,000	40.0	0.431	126	190
15,250 gpm, 1.35 gpm/ft^2		(propeller pumps)				

[a] Turbidity removal.
[b] Softening.

Source: Reprinted with publisher's permission from *J. Am. Water Works Assn., 60*(3):321, © 1968, by the American Water Works Association Inc., 2 Park Ave., New York, N.Y. 10016.

4.2–5 SUGGESTED DESIGN G AND GT VALUES

	G, Sec^{-1}	Gt x 10^3
1. Turbidity/color removal No solids recirculated	50–100 (Taper)	100–150
2. Turbidity/color removal Solids contact reactors (5% volume in suspension)	75–175	125–150
3. Softening Solids contact reactors (10% volume in suspension)	130–200	200–250
4. Softening Ultra high solids contact (20–40% volume in suspension)	250–400	300–400
5. Rapid mix (blending)		
20 sec contact	1,000	
30 sec contact	900	
40 sec contact	790	

Source: Reprinted, with publisher's permission, from *J. Am. Water Works Assn., 60*(3):321, © 1968, by the American Water Works Association, Inc., 2 Park Ave., New York, N.Y. 10016.

4.2–6 SPECIFIC GRAVITY OF MINERAL PRODUCTS OF COAGULATION

Formula	Specific gravity
Al_2O_3	3.97
A1O(OH)	3.3–3.5
$Al_2O_3 \cdot 3H_2O$	2.4
Fe_2O_3	5.26
FeO(OH)	4.05–4.35
$Fe_2O_3 \cdot XH_2O$	2.44–3.60
$Fe(OH)_3$	3.4

Source: Reprinted with publisher's permission from *J. Am. Water Works Assn., 64*(6):382, © 1972, by the American Water Works Association Inc., 2 Park Ave., New York, N.Y. 10016.

4.2–7 SPECIFIC GRAVITY OF MINERALS OF TYPE FOUND NATURALLY SUSPENDED IN WATER

Name	Specific gravity
$CaCO_3$	2.7–2.9
Illite	2.65
SiO_2	2.6–2.9
Kaolinite	2.6–2.7
Montmorillonite	2–3

Source: Reprinted with publisher's permission from *J. Am. Water Works Assn., 64*(6):382, © 1972, by the American Water Works Association Inc., 2 Park Ave., New York, N.Y. 10016.

4.2–8 CHARACTERISTICS OF MUNICIPAL SYSTEMS USING DIATOMITE FILTRATION

Community	Population served	Average daily demand, mgd	Total[a] storage, mg	Daily flow of average daily demand	Period daily use, hr	Design filter rate, gpm/ft²	Filter area, ft²	Average production rate, gpm/ft² period	Brand of pressure filter	Purpose of plant
Altamont, Albany Co.	1,365	0.150	0.74	100	24	1.0	188	0.5	Adams	Fulltime
Menands, Albany Co.	2,300	1.6	1.05	100	24	1.0	1,120	1.0	DeLaval	Fulltime
Peru W.D., Clinton Co.	3,400	0.34	0.3	100	24	1.0	450	1.0	Adams	Auxiliary summer
Sherburne, Chenango Co.	1,700	0.30	0.365	100	18	1.0	300	1.0	BIF	Auxiliary
Tupper Lake, Franklin Co.	6,000	0.8	0.65	50(est.)	12	2.0	620	1.13	BIF	Fulltime summer

[a] Total finished water storage.

Source: Reprinted, with publisher's permission, from *J. Am. Water Works Assn., 59*(7):867, © 1967, by the American Water Works Association, Inc., 2 Park Ave., New York, N.Y. 10016.

4.2–9 DIATOMITE FILTRATION PLANTS FOR POTABLE WATER FILTRATION [a]

Plant location[b]	Year built	Type of service[c]	Source of water	Substance removed[d]	Population served[e]
Alaska					
Anchorage	1955	NH	Ship Creek	Tu	(500)
Ariz.					
Pine,	1962	M	Pine Creek	Tu	350
Tacna	1959	M	irrigation canal	Tu	250
Ark.					
Aerojet General	1955	I	reservoir	Tu	
Calif.					
Almaden	1957	M	stream	Tu	500
Bear Creek Estates	1961	M	Boulder Creek	Tu	30
Clear Lake Oaks	1961	WD	lake	Al	1,100
Coalinga	1959	M	well	O	5,026
Folsom	1951	BR	stream	Tu	
Folsom Lakehills Estates	1961	M	lake	Tu	420–2,800
Goleta	1963	WD	lake	Tu, Ta	20,000
Lake Hemet	1960	WD	stream	Tu	9,600
Julian	1962	M	well	Fe	(500)
Lompoc	1963	M	well	S	(16,800)
Malibu	1951	M	well	Fe, H_2S	(3,500)
Monticeto, office plant	1957, 59	WD	L. Cachuma	Tu	7,000
Monticeto, Barker Pass	1957	WD	L. Cachuma	Tu	total
Riverside	1959	M	lake	Tu	100,500
Saratoga	1959	M	stream	Tu	
Seal Beach	1949	M	well	Ta	
Suisun	1961	M	canal	Tu	
Vacaville	1963	M	canal	Tu	
Ill.					
Bethaltho	1961	M	wells	Fe (3 ppm)	(3,235)
Campbell Hills	1948	M	wells	Fe	(336)
West Dundee	1958	M	well	Fe	(2,530)
Mounds	1948	M	wells	Fe	2,001
Olympia Fields	1960	M	wells	LSS	1,700
Ind.					
(near) Shoals	1955	I	off channel reservoir	Tu	
(near) Avoca	1955	CC	spring	Tu	
Iowa					
Charles City		M	well	Sa	
Me.					
Winterport	1959	M	stream fed reservoir	Tu	(900)

[a] Recent installations not recorded in the table are located at Raybrook State Park, N.Y., and Santa Ana Mountain Valley Water District, Cal.

[b] All plants were approved by the state health department except those at Seal Beach, Folsom, and Riverside, Cal.; Worthington, Minn.; and Iriquois School, N.Y. Filtration rate–1.0 normal design (1.6) maximum.

[c] Meanings of initials are: NH, native hospital; M, municipal; I, industrial; WD, water district; BR, U.S. Bureau of Reclamation; CC, church camp; PU, private utility; P, private; SP, state park; and CCl, country club.

4.2–9 DIATOMITE FILTRATION PLANTS FOR POTABLE WATER FILTRATION[a] *(Continued)*

Approximate production,[e] mgd	Design rate,[e] gpm/ft²	Plant capacity, gpm	Make and type of filter[f]	Remarks
0.015	2.0		Elgin-P	
	2.0		Elgin-P	standby only
	2.0	900	Water cond.	industrial plant
0.26	1.6 (2.5)	185	Delaval–P	limited upstream development[h]
0.2	1.69(3.6)	130	Delaval–P	8 hr operation
0.8	1.6	550	Delaval–P	potential algae problem
	1.0	45	BIF-V	precedes ionics desalting unit
			BIF-P	abandoned
1.0	1.6	750–900	Delaval-P	coagulation before filtration
1.25	1.6	900–1,800	Delaval-P	
1.25	1.2(1.6)	900	Delaval-P	
1.6	1.6(2.1)	100	Delaval-P	
7.4(9.3)	1.0(1.3)	5,140–6,460	BIF-V	automatic
1.0	1.8	450	Bowser-P	abandoned
1.0	0.75(1.50)	500	Sparkler-P	
	1.0	1,000	US-P	
5–6	6.0	11,000	US-P	standby service, turbidity <5
5	1.6(2.5)	900–3,600	Delaval–P	fully automatic[h]
			Olson-P	abandoned
1.25	1.6	900	Delaval-P	h
5	1.6(2.5)	900–3,600	Delaval-P	
0.53	0.92	368	Sparkler-P	supplement sand filters[i]
0.07	1.0	50	Infilco-P	abandoned (1956)
0.20	1.0	600	Elgin-P	
0.16	2.0	350	Infilco-P	abandoned
0.24	1.89	500	Infilco-P	8 hr operation
				batch treatment to clear well
	2.0	500	Infilco-P	
0.05			BIF-V	108 ft²; turbidity, 0–100

[d] Meanings of symbols other than chemical are: Tu, turbidity; Al, algae; O, organic and amorphous matter; Ta, taste; S, softening; LSS, lime-soda ash softening; Sa, sand; Pl, plankton; C, color.

[e] Parentheses enclose estimated data.

[f] P is for pressure filter; V, vacuum filter.

[h] Alternative source available when turbidity exceeds 50 units.

[i] During peak summer periods.

4.2–9 DIATOMITE FILTRATION PLANTS FOR POTABLE WATER FILTRATION[a] *(Continued)*

Plant location[b]	Year built	Type of service[c]	Source of water	Substance removed[d]	Population served[e]
Belfast	1964			Tu	
Wiscasset	1962		stream	Tu	
Md.					
Perryville	Temporary		stream	Tu	
Mich.					
Auburn	1950	M	L. Huron[k]	Tu	(1,497)
Aus Gres	1950	M	L. Huron[k]	Tu	(581)
Bangor-Monitor	1951, 56	M	L. Huron[k]	Tu	1,000
Belleville	1954	M	L. Huron[k]	Tu	(1,921)
Carrollton	1950, 54	M	L. Huron[k]	Tu	3,500
Delta College	1961	M	L. Huron[k]	Tu	
Goetz Waterline	1960	M	L. Huron[k]	Tu	450
Kawkawlin	1952	M	L. Huron[k]	Tu	700
Linwood	1955	M	L. Huron[k]	Tu	500
Saginaw Twp	1954, 56, 61	M	L. Huron[k]	Tu	10,000
Minn.					
Elmore		M	well	Fe	(1,078)
Worthington	1955	I	L. Okabena	Tu, Pl	7,900
N.H.					
Hudson		M	pond	Tu	
N.M.					
Santa Rosa	1950, 54	PU	river	Tu	2,000
N.Y.					
Altamont	1961	M	reservoir	Tu	(1,365)
Canajoharie			lake	Tu	(2,681)
Chaumont	1963	M	lake	Tu,Al,C	(523)
Cherry Valley	1949	M	lake	Tu	760
Copenhagen	1954	M	Deer R.	Tu	(700)
Evans Mills		I	creek	Tu	
Gasport	1949	M	creek	Tu,Fe,M	(800)
Greenwood Lake	1954	M	lake	Tu,Al	(819–4,000)
Hudson	1959	M	reservoir and creek	Tu,Fe,C	(11,000)
Illion	1961	M	reservoir	Tu,Al	(10,300)
Iriquois School	1954	P	reservoir	Tu	
Lyndonville	1955	M	lake	Tu,Al	755
Massina	1963	M	St. Lawrence R.	Tu	16,000
Menands	1962	M	lake, Troy City	Tu	(2,314)
Montrose	1955	WD	Catskill Aqueduct	Tu	14,000
Mount Morris	1948	P	river	Tu	
Newark	1962	M	lake	Tu	(13,000)
Newcomb	1963	M	Hudson R.	Tu	
Newcastle Wtr. Co.	1960	WD	lake	Tu,Al,C	800
Palmyra	1956	M	L. Canandaigua	Tu	3,100
Peekskill	1961	M	lake-reservoir	Tu	18,000
Peru	1959	M	lake-reservoir	Tu,C	(900)

[a] Recent installations not recorded in the table are located at Raybrook State Park, N.Y., and Santa Ana Mountain Valley Water District, Cal.

[b] All plants were approved by the state health department except those at Seal Beach, Folsom, and Riverside, Cal.; Worthington, Minn.; and Iriquois School, N.Y. Filtration rate–1.0 normal design (1.6) maximum.

[c] Meanings of initials are: NH, native hospital; M, municipal; I, industrial; WD, water district; BR, U.S. Bureau of Reclamation; CC, church camp; PU, private utility; P, private; SP, state park; and CCl, country club.

4.2–9 DIATOMITE FILTRATION PLANTS FOR POTABLE WATER FILTRATION[a] *(Continued)*

Approximate production,[e] mgd	Design rate,[e] gpm/ft²	Plant capacity, gpm	Make and type of filter[f]	Remarks
0.35			BIF-V	temporary, abandoned
0.045	g	500	Sparkler-P	
0.025	g		Refinite-P	
0.175	1.0	425	BIF-P	
	1.06		BIF-P	abandoned (1963)
0.20	1.6	600	Refinite-P	
0.03–0.04	0.76	250	BIF-V	
0.03	g	170	Refinite-P	
0.05	g	100–700	Refinite-P	
0.015	g	50	Refinite-P	
0.45–1.36	1.5–1.0	2,200	Refinite-P	
0.36		250	Sparkler-V	
0.6	1.0	454	Gen. Filter-P	municipally owned
			BIF-V	abandoned
0.20		200	BIF-P	abandoned
0.15–0.175	1.0	190	R.P.Adams-P	
			BIF-P	abandoned
0.135	0.7(1.2)	270	BIF-V	incomplete, post-micro-strainer
0.25	0.925		BIF-P	abandoned
0.12	0.83	250	Sparkler-P	8 hr operation
			Sparkler-P	industrial only
0.5		30	BIF-P	standby only
	2.0	500	Refinite-P	auxiliary only
2.0	0.975	1,390	R.P.Adams-P	1,428 ft²; turbidity, 200
0.57	1.0(2.0)	466	R.P.Adams-P	j
			R.P.Adams-P	school
0.164	0.9	126	BIF-P	supplements sand filter
5.0	1.0	3,600	R.P.Adams-P	incomplete
0.8–1.6	1.0	1,120	Delaval-P	turbidity 6–60
2.0	0.5–2.0	500–3,000	Waterite-V	1,500 ft²
	1.0		R.P.Adams-P	U.S. Corps of Engineers
1.0	1.0–1.5	720	R.P.Adams-P	incomplete auxiliary
0.4	1.0	280	R.P.Adams-P	incomplete
0.05	0.5	35	Bowser-P	used on pump suction
0.3(0.19–0.55)	1.0	400	Sparkler-P	avg turbidity, 1.6; range, 0–100
1.0	1.0	700	BIF-V	
0.3	0.75	100–225	R.P.Adams-P	auxiliary

[d] Meanings of symbols other than chemical are: Tu, turbidity; Al, algae; O, organic and amorphous matter; Ta, taste; S, softening; LSS, lime-soda ash softening; Sa, sand; Pl, plankton; C, color.

[e] Parentheses enclose estimated data.

[f] P is for pressure filter; V, vacuum filter.

[g] System demand.

[j] Post-microstaining, presoftening auxiliary to slow sand filters.

[k] Midland-Saginaw pipeline.

4.2–9 DIATOMITE FILTRATION PLANTS FOR POTABLE WATER FILTRATION[a] *(Continued)*

Plant location[b]	Year built	Type of service[c]	Source of water	Substance removed[d]	Population served[e]
Sherburne	1953	M	stream and well	Tu,Fe,C	1,600
Tupper Lake	1952	M	Tupper L.	Tu,Al,Fe,C	5,440
Willsboro	1952	M	L. Champlain	Tu	838
N.C.					
Durham		M	lake	Tu	
Ohio					
Napoleon	1962	I	river	Tu	
Okla.					
Quarry Isle State Pk.	1955	SP	L. Wister	Tu	
Osage Hills	1960	SP	lake	Tu	
Ont.					
Marmora			Crowe R.	Tu,C	1,200
Ore.					
Estacata	1955	M	Clackamas R.	Tu	(950)
Yoncalla	1954	M	stream	Tu	(626)
Pa.					
Bedford	1955	M	river	Tu	(3,600)
Emporium	1956	PU	creek	Tu	5,000
Troy	1964	M	spring-reservoir	Tu,T	
S.C.					
Columbia	1962	CCl	streamfed lake	Tu,Al,Fe	
Va.					
Norton	1954	M	abandoned mine	Fe	(4,315)
W.Va.					
Opequan	1962	WD	quarry	Tu	2,000

[a] Recent installations not recorded in the table are located at Raybrook State Park, N.Y., and Santa Ana Mountain Valley Water District, Cal.

[b] All plants were approved by the state health department except those at Seal Beach, Folsom, and Riverside, Cal.; Worthington, Minn.; and Iriquois school, N.Y. Filtration rate–1.0 normal design (1.6) maximum.

[c] Meanings of initials are: NH, native hospital; M, municipal; I, industrial; WD, water district; BR, U.S. Bureau of Reclamation; CC, church camp; PU, private utility; P, private; SP, state park; and CCl, country club.

[d] Meanings of symbols other than chemical are: Tu, turbidity; Al, algae; O, organic and amorphous matter; Ta, taste; S, softening; LSS, lime-soda ash softening; Sa, sand; Pl, plankton; C, color.

4.2–9 DIATOMITE FILTRATION PLANTS FOR POTABLE WATER FILTRATION[a] *(Continued)*

Approximate production,[e] mgd	Design rate,[e] gpm/ft²	Plant capacity, gpm	Make and type of filter[f]	Remarks
0.44	2.0	624	BIF-P	
0.9	2.0	624	BIF-P	auxiliary only
0.1	0.9	230	BIF-P	
0.14			Sparkler-P	experimental
0.7	1.6	470	Delaval-P	Campbell Soup
0.050		75	Lon Scott-P	state recreation area
			Lon Scott-P	state recreation area
0.044–0.25			Sparkler-V	built by Ontario Water Resources Commission; 7 day runs
	1.0	400	Sparkler-P	unsuccessful
0.36	1.0	200	BIF-P	successful
	1.0	400	Sparkler-P	auxiliary
	0.94(1.9)	750	Sparkler-P	auxiliary
0.200	0.6	280	Vacuumite	
0.014	0.5	100	BIF-V	
			Sparkler-P	emergency (1954), abandoned
0.1	2.0	250	Infilco-P	low turbidity water

[e] Parentheses enclose estimated data.
[f] P is for pressure filter; V, vacuum filter.

Source: Reprinted with publisher's permission from *J. Am. Water Works Assn.*, 67(2):157, © 1965 by the American Water Works Association, Inc., 2 Park Ave., New York, N.Y. 10016.

4.2–10 CHARACTERISTICS OF RAW AND FILTERED WATER

Characteristic	Raw water	Filtered water
Turbidity, Jackson units	24–160	0–0.3
Color, color units	15–45	0–7
Total plankton/ml	610–2,100	10–30
Bacteria, MPN	230–2,000	0–38
Bacteria, MPN[a]	460–24,000	0

[a] Filtrate prechlorinated (average 1.0 ppm total available chlorine).

Source: Reprinted with publisher's permission from *J. Am. Water Works Assn., 56*(8):1047, © 1964, by the American Water Works Association Inc., 2 Park Ave., New York, N.Y. 10016.

4.2–11 CHARACTERISTICS OF FILTER SYSTEMS

Characteristic	Sand filter	Pressure filter	Vacuum filter
Operating time, days	135	110	26
Rate, gpm/ft²	2.2	1.84	1.17
Water filtered, mil gal	16.7	15.2	2.7
Filter cycles	111	139(33[a])	80[a]
Terminal pressure, ft	7.5	110	20
Backwash water, gal	3,500	30	300

[a] Combination of pressure and vacuum filter tests.

Source: Reprinted with publisher's permission from *J. Am. Water Works Assn., 56*(8):1047, © 1964, by the American Water Works Association Inc., 2 Park Ave., New York, N.Y. 10016.

4.2–12 TYPICAL THREE MEDIA HIGH RATE FILTRATION PLANT PERFORMANCE

Location	Turbidity: Applied, JTU	Turbidity: Filtered, JTU	Filter rate, gpm/ft² [a]	Filter runs, hr	Back-wash water,%
Regina, Sask.	0.5	0.1	4	30	2
Siler City, N.C.	1	0.2	5	30	2
Pasadena, Texas (industrial)	3	0.5	6	40	1
Corvallis, Ore.	3	0.2	3	50	1
Ocaquan, Va.	1	0.1	4	48	1
Samoa, Calif. (industrial)	10	0.5	5	18	2
Fort Lauderdale, Fla.	1	0.1	4	80	1
Knoxville, Tenn.	1	0.2	5[b]	60	1
Albany, Ore.	2	0.2	5	20	2
Longview, Texas	4	0.2	3	48	2
Paris, Texas	10	0.3	3	50	2
Newport, Ore.	3	0.2	4	20	3
Norristown, Pa.	6	0.3	4	45	1
Springfield, Mo.	1.5	0.2	4½	24	1
Winnetka, Ill.	2	0.3	5	15	2
Stanton, Del.	2	0.2	4	30	3

[a] Typical summer peak daily flow rate. Hourly peaks are generally in the range of 4–6 gpm/ft².

[b] Under test. Some filters are out of service so that rate could be increased on remaining filters.

Source: Reprinted with publisher's permission from *J. Am. Water Works Assn., 64*(3):203, © 1972, by the American Water Works Association Inc., 2 Park Ave., New York, N.Y. 10016.

4.2–13 OPERATING RESULTS OF MIXED MEDIA AND SAND FILTERS

	Mixed media filter		Sand filter	
	Max	Avg	Max	Avg
Applied water turbidity, ppm	3	2	3	2
Filtration rate, gpm/ft²	5	3.5	3.4	2.0
Filter effluent turbidity, ppm	0.3	0.1	0.2	0.1
Head loss at time of wash, ft	2.8	2.2	4.6	3.0
Length of runs, hr	160	60	62	42
Wash water used, %		1.6		2.1

Source: Reprinted with publisher's permission from *J. Am. Water Works Assn., 60*(12):1377, © 1968, by the American Water Works Association Inc., 2 Park Ave., New York, N.Y. 10016.

4.2–14 PHYSICAL CHARACTERISTICS OF FILTRATION MEDIA

Physical parameters	Poly-styrene	Anthra-cite	Garnet	Mag-netite	Crushed flint sand	Crushed flint sand	Crushed flint sand	Quarry sand	Quarry sand
Sieve size–mm	3.175 to 2.057	1.676 to 1.405	0.599 to 0.500	0.500 to 0.422	0.500 to 0.422	0.599 to 0.500	0.853 to 0.699	1.676 to 1.405	2.411 to 2.057
Fall velocity–cm/sec	3.39	6.35	9.40	10.95	5.00	6.54	8.10	16.85	20.90
Hydraulic diameter–mm	2.50	1.14	0.467	0.415	0.38	0.435	0.60	1.165	1.36
Sphericity	1.00	0.745	0.854	0.90	0.83	0.80	0.78	0.765	0.62
Density–gm/cu cm	1.04	1.40	3.83	4.90	2.65	2.65	2.65	2.65	2.65
Porosity	0.35	0.425	0.47	0.42	0.464	0.464	0.464	0.39	0.39

Source: Reprinted, with publisher's permission from *J. Am. Water Works Assn.*, *61*(10):504, © 1969 by the American Water Works Association, Inc., 2 Park Ave., New York, N.Y. 10016.

4.2–15 TYPICAL RESULTS OF UPFLOW AND DOWNWARD FLOW FILTRATION

Luton, East Hyde Sewage Works, Great Britain

	Feed		Upflow filter effluent				Downward flow filter			
	BOD,	SS,[a]	BOD		SS[a]		BOD		SS[a]	
Month	mg/l	mg/l	mg/l	Removal, %	mg/l	Removal, %	mg/l	Removal, %	mg/l	Removal, %
January 1966[b]	5.3	12.8	2.6	51	5.7	55.5	3.1	41.5	7.5	41.4
February 1966[c]	10.4	28.5	2.6	75	5.3	81.5	4.2	59.6	9.5	66.6
March 1966[c]	9.8	34.8	2.5	74	5	80.5	3.1	68.5	8.5	75.5

[a] SS represents suspended solids.
[b] Upflow filter was operated at 4.17 gpm/ft², downward flow at 4 gpm/ft².
[c] Upflow and downward flow filters were operated at 2.83 gpm/ft².

Source: A.E. Naylor *et al.*, *Water Pollut. Control* (Brit.), *66*:309, 1967.

4.2–16 FILTER BACKWASH CHARACTERISTICS

Plant	Treatment	BOD (5 day), mg/l	COD, mg/l	pH	Total solids, mg/l	Volatile solids, mg/l	Total suspended solids, mg/l	Volatile suspended solids, mg/l
A	Alum Coagulation and sedimentation	4.2	28	7.8	121	44	47	31
B	Alum Coagulation, clarifier	3.7	75	7.2	378	115	104	53
C	Alum Coagulation, clarifier	2.8	160	7.8	166	45	75	40
D	Alum Coagulation, clarifier	1.8	–	–	–	–	100	60
F	Iron and manganese removal	–	–	6.9	1,487	343	1,370	400
H[1,2,a]					360			
I[1,2,b]					340			
K[1,3,d]					390			
L[1,4,e]							800[f]	
M[1,4,g]							120[f]	
N[1,5,h]					370			

[a] Shoremont Plant, Rochester, N.Y.
[b] Wolcott, N.Y.
[c] Ithaca, N.Y.
[d] Colorado Springs, Colo. (Plant No. 1)
[e] Detroit, Mich. (Water Works Park)
[f] Coliforms MPN/100 ml = 0
[g] Detroit, Mich. (Springwells)
[h] Cornell, N.Y.

Source: H.B. Russelmann, *Water Sewage Works, 115*(R.N.):R64, 1968.

REFERENCES

1. E.R. Sutherland, *J. Am. Water Works Assn., 61*(4):186, 1969.
2. *Waste Alum Sludge Characteristics and Treatment,* O'Brien and Gere, Consulting Engineers, Research Report No. 15, New York State Department of Health, Albany, 1966.
3. D.P. Proudfit, *J. Am. Water Works Assn., 60*(6):674, 1968.
4. W.W. Aultman *et al., J. Am. Water Works Assn., 58*(9):1102, 1966.
5. C.D. Gates and R.F. McDermott, *Characteristics of and Methodology for Measuring Water Filtration Plant Wastes,* Research Report No. 14, New York State Department of Health, Albany, 1966.

4.2–17 ANALYSES OF BACKWASH AND RECOVERED WATERS

Colorado Springs Purification Plant 1

Component	Backwash water	Recovered water
Turbidity, JTU	66.0	2.3
Iron, mg/l	1.68	0.09
Bicarbonate, mg/l	33.0	31.0
Sulfate, mg/l	90.0	30.0
Total solids, mg/l	390.0	82.0
Hardness as $CaCO_3$	40.0	42.0
Aluminum, mg/l	1.68	0.33

Source: Reprinted, with publisher's permission, from *J. Am. Water Works Assn., 60*(6):674, © 1968, by the American Water Works Association, Inc., 2 Park Ave., New York, N.Y. 10016.

4.2–18 ANALYSIS OF BACKWASH WATER AT THE CONCLUSION OF A FILTER RUN FOR IRON REMOVAL

Filter depth, in.	Ferrous iron, %	COD, mg/l	Org N, mg/l	Bacteria count, no./ml
3	19.0	17	1.2	6,300
9	16.7	25	1.8	11,300
18	9.8	33	2.1	12,800
30	9.1	42	3.2	16,200

Source: Reprinted, with publisher's permission, from *J. Am. Water Works Assn., 59*(7):878, © 1967, by the American Water Works Association, Inc., 2 Park Ave., New York, N.Y. 10016.

4.2–19 WATER TREATMENT PLANT WASTE PRODUCTION

	Sedimentation basin sludge		Filter backwash	
	Accumulation per mil gal of raw water processed			
Plant	Waste volume, gal	Solids volume, lb	Waste volume, gal	Solids volume, lb
Rochester (Shoremont), N.Y.[1]	1,230	45	21,200	24
Wolcott, N.Y.[1]	2,660	94	60,500	46
Detroit, Mich.				
(Water Works Park)[2]				52
(Springwells)[2]			<20,000	
St. Paul, Minn.[3]	6,500	2,300	10,000	
Ithaca, N.Y.	2,270	159		68
Cornell, N.Y.				
(Filter No. 1)[4]			13,800	48
(Filter No. 2)[4]			9,300	47

Note: All plants use coagulation, sedimentation, and rapid sand filtration, but major variations in processes exist.

Source: Reprinted, with publisher's permission, from *J. Am. Water Works Association., 61*(4):186, © 1969, by the American Water Works Association, Inc., 2 Park Ave., New York, N.Y. 10016.

REFERENCES

1. O'Brien and Gere, Consulting Engineers, *Waste Alum Sludge Characteristics and Treatment,* Res. Rpt. 15, New York State Department of Health, Albany, 1966.
2. W.W. Aultman *et al., J. Am. Water Works Assn., 58*(9):1102, 1966.
3. D.P. Proudfit, *J. Am. Water Works Assn., 60*(6):674, 1968.
4. C.D. Gates and R.F. McDermott, *Characteristics of and Methodology for Measuring Water Filtration Plant Wastes,* Res. Rpt. 14, New York State Department of Health, Albany, 1966.

4.2–20 FILTRATION PROCESS WASTES

Process wastes item	Diatomite filtration	Conventional filtration	
Origin of wastes	Filter backwash	Sedimentation basin sludge	Filter backwash
Typical magnitude of wastes			
Volume, percent of throughout	0.1–0.5	0.05–0.1	0.5–3
Suspended solids, concentration, range, mg/l	10,000–15,000	5,000–50,000	150–400
Suspended solids, weight, lbs/million gals, throughput	100–400	100–250	10–50
Typical characteristics			
Settleability	Filter aid, good Entrapped turbidity, varies from poor to good	Good	Fair
Putrescibility	Negligible or minor	Varies with nature of solids removed and whether or not sludge removal is continuous or intermittent	Negligible
Abrasiveness	Pronounced	Usually none	None
Porosity	Pronounced	None	None
Drainability	Good	Poor to fair	Poor to fair
Form of Solids	Rigid	Amorphous	Amorphous
Preferred means of disposal			
Wastes as produced	To sanitary sewer where allowable or else to dewatering facility	To sanitary sewer where allowable or else to dewatering facility	To settling basin for reclamation of decant or direct to supply reservoir
Dewatered sludge	To landfill, reclamation as a by-product, or reuse as a filter aid	To landfill; in limited cases, alum recovery	Combine with sedimentation basin sludge
Decant or filtrate	Reclaim as raw water where economical or discharge clear liquid to water course or ground water	Reclaim where economical; percolate or discharge clear liquor to water course	Recover for useful purposes where economical or discharge clear liquid to watercourse

Source: Reprinted, with publisher's permission, from *J. Am. Water Works Assn., 62*(8):507, © 1970, by the American Water Works Association, Inc., 2 Park Ave., New York, N.Y. 10016.

4.2–21 COMPARISON OF WATERWORKS SLUDGE TREATMENT METHODS

	Solid phase			Liquid phase		
	Presettling	Maximum total solids, %	Suitable for landfill	Suspended solids, mg/l	BOD, mg/l	COD, mg/l
Drying beds	Yes		Yes	200	60	170
Vacuum filters		20+				
Wolcott	No	20+	Yes	–	–	–
Shoremont	Yes	18	Yes	30	83	85
Centrifuge	Yes	10–12	No	25–30	–	–
Lagoon	No	10–14	No	24	16	5

Source: Reprinted, with publisher's permission, from *J. Am. Water Works Assn., 60*(7):319, © 1968, by the American Water Works Association, Inc., 2 Park Ave., New York, N.Y. 10016.

4.2–22 CENTRIFUGE DATA FOR SELECTED INSTALLATIONS

Installation	Average flow, mgd	No. of centrifuges	Feed, % solids	Concentrate, % solids
Austin, Texas	30	1	1	0.5
Dayton, Ohio	61.2	3	15–25	5.5
Lansing, Mich.	23	1	20	3
Miami, Fla.	81	2	40	5–10

Source: Reprinted, with publisher's permission, from *J. Am. Water Works Assn., 61*(12):682, © 1969, by the American Water Works Association, Inc., 2 Park Ave., New York, N.Y. 10016.

4.2–23 COAGULANT DOSAGE OF ELECTROLYTES

To Coagulate Positive Colloids			To Coagulate Negative Colloids		
Electrolyte	Anion valence	Minimum concentration required, mmol/l	Electrolyte	Cation valence	Minimum concentration required, mmol/l
KCl	1	103	NaCl	1	51
KBr	1	138	KNO_3	1	50
KNO_3	1	131	K_2SO_4	1	63
K_2CrO_3	2	0.325	$MgSO_4$	2	0.81
K_2SO_4	2	0.219	$ZnCl_2$	2	0.68
$K_3Fe(Cn)_6$	3	0.096	$BaCl_2$	2	0.69
			$AlCl_3$	3	0.09

Source: N.L. Nemerow, *Industrial Waste Treatment*, Addison-Wesley, Reading, Mass., 1963.

4.2–24 SCREENING FOR WATER SUPPLIES

Source of water	Water usage	Purpose of treatment by screening or straining	Appropriate fabric, meshes/in.	Rate of treatment, gpd/ft^2
River or lake	Pumpage only (to be treated further by sedimentation, etc.)	Protection of pumping plant	2 x 2	Up to 10^6 [a]
	Pumpage only, direct to fine screens or sand filters	Protection of treatment works	2 x 2 or perhaps coarser followed by 25 x 25 or 50 x 50	$5 \times 10^5 - 10^6$ [a]
	Pumpage only, direct to fine screens or sand filters	Relief of loading of other treatment works but where solids of microscopic size are not of much significance	As above or finer down to 120 x 120 according to suspended solids content or character	$2 \times 10^5 - 4 \times 10^5$ [a]
	Pumpage only, direct to fine screens or sand filters	As above, but where phytoplankton or zooplankton are significant; also in special cases where water is of good quality except for suspended solids content largely removable by straining	Special microfabrics (Mark II adequate for zooplankton, Mark I or 0 for phytoplankton); or very fine mesh fabrics, e.g., 200 x 200	$2 \times 10^4 - 5 \times 10^4$ [a]
Town mains	Industrial or other uses	Polishing treatment for suspended solids or for bacteria	Special pads or fabrics	Very low, perhaps up to 10^3

[a] These ratings are for screens provided with automatic back washing (cleaning).

Note: Ground water sources normally do not require screening; growths of so-called sulfur or iron bacteria may develop in some cases, but usually cannot be successfully treated by screening.

Source: L.D., Bowen, J.W. Lovell, Jr., and D.K.B. Thistlewayte, 3rd Int. Conf. Water Pollut. Res., Munich, 1966. © Water Pollution Control Federation Washington, D.C., 1966.

4.2–25 CHEMICAL DOSAGE EQUIVALENTS

Table A

	1 gpg equals 17.1 ppm or 1 ppm equals 0.0573 gpg						1 gpg equals 142.9 lb/mil gal					
1 ppm	HCO_3	$CaCO_3$	Al_2O_3	$Al_2(SO_4)_3 \cdot 18H_2O$	NH_4OH	$Ca(HCO_3)_2$	$CaSO_4$	CO_2	CaO	NaOH	Cl_2	$FeCl_3$
Alkalinity–HCO_3	1.000	0.822	0.279	1.824	0.575	1.325	1.114	0.181	0.460	0.656	0.582	0.890
Alkalinity–$CaCO_3$	1.218	1.000	0.339	2.219	0.702	1.619	1.360	0.440	0.560	0.800	0.709	1.080
Aluminum oxide–Al_2O_3	3.588	2.945	1.000	6.537	2.064	4.742	4.006	2.589	1.650	2.354	2.085	1.592
Aluminum sulfate–$Al_2(SO_4)_3 \cdot 18H_2O$	0.540	0.451	0.153	1.000	0.316	0.730	0.613	0.198	0.253	0.361	0.319	0.487
Ammonium hydroxide–NH_4OH	1.738	1.424	0.484	3.156	1.000	2.309	1.937	0.313	0.800	1.137	1.011	1.538
Calcium bicarbonate–$Ca(HCO_3)_2$	0.752	0.618	0.209	1.370	0.433	1.000	0.840	0.271	0.346	0.494	0.437	0.667
Calcium sulfate–$CaSO_4$	0.897	0.735	0.250	1.631	0.516	1.191	1.000	0.323	0.412	0.588	0.521	0.792
Carbon dioxide–CO_2	5.545	2.275	0.386	2.524	3.191	3.684	3.095	1.000	1.275	1.820	3.223	2.459
Caustic lime–CaO	2.174	1.784	0.606	3.959	1.251	2.889	2.429	0.784	1.000	1.428	1.264	1.976
Caustic soda–NaOH	1.525	1.250	0.425	2.773	0.877	2.024	1.700	0.549	0.700	1.000	0.886	1.350
Chlorine–Cl_2	1.720	1.409	0.479	3.136	1.009	2.286	1.918	0.310	0.791	1.128	1.000	1.551
Ferric chloride–$FeCl_3$	1.129	0.925	0.314	2.054	0.656	1.499	1.259	0.407	0.519	0.740	0.656	1.000
Ferric oxide–Fe_2O_3	1.143	1.880	0.639	4.173	0.639	3.028	2.558	1.648	1.054	1.503	0.664	2.031
Ferrous sulfate–$FeSO_4 \cdot 7H_2O$	0.439	0.360	0.183	0.799	0.252	0.583	0.490	0.158	0.202	0.288	0.252	0.584
Ferric sulfate–$Fe_2(SO_4)_3$	0.901	0.750	0.255	1.665	0.525	1.209	1.021	0.660	0.421	0.600	0.531	0.811
Fluosilicic acid–H_2SiF_6	0.847	0.695	0.236	4.620	0.487	1.125	0.944	0.153	0.389	0.555	0.492	0.749
Hydrated lime–$Ca(OH)_2$	1.647	1.351	0.458	2.998	0.948	2.188	1.838	0.594	0.757	1.081	0.957	1.460
Magnesium bicarbonate–$Mg(HCO_3)_2$	0.834	0.684	0.235	1.518	0.487	1.108	0.930	0.301	0.383	0.547	0.484	0.739
Magnesium carbonate–$MgCO_3$	1.447	1.187	0.403	2.633	0.833	1.922	1.614	0.522	0.665	0.950	0.841	1.282
Magnesium hydroxide–$Mg(OH)_2$	2.092	1.715	0.583	3.807	1.204	2.762	2.333	0.756	0.961	1.371	1.216	0.927
Magnesium sulfate–$MgSO_4$	1.013	0.831	0.282	1.845	0.583	1.346	1.131	0.365	0.466	0.665	0.588	0.898
Manganese dioxide–MnO_2	0.282	0.231	0.787	0.514	0.162	0.376	0.315	0.509	1.299	1.852	0.164	2.500
Soda ash–Na_2CO_3	1.151	0.943	0.326	2.094	0.662	1.528	1.283	0.415	0.529	0.755	0.745	1.019
Sodium metaphosphate–$NaPO_3$	0.500	0.490	0.167	1.089	0.344	0.794	0.667	0.108	0.275	0.392	0.347	0.536

4.2–25 CHEMICAL DOSAGE EQUIVALENTS (*Continued*)

Table B

1 gpg equals 17.1 ppm or 1 ppm equals 0.0573 gpg	1 gpg equals 142.9 lb/mil gal					1 ppm equals 8.34 lb/mil gal						
1 ppm	Fe_2O_3	$FeSO_4 \cdot 7H_2O$	$Fe_2(SO_4)_3$	H_2SiF_6	$Ca(OH)_2$	$Mg(HCO_3)_2$	$MgCO_3$	$Mg(OH)_2$	$MgSO_4$	MnO_2	Na_2CO_3	$NaPO_3$
Alkalinity–HCO_3	0.873	2.280	1.093	1.181	0.607	1.198	0.690	0.478	0.987	0.327	0.869	1.672
Alkalinity–$CaCO_3$	0.531	2.777	1.331	1.441	0.740	1.462	0.843	0.582	1.203	0.432	1.060	1.002
Aluminum oxide–Al_2O_3	1.566	5.454	3.931	8.476	2.180	4.247	2.481	1.723	3.542	1.270	3.119	6.000
Aluminum sulfate–$Al_2(SO_4)_3 \cdot 18H_2O$	0.239	1.251	0.600	1.297	0.334	0.659	0.380	0.263	0.542	1.948	0.478	0.918
Ammonium hydroxide–NH_4OH	1.519	3.960	1.900	2.055	1.056	2.084	1.202	0.831	1.717	0.615	1.523	2.704
Calcium bicarbonate–$Ca(HCO_3)_2$	0.328	1.715	0.822	0.889	0.457	0.903	0.520	0.360	0.743	2.603	0.654	1.258
Calcium sulfate–$CaSO_4$	0.391	2.041	0.979	1.060	0.544	1.075	0.620	0.429	0.884	0.321	0.779	1.500
Carbon dioxide–CO_2	0.605	3.159	1.514	6.550	1.684	1.663	1.917	1.326	2.737	1.963	2.411	9.045
Caustic lime–CaO	0.949	4.995	2.376	2.569	1.321	2.609	1.504	1.043	2.147	0.759	1.891	3.636
Caustic soda–NaOH	0.665	3.470	1.666	1.802	0.925	1.827	1.035	0.729	1.503	0.540	1.325	2.550
Chlorine–Cl_2	1.505	3.921	1.880	2.032	1.045	2.064	1.189	0.822	1.698	0.609	1.495	2.877
Ferric chloride–$FeCl_3$	0.492	1.713	1.233	1.335	0.685	1.353	0.779	0.539	1.113	0.400	0.981	1.889
Ferric oxide–Fe_2O_3	1.000	3.482	2.504	1.352	1.392	2.712	1.584	1.096	2.262	0.405	1.991	1.912
Ferrous sulfate–$FeSO_4 \cdot 7H_2O$	0.287	1.000	0.719	0.518	0.267	0.525	0.303	0.210	0.433	0.155	0.382	0.737
Ferric sulfate–$Fe_2(SO_4)_3$	0.399	1.390	1.000	1.622	0.556	1.083	0.635	0.437	0.903	0.324	0.795	1.530
Fluosilicic acid–H_2SiF_6	0.742	0.193	0.925	1.000	0.514	1.015	0.585	0.404	0.836	0.300	0.736	1.416
Hydrated lime–$Ca(OH)_2$	0.718	3.751	1.799	1.944	1.000	1.975	1.138	0.787	1.625	0.583	1.432	2.753
Magnesium bicarbonate–$Mg(HCO_3)_2$	0.369	1.899	0.923	0.985	0.506	1.000	0.576	0.404	0.823	0.295	0.725	1.394
Magnesium carbonate–$MgCO_3$	0.631	3.295	1.580	1.709	0.878	1.735	1.000	0.692	1.427	0.512	1.258	2.420
Magnesium hydroxide–$Mg(OH)_2$	0.912	1.588	2.285	2.472	1.270	2.474	1.445	1.000	2.063	0.724	1.817	3.585
Magnesium sulfate–$MgSO_4$	0.442	2.308	1.107	1.030	0.615	1.215	0.701	0.484	1.000	0.358	0.881	1.694
Manganese dioxide–MnO_2	2.469	0.643	3.125	3.336	1.715	3.386	1.951	1.349	2.787	1.000	2.454	4.606
Soda ash–Na_2CO_3	0.502	2.620	1.257	1.359	0.698	1.379	0.795	0.550	1.135	0.407	1.000	1.924
Sodium metaphosphate–$NaPO_3$	0.529	1.362	0.653	0.706	0.363	0.717	0.415	0.281	0.590	0.212	0.520	1.000

Source: E.S. Hopkins and E.L. Bean, *Water Purification Control,* 4th ed., Williams and Wilkins, Baltimore, 1966, p. 319.

4.3 IRON AND MANGANESE REMOVAL

4.3–1 RELATIONSHIP OF pH VALUE TO THE MAXIMUM PRECIPITATION OF MANGANESE AND IRON[a]

	Manganese				Iron			
pH	Residual, mg/l	Precipitated, mg/l	Amount, mol	Percentage	Residual, mg/l	Precipitated, mg/l	Amount, mol	Percentage
7.0	5.49	0.00	0.000000	0.0	5.59	0.00	0.000000	0.0
8.0	5.49	0.00	0.000000	0.0	1.53	4.06	0.000073	72.6
8.4					0.95	4.64	0.000083	83.0
8.8	5.49	0.00	0.000000	0.0	0.47	5.12	0.000093	91.6
8.9	3.00	2.49	0.000045	45.4				
9.0	1.90	3.59	0.000065	65.4				
9.2	1.00	4.49	0.000081	81.7				
9.3	0.28	5.21	0.000094	94.9				
9.4	0.11	5.38	0.000098	98.0	0.06	5.53	0.000099	98.9
9.6	0.03	5.46	0.000099	99.4	0.00	5.59	0.000100	100.0
9.8	0.00	5.49	0.000100	100.0				

[a] $4Fe(HCO_3)_2 + 8Ca(OH)_2 \rightarrow 2F3_2O_3 \cdot 3H_2O + 8CaCO_3 + 6H_2O + O_2$
$2Mn(HCO_3)_2 + 4Ca(OH)_2 + 2O_2 \rightarrow 2MnO_2 \cdot H_2O + 4\ CaCO_3 + 4H_2O$

Source: E.S. Hopkins and E.L. Bean, *Water Purification Control,* 4th ed., Williams & Wilkins, Baltimore, 1966.

4.3–2 EFFECTS OF POLYELECTROLYTES USED WITH PERMANGANATE TREATMENT

$KMnO_4$ treatment	Duration of filter runs, hr	Wash water used, %	Mn removed by basins, %	Fe removed by basins, %	Period of time used, yr
Without polyelectrolytes	60	2	71.3	89.4	4
With polyelectrolyte[a]	67	1.9	74.2	91.4	3
With activated silica	73	1.8	74.7	91.6	4

[a] Hagan's coagulant aid HCA-18.

Source: Reprinted, with publisher's permission, from *J. Am. Water Works Assn., 52*(2):219, © 1960, by the American Water Works Association, Inc., 2 Park Ave., New York, N.Y. 10016.

4.3–3 EFFICIENCY OF ION EXCHANGE TREATMENT OF WATER

Camden, N.J. Plant

Constituent	Raw well water concentration, ppm	Treated water concentration, ppm
Eight Months of Operation[a]		
Hardness	96	2
Sodium	49	224
Potassium	6.5	3
HCO_3	62	130
Free CO_2	95	50
Total iron	37 (as Fe)	0.5 (as Fe)
Manganese	1.7 (as Mn)	0.1 (as Mn)
pH	6.1	6.7
Fifteen Months of Operation[b]		
Hardness	86	2
HCO_3	50	138
Total iron	53 (as Fe)	0.3 (as Fe)
Total manganese	1.0 (as Mn)	0.0 (as Mn)
pH	5.7	6.4
Thirty-six Months of Operation[a]		
Hardness	76	2
Sodium	74	280
HCO_3	15	132
Free CO_2	80	50
Total iron	52 (as Fe)	0.1 (as Fe)
Total manganese	1.3 (as Mn)	0.0 (as Mn)
pH	5.6	6.5

[a] Analyses by Permutit Co.
[b] Analyses by Betz Labs. Inc., Philadelphia, Pa.

Source: Reprinted, with publisher's permission, from *J. Am. Water Works Assn., 55*(6):742, © 1963, by the American Water Works Association Inc., 2 Park Ave., New York, N.Y. 10016.

4.3–4 CHARACTERISTICS OF RAW, SETTLED, AND FILTERED WATER

Iron Removal Process

Depth, in.	Time, hr	DO, mg/l	pH	ORP, mv	Alkalinity, mg/l as $CaCO_3$	Total hardness, mg/l as $CaCO_3$	Ca hardness, mg/l as $CaCO_3$	Organic N, mg/l	NH_4-N, mg/l	COD, mg/l	Bacteria count, no./ml
Raw											
	2	0	7.25	+330	440	343	180	3.5	3.2	34	160
	6	0	7.30	+322	–	–	–	–	–	–	–
	12	0	7.31	+317	–	–	–	3.8	2.9	32	280
	20	0	7.27	+328	–	–	–	–	–	–	–
	24	0	7.28	+325	428	350	184	–	–	–	–
	32	0	7.22	+328	–	–	–	2.9	3.0	35	210
	40	0	7.31	+332	–	–	–	–	–	–	–
	48	0	7.32	+330	434	338	172	3.1	3.1	35	120
Aerated and Settled											
	2	7.2	7.75	+440	424	328	171	3.6	3.2	25	1,700
	6	7.1	7.68	+432	–	–	–	–	–	–	–
	12	6.8	7.70	+428	–	–	–	3.7	3.0	22	2,200
	20	6.8	7.62	+445	–	–	–	–	–	–	–
	24	6.7	7.58	+418	396	331	173	–	–	–	–
	32	7.3	7.56	+429	–	–	–	3.1	2.8	29	3,100
	40	6.9	7.67	+413	–	–	–	–	–	–	–
	48	7.7	7.55	+438	418	317	161	3.0	2.9	31	2,600
Filtered											
3	2	6.5	7.80	+442	410	331	172	2.7	3.2	20.1	1,730
	12	6.3	7.70	+452	–	–	–	2.5	3.0	22.0	2,800
	20	6.4	7.62	+453	–	–	–	–	–	–	–
	24	6.2	7.55	+445	388	342	181	–	–	–	–
	32	6.5	7.58	+457	–	–	–	3.5	3.2	28.0	7,200
	40	6.8	7.65	+462	–	–	–	–	–	–	–
	48	7.1	7.53	+453	412	327	165	4.1	3.1	32.0	7,800
9	2	6.3	7.70	+440	410	315	162	1.2	2.8	7.0	560
	12	6.4	7.71	+458	–	–	–	1.3	–	18.2	1,250
	20	6.0	7.63	+453	–	–	–	–	2.8	–	–
	24	6.1	7.54	+455	386	324	178	–	–	–	–
	32	6.2	7.41	+463	–	–	–	2.0	3.2	20.8	1,800
	40	6.3	7.53	+453	–	–	–	–	–	–	–
	48	6.4	7.58	+453	410	306	163	–	3.1	21.3	2,300
18	2	6.2	7.65	+438	408	312	158	1.0	2.8	6.5	430
	12	6.1	7.62	+465	–	–	–	1.2	2.9	17.2	1,120
	20	6.0	7.58	+468	–	–	–	–	–	–	–
	24	5.9	7.53	+472	386	318	165	1.8	3.1	20.2	1,700
	32	6.0	7.43	+461	–	–	–	–	–	–	–
	40	6.1	7.50	+478	–	–	–	–	–	–	–
	48	5.8	7.57	+474	410	304	163	1.9	3.2	21.1	1,850
30	2	6.1	7.53	+448	402	310	161	nil	3.0	6.4	440
	12	6.0	7.69	+463	–	–	–	1.1	–	8.7	960
	20	5.3	7.61	+478	–	–	–	–	–	–	–
	24	5.8	7.59	+463	384	310	175	1.3	2.8	17.0	1,130
	32	5.9	7.43	+458	–	–	–	–	–	–	–
	40	5.7	7.58	+468	–	–	–	–	–	–	–
	48	5.8	7.60	+460	408	296	160	1.7	3.0	20.0	1,620

Source: Reprinted, with publisher's permission, from *J. Am. Water Works Assn., 59*(7):878, © 1967, by the American Water Works Association, Inc., 2 Park Ave., New York, N.Y. 10016.

4.3–5 IRON AND MANGANESE SLUDGE TREATMENT

Process	Sludge form	Sludge disposal	Comments
Oxidation with or without chemical assist. Slow sand filtration	Retained on filter media	Physical-mechanical removal from filter media, hauled to disposal site	Primarily used in older plants. Considered ultimate disposal; no further study needed
Coagulation and clarification. Rapid sand filtration	Clarifier sludge; retained on filter media	Removed with clarifier or filter backwash sludge	Usually combined with coagulation or softening sludges. Additional disposal study needed
Oxidation with chemicals. Pressure or gravity filtration	Retained on filter media	Removed with filter backwash sludge	Disposal problem same as for other filter backwash sludge
Aeration with pH adjustment. With or without chemicals. Detention. Pressure or gravity filtration	Retained on filter media	Removed with filter backwash sludge	Disposal problem same as for other filter backwash sludge
Manganese zeolite	Precipitation on filter, generally anthracite	Removed with backwash sludge. Regenerant brines removed with rinse water	Additional study needed on effects of brines on receiving vehicle
Ion exchange	Insignificant, predominantly and regenerant brines	Removed with rinse water	Additional study needed on effect of brines on receiving vehicle

Source: Reprinted, with publisher's permission, from *J. Am. Water Works Assn., 61*(10):542, © 1969, by the American Water Works Association, Inc., 2 Park Ave., New York, N.Y. 10016.

4.3–6 EQUILIBRIA DESCRIBING Fe(II) SOLUBILITY

Reaction	Equilibrium constant at 25°C	Ref.
$Fe(OH)_2(s) = Fe^{+2} + 2OH^-$	8×10^{-16}	1
$Fe(OH)_2(s) = FeOH^+ + OH^-$	4×10^{-10}	1
$Fe(OH)_2(s) + OH^- = Fe(OH)_3^-$	8.3×10^{-6}	2
$FeCO_3(s) = Fe^{+2} + CO_3^{-2}$	2.1×10^{-11} [a]	3
$H_2CO_3 = H^+ + HCO_3^-$	4.2×10^{-7}	3
$HCO_3^- = H^+ + CO_3^{-2}$	4.8×10^{-11}	3

[a] The constant for $FeCO_3$ is in the range $(4.8–6.3) \times 10^{-11}$

Source: Reprinted, with publisher's permission, from *J. Am. Water Works Assn., 62*(3):198, © 1970 by the American Water Works Association, Inc., 2 Park Ave., New York, N.Y. 10016.

REFERENCES

1. D.L. Leussing and I.M., Kolthoff, *J. Am. Chem. Soc., 75*:2476, 1953.
2. K.N. Grayer and L. Woontner, *J. Phys. Chem., 60*:1509, 1956.
3. W.E. Latimer, *The Oxidation States of the Elements and Their Potentials in Aqueous Solutions,* 2nd ed., Prentice-Hall, Inc., Englewood Cliffs, N.J., 1952.

4.3–7 FORMS OF IRON IN SOLUTION

Equilibrium[a]	Equilibrium constant
$Fe(OH)_2(aq) = FeOH^+ + OH^-$	2×10^{-5}
$FeOH^+ = Fe^{++} + OH^-$	4.5×10^{-6}
$Fe(OH)_2(c) = Fe^{++} + 2OH^-$	1.8×10^{-15}
$Fe(OH)_2(c) = FeOH^+ + OH^-$	4×10^{-10}
$Fe(OH)_2(c) = FeO_2H^- + H^+$	5×10^{-19}
$Fe(OH)_3(aq) = Fe(OH)_2^+ + OH^-$	2.5×10^{-8}
$Fe(OH)_2^+ = FeOH^{++} + OH^-$	4.5×10^{-10}
$FeOH^{++} = Fe^{+++} + OH^-$	2.7×10^{-12}
$Fe(OH)_3(c) = Fe^{+++} + 3OH^-$	6×10^{-38}
	1×10^{-38}
	3.2×10^{-36}
	1×10^{-42}
$Fe(OH)_3(c) = Fe(OH)_2^+ + OH^-$	5.3×10^{-17}
$Fe(OH)_3(aq) = Fe^{+++} + 3OH^-$	4.0×10^{-29}
$Fe^{+++} + Cl = FeCl^{++}$	33
$FeCl_2^+ + Cl = FeCL_3(aq)$	0.1

[a] Dissolved species are indicated by (aq), the solid phase, by (c).

Source: Reprinted, with publisher's permission, from *J. Am. Water Works Assn., 57*(4):458, © 1965 by the American Water Works Association, Inc., 2 Park Ave., New York, N.Y. 10016.

REFERENCE

1. J.D. Hem and W.H. Cropper, U.S. Geological Survey Water Supply Paper No. 1459-A, 1959.

4.3–8 IRON (III) HYDROLYSIS REACTIONS

Equilibrium[a]	Log K
$Fe^{3+} + H_2O \rightleftharpoons Fe(OH)^{2+} + H^+$	– 3.05
$Fe^{3+} + 2H_2O \rightleftharpoons Fe(OH)_2^{+} + 2H^+$	– 6.31
$Fe^{3+} + 3H_2O \rightleftharpoons Fe(OH)_3(s) + 3H^+$	– 3.55
$Fe^{3+} + 4H_2O \rightleftharpoons Fe(OH)_4^{-} + 4H^+$	– 18.0
$2Fe^{3+} + 2H_2O \rightleftharpoons Fe_2(OH)_2^{4+} + 2H^+$	– 2.96
$2Fe^{3+} + 3H_2O \rightleftharpoons Fe_2(OH)_3^{3+} + 3H^+$	–
$2Fe^{3+} + 5H_2O \rightleftharpoons Fe_2(OH)_5^{+} + 5H^+$	–
$3Fe^{3+} + 2H_2O \rightleftharpoons Fe_3(OH)_2^{7+} + 2H^+$	–
$3Fe^{3+} + 4H_2O \rightleftharpoons Fe_3(OH)_4^{5+} + 4H^+$	– 5.77

[a] The coordinated water molecules have been omitted for clarity.

Source: J.D. Hem, U.S. Geological Survey Water Supply Paper No. 1667A, 1963.

4.3–9 MANGANESE EQUILIBRIUM CONSTANTS

Reaction[a]	Equilibrium constant
$Mn(OH)_2$ [c] $+ H^+ \rightleftharpoons MnOH^+ + H_2O$	4.9×10^4
$MnOH^+ + H^+ \rightleftharpoons Mn^{+2} + H_2O$	4.2×10^{10}
$Mn(OH)_2$ [c] $\rightleftharpoons HMnO_2^- + H^+$	8.7×10^{-20}
$Mn^{+2} + HCO_3 \rightleftharpoons MnHCO_3^+$	63
$Mn^{+2} + SO_4^{-2} \rightleftharpoons MnSO_4$ [aq]	1.9×10^2
$MnS + 2H^+ \rightleftharpoons H_2S + Mn^{+2}$	4.14×10^7
$H_2CO_3^- \rightleftharpoons H^+ + HCO_3^-$	4.16×10^{-7}
$HCO_3 \rightleftharpoons H^+ + CO_3^{-2}$	4.84×10^{-11}
$H_2S \rightleftharpoons H^+ + HS^-$	1.1×10^{-7}
$MnCO_3$ [c] [rhodochrosite] $+ 2H_2O \rightleftharpoons Mn(OH)_2$ [c] $+ 2H^+ + CO_3^{-2}$	7.4×10^{-27}
$MnCO_3$ [c] [ppt.] $+ 2H_2O \rightleftharpoons Mn(OH)_2$ [c] $+ 2H^+ + CO_3^{-2}$	2.3×10^{-26}
$MnHCO_3^+ + H^+ \rightleftharpoons Mn^{+2} + H_2CO_3$ [aq]	4.2×10^4

[a] The letter [c] designates crystalline; [aq] represents dissolved; [c] and [ppt] denotes solids.

Source: J.D. Hem, U.S. Geological Survey Water Supply Paper No. 1667A, 1963.

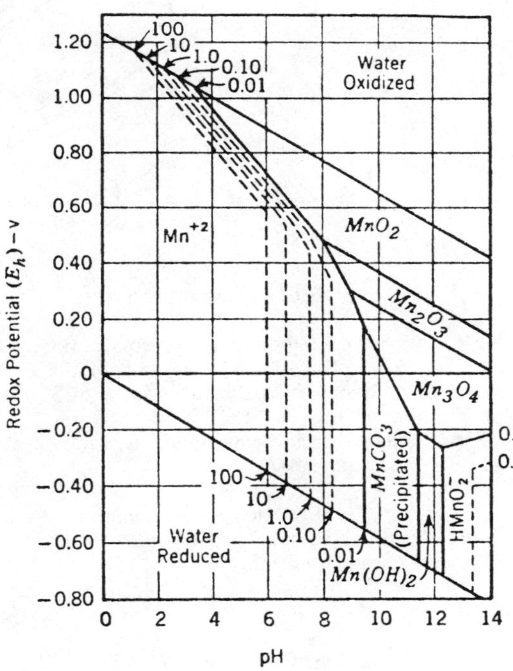

4.3–10 Stability Field Diagram for Manganese.

Note: Dissolved manganese activity is represented by the broken curves. The bicarbonate species activity is 100 ppm as HCO_3^{-4}. Italics indicate the crystalline form.

Source: Reprinted, with publisher's permission from *J. Am Water Works Assn.*, *57*(5):655, © 1965, by the American Water Works Association, Inc., 2 Park Ave., New York, N.Y. 10016.

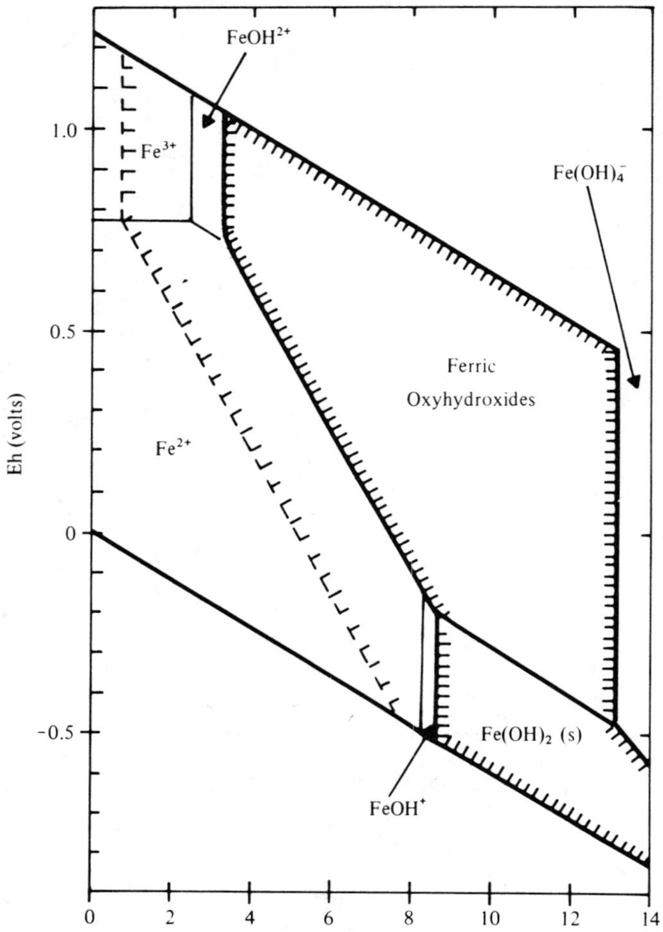

4.3–11 Stability Diagram for Iron.

Note: Eh-pH relations between solids and predominant ionic species in the system Fe-H_2O-O_2. The position of the Fe^{2+}-ferric oxyhydroxide boundary is shown for pK's of 37.1 and 44. The position of the Fe^{3+}-$FeOH^{2+}$ boundary is based on Milburn.[1] Other thermochemical data used to construct the figure are from Langmuir.[2]

Source: D. Langmuir in J.D. Hem, Nonequilibrium Systems in Natural Water Chemistry, *Advances in Chemistry Series No. 106,* Chemical Society, Washington, D.C., 1971, p. 215.

REFERENCES

1. R.M. Milburn, *J. Am. Chem. Soc., 79*:537, 1957.
2. D. Langmuir, U.S. Geological Survey Paper No. 650-B, 1969.

4.4 REMOVAL OF COLOR, ODOR, AND TASTE

4.4–1 ASSOCIATION BETWEEN TURBIDITY AND COLOR REMOVAL

		Filtered water	
Coagulant (alum) dosage, mg/l	**Floc appearance**	**Turbidity, Jackson units**	**Color, standard units**
0	–	1	80
15	None	2.2	20
20	Small	0.75	7
25	Large	0.35	6
30	Large	0.20	3

Source: Reprinted, with publisher's permission, from *J. Am. Water Works Assn., 60*(2):165, © 1968, by the American Water Works Association, Inc., 2 Park Ave., New York, N.Y. 10016.

4.4–2 COMPOSITION OF EXTRACTED COLOR BODY MATERIAL

Constituent	Percentage
Ash (salts)	72.82
Carbon	14
Nitrogen	0.6
Hydrogen	4
Oxygen	8.6 (by difference)

Source: Reprinted, with publisher's permission, from *J. Am. Water Works Assn., 60*(3):357, © 1968, by the American Water Works Association, Inc., 2 Park Ave., New York, N.Y. 10016.

4.4–3 COMPOSITION OF COLOR BODY MATERIAL LOST BY IGNITION

Element	Calculated, %	Literature values[a], %
Carbon	51.8	45–65
Nitrogen	2.2	3–6
Hydrogen	14.8	5 (approximate)
Oxygen	31.8	30–48

[a] For soil organic matter.

Source: Reprinted, with publisher's permission, from *J. Am. Water Works Assn., 60*(3):357, © 1968, by the American Water Works Association, Inc., 2 Park Ave., New York N.Y. 10016.

4.4–4 APPROXIMATE COLOR BODY FRACTIONS FROM DIALYSIS[a]

Fraction	Portion of freeze-dried extract, %	Portion of material lost on ignition, %
Precipitate	7	26
Nondialyzable	4	15
Dialyzable	16 (by difference)	59

[a] Average pore diameter of tubing used = 4.8 nm.

Source: Reprinted, with publisher's permission, from *J. Am. Water Works Assn., 60*(3):357, © 1968, by the American Water Works Association, Inc., 2 Park Ave., New York, N.Y. 10016.

4.4–5 FRACTIONAL MAKEUP OF FREEZE–DRIED COLOR BODY MATERIAL

Fraction	Percentage
Insoluble water	8.35
Ether soluble	0.45
Humic acid	0.74
Fulvic acid (by difference)	90.46

Source: Reprinted, with publisher's permission, from *J. Am. Water Works Assn., 60*(3):357, © 1968, by the American Water Works Association, Inc., 2 Park Ave., New York, N.Y. 10016.

4.4–6 CARBON AND NITROGEN IN VARIOUS COLOR BODY FRACTIONS BY WEIGHT

	Percentage			
Fraction	Carbon	Total carbon	Nitrogen	Total nitrogen
Freeze-dried material	14	100	0.6	100
Insoluble water	25.2	15	1.5	20.9
Ether soluble	62.7	2	1.4	0.01
Humic acid	25.2	1.4	1.3	0.02
Fulvic acid		81.6[a]		79.1[a]
Insoluble pH 1	30.1	27.4	1.8	27.3
Insoluble, hydrolysis	34.2	0.9	0.7	4.3
Nondialyzed	32.4	26.3	1.8	34

[a] Obtained by difference, because sample not homogeneous.

Source: Reprinted, with publisher's permission, from *J. Am. Water Works Assn., 60*(3):357, © 1968, by the American Water Works Association, Inc., 2 Park Ave., New York, N.Y. 10016.

4.4–7 INORGANIC COMPOSITION OF FREEZE–DRIED MATERIAL

Substance	Percentage
Silicon dioxide	3.36
Calcium	7.74
Copper	less than 0.0138
Manganese	0.000552
Magnesium	2.41
Iron	0.254
Zinc	0.099

Source: Reprinted, with publisher's permission, from *J. Am. Water Works Assn., 60*(3):357, © 1968, by the American Water Works Association, Inc., 2 Park Ave., New York, N.Y. 10016.

4.4–8 COLOR REDUCTION WITH OZONE

No. and types of waters tested	O_2 absorbed before ozination, ppm	Color before ozination, units	Reduction in color, %	Residual ozone, ppm
Eight raw waters	2–6 (3.06)[a]	10–79 (28)[a]	25–37 (32)[a]	0–0.08 (0.03)[a]
Six treated public supplies	1.4–3.7 (2.11)	12–44 (19)	12–50 (36)	0–0.10 (0.05)
Eight treated public supplies	0.10–1.30 (0.76)	5.5–8.5 (7)	23–71 (49)	0.01–0.23 (0.11)

[a] Figures in parentheses represent averages.

Source: Reprinted, with publisher's permission, from *J. Am. Water Works Assn., 57*(9):1167, © 1965, by the American Water Works Association, Inc., 2 Park Ave., New York, N.Y. 10016.

4.4–9 EFFECT OF FILTER MEMBRANE PORE SIZE ON APPARENT COLOR OF WATER

Membrane pore size	Color after passage, in APHA units
None (unfiltered water)	55
3 μ	45
1.2 μ	45
0.8 μ	42
0.45 μ	41
0.22 μ	41
100 mμ	40
50 mμ	38
10 mμ	38

Source: Reprinted, with publisher's permission, from *J. Am. Water Works Assn., 60*(3):357, © 1968, by the American Water Works Association, Inc., 2 Park Ave., New York, N.Y. 10016.

4.4–10 MUCIDONE AND ODOR REDUCTION BY CARBON

Carbon, ppm	Percentage of mucidone removed	Threshold odor
0	0	300
10	51	100
20	82	50
25	>90	2

Note: Mucidone concentration, 1 ppm. Mucidone is a chemical substance with a characteristic musty odor and odor threshold number of 3.0×10^8, isolated in high purity from laboratory cultures of actinomycetes taken from the Cedar River near Cedar Rapids, Iowa. Coined from "mucid" (Latin), musty, and "one," denoting a ketone.

Source: Reprinted, with publisher's permission, from *J. Am. Water Works Assn., 59*(10):1320, © 1967, by the American Water Works Association, Inc., 2 Park Ave., New York, N.Y. 10016.

4.4–11 EFFECT OF pH ON REMOVAL OF MUCIDONE BY CARBON

pH	Percentage of mucidone removed
4	45
6	32
7	45
8	36
9	36

Note: Carbon dosage, 10 ppm; contact time, 30 min at 25°C; mucidone concentration, 1 ppm; threshold odor number, 300.

Source: Reprinted, with publisher's permission, from *J. Am. Water Works Assn.*, *59*(10):1320, © 1967, by the American Water Works Association, Inc., 2 Park Ave., New York, N.Y. 10016.

4.4–12 EFFECT OF CONTACT TIME ON REMOVAL OF MUCIDONE BY CARBON

Carbon, ppm	Contact time, min	Percentage of mucidone removed
10	10	47
10	20	47
10	30	51
20	10	79
20	20	81
20	30	82
25	10	83
25	20	90
25	30	>90

Note: pH, 7; mucidone concentration, 1 ppm; threshold odor number, 300.

Source: Reprinted, with publisher's permission, from *J. Am. Water Works Assn.*, *59*(10):1320, © 1967, by the American Water Works Association, Inc., 2 Park Ave., New York, N.Y. 10016.

4.4–13 EFFECT OF POTASSIUM PERMANGANATE ON MUCIDONE AND ODOR

$KMnO_4$, ppm	pH	Contact time, hr	Percentage of mucidone oxidized[a]	Threshold odor
0	7	0.5	0	300
4	7	0.5	48	260
10	7	0.5	>90	285
50	7	0.5	>90	340
4	7	1.0	52	400
4	7	4.0	66	250
10	7	1.0	>90	375
10	7	20.0	>90	200
10	4	4.0	>90	250
100	4	4.0	>90	285
10	7	4.0	>90	285
100	7	4.0	>90	200
10	10	4.0	>90	250
100	10	4.0	>90	285

[a] This oxidation is due to the initial attack on the carbon-carbon double bond and is not reflective of complete oxidation.

Source: Reprinted, with publisher's permission, from *J. Am. Water Works Assn.*, *59*(10):1320, © 1967, by the American Water Works Association, Inc., 2 Park Ave., New York, N.Y. 10016.

4.4–14 EFFECT OF CHLORINE ON MUCIDONE AND ODOR

Chlorine, ppm	pH	Percentage of mucidone oxidized[a]	Threshold odor
10	4	48	250
20	4	>90	220
10	7	6	285
20	7	14	250
10	10	8	300
20	10	15	285

[a] This oxidation is due to the initial attack on the carbon-carbon double bond and is not reflective of complete oxidation.

Note: Contact time, 1 hr. The threshold odors of nonchlorinated samples held for 1 hr at pH values of 4, 7, and 10 were 300, 285, and 285, respectively.

Source: Reprinted, with publisher's permission, from *J. Am. Water Works Assn.*, *59*(10):1320, © 1967, by the American Water Works Association, Inc., 2 Park Ave., New York, N.Y. 10016.

4.4–15 TASTE THRESHOLDS FOR METALS IN WATER

	Ion concentration, ppm			
	Threshold frequency in distilled water		Threshold frequency in spring water	
Metallic constituent[a]	5%	50%	5%	50%
Zinc sulfate	4.3	18	6.8	27
Zinc nitrate	5.2	22		
Zinc chloride	6.3	25	8.6	33
Cupric chloride	2.6	6.6	5.0	13
Ferrous sulfate	0.04	3.4	0.12	1.8
Hydrous ferric oxide	0.7	8.8		
Manganous sulfate	3.6	45		

[a] Threshold values are for the metal ion, not the salt.

Source: Reprinted, with publisher's permission, from *J. Am. Water Works Assn., 52*(5):660, © 1960, by the American Water Works Association, Inc., 2 Park Ave., New York, N.Y. 10016.

4.4–16 REMOVAL OF ACTINOMYCETE ODOR BY ACTIVATED CARBON

Carbon Sample[a]	Threshold odor number	Threshold odor number, % decrease
None (Control)	1,400	
A	8	99.4
B	12	99.1
C	13	99.1
D	8	99.4
E	8	99.4
F	16	98.9

[a] Concentration of each sample 10 ppm. All samples were of Aqua Nuchar, designated respectively as follows: CEE-A, CEE-N, C-190-A, C-190-N, C-145-A, C-145-N.

Source: Reprinted, with publisher's permission, from *J. Am. Water Works Assn., 55*(2):169, © 1963, by the American Water Works Association, Inc., 2 Park Ave., New York, N.Y. 10016.

4.4–17 EFFECT OF CHLORINATION ON ACTINOMYCETE ODOR

Trial 1		Trial 2	
Chlorine added, ppm	Threshold odor number	Chlorine added, ppm	Threshold odor number
0	200	0	400
0.5	70	5.0	216
2.5	80	10.0	80
5.0	75	20.0	200
12.0	80	50.0	97

Source: Reprinted, with publisher's permission, from *J. Am. Water Works Assn., 55*(2):169, © 1963, by the American Water Works Association, Inc., 2 Park Ave., New York, N.Y. 10016.

4.4–18 COMPARISON OF COMMERCIAL ACTIVATED CARBONS FOR BUTANOL ADSORPTION

Carbon	Specifications	Surface area per gram, m^2	Butanol-carbon ratio at break-through[e]	Type of break-through	Relative adsorptive capacity[d]
A	12 x 30	1,164	0.101[c]	sharp[c]	1
B	+30 mesh	995	0.131	sharp	2
C	12 x 30	935	0.060	gradual	3
D	10 x 30	895	0.069	sharp	5
E	10 x 30	830	0.056	gradual	4
F	8 x 20	796	b	immediate	7
G	12 x 20	709	a	immediate	8
H	20 x 40	572	0.028	sharp	9
I	12 x 30	548	a	immediate	10
J	10 x 20	540	0.039	gradual	6

[a] Does not retain more than 99.2%.
[b] Does not retain more than 99.6%.
[c] Extended, low butanol leakage after breakthrough, then sharp dropoff at saturation.
[d] With 60% butanol.
[e] Retention of less than 99.8% of inlet butanol.

Note: Inlet concentration – 3,240 ppm.

Source: Reprinted, with publisher's permission, from *J. Am. Water Works Assn.*, *56*(1):92, © 1964, by the American Water Works Association, Inc., 2 Park Ave., New York, N.Y. 10016.

4.4–19 PHYSICAL PROPERTIES OF COAL-BASED GRANULAR CARBON[a]

	8 x 30 mesh carbon	14 x 40 mesh carbon
Total surface area,[b] m^2/g	800–900	800–900
Bed density,[c] lb/ft^3	30	30
Particle density,[d] g/cm^3	1.4–1.5	1.4–1.5
Effective size, mm	0.80–0.90	0.55–0.65
Uniformity coefficient	⩽1.9	⩽1.7

[a] Type MWT; manufactured by the Pittsburgh Activated Carbon Co., Pittsburgh, Pa.
[b] Measured by the N_2, BET method.
[c] Backwashed and drained.
[d] Wetted in water.

Source: Reprinted, with publisher's permission, from *J. Am. Water Works Assn.*, *57*(11):1440, © 1965, by the American Water Works Association, Inc., 2 Park Ave., New York, N.Y. 10016.

4.4–20 EFFECT OF ACTIVATED CARBON ON HERBICIDES AND INSECTICIDES

Pesticide	Concentration treated with carbon, ppm	Carbon dosage used, ppm	Threshold odor values		Threshold odor units removed per ppm carbon	Concentration after treatment	
			Before	After		Calculated from odor reduction, units	Determined by chemical analysis, ppm
Parathion	10	10	50	4	4.6	0.8	2.6
BHC-37, gamma	25	5	70	6	12.6	0.22	0.08
Malathion, 50%	2	10	50	4	4.6	0.08	0.25
2,4-D, 23.5%	6	20	50	3	2.4	0.085	1.38
2,4-D, 11.7%	1	10	70	2	6.8	0.005	L[a]
Chlordane, 6%	50	10	50	1.4	4.9	0.084	L[a]
DDT, 50%	5	2	70	4	33	0.15	L[a]

[a] Concentrations were too low to determine by available test methods.

Source: Reprinted, with publisher's permission, from *J. Am. Water Works Assn.*, *57*(8):1016, © 1965, by the American Water Works Association, Inc., 2 Park Ave., New York, N.Y. 10016.

4.4–21 ACTIVATED CARBON REQUIRED TO REDUCE PESTICIDE ODOR TO PALATABLE LEVEL

Substance	Concentration of substance, ppm	Threshold odor number	Carbon dosage, ppm	Threshold odor number after carbon
Parathion	10	50	20	4
37 gamma BHC	25	70	15	1.4
Malathion	2	50	20	4
2,4-D	6	50	40	3
DDT	5	70	4	3

Source: Reprinted, with publisher's permission, from *J. Am. Water Works Assn.*, *64*(5):309, © 1972, by the American Water Works Association, Inc., 2 Park Ave., New York, N.Y. 10016.

4.4–22 LOSS OF PESTICIDE FROM RAW RIVER WATER CAUSED BY ADSORPTION AND SEDIMENTATION

	Raw feed		Settling tank effluent	
Pesticide	Turbidity, units	Quantity of pesticide, ppb	Turbidity, units	Quantity of pesticide, ppb
Lindane	13	7.1	13	7.9
Dieldrin	46	8.8	42	6.9
DDT	12	8.0	12	6.3
Parathion	5	9.6	5	7.2
2,4,5-T ester	30	8.3	30	5.6

Source: Reprinted, with publisher's permission, from *J. Am. Water Works Assn.*, *57*(2):181, © 1965, by the American Water Works Association, Inc., 2 Park Ave., New York, N.Y. 10016.

4.4–23 PESTICIDES REMOVED BY CONVENTIONAL WATER TREATMENT

Percent

	Coagulation and filtration				Softening[a]
	Load, ppb				
Pesticide	1	5	10	25	10
Lindane	<10	<10	<10		<10
Dieldrin	55	55	55		<10
DDT			98	97	60
Parathion			20		10
2,4,5-T ester			63		
Endrin	35		35		

[a] Total hardness as $CaCO_3$ reduced from 260–33 ppm and pH increased from 7.6–10.4.

Source: Reprinted, with publisher's permission, from *J. Am. Water Works Assn.*, *57*(2):181, © 1965, by the American Water Works Association, Inc., 2 Park Ave., New York, N.Y. 10016.

4.4–24 ADSORPTION EFFICIENCY FOR CHEMICAL CONTAMINANTS AT SEVERAL CONTACT TIMES

	Carbon contact, gpm/ft³			
	2	1	0.75	0.5
Contaminant	Concentration in effluent, ppb			
Bis(2-chloroethyl) ether	94	44	–[a]	–
Methyl benzyl alcohol	62	13	–	–
2-ethyl hexanol	57	20	–	–
Bis(2-chloroisopropyl) ether	26	10	–	–
Ethylbenzene	20	18	15	5
Isophorane	12	9	–	–
Acetophenone	11	–	–	–
Total	282	114	15	5

[a] Not detectable.

Note: Pilot columns were 4 in. diameter, 5, 10, 15, and 20 ft deep; flow rate was 10 gpm ft². After 41 days of continuous filter service, total CCE in influent was 1,000 ppb and total CCE in effluent was 125 ppb.

Source: Reprinted, with publisher's permission, from *J. Am. Water Works Assn.*, *57*(11):1440, © 1965, by the American Water Works Association, Inc., 2 Park Ave., New York, N.Y. 10016.

4.4–25 IMPURITY REMOVAL BY ADSORPTION-FILTRATION CARBON BEDS[a] AND BY SAND MEDIUM

Impurity	Impurity in influent	Impurity in carbon bed effluent	Impurity in sand medium effluent
Threshold odor number	35–140	0–4[b]	35–70[b]
Color, units	4–14	0–2	1–2
Mn, ppm	0.066–0.15	0.008–0.017	0.008–0.017
Fe, ppm	0.2–0.37	0.006–0.025	0.012–0.087
Turbidity, ppm	0.45–1.4	0.07–0.15	0.10–0.25
Chlorine, ppm	1.4–2.8	0–0.25	1.4–2.8

[a] 24 in. deep, with flow rate of 2 gpm/ft^2.
[b] Odor samples taken over a 60 day period; other samples, over a 150 day period.

Source: Reprinted, with publisher's permission, from *J. Am. Water Works Assn.*, *57*(11):1440, © 1965, by the American Water Works Association, Inc., 2 Park Ave., New York, N.Y. 10016.

4.4–26 REMOVAL OF ORGANIC CONTAMINANTS FROM KANAWHA RIVER WATER

	Nov. 13, 1963		Nov. 21, 1963		
Contaminant	Raw kanawha, ppb	Sand filtered,[a] ppb	Raw kanawha, ppb	After aeration, ppb	Carbon filtered, ppb
Ethylbenzene	11	11	ND	ND	ND
Styrene	18	ND	ND	ND	ND
Bis (2-chloroethyl) ether	55	10	154	104	ND
2-Ethyl hexanol	110	8	ND	7	ND
Bis (2-chloroisopropyl) ether	24	ND	48	23	ND
a-Methyl benzyl alcohol	ND[b]	ND	26	12	ND
Acetophenone	19	ND	29	17	ND
Isophorone	90	ND	25	16	ND
Tetralin	ND	ND	57	103	ND
	–	–	–	–	–
Total	327	29	361	296	
Threshold odor number (25°C)	128	6	237	181	<1

[a] Sample dechlorinated before analysis.
[b] ND–value below detectable concentration.

Source: Reprinted, with publisher's permission, from *J. Am. Water Works Assn.*, *57*(5):663, © 1965, by the American Water Works Association, Inc., 2 Park Ave., New York, N.Y. 10016.

4.4–27 INFLUENCE OF CONTACT TIME ON PASSAGE OF SPECIFIC CONTAMINANTS

Activated carbon treatment of River Water

	Contact time, min			
	1.9	3.8	5.6	7.5
	Depth, ft			
	5	10	15	20
Contaminant	Concentration,[a] ppb			
Ethylbenzene	20	18	15	5
Bis (2-chloroethyl) ether	94	44	ND	ND
2-Ethyl hexanol	57	20	ND	ND
Bis (2-chloro-isopropyl ether	26	10	ND	ND
a-Methyl benzyl alcohol	62	13	ND	ND
Acetophenone	11	ND[b]	ND	ND
Isophorone	12	9	ND	ND
Total	285	114	15	5
Threshold odor number (25°C)	32	25	5	2

[a] Sampled Dec. 30, 1963.
[b] ND–values below detectable concentration.

Source: Reprinted, with publisher's permission, from *J. Am. Water Works Assn., 57*(5):663, © 1965, by the American Water Works Association, Inc., 2 Park Ave., New York, N.Y. 10016.

4.4–28 REACTION OF POTASSIUM PERMANGANATE WITH PURE ORGANIC COMPOUNDS IN DISTILLED WATER

Activated Carbon Treatment of River Water

	Reaction[a]	
Compound	pH 7	pH 10
Saturated, 3-carbon		
propionic acid	–	–
propanol	–	–
propionaldehyde	+	+
propylamine	–	+
ethyl formate	–	+
acetone	–	–
glycerol	–	–
lactic acid	–	–
methyl acetate	–	–
alanine	–	+
pyruvic acid	+	+
Unsaturated, 3-carbon		
acrolein	+	+
allylamine	+	+
acrylic acid	+	+
allyl alcohol	+	+
Aromatic		
benzene	–	–
benzaldehyde	+	+
phenol	+	+
aniline	+	+
benzoic acid	–	–
benzyl alcohol	+	+
Miscellaneous		
octyl alcohol	–	–
caproic acid	–	–
ABS	–	–
2-butanone	–	+
starch	–	–
creatine	–	–

[a] Plus signs indicate positive reactions, minus signs negative reactions.

Note: Reaction between 50 mg/l organic compound and 3.61 mg/l $KMnO_4$. Use of more than 10 mg/l $KMnO_4$ in 6 hr was considered a positive reaction.

Source: Reprinted, with publisher's permission, from *J. Am. Water Works Assn., 57*(4):472, © 1965, by the American Water Works Association, Inc., 2 Park Ave., New York, N.Y. 10016.

4.4–29 EFFECTS ON pH OF PERMANGANATE OXIDATION OF PURE COMPOUNDS

Compound	pH Before reaction	pH After reaction
Saturated, 3-carbon		
propylamine	10.3	9.7
propionaldehyde	7.3; 10.2	5.2; 8.6
ethyl formate	10.2	9.7
alanine	10.2	9.9
pyruvic acid	10.1; 7.0	9.9; 7.2
Aromatic		
benzaldehyde	10.0; 7.0	10.0; 5.2
aniline	6.5; 10.1	9.1; 9.7
benzyl alcohol	6.2; 10.0	7.4; 9.9
phenol	7.0; 10.2	7.6; 9.4
Unsaturated, 3-carbon		
acrylic acid	7.0; 10.0	7.0; 10.0
allyl alcohol	7.3; 10.0	7.4; 9.9
allylamine	8.0; 10.0	7.7; 9.0
acrolein	7.0; 10.0	6.7; 9.0

Source: Reprinted, with publisher's permission, from *J. Am. Water Works Assn.*, 57(4):472, © 1965, by the American Water Works Association, Inc., 2 Park Ave., New York, N.Y. 10016.

4.4–30 OXIDATION OF PURE COMPOUNDS WITH POTASSIUM PERMANGANATE

Compound	Potassium permanganate used[a] mg/l: High pH	Neutral pH
111 mg/l $KMnO_4$ added		
Propylamine	32.4	
Propionaldehyde	80.1	84.5
Ethyl formate	71.1	
Alanine	24.1	
Pyruvic acid	42.5	25.3
Benzaldehyde	48.2	54.6
Benzyl alcohol	23.3	20.8
506 mg/l $KMnO_4$ added		
Aniline	493	484
Phenol	506	506
Acrylic acid	166	194
Allyl alcohol	184	194
Allylamine	404	369
Acrolein	341	328

[a] After 2 hr reaction.

Source: Reprinted, with publisher's permission, from *J. Am. Water Works Assn.*, 57(4):472, © 1965, by the American Water Works Association, Inc., 2 Park Ave., New York, N.Y. 10016.

4.5 *DISINFECTION*

The following section deals with data on disinfection of water entering municipal supply systems. See also Section 6 for related data on biological control measures for water in storage.

4.5–1 DISINFECTANTS USED IN PLANTS SUPPLYING COMMUNITIES OF 25,000 OR MORE

Table A

	1958[1]				1960[2]			
	Plants		Population served		Plants		Population served	
Disinfecting agent	No.	Percentage of total	Estimated no.	Percentage of total	No.	Percentage of total	Estimated no.	Percentage of total
Liquid chlorine	425	68.3	55,383,065	63.2	464	67.8	62,060,010	66.9
Chlorine and ammonia	136	21.9	24,399,390	27.9	142	20.8	22,025,339	23.8
Hypochlorites	14	2.2	473,285	0.5	18	2.6	834,383	0.9
Liquid chlorine and hypochlorites	15	2.4	3,445,840	3.9	15	2.2	3,178,083	3.4
All others	32	5.2	3,898,270	4.5	45	6.6	4,614,365	5.0
Total	622	100.0	87,599,850	100.0	684	100.0	92,712,180	100.0

Table B

	1962[3]				1964[4]			
	Plants		Population served		Plants		Population served	
Disinfecting agent	No.	Percentage of total	Estimated no.	Percentage of total	No.	Percentage of total	Estimated no.	Percentage of total
Liquid chlorine	560	71.5	65,348,008	67.7	806	75.0	77,872,550	73.2
Chlorine and ammonia	144	18.4	23,602,484	24.4	156	14.5	23,874,225	22.4
Hypochlorites	25	3.2	496,837	0.5	79	7.4	1,361,375	1.3
Liquid chlorine and hypochlorites	20	2.6	3,792,455	3.9	31	2.9	3,187,409	3.0
All others	34	4.3	3,370,302	3.5	4	0.4	183,825	0.2
Total	783	100.0	96,610,086	100.0	1,074	100.0	106,399,884	100.0

Source: Reprinted, with publisher's permission, from *J. Am. Water Works Assn.*, *57*(11):1360, © 1965, by the American Water Works Association, Inc., 2 Park Ave., New York, N.Y., 10016.

REFERENCES

1. K.H. Jenkins, *J. Am. Water Works Assn.*, *53*(1):31, 1961.
2. K.H. Jenkins and H.W. Owen, *J. Am. Water Works Assn.*, *55*(1):35, 1963.
3. K.H. Jenkins, *J. Am. Water Works Assn.*, *55*(12):1485, 1963.
4. K.H. Jenkins, *J. Am. Water Works Assn.*, *57*(11):1360, 1965.

4.5–2 INCREASE IN USE OF WATER CHLORINATION IN THE U.S.[a]

Year	Water treatment facilities: Total no.	Water treatment facilities: No. using chlorination	Population served, 1,000,000's: Total	Population served, 1,000,000's: With chlorinated water	Ref.
1910	557+	18+	13.91+	1.18+	1
1915	830	340	22.74	15.58	1
1930	3,015	2,917[b]			1, 2
1940	5,372	4,650	74.31	63.50[c]	2, 3
1948	6,986	6,137	83.25	80.63	2

[a] Data are not strictly comparable because of slight variation in statistical reporting and analytic procedures.
[b] Data appear to be questionable because of the unlikely high percentage of available treatment facilities employing chlorination.
[c] Estimated value.

Source: Reprinted, with publisher's permission, from *J. Am. Water Works Assn., 52*(11):1416, © 1960, by the American Water Works Association, Inc., 2 Park Ave., New York, N.Y. 10016.

REFERENCES

1. R. Porges, *Publ. Works, 89*:172, 1958.
2. J.R. Thoman, *Statistical Summary of Water Supply and Treatment Practices in the United States,* U.S. Public Health Service, 1953.
3. S.R. Weibel, *Publ. Health Reports, 57*:45, 1942.

4.5–3 RECOMMENDATIONS OR REQUIREMENTS FOR MINIMUM POSTCHLORINATION OF SURFACE SUPPLIES

Authority	Treatment specified
USPHS	0.2 ppm free residual after 20 min contact, or 2.0 ppm combined residual after 3 hr contact; 0.05–0.10 ppm free residual or 1.0–2.0 ppm combined residual maintained
Ten State Standards	As necessary for bacteriologic safety
Connecticut	0.2 ppm free residual after 10 min contact; 0.5 ppm combined max residual
Florida	Residual[a] maintained
Iowa	30 min contact
Kansas	30 min contact (max MPN, 20,000)
Kentucky	30 min contact
Massachusetts	0.2 ppm free residual
Minnesota	0.4 ppm residual[b] after 3 hr maintained
Nebraska	0.05 ppm free residual maintained
Nevada	1 ppm residual[b] after 20 min contact
New Mexico	0.2–0.4 ppm residual[b] maintained
Texas	0.20 ppm free residual after 20 min contact or >0.20 ppm combined residual after 3 hr contact maintained
Washington	0.2–0.3 ppm residual[b] maintained
Wyoming	0.1 ppm residual[b] maintained

[a] Amount and type not specified.
[b] Type not specified.

Source: Reprinted, with publisher's permission, from *J. Am. Water Works Assn., 52*(11):1416, © 1960, by the American Water Works Association, Inc., 2 Park Ave., New York, N.Y. 10016.

4.5–4 STATE PRACTICES ON REPAIRED AND NEW WATER MAIN DISINFECTION

Recommended or required (specified)	Recommended or required (implied[a])	Not specified or implied	Status unknown
Alabama[b]	Illinois[b]	Colorado	Alaska
Arkansas	Indiana[b]	Georgia	Arizona[d]
California[c]	Iowa[cd]	New Mexico	Maine
Connecticut	Michigan[cd]	Wisconsin	Maryland[d]
Delaware[c]	Minnesota[c]		Massachusetts[d]
Florida	Missouri[c]		Mississippi
Hawaii	Montana[cd]		New Hampshire
Idaho[cd]	Ohio[cd]		Rhode Island[d]
Kansas[e]	Oklahoma[c]		Tennessee
Kentucky[f]	Pennsylvania[c]		Vermont
Louisiana	South Dakota[c]		Virginia
Nebraska	Utah[c]		
Nevada[dg]			
New Jersey[c]			
New York			
North Carolina[b]			
North Dakota[d]			
Oregon[c]			
South Carolina			
Texas[h]			
Washington			
West Virginia[b]			
Wyoming[i]			

[a] As supporter of Ten State Standards.
[b] $\geqq$ 50 ppm residual for 24 hr minimum contact.
[c] Am. Water Works Assn. recommendations: $\geqq$ 10 ppm residual after 24 hr contact, or 50–100 ppm residual for shorter contact periods (employed in at least seventeen states).
[d] No published standards.
[e] $\geqq$ 100 ppm residual for 24 hr contact; $\geqq$ 25 ppm residual after 24 hr contact.
[f] $\geqq$ 50 ppm residual for 24 hr contact; $\geqq$ 25 ppm residual after 24 hr contact.
[g] $\geqq$ 5 ppm residual after 3 hr contact.
[h] 40–60 ppm minimum residual after 6 hr contact.
[i] 40–50 ppm residual for 3 hr minimum contact; > 5 ppm residual after 3 hr contact.

Source: Reprinted, with publisher's permission, from *J. Am. Water Works Assn., 52*(11):1416, © 1960, by the American Water Works Association, Inc., 2 Park Ave., New York, N.Y. 10016.

4.5–5 STATE PRACTICE REGARDING POSTCHLORINATION OF SURFACE SUPPLIES

Recommended or required (specified)	Recommended or required (implied[a])	Not specified or implied[b]	Status unknown
Alabama	Delaware	Arkansas	Alaska
Florida	Illinois	California	Arizona[c]
Louisiana	Indiana	Connecticut	Maine[c]
Massachusetts[c]	Iowa[c]	Colorado	Maryland[c]
New Jersey	Kansas	Georgia	Mississippi
New Mexico	Michigan[c]	Hawaii	New Hampshire[c]
North Carolina	Minnesota	Idaho[c]	Rhode Island[c]
North Dakota[c]	Missouri	Kentucky	Virginia[c]
South Carolina	Montana[c]	Nebraska	
Tennessee[c]	New York	Nevada[c]	
Texas	Ohio[c]	Oregon	
Washington	Oklahoma[c]	Vermont[c]	
Wisconsin	Pennsylvania		
Wyoming	South Dakota[c]		
	Utah		
	West Virginia		

[a] As supporter of Ten State Standards.
[b] Presumably, postchlorination is used when needed to comply with USPHS *Drinking Water Standards.*
[c] No published standards.
Source: Reprinted, with publisher's permission, from *J. Am. Water Works Assn., 52*(11):1416, © 1960, by the American Water Works Association, Inc., 2 Park Ave., New York, N.Y. 10016.

4.5–6 BACTERIOLOGICAL QUALITY OF DISTRIBUTED WATER RELATED TO CHLORINATION PRACTICE

Type of system	Chlorination practiced	No. systems	Systems exceeding coliform limit, %	Systems containing fecal coliforms, %
Surface water source	No	11	64	64
	Yes	80	7[a]	7[a]
Mixed ground and surface water source	No	4	100	100
	Yes	50	14[a]	0
Spring source	No	62	39	32
	Yes	12	17[a]	8[a]
Well source	No	394	8	2
	Yes	204	5[a]	>1[a]
Mixed spring and well source	No	27	41	33
	Yes	11	27[a]	27[a]
All types	No	98	16	7
	Yes	456	8[a]	4[a]

[a] Almost all of these systems have no chlorine tested in the distribution system.

Source: Reprinted, with publisher's permission, from *J. Am. Water Works Assn., 62*(11):670, © 1970, by the American Water Works Association, Inc., 2 Park Ave., New York, N.Y. 10016.

4.5–7 INFLUENCE OF CHLORINE RESIDUAL ON BACTERIOLOGICAL QUALITY OF DISTRIBUTED WATER

Type of system	Samples showing positive coliforms, %: Unchlorinated systems	All chlorinated systems	Chlorinated systems that maintained a chlorine residual
Surface water source	93	6	2
Mixed ground and surface water source	54	9	6
Spring source	36	9	0
Well source	11	3	0
Mixed spring and well source	43	26	0
All types	17	6	2

Source: Reprinted, with publisher's permission, from *J. Am. Water Works Assn., 62*(11):670, © 1970, by the American Water Works Association, Inc., 2 Park Ave., New York, N.Y. 10016.

4.5–8 SOURCES OF DATA FOR DISINFECTION TIME-CHLORINE CONCENTRATION STUDY

Organism	Temperature range, °C	pH	Ref.
Esch. coli	2–5	7, 8.5	1
Esch. coli	20–25	7, 8.5	1
Salm. typhosa	20–25	7, 8.5	1
Sh. dysenteriae	20–25	7.0	1
Poliomyelitis	20–30	6.85–9.25	2
Poliomyelitis	0	7, 8.5	3
Coxsackie A_2	3–6	7, 9	4
Coxsackie A_2	27–29	7, 9	4
E. histolytica	20–25	7, 8	5
E. histolytica	20–25	7, 8	6
B. anthracis	3	7.2, 8.6	7
B. anthracis	22	7.2, 8.6	7
P. tularensis	15.5–18.5	7.3	8
Hepatitis	room	6.4–6.9	9

Source: Reprinted, with publisher's permission, from *J. Am. Water Works Assn., 54*(11):1379, © 1962, by the American Water Works Association, Inc., 2 Park Ave., New York, N.Y. 10016.

REFERENCES

1. C.T. Butterfield *et al., Public Health Repts., 58*:1837, 1943.
2. S.G. Lensen, M. Rhian, and M.R. Stebbins, *Am. J. Publ. Health, 37*:869, 1947.
3. S.J. Weidenkopf, *Virology, 5*:56, 1958.
4. N.A. Clarke and P.W. Kabler, *Am. J. Hyg., 59*:119, 1954.
5. G.M. Fair, J.C. Morris, and S.L. Chang, *J. New Engl. Water Works Assn., 61*:285, 1947.
6. W.B. Snow, *J. Am. Water Works Assn., 48*:1510, 1956.
7. R.A. Brazis *et al.,* Special Report to Department of the Navy Bureau of Yards and Docks – Sporicidal Action of Free Available Chlorine, R.A. Taft Sanitary Engineering Center, Cincinnati 1957.
8. H.B. Foote *et al., J. Am. Water Works Assn., 35*:902, 1943.
9. J.R. Neffe *et al., Am. J. Publ. Health 37*:365, 1947.

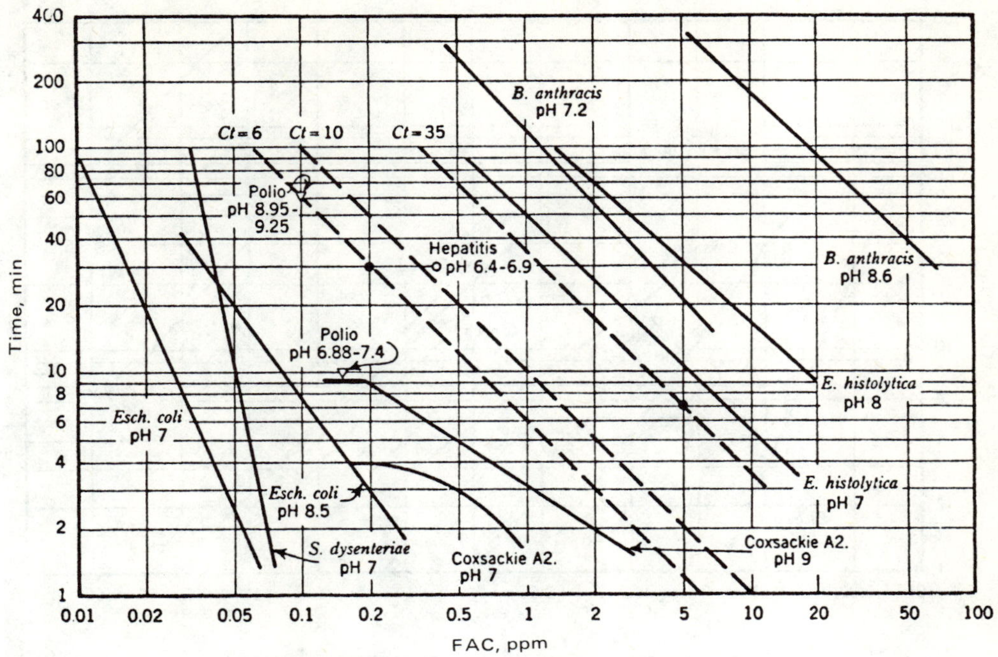

4.5–9 Disinfection Versus FAC[a] Residuals.

[a] FAC–free available chlorine.

Note: Time scale is for 99.5–100 percent kill. Temperature was in the range of 20–29°C, with pH as indicated.

Source: Reprinted, with publisher's permission, from *J. Am. Water Works Assn.*, *54*(11):1379, © 1962, by the American Water Works Association, Inc., 2 Park Ave., New York, N.Y. 10016.

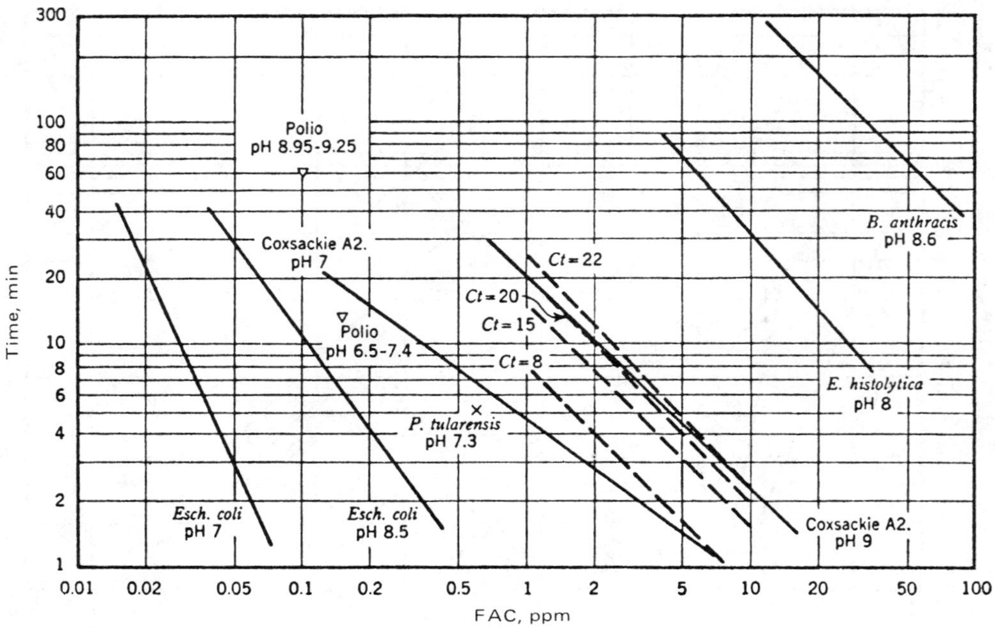

4.5–10 Disinfection Versus FAC Residuals.

Note: Time scale is for 99.6–100 percent kill. Temperature was 10°C, with pH as indicated.

Source: Reprinted, with publisher's permission, from *J. Am. Water Works Assn., 54*(11):1379, © 1962, by the American Water Works Association, Inc., 2 Park Ave., New York, N.Y. 10016.

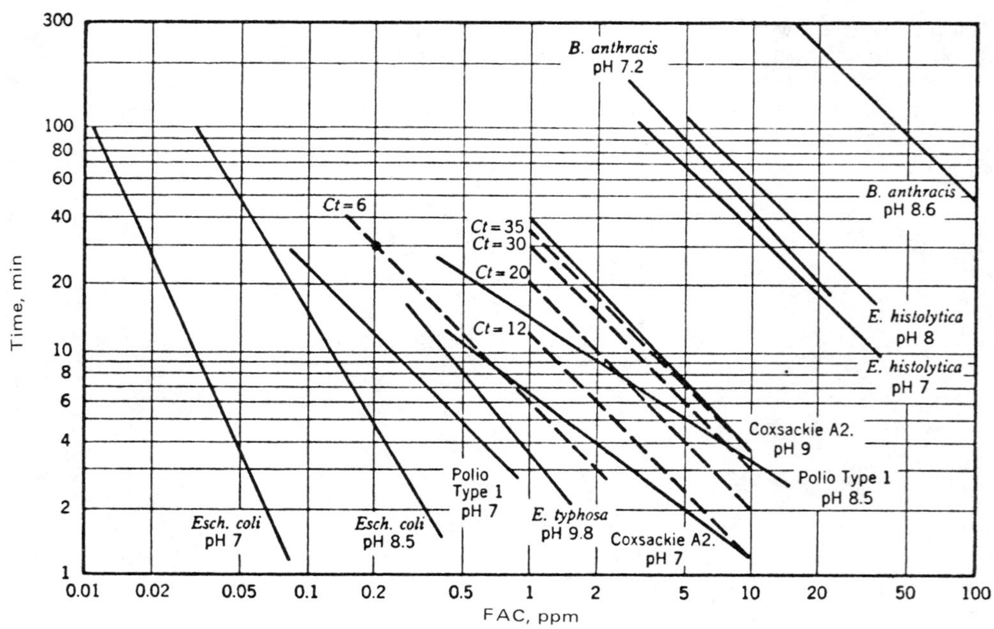

4.5–11 Disinfection Versus FAC Residuals.

Note: Time scale is for 99.6–100 percent kill. Temperature was in the range of 0–5°C, with pH as indicated.

Source: Reprinted, with publisher's permission, from *J. Am. Water Works Assn., 54*(11):1379, © 1962, by the American Water Works Association, Inc., 2 Park Ave., New York, N.Y. 10016.

4.5–12 EFFECT OF CHLORINE ON VIABILITY OF ACTINOMYCETE SPORE SUSPENSIONS[a]

	Viable spores, no./ml		
Culture	Control	30 sec exposure[b]	Reduction, %
S. griseoluteus[c]	170,000	120	99.9
S. lavendulae	710,000	150	99.9
Actinomycete S-5	180,000	760	99.6
Actinomycete S-6	43,000	1,900	95.6

[a] Spore suspension in pH 7.5 $\frac{1}{75}$ *M* phosphate buffer.
[b] Total available chlorine, 1 ppm.
[c] S – *Streptomyces griseoluteus.*

Source: Reprinted, with publisher's permission, from *J. Am. Water Works Assn., 55*(2):169, © 1962, by the American Water Works Association, Inc., 2 Park Ave., New York, N.Y. 10016.

4.5–13 ODOR PRODUCTION BY *STREPTOMYCES GRISEOLUTEUS* IM 3718 IN SOIL PERFUSION UNITS[a]

	Control	Inoculated
Sterilized Units		
Threshold odor number	70	1,200
Odor qualities	Stale	Earthy, musty
Unsterilized units		
Threshold odor number	100	470
Odor qualities	Stale	Earthy, musty

[a] The pH of collected soil was 7.4.

Source: Reprinted, with publisher's permission, from *J. Am. Water Works Assn., 55*(2):169, © 1962, by the American Water Works Association, Inc., 2 Park Ave., New York, N.Y. 10016.

4.5–14 COMPARISON OF LETHAL ACTIVITIES OF CHLORAMINES

	Killing in 10 min, %			Albuminoid nitrogen reduction,[1]	Total nitrogen reduction,[1]
Chloramine	pH=4.0	pH=7.0	pH > 9.0	%	%
Ammonia	99.9	99.9	96.0		90±
Glycine	99.9	<39	<39	90+	50+
Methionine	40.0	<39	50±	80–90	30–40
Taurine	81.5	67.6	99.9	–10	–10
Gelatin	<39	54.9	99.0	20–30	10–20

Source: T.H. Feng, *J. Water Pollut. Control Fed., 38*(4):614, 1966.

REFERENCE

1. M.J. Taras, *J. Am. Water Works Assn., 45*(1):47, 1953.

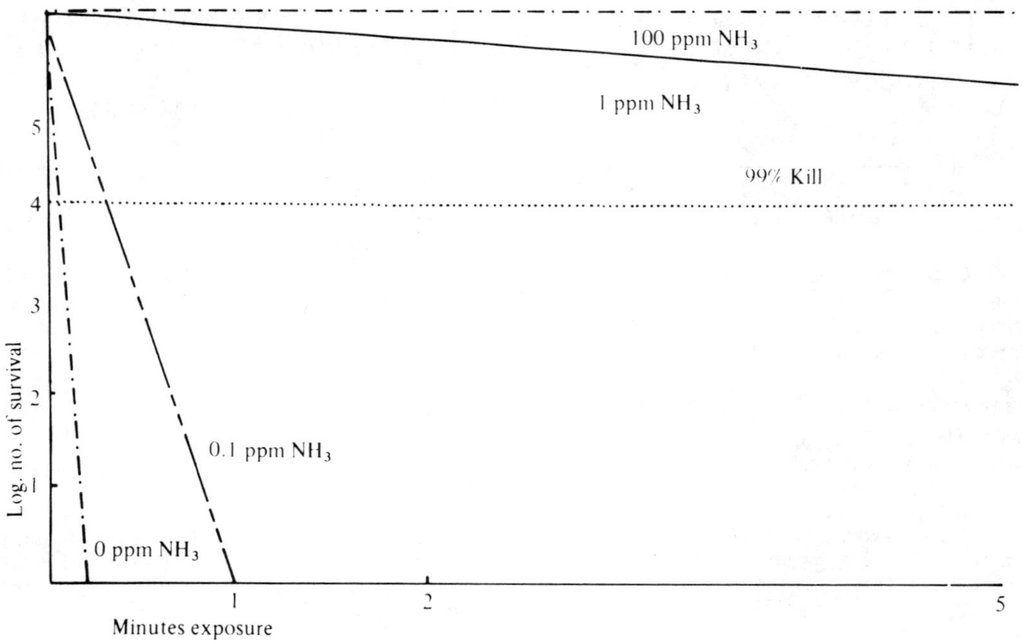

4.5–15 The Effect of Ammonia upon the Germicidal Action of Chlorine.

Note: 0.5 ppm free chlorine–Ammonia as $(NH_4)_2SO_4$; pH–7.0; temperature, 21°C; test organism, *Escherichia coli.*

Source: W.L. Mallman, *Crit. Rev. Environ. Control, 1*(2):221, 1970.

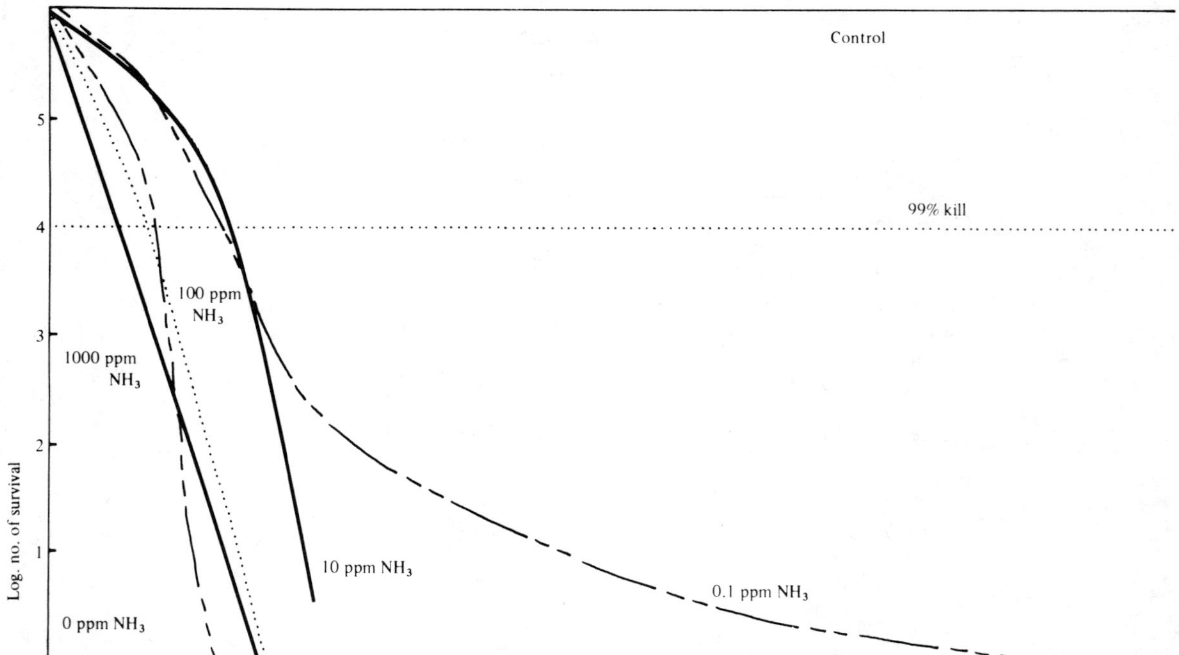

4.5–16 The Effect of Ammonia upon the Germicidal Action of Bromine.

Note: 0.5 ppm free bromine–Ammonia as $(NH_4)_2SO_4$; pH–7.0; temperature, 21°C; test organism, *Escherichia coli.*

Source: W.L. Mallman, *Crit. Rev. Environ. Control, 1*(2):221, 1970.

4.5–17 INACTIVATION OF VIRUS AT pH 7 AND 22°C

Virus	Minimum dose to inactivate TCD_{50},[a] ppm				
	1 min		10 min		
	Cl	Br	Cl	Br	I
Polio I, virulent	2.0	2.0	0.5	0.5	0.5
attenuated			0.2	0.3	0.2
Polio II, virulent	0.3	0.3	0.3	0.3	0.5
attenuated			0.4	0.2	0.2
Polio III, virulent	1.2	0.4	0.3	0.4	0.6
attenuated			0.2	0.3	0.3
Coxsackie B4	1.4	1.4	0.3	0.3	0.4
Echo 2	1.2	0.3	0.3	0.3	0.5
Echo 9	1.2	0.8	0.3	0.2	0.3
Parainfluenza I			0.4	0.4	0.4

[a] Tissue culture death.

Source: W.L. Mallman, *Crit. Rev. Environ. Control,* *1*(2):221, 1970.

4.5–18 COMPARATIVE GERMICIDAL ACTIVITY OF BROMINE AND CHLORINE ON VARIOUS ORGANISMS IN WATER SAMPLES

Free and combined bromine, ppm	Combined chlorine, ppm	No. samples	Number of water samples yielding					
			Coli-forms	*E. coli*	Staphylococcus	*Staphylococcus aureus* coagulase + ve	*Strept. viridans*	Entero-cocci
Bromination								
0–0.99	0.34	84	21	8	71	23	28	23
1–1.99	0.38	31	10	1	27	9	6	9
2–3.99	0.38	24	4	0	8	4	8	4
4.0–5.99	0.33	3	0	0	2	1	2	1
0–5.99	0.44–0.48	142	35	9	108	37	44	37
Chlorination								
Free Cl, ppm								
0–0.09	0.48	4	1	1	4	0	1	2
0.1–0.19	1.87	7	2	2	7	1	2	4
0.2–0.29	1.10	4	1	1	4	2	2	0
0.3–0.39	1.1	1	0	0	1	1	0	0

Source: W.L. Mallman, *Crit. Rev. Environ. Control, 1*(2):221, 1970.

4.5–19 CONCENTRATIONS OF I_2 AND HOI NEEDED FOR POLIOVIRUS TYPE I AND *E. HISTOLYTICA*

Species	Poliovirus Type I, mg/l	*E. histolytica,* mg/l
Iodine	20	2.5
Hypoiodous acid	0.45	4

Source: Reprinted, with publisher's permission, from *J. Am. Water Works Assn., 62*(2):80, © 1970, by the American Water Works Association, Inc., 2 Park Ave., New York, N.Y. 10016.

4.5–20 COMPOSITION OF HALAZONE TABLET

Component	Amount, mg
Halazone	5.30
Soda ash, dried	5.18
Boric acid	11.92
Sodium chloride	114.00
Weight/tablet	136.40

Source: Reprinted, with publisher's permission, from *J. Am. Water Works Assn., 62*(2):80, © 1970 by the American Water Works Association, Inc., 2 Park Ave., New York, N.Y. 10016.

4.5–21 COMPOSITION OF GLOBALINE TABLET

Component	Amount, mg
Tetraglycine hydro-periodide	19.3–21
Disodium dihydrogen pyrophosphate ($Na_2H_2P_2O_7$)	82.5–92.3
Talc	Not more than 6
Weight/tablet	110–120

Source: Reprinted, with publisher's permission, from *J. Am. Water Works Assn., 62*(2):80, © 1970 by the American Water Works Association, Inc., 2 Park Ave., New York, N.Y. 10016.

4.5–22 NORMAL CYSTICIDAL DOSE AT WHICH "OBJECTIONABLE THRESHOLD" IS REACHED

Compound	Percentage					
pH	4	5	6	7	8	9
Halazone	50	40	25	25	25	25
Globaline	–	200	–	200	–	–

Source: Reprinted, with publisher's permission, from *J. Am. Water Works Assn., 62*(2):80, © 1970 by the American Water Works Association, Inc., 2 Park Ave., New York, N.Y. 10016.

4.5–23 EFFECTIVENESS OF GLOBALINE

Kind of water	Temperature, °C	Contact time, min	pH Initial	pH Final	Cysticidal dose, tablet/qt	Cysticidal residual, I_2, ppm
Tap	3	25	8.0 to 9.0	6.5	1	7.5
Tap	10	15	8.0 to 9.0	6.6	1	6.9
Tap	23	10	8.0 to 9.0	7.3	1	7.5
Tap	28	5	8.0 to 9.0	6.65	1	7.5
Tea infusion	23	5	7.2	6.4	2	8.7

Source: Reprinted, with publisher's permission, from *J. Am. Water Works Assn.*, *62*(2):80, © 1970 by the American Water Works Association, Inc., 2 Park Ave., New York, N.Y. 10016.

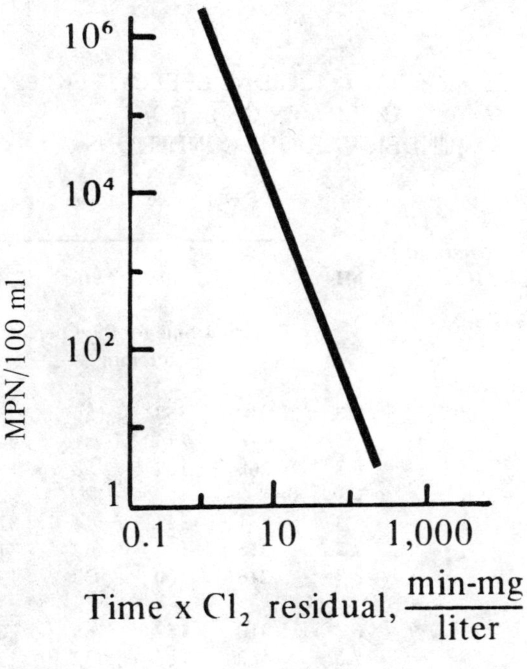

4.5–24 Effect of Time and Concentration on Survival of *E. coli* in Wastewater Using Chlorine as Disinfectant.

Source: Metcalf and Eddy, Inc., *Wastewater Engineering*, McGraw-Hill, New York, 1972.

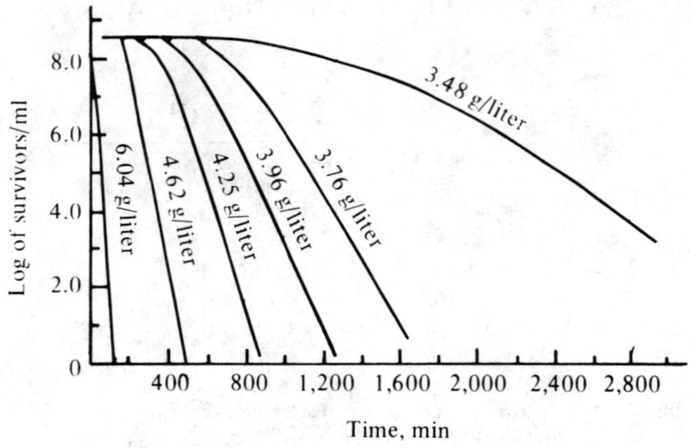

4.5–25 Effect of Time and Concentration on Survival of *E. Coli* Using Phenol as Disinfectant at 35°C.

Source: R.C. Jordan and S.E. Jacobs, *J. Hygiene, 43*:297, 1944.

4.5–26 BACTERICIDAL EFFECTIVENESS OF SILVER ON *E. COLI* UNDER VARIOUS CONDITIONS

Temperature, °C	pH	Silver concentration, ppb 10	30	90	270
		Required time for 99.9 percent kill, min			
5	6.3	1,010	837	156	53
	7.5	466	214	81	34
	8.7	268	109	58	18
15	6.3	831	344	144	32
	7.5	316	177	63	21
	8.7	216	100	38	13
25	6.3	1,210	152	68	20
	7.5	423	86	32	13
	8.7	203	40	20	8

Source: K. Wuhrmann and F. Zobrist, *Schweiz. Z. Hydrol.* (Switzerland), *20*:218, 1958.

4.5–27 EFFICIENCY OF $AgNO_3$ NEUTRALIZATION

Ag, ppm[a]	Neutralizer type	Percent of bacteria surviving after: 0 hr NA[b]	0 hr NA + T[c]	2 hr NA[b]	2 hr NA + T[c]	4 hr NA[b]	4 hr NA + T[c]	6 hr NA[b]	6 hr NA + T[c]
10.0	Sodium sulfite[d]	75.8	100.0	17.9	55.9	13.9	43.1	1.29	24.0
5.0	Sodium sulfite	91.9	100.0	18.1	40.4	17.9	30.6	9.90	26.5
1.0	Sodium sulfite	100.0	100.0	70.0	78.9	52.2	76.4	40.4	69.4
10.0	Thiosulfate-thioglycollate[e]			100.0	98.8	92.5	94.7	95.0	87.1
5.0	Thiosulfate-thioglycollate			99.4	100.0	99.0	100.0	100.0	91.5
1.0	Thiosulfate-thioglycollate			100.0	92.3	93.8	100.0	97.1	90.4
10.0	Sodium sulfite and glycerine[f]	81.0	100.0	4.53	24.7	0	20.6	0	14.8
5.0	Sodium sulfite and glycerine			12.3	43.0	3.60	27.6	4.96	21.8
1.0	Sodium sulfite and glycerine			60.7	87.2	42.0	65.2	38.8	59.0
20.0	Sodium thiosulfate[g]	0	98.2	0	84.0	0	98.0	0	100.0
10.0	Sodium thiosulfate	25.0	92.4	52.2	100.0	60.3	97.3	5.88	100.0
5.0	Sodium thiosulfate	93.6	72.0	97.3	86.9	100.0	98.7	49.9	100.0

[a] Concentration of Ag in 1 ml aliquot added to 9 ml of neutralizer.
[b] Plain nutrient agar.
[c] Nutrient agar containing 500 ppm sodium thioglycollate.
[d] 1,000 ppm Na_2SO_3.
[e] 500 ppm sodium-thioglycollate and 730 ppm thiosulfate mixture.
[f] 2,000 ppm glycerine and 1,000 ppm Na_2SO_3.
[g] 730 ppm $Na_2S_2O_3 \cdot 5H_2O$.

Source: Reprinted, with publisher's permission, from *J. Am. Water Works Assn., 54*(2):208, © 1962 by the American Water Works Association, Inc., 2 Park Ave., New York, N.Y. 10016.

4.5–28 GERMICIDAL ACTION OF SILVER NITRATE

Test water	Ag^{100m}, ppb[b]	Average percent of bacteria surviving after[a]: 1 hr	2 hr	3 hr	4 hr
Phosphate	89	40.1	4.5	0.30	0.028
Distilled	90	22.0	1.22	0.08	0.005

[a] Average for two experiments.
[b] Silver present at the start of an experiment. In the distilled water an average of 13.1 percent of the Ag present was lost to the glass; in phosphate water the corresponding loss was 3.3 percent.

Source: Reprinted, with publisher's permission, from *J. Am. Water Works Assn., 54*(2):208, © 1962 by the American Water Works Association, Inc., 2 Park Ave., New York, N.Y. 10016.

4.5–29 RELATIONSHIP OF BACTERIAL DENSITY TO GERMICIDAL EFFICIENCY

pH 7.0[a]

Germicide	Germicide concentration, ppm	Initial bacteria/ ml	Percent surviving after 2 hr	4 hr	6 hr
Control	0.0	1,790	83.8	82.7	87.2
Control	0.0	1,800,000	99.1	92.9	94.0
$AgNO_3$	Unknown[b]	1,680	44.6	33.0	20.2
$AgNO_3$	Unknown[b]	1,700,000	44.8	28.5	28.7
Control	0.0	27,800	100.0	100.0	100.0
Control	0.0	1,400,000	100.0	91.4	92.1
$AgNO_3$	10[c]	28,300	66.4	52.3	6.36
$AgNO_3$	10[c]	1,400,000	69.8	50.7	4.62

[a] Representative experiments. Parallel tests made at both bacterial densities.
[b] Silver obtained by desorption. No neutralizer used.
[c] Silver initially added. Neutralizer used.

Source: Reprinted, with publisher's permission, from *J. Am. Water Works Assn., 54*(2):208, © 1962 by the American Water Works Association, Inc., 2 Park Ave., New York, N.Y. 10016.

4.5–30 EFFECT OF pH ON GERMICIDAL EFFICIENCY OF $AgNO_3$[a]

Ag, ppb[b]	pH	Percent of bacteria surviving after 15 min	30 min	45 min	60 min	75 min	90 min	105 min	120 min
38	6.2	86.8	65.8	58.4	35.9	49.7	31.3	17.3	14.5
38	7.5	75.8	34.7	10.4	1.93	0.84	0.03	0.02	0

[a] PO_4 concentration constant at $\frac{1}{1,500}$ *M*.
[b] 40 ppb added. Value shown is amount determined by rhodanine test at end of experiment.

Source: Reprinted, with publisher's permission, from *J. Am. Water Works Assn., 54*(2):208, © 1962 by the American Water Works Association, Inc., 2 Park Ave., New York, N.Y. 10016.

4.5–31 EFFECT OF LIGHT ON GERMICIDAL EFFICIENCY OF $AgNO_3$

Distilled Water at pH 6.1[a]

Light status	Ag, ppb[b]	Percent of bacteria surviving after 15 min	30 min	1 hr	2 hr	3 hr	4 hr	5 hr	6 hr
Light	35	84.7	79.4	62.9	11.2	3.72	1.42	0.64	0.53
Dark	35	100.0	74.1	57.2	6.27	7.28	2.41	1.25	0.03

[a] Representative experiment.
[b] Amount of Ag initially added. No final analysis.

Source: Reprinted, with publisher's permission, from *J. Am. Water Works Assn.*, *54*(2):208, © 1962 by the American Water Works Association, Inc., 2 Park Ave., New York, N.Y. 10016.

4.5–32 COMPARISON OF SILVER NITRATE WITH SILVER FROM A FILTER DEVICE

Time, min	Percent[a] of bacteria surviving in distilled water at pH 6.1 with 40 ppb[b] Ag: $AgNO_3$	F[c]
30	94.4	93.1
60	78.9	77.7
90	63.3	63.5
120	52.4	47.3
150	36.0	30.3
180	12.7	18.4

[a] Average of 3 experiments. Three replicate flasks per experiment were used with each type of silver.
[b] Determined by the rhodanine test at start of experiment.
[c] Silver from silver-yielding filter device.

Source: Reprinted, with publisher's permission, from *J. Am. Water Works Assn.*, *54*:(2):208, © 1962 by the American Water Works Association, Inc., 2 Park Ave., New York, N.Y. 10016.

4.5–33 OZONE DOSAGE REQUIREMENTS FOR VARIOUS WATER QUALITIES

Water quality	Approximate gram dosage, mg/l
Clear pure ground water	1/2–1
Treated filtered surface water	2 –3
Same as above but from a heavily polluted source	2.5–5
Clear colored water	3 –6

Source: Reprinted, with publisher's permission, from *J. Am. Water Works Assn.*, *48*:1316, © 1956 by the American Water Works Association, Inc., 2 Park Ave., New York, N.Y. 10016.

4.5–34 REMOVAL OF COLIFORM ORGANISMS BY SLOW SAND FILTRATION AND PASTEURIZATION[a]

1958

	Max treated water temperature, °F	Retention time at max temperature, sec	Density in raw water, MPN/100 ml	Density in clear well, MPN/100 ml
Number of observations (January 22–March 24)	12	12	12	12
Avg	–	2	7,900[b]	515[b]
Max	178	2	46,000	2,300
Min	162	2	1,400	33
Number of observations (April 7–May 26)	10	10	10	10
Avg	–	15	6,600[b]	41[b]
Max	180	15	33,000	220
Min	161	15	460	4.5

[a] No coliform organisms were found in any of five 10-ml portions per sample examined after pasteurization.
[b] Median.

Source: Reprinted, with publisher's permission, from *J. Am. Water Works Assn., 52*(2):247, © 1960, by the American Water Works Association, Inc., 2 Park Ave., New York, N.Y. 10016.

4.5–35 RELATIVE RESISTANCE OF MICROORGANISMS TO STERILIZATION BY MOIST HEAT

Organism	Relative resistance
E. coli	1
Bacterial spores	3×10^6
Mold spores	2–10
Viruses and bacteriophages	1–5

Diphtheria organisms 130°F for 30 min
Tubercule bacilli 139°F for 30 min

Source: O. Rahn, *Bacteriol. Rev.,* 9:1, 1945.

4.5–36 RESISTANCE OF SOME MICROORGANISMS

Microorganisms	Type of radiation	D_dK rad
E. coli	Beta	10–15
E. coli	Gamma	15–20
E. coli	X-ray	5
A. aerogenes	X-ray	3
Ps. geniculata	Gamma	5
S. aureus	Gamma	20

Source: L.E. Brownell, *Radiation Uses in Industry and Science,* U.S. Atomic Energy Commission, 1961.

4.5–37 RADIATION EFFECTS ON *E. COLI* BY GAMMA RADIATION

Radiation dose, Mrep	Population reduction, %	*E. coli* count
0	0	1,976,250,000
0.071	99.8	1,600,000
0.149	99.99	500
0.205	100	0

Source: L.E. Brownell, *Radiation Uses in Industry and Science,* U.S. Atomic Energy Commission, 1961.

4.5–38 SURVIVAL OF *ESCHERICHIA COLI* AND *STREPTOCOCCUS FAECALIS* EXPOSED TO ULTRAVIOLET IRRADIATION

Minimum Exposure Position

UV exposure, sec	Count	Survival, %[a]
	Escherichia coli	
0	1.9×10^7	100
20	6.4×10^3	0.03
40	1.7×10^2	0.0009
60	2.0×10^1	0.0001
90	3.0×10^0	0.000016
	Streptococcus faecalis	
0	8.9×10^6	100
20	1.0×10^4	0.115
40	8.7×10^2	0.0097
60	1.9×10^1	0.0002

[a] Percentage of control group count (no UV exposure) to survive.

Source: Reprinted, with publisher's permission, from *J. Am. Water Works Assn., 57*(4):500, © 1965, by the American Water Works Association, Inc., 2 Park Ave., New York, N.Y. 10016.

4.5–39 COEFFICIENTS OF ULTRAVIOLET ABSORPTION[1] FOR VARIOUS WATERS

(λ = 2,537 Å)

Source	a/cm (coefficient of absorption per centimeter)
Distilled	0.008
Swimming pool	0.031
Cleveland tap water (1944)	0.050
Drilled well	0.056
Fish pool	0.070
Lake Erie (1944)	0.083
Concrete cistern	0.297

for use in
$I = I_o e^{-ax}$
I = effective solution intensity
I_o = initial intensity
a = coefficient of absorption
x = distance between points where the intensities are I and I_o

REFERENCE

1. M. Luckiesh and L.L. Holladay, "Disinfecting Water by Means of Germicidal Lamps," *Gen. Elect. Rev., 47*:45, 1944.

4.5–40 EFFECT OF pH ON THE HYDROLYSIS OF CHLORINE

Percentage Distribution of Chlorine in Solution

pH	Molecular chlorine	Hypochlorous acid	Hypochlorite ion
4	0.5	99.5	0
5	0	99.5	0.5
6	0	96.5	3.5
7	0	72.5	27.5
8	0	21.5	78.5
9	0	1.0	99.0

Source: Reprinted, with publisher's permission, from *J. Am. Water Works Assn., 60*(1):69, © 1968 by the American Water Works Association, Inc., 2 Park Ave., New York, N.Y. 10016.

4.5–41 EFFECT OF pH ON THE HYDROLYSIS OF IODINE

pH	Content of residual, %[a,b] I_2	HIO	IO^-
5	99	1	0
6	90	10	0
7	52	48	0
8	12	88	0.005

[a] Total iodine residual 0.5 ppm.
[b] Reactions:

$I_2 + H_2O \rightleftharpoons HIO + H^+ + I^-$ (1)

$$\frac{[HIO]\ [M]^+\ [I]^-}{[I_2]} = K_h \quad (2)$$

$K_h = 3 \times 10^{-13}$ at 25°C

$HIO \rightleftharpoons H^+ + IO^-$ (3)

$$\frac{[H^+]\ [IO^-]}{[HIO]} = K_a = 4.5 \times 10^{-13} \quad (4)$$

Source: Reprinted, with publisher's permission, from *J. Am. Water Works Assn., 57*(11):1401, © 1965, by the American Water Works Association, Inc., 2 Park Ave., New York, N.Y. 10016.

4.5–42 OXIDATION REDUCTION REACTIONS OF CHLORINE AND ITS OXIDES

Couple	Reaction	Redox potential, v
Hypochlorous acid-chloride	$HOCl+H+2e^-=Cl^-+H_2O$	E_0=+1.49[a,1]
Hypochlorite-chloride	$ClO^-+2H_2O+2e^-=Cl^-+2OH^-$	+0.94[a,1] +0.89[2,3] +0.95[b,4] +1.2[c,4]
Chlorous acid-chloride	$HClO_2+3H^++4e^-=Cl+2H_2O$	+1.57[1]
Chlorite-chloride	$ClO_2^-+2H_2O+4e^-=2Cl^-+4OH^-$	+0.77[1] +0.79[d,4] +0.66[e,4]
Chlorine-chloride	$Cl_2+2e^-=2Cl^-$	+1.3583[1]
Iodine-iodide	$I_2+2e^-=2I^-$	+0.5356[1]
Chlorine dioxide-chlorite	$ClO_2+e^-=ClO_2^-$	+0.950[b,5] +1.15[6]

[a] There is a minor variance depending upon source of information.
[b] Available chlorine at pH 10, 1 g/l.
[c] Available chlorine at pH 7, 1 g/l.
[d] Available chlorine at pH 4, 1 g/l.
[e] Available chlorine at pH 9, 1 g/l.

Source: Reprinted, with publisher's permission, from *J. Am. Water Works Assn., 56*(5):607, © 1964, by the American Water Works Association, Inc., 2 Park Ave., New York, N.Y. 10016.

REFERENCES

1. *Handbook of Chemistry and Physics,* 42nd ed., Chemical Rubber Co., Cleveland, 1960.
2. *Graphic Chemical Prediction,* Graphic Calculator Co., Chicago.
3. J.V. QuaQliano, *Chemistry,* Prentice Hall, Englewood Cliffs, N.J., 1958.
4. M.C. Taylor *et al., Ind. Eng. Chem., 32*:899, 1940.
5. M.B. Sneed, *Comprehensive Inorganic Chemistry, The Halogens,* vol. 3, Van Nostrand, New York, 1954.
6. W.M. Latimer, *Oxidation Potentials,* Prentice Hall, Englewood Cliffs, N.J., 1938.

4.6 SOFTENING AND BOILER WATER TREATMENT

4.6–1 RAW WATER AND ANTICIPATED ANALYSES AFTER TREATMENT

Substance	Symbol	1*	2	3	4	5	6	7
		ppm as $CaCO_3$						
Cations								
Calcium (Hardness)	Ca^{++}	100	35	58	1	–	–	100
Magnesium (Hardness)	Mg^{++}	100	58	7	1	–	–	100
Sodium	Na^{+}	100	100	85	298	–	–	100
Hydrogen Acidity	H^{+}	0	0	0	0	–	–	0
Total Cations		300	193	150	300	5	5	300
Anions								
Bicarbonate (Alkalinity)	HCO_3^{-}	150	0	0	150	–	–	150
Carbonate (Alkalinity)	CO_3^{--}	0	35	21	0	–	–	0
Hydroxide (Alkalinity)	OH^{-}	0	0	0	0	–	–	0
Phosphate (Alkalinity)	PO_4^{---}	0	0	0	0	–	–	0
Chloride	Cl^{-}	75	79	64	75	–	–	75
Sulfate	SO_4^{--}	75	79	63	75	–	–	75
Nitrate	NO_3^{-}	0	0	0	0	–	–	0
Total Anions		300	193	150	300	5	5	300
Total hardness		200	93	65	2	–	–	200
Alkalinity A (Methyl orange)		150	35	23	150	–	–	150
Alkalinity B (Phenolphthalein)		0	17	14	0	–	–	0
Noncarbonate hardness		50	58	55	0	–	–	50
Sodium alkalinity		0	0	0	150	–	–	0
		ppm as Substance						
Carbon dioxide as CO_2		30	0	0	30	5–10	0	30[d]
Silica as SiO_2		15	15[a]	5[b]	15	15	0.20	15
Iron and Manganese as Mn & Fe		10	0.2[b]	0.2[b]	0.2	0.2	0.2	0.2[c]
Turbidity		50	0–2[b]	0–2[a]	0–2[c]	0–2[c]	0–2[c]	0–2[c]
Color		10	10[a]	10[a]	10	10	10	10
Total solids (Cations + SiO_2)		315	208	155	315	20	5	315

[a] Some reduction will occur.
[b] Filtered effluent.
[c] With proper pretreatment.
[d] Affected by pH adjustment.

Note: Ion exchange processes assume that the water was adequately pretreated.

1.* Raw water
2. After cold lime softening and filtration
3. After hot process softening and filtration
4. Ion exchange softening
5. Two step demineralization (weak base anion exchange) and degasification
6. Two step demineralization (strong base anion resin) and degasification
7. Aeration and filtration

Source: Reprinted, with publisher's permission, from *J. Am. Water Works Assn.*, *62*(9):585, © 1970, by the American Water Works Association, Inc., 2 Park Ave., New York, N.Y. 10016.

4.6–2 MAJOR CONSTITUENTS IN FINISHED WATER BEFORE AND AFTER PASSAGE THROUGH A HOME WATER SOFTENER

Constituent	Unsoftened, mg/l	Softened, mg/l
Hardness	170	<1
pH	8.2	8.3
Sodium	23	93
Potassium	3.1	1.0
Calcium	47	<1
Magnesium	13	<1
Alkalinity	48	44
Chloride	44	43
Sulfate	100	100
TOC	1.6	1.4

Source: J.F. Kopp, in *Trace Substances in Environmental Health,* D.D. Hemphill, Ed., vol. 3, University of Missouri, Columbia, 1969.

4.6–3 TRACE CONSTITUENTS IN FINISHED WATER BEFORE AND AFTER PASSAGE THROUGH A HOME WATER SOFTENER

			After 24 hr contact with copper tubing	
Element	Unsoftened	Softened	Unsoftened	Softened
Zinc	<15	<15	290	100
Cadmium	<15	<15	<15	<15
Arsenic	<70	<70	<70	<70
Boron	140	140	77	137
Phosphorus	<70	<70	<70	<70
Iron	79	< 8	77	< 8
Molybdenum	<30	<30	<30	<30
Manganese	< 7.5	< 7.5	< 7.5	< 7.5
Aluminum	<30	<30	<30	<30
Beryllium	< 0.15	< 0.15	< 0.15	< 0.15
Copper	37	< 8	245	126
Silver	< 1.5	< 1.5	< 1.5	< 1.5
Nickel	<15	<15	<15	<15
Cobalt	<15	<15	<15	<15
Lead	<30	<30	3,780	2,250
Chromium	< 8	< 8	< 8	< 8
Vanadium	<30	<30	<30	<30
Barium	29	< 2	2	< 2
Strontium	93	< 2	129	< 2

Source: J.F. Kopp, in *Trace Substances in Environmental Health,* D.D. Hemphill, Ed., vol. 3, University of Missouri, Columbia, 1969.

4.6–4 WASTE BRINES FROM ION EXCHANGE SOFTENERS

Constituent	mg/l[1]	mg/l[2]
Calcium	1,720	3,000–6,000
Magnesium	600	1,000–2,000
Sodium and potassium	3,325	2,000–5,000[a]
Chloride	9,600	9,000–22,000
Total dissolved solids	15,656	
Total hardness	7,762	

[a] Sodium.

REFERENCES

1. H.A. Faber, *J. Am. Water Works Assn., 61*(10):542, 1969.
2. J.W. Krasaukas and L. Streicher, *J. Am. Water Works Assn., 61*(11):621, 1969.

4.6–5 COMPARISON OF SODIUM CONCENTRATION IN WELL WATER OBTAINED BEFORE AND AFTER IT HAS BEEN THROUGH THE SOFTENER

Sodium concentration, mg/l	No. of samples before water went through softener	No. of samples after water went through softener
0–99	48	18
100–199	32	30
200–299	1	15
300–399	5	11
400–499	1	2
500–599	3	7
600–699	0	0
700–799	0	2
800–899	0	3
900–999	0	1
1,000	0	1

Source: G.E. Garrison and O.L. Ader, *Arch. Environ. Health, 13*(11):55, © 1966, American Medical Association.

4.6–6 COMPARISON OF SODIUM CONCENTRATION OF SOFTENED WATER EARLY AND LATE IN THE VARIOUS LENGTH REGENERATION CYCLES

Type of water softener	Length of regeneration cycle and no. of softeners in each group	Average sodium concentration early in cycle, mg/l	Average sodium concentration late in cycle, mg/l
Automatic regeneration models	2 days, 2 softeners	358	361
	3 days, 8 softeners	325	280
	4 days, 4 softeners	251	149
	6 days, 12 softeners	192	147
	7 days, 1 softener	570	530
	12 days, 1 softener	845	62
Manual regeneration models	12 days, 1 softener	240	82
	21 days, 1 softener	1,000	850

Source: G.E. Garrison and O.L. Ader, *Arch. Environ. Health, 13*(11):55, © 1966, American Medical Association.

4.6–7 AMOUNT OF SLUDGE PRODUCED BY LIME HARDNESS REMOVAL

City	Acre ft of 50% sludge/yr/mgd/ 100 ppm hardness removed
Cedar Rapids, Iowa	0.62
Columbus, Ohio	0.52
Sandusky, Ohio	0.66
Lansing, Mich.	0.65
Hinsdale, Ill.	0.45

Source: Reprinted, with publisher's permission, from *J. Am. Water Works Assn., 53*(9):1169, © 1961 by the American Water Works Association, Inc., 2 Park Ave., New York, N.Y. 10016.

4.6–8 EFFECT OF DIFFERENT SLURRIES ON MAGNESIUM REDUCTION

Meramec River Water[a]

Lime dose	Slurry, % by weight	Residual magnesium (as $CaCO_3$), ppm[b]: Slurry from lime and activated silica	Slurry from lime only
Alkalinity equivalent plus 34.24 ppm	1.0	50	50
	2.0	46	46
	3.0	40	38
	4.5	36	30
Alkalinity equivalent plus 17.12 ppm	1.0	60	56
	2.5	52	50
	5.0	44	40
	6.5	44	36

[a] Contact time with the slurries was 1 hr.
[b] Raw water contained 76 ppm magnesium.

Source: Reprinted, with publisher's permission, from *J. Am. Water Works Assn., 52*(1):106, © 1960, by the American Water Works Association, Inc., 2 Park Ave., New York, N.Y. 10016.

4.6–9 CHEMICAL REACTIONS IN COLD LIME, LIME SODA, AND CALCIUM CHLORIDE SOFTENING PROCESSES

	CO_2	+ $Ca(OH)_2$	= $CaCO_3$	+ H_2O	
	Free Carbon Dioxide	+ Calcium Hydroxide	= Calcium Carbonate	+ Water	
	*$Ca(HCO_3)_2$	+ $Ca(OH)_2$	= $2CaCO_3$	+ $2H_2O$	
	Calcium Bicarbonate	+ Calcium Hydroxide	= Calcium Carbonate	+ Water	
	*$CaSO_4$	+ Na_2CO_3	= $CaCO_3$	+ Na_2SO_4	
	Calcium Sulfate	+ Sodium Carbonate	= Calcium Carbonate	+ Sodium Sulfate	
	*$CaCl_2$	+ Na_2CO_3	= $CaCO_3$	+ $2NaCl$	
	Calcium Chloride	+ Sodum Carbonate	= Calcium Carbonate	+ Sodium Chloride	
	*$Ca(NO_3)_2$	+ Na_2CO_3	= $CaCO_3$	+ $2NaNO_3$	
	Calcium Nitrate	+ Sodium Carbonate	= Calcium Carbonate	+ Sodium Nitrate	
	*$Mg(HCO_3)_2$	+ $2Ca(OH)_2$	= $Mg(OH)_2$	+ $2CaCO_3$	+ $2H_2O$
	Magnesium Bicarbonate	+ Calcium Hydroxide	= Magnesium Hydroxide	+ Calcium Carbonate	+ Water
*$MgSO_4$	+ Na_2CO_3	+ $Ca(OH)_2$	= $Mg(OH)_2$	+ $CaCO_3$	+ Na_2SO_4
Magnesium Sulfate	+ Sodium Carbonate	+ Calcium Hydroxide	= Magnesium Hydroxide	+ Calcium Carbonate	+ Sodium Sulfate
*$MgCl_2$	+ Na_2CO_3	+ $Ca(OH)_2$	= $Mg(OH)_2$	+ $CaCO_3$	+ $2NaCl$
Magnesium Chloride	+ Sodium Carbonate	+ Calcium Hydroxide	= Magnesium Hydroxide	+ Calcium Carbonate	+ Sodium Chloride
*$Mg(NO_3)_2$	+ Na_2CO_3	+ $Ca(OH)_2$	= $Mg(OH)_2$	+ $CaCO_3$	+ $2NaNO_3$
Magnesium Nitrate	+ Sodium Carbonate	+ Calcium Hydroxide	= Magnesium Hydroxide	+ Calcium Carbonate	+ Sodium Nitrate
	*$Ca(OH)_2$	+ Na_2CO_3	= $CaCO_3$	+ 2 NaOH	
	+ Calcium Hydroxide	+ Sodium Carbonate	= Calcium Carbonate	+ Sodium Hydroxide	
H_2SO_4	+ $Ca(OH)_2$	+ Na_2CO_3	= $CaCO_3$	+ Na_2SO_4	+ $2H_2O$
Sulfuric Acid	+ Calcium Hydroxide	+ Sodium Carbonate	= Calcium Carbonate	+ Sodium Sulfate	+ Water
	Na_2CO_3	+ $CaCl_2$	= $CaCO_3$	+ $2NaCl$	
	Sodium Carbonate	+ Calcium Chloride	= Calcium Carbonate	+ Sodium Chloride	
	$MgCO_3$	+ $CaCl_2$	= $CaCO_3$	+ $MgCl_2$	
	Magnesium Carbonate	+ Calcium Chloride	= Calcium Carbonate	+ Magnesium Chloride	

*Hot lime soda reactions also.

4.6–9 CHEMICAL REACTIONS IN COLD LIME, LIME SODA, AND CALCIUM CHLORIDE SOFTENING PROCESSES *(Continued)*

	Na_2CO_3 Sodium Carbonate	+ $CaSO_4$ + Calcium Sulfate	= $CaCO_3$ = Calcium Carbonate	+ $NaSO_4$ + Sodium Sulfate	
	$MgCO_3$ Magnesium Carbonate	+ $CaSO_4$ + Calcium Sulfate	= $CaCO_3$ = Calcium Carbonate	+ $MgSO_4$ + Magnesium Sulfate	
	$Al_2(SO_4)_3$ Aluminum Sulfate	+ $3Ca(HCO_3)_2$ + Calcium Bicarbonate	= $2Al(OH)_3$ = Aluminum Hydroxide	+ $3CaSO_4$ + Calcium Sulfate	+ $6CO_2$ + Carbon Dioxide
$Al_2(SO_4)_3$ Aluminum Sulfate	+ $3Na_2CO_3$ + Sodium Carbonate	+ $3Ca(OH)_2$ + Calcium Hydroxide	= $2Al(OH)_3$ = Aluminum Hydroxide	+ $3Na_2SO_4$ + Sodium Sulfate	+ $3CaCO_3$ + Calcium Carbonate
	$Fe_2(SO_4)_3$ Ferric Sulfate	+ $3Ca(HCO_3)_2$ + Calcium Bicarbonate	= $2Fe(OH)_3$ = Ferric Hydroxide	+ $3CaSO_4$ + Calcium Sulfate	+ $6CO_2$ + Carbon Dioxide
$Fe_2(SO_4)_3$ Ferric Sulfate	+ $3Na_2CO_3$ + Sodium Carbonate	+ $3Ca(OH)_2$ + Calcium Hydroxide	= $2Fe(OH)_3$ = Ferric Hydroxide	+ $3Na_2SO_4$ + Sodium Sulfate	+ $3CaCO_3$ + Calcium Carbonate

* Hot Lime Soda Reactions also.

Source: *Data Book*, The Permutit Co., New York, 1958, p. 56.

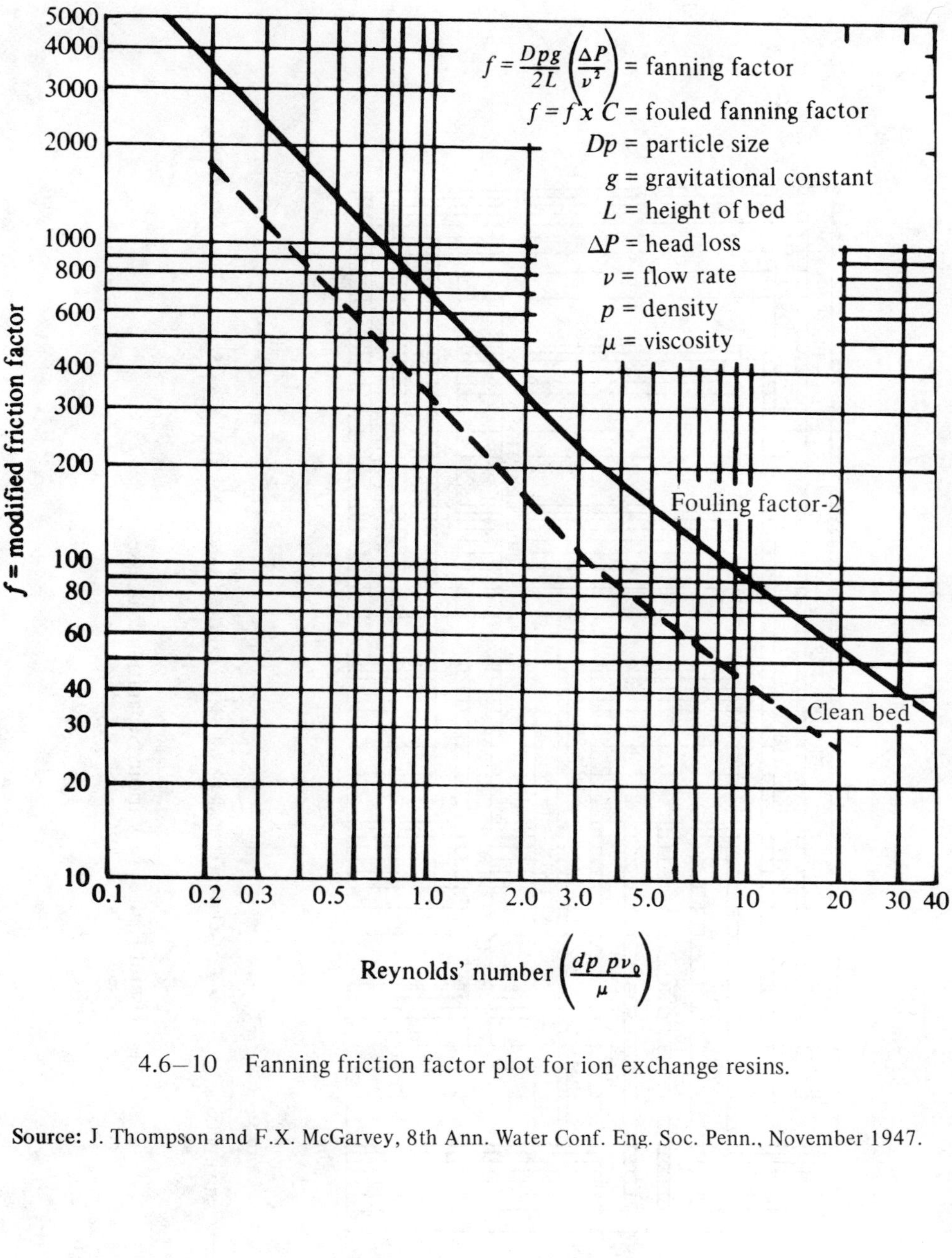

4.6–10 Fanning friction factor plot for ion exchange resins.

Source: J. Thompson and F.X. McGarvey, 8th Ann. Water Conf. Eng. Soc. Penn., November 1947.

4.7 MISCELLANEOUS AND NONCONVENTIONAL PROCESSES

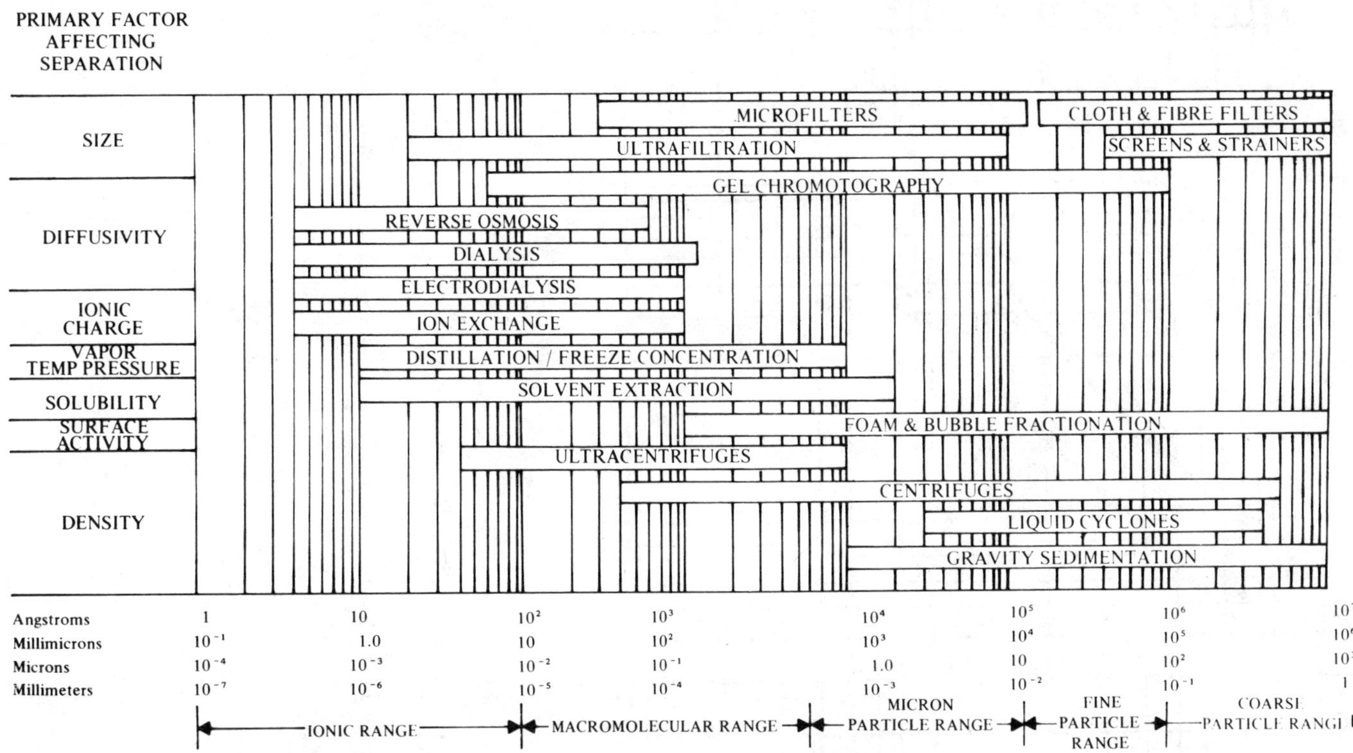

4.7–1 Useful Ranges of Various Separation Processes.

Source: Dorr Oliver, 1969.

4.7–2 EFFECT OF CORROSION INHIBITORS ON IRON CONTENT IN DISTRIBUTION SYSTEM

	Iron concentration, ppm		
Test location	With sodium hexameta-phosphate[a]	With bimetallic glassy phosphate[b]	Reduction, %
	In Distribution System		
1	0.31	0.124	60
2	0.154	0.133	13
3	0.35	0.136	61
4	no data	0.266	–
	Pickup Through System[c]		
1	0.192	0.014	92
2	0.036	0.023	36
3	0.232	0.056	75

[a] Average of 3 months of test data; 15 lb of inhibitor used per day.
[b] Average of 5 months of test data; 3 lb of inhibitor used per day.
[c] Data were obtained by subtracting the iron content of water leaving the treatment plant from the total iron content at the test location.

Source: Reprinted, with publisher's permission, from *57*(6):783, © 1965, by the American Water Works Association, Inc., 2 Park Ave., New York, N.Y. 10016.

4.7–3 PERCENT OF POPULATION USING FLUORIDATED WATER

States and Puerto Rico, 1970

State or other area	Total[a]	Using public water[b]	State or other area	Total[a]	Using public water[b]
Total	44.3	57.3	Mo.	45.0	61.2
Ala.	25.5	40.2	Mont.	18.9	28.0
Alas.	43.5	78.8	Nebr.	47.2	65.7
Ariz.	17.3	19.4	Nev.	3.2	3.8
Ark.	35.7	65.9	N.H.	11.1	16.9
Cal.	14.7	15.8			
			N.J.	12.7	14.5
Colo.	72.0	87.4	N.M.	38.8	53.1
Conn.	72.2	90.4	N.Y.	66.1	75.4
Del.	40.0	54.1	N.C.	37.2	73.5
D.C.	100.0	100.0	N.D.	47.4	89.6
Fla.	26.9	34.8	Ohio	37.6	48.4
			Okla.	55.2	73.0
Ga.	47.0	67.7	Ore.	15.8	20.8
Haw.	12.7	13.1	Pa.	40.0	49.8
Idaho	16.9	25.6	R.I.	80.3	89.8
Ill.	85.8	95.0			
Ind.	57.6	87.3	S.C.	35.6	63.4
			S.D.	50.4	86.1
Iowa	53.6	78.4	Tenn.	43.6	66.9
Kans.	44.5	58.3	Texas	48.9	60.7
Kent.	45.3	79.2	Utah	2.4	2.6
La.	7.5	10.0			
Maine	34.6	57.7	Vt.	26.3	45.1
			Va.	59.8	93.6
Md.	76.2	95.0	Wash.	37.6	45.7
Mass.	11.9	12.8	W.V.	50.2	81.3
Mich.	62.9	89.8	Wisc.	60.8	94.0
Minn.	71.9	95.0			
Miss.	21.4	42.6	Wyo.	29.4	39.3
			P.R.	66.7	92.3

[a] Estimates based on Bureau of the Census, *Current Population Reports* (Series P-25, No. 456) and press release, *1970 Census Final National and State Totals and Congressional Reapportionment Statistics.*
[b] *U.S. Census of Housing: 1960,* Series HC (1), Nos. 1–52. Includes adjustments for population growth based on Bureau of the Census estimates.

Note: As of December 31. Covers both natural and controlled fluoridation.

Source: *Fluoridation Census,* Public Health Service, Dept. of Health, Education, and Welfare, 1970.

4.7–4 DEFLUORIDATION PLANTS

Southern California

Place	Date installed	Defluoridation media	Units: No.	Units: Capacity, gpm	Regeneration chemicals	Remarks
Fort Irwin	1954	bone char	2	33 (each)	caustic soda, phosphoric acid	serves base residential area
Desert Center (State Division of Highways Maintenance)	1955	bone char	1	35	caustic soda, sodium bisulphate	water piped to each residence
Elsinore	1960	activated alumina	2	250 (each)	caustic soda, sulfuric acid	defluoridated water could be served as entire supply, but low fluoride supply now in use and plant inoperative
Apple Valley (Youngtowne Water Co.)	1961	tricalcium phosphate	1	100	caustic soda, carbon dioxide	treats entire well supply
Chocolate Mountain	1961	tricalcium phosphate	1	100	caustic soda, carbon dioxide	defluoridated water could be served as entire supply, but low fluoride supply now in use and plant inoperative
Twentynine Palms (W.D. Fulton)	1959	bone meal derivative	1	13	caustic soda, hydrochloric acid	treats water hauled in tank truck
Death Valley Junction (State Division of Highways Maintenance Station)	to be installed	mixed bed demineralizer; sulfonated polystyrene and quaternary amine polystyrene	1	15	caustic soda, sulfuric acid	demineralization plant; water piped to each residence
Twentynine Palms (County Water District)	1961	bone char	1	3	not regenerated	water available at one tap
Panamint Springs (State Division of Highways maintenance Station)	1951	bone char	3	–	not regenerated	one unit at each residence
Tecopa (Francis Water System)	to be installed	bone char	1	3	not regenerated	water available at one tap
Death Valley National Monument (National Park Service Residential Area)	1955	tricalcium phosphate	1	4½	not regenerated	water available at one tap

Source: Reprinted, with publisher's permission, from *57*(2):245, © 1965, by the American Water Works Association, Inc., 2 Park Ave., New York, N.Y. 10016.

4.7–5 SUMMARY OF PARTICULATE CHARACTERISTICS AND BEHAVIOR DURING MICROSTRAINING

1960–61

	Volume of particulates, μl/l		Mean volume particle in un-treated water	Particu-late re-moval volume, %
	Un-treated	Micro-strained		
Avg	13.7	1.7	37	64
Coefficient of variation	2.1	1.3		
Maximum	112.0	8.1	160	95
Minimum	0.4	0.3	8	-20

Source: Reprinted, with publisher's permission, from *55*(5):606, © 1962, by the American Water Works Association, Inc., 2 Park Ave., New York, N.Y. 10016.

4.7–6 COMPOSITION OF DEBRIS PARTICLES

	Particulates, %		
Particle composition	Influent	Effluent	Waste water
Predominantly mineral	17	10	8
Mineral and organic	38	14	62
Predominantly organic	45	76	30

Source: Reprinted, with publisher's permission, from *J. Am. Water Works Assn.*, *55*(5):606, © 1962, by the American Water Works Association, Inc., 2 Park Ave., New York, N.Y. 10016.

4.7–7 SIZE OF DEBRIS PARTICLES

Particle size, μ	Influent summation, %	Effluent summation, %	Waste summation, %
2	5.3	25.0	1.6
3	16.9	41.3	3.2
5	31.6	57.6	11.1
7	47.4	72.6	22.2
10	64.3	85.1	34.9
15	79.0	92.6	54.0
20	90.6	98.8	74.6
30	96.9	100.0	87.3
50	90.0		95.2
70	100.0		98.4
100			100.0

Source: Reprinted, with publisher's permission, from *J. Am. Water Works Assn.*, *55*(5):606, © 1962, by the American Water Works Association, Inc., 2 Park Ave., New York, N.Y. 10016.

4.7–8 REMOVAL OF MICROORGANISMS BY MICROSTRAINING

Summer, 1963

Date of sampling and class of material	Concentration of microorganisms[a] Raw water	Micro-strained water	Removal, %
June 17			
Diatomaceae	330	10	97
Chlorophyceae	4	0	100
Rotifera	20	20	0
Amorphous	140	148	–6
July 8			
Diatomaceae	398	100	75
Chlorophyceae	68	12	82
Cyanophyceae	4	0	100
Protozoa	4	0	100
Amorphous	442	156	65
July 15			
Diatomaceae	653	70	89
Chlorophyceae	16	2	88
Protozoa	12	0	100
Rotifera	60	0	100
Amorphous	860	206	76
July 29			
Diatomaceae	96	0	100
Chlorophyceae	12	0	100
Cyanophyceae	2	0	100
Rotifera	30	0	100
Amorphous	650	470	28
August 7			
Diatomaceae	1,206	90	93
Chlorophyceae	50	0	100
Protozoa	16	0	100
Rotifera	30	0	100
Amorphous	950	400	58
August 26			
Diatomaceae	150	0	100
Chlorophyceae	4	0	100
Rotifera	80	0	100
Amorphous	464	176	62
September 6			
Diatomaceae	160	10	94
Chlorophyceae	8	0	100
Amorphous	326	220	33

[a] Area standard units per milliliter.

Source: Reprinted, with publisher's permission, from *J. Am. Water Works Assn.*, 57(11):1422, © 1965, by the American Water Works Association, Inc., 2 Park Ave., New York, N.Y. 10016.

4.7–9 MICROSTRAINER SLUDGE TREATMENT

Installation	Raw water	Special provisions	Sludge disposal
Fontana System, San Gabriel Water Co.	Lytle Creek, 0–3 ppm turbidity	Bypass screens when turbidity is greater than 25 ppm. Treat screens with hypochlorite and ultraviolet radiation	1.6% waste water, no information on disposal
Kenosha, Wisc.	Lake Michigan, 5–15 ppm turbidity	Treat screens with hypochlorite each 6 wk, oxidizing agent each 4 mo. to remove iron bacteria	4.1% washwater pumped, 2.7% wasted, returned to Lake Michigan
Denver, Colo.	Surface-Marston Lake, 3–5 ppm avg turbidity	Add Cl_2 to waste. Treat screens with hypochlorite	1% waste, returned to Lake Michigan
Colorado Springs, Colo.	Snow melt, less than 3 JTU turbidity	Microstrainers used only for peak loads	Returned to Monument Creek downstream from plant
Boulder, Colo.	Snow melt		To sludge ponds, then drained off to Boulder Creek
Baker Metro, Colo.			To sewage treatment plant
Palisade, Colo.			To irrigation ditch and used for irrigation
Glenwood Springs, Colo.			Returned to No Name Creek
Loveland, Colo.			Returned to river
Estes Park, Colo.			To sludge pond, supernatant drained off to sewage treatment plant
Grand Junction, Colo.			To irrigation ditch
Durango, Colo.			To washwater pond and water reclaimed
Golden, Colo.			Returned to Clear Creek
Ft. Collins, Colo.		Three units, third unit for reclaiming waste water from others	No data
Greeley, Colo.			Returned to Boyd Lake
Fort Collings-Loveland W.D., Colo.			To irrigation ditch
Little Thompson Valley W.D., Colo.			Returned to Dry Creek
Evergreen, Colo.			Returned to Bear Creek
Danvers, Mass.	Middleton Pond, 4.0 ppm avg turbidity		1–5% net waste water. No information on disposal

4.7–9 MICROSTRAINER SLUDGE TREATMENT (*Continued*)

Installation	Raw water	Special provisions	Sludge disposal
Belleville, Ont.	Lake Ontario	Treat screens with sodium hypochlorite and ultraviolet radiation	1.5–2.7% net waste water
Haweswater, Great Britain	Low turdidity	Pilot plant	No information on waste water. Reference does not consider this washwater as a sludge disposal problem
Turret Water Scheme of the Loch Turret Water Board	Low turbidity		Returned to Turret Burn. Reference considers waste water beneficial to animal life.

Source: Reprinted, with publisher's permission, from *J. Am. Water Works Assn.*, *61*(10):542, © 1969, by the American Water Works Association, Inc., 2 Park Ave., New York, N.Y. 10016.

4.7–10 RESULTS OF FOAM SEPARATION WATER TREATMENT

Location of sample	Initial optical density	Total surfactant dose, mg/l	Residual optical density	Collapsed foam volume,[b] ml	Residual surfactant concentration, mg/l
Des Plaines River, North	0.111	20	0.016	8	1.4
	0.226[a]	40	0.009	22	0.9
Calumet River	0.083	20	0.003	25	1.6
Skokie Lagoons	0.109	30	0.005	17	–
Des Plaines River, South	0.141	30	0.014	22	–
Sanitary Canal	0.143	20	0.012	10	1.3
Vic Bien Hoa, Viet Nam	0.620	20	0.005	10	1.2
Song Dong Nai River, Viet Nam	0.086	10	0.006	2	1.2

[a] Sediment from river bottom added.
[b] Feed volume = 2,000 ml.

Source: Reprinted, with publisher's permission, from *J. Am. Water Works Assn.*, *59*(7):859, © 1967, by the American Water Works Association, Inc., 2 Park Ave., New York, N.Y. 10016.

4.7–11 COMMERCIAL DESALTING TECHNIQUES

Technique	Typical concentrations of total dissolved solids, mg/l[a]		Comments
	Feed water,[a] max	Product water,[a] min	
Distillation	40,000	5–10	Well developed, widely used, in unit sizes up to 6 mgd. Substantial improvements still possible
Ion exchange	2,000	100–500	Well developed for industrial desalting
Crystallization (includes freezing)	40,000	300–500	Recently developed in sizes up to 100 kgd. No utility plants yet in operation
Electrodialysis	5,000	300–500	Well developed. Many small plants in the 1–10 kgd range; a few up to 2 mgd
Reverse osmosis (one stage)	6,000	300–500	Recently developed. One municipal operation; 2 plants now on order but not yet operating

[a] Maximum feed water concentrations and minimum ranges of product water concentration may be considerably surpassed in special industrial applications.

Source: Reprinted, with publisher's permission, from *J. Am. Water Works Assn., 64*(6):367, © 1972, by the American Water Works Association, Inc., 2 Park Ave., New York, N.Y. 10016.

4.7–12 DESALTING PROCESSES

Process	Application	
	Sea water	Brackish
Distillation		
Multistage flash (MSF)	x	x
Multiple effect, long-tube vertical (LTV)	x	x
Combination MSF and LTV	x	x
Vapor compression	x	x
Vapor reheat	x	x
Solar	x	x
Membrane processes		
Electrodialysis	?	x
Reverse osmosis	?	x
Dynamic membrane	?	x
Freezing processes		
Vapor freezing–vapor compression	x	x
Direct contact	x	x
Hydrate	x	x
Others		
Ion exchange		x
Solvent extraction		x
Electrochemical demineralization		x

Source: Reprinted, with publisher's permission, from *J. Am. Water Works Assn., 60*(8):869, © 1968, by the American Water Works Association, Inc., 2 Park Ave., New York, N.Y. 10016.

4.7–13 ENERGY REQUIREMENT FOR COMBINATION NUCLEAR REACTOR–DISTILLATION DESALINATION PLANT

	Energy required, Btu	Total energy requirement, Btu
Separate plants		
Electricity, 1 kWh	10,377	24,237
Water, 147 lb	13,860	
Combination plant		
Electricity, kWh		
Water, 147 lb	17,273	17,273
Difference		6,964

Source: Reprinted, with publisher's permission, from *J. Am. Water Works Assn., 60*(8):869, © 1968, by the American Water Works Association, Inc., 2 Park Ave., New York, N.Y. 10016.

4.7–14 ION REMOVAL SELECTIVITY IN ELECTRODIALYSIS

	Fraction of ion removed / Fraction of all ions removed	
Ion	**Avg.**	**Range**
Bicarbonate	0.77	0.63–0.95
Chloride	1.25	1.07–1.51
Nitrate	1.22	1.02–1.48
Phosphate	0.72	0.25–1.03
Sulfate	1.11	0.21–2.00
Ammonium	1.15	1.05–1.3
Calcium	1.25	1.01–1.48
Magnesium	0.98	0.71–1.35
Potassium	1.14	0.77–1.35
Sodium	0.79	0.52–0.99
Calcium and magnesium	1.13	1.05–1.28

Source: J.M. Cohen, *Adv. Waste Treat. Water Reuse Symp.,* Session 5, Dallas, January 12–14, 1971.

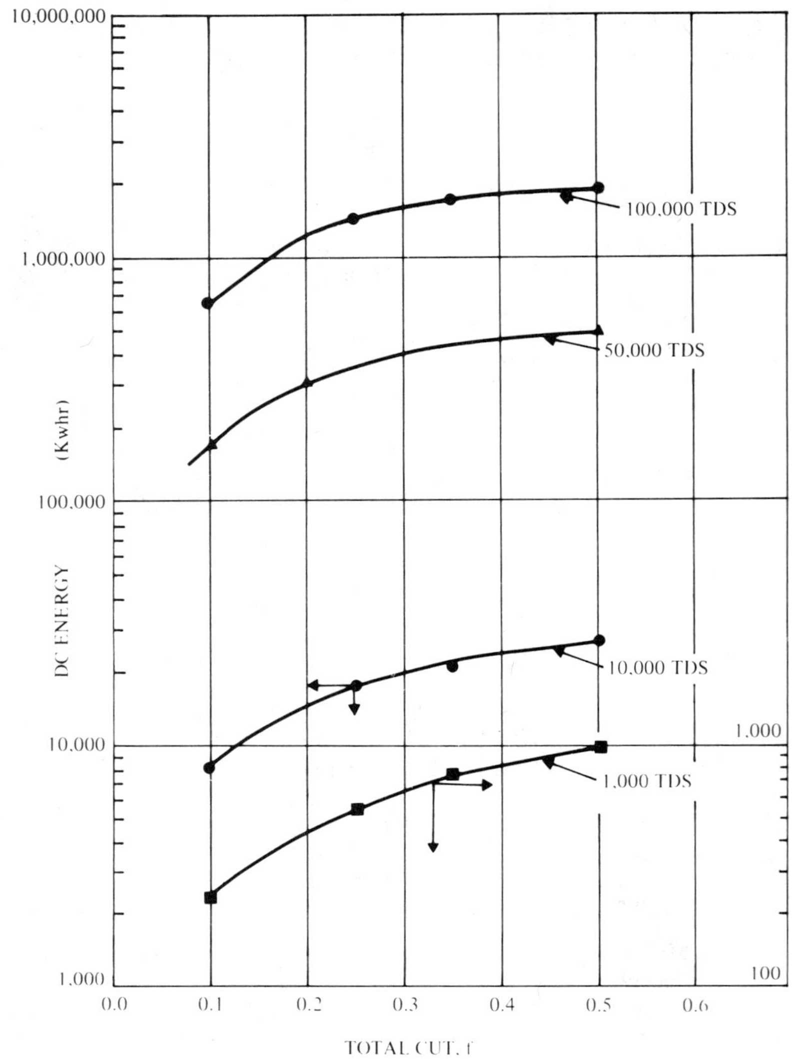

4.7–15 DC Energy Required for a Desired TDS Removal by Electrodialysis.

Source: *Electrodialysis in Advanced Waste Treatment,* Federal Water Pollution Control Administration, 1967.

4.7.1 REVERSE OSMOSIS

4.7–16 TYPICAL WATER ANALYSIS OF 50,000 GAL/DAY PILOT PLANT

	Concentration, mg/l		
	Feed	Brine	Product
Cations			
Sodium	900	3,700	106
Calcium	357	1,500	5.6
Magnesium	221	1,020	3.4
Potassium	26	103	3.8
Iron	7	27	0.1
Manganese	4.1	17	0.0
Anions			
Chloride	2,260	8,900	180
Sulfate	612	2,580	2.9
Bicarbonate	61	18	1.8
Silica	35	58	6.4
Fluoride	1.1	3.6	0.1
Nitrate	0.9	3	0.0
Boron	0.3	0.5	0.2
Total alkalinity ($CaCO_3$)	50	15	1.5
Total hardness ($CaCO_3$)	1,800	7,940	28
Dissolved solids[a]	4,490	17,620	300
Conductivity, μmho/cm at 20°C	7,150	23,000	685
pH	6.0	4.6	6.0

[a] Residue upon drying at 180°C.

Source: E.S. Perry and C.J. Van Oss, *Progress in Separation and Purification*, Wiley-Interscience, New York, 1970.

4.7–17 FEED AND PRODUCT WATER ANALYSIS

Reverse Osmosis Field Test at Coalinga, California

	Feed water, ppm	Product water, ppm
Sulfate	1,227.2	48.1
Sodium	543.0	62.0
Chloride	266.0	78.0
Bicarbonate	158.6	36.6
Calcium	116.6	11.6
Magnesium	75.1	6.7
Silica	35.0	12.0
Iron and alumina (R_2O_3)	12.2	2.4
Boron	2.5	1.5
Iron	0.04	0.04
Total dissolved solids	2,417.6	265.6
Total hardness (as $CaCO_3$)	604.3	56.9
pH	7.6	7.5

Source: E.S. Perry and C.J. Van Oss, *Progress in Separation and Purification*, Wiley-Interscience, New York, 1970.

4.7–18 FIELD TEST RESULTS OF HOLLOW FIBER DESALINATION UNIT

Firebaugh, California

	Run 1		Run 2 (with acidification)	
Ion	Feed, ppm	Product, ppm	Feed, ppm	Product, ppm
SO_4^{2-}	3,265	36	3,515	101
Na^+	1,410	329	1,665	200
Cl^-	449	372	450	212
Ca^{2+}	328	18.9	336	7.7
HCO_3^-	295	165	49.0	25.6
Mg^{2+}	167	7.3	173	3.2
SiO_2	47.1	32.8	47.3	32.5
NO_3^- (as N)	17.1	11.4	9.7	7.9
F^-	0.2	<0.1	0.2	<0.1
K^+	3.4	0.8	3.6	0.5
B	11.0	11.6	12.0	13.0
Fe^{2+}	<0.1	0.0	0.3	0.1
Total alkalinity	242	135	40.2	21.0
Total hardness	3,150	101	1,692	44.0
Total dissolved solids	5,926	958	6,000	654
pH	7.4	7.3	5.8	5.6

Source: E.S. Perry and C.J. Van Oss, *Progress in Separation and Purification,* Wiley-Interscience, New York, 1970.

4.7–19 SUMMARY OF REVERSE OSMOSIS RESULTS FOR REJECTION OF VARIOUS SOLUTES

Composition of solution	Membrane constant, 10^{-5} g/cm^2 sec atm	Water flux, gal/ft^2 day	Rejection, %
Anisotropic Membranes Annealed at 85°C; Applied Pressure, 1,500 psi			
1% NaCl	0.96 ± 0.03	19.1 ± 0.6	98.39 ± 0.07
0.95% $NaHCO_3$	0.90 ± 0.03	18.5 ± 0.5	99.27 ± 0.04
1% NH_4Cl	0.92 ± 0.05	18.3 ± 1.0	97.1 ± 0.1
1% $NaNO_3$	0.98 ± 0.02	20.1 ± 0.4	96.4 ± 0.0
98 ppm H_3BO_3	0.94 ± 0.04	20.2 ± 0.9	56 ± 5
138 ppm urea	0.97 ± 0.02	20.9 ± 0.4	45 ± 1
940 ppm dextrose	1.15 ± 0.06	24.6 ± 1.2	99.74 ± 0.09
0.98% NaH_2PO_4	1.13 ± 0.06	23.4 ± 1.2	99.93 ± 0.03
106 ppm NH_4OH	1.07 ± 0.05	23.0 ± 1.1	25 ± 4
1.5% Na_2HPO_4	0.95 ± 0.05	19.1 ± 0.9	99.984 ± 0.003
33 ppm 2,4-dichlorophenoxy acetic acid (2,4-D)	0.87 ± 0.02	18.7 ± 0.4	92.8 ± 0.0
1% NaCl	0.85 ± 0.02	17.2 ± 0.3	98.21 ± 0.21
96 ppm 2,4-dichlorophenol	0.59 ± 0.01	12.7 ± 0.2	–34 ± 2
1% NaCl	0.65 ± 0.01	13.1 ± 0.2	98.08 ± 0.17
Anisotropic Membranes Annealed at 70°C; Applied Pressure, 800 psi			
1% NaCl	2.59 ± 0.13	25.7 ± 1.3	95.1 ± 0.1
1% $NaHCO_3$	2.65 ± 0.15	27.6 ± 1.5	97.9 ± 0.1
1% NH_4Cl	2.68 ± 0.13	26.5 ± 1.3	91.0 ± 0.2
1.15% NaH_2PO_4	2.68 ± 0.13	28.0 ± 1.9	99.82 ± 0.07
155 ppm urea	2.96 ± 0.17	34.0 ± 1.9	27 ± 2
0.85% $NaNO_3$	2.77 ± 0.15	29.5 ± 1.5	88.5 ± 0.8
100 ppm H_3BO_3	2.99 ± 0.11	34.4 ± 1.3	34.5 ± 0.7
1,120 ppm dextrose	3.15 ± 0.04	36.1 ± 0.5	99.38 ± 0.25
1.18% Na_2HPO_4	2.78 ± 0.06	29.0 ± 0.6	99.94 ± 0.0
90 ppm NH_4OH	2.85 ± 0.06	32.8 ± 0.7	7.4 ± 0.7
1% NaCl	2.75 ± 0.05	27.1 ± 0.6	95.3 ± 0.2

Source: H.K. Lonsdale *et al.*, U.S. Office of Saline Water Research and Development Progress Report, 447, 1969.

4.7–20 SELECTIVITY OF MODIFIED CELLULOSE ACETATE MEMBRANES TO SEVERAL SOLUTES

Solute	Concentration, wt %	Rejection, %	Membrane constant, 10^{-5} g/cm^2 sec atm	Ref.
NaCl	5.25	98.1	1.21	1
NaBr	5.25	98.0	0.88	1
KCl	5.25	95.8	1.02	1
$NaClO_4$	5.25	86.3	1.07	1
NH_4ClO_4	5.25	77.4	0.98	1
NH_4NO_3	0.25	80.3	1.30	1
$CaCl_2$	5.25	99.1	0.98	1
Na_2SO_4	5.25	99.3	0.74	1
HNO_3	0.25	51.3	1.26	1
Sodium lauryl sulfate	0.172	98.2	1.77	1
Sucrose	5.25	99.7	1.58	1
Tetramethyl ammonium chloride	5.25	99.6	1.16	1
Tetraethyl ammonium chloride	5.25	99.4	1.26	1
Phenol	0.0080	-22	0.91	2
Dextrose	0.094	99.74	1.15	3
Na_2HPO_4	1.5	99.98	0.95	3
NH_4OH	0.0106	25	1.07	3
2,4-dichlorophenoxy acetic acid (2,4-D)	0.0035	92.8	0.87	3

Source: E.S. Perry and C.J. Van Oss, *Progress in Separation and Purification,* Wiley-Interscience, New York, 1970.

REFERENCES

1. R. Blunk, Univ. Calif. Los Angeles, Department of Engineering Report 64-28, 1964.
2. H.K. Lonsdale, U. Merten, and M. Tagami, *J. Appl. Polymer Sci., 11*:1807, 1967.
3. H.K. Lonsdale *et al.,* Office of Saline Water Research and Development Progress Report 447, 1969.

4.7–21 CHARACTERISTICS OF WATER AND SALT PERMEABILITY OF SELECTED NYLON FILMS AS MEASURED IN DIRECT OSMOSIS AT 25°C

Material	Film	Thickness, μ	Water permeability g/cm sec	Salt permeability cm²/sec
Du Pont Zytel-31 nylon (polyamide 6/10)	Extruded	29	5.6×10^{-9}	$<1.3 \times 10^{-12}$
Du Pont Fe-2355 nylon copolymer	Extruded	22	2.2×10^{-8}	9.2×10^{-11}
Du Pont Zytel-42	Extruded	24	7.2×10^{-9}	7.9×10^{-11}
Du Pont Zytel-2353	Extruded	50	6.9×10^{-9}	1.3×10^{-10}
Du Pont Zytel-2380	Extruded	45	3.7×10^{-9}	1.1×10^{-10}
Du Pont Zytel-61 nylon copolymer	Ethanol cast	52	1.2×10^{-8}	1.6×10^{-10}
Du Pont Zytel-61-P nylon (Zytel-61 with plasticizer)	Ethanol cast	25	3.2×10^{-8}	3.5×10^{-10}
Belding Corticelli nylon 809 (alkoxy-alkyl substituted nylon 6/6)	Ethanol cast	85	5.0×10^{-7}	1.26×10^{-8}
Belding Corticelli nylon 819 (alkoxy-alkyl substituted nylon 6/6)	Ethanol cast	105 26	5.2×10^{-7} 4.4×10^{-7}	1.1×10^{-8} 1.0×10^{-8}
Belding Corticelli nylon 829 (alkoxy-alkyl substituted nylon 6/6)	Ethanol cast	85	6.8×10^{-7}	1.6×10^{-8}

Source: E.S. Perry and C.J. Van Oss, *Progress in Separation and Purification,* Wiley-Interscience, New York, 1970.

4.7–22 COMMERCIALLY AVAILABLE ULTRAFILTRATION AND REVERSE OSMOSIS MEMBRANES

Membrane	Manufacturer	Type	Chemical composition	Net pressure, psi	Water flux, gal/ft² day	Solute	Rejection, %	Principal use
Loeb-Sourirajan, 85°C annealed	Several	Anisotropic	Cellulose acetate	1000	20	NaCl	97–98	Water desalination
Loeb-Sourirajan, 70°C annealed	Several	Anisotropic	Cellulose acetate	400	20	NaCl	90	Water desalination
Loeb-Sourirajan, unannealed	Several	Anisotropic	Cellulose acetate	150	20	NaCl	25	Water desalination
Gel cellophane	Du Pont, Union Carbide	Homogeneous	Cellulose	100	1.5	Sucrose	15	Concentration and demineralization of proteins and organic solutes
P.E.M.	Gelman	Isotropic	Cellulose triacetate	100	8.5	10,000-MW dextran 67,000-MW BSA[a]	0 100	
P	Schleicher and Schuell	Homogeneous	Cellulosic	100	35	10,000-MW dextran 67,000-MW BSA	0 100	
Diaflo UM-3	Amicon	Anisotropic	Polyelectrolyte complex	100	25	Sucrose	90	Concentration and demineralization of proteins and organic solutes
Diaflo UM-2	Amicon	Anisotropic	Polyelectrolyte complex	100	60	Sucrose	50	Concentration and demineralization of proteins and organic solutes
Diaflo UM-1	Amicon	Anisotropic	Polyelectrolyte complex	100	175	6000-MW PEG[b]	80	Protein concentration, purification
Pellicon PSED	Millipore	Anisotropic	Cellulose ester	100	40	20,000-MW trypsin	50	Concentration, purification of aqueous solutions
Pellicon PSAC	Millipore	Anisotropic	Cellulose ester	100	120	Sucrose	40–60	Concentration, purification of aqueous solutions
HFA 100	Abcor		Cellulosic	100	25	10,000-MW dextran	100	Macromolecular concentration, purification
HFA 300	Abcor		Cellulosic	100	600	13,500-MW cytochrome	20	Macromolecular concentration, purification
Diaflo XM-50	Amicon	Anisotropic	Substituted olefin	100	350	67,000-MW BSA	100	Macromolecular concentration and removal
Diaflo XM-100	Amicon		Substituted olefin	100	1,050	160,000-MW gamma globulin	0	Macromolecular fractionation

[a] BSA – bovine serum albumin.
[b] PEG – polyethylene glycol.

4.7–22 COMMERCIALLY AVAILABLE ULTRAFILTRATION AND REVERSE OSMOSIS MEMBRANES *(Continued)*

Membrane	Manufacturer	Type	Chemical composition	Net pressure, psi	Water flux, gal/ft^2 day	Solute	Rejection, %	Principal use
							100	
Diaflo PM-10	Amicon		Aromatic polymer	100	1,050	1400-MW bacitracin	70	Protein concentration, purification
						35,000-MW pepsin	100	
Diaflo PM-30	Amicon		Aromatic polymer	100	2,800	110,000-MW dextran	60	Macromolecular concentration, purification
VF	Millipore	Isotropic	Cellulose ester	100	1,000	250-Å pores		
HA	Millipore	Isotropic	Cellulose ester	100	15,000	4500-Å pores		Bacteria removal
Nuclepore	General Electric	Isotropic	Polycarbonate	14	34,000	0.5 ± 0.06 μ pores		

[a] BSA – bovine serum albumin.
[b] PEG – polyethylene glycol.

Source: R.E. Lacey and S. Hoeb, *Industrial Processing with Membranes,* Wiley-Interscience, New York, 1972.

REFERENCE

1. A.S. Michaels, in *Progress in Separation and Purification,* E.S. Perry, Ed., Wiley-Interscience, New York, 1968, p. 297.

4.7–23 LIFETIME OF CELLULOSE ACETATE MEMBRANES: TIME FOR NaCl PERMEATION CONSTANT TO INCREASE BY 20, 50, 100, and 400% at 23°C

pH	Hydrolysis rate constant, k_1, sec^{-1}	20% Δt	50% Δt	100% Δt	400% Δt
2	4.8×10^{-9}	15d[a]	33d	42d	130d
3	6×10^{-10}	120d	270d	1.2y	2.8y
4	2×10^{-10}	1.0y[b]	2.2y	3.7y	8.2y
5	1.8×10^{-10}	1.1y	2.4y	4.1y	9.5y
6	3×10^{-10}	180d	1.4y	2.5y	5.7y
7	1.3×10^{-9}	55d	120d	210d	1.3y
8	1.3×10^{-8}	5.5d	12d	21d	49d
9	1.3×10^{-7}	13h[c]	1.2d	2.1d	4.9d
10	1.3×10^{-6}	1.3h	2.9h	5.0y	12h

[a] d – days.
[b] y – years.
[c] h – hours.

Source: E.S. Perry and C.J. Van Oss, *Progress in Separation and Purification,* Wiley-Interscience, New York, 1970.

4.7–24 PRODUCTIVITY OF REVERSE OSMOSIS MEMBRANE MODULES

	ft^2 membrane / ft^3 equipment	Flux, gal/day ft^2	Productivity, gal/day ft^3
Tubular	20	32	640
Spiral wound	250	32	8,000
Hollow fiber (nylon)	5,400	1	5,400
Hollow fiber (cellulose acetate)	2,500	10	25,000

Source: F.M. Middleton *et al., Adv. Waste Treat. Water Reuse Symp.,* Session 4, Dallas, January 12–14, 1971.

Section 5

Water Distribution

5.1 Water Supply Requirements
Pages 521–531

5.2 Plumbing System Standards
Pages 532–545

5.3 Storage
Pages 546–553

5. WATER DISTRIBUTION

5.1 WATER SUPPLY REQUIREMENTS

See also Section 2. Needs.

5.1–1 ESTIMATED WATER CONSUMPTION AT DIFFERENT TYPES OF ESTABLISHMENTS

Type of establishment	gal/cap day
Dwelling units, residential	
Private dwellings on individual wells or metered supply	50–75
Apartment houses on individual wells	75–100
Private dwellings on public water supply, unmetered	100–200
Apartment houses on public water supply, unmetered	100–200
Subdivision dwelling on individual well, or metered supply, per bedroom	150
Subdivision dwelling on public water supply, unmetered, per bedroom	200
Dwelling units, treatment	
Hotels	50–100
Boarding houses	50
Lodging houses and tourist homes	40
Motels, without kitchens, per unit	100–150
Camps	
Pioneer type	25
Children's, central toilet and bath	40–50
Day, no meals	15
Luxury, private bath	100–150[1]
Labor	35–50
Trailer with private toilet and bath, per unit (2½ persons)[a]	125–150
Restaurants (including toilet)	
Average	7–10
Kitchen wastes only	2½–3
Short order	4
Short order, paper service	1–2
Bars and cocktail lounges	2
Average type, per seat	35
Average type, 24 hr, per seat	50
Tavern, per seat	20
Service area, per counter seat (toll road)	350
Service area, per table seat (toll road)	150
Institutions	
Average type	75–125
Hospitals	150–250
Schools	
Day, with no cafeteria, gymnasium, or showers	15[1]
Day, with cafeteria or lunch room, no gymnasiums or showers	10–20[1]
Day, with cafeteria and showers	15–25[1]
Boarding	75–100[1]
Theatres	
Indoor, per seat, two showings per day	3
Outdoor, including food stand, per car (3 1/3 persons)	3–5
Automobile service stations	
Per vehicle served	10
Per set of pumps	500
Stores	
First 25 ft frontage	450
Each additional 25 ft frontage	400
Country clubs	
Resident type	100
Transient type, serving meals	17–25
Offices	10–15
Factories, sanitary wastes, per shift	15–35
Self-service laundry, per machine	250–500
Bowling alleys, per alley	200
Swimming pools and beaches, toilet and shower	10–15
Picnic parks, with flush toilets	5–10
Fairgrounds (based on daily attendance)	1
Assembly halls, per seat	2
Airport, per passenger	3–5

[a] Add 125 gallons per trailer space for lawn sprinkling, car washing, leakage, etc.

Note: Water under pressure, flush toilets, and wash basins are assumed provided unless otherwise indicated. These figures are offered as a guide; they should not be used blindly. Add for any continuous flows and industrial usages. Figures are flows per capita per day, unless otherwise stated.

Source: J.A. Salvato, *Public Works, 91*(5), 1960.

REFERENCE

1. *Report of Public Health Service Committee on Plumbing Standards,* U.S. Department of Health, Education, and Welfare, Public Health Service, 1963.

5.1–2 MINIMUM NUMBER OF PLUMBING FIXTURES

Type of building occupancy	Type of fixture: Water closets		Urinals		Lavatories		Bathtubs or showers	Drinking fountains	Other fixtures
	No. persons	No. fixtures	No. persons	No. fixtures	No. persons	No. fixtures			
Assembly – places of worship	150 women	1	300 men[a]	1	1			1	
	300 men	1							
Assembly – other than places of worship (auditoriums, theaters, convention halls)	1–100	1	1–200	1	1–200	1		1 for each 300 persons	1 slop sink
	101–200	2	201–400	2	201–400	2			
	201–400	3	401–600	3	401–750	3			
	Over 400, add 1 fixture for each additional 500 men and 1 for each 300 women		Over 600, add 1 fixture for each 300 men[a]		Over 750, add 1 fixture for each 500 persons				
Dormitories–school or labor, also institutional	Men: 1 for each 10 persons Women: 1 for each 8 persons.		1 for each 25 men. Over 150, add 1 fixture for each 50 men[a]		1 for each 12 persons. (Separate dental lavatories should be provided in community toilet rooms. A ratio of 1 dental lavatory to each 50 persons is recommended.)		1 for each 8 persons. For women's dormitories, additional bathtubs should be installed at the ratio of 1 for each 30 women. Over 150 persons add 1 fixture for each 20 persons	1 for each 75 persons	Laundry trays, 1 for each 50 persons Slop sinks, 1 for each 100 persons
Dwellings – one and two family	1 for each dwelling unit				1 for each dwelling unit		1 for each dwelling unit		Kitchen sink, 1 for each dwelling unit
Dwellings – multiple or apartment	1 for each dwelling unit or apartment				1 for each dwelling unit or apartment		1 for each dwelling unit or apartment		Kitchen sink, 1 for each dwelling unit or apartment. For apartments or multiple dwelling units in excess of 10 apartments or units, 1 double laundry tray for each 10 units or 1 automatic laundry washing machine for each 20 units

[a] Where urinals are provided for the women, the same number shall be provided as for men.

5.1–2 MINIMUM NUMBER OF PLUMBING FIXTURES (*Continued*)

Type of building occupancy	Type of fixture: Water closets		Urinals		Lavatories		Bathtubs or showers	Drinking fountains	Other fixtures
Industrial – factories, warehouses, foundries, and similar establishments	No. each sex	No. fixtures	Where more than 10 men are employed:		No. persons	No. fixtures	1 shower for each 15 persons exposed to excessive heat or to occupational hazard from poisonous, infectious, or irritating material	1 for each 75 persons.	
	1–10	1	No. men	No. urinals	1–100	1–10			
	11–25	2	11–30	1	Over 100	1–15			
	26–50	3	31–80	2					
	51–75	4	81–160	3					
	76–100	5	161–240	4					
	1 fixture for each additional 30 employees.								
Institutional – other than hospitals or penal institutions (on each occupied story)	1 for each 25 men 1 for each 20 women		1 for each 50 men[a]		1 for each 10 persons		1 for each 10 persons	1 for each 50 persons	
Hospitals	1				1		1		1 slop sink per floor
Individual room wards	1 for each 8 patients				1 for each 10 patients		1 for each 20 patients	1 for each 100 patients	
Waiting rooms	1				1				
Employees	Same as public		Same as public		Same as public			Same as public	
Penal institutions	1 in each cell		1 in each exercise room		1 in each cell		1 on each cell block floor	1 on each cell block floor	1 slop sink per floor
Prisoners	1 in each exercise room				1 in each exercise area			1 in each exercise area	
Employees	Same as public		Same as public		Same as public			Same as public	
Public buildings, offices, business, mercantile, storage, and institutional employees	No. each sex	No. fixtures	Urinals may be provided in men's[a] toilet rooms in lieu of water closets but for not more than 1/3 of the required number of water closets		No. employees	No. fixtures		1 for each 75 persons	1 slop sink per floor
	1–15	1			1–15	1			
	16–35	2			16–35	2			
	36–55	3			36–60	3			
	56–80	4			61–90	4			
	81–110	5			91–125	5			
	111–150	6			1 fixture for each additional 45 persons				
	1 fixture for each additional 40 employees								

5.1–2 MINIMUM NUMBER OF PLUMBING FIXTURES (*Continued*)

Type of building occupancy	Type of fixture: Water closets		Urinals	Lavatories	Bathtubs or showers	Drinking fountains	Other fixtures
	Boys	Girls					
Schools							1 slop sink per floor
Elementary	1/40	1/35	1/30 boys	1/30 pupils	In gym or pool shower rooms, 1/5 pupils of a class	1/100 pupils but at least 1 per floor	
Secondary	1/75	1/45	1/30 boys	1/30 pupils			
Working men, temporary facilities	1/30 working men		1/30 working men	1/30 working men		1 fixture or equivalent for each 100 working men	

Source: *Report of Public Health Service Committee on Plumbing Standards,* U.S. Department of Health, Education, and Welfare, Public Health Service, 1963.

5.1–3 SUPPLY DEMAND FOR VARIOUS LOADS IN SUPPLY FIXTURE UNITS

Load	Supply demand: Flush valve water closets predominate, gpm	Supply demand: Tank water closets predominate, gpm	Load	Supply demand: Flush valve water closets predominate, gpm	Supply demand: Tank water closets predominate, gpm
Supply fixture units			Supply fixture units		
5	22	4	500	143	124
10	27	8	600	157	143
20	35	14	660	162	152
30	42	20	700	170	161
40	46	24	800	183	178
50	51	28	850	189	185
60	54	32	900	197	195
88	64	40	1,000	208	208
124	74	48	1,060	216	216
160	81	56	1,280	243	243
236	98	72	1,510	270	270
300	108	85	1,990	324	324
400	127	106	2,480	378	378
470	135	118	2,990	432	432

Source: *Report of Public Health Service Committee on Plumbing Standards,* U.S. Department of Health, Education, and Welfare, Public Health Service, 1963.

5.1–4 COMBINED HOT AND COLD WATER DEMAND[a]

Fixture	No. fixture units: Private use	No. fixture units: Public use	Supply control	Ref.
Automatic clothes washer (2 in. standpipe)	1.5–3	Depends on size		3,8
Bar sink	1	2		1
Bathroom group[b]	8		Flushometer valve closet	3,5,6,7
Bathroom group[b]	6		Flush valve closet	3,5,6,7
Bathtub (with or without shower over)	2	4	Faucet	1,2,3,5,6,7
Bed pan washer		10		5
Bidet	3	4		5
Combination fixture	3			5,6,7
Combination sink and tray with food disposal unit	4			3
Combination sink and tray with one 1½ in. trap	2			3
Combination sink and tray with separate 1½ in. trap	3			3
Dental lavatory	1	2		3,5
Dental unit or cuspidor	·	1		1,3,5
Dishwashing machine	1–1.5	4–6	0.5 in. outlet	2,8
Drinking fountain (each head)	½	1		1,3,5
Electric water cooler	1			5
Floor drains with 2 in. waste	3			3
Garden hose	10			4

[a] One fixture unit is equivalent to a flow of 7.5 gpm.
[b] A bathroom group consists of a water closet, lavatory, and bathtub or shower stall.

5.1–4 COMBINED HOT AND COLD WATER DEMAND[a] (*Continued*)

Fixture	No. fixture units: Private use	Public use	Supply control	Ref.
Hose bibb or sill cock (standard type)	2–4[c]	4–8[c]		8
House trailer (each)	6	6		1
Kitchen sink, domestic	2	4	Faucet	2,3
Kitchen sink, domestic, with food waste grinder	2			3
Laundry tray or clothes washer (each pair of faucets), 1 or 2 tubs	2	4		1,3,5
Laundry tray, 1 to 3 tubs	3	3	Faucet	2,6,7
Lavatory	1	2	Faucet	1,2,5,6,7
Lavatory, barber, beauty parlor		3		5
Lavatory, surgeons		3		5
Lawn sprinklers (standard type, each head)	1	1		1
Shower (each head)	1½–2	3–4	Mixing valve	2,3,5,6,7,8
Sink, flushing rim		10	Flushometer	1,5
Sink, flushing rim		6	Flush valve	3
Sink, pantry	1			4
Sink, pot or scullery		5		5
Sink, service (P-trap)	2			3
Sink, service (trap standard)	3			3
Sink, slop		1		4
Sink, washup, circular spray		4		1
Sink, washup, each set of faucets		2		1,3,5
Urinal, pedestal or similar type		10	Flushometer	1,5,6,7
Urinal, pedestal, siphon jet blowout		6		3
Urinal, stall, washout		3–5		1,3,5,6,7
Urinal, trough (each 6 ft section)		2		3
Urinal, wall, lip		3–5		1,3,5,6,7
Urinal, wall or stall		3	Flush tank valve	1,5
Water closet	3	5	Flush tank	1,5,6,7
Water closet	4–6	10	Flushometer	1,5,6,7,8
Water supply outlets for items not listed				
3/8 in.	1	2		1
½ in.	2	4		1
3/4 in.	3	6		1
1 in.	6	10		1
Unlisted fixture drain or trap size				
1¼ in.	1	1		3
1½ in.	2	2		3
2 in.	3	3		3
2½ in.	4	4		3
3 in.	5	5		3
4 in.	6	6		3

[a] One fixture unit is equivalent to a flow of 7.5 gpm.
[c] Function of size: 3/4 in. 2 fixture units (private), 4 fixture units (public); 1 in. 4 fixture units (private), 8 fixture units (public).

REFERENCES

1. N.S. Moodhe, *J. Am. Water Works Assn.*, *59*(1):43, 1967.
2. H.V. Aldrich, *J. Am. Water Works Assn.*, *53*(3):267, 1961.
3. *Report of Public Health Service Committee on Plumbing Standards,* U.S. Department of Health, Education, and Welfare, Public Health Service, Washington, D.C., 1963.
4. H.E. Babbitt, *Plumbing,* 2nd ed., McGraw-Hill Book Company, Inc., New York, 1950.
5. *The Uniform Plumbing Code for Housing,* Technical Paper 6, Housing and Home Finance Agency, 1948.
6. *Plumbing Manual,* Report BMS66, National Bureau of Standards, 1940.
7. *National Plumbing Code,* ASA A40.8, 1955.
8. A.M. Cunningham, *J. Am. Water Works Assn.*, *64*(1):3, 1972.

5.1–5 WATER DEMANDS OF RESIDENTIAL AND COMMERCIAL ESTABLISHMENTS

	Maximum demand rates, gpm	
Units	**Hourly average**	**Instantaneous**
Apartments	0.64	0.84
Motels	0.83	1.65
Mobile parks	0.77	0.87
Car washes	1.56	2.82
Hotels	0.43	0.66

Source: Reprinted, with publisher's permission, from *J. Am. Water Works Assn.*, *64*(1):3, © 1972 by the American Water Works Association, Inc., 2 Park Ave., New York, N.Y. 10016.

5.1–6 MINIMUM FLOW PRESSURE AND FLOW RATES

Location	Flow pressure, psi	Flow rate, gpm
Ordinary basin faucet	8	2.0
Self-closing basin faucet	8	2.5
Sink faucet, 3/8 in.	8	4.5
Sink faucet, ½ in.	8	4.5
Bathtub faucet	8	6.0
Laundry tub cock, ½ in.	8	5.0
Shower	8	5.0
Ballcock for closet	8	3.0
Flush valve for closet	15	15.35
Flushometer valve for urinal	15	15.0[a]
Drinking fountains	15	0.75
Sill cock–wall hydrant	10	5.0

[a] Wide range due to variation in design and type of closet flush valves.

Source: *Report of Public Health Service Committee on Plumbing Standards,* U.S. Department of Health, Education, and Welfare, Public Health Service, 1963.

5.1–7 DEMAND CONSTANT–PEAK DEMAND

Demand constant[a]	Demand, gpm: Flush tank	Demand, gpm: Flush valve	Demand Constant[a]	Demand, gpm: Flush tank	Demand, gpm: Flush valve
10	12.00	30.00	400	105.0	126.0
15	15.00	33.00	450	115.0	133.0
20	18.00	36.00	500	125.0	140.0
25	20.00	39.00	600	144.0	154.3
30	22.00	42.00	700	163[b]	
35	24.50	44.00	800	174	
40	26.50	46.00	900	185	
45	28.50	48.00	1,000	197	
50	30.50	50.00	1,100	208	
55	32.50	52.00	1,200	219	
60	34.50	54.00	1,300	231	
65	35.87	56.00	1,400	242	
70	37.24	58.00	1,500	253	
75	38.62	60.00	1,600	264	
80	40.00	61.50	1,700	276	
85	41.25	63.00	1,800	287	
90	42.50	64.50	1,900	298	
95	43.75	66.00	2,000	310	
100	45.00	67.50	2,100	321	
105	46.00	68.75	2,200	332	
110	47.00	70.00	2,300	344	
115	48.00	71.25	2,400	355	
120	49.00	72.50	2,500	366	
125	50.00	73.70	2,600	377	
130	51.00	75.00	2,700	389	
135	52.00	76.20	2,800	400	
140	53.00	77.40	2,900	411	
145	54.00	78.70	3,000	422	
150	55.00	79.90	3,100	433	
160	57.00	82.40	3,200	445	
170	59.00	84.90	3,300	456	
180	61.00	87.40	3,400	467	
190	63.00	89.80	3,500	478	
200	65.00	92.20	3,600	490	
210	67.50	93.80	3,700	501	
220	69.00	95.30	3,800	512	
230	70.5	96.90	3,900	524	
240	73.00	98.40	4,000	535	
250	75.00	100.0	4,100	546	
300	85.00	110.0	4,200	558	
350	95.00	119.0	4,300	569	

[a] See 5.1–4.
[b] The following will be the same for either the tank or valve type.

5.1–8 OPTIMUM FLOW RATES FOR RESIDENTIAL WATER DEMANDING FIXTURES

		Authority, gpm					
Fixture	**Residual pressure,[1] psi**	**Mamrelli[2]**	**Arnold[3]**	**Mathews[1]**	**Moodhe[4]**	**Basic Water Works Manual[5]**	**Compiler's measurements to nearest whole number**
Bathtub	5		4	6	8–10	4	7
Shower	12		3	5	5–8	3	4
Wash basin	8		3	3	5–8	3	3
Toilet (tank)	15		4	3	3–5	4	4
Kitchen sink	5		4		8–10	4	3
Dishwasher		2–5				3–5	
Disposal		2				2–3	
Clothes washer		7	5			3–7	
Laundry tub			5	5	8–10	5	
Hose bibb[a]	30		6	5	5–10	5	9[b]

[a] With 50 ft garden hose and one garden hose.
[b] With 50 ft of 0.75 in. garden hose; free discharge, no sprinkler.

Source: Reprinted, with publisher's permission, from *J. Am. Water Works Assn., 62*(2):85, © 1970 by the American Water Works Association, Inc., 2 Park Ave., New York, N.Y. 10016.

REFERENCES

1. C.M. Mathews, *J. Am. Water Works Assn., 49*(9):1191, 1957.
2. E.S. Mamrelli, *J. Am. Water Works Assn., 47*(4):398, 1955.
3. G.E. Arnold, *J. Am. Water Works Assn., 48*(6):744, 1956.
4. N.S. Moodhe, *J. Am. Water Works Assn., 59*(1):43, 1967.
5. *Basic Water Works Manual,* Price Brothers Co., Dayton, Ohio, 1958.

5.1–9 RECOMMENDED CAPACITIES OF HOT WATER HEATING EQUIPMENT[1,2]

Gallons Per Hour Per Fixture

Fixture	Apartment house	Club	Gymnasium	Hospital	Hotel	Industrial plant	Laundry	Office building	Public bath	Private residence	School	YMCA
Private lavatory	3	3	3	3	3	3	3	3	3	3	3	3
Public lavatory	5	8	10	8	10	15	10	8	15		18	10
Bathtub	15	15	30	15	15	30			45	15		30
Dishwasher	15	30		30	30	30				15	15	
Foot basin	3	3	12	3	3	12				3	3	12
Kitchen sink	10	20		20	20	20				10	10	20
Laundry stationary tubs	25	35		35	35		42			25		35
Laundry revolving tubs	75	75		100	150		75 to 100		100	75		100
Pantry sink	10	20		20	20					10	20	20
Shower	100	200	300	100	100	300			300	100	300	200
Slop sink	20	20		20	30	20	10	15	15	15	20	20
Dishwasher			200 gal per hr at 180° per 500 people									
Dishwasher[a]			300 gal per hr at 180° per 500 people									
Total water for all fixtures likely to be drawn at one time	35	60	80	75	60	90	100	20	100	50	25	
Percent heater capacity[3]	20	50	80	60	50	90	100	15	100	50	25	
Storage capacity in percent of	100	75	50	50	25	50	25	100	50	100		
maximum heating capacity[2]	35	40	40	45	45	50	50	40	50	30	40	

Apartments: One kitchen and one bath

Number of families	up to 25	25–50	50–75	70–100	over 100
Gallons per hour per family	35	30	35	20	15

Swimming pools
Fill in 24 hr.
Refilter every 24 hr.
50,000 gal pool good for 100 persons.

[a] Computed at final temperature of 150°F.

Source: H.E. Babbitt, *Plumbing*, McGraw-Hill, New York, 1960, p. 178.

REFERENCES

1. A. Buenger, *J. Am. Soc. Heating Ventilating Engrs.*, *26*:701, 1920.
2. F.M. Reiter, *Ind. Gas*, November 1954, p. 8.
3. P. West, *J. Am. Soc. Heating Ventilating Engrs.*, *27*:253, 1921.

5.1–10 DATA USED IN STUDY OF STORAGE REQUIREMENTS AND POPULATION DENSITIES

Zone designation	Type of home	Lot size, ft^2	Density: Dwelling units per acre[a]	Density: Persons per acre	Maximum demands, gpm/acre: Hourly	Maximum demands, gpm/acre: Daily
R-G	group house	2,000– 2,400	9.0	30.6	3.8	2.9
R-6	detached dwelling	5,000– 7,500	4.2	16.8	1.8	0.7
R-10	detached dwelling	9,000–12,000	3.0	10.50	2.2	1.2
R-20	detached dwelling	15,000–25,000	1.6	7.46	2.9	1.3
R-40	detached dwelling	40,000 and over	0.8	3.33	2.7	1.3

[a] Adjusted by taking into account such areas as streets, public buildings, parks, institutions, and drainage courses.

Source: Reprinted, with publisher's permission, from *J. Am. Water Works Assn., 53*(10):1251, © 1961 by the American Water Works Association, Inc., 2 Park Ave., New York, N.Y. 10016.

5.1–11 COMPARISON OF MAXIMUM DAILY AND HOURLY SERVICE DEMANDS

Lot Type	No. observations	Max day avg rate, gpd/service	Peak hour, gpd/service	Ratio of peak hour to max day	Ratio of peak hour to avg day	Max day, gal/cap day	Avg day, gal/cap day
R-G	63	327	625	1.91:1	3.84:1	93	48
R-6	51	228	605	2.67:1	3.30:1	64	51
R-10	67	556	1,050	1.89:1	4.63:1	163	67
R-20	70	1,170	2,630	2.25:1	7.90:1	278	80
R-40	61	2,320	5,400	2.33:1	10.30:1	564	117

Source: Reprinted, with publisher's permission, from *J. Am. Water Works Assn., 53*(10):1251, © 1961 by the American Water Works Association, Inc., 2 Park Ave., New York, N.Y. 10016.

5.2 PLUMBING SYSTEM STANDARDS

5.2–1 CROSS CONNECTIONS WHERE PROTECTIVE DEVICES ARE REQUIRED AND CRITICAL LEVEL (C-L) SETTINGS FOR BACKFLOW PREVENTERS[a]

Fixture or equipment	Method of installation
Aspirators and ejectors	C-L at least 6 in. above flood level of receptacle
Dental units	On models without built-in vacuum breakers – C-L at least 6 in. above flood level rim of bowl
Dishwashing machines	C-L at least 6 in. above flood level of machine. Install on both hot and cold water supply lines
Flushometers (closet and urinal)	C-L at least 6 in. above top of fixture supplied
Garbage can cleaning machine	C-L at least 6 in. above flood level of machine. Install on both hot and cold water supply lines
Hose outlets	C-L at least 6 in. above highest point on hose line
Laundry machines	C-L at least 6 in. above flood level of machine. Install on both hot and cold water supply lines
Lawn sprinklers	C-L at least 12 in. above highest sprinkler or discharge outlet
Steam tables	C-L at least 6 in. above flood level
Tank and vats	C-L at least 6 in. above flood level rim or line
Trough urinals	C-L at least 30 in. above perforated flush pipe
Flush tanks	Equip with approval ball cock. Where ball cocks touch tank water equip with vacuum breaker at least 1 in. above overflow outlets. Where ball cock does not touch tank water install ball cock outlet at least 1 in. above overflow outlet or provide vacuum breaker as specified above
Hose bibs (where aspirators or ejectors could be connected)	C-L at least 6 in. above flood level of receptacle served

[a] Critical level (C-L) is defined as the level to which the backflow preventer (vacuum breaker) may be submerged before backflow will occur. Where the C-L is not shown on the preventer, the bottom of the device shall be taken as the C-L.

Source: *Report of Public Health Service Committee on Plumbing Standards,* U.S. Department of Health, Education, and Welfare, Public Health Service, 1963.

5.2–2 SIZES OF OVERFLOW PIPES FOR WATER SUPPLY TANKS

Maximum capacity of water supply line to tank, gpm	Diameter of overflow pipe, in. ID
0–50	2
50–100	2½
100–200	3
200–400	4
400–700	5
700–1,000	6
Over 1,000	8

Source: *Report of Public Health Service Committee on Plumbing Standards,* U.S. Department of Health, Education, and Welfare, Public Health Service, 1963.

5.2–3 SIZE OF DRAIN PIPES FOR WATER TANKS

Tank capacity, gal	Drain pipe, in.
up to 750	1
751–1,500	1½
1,501–3,000	2
3,001–5,000	2½
5,001–7,500	3
Over 7,500	4

Source: *Report of Public Health Service Committee on Plumbing Standards,* U.S. Department of Health, Education, and Welfare, Public Health Service, 1963.

5.2–4 MINIMUM SIZES OF FIXTURE WATER SUPPLY LINES

Type of fixture or device	Nominal pipe size, in.
Bathtubs	1/2
Combination sink and tray	1/2
Drinking fountain	3/8
Dishwasher (domestic)	1/2
Kitchen sink	
residential	1/2
commercial	3/4
Lavatory	3/8
Laundry tray; 1, 2, or 3 compartments	1/2
Shower (single head)	1/2
Sink (service, slop)	1/2
Sinks flushing rim	3/4
Urinal (flush tank)	1/2
Urinal (direct flush valve)	3/4
Water closet (tank type)	3/8
Water closet (flush valve type)	1
Hose bibs	1/2
Wall hydrant	1/2

Source: *Report of Public Health Service Committee on Plumbing Standards,* U.S. Department of Health, Education, and Welfare, Public Health Service, 1963.

5.2–5 CAPACITY OF GREASE INTERCEPTORS

Flow-through rating, gpm	Grease retention capacity, lb
4	8
6	12
7	14
9	18
10	20
12	24
14	28
15	30
18	36
20	40
25	50
35	70
50	100

Source: *Report of Public Health Service Committee on Plumbing Standards,* U.S. Department of Health, Education, and Welfare, Public Health Service, 1963.

5.2–6 FIXTURES AND THEIR WATER SUPPLY PROTECTION

Fixtures	Type of protection[a]	Remarks
Aspirators		
Laboratory	Vacuum breaker	
Portable	Vacuum breaker	
Vacuum system	Vacuum breaker	
Bedpan		
Washers	Vacuum breaker	
Washer hose	Vacuum breaker	Locate 5 ft above floor
Boiling type sterilizer	Air gap	Not less than twice the effective opening of the water supply
Exhaust condenser	Vacuum breaker	
Flush floor drain	Vacuum breaker	
Hose connection	Vacuum breaker	Locate 6 ft above floor
Pressure instrument washer-sterilizer	Vacuum breaker	
Pressure sterilizer	Vacuum breaker	
Vacuum systems		
Cleaning	Air gap or vacuum breaker	
Fluid suction	Air gap or vacuum breaker	

[a] Where vacuum breakers are used, they shall be installed after the last control valve.

Source: *Report of Public Health Service Committee on Plumbing Standards,* U.S. Department of Health, Education, and Welfare, Public Health Service, 1963.

5.2–7 SIZE OF FIXTURE TRAPS FOR DIFFERENT TYPE PLUMBING FIXTURES

Plumbing fixture	Trap size, in.
Bathtub (with or without overhead shower)	1½
Bidet	1½
Combination sink and wash (laundry) tray	1½
Combination sink and wash (laundry) tray with food waste grinder unit	1½[a]
Dental unit or cuspidor	1¼
Dental lavatory	1¼
Drinking fountain	1
Dishwasher, commercial	2
Dishwasher, domestic	1½
Floor drain	2
Food waste grinder unit	1½
Kitchen sink, domestic	1½
Kitchen sink, domestic with food waste grinder unit	1½
Lavatory, common	1¼
Lavatory (barber shop, beauty parlor, or surgeon's)	1½
Lavatory, multiple type (wash fountain or waste sink)	1½
Laundry tray (1 or 2 compartments)	1½
Shower stall	1½
Sink (surgeon's)	1½
Sink (flushing rim type, flush valve supplies)	3
Sink (service type with floor outlet trap standard)	3
Sink (service trap with P trap)	2
Sink, commercial (pot, scullery, or similar type)	2
Sink, commercial (with food waste grinder unit)	2
Urinal (pedestal, syphon jet, or blowout type, flush valve supplied	3
Urinal (stall type)	2
Urinal (wall lip type, flush tank supplied)	1½
Water closet	3

[a] Separate trap required for wash tray and separate trap required for sink compartment with food waste grinder unit.

Source: *Report of Public Health Service Committee on Plumbing Standards,* U.S. Department of Health, Education, and Welfare, Public Health Service, 1963.

5.2–8 MINIMUM AIR GAPS FOR PLUMBING FIXTURES

Fixture	Minimum air gap: When not affected by near wall, in.[a]	Minimum air gap: When affected by near wall, in.[b]
Lavatories and other fixtures with effective opening not greater than ½ in. diameter	1.0	1.50
Sink, laundry tray, gooseneck bath faucets, and other fixtures with effective openings not greater than ¾ in. diameter	1.5	2.25
Over rim bath fillers and other fixtures with effective openings not greater than 1 in. diameter	2.0	3.0
Drinking water fountains-single orifice 7/16 (0.437) in. diameter or multiple orifices having total area of 0.150 in.^{2} (area of circle 7/16 in. diameter)	1.0	1.50
Effective openings greater than 1 in.	2x diameter of effective opening	3x diameter of effective opening

[a] Side walls, ribs, or similar obstructions do not affect air gaps when spaced from inside edge of spout opening a distance greater than 3x the diameter of the effective opening for a single wall, or a distance greater than 4x the diameter of the effective opening for two intersecting walls.

[b] Vertical walls, ribs, or similar obstructions extending from the water surface to or above the horizontal plane of the spout opening require a greater air gap when spaced closer to the nearest inside edge of spout opening than specified in a, above. The effect of three such vertical walls or ribs has not been determined. In such cases the air gap shall be measured from the top of the wall.

Source: *Report of Public Health Service Committee on Plumbing Standards,* U.S. Department of Health, Education, and Welfare, Public Health Service, 1963.

5.2–9 DISTANCE OF FIXTURE TRAP FROM VENT

Size of fixture drain, in.	Distance from trap to vent, ft
1¼	2.5
1½	3.5
2	5
3	6
4	10

Source: *Report of Public Health Service Committee on Plumbing Standards,* U.S. Department of Health, Education, and Welfare, Public Health Service, 1963.

5.2–10 MINIMUM DISTANCES FROM SOURCE OF CONTAMINATION TO INDIVIDUAL WATER SUPPLIES AND PUMP SUCTION LINES

Source of contamination	Distance, ft[a]
Sewer	50
Septic tanks	50
Subsurface pits	50
Pasture	100
Sewer of cast iron (leaded or mechanical joints)	10
Subsurface disposal fields	100[b]
Seepage pits	100
Cesspools	150
Barnyards	100
Farm site	25
Pumphouse floor drain of cast iron draining to ground surface	2

[a] These distances constitute minimum separation and shall be increased in areas of creviced rock or limestone or where the direction of movement of the ground water is from source of contamination toward the well.

[b] When approved by the authority having jurisdiction and under special soil conditions, this distance may be reduced. However, a minimum separation of not less than 50 feet shall be maintained.

Source: *Report of Public Health Service Committee on Plumbing Standards,* U.S. Department of Health, Education, and Welfare, Public Health Service, 1963.

5.2–11 EFFECT OF SALINITY ON HOUSEHOLD FACILITIES

Item	Cost per household, $	Life: With 600 ppm water, yr	Life: With 800 ppm water, yr	Annual cost per household: With 600 ppm water, $	Annual cost per household: With 800 ppm water, $	Difference in cost due to increased salinity, $
Water piping	250	22	19	11.40	13.20	1.80
Waste water pipe	450	42.5	38	10.60	11.80	1.20
Water heaters	110	10	9	11.00	12.20	1.20
Faucets	165	9.5	9.25	17.40	17.80	0.40
Toilet flushing equipment	20	7	6	2.90	3.30	0.40
Garbage disposal units	8	7.25	6.25	1.10	1.30	0.20
Washing equipment	120	9	8.25	13.30	14.50	1.20
Cost per household	1,123					6.40

Source: Reprinted, with publisher's permission, from *J. Am. Water Works Assn., 62*(2):106, © 1968 by the American Water Works Association, Inc., 2 Park Ave., New York, N.Y. 10016.

REFERENCE

1. W.L. Patterson and R.F. Banker, *J. Am. Water Works Assn., 60*(9):1060, 1968.

5.2–12 SUMMARY OF WATER QUALITY CHARACTERISTICS AND ESTIMATED AVERAGE LIVES OF FACILITIES

				Characteristics of distributed water, ppm					Estimated average service lives, yr.						
State	Population served	Source of supply	Treatment[a]	TDS	Hardness[b]	Sulfate	Chloride	Dissolved oxygen[c]	Galvanized iron water piping	Waste-water piping	Water heaters	Faucets	Toilet flushing mechanisms	Garbage grinders	Washing equipment
Arizona	35,200	Ground[d]	None	818	309	59	277	Low	10	–	8	7	5	5	7
	30,000	Surface	P	1,787	640	440	575	High	15	–	7	13	3	6	5
Colorado	640,000	Surface	P	118	68	33	15	High	25	50	10	8	10	9	–
	9,000	Ground	D	1,950	960	965	45	Low	20	20	6	11	3	6	–
	3,400	Ground	D	2,850	1,160	1,520	100	Low	20	30	10	8	2	–	–
	5,000	Surface[e]	P	1,000	435	430	30	High	10	–	5	7	–	–	–
Kansas	14,500	Ground	DK	843	358	130	273	Low	15	–	10	6	7	5	10
	20,000	Surface	PH	150	90	41	19	High	35	–	13	–	–	–	–
	2,500	Ground	DK	1,807	1,312	1,017	47	Low	15	–	6	7	5	5	8
	2,200	Surface	P	1,606	789	706	367	High	20	–	8	8	4	9	9
	15,300	Ground	D	255	164	20	11	Low	–	–	15	–	12	–	–
	4,800	Surface	PH	130	96	24	11	High	40	–	12	–	–	–	–
	255,000	Ground[f]	APKH	304	95	67	56	High	35	–	13	12	–	–	–
Missouri	650,000	Surface	PKH	258	85	126	22	High	25	–	15	15	8	–	–
	8,500	Ground	ADK	1,374	362	79	562	High	7	45	4	3	3	3	–
	1,600	Ground	A	1,932	414	81	834	High	6	–	3	7	3	–	–
	385	Ground[d]	None	5,125	380	1,650	1,152	Low	25	–	12	–	3	–	–
Montana	1,100	Ground	D	1,770	330	660	90	Low	10	–	10	7	7	10	8
	1,200	Ground[g]	D	2,370	30	1,247	58	Low	17	–	12	–	–	9	8
New Mexico	28,000	Ground[h]	D	1,320	659	244	161	Low	15	20	8	6	5	–	–
	15,000	Ground	D	1,352	216	132	26	Low	25	25	5	6	2	5	–
	45,000	Ground[i]	None	1,447	709	502	270	Low	15	30	8	6	2	5	–
	2,500	Surface[j]	P	1,062	690	616	8	High	25	–	–	13	5	–	–

[a] P – Purification; H – Softening; A – Aeration; D – Disinfection; K – Corrosion control or stabilization.
[b] Total hardness as calcium carbonate.
[c] High dissolved oxygen – natural or mechanically aerated water.
Low dissolved oxygen – nonaerated ground water.
[d] New surface source of supply and treatment plant into operation in 1966; characteristics shown are for the old supply.
[e] Primary supply from surface source; supplemental supply from ground source; characteristics shown are estimated averages.
[f] Additional supply from surface source recently developed is not included.
[g] New source of supply obtained in 1963; characteristics shown are for old supply.
[h] New source of supply obtained in 1962; characteristics shown are for old supply.
[i] Office of Saline Water demineralization plant provides a small percentage of supply.
[j] New ground source of supply obtained in 1964; characteristics shown are old supply.

5.2–12 SUMMARY OF WATER QUALITY CHARACTERISTICS AND ESTIMATED AVERAGE LIVES OF FACILITIES (*Continued*)

State	Population served	Source of supply	Treatment[a]	Characteristics of distributed water, ppm: TDS	Hardness[b]	Sulfate	Chloride	Dissolved oxygen[c]	Estimated average service lives, yr: Galvanized iron water piping	Wastewater piping	Water heaters	Faucets	Toilet flushing mechanisms	Garbage grinders	Washing equipment
Oklahoma	18,000	Surface	P	1,230	606	522	215	High	25	25	6	8	4	–	–
	16,000	Surface	P	1,180	748	529	79	High	13	30	9	6	2	5	–
South Dakota	1,440	Ground	k	2,553	64	1,179	325	Low	–	–	–	–	2	–	–
	520	Ground	None	2,973	58	1,227	432	Low	–	–	–	–	–	–	–
	1,600	Ground	DHK	2,238	56	1,324	78	Low	–	–	10	–	2	5	7
	1,950	Ground	D	1,108	38	185	665	Low	–	–	–	–	–	–	
	1,060	Ground	D	2,188	212	1,228	125	Low	–	–	–	–	–	–	–
	2,500	Ground[l]	DH	1,510	165	783	14	Low	–	–	5	–	–	–	–
Texas	14,500	Ground	AD	1,560	64	7	475	High	13	25	7	–	3	2	7
	30,000	Ground[m]	D[m]	470	99	106	100	Low	35	45	15	7	7	–	–
	6,000	Ground	D	1,710	690	465	452	Low	4	25	5	4	2	4	–
	2,900	Ground	AD	1,579	18	64	545	High	20	25	4	4	2	7	5
	5,000	Ground	D	1,810	287	158	574	Low	10	–	5	–	2	–	–
Utah	9,500	Ground	None	1,168	142	216	300	Low	10	–	7	8	–	8	7
Wyoming	3,600	Ground	None	1,000	580	380	20	Low	25	–	10	14	–	9	6

[a] P – Purification; H – Softening; A – Aeration; D – Disinfection; K – Corrosion control or stabilization.
[b] Total hardness as calcium carbonate.
[c] High dissolved oxygen – natural or mechanically aerated water. Low dissolved oxygen – nonaerated ground water.
[k] Fluoride reduction only.
[l] Office of Saline Water demineralization plant has provided most of supply since 1962; characteristics shown are representative of water distributed during the period of 1947 to 1962
[m] Approximately 85 percent of supply is from ground source with disinfection the only treatment; remaining 15 percent of supply is from surface source with purification, stabilization and softening. Characteristics shown are estimated average.

Source: Reprinted, with publisher's permission, from *J. Am. Water Works Assn.*, *60*(9):1060, © 1968 by the American Water Works Association, Inc., 2 Park Ave., New York, N.Y. 10016.

5.2–13 LANGELIER INDEX AND IRON PICKUP IN DISTRIBUTION SYSTEMS

Location	Langelier Index (pH-pHs) at 60°F	Ryznar[1] Index (2pHs-pH)	Average iron pickup in system, ppm	Maximum iron in system test, ppm
Alexandria, Va.	−2.05	11.10	0.17	0.60
Ashtabula, Ohio	−0.71	8.80	0.15	0.75
Ashtabula, Ohio	−0.59	8.48	0.08	0.30
Bernardsville, N.J.	−0.60	9.50	0	0.09
Chattanooga, Tenn.	−1.24	10.18	0	0.04
Clinton, Iowa	0.08	7.34	0.08	0.15
Davenport, Iowa	−1.01	9.12	0.06	0.10
East St. Louis, Ill.	−0.31	8.02	0.01	0.08
East St. Louis, Ill.	−0.17	7.74	0.03	0.25
Ellwood City, Pa.	−0.13	8.20		
Huntington, W. Va.	0.32	8.26	0	0.10
Monmouth, N.J.	−1.18	10.62	0.03	0.14
Monongahela, Pa.	−0.14	8.70	0	0.12
Muncie, Ind.	0.20	7.30	0.05	0.18
Newark, N.J.	−2.40	11.70		
New Castle, Pa.	−0.57	8.94	0.04	0.07
Noroton, Conn.	−2.38	11.46		
Old Dominion, Va.	−1.86	11.82	0.10	0.20
St. Joseph, Mo.	0	7.70		
St. Joseph, Mo.	0.29	7.22	0	trace
South Pittsburgh, Pa.	−1.00	9.70	0.04	0.05
Uniontown, Pa.	−3.60	3.60	0.24	0.35
Uniontown, Pa.	−2.09	11.09	0.22	0.65
White Deer, Pa.	−2.97	12.94	0.72	1.68

Source: Reprinted, with publisher's permission, from *J. Am. Water Works Assn., 53*(12):1461,© 1961 by the American Water Works Association, Inc., 2 Park Ave., New York, N.Y. 10016.

REFERENCE

1. J. W. Ryznar, *J. Am. Water Works Assn., 36*(4):472, 1944.

5.2–14 BURSTING PRESSURES AND NUMBER OF PERFORATIONS OF CAST IRON PIPE SPECIMENS

Soil	Exposure time, yr	Specimen[a]	Bursting pressure, psi	No. perforations after cleaning
Lakes Charles clay (El Vista, Texas)	7.2	a	350.4	2
		b	500+	1
	8.7	a	500+	3
	10.9	a	500+	1
		b	500+	5
Muck (New Orleans, La.)	7.2	a	500+	1
	10.9	a	500+	2
Docas clay (Cholame, Cal.)	7.2	a	500+	6
		b	500+	5
	8.8	a	425	4
	11.0	a	500+	2
		b	500+	10
Mohave fine gravelly loam (Phoenix, Ariz.)	7.2	a	450	5
		b	500+	2
	8.8	a	500+	1
	11.0	a	500+	1
Merced silt loam (Buttonwillow, Cal.)	7.2	a	500+	5
		b	500+	4
	11.0	a	500+	12
		b	50	16
		c	500+	11

[a] Specimen letter designations indicate that the test results are for the same section of pipe.

Source: Reprinted, with publisher's permission, from *J. Am. Water Works Assn.*, *56*(9):1129,© 1964 by the American Water Works Association, Inc., 2 Park Ave., New York, N.Y. 10016.

5.2–15 CORROSION OF DUCTILE CAST IRON IN DIFFERENT SOILS

Soil	Location	Resistivity at 60°F, ohm-cm	pH	Aeration[c]	Exposure time, yr	Loss in weight,[a] oz/ft²	Maximum pit depth, mil
Hagerstown loam	Loch Raven, Md.	5,200	5.8	G	1.0	1.3	49
					2.1	2.0	68
					4.1	2.2	81
Sagemoor sandy loam	Toppenish, Wash.	400	8.8	G	1.0	0.3	45
					2.0	1.7	66
					4.0	1.1	60
Clay	Cape May, N.J.	300	4.0	P	1.0	2.4	61
					2.0	4.7	108
					5.0	5.9	114+[b]
Lakewood sand	Wildwood, N.J.	30,000	7.3	G	1.0	0.7	50
					2.0	1.4	44
					4.0	2.1	43
Coastal sand	Wildwood, N.J.	55	7.1	P	1.0	1.8	35
					2.0	3.8	49
Tidal marsh	Patuxent, Md.	300	7.1	P	0.9	2.7	63
					1.8	3.4	56
					3.9	10.3	142+[b]

[a] Average of four specimens.
[b] One specimen perforated by corrosion. Pipe wall thickness, 250 mil.
[c] G – good; P – poor aeration.

Source: Reprinted, with publisher's permission, from *J. Am. Water Works Assn.*, *56*(9):1129, © 1964 by the American Water Works Association, Inc., 2 Park Ave., New York, N.Y. 10016.

5.2–16 CORROSION OF PLAIN CAST IRON BURIED IN SOILS WITH PROPERTIES SHOWN[a]

Soil type	Test site location	Exposure time, yr	Weight loss, oz/ft^2	Maximum pit depth, mils	Soil aeration[d]	Resistivity at 60°F, ohm-cm	pH	Total acidity[c]	Composition of water extract mg equivalents/100 g of soil: HCO_3	Cl	SO_4
Cecil clay loam	Atlanta, Ga.	14	3.3	138	G	17,800	4.8	5.1			
Hagerstown loam	Loch Raven, Md.	14	3.6	149	G	5,200	5.8	10.9			
Chino silt loam	Wilmington, Cal.	14	9.0	146	G	148	8.0	A	1.30	6.05	16.90
Susquehanna clay	Meridian, Miss.	14	8.5	140	F	6,900	4.5	12.0			
Docas clay	Cholame, Cal.	9	36+	H[b]	F	62	7.5	A	0.89	28.80	0.26
Mohave fine gravelly loam	Phoenix, Ariz.	14	28+	H	F	232	8.0	A	0.73	2.77	2.97
Merced silt loam	Buttonwillow, Cal.	7	35+	H	F	278	9.4	A	1.87	1.12	5.57
Rifle peat	Plymouth, Ohio	9	20.8	H	P	218	2.6	297.4	0.00	0.00	56.70
Sharkey clay	New Orleans, La.	14	12.3	98	P	943	6.8	4.9	0.71	0.10	0.91
Acadia clay	Spindletop, Texas	5	18.4	H	P	190	6.2	13.2	0.56	5.75	22.0
Lake Charles clay	El Vista, Texas	7	22.0	H	VP	406	7.1	5.1	0.80	1.59	3.04
Muck	New Orleans, La.	5	12.1	H	VP	712	4.8	15.0	0.00	0.47	2.54
Carlisle muck	Kalamazoo, Mich.	14	6.8	70	VP	1,660	5.6	12.6	0.00	3.47	1.04
Tidal marsh	Charleston, S.C.	14	13.1	156	VP	84	6.9	14.6	0.00	12.70	36.60
Cinders	Milwaukee, Wisc.	5	48+	H	VP	455	7.6	A	0.55	0.08	2.89

[a] For the more corrosive soils, weight losses and pit depths are given for the earliest time at which perforation of the pipe wall was observed.
[b] H indicates perforation of the pipe wall by corrosion. Thickness of pipe, 250 mil.
[c] A is for alkaline reaction.
[d] G is for good, F, fair; P, poor; and VP, very poor aeration.

Source: Reprinted, with publisher's permission, from *J. Am. Water Works Assn., 56*(9):1129, © 1964 by the American Water Works Association, Inc., 2 Park Ave., New York, N.Y. 10016.

5.2–17 RELATIONSHIP BETWEEN CORROSION AND SOIL RESISTIVITY

Soil resistivity, ohm-cm	Corrosiveness
<700	very corrosive
700–2,000	corrosive
2,000–5,000	moderately corrosive
>5,000	mildly to noncorrosive

Source: Reprinted, with publisher's permission, from *J. Am. Water Works Assn., 56*(9):1129, © 1964 by the American Water Works Association, Inc., 2 Park Ave., New York, N.Y. 10016.

5.2–18 CURRENT REQUIRED FOR TIME SPECIFIED TO THAW STEEL PIPE

Approximate time, min	Current, amp					
	½ in.	¾ in.	1 in.	1¼ in.	1½ in.	2 in.
3–4	200	270	400			
6–8	150	200	300	400		
8–12	125	170	250	340	400	
12–16	100	135	200	270	320	440
25–30	75	100	150	200	240	330
60	50	67	100	135	160	220

Source: Reprinted, with publisher's permission, from *J. Am. Water Works Assn., 55*(5):602, © 1963 by the American Water Works Association, Inc., 2 Park Ave., New York, N.Y. 10016.

5.2–19 EQUIPMENT CAUSING POTENTIAL PLUMBING HAZARDS

The following is a list of types of water-using equipment creating plumbing hazards, one or more of which may be found in premises supplied with potable water:

Air conditioning systems
Aspirators (including mortuary)
Blueprint machines
Car washing equipment
Chemical and chemical solution lines
Coffee urns
Compressors
Cooling water systems and towers
Dental cuspidors
Display fountains
Ejectors (steam or water)
Fish ponds
Garbage can washing equipment
Garden spray aspirators
Hospital and laboratory equipment – autoclaves, bedpan washers, instrument sterilizers, pipet washers, and x-ray processing equipment
Hydraulic applicators for fertilizer and other farm chemicals
Hydraulic lifts
Industrial process fluid lines, tanks, and vats
"Industrialized" water lines
Irrigation systems under pressure
Lawn sprinklers
Laundry equipment
Mechanical equipment and engines (for cooling)
Plumbing fixtures – flushometer valves on toilets, below rim inlets to basins and tubs, washing machines, dishwashers, and garbage grinders
Pressure cookers and food processing equipment
Process water recirculating system
Sewer flush tanks
Slop sinks and utility hoppers
Soda fountains
Steam boilers
Steam tables and other kitchen equipment
Swimming pools
Tanks and vats
Vacuum systems
Water-using mechanical equipment
Water troughs

Source: Reprinted, with publisher's permission, from *J. Am. Water Works Assn., 58*(4):423, © 1966 by the American Water Works Association, Inc., 2 Park Ave., New York, N.Y. 10016.

5.2–20 INDUSTRIES IN WHICH PLUMBING HAZARDS CAN OCCUR

Aircraft and missile plants
Automotive plants
Beverage bottling plants
Breweries
Buildings (multistory)
Canneries, packing houses, and reduction plants
Chemical plants
Dairies and cold storage plants
Fertilizer manufacturing plants
Film laboratories
Hospitals, medical buildings, mortuaries, morgues, sanitariums, and nursing homes
Laundries and dye works
Metal manufacturing, cleaning, plating, processing, and fabricating plants
Motion picture studios
Oil and gas production facilities
Paper and paper products plants
Plating plants
Power plants
Rubber plants
Schools and colleges with laboratories
Sewage treatment plants, and sewage and storm water pumping stations
Water front facilities and industries
Water treatment plants

Source: Reprinted, with publisher's permission, from *J. Am. Water Works Assn., 58*(4):421, © 1966 by the American Water Works Association, Inc., 2 Park Ave., New York, N.Y. 10016.

5.2–21 POTENTIAL CROSS CONNECTIONS WITH CONSUMER WATER SYSTEM

Potential cross connections exist between the consumer's water system and the following:

1. Sewage pumps used for priming, cleaning, flushing, or unclogging
2. Water operated sewage sump ejectors
3. Sewer lines used for disposing of filter or softener backwash water or water from cooling systems, or for providing a quick drain for building lines, or for flushing or blowing out obstructions in a sewer line
4. Water cooled equipment that may be sewer connected, such as heat exchangers, compressors, and air-conditioning equipment.

Source: *Report of Public Health Service Committee on Plumbing Standards,* U.S. Department of Health, Education, and Welfare, Public Health Service, 1963.

5.3 STORAGE

Reservoir problems including quality changes, artificial aeration, and destratification. See also Section 6., Biological Control.

5.3–1 CHANGES IN WATER QUALITY RESULTING FROM STORAGE

	Average of samples			Maximum value		
	River water	Reservoir water	Percentage reduction	River water	Reservoir water	Percentage reduction
Nitrogen						
Ammonia	1.7	1.2	30	3.6	1.6	55
Albuminoid	0.47	0.22	53	1.67	0.78	53
$KMnO_4$ demand	55.8	50.6	9	107.5	66.4	38
Color, pt. standard	95	88	7	240	130	46
Odor						
20°C	20.2	10.1	50	70	18	74
60°C	45.6	21.0	54	300	30	90
Bacteria						
No. colonies, 20°C	11,800	738	94			
No. colonies, 37°C	3,590	245	93			
Coliforms	88	10.1	89	150	50	67
Esch. coli	56	4.4	92	183	10.4	94
Oxygen	11.5	9.8				
Iron	1.3	1.3				
Sulfates	15.3	16.2				
Nitrates	3.9	5.4				
Nitrites	0.07	0.08				
Total solids	129	132				
Fixed solids	89	92				

Source: Reprinted, with publisher's permission, from *J. Am. Water Works Assn.*, *59*(6):680, © 1967 by the American Water Works Association, Inc., 2 Park Ave., New York, N.Y. 10016.

5.3–2 OXYGEN UPTAKE FOR VARIOUS RESERVOIR AERATION SYSTEMS

Method	Oxygen uptake related to the oxygen blown, %	Remarks
Air bubble method Wahnbach-Reservoir[1]	25	O_2 saturation, 10–20%
Aeration and mixing limited to the hypolimnion Wahnbach Reservoir system[2]	50	O_2 saturation, 10%
Air bubble method Baldeney lake[3]	5	Water depth, 17–20 ft. O_2 saturation, 68%
Lake Pfaffiker system[4,5]	45	O_2 saturation, 10%
River impoundment aeration according to R.v. Wolff[6,7]	17 (average)	O_2 saturation, 20%. Water temperature 10–25°C
River impoundment aeration according to Wagner-Voith[8]	15–20	Average result of numerous tests in the power station in Poppenweiler/Neckar. Water temperature 17–18°C, O_2 saturation, 40%
	50	O_2 saturation, 10%
Medium bubble aeration in activated sludge plants[9]	2.7 per 3.3 ft water depth	From tests with tap water O_2-saturation, 10%

Source: Reprinted, with publisher's permission, from *J. Am. Water Works Assn.*, *59*(8):943, © 1967 by the American Water Works Association, Inc., 2 Park Ave., New York, N.Y. 10016.

REFERENCES

1. H. Bernhardt, *Jahrbuch "Vom Wasser," 30*:11, 1963.
2. H. Bernhardt, *Archiv fur Hydrobiologie, 63*:404, 1967.
3. K.R. Imhoff, *83*:40, 1965. Vulkan-Verlag Dr. W. Classen, Essen.
4. G. Muller, *Schweizer Verkehrs- und Industrierevue,* Juliheft 5, 1963.
5. H. Ambuhl, *Verband Schweizer Abwasserfachleute,* Verbandsbericht Nr. 77/3, 1963.
6. M. Eckoldt, *Deutsche Gewasserkundliche Mitteilungen, 10*:1, 1966.
7. R.v. Wolff, Deutsches Bundespatent 1,153,725.
8. H. Wagner, *Deutsche Gewasserkundliche Mittelungen, 2*:73, 1958.
9. E. Knop, W. Bischofsberger, and V. Stalmann, Versuche mit verschiedenen Beluftungssytemen im technischen Massstab. Teil 2. Vulkan-Verlag Dr. W. Classen, Essen.

5.3–3 OXYGENATION CAPACITY FOR VARIOUS AERATION SYSTEMS

Method	Oxygenation efficiency, lb O_2 transferred/kWh
Diffused air Wahnbach Reservoir[1]	1.2
Hypolimnion mixing Wahnbach Reservoir[2]	2.1
Diffused air Baldeney Lake[3]	0.33[a]
Lake Pfaffiker[4,5]	4.3
Turbine aeration[6,7]	1.3–5.6 2.4 (average)
Turbine aeration[8]	4.3–5.6 (O_2 saturation below 10%)
Activated sludge[9]	3.1 (coarse bubble) 5.6 (fine bubble)
Aero-Hydraulics gun[10]	1.4
Boltz Lake Mechanical pump[11]	1.0

[a] The low oxygenation efficiency of this installation was caused by excessive pressure that could not be changed during the experiment.

Source: Reprinted, with publisher's permission, from *J. Am. Water Works Assn., 59*(8):943, © 1967 by the American Water Works Association, Inc., 2 Park Ave., New York, N.Y. 10016.

1. H. Bernhardt, *Jahrbuch "Vom Wasser," 30*:11, 1963.
2. H. Bernhardt, *Archiv fur Hydrobiologie, 63*:404, 1967.
3. K.R. Imhoff, *83*:40, 1965. Vulkan-Verlag Dr. W. Classen, Essen.
4. G.Muller, *Schweizer Verkehrs- und Industrierevue, Juliheft 5*, 1963.
5. H. Ambuhl, *Verband Schweizer Abwasser-fachleute, Verbandsbericht Nr.* 77/3, 1963.
6. M. Eckoldt, *Deutsche Gewasserkundliche Mitteilungen, 10*:1, 1966.
7. R.v. Wolff, Deutsches Bundespatent 1,153 725.
8. H. Wagner, *Deutsche Gewasserkundliche Mitteilungen, 2*:73, 1958.
9. E. Knop, W. Bischofsberger, and V. Stalmann, Versuche mit verschiedenen Beluftungssystemen im technischen Massstab. Teil 2. Vulkan-Verlag Dr. W. Classen, Essen.
10. J.G. Bryan, *J. Brit. Waterworks Assn., 46*:546, 1964.
11. J.M. Symons *et al.*, Management and Measurement of "DO" in Impoundments, *J. San. Eng. Div., Proc. Am. Soc. Civ. Eng.*

5.3–4 OXYGENATION CAPACITY OF REAERATION DEVICES

Type	Oxygenation capacity,[a] lb DO/kWh
Turbine[1]	2.7–4.0
Turbine[2]	5.0
Turbine[3]	3.4
Floating turbine[4]	2.0–2.4 – low river flow 5.4–6.0 – high river flow
DO aerator[5]	3.3
Diffused air activated sludge	3.1–3.3

[a] These values may be used as a guide, but are influenced by test conditions.

Source: Reprinted, with publisher's permission, from *J. Am. Water Works Assn., 59*(10):1268, © 1967 by the American Water Works Association, Inc., 2 Park Ave., New York, N.Y. 10016.

REFERENCES

1. C. Bueltman *et al., Advances in Biological Waste Treatment,* Eckenfelder and McCabe, Eds., Pergamon Press, London, 1963, p. 175.
2. R.F. Weston, in *Proc. 16th Industrial Waste Conf.,* Lafayette, Ind., May 2–4, 1961, Engineering Extension Series No. 109, Engineering Bulletin, Purdue University, *46*:505, 1962.
3. R.L. Laws and O.B. Burns, Jr., *Proc. 14th Industrial Waste Conf.,* Lafayette, Ind., May 5–7, 1959, Engineering Extension Series No. 104, Engineering Bulletin, Purdue University, *44*:633, 1960.
4. A.J. Kaplovsky, W.R. Walters, and B. Sosewitz, *J. Water Pollut. Control Fed., 36*:463, 1964.
5. T.W. Lesperance, *Advances in Biological Waste Treatment,* Eckenfelder and McCabe, Eds., Pergamon Press, London, 1963, p. 169.
6. A.L. Downing and A.G. Boon, *Advances in Biological Waste Treatment,* Eckenfelder and McCabe, Eds., Pergamon Press, London, 1963, p. 131.

5.3–5 COMPARISON OF VARIOUS DESTRATIFICATION EFFICIENCIES

Lake	Stability: Before aeration, kWh	Stability: After destratification, kWh	Stability decrease, kWh	Total energy used, kWh	Energy used per unit lake volume, kWh/acre ft	Destratification efficiency, %[d] column
Wahnbach Reservoir[c] 530 acres, 33,740 acre ft, Max depth, 141 ft	400	0	400	104,880	3.1	0.38
Boltz Lake[b] 96 acres, 2,400 acre ft, Max depth 60 ft[1-3]	44.4	11.8	32.6–13[a] = 19.6	14,300	5.9	0.14
Lake Wohlford[c] 130 acres, 2,500 acre ft, Max depth 50 ft[4,5]	14.7	1.1	13.6	2,910	1.2	0.47

[a] Stability decrease because of natural partial circulation.
[b] Using a pump.
[c] Blowing in air on the bottom of the lake.
[d] $\frac{\text{Stability decrease}}{\text{Total energy used}}$.

Source: Reprinted, with publisher's permission, from *J. Am. Water Works Assn., 59*(8):943, © 1967 by the American Water Works Association, Inc., 2 Park Ave., New York, N.Y. 10016.

REFERENCES

1. J.M. Symons *et al.,* Management and Measurement of "DO" in Impoundments, *J. San. Eng. Div., Proc. Am. Soc. Civ. Eng.*
2. W.H. Irwin, J.M. Symons, and G.G. Robeck, *J. San. Eng. Div., Proc. Am. Soc. Civ. Eng., 92*SA-6, 1966.
3. J.M. Symons, W.H. Irwin, and G.G. Robeck, *J. San Eng. Div., Proc. Am. Soc. Civ. Eng., 93*:1, SA-2, 1967.
4. G.E. Koberg, Elimination of Thermal Stratification in Reservoirs and the Resulting Benefits. Geological Survey Water Supply, Paper 1809–M, U.S. Govt. Printing Office, 1965.
5. M.E. Ford, Jr., *J. Am. Water Works Assn., 55*:267, 1963.

5.3–6 WATER SUPPLY INSTALLATIONS OF ARTIFICIAL DESTRATIFICATION FOR QUALITY CONTROL

Owner	Name of reservoir	Problems	Dates of operations	Success
City of San Diego, Cal.	El Capitan[a]	H_2S, Mn, high Cl_2 demand, high evaporation	6/10/65 to date	Mostly
City of Santa Barbara, Cal.	Lake Cachuma[a]	Taste and odor (may have occurred in Tecolote Tunnel, not in reservoir)	1968	No improvement, but DO maintained above 2.5 mg/l
Helix Irrigation District, La Mesa, Cal.	Lake Jennings[a]	H_2S, Fe, Mn	1964 to date	Yes
City of Santa Cruz, Cal.	Loch Lomond[a]	Mn, Fe (slight)	6/67 to date	Yes
Escondido Mutual Water Company, Escondido, Cal.	Lake Wohlford[a]	H_2S, high Cl_2 demand	1962 to date	Yes
Municipal Authority of Westmoreland County, Greensburg, Pa.	Unity[e]	Taste and odor	1969 to date	Yes
Laurel Management Company, Johnstown, Pa.	Laurel Run[e]	Taste and odor	6/65 to date	Yes
	Millcreek[e]	Excessive bottom sludge	6/65 to date	Yes
Highland Sewer and Water Authority, Highland, Pa.	Lloydell[e]	Taste and odor	1966 to date	Yes
East Bay Municipal Utility District Oakland, Cal.	Lafayette[a]	Taste and odor	12/67 to date	Yes
California American Water Co., Chula Vista, Cal.	San Clemente[e]	H_2S, Fe, Mn, taste and odor, high Cl_2 demand	8/4/69 to date	Yes
Ventura River Municipal Water District, Oakview, Cal.	Casitas[a]	Reduced volume of usable water, Mn, H_2S, excess algae	6/20–11/1/68 4/2–10/17/69	Yes Yes
Wahnbachtalsperrenverband Siegburg, Federal Republic of Germany	Wahnbach[b]	Fe, Mn, Blue-green algae, dissolution dissolution of phosphate	1966 to date	Yes
Metropolitan Water Board, London, U.K.	Queen Elizabeth II[c]	Reduced volume of usable water, excess algae	1963 to date	Yes
	King George VI[d]	Reduced volume of usable water, excess algae	1965 to date	Yes
Kalutuwawa, Ceylon	No name[a]	Air binding in filters, light floc	–	Yes
Cameron Highlands, Malaysia	No name[a]	H_2S	–	Yes

[a] Diffused air system.
[b] Hypolimnion reaeration only.
[c] Subsurface jets.
[d] Mechanical pump.
[e] Air-Aqua system.

Source: Reprinted, with publisher's permission, from *J. Am. Water Works Assn., 62*(5):323, © 1970 by the American Water Works Association, Inc., 2 Park Ave., New York, N.Y. 10016.

5.3–7 SUMMARY OF PERFORMANCE FACTORS DURING ARTIFICIAL DESTRATIFICATION

Lake	Dates of mixing	Method	Volume, acre ft	Surface area, acres	Max. depth, ft	Oxygenation capacity,[a] lb DO/kWh	Destratification efficiency,[a] %
Vesuvius	9/3–9/17/64	Mechanical pump	1,260	100	30	Unknown	0.4
Boltz	8/6–9/10/65	Mechanical pump	2,900	96	62	1.3	0.2
Boltz	6/2–6/7	Diffused air pump	2,900	96	62	1.6	1.5
(Test Lake I)	6/21–6/24					4.3	1.4
	7/15–7/20					3.3	1.5
	8/18–8/23					–	0.6
	1966						
Falmouth	(5/16–5/19,	Diffused air pump	4,600	225	42		
(Test Lake II)	5/20–5/23)					0.0	0.2
	6/10–6/15					1.4	0.9
	7/1–7/3					3.0	0.9
	7/26–8/3					4.2	0.8
	9/8–9/13					3.9	0.3
	1966						

[a] These values are intended as a rough guide only and are influenced by many factors.

Source: Reprinted, with publisher's permission, from *J. Am. Water Works Assn.,59*(10):1262, © 1967, by the American Water Works Association, Inc., 2 Park Ave., New York, N.Y. 10016.

REFERENCE

1. J.M. Symons *et al., J. Am. Water Works Assn., 62*(5):323, 1970.

5.3–8 EVAPORATION REDUCTION IN RESERVOIRS USING MONOMOLECULAR FILMS

Results of Field Trials

Location	Date	Water surface area, hectares	Evaporation reduction, %
Australia	1954–55	0.4–9	15–40
Australia	1956–57	372	37
East Africa	1956	0.4–2.6	25–30
East Africa	1957	52	12
United States	1956	1.6	4–18
United States	1957–58	1.0	22–43
United States	1958	1,000	9
United States	1959	0.1	25
United States	1959–63	0.1	0–23
United States	1960	16	19
United States	1960	400	14
United States	1961	960	8
Spain	1957	12	35
Spain	1958	12	31

Source: United Nations Educational, Scientific, and Cultural Organization, 1965.

5.3–9 COMPARISON OF DESTRATIFICATION EFFICIENCIES

Lake reference	Dates of mixing	Method	Volume, acre ft	Change in stability, Uncorrected (U) or corrected (C), kWh	Total energy input, kWh	Destratification efficiency, DE[c] (calculated over entire mixing period), %	Total energy input/unit volume/mixing[c] kWh/acre ft
Boltz Lake	6/2–6/7,	Diffused air pump	2,930	27 (C)	1,760	1.5	0.6
	6/21–6/24,			23 (C)	1,680	1.4	0.6
	7/15–7/20,			33 (C)	2,200	1.5	0.7
	8/18–8/23,			16 (C)	2,860	0.6	1.0
	1966					1.2[b]	0.7[b]
Falmouth Lake	(5/16–5/19,	Diffused air pump					
	5/20–5/23)		4,600	7 (C)	3,285	0.2	0.7
	6/10–6/15,			24 (C)	2,740	0.9	0.6
	7/1–7/8,			35 (C)	3,740	0.9	0.8
	7/26–8/3,			28 (C)	3,550	0.8	0.8
	9/8–9/13,			8 (C)	2,540	0.3	0.5
	1966					0.6[b]	0.7[b]
Vesuvius Lake[1]	9/3–9/17/64	Mechanical pump	1,260	6.7(U)	1,862	0.4[a]	1.5
Boltz Lake[2,3]	8/6–9/10/65	Mechanical pump	2,930	28 (C)	14,300	0.2	4.9
King George IV Reservoir[4]	7/13–8/16/66	Mechanical pump	18,375	23.8(U)	32,640	0.1[a]	1.8
Lake Wolford[5]	4/18–4/25/62	Diffused air pump	2,500	14.8(U)	2,570	0.6[a]	1.0
Wahnbach Reservoir[6]	6/9–9/9/64	Diffused air pump	33,740	400 (U)	104,800	0.4[a]	3.1
Cox Hollow Lake[7]	6/30–7/18/66	Aero-Hydraulic Gun	1,190	7.9(U)	2,140	0.4[a]	1.8
						0.5 (avg)	2 (avg)

[a] These data may be somewhat in error as the change in stability was not corrected for any natural change in stability that may have occurred during the mixing operation.
[b] Average for all mixings in 1966.
[c] These data are intended as a rough guide only and are influenced by many factors: shape of lake, time of year, meteorologic conditions (temperature, sunlight, wind) quality requirements, circulation requirements, and so forth.

Source: Reprinted, with publisher's permission, from *J. Am. Water Works Assn., 59*(10):1268, © 1967 by the American Water Works Association, Inc., 2 Park Ave., New York, N.Y. 10016.

REFERENCES

1. W.H., Irwin, J.M. Symons, and G.G. Robeck, *J. San. Eng. Div., Am. Soc. Civ. Eng., 92*:SA-6, 1966.
2. J.M. Symons, W.H. Irwin, and G.G. Robeck, *J. San. Eng. Div., Am. Soc. Civ. Eng., 93*:SA-2, 1967.
3. J.M. Symons *et al., J. San. Eng. Div., Am. Soc. Civ. Eng., 93*:SA-6, 1967.
4. J.E. Ridley, P. Cooley, and J.A.P. Steel, Control of Thermal Stratification in Thames Valley Reservoir *Proc. Society for Water Treatment and Examn*
4. J.E. Ridley, P. Cooley, and J.A.P. Steel, Control of Thermal Stratification in Thames Valley Reservoir. *Proc. Society for Water Treatment and Examination, 15*:225, 1966.
5. M.E. Ford, Jr. Air Injection for Control of Reservoir Limnology. *J. Am. Water Works Assn., 55*:267, 1963.
6. H. Bernhardt, *J. Am. Water Works Assn., 59*:943, 1967.
7. T.L. Wirth, and R.C. Dunst, *Limnological Changes Resulting from Artificial Destratification and Aeration of an Impoundment.* Presented at 28th Midwest Fish and Wildlife Conference, Dec. 11–14, Chicago, 1966.

Section 6

Biological Control in Water Supply Systems

6. Introduction
Pages 555–565

6.1 Algae
Pages 566–571

6.2 Bacteria
Pages 572–579

6.3 Viruses
Pages 580–595

6. BIOLOGICAL CONTROL IN WATER SUPPLY SYSTEMS

6–1 SURVEY OF WATER SUPPLY SYSTEMS FOR BIOLOGICAL CONTROL PROBLEMS

Organism creating problem	No. managers reporting problem
Algae	164
Iron bacteria	54
Pond weeds	31
Midge larvae	17
Copepods	10
Asiatic clams	5
Nematodes	3
Manganese bacteria	3
Sulfur bacteria	3
Actinomycetes	2
Dead fish	2
Mosquito larvae	2
Fungi	2
Bryozoa	1
Water milfoil	1
Frogs	1

Type of problem	No. managers reporting problem
Taste and odors	162
Clogging of filters	64
Organism in distribution system	50
Increased reservoir maintenance	45
Reduced flows in distribution system	6
Solids or color in finished water	5

Origin of problem	No. reports indicating origin
Supplying lake or reservoir	149
Supplying river	45
Supplying well	40
In distribution system	29
In clear well	6
Supply canal	4
Finish water storage reservoir	2

Nonchemical control method	No. managers using method
Mechanical cleaning	46
Flushing of system	14
Redesign of system	10
Microstraining	9
Screening	5
Aeration of reservoir	3
Adjustment of filter rates	1
Cutting of pond weeds	1

Control chemicals	No. managers using chemical control
Chlorine	85
Copper sulfate	84
Carbon	44

Source: Reprinted, with publisher's permission, from *J. Am. Water Works Assn.*, *62*(8):520, © 1970 by the American Water Works Association, Inc., 2 Park Ave., New York, N.Y. 10016.

6–2 DISTRIBUTION OF FREE-LIVING NEMATODES AND AMEBAS IN WATER SUPPLIES

Organism[a]	No. supplies[b] containing organism	Location
Nematodes		
Aphelenchus	3	W.Va., Tenn., Texas
Cephalobus	2	Kans., Ind.
Cheilobus quadrilabiatus[c]	1	Wash.
Diplogaster nudicapitatus	1	W.Va.
Diplogaster	2	Ill., Pa.
Dorylaimus	1	Md.
Monhystera	4	W.Va., Mass., N.Y., Texas
Rhabditis	3	Pa., Md., Texas
Turbatrix aceti	2	Ga., Tenn.
Unidentified[d]	3	Cal., La., Colo.
Amebas		
Naegleria gruberi	2	W.Va., N.Y.
Hartmannella glebae	2	Kans., Texas
Hartmannella agricola	1	Texas
Hartmannella rhysodes	1	Tenn.
Unidentified	2	Mass., Md.

[a] Species included where known.
[b] All are river sources except for the Indiana source, which is a lake.
[c] Species identification uncertain.
[d] Specimens lost in preparation for staining.

Source: Reprinted, with publisher's permission, from *J. Am. Water Works Assn., 52*(5):613, © 1960 by the American Water Works Association, Inc., 2 Park Ave., New York, N.Y. 10016.

6–3 PREDOMINANT ORGANISMS AS MEASURED BY BIOMASS

Lafayette Reservoir

Week	1967	1968
27	Ceratium[a]	Pandorina[b]
28	Ceratium	Stephanodiscus[c]
29	Ceratium	Dinobryon[a]
30	Ceratium	Dinobryon
31	Pandorina[b]	Dinobryon
32	Anacystis[d]	Dinobryon
33	–	Cyclotella[c]
34	Pandorina[b]	Cyclotella
35	Pandorina	Pandorina[b]
36	Pandorina	Anacystis[a]
37	Pandorina	Melosira[c]
38	Pandorina	–
39	Pandorina	Ceratium[a]
40	Dinobryon[a]	–
41	Oocystis[b]	Stephanodiscus[c]
42	Carteria[b]	Stephanodiscus
43	Oocystis[b]	Stephanodiscus
44	Carteria	Anabaena[d]

[a] Protozoa.
[b] Chlorophycene.
[c] Diatomaceae.
[d] Cyanophyceae.

Source: Reprinted, with publisher's permission, from *J. Am. Water Works Assn., 62*(11):711, © 1970 by the American Water Works Association, Inc., 2 Park Ave., New York, N.Y. 10016.

6–4 APPROXIMATE DIAMETERS OF BACTERIA, VIRUSES, DETERGENTS, AND OTHER PARTICLES

Item	Diameter, mμ
Bacteria (coliform organisms, typhoid)	1,000
Poliomyelitis virus (in feces)	30
ECHO virus (in feces)	18
Coxsackie virus (in feces)	18
Hemoglobin molecule (in blood)	3
ABS molecule (in detergents)	1
Uranin molecule	1
Water molecule	0.4
Hepatitis Virus A (in feces)	a

[a] Size not known.

Source: Reprinted, with publisher's permission, from *J. Am. Water Works Assn., 53*(3):303, © 1961 by the American Water Works Association, Inc., 2 Park Ave., New York, N.Y. 10016.

6–5 BIOLOGICALLY MEDIATED REACTIONS AFFECTING pH IN NATURAL WATER SYSTEMS

Process	Reaction	Effect on pH
Photosynthesis	$6CO_2+6H_2O \rightarrow C_6H_{12}O_6+6O_2$	increase
Respiration	$C_6H_{12}O_6+6O_2 \rightarrow 6CO_2+6H_2O$	decrease
Methane fermentation	$C_6H_{12}O_6+3CO_2 \rightarrow 3CH_4+6CO_2$	decrease
Nitrification	$NH_4^++2O_2 \rightarrow NO_3^-+H_2O+2H^+$	decrease
Denitrification	$5C_6H_{12}O_6+24NO_3^-+24H^+ \rightarrow 30CO_2+12N_2+42H_2O$	increase
Sulfide oxidation	$HS^- + 2O_2 \rightarrow SO_4^{--} + H^+$	decrease
Sulfate reduction	$C_6H_{12}O_6+3SO_4^{--}+3H^+ \rightarrow 6CO_2+3HS^-+6H_2O$	increase

Source: Reprinted, with publisher's permission, from *J. Am. Water Works Assn., 55*(12):1553, © 1963 by the American Water Works Association, Inc., 2 Park Ave., New York, N.Y. 10016.

6–6 CHEMICALS REQUIRED FOR TREATMENT OF DIFFERENT GENERA

Organisms and Odor[a]	Copper sulfate, ppm	Copper sulfate, lb/mil gal	Chlorine, ppm	Chlorine dioxide, ppm[1]
Diatomaceae				
Asterionella[d]	0.12[b]–0.20[b]	1.0–1.7	0.5[b]–1.0[b]	
Aromatic, geranium, fishy				
Cyclotella[cd]			1.0	0.4
Faintly aromatic				
Diatoma[c]				
Faintly aromatic				
Fragilaria[c]	0.25	2.1		
Melosira[cde]	0.20[b]	1.7	2.0	0.8
Meridion				
Aromatic				
Navicula[c]	0.07	0.6		0.2
Nitzchia[e]	0.50	4.2		
Synedra[cd]	0.36[b]–0.50[b]	3.0–4.2	1.0	0.7
Earthy				
Stephanodiscus	0.33	2.8		
Tabellaria[d]	0.12[b]–0.50[b]	1.0–4.2	0.5[b]–1.0	
Aromatic, geranium, fishy				
Chlorophyceae				
Cladophora[c]	0.50	4.2		1.2
Closterium	0.17	1.4		1.4
Coelastrum[c]	0.05–0.33	0.4–2.8	1.0	
Conferva[c]	0.25	2.1		
Desmidium	2.00	16.6		
Dictyosphaerium[d]			0.5[b]–1.0	1.2
Grassy, nasturtium, fishy				
Draparnaldia	0.33	2.8		1.6
Eudorina	2.00–10.00	16.6–83.0		
Faintly fishy				
Enteromorpha	0.50	4.2		
Gloeocystis	0.50	4.2		
Offensive				
Hydrodictyon[c]	0.10	0.08		
Very offensive				
Microspora	0.40	3.3		
Palmella	2.00	16.6		2.0
Pandorina	2.00–10.00	16.6–83.0		3.8
Faintly fishy				
Protococcus			1.0	
Raphidium	1.00	8.3		
Scenedesmus	1.00[b]	8.3		1.6
Spirogyra[c]	0.12	1.0	0.7–1.5	0.3
Staurastrum	1.50	12.5		0.8
Grassy				
Tetrastrum			1.0	
Ulothrix	0.20[b]	1.7		0.3

[a] The organisms where odors are expressed have been in sufficient quantity at one time or another to cause the characteristic odor.
[b] Dosage successful in New York City's supplies, or personal knowledge.
[c] These organisms have caused trouble other than odor.
[d] These organisms have been affected by chlorine sometimes producing characteristic odor and in many cases controlled by dosage ranging from 0.3–3 ppm, depending largely on amount of organisms.
[e] Controlled by excess caustic lime, 5 ppm, in the case of diatoms (48 hr).

6–6 CHEMICALS REQUIRED FOR TREATMENT OF DIFFERENT GENERA *(Continued)*

Organisms and Odor[a]	Copper sulfate ppm	Copper sulfate lb/mil gal	Chlorine, ppm	Chlorine dioxide, ppm[1]
Volvox[d]	0.25	2.1	0.3[b]–1.0	
Fishy				
Zygnema	0.50	4.2		
Cyanophyceae				
Anabaena[d]	0.12[b]–0.48	1.0–4.0	0.5[b]–1.0	1.2
Moldy, grassy, vile				
Aphanizomeon[d]	0.12–0.50[b]	1.0–4.2	0.5[b]–1.0[b]	
Moldy, grassy, vile				
Clathrocystis	0.12[b]–0.25[b]	1.0–2.1	0.5[b]–1.0	
Sweet grassy, vile				
Coelosphaerium[d]	0.20[b]–0.33	1.7–2.8	0.5[b]–1.0	0.6
Sweet grassy				
Cylindrospermum	0.12	1.0		
Grassy				
Gloeocapsa (red)	0.24[b]	2.0		0.4
Microcystis	0.20	1.7		0.5
Oscillaria[c]	0.20–0.50[b]	1.7–4.2	1.1	0.5
Rivularia				
Moldy, grassy				
Protozoa				
Bursaria				
Irish moss, salt marsh, fishy				
Ceratium[cd]	0.24[b]–0.33	2.0[b]–2.8	0.3[b]–1.0	0.5
Fishy, vile (rusty brown color)				
Chlamydomonas	0.36[b]–1.0	3.0–8.3		1.6
Cryptomonas	0.50	4.2		
Candied violets				
Dinobryon[d]	0.18[b]	1.5	0.3[b]–1.0	1.5
Aromatic, violets, fishy				
Endamoeba histolytica (cysts)			25.0–100.0	
Euglena	0.50	4.2	1.0	0.8
Glenodinium	0.50[b]	4.2		
Fishy				
Mallomonas	0.50	4.2		
Aromatic, violets, fishy				
Peridinium	0.50[b]–2.00	4.2–16.6		
Fishy, like clamshells				
Synura[d]	0.12[b]–0.25[b]	1.0–2.1	0.3[b]–1.0	0.4
Cucumber, muskmelon, fishy, bitter taste				
Uroglena[d]	0.05[b]–0.20[b]	0.4–1.6	0.3[b]–1.0	
Fishy, oily, cod-liver oil taste				
Rotifer				
Stentor	0.24[b]	2.0		
Crustacea				
Cyclops[cde]			1.0–3.0	
Daphnia[cde]	2.00	16.6	1.0–3.0	

[a] The organisms where odors are expressed have been in sufficient quantity at one time or another to cause the characteristic odor.

[b] Dosage successful in New York City's supplies, or personal knowledge.

[c] These organisms have caused trouble other than odor.

[d] These organisms have been affected by chlorine sometimes producing characteristic odor and in many cases controlled by dosage ranging from 0.3–3 ppm, depending largely on amount of organisms.

[e] Controlled by excess caustic lime, 5 ppm, in the case of diatoms (48 hr).

6–6 CHEMICALS REQUIRED FOR TREATMENT OF DIFFERENT GENERA *(Continued)*

Organisms and Odor[a]	Copper sulfate ppm	Copper sulfate lb/mil gal	Chlorine, ppm	Chlorine dioxide, ppm[1]
Schizomycetes				
Beggiatoa[c]	5.00	41.5		
Very offensive decayed				
Cadothrix	0.20	1.7		
Crenothrix[cd]	0.33–0.5[b]	2.8–4.2	0.5[b]	
Very offensive decayed, medicinal with chlorine				
Leptothrix[c]				
Sphaerotilis, natans[c]	0.40	3.3		
Very offensive decayed				
Spirophyllum[c]			0.25[b]	
Thiothrix[cd]			0.5[b]–1.0	
Fungus				
Achlya			0.6[b]	
Leptomitus	0.40	3.3		
Saprolegnia[c]	0.18	1.5		
Miscellaneous				
Chara[c]	0.10–0.50	0.8–4.2		
Nitella, flexilis[c]	0.10–0.18[b]	0.8–1.5		
Objectionable				
Potamogeton[c]	0.30–0.80	2.5–6.7		
Nais[c]			1.0[b]	
Blood worm, chironomus[c]			15.0–50.0	
Blood worm, gnats			3.0	

[a] The organisms where odors are expressed have been in sufficient quantity at one time or another to cause the characteristic odor.
[b] Dosage successful in New York City's supplies, or personal knowledge.
[c] These organisms have caused trouble other than odor.
[d] These organisms have been affected by chlorine sometimes producing characteristic odor and in many cases controlled by dosage ranging from 0.3–3 ppm, depending largely on amount of organisms.

Source: F.E. Hale, *Use of Copper Sulfate in Control of Microscopic Organisms,* Phelps Dodge Refining Corp., p. 27.

REFERENCE

1. E.S. Hopkins and E.L. Bean, *Water Purification Control,* Williams and Wilkins, Baltimore, 1966.

6–7 TOLERANCE TO SEVERAL CHEMICALS OF LIVE ORGANISMS FOUND IN POTABLE WATER

Retention Period up to 24 hr

SYMBOLS

Unaffected, blank.
Stunned, subsequently recovered, *.
Killed, +.

	Chemical, ppm															
										Copper chloramine						
	Chloramine				Pyrethrins					4 copper as Cu ion, 0.5 chlorine, 0.15 ammonia	4 copper as Cu ion, 1 chlorine, 0.15 ammonia	Chlorine			
Organism	**0.4**	**0.5**	**1.0**	**2.0**	**0.0005**	**0.001**	**0.01**	**0.1**	**1.0**			**0.4**	**0.5**	**1.0**	**2.0**
Asellus				*	*	*	+	+	+		*				
Nais	*	*	+	+						*	+	*	*	+	+
Cyclops			*	+							+				+
Ostracoda				*					*						*
Cladocera									*		+				*
Snails										*	*				

Source: Reprinted, with publisher's permission, from *J. Am. Water Works Assn., 60*(2):228, © 1968 by the American Water Works Association, Inc., 2 Park Ave., New York, N.Y. 10016.

6–8 MEAN TOLERANCE LIMITS OF *DAPHNIA MAGNA* IN CRUDE OIL, EMULSIFIERS *F* AND *W*, AND CRUDE OIL PLUS EMULSIFIERS *F* AND *W* IN AN ARTIFICIAL AND A NATURAL TEST MEDIUM

Materials tested	24 hr TL_m, mg/l	48 hr TL_m, mg/l	72 hr TL_m, mg/l
Reference Dilution Water As Test Medium			
Crude oil	10,000.0	4,613.3	752.2
F[a]	694.0	290.0	204.0
W[b]	13.7	2.9	1.5
Crude oil plus *F*	370.3	275.0	102.0
Crude oil plus *W*	24.4	10.7	9.1
University Lake Water As Test Medium			
Crude oil	2,110.0	1,095.0	750.0
F	590.0	357.0	240.0
W	5.5	3.0	1.7
Crude oil plus *F*	455.3	207.3	115.0
Crude oil plus *W*	2.3	1.1	0.9

[a] Emulsifier *F* is Conco LCP-12 from the Conco Chemical Corp., Dallas, Texas.
[b] Emulsifier *W* is Gamosol Solvent "D" from the Gamlen Chemical Co., East Paterson, N.J.

Source: B.F. Dowden, *J. Water Pollut. Control Fed.*, *34*(10):1010, 1962.

6–9 NEMATODE COUNTS IN WATER SAMPLES

Sample no.	Counts, no./gal
1 Liter Grab Samples	
1	4,560
2	2,480
3	4,380
4	3,980
5	4,780
6	3,640
7	1,500
8	2,340
9	1,800
10	4,300
Avg	3,380
10 Liter Composites	
1	3,740
2	3,540
3	3,300
Avg	3,300

Source: Reprinted, with publisher's permission, from *J. Am. Water Works Assn.*, *56*(1):73, © 1964 by the American Water Works Association, Inc., 2 Park Ave., New York, N.Y. 10016.

6–10 NEMATODE CONCENTRATIONS FROM SEWAGE TREATMENT PLANT OUTFALLS

Various Points Upstream and Downstream

City and stream	Sampling point[a]	Nematode concentration, no./gal	Distance from outfall,[b] yd	Type of waste treatment
Danville, Ill.	A	0	−880	
Vermilion R.	B	0	−200	
	C	83	0	AS[c]
	D	79	+200	
Decatur, Ill.	A	0	−2,200	
Sangamon R.	B	0	−1,760	
	C	931	0	AS
	D	23	+50	
Westville, Ill.	C	749	0	
Grape Creek	D	193	+100	TF[d]
Paris, Ill.	A	0	−880	
Sugar Creek	C	106	0	
	D	49	+75	TF
	E	23	+200	
Bloomington, Ill.	A	0	−500	
Sugar and Salt	C	1,983	0	
Creek	D	469	+200	TF
	E	170	+880	

[a] A, water treatment plant intake; B, upstream from sewage treatment plant outfall; C, sewage treatment plant effluent; D, downstream from sewage treatment plant outfall; E, downstream from D.
[b] Minus (−) sign indicates upstream and plus (+) sign downstream from sewage treatment plant outfall.
[c] Activated sludge.
[d] Trickling filter.

Source: Reprinted, with publisher's permission, from *J. Am. Water Works Assn.*, *56*(1):73, © 1964 by the American Water Works Association, Inc., 2 Park Ave., New York, N.Y. 10016.

6–11 THRESHOLD ODOR CONCENTRATIONS OF FISH POISON FORMULATIONS AND COMPONENTS

Material[a]	Threshold odor concentration, mg/l	95% confidence limits	
		Upper	Lower
Formulation A	0.020	0.028	0.018
Formulation B	0.023	0.033	0.015
Formulation C	0.007	0.009	0.005
Formulation D	0.018	0.025	0.011
Cube powder	30.6	41.6	24.0
Rotenone	13.8	23.1	9.6
Toxaphene	0.0052	0.010	0.0022
Sulfoxide	0.091	0.17	0.05
Aerosol OT	14.6	23.1	10.3
Solvent 1	0.016	0.021	0.004
Solvent 2	13.9	20.5	8.1
Solvent 3	0.090	0.143	0.057
Deodorized kerosene	0.082	0.106	0.057

[a] Formulation A – rotenone, rotenoids, toxaphene, solvent, emulsifier, clay.
Formulation B – rotenone, rotenoids, solvent, emulsifier.
Formulation C – rotenone, rotenoids, synergist, solvent, emulsifier.
Formulation D – rotenone, rotenoids, solvent, emulsifier.
Rotenone – rotenone.
Toxaphene – toxaphenol.
Sulfoxide – synergist.
Aerosol OT – emulsifier.
Solvent 1 – solvent.
Solvent 2 – solvent.
Solvent 3 – solvent.
Deodorized kerosene – solvent.

Source: Reprinted, with publisher's permission, from *J. Am. Water Works Assn., 53*(1):49, © 1961, by the American Water Works Association, Inc., 2 Park Ave., New York, N.Y. 10016.

6–12 LIMITING SAFE DOSAGE OF COPPER SULFATE FOR FISH

Fish	ppm	lb/mil gal
Trout	0.14	1.2
Carp	0.33	2.8
Suckers	0.33	2.8
Catfish	0.40	3.5
Pickerel	0.40	3.5
Gold Fish	0.50	4.2
Perch	0.67	5.5
Sunfish	1.35	11.1
Black Bass	2.00	16.6

Source: F.E. Hale, *Use of Copper Sulfate in Control of Microscopic Organisms,* Phelps Dodge Refining Corp., New York, p. 20.

6–13 SUMMARY OF DATA ON TREATMENT OF WATER CONTAINING FISH POISONS

Poison	Initial concentration, ppm	Final concentration,[b] ppm	Carbon a	Carbon b	Chlorine contact time, min 15	60	180	Chlorine dioxide
					Concentration added, ppm			
Rotenone	0.1	0.003	d	6.6	41.5	30.0	11.5	0.50–0.88
Toxaphene	0.1	0.007	5.0	5.0	c	c	c	c
Cube powder	2.0	0.03	d	9.8				
Formulation A[e]	2.0	0.05	36	8.8	c	c	c	
Formulation B[f]	2.0	0.03	78	26.4	32	21	15	
Formulation C[g]	2.0	0.03	85	21.6	34	20	12	
Formulation D[h]	2.0	0.05	62	16.7	38	21	10	
Solvent 1	2.0		50	i				
Deodorized kerosene	2.0		52	i				

[a] Carbon required to reduce threshold odor number to 1.
[b] Carbon required to reduce poison concentration to half of 96-hr TLm (median toxicity limit) for fathead minnows.
[c] Both treatment shown and alum treatment ineffective.
[d] Material nonodorous in the concentrations normally used.
[e] Rotenone, rotenoids, toxaphene, solvent, emulsifier, and clay.
[f] Rotenone, rotenoids, solvent, and emulsifier.
[g] Rotenone, rotenoids, synergist, solvent, and emulsifier.
[h] Rotenone, rotenoids, solvent, and emulsifier.
[i] Material is nontoxic to fish in the concentration normally used.

Source: Reprinted with publisher's permission, from *J. Am. Water Works Assn., 56*(3):267, © 1964, by the American Water Works Association, Inc., 2 Park Ave., New York, N.Y. 10016.

REFERENCES
1. J.M. Cohen, *et al., J. Am. Water Works Assn., 53*(2):233, 1961.
2. J.M. Cohen, *et al., J. Am. Water Works assn., 53*(1):49, 1961.
3. J.M. Cohen, *et al., J. Am. Water Works Assn., 52*(12):1551, 1960.

6–14 THERMAL DEATH CONDITIONS

Organism	Temperature °C	°F	Time
Salmonella typhosa	55–60	131–140	30 min
	60	140	20 min
Salmonella sp.	55	131	60 min
Shigella sp.	55	131	60 min
Escherichia coli	55	131	60 min
	60	140	15–20 min
Entamoeba histolytica cysts	45	113	few min
Taenia saginata	55	131	few min
Trichinella spiralis larvae	55	131	few min
	60	140	few sec
Brucella abortus	62–63	144–145	3 min
Micrococcus pyogenes var aurens	50	122	10 min
Steptococcus pyogenes	54	130	10 min
Mycobacterium tuberculosis var hominis	66	151	15–20 min
	67	153	few min
Corynebacterium diphtheriae	55	131	45 min
Ascaris lumbricoides eggs	50	122	60 min

Source: ,R. Rickles, *Pollution Control,* Noyes Development Corp., Park Ridge, N.J., 1965, p. 143.

REFERENCES

1. E. Jordon and W. Burrows, *Textbook of Bacteriology,* 14th ed., Saunders, Philadelphia, 1946.
2. K.F. Maxcy, *Rosenau's Preventive Medicine and Hygiene,* 7th ed., New York, 1951.
3. I.S. Perold, *Public Health, 15*:70, 1950.

6.1 ALGAE

See also Section 4 of Volume IV, Nutrients, Lakes, Eutrophication.

6.1–1 CHEMICAL COMPOSITION OF SOME ALGAE FROM PONDS AND LAKES

Southeastern U.S.

Table A

Analysis	Chara	Pithophora	Spirogyra	Giant Spirogyra	Rhizoclonium	Oedogonium	Mougeotis
Ash, %	43.4	27.77	13.06	13.86	17.36	12.69	14.54
C, %	29.3	35.38	42.40	41.16	39.10	40.84	40.74
N, %	2.46	2.57	3.01	2.35	3.46	2.64	1.77
P, %	0.25	0.30	0.20	0.23	0.43	0.08	0.25
S, %	0.55	1.42	0.27	0.24	0.27	0.15	0.36
Ca, %	8.03	3.82	0.57	0.84	0.52	0.44	1.68
Mg, %	0.92	0.20	0.45	0.30	0.21	0.16	0.57
K, %	2.35	3.06	0.92	0.99	1.90	3.03	1.20
Na, %	0.13	0.07	1.42	1.43	0.09	0.06	0.49
Fe, mg/l	2,520	2,836	1,368	1,793	1,820	1,645	60
Mn, mg/l	2,926	829	1,641	1,658	1,687	1,729	1,080
Zn, mg/l	89	29	72	46	89	119	520
Cu, mg/l	19	23	47	34	75	75	143
B, mg/l	6.7	65	4.2	4.3	1.8	8.1	8

Table B

Analysis	Anabaena	Cladophora	Euglena	Hydrodictyon	Microcystis	Lyngbya	Nitella	Aphanizomenon
Ash, %	5.19	23.38	4.12	17.94	6.2	17.20	19.11	7.21
C, %	49.70	35.27	48.14	39.96	46.46	40.23	38.43	47.65
N, %	9.43	2.30	5.14	3.87	8.08	5.01	2.70	8.57
P, %	0.77	0.56	0.67	0.24	0.68	0.31	0.23	1.17
S, %	0.53	1.58	0.19	1.41	0.27	0.28	0.34	1.18
Ca, %	0.36	1.69	0.05	0.69	0.53	0.45	1.89	0.73
Mg, %	0.42	0.23	0.07	0.17	0.17	0.14	0.95	0.21
K, %	1.20	6.08	0.34	4.21	0.79	0.42	3.73	0.68
Na, %	0.18	0.18	0.02	0.38	0.04	0.06	0.28	0.19
Fe, mg/l	80	2,300	1,545	1,963	322	3,866	2,180	833
Mn, mg/l	800	1,040	240	1,373	2,751	5,230	2,388	167
Zn, mg/l	–	10	73	129	48	171	240	120
Cu, mg/l	70	190	290	114	37	101	39	187
B, mg/l	–	84.6	3.8	–	3.6	112	9.8	–

Source: *Water Quality Criteria,* U.S. Department of the Interior, Fed. Water Pollut. Control Adm., April 1968, p. 55. From personal communication of Dr. J.M. Lawrence, Auburn University, Auburn, Ala.

6.1–2 ANALYSES OF VARIOUS ALGAE

Organism	Substance, %					
	Nitrogen	Phosphorus	Calcium	Magnesium	Silicon	R_2O_3[b]
Blue-green						
Microcystis	8.35[a]	0.52	0.66	0.38	0.06	0.84
	8.60					
	9.27					
	6.32					
Microcystis aeruginosa[2]	4.0–9.4[c]	0.12–0.67[c]				
	5.64–8.68[d]	0.52–0.98[d]				
Anabaena	8.27	0.51	1.02	0.42	0.44	1.27
Anabaena and coelosphaerium	8.35	–	0.58	0.38	0.13	–
Volvox	7.61	1.09	0.78	0.56	0.11	0.80
Aphanizomenon	9.30	–	–	–	–	–
Green						
Spirogyra	3.47	–	–	–	–	–
Cladophora	2.77	0.14	2.32	0.97	3.30	1.80

[a] Values on specimens from three different lakes.
[b] $Al_2O_3 + Fe_2O_3$ combined, no differentiation.
[c] Laboratory studies.
[d] Seven samples from three lakes.

REFERENCES

1. E.A. Birge and C. Juday, *Wisc. Geol. Nat. Hist. Survey*, Bull. 64, Madison, 1922.
2. G.C. Gerloff and F. Skoog, *Ecology*, *35*:348, 1954.

6.1–3 TOXIC RESPONSE OF FREEZE-DRIED CELLS

From 24 Waterblooms of Blue-Green Algae[a]

Dominant species	Response			
	Very fast deaths (1–10 min), %	Fast deaths (1–2 hr), %	No deaths or slow deaths (4–48 hr), %	Total in test group
Microcystis aeruginosa		70	30	10
Anabaena flos-aquae	62		38	8
Aphanizomenon flos-aquae			100	1
50 percent *Aphanizomenon*, 50 percent *Microcystis*		100		4
Lyngbya birgei			100	1

[a] Intraperitoneal injection, 20 g male mice, dosage range 40-640 mg/p kg.[1]

Source: Reprinted, with publisher's permission, from *J. Am. Water Works Assn.*, *56*(11):1481, © 1964 by the American Water Works Association, Inc., 2 Park Ave., New York, N.Y. 10016.

REFERENCE

1. P.R. Gorham, *Algae and Man*, Plenum Press, New York, 1964, p.307.

6.1–4 TOXIC RESPONSE OF FREEZE-DRIED CELLS AND CULTURE FILTRATES

From 59 Strains of Blue-Green Algae[a]

Species	Response			Total in test group
	Very fast deaths (1–10 min), %	Fast deaths (1–2 hr), %	No deaths or slow deaths (4–48 hr), %	
Microcystis aeruginosa		29	71	28
Anabaena flos-aquae	57		43	14
Aphanizomenon flos-aquae			100	5
Gloeotrichia echinulata			100	1
Coelosphaerium kützingianum			100	1
Other planktonic species			100	10

[a] Intraperitoneal injection, 20 g male mice, dosage range 40–640 mg/p kg.[1]

Source: Reprinted, with publisher's permission, from *J. Am. Water Works Assn., 56*(11):1481,

REFERENCE

1. P.R. Gorham, *Algae and Man,* Plenum Press, New York, 1964, p. 307.

6.1–5 DATA ON ALGAE REMOVAL AND WATER QUALITY OF SUPERNATANT OBTAINED IN PILOT-PLANT STUDIES ON USE OF ALUM AS A FLOCCULENT

Item	Alum dosage, mg/l				
	60	75	90	105	120
Solids					
Total suspended, mg/l	21	23	15	15	8
Removal, %[a]	86	85	89	91	94
Volatile suspended, mg/l	13	11	8	10	6
Total dissolved, mg/l	479	431	467	434	506
Removal, %[a]	0	15	0	17	15
Volatile dissolved, mg/l	103	121	151	116	88
Algal cells, 10^4/ml	30	0.3	0	0	0
Removal, %[a]	85	99	100	100	100
Coliforms, per ml	–	200	150	127	95
Removal, %[a]	–	98	99.9	99.9	99.9
Optical density, 380 mμ	0.094	0.085	0.062	0.021	0.021
Total nitrogen, mg/l[b]	2.1	6.0	6.0	5.1	4.3
Removal, %[a]	75	70	75	91	92
NH_4-nitrogen, mg/l	0	3.1	4.2	3.7	2.8
$PO_4^{\equiv}$, mg/l	4	3	0	0	0
$SO_4^{=}$, mg/l	–	175	190	201	167
Cl^-, mg/l	–	–	–	44	51
Na^+, mg/l	–	–	–	54	57
Alkalinity, mg/l	108	103	24	–	–
Hardness, mg/l as $CaCO_3$	203	–	–	150	124
Conductivity, mmho	–	–	–	0.7	0.69
BOD, mg/l	7	5	2	5.5	4
pH	6.5	6.3	6.5	6.3	6.5

[a] Removal based on concentration in pond effluent.
[b] Amount contained in dissolved and suspended solids combined.

Source: C.G. Golueke and W.J. Oswald, *J. Water Pollut. Control Fed., 37*(4):471, 1965.

6.1–6 REMOVAL OF ALGAE BY WATER TREATMENT PROCESSES

Type of algae	Sedimentation with alum, softening[a]						Sedimentation with lime[b]			All operations[c]		
	Raw water, organisms/ml	Sedimentation basin effluent,[a] organisms/ml	Removal, %	Softener effluent, organisms/ml	Removal, %	Overall removal, %	Raw water, organisms/ml	Sedimentation basin effluent,[a] organisms/ml	Removal, %	Plant influent, organisms/ml	Plant effluent, organisms/ml	Removal, %
Blue-green algae												
Aphanizomenon	96	67	28	5	93	95	27	7	73	24	0	100
Oscillatoria	32	5	84	4	20	88	5	1	80	6	1	83
Diatoms												
Cyclotella	530	310	42	7	98	99	9,230	187	98	3,407	3	>99
Fragilaria	91	52	43	5	90	95	580	42	93	235	1	>99
Melosira	295	190	36	30	84	90	501	2	99	300	0.3	>99
Navicula	20	5	75	0	100	100	37	2	95	32	1	97
Stephanodiscus	24	20	17	0	100	100	575	0	100	230	0	100
Others	7	5	29	2	60	72	43	1	98	95	0	100
Green algae												
Ankistrodesmus	11	25	127 increase	1	96	91	21	1	95	13	0.2	>98
Chlorella	249	106	57	26	75	90	2,300	78	97	920	5	>99
Scenedesmus	4	9	125 increase	1	89	75	69	7	90	34	0.3	>99
Others	5	7	40 increase	0	100	100	43	2	95	30	0.4	>98
Flagellates												
Chlamydomonas	160	162	1 increase	1	99	99	85	11	88	62	0.6	99
Chloromonad	28	21	25	0	100	100						
Euglena	10	13	30 increase	2	85	80	75	4	95	40	2	95
Others	40	20	50	0	100	100	5	0	100	36	0	100
Totals	1,599	1,017	37	84	92	95	13,596	345	97	5,466	14.9	>99

[a] Four samplings averaged.
[b] Three samplings averaged.
[c] Ten samplings averaged.

Source: Reprinted, with publisher's permission, from *J. Am. Water Works Assn., 61*(6):289, © 1969, by the American Water Works Association, Inc., 2 Park Ave., New York, N.Y. 10016.

6.1–7 EXTENT OF ALGAE REMOVAL AND EFFECT OF FLOCCULATION ON EFFLUENT QUALITY

Field Scale Experiments

		Time of day							
		2400 (Midnight) Effluent				0600 Effluent			
Item	Influent[a]	Pond	Removal, %	Flocculator	Removal, %	Pond	Removal, %	Flocculator	Removal, %
Suspended solids, mg/l									
Total	100	199	+[b]	13	87	121	+	0	100
Volatile	88	122	+	7	92	96	+	0	100
Dissolved solids, mg/l									
Total	900	774	14	777	14	777	14	836	7
Volatile	202	164	19	129	36	164	19	211	1
Optical density, 380 mμ	–	0.512		0.029		0.355	–	0.015	
Total nitrogen, mg/l	45	32	29	16	64	29	36	19	58
Ammonia nitrogen, mg/l	29	3	90	3	90	3	90	4	86
$PO_4^{\equiv}$, mg/l	20	10	50	1	95	9	55	2	90
$SO_4^{=}$, mg/l	186	165	11	289	+	181	3	291	+
Cl^-, mg/l	143	158	+	157	+	159	+	156	+
Na^+, mg/l	135	160	+	160	+	161	+	160	+
K^+, mg/l	10	7	30	7	30	8	20	7	30
Mg^{++}, mg/l	59	37	37	32	46	39	34	40	32
Ca^{++}, mg/l	65	31	52	31	52	30	54	27	58
Hardness, mg/l as $CaCO_2$	405	229	44	210	48	238	41	234	42
BOD, mg/l	–	23		1		11		1	
pH	7.8	9.8	+	7.1	–[c]	9.7	+	6.6	–

[a] Sewage influent to algae pond.
[b] Increase.
[c] Decrease.

Source: C.G. Golueke and W.J. Oswald, *J. Water Pollut. Control Fed.*, *37*(4):471, 1965.

6.1–8 CLASSIFICATION OF ALGAL PRODUCT QUALITY ACCORDING TO PRODUCTION METHOD

Grade criteria	Grade code	Production method	Quality characteristic
Growth	1000[a]	Grown sterile in inorganic media	Pure algae
	2000	Grown unsterile in inorganic media	Contains some bacteria
	3000	Grown unsterile in organic media	Contains many bacteria
	4000	Grown in sewage	May contain pathogens
	5000	Grown in land drainage	May contain toxic substances
Primary concentration	0100	Concentrated by centrifuge	Contains no additives
	0200	Concentrated by flocculation and sedimentation (organic reagents)	Contains organic coagulants
	0300	Concentrated by flocculation and sedimentation (inorganic reagents)	Contains lime, alum, ammonia, and/or phosphates
	0400	Concentrated by autoflocculation	Contains ashed compounds of $CaCO_3$, $CaSO_4$, $Mg(OH)_2$, etc
Secondary concentration	0010	Gravity filter	Contains no additives
	0020	Sand bed	Contains sand, dust, etc
	0030	Solid bowl centrifuge	Contains no additives
Drying method, sterility, digestibility	0001	Drum or spray dried	Sterile, digestible
	0002	Air dried, sterilized	Sterile, low digestibility
	0003	Air dried, unsterilized	Unsterile, low digestibility
	0004	Sand bed or sun dried	Unsterile, low digestibility

[a] In this notation the 0 indicates an unspecified process.

Source: C.G. Golueke and W.J. Oswald, *J. Water Pollut. Control Fed.*, *37*(4):471, 1965.

6.1–9 COMPARISON OF ALGAL REMOVAL IN SAND AND ANTHRAFILT FILTERS

Depth below filter surface, in.	Sand filter, organisms/ml	Anthrafilt filter, organisms/ml
2	380	359
4	468	594
6	271	25
12	25	140
24	22	37
36	4	10

Source: Reprinted, with publisher's permission, from *J. Am. Water Works Assn.*, *61*:259, © 1969 by the American Water Works Association, Inc., 2 Park Ave., New York, N.Y. 10016.

6.2 BACTERIA

6.2–1 BACTERIAL CONTENT OF VARIOUS TYPES OF WATER

	Coliform organisms	*Salmonella paratyphi*
Type of water	**Concentration, No./l**	
Domestic sewage	500,000,000[a]	500[a]
Sewage after biologic treatment	5,000,000[a]	5[a]
Polluted river	50,000[b]	0.05[b]
Potable	0.5[b]	0[b]

[a] London water utility laboratories.
[b] Research by author.

Source: Reprinted, with publisher's permission, from *J. Am. Water Works Assn., 55*(7):917, © 1963 by the American Water Works Association, Inc., 2 Park Ave., New York, N.Y. 10016.

6.2–2 COLIFORM DISCHARGE FACTORS FOR DOMESTIC SEWAGE

Season	*E. coli*/person day
Summer	4×10^{11}
Winter	1.33×10^{4}

Source: H.A. Thomas, Jr., *Sew. Ind. Wastes, 23*(1):1, 1951.

6.2–3 FECAL STREPTOCOCCUS DISTRIBUTIONS IN WARM-BLOODED ANIMAL FECES

		Percent occurrence			
Fecal source	**Total strains examination, no.**	Enterococci, %	*S. bovis S. equinus,* %	Atypical *S. faecalis,* %	*S. faecalis liquefaciens,* %
Human	1,067	73.8	None	None	26.2
Animal pets					
Cat	268	89.9	1.5	2.2	6.3
Dog	585	44.1	32.0	14.4	9.6
Rodents	539	47.3	17.1	0.4	35.3
Livestock					
Cow	438	29.7	66.2	None	4.1
Pig	296	78.7	18.9	None	2.4
Sheep	321	38.9	42.1	None	19.0
Poultry					
Duck	328	51.2	48.8	None	None
Chicken	275	77.1	1.1	None	21.8
Turkey	317	76.7	1.6	None	21.8

Source: Reprinted, with publisher's permission, from *J. Am. Water Works Assn., 62*(2):113, © 1970 by the American Water Works Association, Inc., 2 Park Ave., New York, N.Y. 10016.

6.2–4 DISTRIBUTION OF FECAL STREPTOCOCCI IN PRAIRIE WATERSHEDS, RECREATIONAL WATERS, AND PRIVATE WELL SUPPLIES

	Densities/100 ml[a]				Occurrence			
Water source	Fecal coliforms	Fecal streptococci	FC[b]/FS	Total strains examination	Enterococci, %	S. bovis S. equinus, %	Atypical S. faecalis, %	S. faecalis liquefaciens, %
Prairie watersheds								
Cherry Creek, Wyo.	90	83	1.08	17	64.7	None	None	35.3
Saline River, Kan.	95	180	0.53	140	75.7	None	None	24.3
Club River, Idaho	110	160	0.69	34	97.1	2.9	None	None
Clear Creek, Colo.	170	110	1.5	70	95.8	None	None	4.3
Recreational waters								
Lake Mead	2	444	0.0045	50	96.0	None	None	4.0
Lake Moovalaya	9	170	0.053	100	78.0	None	None	22.0
Colorado River	4	256	0.016	100	100.0	None	None	None
Whitman River	32	88	0.36	45	100.0	None	None	None
Merrimack River	100	96	1.04	42	92.9	2.4	None	4.8
Private well supplies								
Gilman	<0.8	8,800	0.000081	25	None	None	None	100.0
Miami	<1.0	18	5.4	19	89.5	5.3	5.3	None
Baas	1.1	32	3.4	15	73.3	None	None	26.7
Kilb	2.8	106	2.6	34	73.5	None	None	26.5
Hopper	5.0	134	3.7	24	100.0	None	None	None

[a] All bacterial densities for prairie watersheds and recreational waters were based on median values. Data for private well supplies were developed from single samples only.
[b] Inserted by the editors.

Source: Reprinted, with publisher's permission, from *J. Am. Water Works Assn., 62*(2):113, © 1970 by the American Water Works Association, Inc., 2 Park Ave., New York, N.Y. 10016.

6.2–5 SOME COLIFORM, *E. COLI*, AND ENTEROCOCCUS INDICES OF WATER FROM WATER CRESS PONDS IN HAWAII

MPN per 100 ml

				Ratio of Enterococcus to	
Water cress Pond	Coliform	*E. Coli*	Enterococcus	Coliform index	*E. coli* index
A	7,000	240	2,400	0.3	10.0
B	2,400	6	240	0.1	40.0
C	2,400	23	240	0.1	10.4
D	2,400	23	2,400	1.0	104.0
E	7,000	23	240	<0.1	10.4
F	2,400	23	700	0.3	30.5
G	2,400	700	7,000	2.9	10.0
H	700	6	700	1.0	116.7

Source: M. Levine, H. Minette, and R.H. Tanimoto, *Proc. 1st Int. Conf. Waste Disp. Marine Environ.*, July 22–25, 1959, Pergamon Press, New York, 1960.

6.2–6 MICROBIOLOGIC FINDINGS AT THREE WADING POOLS

Wading pool	Concentration of chlorine residual, ppm	Coliform organisms, No./100 ml	*Esch. coli*, No./100 ml	*Streptococcus faecalis*, No./100 ml
A	<0.1	4,666	4,666	70
	>0.1	36	3	0
B	<0.1	608	472	81
	>0.1	3	2	6
C	<0.1	46,000	22	1
	>0.1	0	0	0

Source: Reprinted, with publisher's permission, from *J. Am. Water Works Assn., 56*(5):585, © 1964 by the American Water Works Association, Inc., 2 Park Ave., New York, N.Y. 10016.

6.2–7 CORRELATION OF BACTERIOLOGIC QUALITY AND DETERGENT CONTENT

Classification of waters	Total No.	Percentage of total	Number unsafe	Percentage unsafe in category
All waters tested	2,167	100	277	12.8
Free of detergents	1,471	67.9	136	9.2
Detergents present in trace[a] amounts to 10 mg/l	696	32.1	141	20.3
Trace[a] amounts of detergents only	316	14.5	54	17.0
1.0 mg/l detergents	227	10.5	39	17.2
2.0 mg/l detergents	81	3.8	17	20.9
3.0–10 mg/l detergents	72	3.3	31	43.1

[a] Less than 1.0 mg/l.

Source: Reprinted, with publisher's permission, from *J. Am. Water Works Assn., 53*(3):303, © 1961 by the American Water Works Association, Inc., 2 Park Ave., New York, N.Y. 10016.

6.2–8 EVALUATION OF RESULTS WITH MFC MEDIUM

Ohio Waters

Sample	Number of coliform organisms: Total, MPN,[a] Confirmed	Total, MPN,[a] Completed	Total, MF[a]	Fecal, MPN[a]	MFC[a] Count	MFC[a] Verified	Coliform-organism ratios: Total ME / Total MPN	Fecal MF / Fecal MPN
Wells								
Baas	33	17	4.3	2.0	1.3	1.1	0.25	0.65
Kilb	130	33	37	<2.0	2.8	2.8	1.12	Indt.[b]
Gilman	278	221	215	<2.0	<0.8	<0.8	0.97	Indt.
Lakes								
Devou	109	70	36	11	8	8	0.51	0.73
Sharon	221	94	80	<2.0	1.5	1.5	0.85	Indt.
Burnet	1,700	1,100	380	220	300	270	0.35	1.36
Winton	4,600	3,300	3,300	2,300	2,500	2,500	1.00	1.09
Smith	172,000	49,000	35,000	13,000	5,500	5,500	0.71	0.42
Creeks								
4-Mile	490	141	80	49	22	22	0.57	0.45
8-Mile	350	350	160	23	10	10	0.46	0.44
Taylor	490	490	700	490	550	550	1.43	1.12
Huntley	2,300	1,300	450	14	17	17	0.35	1.21
Upper Cluff	7,000	2,210	900	490	370	370	0.41	0.76
Berkshire	22,000	4,900	10,000	790	1,000	1,000	2.04	1.27
9-Mile	10,900	7,900	15,000	330	550	550	1.90	1.64
Lower Cluff	7,900	7,900	5,300	2,300	1,700	1,500	0.67	0.74
Duck	78,000	49,000	86,000	10,900	6,000	5,400	1.76	0.55
Mill	4,900,000	2,210,000	4,800,000	221,000	440,000	400,000	2.17	1.99
Lagoon								
Ludlow	49,000	14,100	12,000	1,720	3,600	2,900	0.85	2.09
Rivers								
Little Miami-								
Kellogg	1,720	1,090	2,100	330	150	150	1.93	0.46
Beechmont	10,900	2,210	3,300	330	290	260	1.49	0.88
Newtown	10,900	10,900	22,000	3,300	2,800	2,800	2.02	0.85
Great Miami	840	310	470	23	58	46	1.52	2.52
Licking	10,900	10,900	16,000	3,300	3,700	3,700	1.47	1.12
Ohio (mi no.)								
453.9	3,300	3,300	1,300	172	200	180	0.39	1.16
461.6	1,720	700	2,300	172	160	160	3.29	0.93
463.0	460	310	260	130	56	50	0.84	0.43
464.1	700	490	1,300	172	140	140	2.65	0.81
465.0	17,200	17,200	29,000	3,300	1,400	1,100	1.69	0.42
469.8	33,000	17,200	44,000	3,300	6,800	6,800	2.56	2.06
472.4	46,000	14,100	45,000	4,600	4,200	4,200	3.19	0.91
475.1	1,300,000	1,300,000	3,100,000	790,000	1,300,000	1,300,000	2.30	1.65
482.0	49,000	49,000	52,000	14,100	10,200	10,200	1.06	0.72
485.8	23,000	13,000	64,000	4,900	4,800	4,300	4.92	0.98
513.3	230	230	440	79	58	58	1.91	0.73
517.6	130	27	56	17	12	12	2.07	0.71
522.1	130	11	22	5	<2	<2	2.00	Indt.
Domestic sewage								
Linwood	13,000,000	7,900,000	19,000,000	4,900,000	5,400,000	5,400,000	2.41	1.10
Beechmont	33,000,000	17,200,000	31,000,000	4,600,000	6,400,000	5,800,000	1.80	1.39
Berkshire	54,200,000	34,800,000	35,000,000	4,900,000	5,000,000	4,500,000	1.01	1.02

[a] MPN – most probable number by multi-tube procedure in liquid medium; MF – membrane filter total coliform-organism procedure; MFC – medium for fecal coliform organisms.
[b] Indeterminate.

Source: Reprinted, with publisher's permission, from *J. Am. Water Works Assn.*, *57*(2):208, © 1965, by the American Water Works Association, Inc., 2 Park Ave., New York, N.Y. 10016.

6.2–9 SULFATE-REDUCING BACTERIA FROM NATURAL SOURCES

Ohio Waters

Source of Sample	MPN of Sulfate-Reducing Bacteria per 100 ml of Sample
Tap water	0.045[a]
Upper Mill Creek	490
Lower Mill Creek	70,000
Little Miami River	11,000
Greater Miami River	33,000
Licking River	33,000
Ohio River	13,000
Sewage effluent	7,900,000
Well No. 1	<2
Well No. 2	220
Well No. 3	11,000

[a] MPN run by filtering 10 liters, 1 liter, and 100 ml through membrane filters.

Source: Reprinted, with publisher's permission, from *J. Am. Water Works Assn., 57*(8):1011, © 1965 by the American Water Works Association, Inc., 2 Park Ave., New York, N.Y. 10016.

6.2–10 OCCURRENCE AND DISTRIBUTION OF NONAEROGENIC COLIFORM STRAINS

44.5°C in Warm-Blooded Animals and Various Environmental Sources

Source	Coliform strains examined, no.	Nonaerogenic strains, total no.	Occurrence of non-aerogenic strains, %
Feces			
Human	6,049	150	2.5
Livestock	2,339	10	0.4
Poultry	1,896	121	6.4
Cat and dog	1,716	4	0.2
Rodent	718	51	7.1
Environmental			
Vegetation	1,477	214	14.5
Insects	1,086	249	22.9
Soil	2,953	601	20.4
Water			
Stormwater	856	330	38.6
Water supply intakes[a]	6,742	803	11.9

[a] Data from fourteen surface water supplies widely distributed throughout the U.S.

Source: Reprinted, with publisher's permission, from *J. Am. Water Works Assn., 62*(2):113, © 1970, by the American Water Works Association, Inc., 2 Park Ave., New York, N.Y. 10016.

6.2–11 FECAL COLIFORM CORRELATIONS WITH *SALMONELLA* OCCURRENCE

Source	Fecal coliform density per 100 ml	Total examinations	*Salmonella* Detection: Positive occurrence Number	Percentage
Fresh	1–200	29	8	27.6
	201–2,000	27	19	85.2
	over 2,000	54	53	98.1
Estuarine water	1–70	184	12	6.5
	71–200	74	21	28.4
	201–2,000	91	40	44.0
	over 2,000	75	45	60.0

Source: Reprinted, with publisher's permission, from *J. Am. Water Works Assn., 62*(2):113, © 1970 by the American Water Works Association, Inc., 2 Park Ave., New York, N.Y. 10016.

6.2–12 FECAL COLIFORM CORRELATIONS WITH *SALMONELLA* OCCURRENCE IN BOTTOM MUDS

Overlying water fecal coliform density per 100 ml	Bottom water sediment interface: Total examinations	Fecal coliform density per 100 cc	*Salmonella* positive occurrence: Number	Percentage
1–200	17	2,500–430,000	4	23.5
Above 200	22	10,000–5,800,000	15	68.2

Source: Reprinted, with publisher's permission, from *J. Am. Water Works Assn., 62*(2):113, © 1970 by the American Water Works Association, Inc., 2 Park Ave., New York, N.Y. 10016.

6.2–13 TYPICAL NUCLEIC ACID CONTENT OF SEVERAL BACTERIAL SPECIES

Percent Dry Weight

Species	Total nucleic acid	RNA	DNA[a]	Ref.
Staphylococcus	11.57	8.75	2.82	1
Eberthella typhosa	12.84	9.12	3.72	1
Escherichia coli	13.90	9.73	4.17	1
(various strains)	13.12	8.72	4.40	1
	14.67	10.43	4.21	1
	12.37	8.59	3.78	1
	13.98	10.09	3.89	1
	15.76	11.47	4.29	1
			5.10	1
	12.39	8.54	3.85	2
			4.20	3
Bacillus aertrycke	8.40	5.40	3.00	1
	10.64	7.00	3.64	1
Proteus vulgaris			3.70	3
Sphaerotilus			1.75[b]	4
Zooglea ramigera			5.41[b]	5

[a] $\frac{\text{mg DNA}}{\text{mg volume dry weight}} \times 100.$

[b] Average of five determinations.

REFERENCES

1. R. Vendrely, "Un Symposium sur les Proteins," Masson et Cie, Paris, 1947.
2. J.M. Webb, *J. Biol. Chem.*, *221*:635, 1956.
3. J.M. Webb and M.B. Levy, *J. Biol. Chem.*, *213*:107, 1955.
4. H. Okrend, "A Nutritional Study of *Spaerotilus natans*," Ph.D. Thesis, Rutgers University, New Brunswick, N.J., 1963.
5. E.J. Genetelli, *J. Water Pollut. Control Fed.*, *39*(10)(Part 2):R32, 1967.

6.2–14 MICROORGANISMS IN SYNTHETIC SEWAGE ADJUSTED TO VARIOUS pH LEVELS

Incubated at 20° for 24 Hours[a]

Final pH of synthetic sewage	Organisms/ml x 10^7 [b]
7.8	21.2
7.4	16.6
5.6	18.5
4.6	6.5
3.5	0.01

[a] Plates were counted after 48 hr incubation at 20°C.
[b] Based on an average of five plates.

Source: T.O. Rogers and H.A. Wilso, *J. Water Pollut. Control Fed., 38*(6):990, 1966.

6.2–15 ORGANISMS, BASED ON COLONY COUNT, IN A MONONGAHELA RIVER WATER SAMPLE

pH 6.50 Plated at Various pH Levels[a]

pH of plating medium	Organisms/ml x 10^3
7.0	920
5.0	108
4.0	10.8
3.5	0.7
3.0	0.5

[a] Incubation 48 hr at 25°C.

Source: T.O. Rogers and H.A. Wilso, *J. Water Pollut. Control Fed., 38*(6):990, 1966.

6.2–16 LONGEVITY OF MICROORGANISMS IN RAW SEWAGE SUSPENDED IN BUFFER SOLUTIONS OF VARIOUS pH

Hours exposure to buffer	Organisms/ml x 10^3 [a]			
	pH 7.0	pH 5.0	pH 4.0	pH 3.5
0	72	54	45	47
4	54	29	11.4	5.4
8	47	25	10.4	5
28	46	25	7.5	1.9
48	20	15	6.6	0.068

[a] The average count of five plates.

Source: T.O. Rogers and H.A. Wilso, *J. Water Pollut. Control Fed., 38*(6):990, 1966.

6.3 VIRUSES

6.3–1 POLIOVIRUS REMOVAL BY STRAIGHT LIME SOFTENING[a]

Initial conditions mg/l as $CaCO_3$				Final conditions				
				Hardness removed				
$Ca(HCO_3)_2$ Hardness	Alkalinity	CO_2	$Ca(OH)_2$ added	mg/l as $CaCO_3$	Percent	$CaCO_3$ precipitated, mg/l	pH	Poliovirus removal, %
0	–	–	16.4	–	–	–	8.9	0[b]
100	106	9.5	117	63.8	63.8	181	9.0	8.9
200	206	10.5	210	140	70.5	359	8.3	43
300	306	11.5	319	218	72.7	537	8.1	70
400	406	12.6	424	301	75.3	725	8.0	81

[a] Average values for two runs.
[b] By definition.

Source: Reprinted, with publisher's permission, from *J. Am. Water Works Assn., 60*(8):939, © 1968 by the American Water Works Association, Inc., 2 Park Ave., New York, N.Y. 10016.

6.3–2 POLIOVIRUS REMOVAL BY PREFORMED CALCIUM CARBONATE

Initial pH	$CaCO_3$ added, mg/l	Ca $(OH)_2$ added, mg/l	Final pH	Poliovirus survival, %
8.5	–	6.7	8.6	100[a]
8.5	167	6.7	8.9	123
8.5	367	6.5	8.9	80.5
8.5	567	6.5	8.9	92.6

[a] By definition.

Source: Reprinted, with publisher's permission, from *J. Am. Water Works Assn., 60*(8):939, © 1968 by the American Water Works Association, Inc., 2 Park Ave., New York, N.Y. 10016.

6.3–3 POLIOVIRUS REMOVAL BY STRAIGHT LIME SOFTENING IN THE PRESENCE OF MAGNESIUM CHLORIDE

Initial conditions, mg/l as $CaCO_3$				Final conditions				
Hardness				Hardness removed mg/l as $CaCO_3$				
$MgCl_2$	$Ca(HCO_3)_2$	Alkalinity	CO_2	Mg	Ca	$CaCO_3$ precipitated mg/l	pH	Poliovirus removal, %
0	300	305	4	0	241	547	9.1	0[a]
10	300	305	4	3.9	232	535	9.0	16
10	300	305	4	2.8	242	555	9.2	24
50	300	305	4	6.0	215	528	8.9	44
50	300	305	4	6.0	231	544	9.2	44

[a] By definition.

Source: Reprinted, with publisher's permission, from *J. Am. Water Works Assn., 60*(8):939, © 1968 by the American Water Works Association, Inc., 2 Park Ave., New York, N.Y. 10016.

6.3–4 POLIOVIRUS REMOVAL BY PRECIPITATION OF MAGNESIUM HYDROXIDE[a]

Initial hardness $MgCl_2$, mg/l as $CaCO_3$	NaOH added, mg/l	$MgCl_2$ hardness removed		Final pH	Poliovirus survival, %
		mg/l as $CaCO_3$	%		
0	5	–	–	7.3	100[b]
100	80	56	56	10.3	0.74
200	160	169	84	10.4	0.39
300	240	277	92.2	10.4	0.12
400	320	381	95.3	10.5	<0.05
500	400	486	97.3	10.6	0.08

[a] Average for two runs.
[b] By definition.

Source: Reprinted, with publisher's permission, from *J. Am. Water Works Assn., 60*(8):939, © 1968 by the American Water Works Association, Inc., 2 Park Ave., New York, N.Y. 10016.

6.3–5 POLIOVIRUS REMOVAL BY PREFORMED MAGNESIUM HYDROXIDE

	Magnesium hydroxide, mg/l				
Initial pH	**Initial dissolved**	**Preformed added**	**Residual dissolved**	**Final pH**	**Poliovirus removal, %**
9.8	20.4	–	20.3	9.8	0[a]
9.9	20.4	133	26.9	10.1	24
9.9	20.4	133	26.9	10.1	20
9.8	20.4	400	32.4	10.3	44
9.8	20.4	400	31.3	10.3	40

[a] By definition.

Source: Reprinted, with publisher's permission, from *J. Am. Water Works Assn.*, *60*(8):939, © 1968 by the American Water Works Association, Inc., 2 Park Ave., New York, N.Y. 10016.

6.3–6 POLIOVIRUS REMOVAL BY EXCESS LIME-SODA ASH SOFTENING

Initial conditions, mg/l as $CaCO_3$						Final conditions				
Hardness						Hardness removed, mg/l as $CaCO_3$				
Total	**$CaCl_2$**	**$MgCl_2$**	**Alkalinity, $NaHCO_3$**	**$Ca(OH)_2$ added**	**Na_2CO_3 added**	**Ca**	**Mg**	**$CaCO_3$ precipitated, mg/l**	**Final pH**	**Poliovirus survival, %**
0	–	–	–	12.7	–	–	–	–	7.3	100[a]
100	50	50	50	167	50	15	24	65	10.7	0.17
200	133	67	100	234	100	76	50	209	10.8	0.10
200	67	133	100	301	100	10	106	77	10.7	0.051
300	200	100	150	323	150	123	80	323	10.8	0.017
300	200	100	150	323	150	137	71	337	11.1	<0.008
400	100	300	200	573	200	13	257	113	10.9	<0.007
400	100	300	200	573	200	16	265	116	11.1	<0.007
400	200	200	200	473	200	105	173	305	11.2	<0.007
400	300	100	200	373	200	204	82	504	11.2	<0.007
500	400	100	250	423	250	279	85	679	11.2	<0.007

[a] By definition.

Source: Reprinted, with publisher's permission, from *J. Am. Water Works Assn.*, *60*(8):939, © 1968 by the American Water Works Association, Inc., 2 Park Ave., New York, N.Y. 10016.

6.3–7 INACTIVATION OF POLIOVIRUS TYPE I BY WATER SOFTENING PRECIPITATION PROCESSES

Initial conditions, mg/l as $CaCO_3$			Final conditions			
Hardness			Hardness removal as $CaCO_3$			
$MgCl_2$	$Ca(HCO_3)_2$	Alkalinity	Mg	Ca	pH	Poliovirus removal, %
Straight lime softening						
	100	106		64	9.0	9
	300	306		218	8.1	70
Precipitation of Mg only with NaOH						
100	0	Minor	56		10.3	99.26
300	0	Minor	277		10.4	99.88
Excess lime-soda ash softening						
67	133	100	50	76	10.8	99.90
100	300	200	82	204	11.2	99.993

Source: Adapted from D.F. Wentworth *et al., J. Am. Water Works Assn., 60*(8):939, 1968. Reprinted, with publisher's permission, from *J. Am. Water Works Assn., 64*(1):31, © 1972, by the American Water Works Association, Inc., 2 Park Ave., New York, N.Y. 10016.

6.3–8 INACTIVATION OF VIRUS BY pH

Virus used	Contact time, min	System	pH	Inactivation, %
T2 Bacteriophage[2]	90	$Ca(OH)_2$ and distilled water	10.5	90
Poliovirus Type 1 (Sabin)[1]	90	$Ca(OH)_2$ and distilled water	11.2	No significant inactivation
Poliovirus Type I (L Sc)[4]	90	$Ca(OH)_2$ and a precipitated and	11.2	98.5
		filtered secondary effluent	10.8	50
Poliovirus Type I (Sabin)[3]	90	$Ca(OH)_2$ + distilled water	11.9	90
		NaOH + distilled water	11.9	90
		KOH & distilled water	12.2	90

Source: Reprinted, with publisher's permission, from *J. Am. Water Works Assn., 64*(1):31, © 1972 by the American Water Works Association, Inc., 2 Park Ave., New York, N.Y. 10016.

REFERENCES

1. D.F. Wentworth *et al., J. Am. Water Works Assn., 60*(8):939, 1968.
2. S.E. Thayer and O.J. Sproul, *J. Am. Water Works Assn., 58*(8):1063, 1966.
3. O.J. Sproul *et al., Salt and Virus Inactivation by Chlorine and High pH,* Natl. Spec. Conf. Disinfect., Am. Soc. Civil Eng., Amherst, Mass., 1970.
4. G. Berg *et al., J. Am. Water Works Assn., 60*(2):193, 1968.

6.3–9 VIRUS REMOVAL BY COAGULATION WITH ALUMINUM SULFATE AND CATIONIC POLYELECTROLYTES[a]

Coagulant				Removals	
$Al_2(SO_4)_3$, mg/l	Polyelectrolyte, mg/l	Clay, mg/l	Virus	Virus, %	Turbidity, %
10	0	50	Polio Type I[1]	86	96
10	1 (Purifloc C32)	50	Polio Type I[1]	81	95
25.7	0	120	Phage T4[2]	98	99
			Phage MS2[2]	99.8	98
25.7	1 (Primafloc C7)	120	Phage T4[2]	99.9	98
			Phage MS2[2]	99.5	98

[a] Poliovirus data adapted from reference 1. T4 and MS2 data adapted from reference 2.

Source: Reprinted, with publisher's permission, from *J. Am. Water Works Assn., 64*(1):31, © 1972 by the American Water Works Association, Inc., 2 Park Ave., New York, N.Y. 10016.

REFERENCES

1. R.T. Thorup *et al., J. Am. Water Works Assn., 62*(2):97, 1970.
2. M. Chaudhuri and R.S. Engelbrecht, *J. Am. Water Works Assn., 62*(9):563, 1970.

6.3–10 EFFECT OF IONS ON REMOVAL OF T2 VIRUS BY PRIMARY POLYELECTROLYTE COAGULATION[a]

Ion species and dosage			Virus removal			Degree of floc formation		
	mg/l	M	Anionic, %	Cationic, %	Nonionic, %	Anionic	Cationic	Nonionic
Na^+	1.6	0.00007	0	32	0	None	Poor	None
	5	0.001	0	68	0	None	Poor	None
	25	0.002	0	86	0	None	Good	None
	126	0.005	28	93	36	Poor	Good	Poor
K^+	2.7	0.00007	0	28	0	None	Poor	None
	5	0.0001	0	71	0	Poor	Fair	None
	25	0.0006	5	84	0	Poor	Good	None
	126	0.003	32	94	27	Poor	Good	Poor
Ca^{++}	1.4	0.00004	0	61	0	None	Fair	None
	5	0.0001	0	85	0	None	Fair	Poor
	25	0.0006	40	93	35	Fair	Good	Poor
	126	0.003	41	95	37	Fair	Good	Poor
Mg^{++}	5	0.0002	5	86	3	Poor	Good	Poor
	25	0.001	38	94	37	Poor	Good	Poor
	126	0.005	41	96	37	Fair	Good	Fair

[a] pH – 7.0 ± 0.2. NaOH, KOH, $Ca(OH)_2$ and $Mg(HCO_3)_2$ respectively, used to adjust; coagulant – as indicated, no aluminum sulfate, 1 mg/l polyelectrolyte; turbidity – 5 mg/l each of bentonite clay, and infusorial earth.

Source: Reprinted, with publisher's permission, from *J. Am. Water Works Assn., 62*(2):97, © 1970 by the American Water Works Association, Inc., 2 Park Ave., New York, N.Y. 10016.

6.3–11 EFFECT OF COAGULANT AIDS ON T2 VIRUS REMOVAL AT LOW ALUMINUM SULFATE CONCENTRATION[a]

Coagulants	Virus removal in presence of 20 mg/l of the indicated cation				Degree of floc formation in presence of 20 mg/l of the indicated cation			
	Na^+, %	K^+, %	Ca^{++}, %	Mg^{++}, %	Na^+	K^+	Ca^{++}	Mg^{++}
Anionic acid +$Al_2(SO_4)_3$ (5 mg/l)	39	41	49	52	Poor	Poor	Poor	Poor
Cationic acid +$Al_2(SO_4)_3$ (5 mg/l)	84	80	94	93	Fair	Fair	Fair	Good
Nonionic acid +$Al_2(SO_4)_3$ (5 mg/l)	63	60	70	73	Fair	Fair	Good	Fair
$Al_2(SO_4)_3$ alone (5 mg/l)	58	56	57	57	Poor	Poor	Poor	Poor

[a] pH – 7.0 ± 0.2; ions contributed by buffer – 12.6 mg/l Na^+; polyelectrolyte dosage – 1 mg/l; turbidity – 5 mg/l each of bentonite clay and infusorial earth.

Source: Reprinted, with publisher's permission, from *J. Am. Water Works Assn., 62*(2):97, © 1970 by the American Water Works Association, Inc., 2 Park Ave., New York, N.Y. 10016.

6.3–12 POLIOVIRUS AND TURBIDITY REMOVAL BY ALUMINUM SULFATE COAGULATION AIDED BY CATIONIC POLYELECTROLYTES[a]

Coagulants[b]	Virus and turbidity removal, % $CaCl_2$ concentration, M			
	0.1 M	0.01 M	0.001 M	0 M
$Al_2(SO_4)_3$	–	90[c] (97)[d]	–	–
$Al_2(SO_4)_3$ + Drewfloc 21	85 (96)	91 (96)	79 (93)	0 (0)
$Al_2(SO_4)_3$ + Reten 210	83 (92)	91 (93)	94 (91)	0 (0)
$Al_2(SO_4)_3$ + Purifloc C32	69 (95)	81 (95)	75 (94)	24 (92)

[a] pH – 6.8:
[b] Dosages – $Al_2(SO_4)_3$ – 10 mg/l (0.029m*M*); Clay – 50 mg/l; Polyelectrolyte – 1 mg/l.
[c] Virus removal.
[d] Turbidity removal.

Source: Reprinted, with publisher's permission, from *J. Am. Water Works Assn., 62*(2):97, © 1970 by the American Water Works Association, Inc., 2 Park Ave., New York, N.Y. 10016.

6.3–13 REMOVAL OF T2 VIRUS DURING AIDED COAGULATION[a]

Coagulants	Polyelectrolyte dosage, mg/l	Virus removal, %
Cationic Aid	0.5	97
+	1.0	98
40 mg/l $Al_2(SO_4)_3$	1.5	98
	2.0	98
Nonionic Aid	0.5	97
+	1.0	96
40 mg/l $Al_2(SO_4)_3$	1.5	96
	2.0	88
Anionic Acid	0.5	97
+	1.0	96
40 mg/l $Al_2(SO_4)_3$	1.5	96
	2.0	96
$Al_2(SO_4)_3$ Alone	–	98
+		
40 mg/l $Al_2(SO_4)_3$		

[a] pH – 7.0 ± 0.2; ions – 126 mg/l Na^+; turbidity – 5 mg/l each of bentonite clay and infusorial earth.

Source: Reprinted, with publisher's permission, from *J. Am. Water Works Assn., 62*(2):97, © 1970 by the American Water Works Association, Inc., 2 Park Ave., New York, N.Y. 10016.

6.3–14 REMOVAL OF BACTERIOPHAGE T4 IN THE PRESENCE OF CALCIUM AND MAGNESIUM

Calcium, mg/l	Magnesium, mg/l	Percent removal
0	0	97.9
25	25	98.0
50	50	97.9

Note: Average input virus concentration: 4.05 x 10^5/ml; pH range: 5.1–5.4 Aluminum sulfate: 50 mg/l; Average turbidity: 12.5 JTU.

Source: Reprinted, with publisher's permission, from *J. Am. Water Works Assn., 62*(9):563, © 1970 by the American Water Works Association, Inc., 2 Park Ave., New York, N.Y. 10016.

6.3–15 REMOVAL OF POLIOVIRUS AND TURBIDITY BY UNAIDED COAGULATION, AND CATIONIC POLYELECTROLYTE AIDED COAGULATION WITH ALUMINUM AND FERRIC SULFATES

Waters with Varying Ionic Concentrations[a]

	CaCl$_2$ Concentration, M									
	0.1		0.01		0.001		0.0001		0	
Substance	**Virus removal, %**	**Turbidity removal, %**	**Virus removal, %**	**Turbidity removal, %**	**Virus removal, %**	**Turbidity removal, %**	**Virus removal, %**	**Turbidity removal, %**	**Virus removal, %**	**Turbidity removal, %**
System – Aluminum Sulfate										
Polyelectrolyte only	6		36		25		10		0	
Clay only	79	91	82	91	13	26	0	2	0	0
$Al_2(SO_4)_3$ only	12		9		27		33		36	
$Al_2(SO_4)_3$ + clay	79	94	86	96	83	93	26	18	11	0
$Al_2(SO_4)_3$ + clay polyelectrolyte	69	95	81	95	75	94	19	93	24	92
System – Ferric Sulfate										
Polyelectrolyte only	6		36		25		10		0	
Clay only	79	91	82	91	13	26	0	2	0	0
$Fe(SO_4)_3$ only	20		20		20		0		0	
$Fe(SO_4)_3$ + clay	85	93	83	93	4	0	0	0	0	0
$Fe(SO_4)_3$ + clay + polyelectrolyte	76	95	75	93	60	83	62	0	35	0

[a] pH – 6.8, adjusted with 1.25–4.0 ml of 0.1 N NaOH: Dosages: Polyelectrolyte (Purifloc C32), 1 mg/l; clay, 50 mg/l; $Al_2(SO_4)_3$, 10 mg/l (0.29 m*M*); $Fe_2(SO_4)_3$, 11.6 mg/l (0.029 m*M*).

Source: Reprinted, with publisher's permission, from *J. Am. Water Works Assn., 62*(2):97, © 1970 by the American Water Works Association, Inc., 2 Park Ave., New York, N.Y. 10016.

6.3–16 REMOVAL OF BACTERIOPHAGE T4 BY COAGULATION AND FLOCCULATION IN THE PRESENCE OF ORGANIC MATTER

		Percent removal	
Organic matter	Concentration	Bacteriophage T4	Turbidity
Egg albumin	0 mg/l	98.0	99.0
	10 mg/l	96.9	98.2
	20 mg/l	96.4	98.2
	30 mg/l	95.3	98.1
	50 mg/l	94.8	95.8
Bovine serum albumin	0 mg/l	98.6	99.1
	10 mg/l	97.7	98.8
	20 mg/l	97.6	98.6
	30 mg/l	96.9	98.4
	50 mg/l	96.0	95.4
Settled wastewater[b]	0 ml/l	98.0	99.0
	200 ml/l	95.7	98.6
Wastewater effluent[b]	0 ml/l	98.0	99.0
	200 ml/l	97.0	98.8

Note: Average input virus concentration: 3.58×10^5/ml; pH range: 5.1–5.4 Aluminum sulfate: 50 mg/l; Average turbidity; 12.5 JTU.[a]

[a] Initial turbidity was 18 JTU when settled wastewater was added.
[b] Characteristics:

	Raw wastewater	Wastewater effluent
5 day BOD, mg/l	181	50
Total suspended solids, mg/l	253	10
Volatile suspended solids, mg/l	178	9

Source: Reprinted, with publisher's permission, from *J. Am. Water Works Assn., 62*(9):563, © 1970 by the American Water Works Association, Inc., 2 Park Ave., New York, N.Y. 10016.

6.3–17 REMOVAL OF BACTERIOPHAGE MS2 BY COAGULATION AND FLOCCULATION IN THE PRESENCE OF WASTEWATER AND WASTEWATER EFFLUENT

	Percent removal	
	Bacteriophage MS2	Turbidity
No settled wastewater or wastewater effluent	99.8	98.1
200 ml/l settled wastewater[b]	89.6	92.1
200 ml/l wastewater effluent[b]	94.0	93.6

Note: Average input virus concentration: 3.87×10^5/ml; pH range: 5.9–6.1 Aluminum sulfate: 50 mg/l; Average turbidity; 12.5 JTU.[a]

[a] Initial turbidity was 19.0 JTU when settled wastewater was added.
[b] Characteristics:

	Raw wastewater	Wastewater effluent
5 day BOD, mg/l	181	50
Total suspended solids, mg/l	253	10
Volatile suspended solids, mg/l	178	9

Source: Reprinted, with publisher's permission, from *J. Am. Water Works Assn., 62*(9):563, © 1970 by the American Water Works Association, Inc., 2 Park Ave., New York, N.Y. 10016.

6.3–18 INACTIVATION OF BACTERIOPHAGES T4 AND MS2 IN THE PRESENCE OF POLYELECTROLYTES (COAGULANT AIDS)

Average input virus concentration: 3.44×10^5/ml

Polyelectrolyte	Concentration, mg/l	pH	Bacteriophage	Percent[b] inactivation
Primafloc C-7 (cation)	0.5–1.5	5.2	T4	81[a]
	0.5	6.0	MS2	85
Catfloc (anion)	0.5–10.0	5.2	T4	77[a]
	1.0	6.0	MS2	82
Primafloc A-10 (cation)	1.0–5.0	5.2	T4	None
	1.0	6.0	MS2	None
Coagulant aid #243	1.0–5.0	5.2	T4	None
(anion)	1.0	6.0	MS2	None

[a] Average value for the concentration range of polyelectrolyte.
[b] Compared against deionized water control.

Source: Reprinted, with publisher's permission, from *J. Am. Water Works Assn.*, *62*(9):563, © 1970 by the American Water Works Association, Inc., 2 Park Ave., New York, N.Y. 10016.

6.3–19 REMOVAL OF BACTERIOPHAGES T4 AND MS2 BY COAGULATION AND FLOCCULATION

Aluminum Sulfate and Cationic Polyelectrolytes as Coagulant Aids

		Experiment 1 Percent removal		Experiment 2 Percent removal	
Cationic polyelectrolyte	Concentration, mg/l	Bacteriophage T4	Turbidity	Bacteriophage MS2	Turbidity
Primafloc C-7	0.00	97.7	99.1	99.8	98.3
	0.50	99.8	98.9	–	–
	1.00	99.9	97.5	99.5	98.3
	1.50	99.9	95.5	–	–
	2.00	99.9	94.3	99.2	96.6
	3.00	–	–	99.4	93.8
Catfloc	0.00	97.8	99.3	99.7	98.6
	0.25	98.4	99.3	–	–
	0.50	99.5	99.4	–	–
	1.00	99.9	98.9	99.7	98.4
	1.50	99.9	98.5	–	–
	2.00	–	–	99.6	97.7
	3.00	–	–	98.9	96.2

Note: Average input virus concentration: 4.49×10^5/ml (Experiment 1); 2.74×10^5/ml (Experiment 2)
Aluminum sulfate: 50 mg/l; Average turbidity: 12.5 JTU; pH range: 5.1–5.4 (Experiment 1); 5.9–6.0 (Experiment 2).

Source: Reprinted, with publisher's permission, from *J. Am. Water Works Assn.*, *62*(9):563, © 1970 by the American Water Works Association, Inc., 2 Park Ave., New York, N.Y. 10016.

6.3–20 REMOVAL OF VIRUSES FROM WATER BY MAGNETIC IRON OXIDE[a]

Virus type	Virus input, pfu/ml	Filtrate, pfu/ml
Coxsackie B-2	55	0
Echo 11	107	0
Polio 3	75	0
Adeno 2	122	0
SV-40	145	0
Influenza A	140	0
Herpes simplex	170	0
Reo 1	80	0
Pool of above viruses	100 (each virus)	0

[a] M.O. 2530 magnetic iron oxide used.

Source: Reprinted, with publisher's permission, from *J. Am. Water Works Assn., 60*(11):1288, © 1968 by the American Water Works Association, Inc., 2 Park Ave., New York, N.Y. 10016.

6.3–21 REMOVAL OF COXSACKIE VIRUS A-9 FROM SEWAGE EFFLUENT, RIVER WATER, AND TAP WATER BY MAGNETIC IRON OXIDE

Suspending menstruum	Virus input, pfu/ml	Virus in filtrate, pfu/ml	Percent virus recov-ered in eluate
Sewage effluent–pH 6	105	1.5	93.5
Ohio River water–pH	137	2.0	70.6
Tap water–pH 7.9	100	0	86

Source: Reprinted, with publisher's permission, from *J. Am. Water Works Assn., 60*(11):1288, © 1968 by the American Water Works Association, Inc., 2 Park Ave., New York, N.Y. 10016.

6.3–22 SUMMARY OF POLIO VIRUS REMOVAL

Process	Removal %
Die-off (under experimental conditions, 20°C)	≈5–10 in 24–48 hr
Activated sludge[1]	90–95
Polluted stream[2,3]	99.9 in 18 days
Filtration or seepage	
Through unsaturated sand at 20–40 gpd/ft²	>99.99
Ground water movement	
Upflow 9 gpd/ft² (3 ft/day)	>99.99
Downflow 12 gpd/ft² (4 ft/day)	>98
Slow sand filter rates–50 gpd/ft²	22–96
Rapid filter rates–2,880–8,640 gpd/ft² (2–6 gpm/ft²)	
Without alum	1–50
With alum	
Without settling	90–99
With settling	>99.7
Flocculation and settling[4] (varies with dose and time)	95–99
Chlorination[5] (varies with dose and contact)	99.99

Source: Reprinted, with publisher's permission, from *J. Am. Water Works Assn., 54*(10):1275, © 1962 by the American Water Works Association, Inc., 2 Park Ave., New York, N.Y. 10016.

REFERENCES

1. N.A. Clarke *et al., Am. J. Publ. Health, 51*:1118, 1961.
2. N.A. Clarke *et al., J. Am. Water Works Assn., 48*(6):677, 1956.
3. A.J. Rhodes *et al., Can. J. Publ. Health., 41*:46, 1950.
4. M.S. Nichol and E. Koepp, *J. Am. Water Works Assn., 53*(3):303, 1961.
5. S.J. Weidenkopf, *Virology, 5*:56, 1958.

6.3–23 REMOVAL OF POLIOVIRUS 1 FROM SECONDARY EFFLUENT BY FLOCCULATION WITH $Ca(OH)_2$ AND RAPID SAND FILTRATION

Test no.	$Ca(OH)_2$ concentration, mg/l	Initial virus concentration, pfu/l	pH of treated effluent	Surviving virus concentration, pfu/l[b]	Virus removal, %	pH after sand filtration	Virus concentration after sand[a] filtration pfu/l	Virus removal, after filtration, %	Overall virus removal, %
1	200	33,333	9.30	2,200	92.3	9.26	397	82.0	99
2	300	51,480	10.21	15,912	69.1	10.10	750	95.3	98.5
3	400	55,000	11.30	1,940	96.5	11.20	<4.6	>99.8	>99.99
4	500	33,333	11.03	505	98.5	11.07	12.5	97.5	99.96
5	500	33,333	11.01	47	99.86	10.93	2.8	94.0	99.99

[a] Filtration rate 2.25 gpm/ft^2 through 8 in. sand.
[b] Virus concentration in flocculated effluent just prior to sand filtration.

Source: Reprinted, with publisher's permission, from *J. Am. Water Works Assn.*, *60*(2):193, © 1968 by the American Water Works Association, Inc., 2 Park Ave., New York, N.Y. 10016.

6.3–24 OZONATION OF POLIOVIRUS TYPE I IN DISTILLED WATER

	Time of sample after addition of viruses, min			
	0	2½	5	10
Ozone, mg/l	0.15	0.10	0.00	0.00
Inactivation, %		50	57	46
Ozone, mg/l	0.75	0.23	0.14	0.10
Inactivation, %		95	96	95
Ozone, mg/l	1.27	0.23	0.18	0.14
Inactivation, %		99.99	99.99+	99.99+

Source: Adapted from T.J. Schaffernoth, *High Level Inactivation of Poliovirus in Biologically Treated Waste Water by Ozonation,* Thesis, University of Maine, Orono, 1970. Reprinted from O.J. Sproul, *J. Am. Water Works Assn., 64*(1):31, 1972.

6.3–25 SABIN POLIOVIRUS 1 DISAPPEARANCE

pH 6.8–7.0

Holding medium	Holding temperature, °C	Average time of disappearance, days
Raw pond water	37	98
Autoclaved pond water	37	91
Raw filtered pond water	37	91
Autoclaved filtered pond water	37	77
Double distilled deionized water	37	91
Phosphate buffered saline	37	91
Raw pond water	20–25	35
Autoclaved pond water	20–25	35
Double distilled deionized water	20–25	42
Phosphate buffered saline	20–25	42

Source: Reprinted, with publisher's permission, from *J. Am. Water Works Assn., 59*(4):491, © 1967 by the American Water Works Association, Inc., 2 Park Ave., New York, N.Y. 10016.

6.3–26 SABIN POLIOVIRUS 1 DISAPPEARANCE

Varying pH, Constant Temperature, 20–25°C

Holding medium	Original pH	Final pH	Time of disappearance, days
Double distilled deionized water	8.0	8.8	35
	7.5	7.8	35
	7.0	7.5	35
	6.5	6.9	35
	6.0	6.9	35
	5.0	5.9	35
Hank's balanced salt solution	8.0	8.1	35
	7.5	7.5	35
	7.0	7.1	35
	6.5	6.6	35
	6.0	6.1	35
	5.0	5.1	35

Source: Reprinted, with publisher's permission, from *J. Am. Water Works Assn., 59*(4):491, © 1967 by the American Water Works Association, Inc., 2 Park Ave., New York, N.Y. 10016.

6.3–27 DISAPPEARANCE OF SABIN POLIOVIRUS 1 IN THE PRESENCE OF MIXED BACTERIAL CULTURE

Varying pH, Constant Temperature, 20–25°C

Holding medium	Bacterial count, no./ml	Initial pH	Final pH	Time of disappearance, days
Double distilled deionized water	2,000	8.0	8.4[a]	28
		7.0	6.9	35
		6.0	6.3[a]	28
		5.0	5.4[a]	28
	20,000	8.0	7.8	14
		7.0	7.1[a]	14
		6.0	6.5[a]	7
		5.0	5.2[a]	7
Hank's balanced salt solution	2,000	8.0	8.1[a]	28
		7.0	7.1[a]	28
		6.0	6.1[a]	28
		5.0	5.2[a]	28
	20,000	8.0	7.9	7
		7.0	6.9	14
		6.0	6.0	7
		5.0	5.1[a]	7

[a] The pH fell during the course of this study but rose to indicated levels by the termination of the experiment. This may have been due to loose fitting screw caps.

Source: Reprinted, with publisher's permission, from *J. Am. Water Works Assn.*, *59*(4):491, © 1967 by the American Water Works Association, Inc., 2 Park Ave., New York, N.Y. 10016.

6.3–28 VIRUS DISAPPEARANCE FROM POND WATERS

Pond no.	pH	Temperature, °C	Virus[b]	First day not recovered
1	7.8	20–25	p	70
		4	p	present at 91 days
5	8.0	20–25	p	84
		4	p	present at 91 days
6	8.1	20–25	p	70
		4	p	present at 91 days
8	8.0	20–25	p	63
		4	p	63
A[a]	8.0	20–25	p	present at 91 days
		4	p	present at 91 days
1	7.8	20–25	c	84
		4	c	84
5	8.0	20–25	c	91
		4	c	91
6	8.1	20–25	c	present at 91 days
		4	c	91
8	8.0	20–25	c	56
		4	c	77
A[a]	8.0	20–25	c	present at 91 days
		4	c	present at 91 days
1	7.8	20–25	e	84
		4	e	84
5	8.0	20–25	e	84
		4	e	91
6	8.1	20–25	e	91
		4	e	91
8	8.0	20–25	e	63
		4	e	63
A[a]	8.0	20–25	e	84
		4	e	present at 91 days

[a] Double distilled deionized water.
[b] p, poliovirus; c, coxsackie; e, echo.

Source: Reprinted, with publisher's permission, from *J. Am. Water Works Assn., 59*(4):491, © 1967 by the American Water Works Association, Inc., 2 Park Ave., New York, N.Y. 10016.

6.3–29 DAY OF DISAPPEARANCE OF T_2 BACTERIOPHAGE AND BACTERIA

Raw and Autoclaved Farm Pond Water

	Pond 5				Pond 6			
	Raw		Autoclaved		Raw		Autoclaved	
Organism	20°C	4°C	20°C	4°C	20°C	4°C	20°C	4°C
Esch. coli B	8	10	>14	>14	5	>14	>14	>14
T_2 phage	12	>14	>14	>14	8	>14	>14	>14
Bacteria other than *Esch. coli*	>14	>14			>14	>14		
T_2 without parent strain	3	7	4	5	5	7	5	10
Esch. coli B without phage	12	>14	>14	>14	10	>14	>14	>14
Bacteria other than *Esch. coli* B, without phage	>14	>14			>14	>14		

Source: Reprinted, with publisher's permission, from *J. Am. Water Works Assn.*, *59*(4):491, © 1967 by the American Water Works Association, Inc., 2 Park Ave., New York, N.Y. 10016.

6.3–30 RATE CONSTANTS FOR ADSORPTION OF BACTERIOPHAGE T_4

Activated Carbon at pH 6.9 and Ionic Strengths of 0.08 and 0.10

		Rate constants	
Test no.	Ionic strength	Forward $k_1 \times 10^{13}$ Ml. Virus particle/sec	Reverse $k_2 \times 10^7$ sec^{-1}
1	0.10	7.2	18.0
2	0.08	8.3	20.5
3	0.08	8.8	22.0
4	0.08	8.3	20.8
5	0.08	8.1	20.3
6	0.10	7.9	19.8
7	0.08	8.6	21.5

Source: J.T. Cookson, Jr. and W.J. North, *Environ. Sci. Technol.*, *1*(1):46, 1967.

Section 7

Recreational Waters

7.1 Swimming and Primary Body Contact
Pages 599–602

7.2 Fishing and Secondary Body Contact
Pages 603–613

7. RECREATIONAL WATERS

7.1 *SWIMMING AND PRIMARY BODY CONTACT*

7.1–1 TESTS OF SWIMMING POOL WATER

Michigan State University

Date of sampling	No. people in pool	Chlorine residual, ppm	Coliform MPN/100 ml	Cocci MPN/100 ml	Total count/1 ml
Women's Intramural Pool					
Nov. 3, 1969	26	0.5	0	2.2	2
Nov. 5	26	0.6	0	0	0
Nov. 7	4	0.6	0	16	4
Nov. 10	26	0.6	0	0	0
Nov. 12	26	0.5	0	5.1	1
Nov. 14	3	0.5	0	16	1
Nov. 17	22	0.5	0	0	5
Nov. 19	4	0.5	0	2.2	17
Nov. 21	12	0.5	0	16	9
Nov. 24	20	0.5	0	0	1
Nov. 26	15	0.4	0	0	2
Women's Gymnasium Pool					
Nov. 3, 1969	29	0.6	0	2.2	0
Nov. 5	22	0.5	0	0	0
Nov. 7	0	0.6	0	2.2	2
Nov. 10	28	0.5	0	0	1
Nov. 12	22	0.5	0	2.2	6.1
Nov. 14	0	0.5	0	0	1
Nov. 17	20	0.5	0	0	4
Nov. 19	19	0.5	0	16	10
Nov. 21	0	0.5	0	5.1	2
Nov. 24	18	0.15	0	2.2	29
Nov. 26	15	0.5	0	0	6

Source: W.L. Mallman in "Public Health Aspects of Beaches and Swimming Pools," *Crit. Rev. Environ. Control, 1*(2):221, 1970.

7.1–2 AVERAGE NUMBER OF INDICATOR ORGANISMS IN POOL WATER

85 Samples

	Average	**Range**
Chlorine, ppm	0.53	0.1–1.0
Total count per 1 ml	8	0–300
Staphylococcus per 100 ml	47	1–658
Coliform per 100 ml	0.8	0–58
Strept. salvarius per 100 ml	1.9	0–52
Enterococci per 100 ml	0.7	0–32
Ps. aeruginosa per 100 ml	0	0
Bathing load per day	90	30–250
pH	7.7	7.4–8.0

Source: M.S. Favero, C.H. Drake, and G.B. Randal, *U.S. Public Health Rep., 79*:61, 1964.

7.1–3 POLLUTION CHARACTERISTICS OF SWIMMING POOL AND SALINE BATHING BEACHES

	Fresh water swimming pools	**Saline bathing beaches**
Water supply	Normally free from contamination	Normally contaminated by soil washings and possibly animal wastes
Primary source of pollution	Respiratory tract and body surfaces of swimmers	Soil washings, animal wastes, diluted raw and treated sewage
Secondary sources of pollution	Soil trackings from immediate vicinity of pool	Respiratory tract and body surfaces of bathers
Effect of dilution and currents	Relatively slight	Considerable
Natural purification	Slight, if any	May be considerable
Effect of artificial purification	Considerable	None
Bacterial findings	Measure of efficiency of filtration and disinfection	Measure of resultant of contamination plus purification processes
Possible bacterial indices	*Salmonella salivarius* *S. pyrogenes* *Neisseria catarrhalis* *N. sicca* Coliform *Escherichia coli* Enterococci *S. fecalis*	Coliform *E. coli* Enterococci (*S. fecalis*)

Source: M. Levine *et al., Proc. 1st Int. Conf. Waste Disp. Marine Environ.,* July 22–25, 1959, Pergamon Press, 1960.

7.1–4 CHEMICAL ANALYSIS OF SWIMMING POOL WATER

University of Florida Pool

Constituent	mg/l
Total dissolved solids	164
Silica, SiO_2	3.4
Iron, Fe^{+++}	0.06
Calcium, Ca^{++}	28
Magnesium, Mg^{++}	10
Sodium and Potassium as Na^+	8.7
Bicarbonate, HCO_3^-	29
Sulfate, SO_4^-	35
Chloride, Cl^-	48
Iodide, I^-	4.9
Fluoride, F^-	0.63
Nitrate, NO_3^-	0.40
Carbonate hardness as $CaCO_3$	35
Noncarbonate hardness as $CaCO_3$	76
Total hardness as $CaCO_3$	111
pH 7.40	

Note: Pool was treated with iodine as a disinfectant.

Source: Reprinted, with publisher's permission, from *J. Am. Water Works Assn., 59*(4):471, © 1967, by the American Water Works Association, Inc., 2 Park Ave., New York, N.Y. 10016.

7.1–5 CONCENTRATION OF IODINE AND IODATE IN IODINATED WATER

	Old pipe, ppm	New pipe, ppm
Iodate	0.14	0.15
Iodide	0.33	0.12
Initial iodide	1.08	1.08
Unaccounted for	0.61	0.81
pH	8.45	8.55

Source: Reprinted, with publisher's permission, from *J. Am. Water Works Assn., 60*(1):69, © 1968, by the American Water Works Association, Inc., 2 Park Ave., New York, N.Y. 10016.

7.1–6 PERCENTAGE OF I_2 RESIDUAL OF 0.5 PPM PRESENT

pH	I_2	Hypoiodous Acid, HIO	Hypoiodite Ion, IO^-
5	99	1	0
6	90	10	0
7	52	48	0
8	12	88	0.005

Source: Reprinted, with publisher's permission, from *J. Am. Water Works Assn., 60*(1):69, © 1968, by the American Water Works Association, Inc., 2 Park Ave., New York, N.Y. 10016.

7.1–7 EFFECTIVENESS OF IODINE AND FREE CHLORINE FOR DISINFECTING SWIMMING POOL WATER

	Iodine	Chlorine
Total number of samples	790	439
Desired residual range, ppm	0.6–0.8	0.4–0.6
Samples confirming for coliforms, percent	2	3
Samples confirming for fecal streptococci, percent	3	2
Average number of staphylococci per 100 ml	3.5[a]	46[b]
Median 48 hr plate count	210	5

[a] Based upon 60 samples.
[b] Based upon 54 samples.

Source: Reprinted, with publisher's permission, from *J. Am. Water Works Assn., 60*(1):69, © 1968, by the American Water Works Association, Inc., 2 Park Ave., New York, N.Y. 10016.

7.1–8 CORRELATION OF LEUCO CRYSTAL VIOLET $HgCl_2$ COLOR TEST FOR RESIDUAL IODINE DISINFECTANT WITH TITRATION RESULTS

University of Florida Swimming Pool

Deviation from titration results mg/l of I_2	No. tests	Percent total tests	Cumulative, percent
0.00	10	14.9	14.9
0.01	15	22.4	37.3
0.02	10	14.9	52.2
0.03	13	19.4	71.6
0.04	9	13.4	85.0
0.05	2	3.0	88.0
0.06	1	1.5	89.5
0.07	2	3.0	92.5
0.08	4	6.0	98.5
0.09	1	1.5	100.0
Total tests	67	100.0	100.0

Source: Reprinted, with publisher's permission, from *J. Am. Water Works Assn., 59*(4):471, © 1967, by the American Water Works Association, Inc., 2 Park Ave., New York, N.Y. 10016.

7.1–9 BACTERIOLOGICAL TESTS ON WADING POOLS

Halogen	Plate count	Coli-aerogenes	*E. coli*	Staph.	Entero-cocci	Diphth-oids
0–0.19 ppm						
Chlorine	71800	54	23	3550	55	475
Iodine	81600	58	11	2360	18	20
Bromine	61000	66	30	4930	33	288
0.2–0.40 ppm						
Chlorine	5280	22	15	2750	8	134
Iodine	7700	10	2	380	3	17
Bromine	3270	29	11	1720	12	15
0.5–1.0 ppm						
Chlorine	1550	1	0	14	0	6
Iodine	2400	0	0	100	2	6
Bromine	1600	0	0	260	0	4

Source: J.R. Brown, D.M. McLean, and M.C. Nixon, *Can. J. Publ. Health, 55*:251, 1964.

7.2 *FISHING AND SECONDARY BODY CONTACT*

7.2–1 ESTIMATES OF RECREATIONAL FISHING

Year	No. of fishermen, 1,000	Man-days of fishing, 1,000	Expenditure, $1,000	Expenditure, dollars[a] Per fishermen	Per man-day of fishing
1954	20,813	397,447	1,914,292	92	4.83
1980	46,518	930,379	4,481,144	96	4.82
2000	62,613	1,299,102	6,257,087	100	4.82

[a] Added by the editors.

Source: R.D. Hoak, *J. Water Pollut. Control Fed.*, *33*(12):1267, 1961.

7.2–2 STATISTICS ON FISHING PERMITS

	Oradell Reservoir			DeForest Lake Reservoir		
	No. permits issued		No. minors fishing per day	No. permits issued		No. minors fishing per day
Year	Adult	Minor		Adult	Minor	
1959	624	469	29	627		
1960	646	299	24	477	260	9
1961	500	515	29	416	217	4
1962	557	361	18	411	177	5

Note: Oradell and DeForest Lake Reservoirs are part of the Hackensack Water Co. and its subsidiary, the Spring Valley Water Works and Supply Co., New Milford, New Jersey.

Source: Reprinted, with publisher's permission, from *J. Am. Water Works Assn.*, *55*(6):803, © 1963, by the American Water Works Association, Inc., 2 Park Ave., New York, N.Y. 10016.

7.2–3 PUBLIC HUNTING LANDS IN THE U.S.

Region	Land per resident hunter, acres
West	121.6
Midwest	5.2
South	3.3
Northeast	3.7

Source: A.J. Beneditti, *J. Am. Water Works Assn.*, *56*(11):1467, 1964. Taken from data in *National Survey of Hunting and Fishing,* U.S. Fish and Wildlife Service, 1960.

7.2–4 VISITORS AT FORREST LAKE AND THOUSAND HILLS STATE PARK

Month	1958	1959	1960
Jan	3,070	1,467	1,714
Feb	5,944	2,984	1,753
Mar	12,599	14,491	3,447
Apr	30,327	22,309	22,268
May	43,678	47,178	44,723
June	52,160	52,054	37,150
July	57,196	57,246	57,572
Aug	59,834	44,641	37,819
Sept	29,968	26,293	27,956
Oct	23,225	7,811	12,645
Nov	9,866	5,529	5,370
Dec	4,266	2,824	3,983
Total	332,033	284,827	256,400

Note: Forrest Lake is located in Thousand Hills State Park, 4 miles west of Kirksville, Mo. The lake has an area of 702 acres and a shoreline of 17 miles.

Source: Reprinted, with publisher's permission, from *J. Am. Water Works Assn., 56*(1):43, © 1964, by the American Water Works Association Inc., 2 Park Ave., New York, N.Y. 10016.

7.2–5 RECREATIONAL PREFERENCE OF VISITORS TO THOUSAND HILLS STATE PARK

Kirksville, Mo., 1960

Preferred activity	Percentage of cars[a]
Motorboating	10.7
Other boating	0.2
Water skiing	3.9
Fishing	13.6
Swimming	16.5
Picnicking	19.3
Sightseeing	21.6
Camping	6.4
Relaxing	3.7
Other	3.3
No preference	1.6

[a] Occupants of 508 cars were interviewed.

Source: Reprinted, with publisher's permission, from *J. Am. Water Works Assn., 56*(1):43, © 1964, by the American Water Works Association, Inc., 2 Park Ave., New York, N.Y. 10016.

7.2–6 SUMMARY OF CREEL CENSUS DATA

Thousand Hills State Park, Kirksville, Mo.

Fishermen on lake	7,947
Male, percent	83.81
Female, percent	16.19
Fishing methods	
From shore, percent	11
From boats, percent	89
Fishermen on lake, daily mean	63.1
Fishermen on lake, weekly mean	441.5
Ratio of fishermen, 8 a.m. – 5 p.m.	1.287:1
Boats on lake, total	3,285
Boats on lake, daily mean	26.06
Boats on lake, weekly mean	182.44
Boat party, mean size	2.154
Mean fishing day, hr	3.75
Catch per fisherman	4.84
Catch, hourly mean	1.29
Time per fish caught, min	46
Fish caught[a]	38,465
Total time spent on lake, hr	29,794

[a] Data cover period, May 24–Sept. 27, 1958.

Source: Reprinted, with publisher's permission, from *J. Am. Water Works Assn., 56*(1):43, © 1964, by the American Water Works Association, Inc., 2 Park Ave., New York, N.Y. 10016.

7.2–7 EXISTING AND POTENTIAL RECREATION AREAS IN THE U.S.

	Campsites, acres/100 population		Picnic areas, acres/100 population	
Region	Developed	Potential	Developed	Potential
West	1.5	18.1	2.1	8.2
Midwest	0.3	2.5	1.8	3.7
South	0.2	3.4	1.2	2.5
Northeast	0.25	0.85	2.25	0.85

Source: Outdoor Recreation Resources Review Comm., *Public Outdoor Recreation Acres – Acreage, Use, Potential,* U.S. Government Printing Office, 1962.

7.2–8 BACTERIOLOGICAL RESULTS AT STATIONS WITH VARYING INTENSITIES OF RECREATIONAL USE

Location	Total coliform				Fecal coliform				Fecal streptococci			
	No. samples	MPN/100 ml			No. samples	MPN/100 ml			No. samples	MPN/100 ml		
		Min	Max	Med		Min	Max	Med		Min	Max	Med
Folsom Reservoir, First Study, May 4–May 15, 1960[a]												
Low Use Area												
Intake	21	2.1	64	15	21	<2.1	2.1	<2.1	21	<2.1	2.1	<2.1
Boat refueling	11	8.6	>700	32	7	<2.1	32	2.1	9	<2.1	32	12
Docks	8	48	240	8.6	4	<2.1	32	6.7	5	<2.1	240	<2.1
Docks	9	7.4	32	15	4	<2.1	<8.6	2.1	4	<2.1	2.1	<2.1
Docks	8	8.6	700	20	3	2.1	32	<4.8	5	<2.1	2.1	<2.1
Docks	10	12	700	20	6	<2.1	32	3.4	9	<2.1	12	2.1
High Use Area												
Shore fishing	9	15	>700	64	5	2.1	64	32	7	2.1	64	15
Supervised swimming	12	12	700	185	6	<2.1	32	6.1	12	<2.1	700	4.5
	13	7.4	>700	210	10	<2.1	240	32	13	<2.1	700	2.1
	11	8.6	700	130	6	2.1	32	20.3	11	<2.1	700	4.8
Boat launching	8	32	700	240	4	4.8	64	12.4	5	<2.1	240	15
	8	15	>700	470	4	4.8	32	5.3	8	4.8	240	15
	9	2.1	700	130	4	4.8	64	18.4	9	<2.1	240	20
Fishing, picnicking	10	25	>700	700	5	2.1	32	32	10	4.8	240	20
Swimming	10	240	>700	700	5	8.6	240	32	10	20	>700	470
	10	240	>700	700	10	2.1	240	64	7	7.4	>700	32
Cackuma Reservoir, First Study, May 20–June 7, 1960[a]												
Low Use Area												
Dam intake	16	<4.5	700	47	14	<2.1	2.1	<2.1	16	<2.1	4.8	<2.1
Dam surface	19	7.4	2,100	64	17	<2.1	12	<2.1	19	<2.1	240	2.1
Near toilet	20	4.8	>700	68	17	<2.1	4.8	<2.1	21	<2.1	>700	2.1
Near park intake	29	4.8	>24,000	500	18	<2.1	23	6	29	<2.1	>700	240
Tower intake surface	19	15	>24,000	620	17	<2.1	1,300	<2.1	19	<2.1	>700	15

[a] Data are for the entire study period.

7.2–8 BACTERIOLOGICAL RESULTS AT STATIONS WITH VARYING INTENSITIES OF RECREATIONAL USE (*Continued*)

Location	Total coliform				Fecal coliform				Fecal streptococci			
	No. samples	MPN/100 ml			No. samples	MPN/100 ml			No. samples	MPN/100 ml		
		Min	Max	Med		Min	Max	Med		Min	Max	Med
Cackuma Reservoir, First Study, May 20–June 7, 1960[a]												
High Use Area												
Shore fishing	23	20	>24,000	620	13	<2.1	62	23	23	<2.1	>7,000	64
	27	2.1	>24,000	240	16	<2.1	7,000	14	27	<2.1	>700	32
	19	15	7,000	620	18	<2.1	230	6	19	<2.1	>700	130
Boat refueling	29	4.8	>2,400	500	16	<2.1	62	<2.1	29	<2.1	700	15
Launching	31	4.8	7,000	230	18	<2.1	62	<2.1	31	<2.1	240	4.8
Docks	27	32	2,400	620	16	<2.1	64	6	27	<2.1	240	15
Near toilet	18	20	>24,000	620	17	<2.1	230	6	18	<2.1	>700	152

[a] Data are for the entire study period.

Source: Reprinted, with publisher's permission, from *J. Am. Water Works Assn.*, *61*(6):297, © 1969, by the American Water Works Association, Inc., 2 Park Ave., New York, N. Y. 10016.

7.2–9 COMPARISON OF UPSTREAM AND DOWNSTREAM INDICATOR ORGANISM DATA

	Sample station	No. samples	Total coliform			Fecal coliform			Fecal streptococci		
			max	min	mean	max	min	mean	max	min	mean
Cedar	C-1	84	2,300	1.1	37	79	0.45	5.6	240	0.2	5.5
River	C-11	75	540	0.2	9.4	49	0.2	0.8	49	0.2	1.6
Green	G-1	87	920	0.45	19	24	0.2	1.4	490	0.2	3.3
River	G-16	83	170	0.2	11	13	0.2	0.6	350	0.2	0.4
Clackamas	CL-1	81	920	0.68	21	79	0.2	1.1	1,600	0.2	7.3
River	CL-9	72	350	0.2	3.3	68	0.2	0.4	920	0.2	2.6

Note: All data are in density/100 ml.

Source: Reprinted, with publisher's permission, from *J. Am. Water Works Assn.*, *62*(7):412, © 1970, by the American Water Works Association, Inc., 2 Park Ave., New York, N.Y. 10016.

7.2–10 CORRELATION COEFFICIENT FOR INDICATOR ORGANISM DENSITIES AT THE MOST DOWNSTREAM SAMPLING STATION

Different Watersheds within Seasons

Watershed	Spring, 1966	Summer, 1966	Fall, 1966	Winter, 1966/67	Spring, 1967	Summer, 1967
Total coliform						
Clackamas-Cedar	–0.15	0.03	–0.31	0.64	0.42	0.03
Clackamas-Green	–0.55	0.22	–0.11	0.29	0.79	–0.20
Cedar-Green	0.35	0.16	–0.31	0.15	0.61	0.11
Fecal coliform						
Clackamas-Cedar	0.26	0.35	0.38	–0.10	–0.03	0.21
Clackamas-Green	0.00	–0.09	0.51	–0.12	0.03	–0.18
Cedar-Green	–0.04	–0.19	0.21	0.47	–0.04	0.55
Fecal streptococci						
Clackamas-Cedar	0.04	0.19	0.35	–0.09	0.18	–0.07
Clackamas-Green	0.30	–0.11	–0.09	–0.17	–0.22	0.07
Cedar-Green	0.69	–0.21	0.74	–0.07	–0.50	0.74

Source: Reprinted, with publisher's permission, from *J. Am. Water Works Assn.*, *62*(7):412, © 1970, by the American Water Works Association, Inc., 2 Park Ave., New York, N.Y. 10016.

7.2–11 STATISTICAL SUMMARY OF BACTERIOLOGICAL DATA

Most Downstream Station for Entire Study

	Total coliforms, density/ 100 ml (MPN)			Fecal coliforms, density/ 100 ml (MPN)			Fecal streptococci, density/ 100 ml (MPN)		
	Cedar	Green	Clackamas	Cedar	Green	Clackamas	Cedar	Green	Clackamas
No. samples	84	87	81	84	86	80	84	84	80
Maximum	2,300	920	920	79	24	79	240	490	1,600
Minimum	1.1	0.45	0.68	0.45	0.18	0.18	0.18	0.18	0.18
Geometric mean	37	19	21	5.6	1.4	1.1	5.5	3.3	7.3
95 percent of the data fall between	4.0–340	1.5–235	2.4–188	0.9–34	0.2–10.1	0.1–8.4	0.5–61	0.3–40	0.4–130

Source: Reprinted, with publisher's permission, from *J. Am. Water Works Assn., 62*(7):412, © 1970, by the American Water Works Association, Inc., 2 Park Ave., New York, N.Y. 10016.

7.2–12 ISOLATION OF PATHOGENIC ENTEROBACTERIACEAE AND VIRUS FROM FLOWTHROUGH SAMPLER

Most Downstream Sampling Station

	Pathogenic enterobacteriaceae			Virus		
	Cedar	Green	Clackamas	Cedar	Green	Clackamas
No. samples examined	61	57	38	59	61	31
No. samples positive	17	22	6	0	0	0
Percent positive	28	39	16	0	0	0

Source: Reprinted, with publisher's permission, from *J. Am. Water Works Assn., 62*(7):412, © 1970, by the American Water Works Association, Inc., 2 Park Ave., New York, N.Y. 10016.

7.2–13 ESTIMATED ANIMAL AND MAXIMUM HUMAN POPULATIONS

Watershed	Max total population, man-days	Date of max population	Estimated total animal population	Water-shed area, mi²	Main-stem river miles	Human population/area, man-days/mi²	Animal population/area number/mi	Human population/river mile, man-days/mi	Animal population/river mile, number/mi
Cedar	95	June 29, 1967	8,200	143	31.6	0.7	57	3	260
Green	295	Sept. 5, 1966	10,700	231	30.5	1.3	46	10	350
Clackamas	3,400	July 4, 1967	27,300	610	77.5	5.7	45	45	350

Source: Reprinted, with publisher's permission, from *J. Am. Water Works Assn.*, 62(7):412, © 1970, by the American Water Works Association, Inc., 2 Park Ave., New York, N.Y. 10016.

7.2–14 MAXIMUM VALUES FOR CHEMICAL DATA DETERMINED BY WET CHEMISTRY METHODS FOR ALL SAMPLES

Most Downstream Sampling Station

	Cedar	Green	Clackamas
No. samples	17	28	54
Constituent		Units,[a] mg/l	
Selenium[b]	2.0(15)	3.0(23)	2.0(49)
Chloride	1.0(14)	1.0(20)	>4.0(16)
MBAS	0.06(16)	0.15(24)	0.2(49)
Cyanide	0.005(16)	0.007(26)	0.007(50)
Fluoride	0.08(14)	0.10(23)	0.11(30)
Nitrate-N	0.3(9)	0.10(26)	0.10(53)
Nitrite-N	0.007(15)	All BDL	All BDL
Ammonia-N	0.2(15)	0.2(27)	0.3(53)
Ortho-phosphorus-P	All BDL	0.10(27)	0.10(48)
Sulfate	2.0(13)	2.0(26)	All BDL
Tannin	0.3(16)	All BDL	0.3(53)

[a] The numbers in parentheses are the numbers of samples that were below the detectable level (BDL).
[b] Units for selenium in μg/l.

Source: Reprinted, with publisher's permission, from *J. Am. Water Works Assn.*, *62*(7):412, © 1970, by the American Water Works Association, Inc., 2 Park Ave., New York, N.Y. 10016.

7.2–15 ARITHMETIC MEAN VALUES FOR RESIDUE DATA

All Samples at the Most Downstream Sampling Station

	Cedar	Green	Clackamas
No. samples	17	28	54
Constituent		Units, mg/l	
Total solids	46.8	48.4	70.1
Volatile total solids	15.4	13.2	16.0
Total dissolved solids	42.3	40.3	47.1
Volatile dissolved solids	13.2	11.8	14.6

Source: Reprinted, with publisher's permission, from *J. Am. Water Works Assn.*, *62*(7):412, © 1970, by the American Water Works Association, Inc., 2 Park Ave., New York, N.Y. 10016.

7.2–16 MAXIMUM VALUES FOR CHEMICAL DATA DETERMINED BY SPECTROGRAPHIC ANALYSIS

Most Downstream Sampling Station

	Cedar	Green	Clackamas
No samples	11	12	19
Constituent		**Units,[a] μg/l**	
Silver	2.5(10)	1.9(10)	3.4(13)
Nickel	All BDL	3.0(10)	17(15)
Cobalt	All BDL	All BDL	12(15)
Lead	34(9)	45(10)	37(12)
Chromium	3.0(8)	2.0(10)	5.0(14)
Vanadium	All BDL	All BDL	21(18)
Barium	7.0(0)	4.0(1)	15(0)
Strontium	20(0)	20(0)	27(0)
Arsenic	All BDL	All BDL	10(18)
Zinc	22(2)	42(2)	60(0)
Cadmium	All BDL	All BDL	7(16)
Boron	20(0)	30(0)	42(0)
Total phosphorus	60(9)	42(8)	76(14)
Iron	68(0)	380(0)	136(1)
Molybdenum	10(10)	3(11)	23(15)
Manganese	4.9(4)	17(2)	50(6)
Aluminum	120(5)	480(2)	220(6)
Beryllium	All BDL	0.01(11)	0.04(18)
Copper	14(3)	30(4)	51(8)

[a] The numbers in parentheses are the numbers of samples that were below the detectable level (BDL).

Source: Reprinted, with publisher's permission, from *J. Am. Water Works Assn., 62*(7):412, © 1970, by the American Water Works Association, Inc., 2 Park Ave., New York, N.Y. 10016.

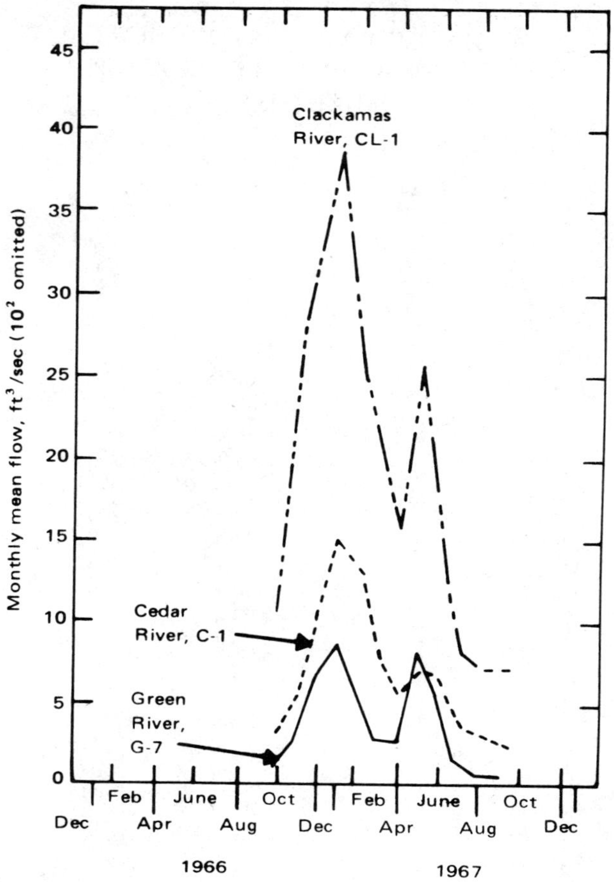

7.2–17 Monthly Mean Flow.

Source: Reprinted, with publisher's permission, from *J. Am. Water Works Assn., 62*(7):412, © 1970, by the American Water Works Association, Inc., 2 Park Ave., New York, N.Y. 10016.

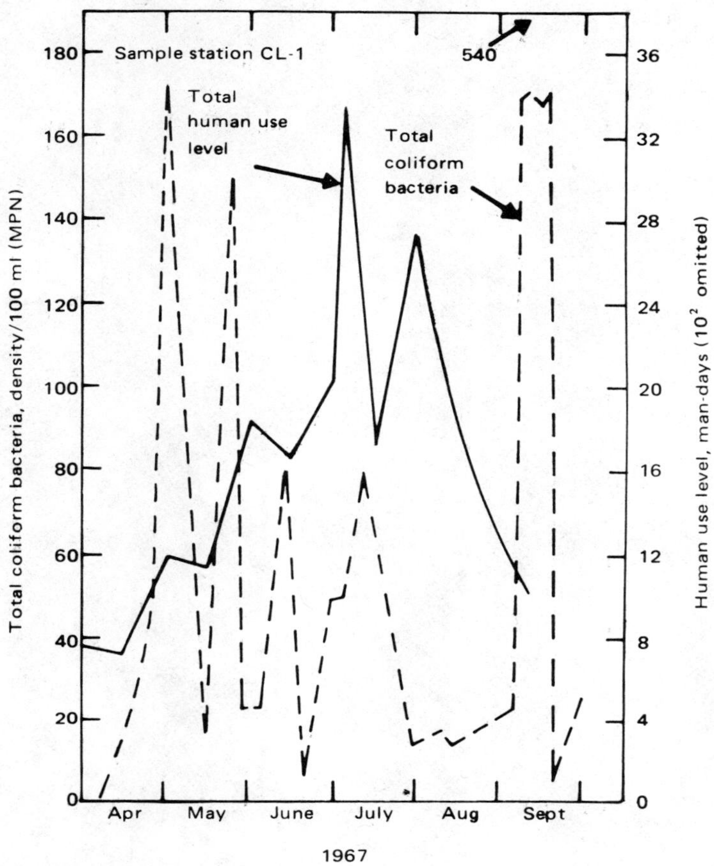

7.2–18 Comparison of Human Use Level and Total Coliform Densities for the Clackamas River.

Source: Reprinted, with publisher's permission, from *J. Am. Water Works Assn.*, *62*(7):412, © 1970, by the American Water Works Association, Inc., 2 Park Ave., New York, N.Y. 10016.

Section 8

Analytical Methods

8. **Introduction**
Pages 617–651

8.1 **Electrometric Methods**
Pages 652–656

8.2 **Spectrometric Methods**
Pages 657–663

8.3 **Oxygen Demand**
Pages 664–693

8. *ANALYTICAL METHODS*

This section is concerned with methods for the examination of water. Recommended methods are briefly summarized. The reader is given definite references to see for details (e.g., *Standard Methods for the Examination of Water and Waste-Water*). Further data are presented on analyses of problems of current importance: nutrients, heavy metals, pesticides, etc. to enable the reader to judge analytical results and then to carry out the analysis himself.

Subsections follow on specific measuring methodologies and on oxygen demand parameters.

8–1 RECOMMENDED METHODS FOR THE ANALYSIS OF SURFACE WATERS, WASTE WATERS AND SLUDGES

Property	Surface water	Waste water	Sludge
Acidity	Same method as for waste water	Sample is titrated to a pH of 8.3 and results reported as mg/l $CaCO_3$. Not applicable to analysis of acid mine drainage water[1]	Supernatant from the settled sludge is tested for acidity either titrimetrically or, if colored, potentiometrically[3]
Alkalinity	Same method as for waste water	Titrated unaltered sample to an electrometrically determined end point of pH 4.5. Suitable for all concentration ranges; large concentrations of weak organic and inorganic acids, and oil and grease interfere[1]	Sludge is allowed to stand and the supernatant liquor tested for alkalinity either titrimetrically, or, if colored, potentiometrically[3]
Alkalinity, total	Automated colorimetric method; addition of alkalinity causes loss of color of the methyl orange indicator buffered at 3.1 pH; absorbance measured at 550 nm. Analytical range: 10–200 mg/l $CaCO_3$. No significant interferences. Also applicable to saline waters[1]	Same method as for surface water	
Aluminum – atomic absorption method[a]	Analytical range: 10–1,000 mg/l using 309.2 nm line; sensitivity: 0.4 mg/l; detection limit: 0.1 mg/l. Other (less sensitive) wavelengths: 308.2, 396.2, and 394.4 nm	Same method as for surface water	
Arsenic – atomic absorption method[a]	Analytical range: 10–100 mg/l using 193.7 nm line; sensitivity: 1.0 mg/l; detec-	Same method as for surface water	

[a] Atomic absorption methods: Free, unexcited metal atoms present in a flame, when the sample is atomized and aspirated into it, absorb light of a suitable wavelength in proportion to the metal concentration in the sample. Detection limits may be extended by sample concentration, solvent extraction, or scale expansion. Use of less sensitive wavelengths allows direct measurement of high concentrations. Sludges may be analyzed by using acid digestion to solubilize the metals.[1]

8–1 RECOMMENDED METHODS FOR THE ANALYSIS OF SURFACE WATERS, WASTE WATERS AND SLUDGES (*Continued*)

Property	Surface water	Waste water	Sludge
Arsenic–atomic absorption method[a] (continued)	tion limit: 0.5 mg/l. Dissolved salts (above 1%) produce positive interference at the 193.7 nm line. High solids burner lowers detection limit		
Arsenic, inorganic		Arsine, AsH_3, is formed by reduction of As and is absorbed in a solution of silver diethyldithiocarbamate in pyridine. The resulting red complex is measured at 535 nm. Lead acetate scrubber removes sulfides from AsH_3. Sample should contain >10 μg/l As. High concentrations of Cr, Co, Cu, Hg, Mo, Ni, and Ag interfere[1]	
BOD	Same method as for waste water	Sample is incubated at 20° for 5 days in dark and BOD calculated from the reduction in dissolved oxygen (DO) concentration. The DO may be determined either by the modified Winkler method or the probe method. No specific standard test for BOD has been selected by the Water Quality Office[1]	
Cadmium – atomic absorption method[a]	Analytical range: 0.1–1.0 mg/l using 228.8 nm line; sensitivity: 0.004 mg/l; detection limit: 0.001 mg/l	Same method as for surface water	
Calcium – atomic absorption method[a]	Analytical range: 1–200 mg/l using 427.7 nm line; sensitivity: 0.07 mg/l; detection limit: 0.003 mg/l. Acid lanthanum chloride removes interference by PO_4^{-3}, SO_4^{-2} and Al^{+3}. More than 1,000 mg/l of Mg suppresses absorption. Up to 500 mg/l each of Na^+, K^+ and NO_3^- do not interfere. Sensitivity at 239.9 nm is 20 mg/l		

[a] Atomic absorption methods: Free, unexcited metal atoms present in a flame, when the sample is atomized and aspirated into it, absorb light of a suitable wavelength in proportion to the metal concentration in the sample. Detection limits may be extended by sample concentration, solvent extraction, or scale expansion. Use of less sensitive wavelengths allows direct measurement of high concentrations. Sludges may be analyzed by using acid digestion to solubilize the metals.[1]

8–1 RECOMMENDED METHODS FOR THE ANALYSIS OF SURFACE WATERS, WASTE WATERS AND SLUDGES (*Continued*)

Property	Surface water	Waste water	Sludge
Chemical Oxygen Demand (COD), high level		Oxidation of organics by refluxing with potassium dichromate in 50% H_2SO_4 solution followed by titrimetric determination of excess dichromate with standard ferrous ammonium sulfate, using orthophenanthroline ferrous complex indicator. Ag_2SO_4 catalyzes the oxidation and Hg_2SO_4 removes chloride interference. Organic carbon concentration should be >15 mg/l. For saline waters with >1,000 mg/l chlorides, and COD <250 mg/l, the results are highly questionable. Chloride correction necessary for >250 mg/l COD[1]	
Chemical Oxygen Demand (COD), low level	Method same as for high level COD test. Analytical range: 5–50 mg/l. Extreme care necessary to prevent sample contamination[1]		
Chloride	Same method as for water method	Acidified sample is titrated with dilute HNO_3 solution using diphenyl-bromophenol blue indicator. Endpoint is indicated by the formation of blue-violet mercury-diphenylcarbazone complex. Preferred concentration range: 200–400 mg Cl/l. H_2O_2 oxidation removes sulfite interference[1]	
	Same method as for water method	Automated ferricyanide method: Thiocyanate ion (SCN) is liberated from $Hg(SCN)_2$ by chloride sequestration of mercury and forms colored ferric thiocyanate; absorbance measured at 480 nm. Analytical range: 1–250 mg Cl/l. No significant interferences[1]	

8–1 RECOMMENDED METHODS FOR THE ANALYSIS OF SURFACE WATERS, WASTE WATERS AND SLUDGES (*Continued*)

Property	Surface water	Waste water	Sludge
Chlorine requirement	Same methods as for waste water	A solution of known chlorine content is added incrementally to a series of sample aliquots. At the end of the stipulated contact time or when the desired result has been achieved, the residual chlorine is measured. Results must refer to the pH, temperature, contact time of chlorination and the methods used for chlorine determination[1]	
Chromium – atomic absorption method[a]		Analytical range: 1–200 mg/l, using 357.9 nm line. Sensitivity: 0.02 mg/l; detection limit: 0.01 mg/l; other (less sensitive) lines: 360.5, 359.3, 425.4, 427.4, and 428.9 nm. Absorption very sensitive to fuel/air ratio. NO-acetylene flame eliminates negative interference by Fe and Ni. MIBK-APDC extraction recommended for low Cr levels. MIBK-Methyl isobutyl ketone APDC-Ammonium pyrrolidine dithiocarbamate	
Color	Visual comparison of sample with platinum-cobalt standards. Analytical range: 0–70 units. Not applicable to highly colored industrial wastes; measures color derived from naturally occurring materials[1]	No recommended terms. Describe by such color terms as seem suitable[3]	
Copper – atomic absorption method[a]	Analytical range: 0.1–10 mg/l, using 324.7 nm line. Sensitivity: 0.04 mg/l; detection limit: 0.005 mg/l other (less sensitive) lines: 326.4, 217.8, 216.5, 218.1, 222.5, 202.4, and 249.2 nm (in order of decreasing sensitivity). For concentration below 0.05 mg/l, use extraction procedure	Same method as for surface water	

[a] Atomic absorption methods: Free, unexcited metal atoms present in a flame, when the sample is atomized and aspirated into it, absorb light of a suitable wavelength in proportion to the metal concentration in the sample. Detection limits may be extended by sample concentration, solvent extraction, or scale expansion. Use of less sensitive wavelengths allows direct measurement of high concentrations. Sludges may be analyzed by using acid digestion to solubilize the metals.[1]

8–1 RECOMMENDED METHODS FOR THE ANALYSIS OF SURFACE WATERS, WASTE WATERS AND SLUDGES (*Continued*)

Property	Surface water	Waste water	Sludge
Cyanide		HCN released from metallic cyanide complex ions by reflux distillation is absorbed in a NaOH scrubber and the resulting cyanide ion is titrated with $AgNO_3$ using *p*-dimethylaminobenzalrhodanine indicator. Applicable to samples containing more than 1 mg CN/l[1]	
	Cyanide reacts first with free chlorine and next with pyridine to form a glutaconic aldehyde which forms a blue dye with 1-phenyl-3-methyl-5-pyrazolone; measure absorbance at 620 nm. Applicable to samples containing <1 mg CN/l; sensitivity ~5 μg CN/l. Sulfides interfere and are removed as PbS. Oxidizing substances interfere and should be removed with ascorbic acid[1]		
Dissolved oxygen	Same method as for waste water	Modified Winkler method: Manganous hydroxide, precipitated by the addition of manganous sulfate and alkaline KI to the sample, is oxidized to manganic hydroxide, $MnO(OH)_2$, by the dissolved oxygen. Upon acidification, $MnO(OH)_2$ is converted to manganic sulfate which releases free iodine (stoichiometrically equivalent to dissolved oxygen) from KI. The released iodine is titrated with sodium thiosulfate, using starch indicator. Reproducibility: ~0.2 ppm DO at 7.5 ppm level. Fe^{+3} interference (up to 200 mg Fe^{+3}/l) is removed by adding fluoride solution. Sodium azide removes NO_2^- interference. Presence of sulfite, thiosulfate, polythionate, free chlorine, OCl^-, suspended solids, oxidizable organics, domestic sewage, biological flocs,	Diluted sludge sample is incubated and BOD measured as the DO consumed. Alum flocculation or copper sulfate-sulfamic acid flocculation is used to clarify the sample before DO measurement[3]

8–1 RECOMMENDED METHODS FOR THE ANALYSIS OF SURFACE WATERS, WASTE WATERS AND SLUDGES (*Continued*)

Property	Surface water	Waste water	Sludge
Dissolved oxygen (*continued*)		and sample color make the azide modification inapplicable. NO_3^- nitrogen and up to 1 mg/l Fe^{-2} ion are permissible[1]	
		Probe method: The current produced by the cathodic reduction of oxygen diffusing across the oxygen-permeable membrane of the probe depends upon the DO content of the sample. Analytical range: 0–20 mg/l DO. Sensitivity: 0.05 mg/l. Dissolved inorganic salts, and reactive gases like Cl_2 and H_2S may interfere[1]	
Fluoride	Same method as for waste water	Loss of color resulting from the reaction of fluoride with the Zirconyl-SPADNS dye is proportional to fluoride concentration; absorbance measured at 570 nm. Analytical range: 0.1–2.5 mg F/l. High concentrations of Al^{+3}, SiO_2 and Cl^- and oily matter interfere[1,3]	
		Fluoride ion reacts with red cerous chelate of alizarin complexone; absorbance measured at 650 nm. Analytical range: 0.05–1.5 mg F/l. Aluminum interferes and can be removed by complexing with 8-hydroxy quinoline. Analysis can be automatized[1]	
		Probe method: Potential developed across a LaF_3 crystal depends upon the F^- ion concentration; potentiometric measurement employs the specific ion fluoride electrode in conjunction with a standard reference electrode. Analytical range: 0.1–1,000 mg F/l. Extreme pH values, Si^{+4}, Fe^{+3} and Al^{+3} interfere. A 5 pH buffer containing a strong chelating agent removes the interferences.[1]	

8–1 RECOMMENDED METHODS FOR THE ANALYSIS OF SURFACE WATERS, WASTE WATERS AND SLUDGES (*Continued*)

Property	Surface water	Waste water	Sludge
Hardness	Same method as for waste water	Calcium and magnesium ions are sequestered by disodium dihydrogen ethylenediamine tetracetate (Na_2 EDTA); chrome black T or calmagite indicator changes from red to blue when sequestering is complete. Preferred hardness: 500 mg/l as $CaCO_3$. Heavy metal interference removed by complexing with cyanide. Titration may be automated[1]	
Hardness, total	Same method as for waste water	Calcium and/or other cations in the sample react with disodium magnesium EDTA forming stable chelates and releasing magnesium which, in conjunction with the free magnesium in sample, forms a red-violet complex with calmagite at pH 10; absorbance measured at 520 nm. Analytical range; 10–400 mg/l as $CaCO_3$ [1]	
Iron – atomic absorption method[a]	Analytical range: 0.1–20 mg/l, using 248.3 nm line. Sensitivity: 0.006 mg/l; detection limit: 0.004 mg/l; absorption strongly dependent upon lamp current. Other (less sensitive) lines: 248.8, 252.2, 271.9, 302.1, 252.7, 272.1, 372.0, 296.7, 386.0, and 344.1 nm (in order of decreasing sensitivity)	Same method as for surface water	
Lead – atomic absorption method[a]	Analytical range: 1–10 mg/l, using 217.0 nm line. Sensitivity: 0.06 mg/l; detection limit: 0.01 mg/l; other (less sensitive) lines: 283.3, 261.4, and 368.3 (in order of decreasing sensitivity). Careful alignment of light beam necessary for good results. For lead concentrations below 0.2 mg/l, extraction at pH 2.8 is recommended	Same method as for surface water	

[a] Atomic absorption methods: Free, unexcited metal atoms present in a flame, when the sample is atomized and aspirated into it, absorb light of a suitable wavelength in proportion to the metal concentration in the sample. Detection limits may be extended by sample concentration, solvent extraction, or scale expansion. Use of less sensitive wavelengths allows direct measurement of high concentrations. Sludges may be analyzed by using acid digestion to solubilize the metals.[1]

8–1 RECOMMENDED METHODS FOR THE ANALYSIS OF SURFACE WATERS, WASTE WATERS AND SLUDGES (*Continued*)

Property	Surface water	Waste water	Sludge
Magnesium – atomic absorption method[a]	Analytical range: 0.01–2 mg/l, using 285.2 nm line. Sensitivity: 0.005 mg/l; detection limit: 0.0005 mg/l; other (less sensitive) lines: 202.5, 702.5 and 279.6 nm (in order of decreasing sensitivity). Sensitivity decreases with increasing lamp current. La removes interference by Al. Na, K and Ca do not interfere at less than 400 mg/l concentrations	Same method as for surface water	
Manganese – atomic absorption method[a]	Analytical range: 0.1–20 mg/l, using 279.5 nm line. Sensitivity: 0.04 mg/l; detection limit: 0.005 mg/l extraction at pH 4.5–5 recommended for <0.01 mg/l of manganese		
Mercury	Flameless atomic absorption at 253.7 nm by Hg vapor. Hg in the sample is converted to Hg^{+2} form by $KMNO_4$ and $K_2S_2O_8$, reduced to elemental state by Sn^{+2} ions and the vapor passed through the light path of an AAS. $KMNO_4$ eliminates S^{-2} interference. Cu at 10 mg/l does not interfere. Free chlorine and volatile organics interfere[1]	Same method as for surface water	
Methylene blue active substances (MBAS)	Same method as for waste water	Anionic surface active materials react with aqueous methylene blue dye to form a blue colored salt extractable with $CHCl_3$; measure absorbance at 650 nm. Analytical range: 0.025–100 mg/l LAS. Insignificant interference from naturally occurring surface active agents; not applicable to brine samples[1,3]	

[a] Atomic absorption methods: Free, unexcited metal atoms present in a flame, when the sample is atomized and aspirated into it, absorb light of a suitable wavelength in proportion to the metal concentration in the sample. Detection limits may be extended by sample concentration, solvent extraction, or scale expansion. Use of less sensitive wavelengths allows direct measurement of high concentrations. Sludges may be analyzed by using acid digestion to solubilize the metals.[1]

8–1 RECOMMENDED METHODS FOR THE ANALYSIS OF SURFACE WATERS, WASTE WATERS AND SLUDGES (*Continued*)

Property	Surface water	Waste water	Sludge
Nitrogen, ammonia	Same methods as for waste water	Borate buffered (pH 9.5) sample is distilled into a solution of H_3BO_3. The NH_3 in the distillate is measured by one of the following: Titration with H_2SO_4 using a mixed methyl-red-methylene blue indicator; analytical range: 1–25 mg/l $NH_3 \cdot N$; volatile alkaline compounds (i.e. hydrazine) interfere[1]	~20 g sludge is diluted to 250 ml with ammonia-free distilled water. The ammonia-nitrogen in solution is determined by distilling into boric acid solution and titrating with 0.02 N H_2SO_4[3]
		Nesselerization with alkaline HgI_2-KI solution; absorbance measured at 425 nm. Analytical range: 0.05–1.0 mg/l NH_3-N; certain ketones, aldehydes and alcohol may interfere[1]	
NH_3-nitrogen		Automated method: NH_3 reacts with alkaline phenol hypochlorite to form indophenol blue; color is intensified by sodium nitroprusside and measured at 650 nm. Analytical range: 0.01–20 mg/l NH_3-N Ca, Mg interference eliminated by EDTA; color intensity is pH dependent[1]	
Nitrogen, total Kjeldahl	Same method as for waste water	Digestion with concentrated H_2SO_4, K_2SO_4 and $HgSO_4$ converts organic nitrogen to NH_4^+ which is then reduced by alkaline thio-sulfate solution to NH_3. The ammonia is distilled and measured as described above. Amines, nitro compounds, hydrazines, oximes, semicarbazones and refractory tertiary amines may not be converted by this digestion[1]	Sludge is dried at 103°C, ground and dried again. A known weight is digested with H_2SO_4, K_2SO_4 and the resulting NH_3 determined by the method used for surface waters and waste waters. Some NH_3 may be lost on drying[3]

8–1 RECOMMENDED METHODS FOR THE ANALYSIS OF SURFACE WATERS, WASTE WATERS AND SLUDGES (*Continued*)

Property	Surface water	Waste water	Sludge
Nitrogen, total Kjeldahl (continued)	Same method as for waste water	Automated method: Automatic digestion with concentrated H_2SO_4, K_2SO_4, and $HgSO_4$ to convert organic nitrogen to NH_4^+ followed by reduction to NH_3 which is then measured by developing the indophenol blue color as described above. (NH_3-nitrogen automated method). Analytical range: 0.05–20 mg/l Fe, Cr and Cu may interfere. Other interferences mentioned in automated NH_3-nitrogen and kjeldahl nitrogen analysis methods[1]	
Nitrogen, nitrate	Same method as for waste water	NO_3^- reacts with brucine sulfate in 13 NH_2SO_4 at 100°C to form yellow-colored complex; absorbance measured at 410 nm; analytical range: 0.1–2 mg NO_3 N/l. Temperature control of reaction is extremely important; strong oxidizing or reducing agents interfere; Fe^{+2}, Fe^{+3} and Mn^{+4} ions at <1 mg/l concentrate give negligible interference; residual chlorine interference eliminated by sodium arsenite[1,2]	
Nitrogen, nitrate and nitrite	Automated Cd reduction: cadmium-copper catalyst reduces NO_3^- to NO_2^- which reacts first with sulfanilamide, and then with N-l naphthyl-ethylene diamine hydrochloride to form an azo dye; measure absorbance at 540 nm. (EPA manual does not recommend this method explicitly for waste waters), analytical range: 0.05–10 mg/l N as NO_3. Oxidizing and reducing agents and metal ions interfere at high concentrations; NH_3 and primary amines may destroy nitrites[1]	Same method as for surface water	

8–1 RECOMMENDED METHODS FOR THE ANALYSIS OF SURFACE WATERS, WASTE WATERS AND SLUDGES (*Continued*)

Property	Surface water	Waste water	Sludge
	Same method as for waste water	Automated hydrazine reduction: NO_3^- is reduced to NO_2^- with hydrazine sulfate and the NO_2^- determined by diazotization-coupling reaction; analytical range 0.05–10 mg/l. PO_4^{-3}, NH_3, Mg^{+2}, Fe^{+3} and ABS do not interfere at up to 30 mg/l; Ca^{+2} at 240 and Cl^- at 30,000 mg/l levels do not interfere: S^{-2} interferes[1]	
Nitrogen, nitrite		The diazonium compound formed by diazotation of sulfanilamide by NO_2^- in acidic aqueous solutions is coupled with N-(1-naphthyl)-ethylenediamine to produce a reddish-purple color measured at 540 nm; analytical range: 0.05–1.0 mg/l NO_2-N strong oxidants and reductants affect NO_2^- concentration. High alkalinity (>600 mg/l) suppresses color formation[1]	
Nitrogen, organic and ammonia	Automated phenolate method: Manual digestion with H_2SO_4 and $K_2S_2O_8$ followed by automated phenol-hypochlorite procedure to determine ammonia (described above). Analytical range: 1–10 mg N/l NO_3^-, NO_2^- nitrogen not measured. No significant interferences[1]		
NTA (tri-sodium salt of nitrilotriacetic acid)	NTA breaks down the blue-color formed by zinc with 2 carboxy-2′hydroxy-5′-sulfoformazylbenzene. Optical density measured at 620 nm. Analytical range: 0.5–10.0 mg/l NTA. Cations complex with NTA and interfere. Not applicable to saline waters[1]		

8–1 RECOMMENDED METHODS FOR THE ANALYSIS OF SURFACE WATERS, WASTE WATERS AND SLUDGES (*Continued*)

Property	Surface water	Waste water	Sludge
NTA (continued)	Automated method: same as above. Analytical range: 0.04–1.0 mg/l and 0.5–10.0 mg/l depending on the manifold system used. Positive interference of 0.5–1.0 mg/l obtains with sewage-type samples. Not applicable to saline waters. Cations complex with NTA and interfere[1]	Same method as for surface water	
Oil and grease		The sample is acidified to pH 3.0 and extracted with hexane using a soxhlet extraction. The solvent is evaporated from the separated extract and the residue weighed. Analytical range: 5–1,000 mg/l extractable material. Not applicable to measurement of light hydrocarbons that volatilize below 80°C[1]	The sample is acidified, dried with magnesium sulfate monohydrate and the grease in the dry sludge extracted with hexane. Results reproducible up to 650 mg/l grease. Empirical method: strict adherence to procedural details necessary[3]
Oil and grease, hydrocarbon and fatty matter content of		Activated alumina adsorbs grease from its solution in hexane. The nonpolar hydrocarbons are later eluted by passing hexane through the column. The alumina retains polar fatty acids and esters during elution. More polar hydrocarbons like complex aromatics and hydrocarbon derivatives of Cl, S, and N are retained by alumina and reported as fatty matter[3]	
Organic carbon, total and dissolved	Same method as for waste water	Catalytic combustion of organic matter at 950°C to CO_2 and H_2O followed by measurement of CO_2 produced by an infrared analyzer. Analytical range: 1–150 mg/l. CO_3^{-2} and HCO_3^{-} in the sample should be removed by acidification[1]	

8–1 RECOMMENDED METHODS FOR THE ANALYSIS OF SURFACE WATERS, WASTE WATERS AND SLUDGES (*Continued*)

Property	Surface water	Waste water	Sludge
pH	Same method as for waste water	Electrometric pH measurement, using either a glass electrode in combination with a reference electrode (i.e., saturated calomel electrode) or a combination electrode. Oil and grease cause sluggish response[1]	Electrometric pH measurement on the liquor that separates on standing. Loss of CO_2 gives high results[3]
Phenolics	Same method as for waste water	Phenolic material reacts with 4-amino-antipyrine in the presence of $K_3Fe(CN)_6$ at a pH of 10 to form reddish-brown antipyrine dye; absorbance measured at 460 nm (when chloroform extraction is used) or 510 nm (without extraction). Analytical range: 5–1,000 μg/l with extraction, 50–5,000 μg/l without extraction. Does not differentiate between different kinds of phenols. Preliminary distillation may be necessary to minimize interferences[1,2]	
Phosphorus, all forms	Same method as for waste water	Ammonium molybdate and potassium antimonyltartrate react in acid medium with orthophosphate ion to form an antimony-phospho-molybdate complex which is reduced to a blue-colored complex by ascorbic acid; absorbance measured at 880 nm. Analytical range: 0.01–0.5 mg/l P. H_2SO_4 hydrolysis or persulfate digestion necessary to convert polyphosphates and organic phosphorus compounds to the orthophosphate form. High Fe^{+3} concentrations interfere by precipitating phosphorus. Spiking with NaCl removed $MgCl_2$ interference[1]	

8–1 RECOMMENDED METHODS FOR THE ANALYSIS OF SURFACE WATERS, WASTE WATERS AND SLUDGES (*Continued*)

Property	Surface water	Waste water	Sludge
Phosphorus, (continued)	Same method as for waste water	Automated single reagent method: Ammonium molybdate and potassium antimonyl-tartrate react in an acid medium with dilute solutions of phosphorus to form an antimony-phosphor-molybdate complex which is reduced to a blue-colored complex by ascorbic acid; absorbance measured at 650 nm using 50 mm tubular flow cell. Analytical range: 0.01–1.0 mg/l P. H_2SO_4 hydrolysis or persulfate digestion necessary to convert polyphosphates and organic phosphorus compounds to the orthophosphate form. High Fe concentrations interfere by precipitating phosphorus. Spiking with NaCl removes $HgCl_2$ interference[1] Automated stannous chloride method: Manual digestion with ammonium persulfate and sulfuric acid converts the phosphorus to the orthophosphate form which forms a heteropoly acid, molybdophosphoric acid, with ammonium molybdate. The acid is reduced by $SnCl_2$ to form molybdenum blue complex. Analytical range: 0.01–1.0 mg P/l. Not applicable to saline waters[1]	
Potassium – atomic absorption method[a]	Atomic absorption at 766.5 nm; analytical range: 0.01–2 mg/l; sensitivity: 0.01 mg/l; detection limit: 0.005 mg/l. Na, Li and Cs enhance absorption; burner height adjustment necessary to eliminate this interference	Same as for surface water	
Residue (see Solids, total)			

[a] Atomic absorption methods: Free, unexcited metal atoms present in a flame, when the sample is atomized and aspirated into it, absorb light of a suitable wavelength in proportion to the metal concentration in the sample. Detection limits may be extended by sample concentration, solvent extraction, or scale expansion. Use of less sensitive wavelengths allows direct measurement of high concentrations. Sludges may be analyzed by using acid digestion to solubilize the metals.[1]

8–1 RECOMMENDED METHODS FOR THE ANALYSIS OF SURFACE WATERS, WASTE WATERS AND SLUDGES (*Continued*)

Property	Surface water	Waste water	Sludge
Selenium	Same test as for waste water	Acid permanganate oxidizes selenium to selnate which is reduced to selenite and reacted with diaminobenzidine to form piazselenol complex. This is extracted into toluene and absorbance measured at 420 nm. Analytical range: 0.003–0.05 mg/l selenium. Temperature, time, and pH control essential for reliable results. EDTA eliminate Fe^{+3} interference; oxidative distillation with $KBr\text{-}N_2SO_4$ and H_2O_2 removes interfering I^- and Br^{-1}	
Silica, dissolved	Same test as for waste water	Dissolved silica reacts with molybdate ion in acidic solution to form a greenish-yellow complex; absorbance measured at 410 nm. Analytical range: 2–25 mg silica/l; precision: ~4.1. Beer's law followed only approximately. Turbidity, color, and chromate ion interfere[1,2]	
Silica		Amino-naphthol sulfonic acid method: For low levels, the silico molybdate complex is reduced by amino naphthol sulfonic acid to form heteropoly blue complex; absorbance measured at 815 nm (preferably) or in the 640–700 nm range. Analytical range: 0–2 mg/l; precision: ±0.10 mg/l in the 0–2 mg/l range. Oxalic acid addition eliminates phosphate interference[1,2]	
Silver – atomic absorption method[a]	Atomic absorption at 328.1 nm; analytical range: 0.1–20 mg/l, sensitivity 0.05 mg/l; detection limit 0.01 mg/l. Higher concentrations may be analyzed by using 338.2 nm	Same as for surface water	

[a] Atomic absorption methods: Free, unexcited metal atoms present in a flame, when the sample is atomized and aspirated into it, absorb light of a suitable wavelength in proportion to the metal concentration in the sample. Detection limits may be extended by sample concentration, solvent extraction, or scale expansion. Use of less sensitive wavelengths allows direct measurement of high concentrations. Sludges may be analyzed by using acid digestion to solubilize the metals.[1]

8–1 RECOMMENDED METHODS FOR THE ANALYSIS OF SURFACE WATERS, WASTE WATERS AND SLUDGES (*Continued*)

Property	Surface water	Waste water	Sludge
Sodium – atomic absorption method[a]	Atomic absorption at 330.2 nm; analytical range 1–200 mg/l; sensitivity 0.003 mg/l; detection limit 0.001 mg/l; 589.0 nm line may be used for higher sensitivity	Same as for surface water	
Solids, filterable (dissolved)	Same test as for waste water	Well-mixed sample is filtered through glass fiber filter. Filtrate is evaporated and dried to constant weight at 180°C. Analytical range: 10–20,000 mg/l. Sulfates and chlorides of calcium and magnesium and high concentrations of bicarbonate require prolonged drying[1]	
Solids, nonfilterable (suspended)	Same test as for waste water	A well-mixed sample is filtered through glass fiber filter and the residue retained on the filter is dried to constant weight at 103–105°C. Analytical range: 20–20,000 mg/l[1]	For activated sludge: filter with Whetman No. 1 or equivalent filter, dry the residue at 103–105°C for 30 min and weigh[3]
Solids, total	Same test as for waste water	Gravimetric; well mixed aliquot evaporated to dryness at 103–105°C and residue weighed. Analytical range: 10–30,000 mg/l[1]	Dry 25–50 g sludge sample at 103°C for 1 hr after evaporating to dryness on a water bath. Longer, possibly overnight, drying at 103° necessary if solid content is >15%[3]
Solids, volatile		The residue obtained from the determination of total, suspended, or dissolved solids is ignited at 550°C in a muffle furnace. Loss of weight on ignition is volatile solids, reported as mg/l. Sources of error: loss of water of crystallization, loss of volatile organic matter prior to combustion, incomplete oxidation of complex organics and decomposition of mineral salts[1]	The residue obtained from the determination of total, suspended, or dissolved solids is ignited at 550°C in a muffle furnace. Loss of weight on ignition is volatile solids, reported as mg/l. Sources of error: loss of water of crystallization, loss of volatile organic matter prior to combustion, incomplete oxidation of complex organics and decomposition of mineral salts[1]

[a] Atomic absorption methods: Free, unexcited metal atoms present in a flame, when the sample is atomized and aspirated into it, absorb light of a suitable wavelength in proportion to the metal concentration in the sample. Detection limits may be extended by sample concentration, solvent extraction, or scale expansion. Use of less sensitive wavelengths allows direct measurement of high concentrations. Sludges may be analyzed by using acid digestion to solubilize the metals.[1]

8–1 RECOMMENDED METHODS FOR THE ANALYSIS OF SURFACE WATERS, WASTE WATERS AND SLUDGES (*Continued*)

Property	Surface water	Waste water	Sludge
Specific conductance (EC)	Same method as for waste water	Measured by use of a self-contained conductivity meter, wheatstone bridge-type, or equivalent. Results reported at 25°C; if necessary, after correction[1]	
Sulfate	Same method as for waste water	Barium chloride converts sulfate to barium sulfate suspension; turbidity measured spectrophotometrically. Preferable concentration not greater than 40 mg SO_4/l[1]	
		Automated chloranilate method: Barium chloranilate reacts with sulfate ion to form barium sulfate and colored acid chloranilate ion, measured at 520 nm. Analytical range: 10–400 mg SO_4/l. Ion exchange removes cation interference[1]	
Sulfide	Same method as for waste water	Acidified sample is aerated with an inert gas to strip the sulfide and collect it in a zinc acetate solution where it reacts with iodine. Unreacted iodine is measured titrimetrically with thiosulfate. Analytical range: $\ngtr$1 mg/l. Test does not measure acid insoluble sulfides. Sulfite, thiosulfate, and hydrosulfite and volatile iodine-consuming substances interfere[1]	
Temperature	Measured with any good grade of mercury filled or dial type centigrade thermometer, or thermistor[1]	Measured with any good grade of mercury filled or dial type centigrade thermometer, or thermistor[1]	Record to the nearest degree centigrade at the time of collection[3]

8–1 RECOMMENDED METHODS FOR THE ANALYSIS OF SURFACE WATERS, WASTE WATERS AND SLUDGES (*Continued*)

Property	Surface water	Waste water	Sludge
Threshold odor	Same method as for waste water	Water sample is diluted with odor-free water until a dilution of the least definitely perceptible odor to the tester is obtained. The dilution ratio is reported as the "Threshold Odor Number." A panel of not less than five persons is required to overcome the variability of a single observer[1]	Record in any suitable descriptive way[3]
Turbidity	Same method as for waste water	Measured by the comparison of the intensity of light scattered by the sample under defined conditions with the intensity of light scattered by standard reference suspensions. Results reported in Jackson Turbidity Units (J.T.U.). Preservation of samples not recommended. Floating debris, settleable coarse sediments, finely divided air bubbles and true color interfere[1]	
Zinc – atomic absorption method[a]	Atomic absorption at 213.9 nm; analytical range; 0.1–2 mg/l; sensitivity; 0.02 mg/l; detection limit; 0.005 mg/l. Silicon interferes at high concentration levels; high flame absorption	Same as for surface water	

[a] Atomic absorption methods: Free, unexcited metal atoms present in a flame, when the sample is atomized and aspirated into it, absorb light of a suitable wavelength in proportion to the metal concentration in the sample. Detection limits may be extended by sample concentration, solvent extraction, or scale expansion. Use of less sensitive wavelengths allows direct measurement of high concentrations. Sludges may be analyzed by using acid digestion to solubilize the metals.[1]

REFERENCES

1. *Methods for Chemical Analysis of Water and Wastes,* Environmental Protection Agency, Water Quality Office, 1971.
2. *Annual Book of ASTM Standards,* Part 23, American Society for Testing and Materials, 1971.
3. *Standard Methods for the Examination of Water and Wastewater,* Am. Publ. Health Assn., 1969.

8–2 ANALYSIS OF NO_3^- IN NATURAL WATER SAMPLES

By Specific Ion Electrode Methods and Brucine Method

Sample	Specific conductance, μmho, 25°C	HCO_3^-, ppm	Cl^-, ppm	NO_3^-, ppm Brucine	Electrode[a]	Electrode[b]
1	558	62	37	122	123	120
2	522	121	47	101	99	99
3	427	48	29	93	95	94
4	415	125	17	89	92	88
5	372	42	26	78	82	81
6	366	41	24	77	81	79
7	336	38	22	73	73	73
8	285	33	19	65	63	62
9	246	27	16	55	56	56
10	352	156	11	43	43	44
11	335	122	9.0	40	40	41
12	297	150	5.8	34	34	35
13	519	187	31	33	34	36
14	340	107	19	28	29	30
15	473	146	27	27	28	29
16	432	168	35	26	26	28
17	327	163	7.8	24	23	25
18	254	126	8.6	24	20	22
19	366	173	11	21	21	24
20	407	164	28	21	21	25
21	218	58	18	19	20	21
22	475	260	16	17	15	19
23	439	217	19	16	14	17
24	284	148	6.8	15	16	18
25	258	143	9.8	14	11	14
26	310	157	10	14	14	17
27	407	194	16	12	10	14
28	330	168	6.8	12	13	16
29	325	171	8.9	12	13	16
30	268	162	2.7	11	12	14
31	190	116	2.4	4.5	3.5	5.6
32	163	81	3.7	4.3	3.7	5.2
33	127	59	1.5	3.8	4.8	6.0
34	239	154	2.5	3.5	2.2	5.0
35	198	85	7.0	3.5	2.5	4.2
36	151	79	1.2	3.2	2.9	4.0
37	60	17	0.8	1.5	1.2	1.5

[a] Includes correction for HCO_3^- and Cl^- interferences and ionic strength activity correction.
[b] Ignores correction.

Source: D. Langmuir and R.L. Jacobson, *Environ. Sci. Technol.*, *4*(10):834, 1970.

8–3 ANALYTICAL METHODS FOR NITROGEN AND PHOSPHORUS IN NATURAL WATERS

Compound	Reagent or principle	Ref.
Nitrate	Chromotropic acid-antimony	1
	Brucine	2
	Cd-amalgam reduction	3
Nitrate	Sulfanilamide and *n*-(l-naphthyl)-ethylenediamine	4
Ammonia	Oxidation to NO_2	5
o-Phosphate	Vanadomolybdophosphoric acid	6,7
	Extraction with benzene/isoamyl alcohol	8
	Ascorbic acid-antimony	9
	Persulfate oxidation	10
Total phosphate	Sulfuric-nitric or perchloric acid	11

Source: G.F. Lee, *Water Sewage Works, 116*:R38, 1969.

REFERENCES

1. P.W. West and T.P. Ramachandran, *Anal. Chim. Acta, 35*:317, 1966.
2. D. Jenkins and L.L. Medsker, *Anal. Chem., 36*:610, 1964.
3. A.W. Morris and J.P. Riley, *Anal. Chim. Acta, 29*:272, 1963.
4. M.B. Shinn, *Ind. Eng. Chem. Anal. Ed., 13*:33, 1941.
5. F.A. Richards and R.A. Kletsch, *Recent Research in the Fields of Hydrosphere, Atmosphere, and Nuclear Geochemistry,* Y. Miyake and T. Koyama, Eds., 1964, p. 65.
6. D.A. Wentz, Master's Thesis, University of Wisconsin, Madison, 1967.
7. R.E. Kitson and M.G. Mellon, *Ind. Eng. Chem. Anal. Ed., 16*:379, 1944.
8. D. Jenkins, *J. Water Pollut. Control Fed., 39*:159, 1967.
9. J. Murphy and J.P. Riley, *Anal. Chim. Acta, 27*:31, 1962.
10. D.W. Menzel and H. Corwin, *Limnol. Oceanogr., 10*:240, 1965.
11. G.F. Lee *et al., Air Water Pollut., 9*:715, 1965.

8–4 PRECISION OF THE BRUCINE NITRATE METHOD

Number of replicates[a]	Procedure	Mean	Range	Standard deviation	Coefficient of variation, %
19	Automated	0.080	0.065–0.088	0.0051	6.4
20	Automated	0.265	0.250–0.277	0.0062	2.3
13	Automated	0.447	0.427–0.460	0.0088	2.0
24[b]	Manual	0.341	0.317–0.378	0.015	4.4
25[b]	Manual	0.552	0.522–0.582	0.018	3.3

[a] Nitrate nitrogen, mg/l.

[b] These results were obtained by the recommended procedure described by the source.

Source: L. Kahn and F.T. Brezenski, *Environ. Sci. Technol., 1*(6):492, 1967.

8–5 VANADOMOLYBDOPHOSPHORIC YELLOW COLOR PROCEDURE DATA

Light path, cm	Sensitivity,[a] mg P/l/%T	Minimum detectable concentration (MDC),[b,c] mg P/l	Precision,[c,d] mgP/l	Working range,[e] mg P/l
1	0.15	0.2	0.02	2–18
5	0.030	0.04	0.004	0.4–3.6
10	0.015	0.02	0.002	0.2–1.8

[a] The sensitivity is defined as the change in concentration which accompanies a 1% change in transmittance at a level of 50% T.

[b] MDC @ (A/S = C) $\sqrt{n}$, where A is the mean absorbance determined from n different solutions of a single known concentration, C, and S is the slope of the standard curve [see Wentz (1967) for derivation of the equation].

[c] Determined from 10.0, 2.00, and 1.00 mg of P/l standards for 1, 5, and 10 cm cells, respectively; $n = 5$.

[d] C is the standard deviation for concentration based on $n - 1$ degrees of freedom and is a measure of the precision of the method.

[e] Determined from the intersections of the 10 and 75% T lines with the standard curves.

Source: D.A. Wentz and G.F. Lee, *Environ. Sci. Technol.*, *3*(8):751, 1969.

8–6 SENSITIVITIES[a] OF METHODS OF MANGANESE ANALYSIS

Analytic method	Sensitivity, ppm	Ref.
Neutron activation[b]	0.0006	1
Amperometric titration	0.003	2
Spectrophotometry	0.001	3
Emission spectroscopy[c]	0.02	4
Flame photometry	0.1	5

[a] For about 10% error and assuming a 50 ml sample available.

[b] Flux of $5 \times 10^{11}/cm^2$ sec.

[c] Copper spark.

Source: Reprinted, with publisher's permission, from *J. Am. Water Works Assn.*, *57*(1):107, © 1965 by the American Water Works Association, Inc., 2 Park Ave., New York, N.Y. 10016.

REFERENCES

1. W.W. Memke, *Science, 121*:177, 1955.
2. W.D. Cooke *et al.*, *Anal. Chem.*, *24*:205, 1952.
3. E.M. Gates and G.H. Ellis, *J. Biol. Chem.*, 168:537, 1947.
4. N.N. Nachtrieb, *Principles and Practice of Spectrochemical Analysis,* McGraw-Hill, New York, 1950, p. 272.
5. Reprint No. R-56, Beckman Instruments, Pasadena, Cal., 1952.

8–7 COCRYSTALLIZATION OF TRACE ELEMENTS FROM AQUEOUS SOLUTIONS WITH ORGANIC REAGENTS

		Trace elements recovered, %				
Reagent	Elements cocrystallized	From laboratory solution	At pH	From sea water	At pH	Ref.
α-Benzoinoxime	Mo	99–100	1.8–5.5	99	2	1
1-Nitroso-2-naphthol	Co	100	1–9	99	8[a]	2
	U	>95	7	99	7–8	3
	Fe	99	2–9			3
	Ce,Zn	>95	7			3
	Zr	97	5			3
Thionalide	Ag	100	3.5–7	92	3.5	4
	In,Hg	98–100	3.5–7			4
	Au	98	0.7			4
	Os	100	7			4
	Ta	98	3.5			4
2-Mercaptobenzimidazole	Au	99	1	99	1	5
	Sn	100	5			5
	Hg,Ag	99	1–5			5
	Ta	95–98	1–5			5
Potassium rhodizonate	Ba	100	5–7	99	7–8[b]	6
	Pu			100	7–8[b]	6
	Ra,Sr	100	5–7			6
Ammonium dipicrylaminate	Cs	⩾98	2–5			7
	Rb	⩾95	2–8			7

[a] Untreated sea water.
[b] 2 volumes of 1.6% K-rhodizonate in distilled water + 1 volume of seawater.

Source: T. Joyner *et al.*, *Environ. Sci. Technol.*, *1*(5):417, 1967.

REFERENCES

1. H.V. Weiss and M.G. Lai, *Talanta, 8*:72, 1961.
2. H.V. Weiss and J.A. Reed, *J. Marine Res., 18*:185, 1960.
3. H.V. Weiss, M.G. Lai, and A. Gillespie, *Anal. Chim. Acta, 25*:550, 1961.
4. M.G. Lai and H.V. Weiss, *Anal. Chem., 34*:1012, 1962.
5. H.W. Weiss and M.G. Lai, *Anal. Chim. Acta, 28*:242, 1963.
6. H.V. Weiss and M.G. Lai, *Anal Chem., 32*:475, 1960.
7. H.V. Weiss and M.G. Lai, *J. Inorg. Nucl. Chem., 17*:366, 1961.

8–8 SOLVENT EXTRACTION SYSTEMS FOR CONCENTRATING TRACE ELEMENTS FROM SEA WATER

Elements analyzed	Original sample	Sequestering agents	Solvent	Method of analysis	Ref.
Ce	Biological matter	Bis(2-ethyl hexyl)-	*n*-Heptane	Radiochemical	1
	Sea water	hydrogen phosphate			
Cr^{VI}	Brines	Dithizone	Methyl isobutyl ketone (MIBK)	Atomic absorption	2
Cu	Sea water	Dithizone	CCl_4	Polarographic	3
	Sea water	Sodium diethyldithiocarbamate (DEDC)	CCl_4	Spectrophotometric	4
	Sea water	2,2′-Diquinolyl neocuproine	*n*-Hexanol	Spectrophotometric	5
	Many trace metals		Isobutyl alcohol	Spectrophotometric	6
Fe	Many trace metals	Diphenylphenanthroline	Isobutyl alcohol	Spectrophotometric	6
Mn^{II}	Many trace metals	2-Thenoyltrifluoroacetone (TTA)	Acetone-benzene	Spectrophotometric	7
Mn	Sea water	1-Nitroso-2-naphthol	$CHCl_3$	Spectrophotometric	5
Mo	Radionuclide mixture	*α*-Benzoinoxime	$CHCl_3$	Radiochemical	8
	Brines	Dithiol	MIBK	Atomic absorption	2
		2-Amino-4-chlorobenzenethiol hydrochloride	$CHCl_3$	Spectrophotometric	9
Ni	Many trace metals	Dimethylglyoxime	$CHCl_3$	Spectrophotometric	6
Pb	Environmental samples	NaI, dithizone	Isopropyl methyl ketone	Radiochemical	10
	Sea water	Diethylammonium diethyldithiocarbamate	$CHCl_3$	Spectrophotometric	5
	Many trace metals	Dithizone	$CHCl_3$	Spectrophotometric	6
Tl^{III}	Biological samples	NH_2 SO_4 and rhodamine-B	Benzene	Spectrophotometric	11
W	Steels	8-Quinolinol	$CHCl_3$	Spectrophotometric	12
V	Many trace metals	Phosphotungstate	Isobutyl alcohol	Spectrophotometric	6
Zn	Rocks	Dithizone	$CHCl_3$	Spectrophotometric	13
	Urine	Dithizone	CCl_4	Spectrophotometric	14
Co,Ni	Natural waters	Ammonium pyrrolidine dithiocarbamate (APDC)	MIBK	Atomic absorption	15
Co,Zn	Sea water	Dithizone	CCl_4	Spectrophotometric	16
Fe,Mn	Sea water	DEDC	MIBK	Atomic absorption	17
Al,Cu,Fe	Alloys	Trifluoroacetyl acetone	$CHCl_3$	Gas-liquid chromatography	18
Cu,Mo,Zn	Sea water	8-Quinolinol	$CHCl_3$	Spectrochemical	19
Cu,Mo,Mn,V	Brines	APDC	MIBK	Atomic absorption	20
Cu,Fe,Mn,Ni	Brines	Cupferron	MIBK	Atomic absorption	2
Fe,Mn,Mo, Zn	Milk	APDC	$CHCl_3$	Emission spectrographic	21
Cu,Co,Ni, Pb,Zn	Tungsten	Dithizone	$CHCl_3$	x-ray spectrographic	22
Ag,Cu,Fe, Mo,Ni,V	Sea water	DEDC	$CHCl_3$	Emission spectrographic	23

Source: T. Joyner *et al.*, *Environ. Sci. Technol.*, *1*(5):417, 1967.

8–8 SOLVENT EXTRACTION SYSTEMS FOR CONCENTRATING TRACE ELEMENTS FROM SEA WATER (*Continued*)

REFERENCES

1. B.L. Hampson, *Analyst, 89*:651, 1964.
2. B. Delaughter, *Atomic Absorption Newsletter, 4*:273, Norwalk, Con., 1965.
3. M.K. Tikhonov and V.K. Zhavoronkina, *Soviet Oceanography, 3*:22, 1960.
4. J.D.H. Strickland and T.R. Parsons, *A Manual of Sea Water Analysis,* Fisheries Research Board Canada, 1960.
5. B.A. Loveridge *et al., Determination of Copper, Chromium, Lead and Manganese in Sea Water,* AERE (Harwell), R-3323, 1960.
6. R.M. Sherwood and F.W. Chapman, *Anal. Chem., 27*:88, 1955.
7. A.K. De, and M.S. Rahaman, *Anal. Chem., 35*: 159, 1963.
8. L. Wish, *Anal. Chem., 34*:625, 1962.
9. G.F. Kirkbright and J.H. Roe, *Talanta, 11*:415, 1964.
10. N.A. Talvitie and W.J. Garcia, *Anal. Chem., 37*:851, 1965.
11. V. Miketukova and J. Kohlicek, *Z. Anal. Chem., 208*:7, 1965.
12. A.R. Eberle, *Anal. Chem., 35*:669, 1963.
13. C. Huffman, H.H. Lipp, and L.F. Rader, *Geochim. Cosmochim. Acta, 27*:209, 1963.
14. J.H.R. Kagi and B.L. Vallee, *Anal. Chem., 30*:1951, 1958.
15. D.C. Burrell, *Atomic Absorption Newsletter, 4*:309, 1965.
16. S.H. Fonselius and F. Koroleff, *Bull. Inst. Oceanogr., Monaco, 61*:1, 1963.
17. T. Joyner and J.S. Finley, *Atomic Absorption Newsletter, 5*:4, 1966.
18. D.F. Schutz and K.K. Turekian, *Geochim. Cosmochim. Acta, 29*:259, 1965.
19. R.R. Brooks, *Cosmochim. Acta, 29*:1369, 1965.
20. H. Malissa and E. Schöffmann, *Mikrochim. Acta, 40*:187, 1955.
21. G.L. Hubbard and J.L. Voth, *Anal. Chem., 35*:1957, 1963.
22. T.K. Zhavoronkina, *Soviet Oceanography, 3*:27, 1960.

8–9 EFFICIENCY OF FREEZE CONCENTRATION OF ORGANIC COMPOUNDS IN DILUTE AQUEOUS SOLUTIONS

Compound[a]	Water source	Initial concentration, mg/l	Concentration factor	Recovery, %
Glucose	Distilled	0.019–0.022	3.33–4.56	80–93
	Distilled	1.0	5.65–6.05	97–102
	Lake	0.12–0.15	3.33–5.27	96–100
Glycine	Distilled	0.017–0.023	4.70–6.25	99–103
	Distilled	0.10	21.8	101
	Distilled	1.0	4.35–5.17	99–102
	Lake	0.12–0.15	5.33–7.43	99–102
		0.10	14.5	106
Phenylalanine	Distilled	0.013–0.017	4.54–6.25	98–106
	Distilled	1.0	3.14–5.75	90–98
	Lake	0.12–0.15	3.48–5.54	96–100
Citric acid	Distilled	0.015–0.020	4.04–5.42	97–113
	Distilled	1.0	4.05–6.04	93–104
	Lake	0.12–0.15	5.33–7.03	91–99
Lindane	Distilled	.0134	5.20–5.42	99–101
	Distilled	.0139	4.85–5.86	88–93

[a] Analytical method in all cases – ^{14}C-tracer.

Source: P.A. Kammerer and G.F. Lee, *Environ. Sci. Technol., 3*(3):276, 1969.

8–10 THRESHOLD ODOR NUMBERS FOR VARIOUS DILUTIONS

Sample volume diluted to 200 ml, ml	Threshold odor number	Sample volume diluted to 200 ml, ml	Threshold odor number
200	1	12	17
140	1.4	8.3	24
100	2	5.7	35
70	3	4	50
50	4	2.8	70
35	6	2	100
25	8	1.4	140
17	12	1.0	200

Source: J.W. Clark and W. Viessman, Jr., *Water Supply and Pollution Control,* 2nd ed., International Textbook, Scranton, Pa., 1971.

8–11 QUALITATIVE DESCRIPTIONS OF ODORS

Code	Nature of odor	Description (such as odors of:)
A	Aromatic (spicy)	Camphor, cloves, lavender, lemon
Ac	Cucumber	*Synura*
B	Balsamic (flowery)	Geranium, violet, vanilla
Bg	Geranium	*Asterionella*
Bn	Nasturtium	*Aphanizomenon*
Bs	Sweetish	*Coelosphaerium*
Bv	Violet	*Mallomonas*
C	Chemical	Industrial wastes or treatment chemicals
Cc	Chlorinous	Free chlorine
Ch	Hydrocarbon	Oil refinery wastes
Cm	Medicinal	Phenol and iodoform
Cs	Sulfuretted	Hydrogen sulfide
D	Disagreeable	(pronounced, unpleasant)
Df	Fishy	*Uroglenopsis, Dinobryon*
Dp	Pigpen	*Anabaena*
Ds	Septic	Stale sewage
E	Earthy	Damp earth
Ep	Peaty	Peat
G	Grassy	Crushed grass
M	Musty	Decomposing straw
Mn	Moldy	Damp cellar
V	Vegetable	Root vegetables

Source: Am. Publ. Health Assn., Am. Water Works Assn., and Water Pollut. Control Fed., *Standard Methods for the Examination of Water and Waste Water,* 13th ed., Am. Publ. Health Assn., 1971, p. 251.

8–12 MPN DATA FOR FIVE 10 ML PORTIONS

Number of tubes giving positive reaction out of 5 of 10 ml each	MPN index	95% confidence limits	
		Lower limit	Upper limit
0	0	0	6.0
1	2.2	0.1	12.6
2	5.1	0.5	19.2
3	9.2	1.6	29.4
4	16.0	3.3	52.9
5	Infinite	8.0	Infinite

Note: MPN Index – most probable number of viable bacterial cells in ml of sample, as assayed by pattern of positive and negative results after incubation of media tubes inoculated with 10 ml, 1 ml, or 0.1 ml of sample.

Source: *International Standards for Drinking Water,* World Health Organization, Geneva, Switzerland, 1963.

8–13 MPN DATA FOR FIVE 10 ML, FIVE 1 ML, AND FIVE 0.1 ML PORTIONS

Number of tubes giving positive reaction out of		MPN index	95% of confidence limits	
5 of 1 ml each	5 of 0.1 ml each		Lower limit	Upper limit
0	1	2	<0.5	7
1	0	2	<0.5	7
2	0	4	<0.5	11
0	0	2	<0.5	7
0	1	4	<0.5	11
1	0	4	<0.5	11
1	1	6	<0.5	15
2	0	6	<0.5	15
0	0	5	<0.5	13
0	1	7	1	17
1	0	7	1	17
1	1	9	2	21
2	0	9	2	21
3	0	12	3	28
0	0	8	1	19
0	1	11	2	25
1	0	11	2	25
1	1	14	4	34
2	0	14	4	34
2	1	17	5	46
3	0	17	5	46
0	0	13	3	31
0	1	17	5	46
1	0	17	5	46
1	1	21	7	63

8–13 MPN DATA FOR FIVE 10 ML, FIVE 1 ML, AND FIVE 0.1 ML PORTIONS (*Continued*)

Number of tubes giving positive reaction out of				95% of conficence limits	
5 of 10 ml each	5 of 1 ml each	5 of 0.1 ml each	MPN index	Lower limit	Upper limit
	1	2	26	9	78
	2	0	22	7	67
	2	1	26	9	78
	3	0	27	9	80
	3	1	33	11	93
	4	0	34	12	96
	0	0	23	7	70
	0	1	31	11	89
	0	2	43	15	114
	1	0	33	11	93
	1	1	46	16	120
	1	2	63	21	154
	2	0	49	17	126
	2	1	70	23	168
	2	2	94	28	219
	3	0	79	25	187
	3	1	109	31	253
	3	2	141	37	343
	3	3	175	44	503
	4	0	130	35	302
5	4	1	172	43	486
5	4	2	221	57	698
5	4	3	278	90	849
5	4	4	345	117	999
5	5	0	240	68	754
5	5	1	348	118	1,005
5	5	2	542	180	1,405
5	5	3	918	303	3,222
5	5	4	1,609	635	5,805

Note: MPN Index – most probable number of viable bacterial cells in ml of sample, as assayed by pattern of positive and negative results after incubation of media tubes inoculated with 10 ml, 1 ml, or 0.1 ml of sample.

Source: *International Standards for Drinking Water,* World Health Organization, Geneva, Switzerland, 1963.

8–14 MPN DATA FOR ONE 50 ML AND FIVE 10 ML PORTIONS

Number of tubes giving positive reaction out of			95% confidence limits	
1 of 50 ml	5 of 10 ml each	MPN index	Lower limit	Upper limit
0	1	1	<0.5	4
0	2	2	<0.5	6
0	3	4	<0.5	11
0	4	5	1	13
1	0	2	<0.5	6
1	1	3	<0.5	9
1	2	6	1	15
1	3	9	2	21
1	4	16	4	40

Note: MPN Index – most probable number of viable bacterial cells in ml of sample, as assayed by pattern of positive and negative results after incubation of media tubes inoculated with 10 ml, 1 ml, or 0.1 ml of sample.

Source: *International Standards for Drinking Water,* World Health Organization, Geneva, Switzerland, 1963.

8–15 MPN DATA FOR ONE 50 ML, FIVE 10 ML, AND FIVE 1 ML PORTIONS

Number of tubes giving positive reaction out of				95% confidence limits	
1 of 50 ml	5 of 10 ml each	5 of 1 ml each	MPN index	Lower limit	Upper limit
0	0	1	1	<0.5	4
0	0	2	2	<0.5	6
0	1	0	1	<0.5	4
0	1	1	2	<0.5	6
0	1	2	3	<0.5	8
0	2	0	2	<0.5	6
0	2	1	3	<0.5	8
0	2	2	4	<0.5	11
0	3	0	3	<0.5	8
0	3	1	5	<0.5	13
0	4	0	5	<0.5	13
1	0	0	1	<0.5	4
1	0	1	3	<0.5	8
1	0	2	4	<0.5	11
1	0	3	6	<0.5	15
1	1	0	3	<0.5	8
1	1	1	5	<0.5	13
1	1	2	7	1	17
1	1	3	9	2	21
1	2	0	5	<0.5	13

8–15 MPN DATA FOR ONE 50 ML, FIVE 10 ML, AND FIVE 1 ML PORTIONS (*Continued*)

Number of tubes giving positive reaction out of				95% confidence limits	
1 of 50 ml	5 of 10 ml each	5 of 1 ml each	MPN index	Lower limit	Upper limit
1	2	1	7	1	17
1	2	2	10	3	23
1	2	3	12	3	28
1	3	0	8	2	19
1	3	1	11	3	26
1	3	2	14	4	34
1	3	3	18	5	53
1	3	4	21	6	66
1	4	0	13	4	31
1	4	1	17	5	47
1	4	2	22	7	69
1	4	3	23	9	85
1	4	4	35	12	101
1	4	5	43	15	117
1	5	0	24	8	75
1	5	1	35	12	101
1	5	2	54	18	138
1	5	3	92	27	217
1	5	4	161	39	>450

Note: MPN Index – most probable number of viable bacterial cells in ml of sample, as assayed by pattern of positive and negative results after incubation of media tubes inoculated with 10 ml, 1 ml, or 0.1 ml of sample.

Source: *International Standards for Drinking Water,* World Health Organization, Geneva, Switzerland, 1963.

8–16 MPN DATA FOR FIVE 50 ML, FIVE 10 ML, AND FIVE 1 ML PORTIONS

Number of tubes giving positive reaction out of				95% confidence limits	
5 of 50 ml each	5 of 10 ml each	5 of 1 ml each	MPN index	Lower limit	Upper limit
0	0	1	1	<0.5	2
0	1	0	1	<0.5	2
0	1	1	1	<0.5	2
0	2	0	1	<0.5	2
0	3	0	1	<0.5	2
1	0	0	1	<0.5	2
1	0	1	1	<0.5	2
1	1	0	1	<0.5	2
1	1	1	1	<0.5	2
1	2	0	1	<0.5	2
1	2	1	2	<0.5	4
1	3	0	2	<0.5	4
2	0	0	1	<0.5	2
2	0	1	1	<0.5	2
2	1	0	1	<0.5	2
2	1	1	2	<0.5	4
2	2	0	2	<0.5	4
2	2	1	2	<0.5	4
2	3	0	2	<0.5	4
2	3	1	3	1	7
2	4	0	3	1	7
3	0	0	2	<0.5	4
3	0	1	2	<0.5	4
3	1	0	2	<0.5	4
3	1	1	2	<0.5	4
3	1	2	3	1	7
3	2	0	3	1	7
3	2	1	3	1	7
3	2	2	4	1	9
3	3	0	3	1	7
3	3	1	4	1	9
3	4	0	4	1	9
3	4	1	4	1	9
4	0	0	2	<0.5	4
4	0	1	3	1	7
4	0	2	3	1	7
4	1	0	3	1	7
4	1	1	4	1	9
4	1	2	4	1	9
4	2	0	4	1	9

8–16 MPN DATA FOR FIVE 50 ML, FIVE 10 ML, AND FIVE 1 ML PORTIONS (*Continued*)

Number of tubes giving positive reaction out of				95% confidence limits	
5 of 50 ml each	**5 of 10 ml each**	**5 of 1 ml each**	**MPN index**	**Lower limit**	**Upper limit**
4	2	1	4	1	9
4	2	2	5	2	12
4	3	0	5	2	12
4	3	1	5	2	12
4	3	2	6	2	14
4	4	0	6	2	14
4	4	1	7	3	17
4	5	0	7	3	17
4	5	1	8	3	19
5	0	0	4	1	9
5	0	1	4	1	9
5	0	2	6	2	14
5	1	0	5	2	12
5	1	1	6	2	14
5	1	2	7	3	17
5	2	0	6	2	14
5	2	1	8	3	19
5	2	2	10	4	23
5	2	3	12	4	28
5	3	0	9	3	21
5	3	1	11	4	26
5	3	2	14	5	34
5	3	3	18	6	53
5	4	0	13	5	31
5	4	1	17	6	47
5	4	2	22	7	70
5	4	3	28	9	85
5	4	4	35	11	101
5	5	0	24	8	75
5	5	1	35	11	101
5	5	2	54	18	140
5	5	3	92	27	218
5	5	4	161	39	424

Note: MPN Index – most probable number of viable bacterial cells in ml of sample, as assayed by pattern of positive and negative results after incubation of media tubes inoculated with 10 ml, 1 ml, or 0.1 ml of sample.

Source: *International Standards for Drinking Water,* World Health Organization, Geneva, Switzerland, 1963.

8–17 MPN DATA FOR THREE 10 ML, THREE 1 ML, AND THREE 0.1 ML PORTIONS

Number of tubes giving positive reaction out of				95% confidence limits	
3 of 10 ml each	3 of 1 ml each	3 of 0.1 ml each	MPN index	Lower limit	Upper limit
0	0	1	3	<0.5	9
0	1	0	3	<0.5	13
1	0	0	4	<0.5	20
1	0	1	7	1	21
1	1	0	7	1	23
1	1	1	11	3	36
1	2	0	11	3	36
2	0	0	9	1	36
2	0	1	14	3	37
2	1	0	15	3	44
2	1	1	20	7	89
2	2	0	21	4	47
2	2	1	28	10	149
3	0	0	23	4	120
3	0	1	39	7	130
3	0	2	64	15	379
3	1	0	43	7	210
3	1	1	75	14	230
3	1	2	120	30	380
3	2	0	93	15	380
3	2	1	150	30	440
3	2	2	210	35	470
3	3	0	240	36	1,300
3	3	1	460	71	2,400
3	3	2	1,100	150	4,800

Note: MPN Index – most probable number of viable bacterial cells in ml of sample, as assayed by pattern of positive and negative results after incubation of media tubes inoculated with 10 ml, 1 ml, or 0.1 ml of sample.

Source: *International Standards for Drinking Water,* World Health Organization, Geneva, Switzerland, 1963.

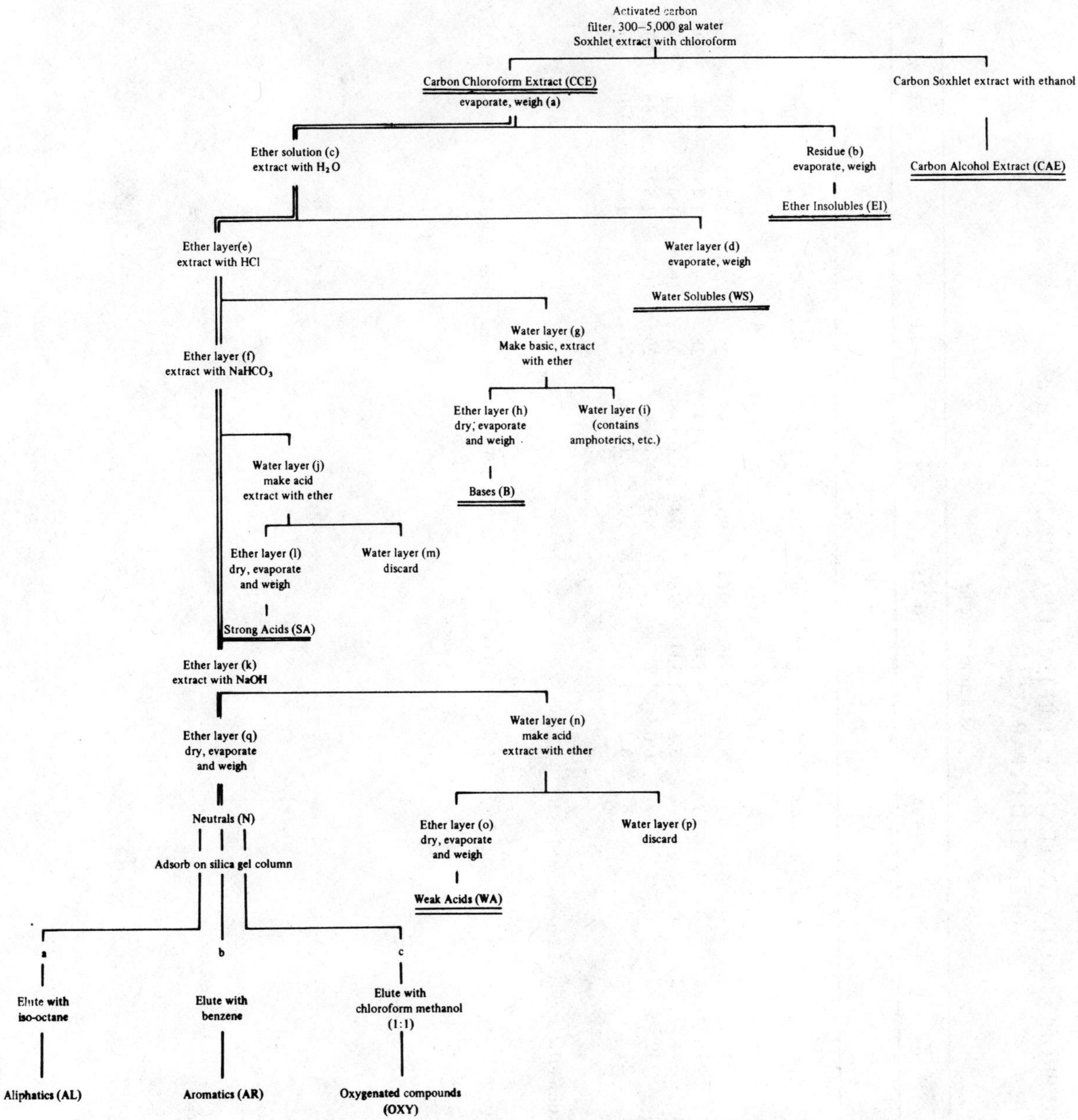

8–18 Flow Scheme for Concentration, Separation, and Identification of Trace Organics in Carbon Adsorption, Extraction, and Chromatography.

General Composition of the Extracts

1. Chloroform extract
   - (a) ether insolubles (0–30%)
     carboxylic acids, ketones, higher alcohols
   - (b) water insolubles
     acids with low boiling points such as hydroxy-acids, keto-acids and keto-alcohols (10–20%)
   - (c) weak acids
     phenols, enols, imides, sulfonamides (5–20%)
   - (d) strong acids
     carboxylic acids such as acetic, benzoic, salicylic or butyric (5–20%)
   - (e) bases
     organic amines such as aniline and pyridine, lower amines (1–2%)
   - (f) neutrals
     hydrocarbons, aldehydes, ketones, esters, ethers (50–70%)
     - (1) aliphatic neutrals: petroleum hydrocarbons, mineral oil type
     - (2) aromatic neutrals: coal tar hydrocarbons such as benzene, toluene, chlorinated hydrocarbon pesticides
     - (3) neutral oxygenated compounds: aldehydes, ketones, esters

2. Alcohol extract
   Contains substances more polar than those removed by chloroform. The synthetic detergents form one major group in this category.

Source: *Handbook of Analytical Toxicology,* I. Sunshine, Ed., CRC Press, Cleveland, 1969, p. 573.

8–19 REPRODUCIBILITY AND EXTRACTION TIME OF CARBON COLUMNS OPERATED IN PARALLEL

		Extraction time, hr																
		0–24					24–35			0–35			35–72			0–72		
		CAE		CCE														
Column	Volume, gal	g	ppb	g	ppb	%[a]	g	ppb	%[a]	g	ppb	%[a]	g	ppb	%[a]	g	ppb	%[a]
1	1,354	1.249	244	1.241	242	84.0	0.096	19	6.6	1.337	261	90.6	0.138	27	9.4	1.475	288	100
2	1,329	1.320	262	1.244	247	84.0	0.094	19	6.5	1.338	266	90.5	0.143	28	9.5	1.481	294	100
3	1,355	1.380	269	1.253	244	83.8	0.081	16	5.5	1.334	260	89.3	0.158	31	10.7	1.492	291	100
4	1,327	1.317	262	1.198	239	83.9	0.085	17	6.0	1.283	256	89.9	0.145	29	10.1	1.428	285	100
Avg	1,341	1.317	259	1.234	243	83.9	0.089	18	6.2	1.323	261	90.1	0.146	29	9.9	1.469	290	100

[a] Percentage of adsorbed material in CCE; extraction efficiency at 72 hr assumed to be 100%.

Source: Reprinted, with publisher's permission, from *J. Am. Water Works Assn.*, *57*(2):215, © 1965 by the American Water Works Association, Inc., 2 Park Ave., New York, N.Y. 10016.

8–20 COMPARISON OF BACTERIAL SPECIES DETECTED IN LUMINOL CHEMILUMINESCENCE SYSTEMS

	Viable cells/ml seeded water	
Organism	Peroxide system[a]	Perborate system[b]
A. aerogenes	7.2 x 10^4	7.2 x 10^4
B. subtilis	2.3 x 10^3	2.6 x 10^3
E. coli	3.4 x 10^3	8.0 x 10^3
P. mirabilis	9.2 x 10^4	7.1 x 10^3
P. fluorescens	1.4 x 10^4	9.3 x 10^3
S. typhimurium	2.1 x 10^5	4.2 x 10^4
S. marcescens	1.9 x 10^4	1.9 x 10^4
S. aureus	2.6 x 10^5	8.0 x 10^4

[a] $Na_4P_2O_7 \cdot 2H_2O_2$.
[b] $NaBO_3 \cdot 4H_2O$.

Note: Luminol (5-amino-2,3-dihydro-1,4-phthalazinedione) chemiluminescence in the presence of sodium perborate ($NaBO_3 \cdot 4H_2O$).

Source: W.S. Oleniacz *et al., Environ. Sci. Technol., 2*(11):1032, 1968.

8–21 THIN LAYER CHROMATOGRAPHY OF CHLORINATED HYDROCARBON PESTICIDES

R_f Values for Silica Gel-G Developed With CCl_4

Pesticide	R_f value	Zone
Methyl parathion	0.05	
Parathion	0.07	
Dieldrin	0.17	II
Endrin	0.20	
Heptachlor epoxide	0.29	
Lindane	0.37	
DDD	0.54	III
γ-Chlordane	0.55	
Heptachlor[a]	0.67	
DDT	0.68	IV
DDE	0.72	
Aldrin	0.73	

[a] Technical grade heptachlor contains approximately 30% γ-chlordane.

Source: *Handbook of Analytical Toxicology,* I. Sunshine, Ed., CRC Press, Cleveland, 1969, p. 576.

8–22 SOLVENT EXTRACTION OF ORGANO–PHOSPHORUS PESTICIDES

Pesticide	Solvent	No. extractions	pH	Mean recovery, %
Parathion	1:1 Hexane and benzene	continuous	–	97
Parathion	Benzene	1	acid	99–100
Malathion	Dichloromethane	3	7.0–8.0	100?
Abate	Chloroform	3	1.0	70
Dipterex and DDVP	Ethyl acetate	2	8.0	94.5

Source: D.C. Abbott *et al., Analyst, 89*:480, 1964.

8–23 THIN LAYER CHROMATOGRAPHY ANALYSIS FOR CHLOROPHENOLS

R_f Values and Sensitivity

Compound	R_f (solvent front)	Sensitivity for chromogens, μg/l	
		$AgNO_3$	Antipyrine
m-Chlorophenol	0.94	0.1	0.5
2,4-Dichlorophenol	0.71	0.1	0.5
2,4,5-Trichlorophenol	0.62	0.1	0.5
2,4,6-Trichlorophenol	0.42	0.1	0.5
Pentachlorophenol	0.09	0.5	–

Source: M.G. Ziegler *et al., Environ. Sci. Technol., 1*(1):65, 1967.

8.1 ELECTROMETRIC METHODS

8.1–1 DETECTION LIMITS OF ELECTROANALYTICAL METHODS OF ANALYSIS

Method	Detection limit, mol/l
AC polarography, chronopotentiometry, potentiometry	$10^{-4}-10^{-5}$
Classical polarography, coulometry at controlled potential, precision null-point potentiometry	$10^{-5}-10^{-6}$
Derivative polarography, square wave polarography, linear sweep voltammetry	$10^{-6}-10^{-7}$
Pulse polarography, amperometry with rotating electrodes	$10^{-7}-10^{-8}$
Anodic stripping with hanging mercury drop electrodes	$10^{-8}-10^{-9}$
Anodic stripping with thin film electrodes or solid electrodes	$10^{-9}-10^{-10}$

Source: K.H. Mancy *et al., Instrumental Analysis for Water Pollution Control,* Ann Arbor Science, Ann Arbor, Mich., 1973.

8.1–2 CHARACTERISTIC STRIPPING PEAK POTENTIAL

Metal	$E_{1/2}$[1]	Sea water, pH 5.8	Distilled water, pH 4.4
Zn	–0.995	–0.98	–0.93
Cd	–0.599	–0.63	–0.53
Pb	–0.396	–0.43	–0.35
Sn(II)	–0.47[a]	–0.43	not observed
Cu	+0.02[b]	–0.15	+0.07

[a] 0.1*N* HCl.
[b] 0.1*N* KNO_3.

Note: E_p in volts vs. saturated calomel electrode.

Source: A. Zirino and M.L. Healy, *Environ. Sci. Technol.,* 6(3):243, 1972.

REFERENCE

1. I.M. Kolthoff and J.J. Lingane, *Polarography,* Interscience, New York, 1961.

8.1–3 POTENTIOMETRIC TITRATION[a] DETERMINATION OF AMMONIUM NITROGEN

Composition of sample	NH_4N, mg/l found in sample when taking end point at various pH: pH 7.5	pH 8.6
Deionized water	0[b]	0[b]
100,000 mg/l of NaCl solution	0	0.8
0.8 mg/l of NH_4N solution	0.7	0.8
3.9 mg/l of NH_4N solution	3.1	4.0
Solution containing 3.9 mg/l of NH_4N and 100 mg/l HCO_3^-	2.1	2.7
Above solution after removal of HCO_3^-	2.7	4.5
Tap water containing <0.1 mg/l of NH_4N	0	0.5
Tap water with 5.3 mg/l of NH_4N added	3.8	5.4
Oilfield brine high in solids[c]	0	5.2
Above brine with 5.3 mg/l of NH_4N added	2.5	9.7
53 mg/l of NH_4N solution	34.2	50.9

[a] The ammonium ion is too weak an acid to be successfully titrated; however, when treated with formaldehyde, hexamethylenetetramine and a strong acid are produced. This strong acid can be titrated with a base using indicators or a potentiometer to determine the end point. The weak hydroxide titrant must be protected from atmospheric carbon dioxide, and a reagent blank must be determined because formaldehyde contains formic acid.

[b] Blank amounting to 0.85 ml (2.1 mg/l of NH_4N) for end point at pH 7.5 and 1.0 ml (2.6 mg/l of NH_4N) for end point at pH 8.6 subtracted from all results.

[c] Sample contained less than 0.4 mg/l of NH_4N, according to Nessler method results without distillation.

Source: A.G. Collins *et al., Environ. Sci. Technol., 3*(3):274, 1969.

8.1–4 COMMERCIALLY AVAILABLE ION-SELECTIVE ELECTRODES

Ion	Membrane	Lower detection limit	Principal interferences
H^+	Glass	10^{-14} M	none below pH 13
Na^+	Glass	10^{-6} M	H^+, Ag^+
K^+	Glass	10^{-6} M	H^+, Ag^+, Na^+
NO_3^-	Liquid	10^{-5} M	I^-, Br^-
Cl^-	Liquid	10^{-5} M	I^-, NO_3^-, Br^-
Ca^{++}	Liquid	10^{-5} M	Zn^{++}, Fe^{++}, Pb^{++}, Cu^{++}
Water hardness	Liquid	10^{-8} M	Zn^{++}, Fe^{++}, Pb^{++}
F^-	Solid	10^{-6} M	OH^-
Cl^-	Solid	5×10^{-5} M	$S^=$, I^-, Br^-, CN^-
Br^-	Solid	5×10^{-6} M	I^-, $S^=$
I^-	Solid	5×10^{-8} M	$S^=$
$S^=$	Solid	10^{-17} M	Hg^{++}
CN^-	Solid	10^{-6} M	$S^=$, I^-
Ag^+	Solid	10^{-17} M	Hg^{++}
Ca^{++}	Solid	10^{-7} M	Ag^+, Cu^{++}, Hg^{++}
Pb^{++}	Solid	10^{-7} M	Ag^+, Cu^{++}, Hg^{++}
Cu^{++}	Solid	10^{-8} M	Ag^+, Hg^{++}, $S^=$

Source: K.H. Mancy *et al., Instrumental Analysis for Water Pollution Control,* Ann Arbor Science, Ann Arbor, Mich., 1973.

8.1–5 SPECIFIC ION ELECTRODES

Ion	Type of electrode	Lower detectable limit, mg/l	Principal interferences	Possible source industries
Bromide	Solid state, heterogeneous[1]	0.4	CN^-, I^-, S^-	Photographic; pharmaceuticals
Cadmium	Solid state	0.01	Ag^+, Hg^{++}, Cu^{++}, Fe^{++}, Pb^{++}	Plating nonferrous alloy; paper; pigments; lubricating oils and greases
Calcium	Liquid ion exchange, heterogeneous[1]	0.4	Zn^{++}, Fe^{++}, Pb^{++}, Cu^{++}, Ni^{++}, Sr^{++}, Mg^{++}, Ba^{++}	Water treatment, softening and ion exchange material; food processing; pharmaceuticals; fertilizers; liquid propellants; explosives
Chloride	Liquid ion exchange, solid state,[1] heterogeneous[1]	0.4	ClO_4^-, I^-, OH^-, NO_3^-, Br^-, OAc^-, HCO_3^-, F^-, $SO_4^=$	Desalination; hydrochloric acid; food; beverages; pharmaceuticals; plastics; glass; pesticides; soaps
	Solid state	1.8	Br^-, CN^-, SCN^-, I^-, NH_3, $S^=$	Desalination; hydrochloric acid; food; beverages; pharmaceuticals; plastics; glass; pesticides; soaps
Copper (II)	Solid state	0.006	Ag^+, Hg^{++}, Fe^{+++}	Plating; printed circuits; ore extractions; mining; algaecides and fungicides; sewage treatment; pharmaceuticals; foods; beverages; paints; pigments
Cyanide	Solid state	0.3	$S^=$, I^-	Plating; metal finishing and extraction; petrochemical
Divalent[1]	Solid state,[1] liquid ion exchange[1]			
Fluoride	Solid state	0.02	OH^-	Etching; glass; ore processing; hydrofluoric acid; electroplating; metal cleaning; plastics; porcelains; fertilizers; pesticides; pharmaceuticals; toothpastes
Fluoroborate (boron)	Liquid ion exchange, heterogeneous[1]	0.11	I^-, HCO_3^-, NO_3^-, F^-, Br^-, OAc^-, OH^-, Cl^-, $SO_4^=$	Plating; alloys; refractories; semiconductors; high-energy fuels; paints; pigments; enamels; glasses
Hydrogen[1]	Glass[1]			
Iodide	Solid state, heterogeneous[1]	0.007	$S^=$, CN^-, $S_2O_3^=$	Salt brine processing; pharmaceuticals
Lead	Solid state	0.02	Ag^+, Hg^{++}, Cu^{++}, Cd^{++}, Fe^{++}	Electroplating; petrochemicals; pesticides; paints; glass; food; nonferrous alloys
Mercury[1]	Solid state[1]			

8.1–5 SPECIFIC ION ELECTRODES (*Continued*)

Ion	Type of electrode	Lower detectable limit, mg/l	Principal interferences	Possible source industries
Nitrate	Liquid ion exchange, heterogeneous[1]	0.6	ClO_4^-, I^-, ClO_3^-, $S^=$, Br^-, NO_2^-, CN^-, HCO_3	Sewage treatment; fertilizers; food; pharmaceuticals; explosives; photographic
Perchlorate	Liquid ion exchange, heterogeneous[1]	1	OH^-, I^-, NO_3^-, $MnO_4^=$, IO_4^-, $Cr_2O_7^=$	Chemical; explosives; solid propellants
Potassium	Liquid ion exchange, glass[1]	2	H^+, NH_4^+, Ag^+, Na^+, Li^+, Cs^+	
Redox	Solid state	Varies	All redox systems	Bleaches; cyanide and chromate
Silver	Solid state, heterogeneous,[1] glass[2]	0.01	Hg^{++}	Photographic
Sodium	Solid state, glass[1]	0.02	Ag^+, H^+, Li^+, K^+	Food; beverages
Sulfide	Solid state, heterogeneous[1]	0.003		Paper pulp; petroleum processing; beer; drilling muds
Thiocyanate	Solid state	0.6	I^-, $S_2O_3^-$, Br^-, Cl^-, NH_3	
Univalent[1]	Glass[1]			
Water hardness	Liquid ion exchange	0.001	Zn^{++}, Fe^{++}, Cu^{++}, Ni^{++}, Ba^{++}, Sr^{++}	

Source: *American Laboratory,* July 1970, p. 20.

REFERENCE

1. R.A. Durst, *American Scientist, 59*(3):353, 1971.

8.1–6 ACCURACY AND PRECISION OF COULOMETRIC TOTAL NITROGEN ANALYSIS

Prepared sample	Sample volume, μl	Prepared standard, N, mg/l	Mean analytical result, N, mg/l	Relative standard deviation, %	No. determinations
Organics in H_2O					
Diethanol amine	8.4	1.14	1.14	1.6	7
Diethanol amine	5.4	11.4	11.1	0.4	6
Aminotriazol	5.6	18.1	14.7	0.6	4
Chloropheniramine	5.6	10.3	10.1	1.6	6
Ethylene diamine	5.8	35.0	35.1	1.8	4
Ethylene diamine	1.8	70.0	71.6	1.4	4
Ethylene diamine	1.6	140.0	143.0	2.5	4
Organics in 0.1*M* NaOH					
Bovine albumine (Recrystallized)	5.6	23.6	23.3	2.5	3
Cystine (NBS standard)	5.6	10.8	10.3	2.7	3
Tryptophan	5.6	10.0	9.8	9.0	3
Valine	5.6	9.50	9.40	8.5	3
Inorganics in H_2O					
Calcium nitrate	5.6	9.5	9.5	2.4	6
Manganous nitrate	5.6	10.0	10.0	2.0	4
Nickel nitrate	5.6	10.7	10.6	2.0	3
Potassium nitrate	5.6	10.5	10.3	2.9	4
Potassium nitrite	5.6	15.0	14.7	2.7	3
Organic Pesticides in Benzene					
Linuron 3-(3,4-dichlorophenyl)-1-methoxy-1-methylurea	5.6	5.3	5.35	3.0	3
EPN (Ethyl P-nitrophenyl thionobenzene-phosphonate)	5.6	10.6	10.2	2.3	5
Organics in Alcohol					
Diethanol amine	5.6	15.8	15.9	1.5	3

Source: R.T. Moore and J.A. McNulty, *Environ. Sci. Technol.*, *3*(8):741, 1969.

8.2 SPECTROMETRIC METHODS

8.2–1 DETECTION LIMITS OF SPECTROMETRIC METHODS OF ANALYSIS

Method	Principle	Detection limit, mol/l
Molecular absorption spectrophotometry	Absorption of radiation by dissolved molecules	$10^{-5}-10^{-6}$
Molecular fluorescence spectrophotometry	Re-emission of radiation absorbed by dissolved molecules	$10^{-7}-10^{-8}$
Atomic absorption spectrophotometry	Absorption of radiation by free atoms	$10^{-6}-10^{-7}$
Atomic fluorescence spectrophotometry	Re-emission of radiation absorbed by free atoms	$10^{-7}-10^{-8}$
Optical and x-ray spectroscopy	Spectral emission analysis by flame, arc or x-ray excitation	$10^{-5}-10^{-6}$
Neutron activation analysis	Nuclear activation by thermal neutrons	$10^{-9}-10^{-10}$

Source: K.H. Mancy *et al.*, *Instrumental Analysis for Water Pollution Control*, Ann Arbor Science, Ann Arbor, Mich., 1973.

8.2–2 METAL IONS DETERMINED BY COLORIMETRY AND MOLECULAR ABSORPTION SPECTROPHOTOMETRY

Metal	Color developing agent	Optimum wave length, mμ	Minimum detectable concentration, μg/l	Possible interferences
Al	"Aluminon"	525	2	F, Ca, chlorine, sulfite polyphosphate
As	Diethyldithiocarbamate	535	1 μg	Co, Ni, Hg, Ag, Pt, Cu, Cr, Mo, Sb
B	Curcumin	540	0.2 μg	Nitrate
Cd	Dithizone	518	0.5 μg	Pb, Zn, Cu
Cu	Cuprethol	435	0.2 (visual)	Bi, Hg, Co, Ni, Ag, etc.
Cr	Diphenylcarbazide	540	10	V, Fe, Mo, Cu, Hg
Fe	Phenanthroline	510	3 μg	Cr, Zn, Co, Cu, Ni, etc. oxidants
Pb	Dithizone	510	2 μg	Heavy metals, organics
Mn	Persulfate	525	5 μg	Br, I
Ni	Heptoxime	445	–	–
Se	Diaminobenzidine	420	1 μg	Fe, I, Br, oxidants
Ag	Dithizone	462 620	0.2 μg	Fe, Cl_2, oxidants
Zn	Dithizone	535	–	Cd, Cu, Pb, Ni

Source: *Standard Methods for the Examination of Water and Waste Water,* 12th ed., Am. Publ. Health Association, New York, 1965.

8.2–3 MOLECULAR ABSORPTION SPECTROPHOTOMETRY FOR METALS IN NATURAL AND WASTE WATERS

Metal	Complexing agent	Solvent extraction	Color of complex	pH range	Suitable wave length, $m\mu$	Useful range, mg/l
Cobalt	Diethyldithio-carbamate	Ethylacetate	Blue	Acidic pH 3.0	367	–
Cadmium	dithizone	Carbon tetrachloride	Red	Alkaline pH 10–12	518	0.1–5
Chromium	1,5 diphenyl carbohydrazide	Butanol	Violet	Acidic pH 2–3	540	0.05–0.5
Copper	Dithizone	Carbon tetrachloride	Violet	Acidic pH 0.5	510	0.04–14
	Diethyldithio-carbamate	Carbon tetrachloride	Yellow-brown	Alkaline pH 9.0	436	0.1–0.8
	Cuprione (2-2′diquinolyl)	Isoamyl alcohol	Purple	pH 5–6	540	–
Iron	*O*-phenanthroline	–	Orange-red	pH 2–9	490	0.01–1.0
	Thioglycollic acid	–	Purple	pH 8–12	540	0.04–1.2
	Tripyridyl	–	Red-purple	pH 9–10	560	0.01–2.0
Lead	Dithizone	Chloroform	Red	pH 7–10	520	–
Mercury	Dithizone	Carbon tetrachloride	Yellow-orange	pH 0–1	500	–
Nickel	Dimethylyoxime	–	Reddish-brown	Neutral	465	–
Zinc	Dithizone	Chloroform	Purple-red	pH 4–5.5	530	0.1–1.0
	Zincom	–	Blue	pH 9.0	620	0.1–2.4

Source: K.H. Mancy *et al., Instrumental Analysis for Water Pollution Control,* Ann Arbor Science, Ann Arbor, Mich., 1973.

8.2–4 SPECTROFLUOROMETRIC METHODS FOR CERTAIN INORGANIC IONS

Determination	Reagent	Conditions	Maxima: Absorption, mμ	Maxima: Fluorescence, mμ	Sensitivity, μg/l	Interferences
Al	Alizarin Garnet R(CI 168)	pH 4.6	470	500	0.007	Be, Co, Cr, Cu, F^-, Fe, NO_3^-, Ni, PO_4^{-3}, Th, Zr
Mg	Bissalicylidene diaminobenzofuran	pH 10.5	475	545	0.002	Mn
Mo	Carminic acid	pH 5.2	560	590	0.1	
Cd	2-(o-hydroxyphenyl) benzoxazole	Precipitated, dissolved in HOAc	365	Blue	2	NH_3
Cu(II)	Tetrachlorotetraiodofluorescein *o*-phenanthroline	pH ≅ 7	560	570	0.001	Cyanide
F^-	Al complex of alizarin garnet R or Eriochrome red B	pH 4.6 (quenched reaction)	470	590	0.001	Be, Co, Cr, Cu, Ni, PO_4^{-3}, Tl
Be	l-amino-4-hydroxyanthraquinone	0.02 M NaOH	540	620	0.2	CrO_4, Li
	2,3-hydroxynaphthoic acid	pH 7.5	380	460	0.0002	Sc, Cr(III), Bi(III), Ce, Sn(IV), Th(IV), Fe
	8-hydroxyquinaldine	($CHCl_3$)pH 8.0	–	–	0.001	Al, Bi, Cd, Cr, Cu, Fe, In, Sn, Ti, Zn

Source: K.H. Mancy *et al., Instrumental Analysis for Water Pollution Control,* Ann Arbor Science, Ann Arbor, Mich., 1973.

14:12

8.2–5 MINIMUM DETECTABLE CONCENTRATIONS[a] FOR ATOMIC ABSORPTION AND EMISSION SPECTROGRAPHY

Element	Minimum for atomic absorption,[b] mg/l	Minimum for emission spectrography, mg/l: Evaporation method[c]	Minimum for emission spectrography, mg/l: Extraction method[d]
Arsenic	0.2–2	0.100	0.020
Barium	0.02–0.33	0.002	Not extracted
Boron	0.1(by flame emission)–6.0	0.010	Not extracted
Cadmium	0.005–0.012	0.020	0.005
Chromium	0.005–0.1	0.010	0.0025
Copper	0.005–0.1	0.010	0.0025
Iron	0.03–0.1	0.010	0.0025
Lead	0.03–0.08	0.040	0.010
Manganese	0.002–0.05	0.010	0.0025
Selenium	0.5–2.0	–	–
Silver	0.01–0.05	0.002	0.0005
Uranium	30–112		
Zinc	0.0012–0.01	0.020	0.005

[a] Smallest concentration of material that can be distinguished from background noise or interference.
[b] Manufacturers' data; range of four systems.
[c] Based on 400 mg/l total dissolved solids (TDS) in sample.
[d] Used if TDS in sample is so large as to render detection limits meaningless.

Source: Reprinted, with publisher's permission, from *J. Am. Water Works Assn., 62*(3):172, © 1970 by the American Water Works Association, Inc., 2 Park Ave., New York, N.Y. 10016.

REFERENCE

1. R.C. Kroner, Federal Water Pollution Control Administration, Cincinnati, Ohio.

8.2–6 ATOMIC ABSORPTION DETECTION LIMITS

Metal	Detection limit[a]	Analytical wave length	Suggested resolution
Silver	0.005	3281	7A
Aluminum[b]	0.1	3093	2A
Arsenic[c]	0.1	1937	7A
Boron[b]	6.0	2497	7A
Barium[b]	0.05	5536	4A
Beryllium[b]	0.002	2349	20A
Bismuth	0.05	2231	2A
Calcium	0.002	4227	13A
Cadmium[c]	0.001	2288	7A
Cobalt	0.005	2407	2A
Chromium	0.005	3579	2A
Copper	0.005	3247	7A
Iron	0.005	2483	2A
Mercury	0.5	2537	20A
Potassium	0.005	7665	13A
Lanthanum[b]	2.0	3928	4A
Lithium	0.005	6708	40A
Magnesium	0.0003	2852	20A
Manganese	0.002	2795	7A
Molybdenum	0.03	3133	2A
Sodium	0.002	5890	4A
Nickel	0.005	2320	2A
Lead	0.03	2833	7A
Antimony	0.1	2175	2A
Selenium[c]	0.1	1961	20A
Silicon[b]	0.1	2516	2A
Tin[c]	0.02	2246	7A
Tellurium	0.1	2143	7A
Titanium[b]	0.1	3643	2A
Thallium	0.025	2768	20A
Vanadium[b]	0.02	3184	7A
Tungsten[b]	3.0	4008	2A
Zinc	0.002	2138	20A
Zirconium[b]	5.0	3601	2A

[a] The detection limit is given by the metal concentration in ppm which gives a signal twice the size of the peak-to-peak variability of the background.
[b] Nitrous oxide flame required.
[c] Indicates use of argon-hydrogen flame.

Source: K.H. Mancy *et al., Instrumental Analysis for Water Pollution Control,* Ann Arbor Science, Ann Arbor, Mich., 1973.

8.2–7 COMPARISON OF DETECTION LIMITS FOR SEVERAL METALS WITH ATOMIC ABSORPTION SPECTROPHOTOMETRY

Directly in Water and Using Solvent Extraction

	Detection limits, μg/l			
		Solvent extraction		
Metal	**Direct in H_2O[1]**	**Ref. 2**	**Ref. 3**	**Ref. 4**
Fe	5	2–3		0.2
Cu	5	0.2		
Zn	2	0.3		
Pb	30	5	0.8	
Cd	1	0.7		
Co	5		0.4	
Ni	5		0.4	1.0

Source: K.H. Mancy *et al., Instrumental Analysis for Water Pollution Control,* Ann Arbor Science, Ann Arbor, Mich., 1973.

REFERENCES

1. H.L. Kahn, in *Trace Inorganics in Water,* R.F. Gould, Ed., Adv. Chem. Series No. 73, Am. Chem. Soc., 1968.
2. J.A. Platte, in *Trace Inorganics in Water,* R.F. Gould, Ed., Adv. Chem. Series No. 73, Am. Chem. Soc., 1965.
3. M.J. Fishman and M.R. Midgett, in *Trace Inorganics in Water,* R.F. Gould, Ed., Adv. Chem. Series No. 73, Am. Chem. Soc., 1968.
4. J.L. Jones and R.D. Eddy, *Anal. Chim. Acta, 43*:165, 1968.

8.2–8 ELIMINATION OF INTERFERENCE DUE TO CALCIUM CARBONATE PRECIPITATION IN ATOMIC ABSORPTION SPECTROPHOTOMETRY ANALYSIS FOR CALCIUM

Apparent Calcium Concentration for Sample Pretreatments, mg/l

Sample source (Wisconsin)	Direct analysis	50% EDTA	1% Lanthanum	pH adjusted to 2.4
Black Earth Creek	49.0			52.0
Comstock Lake	19.1			23.5
Big Twin Lake	21.5			23.0
Fox Lake	31.3			38.8
Lake Katrine	6.0			5.3
Fox Lake	33.5	39.5		40.0
Cox Hollow Lake	22.0	26.0		26.5
Parker Lake	16.0	19.0		19.5
Lake Mendota	23.5	26.0		26.0
Little Cedar Lake			30.0	30.0
Cox Hollow Lake			24.0	23.5
Lake Redstone			25.5	26.0
Lake Mendota			28.0	27.0

Source: E.M. Bentley and G.F. Lee, *Environ. Sci. Technol.*, *1*(9):721, 1967.

8.2–9 REPRESENTATIVE RESULTS OF Hg ANALYSES IN EDIBLE FISH TISSUE

Method	Results, ppm
Sample 1	
Atomic absorption[a]	0.60[1]
Atomic absorption[a]	1.0[1]
Atomic absorption[a]	0.85[1]
X-ray fluorescence	0.40[1]
Destructive NAA[b]	0.87[1]
This method[c]	0.86 (10% standard deviation)
Sample 2	
This method[c]	0.58 (10% standard deviation)
National Bureau of Standards (nondestructive) NAA[2]	0.54
Average results of 15 labs (various methods)[1]	0.51 (range, 0.19–1.2)

[a] Atomic absorption results of three different laboratories.
[b] NAA – neutron activation analysis.
[c] A simple rapid neutron activation method for mercury in fish tissue. The technique employs separation of the activated mercury as an $HgCl_4{}^{2-}$ complex on an anion-exchange resin and direct counting of the resin.

Source: J.M. Rottschafer *et al.*, *Environ. Sci. Technol.*, *5*(4):336, 1971.

REFERENCES

1. R. Grieg, Bureau of Commercial Fisheries, Department of the Interior, Ann Arbor, Mich., personal communication, 1970.
2. P.D. LaFleur, National Bureau of Standards, personal communication, 1970.

8.3 OXYGEN DEMAND

The following section deals with concepts of oxygen demand and other gross measures of organic content. Of major importance are BOD (biological – or biochemical – oxygen demand), COD (chemical oxygen demand), and TOC (total organic carbon).

8.3–1 RELATIONSHIP BETWEEN COD AND TOC FOR ORGANIC OMPOUNDS

Substance	COD/TOC (calculated)	COD/TOC (measured)
Acetone	3.56	2.44
Ethanol	4.00	3.35
Phenol	3.12	2.96
Benzene	3.34	0.84
Pyridine	3.33	Nil
Salicylic acid	2.86	2.83
Methanol	4.00	3.89
Benzoic acid	2.86	2.90
Sucrose	2.67	2.44

Source: D.L. Ford, "Application of the Total Carbon Analyser for Industrial Wastewater Evaluation," *Proc. 23rd Ind. Waste Conf.,* Purdue University, Lafayette, Ind., 1968.

8.3–2 BOD MEASURABLE WITH VARIOUS DILUTIONS OF SAMPLES

Using percent mixtures		By direct pipetting into 300 ml bottles	
% mixture	**Range of BOD**	**ml**	**Range of BOD**
0.01	20,000–70,000	0.02	30,000–105,000
0.02	10,000–35,000	0.05	12,000–42,000
0.05	4,000–14,000	0.10	6,000–21,000
0.1	2,000–7,000	0.20	3,000–10,500
0.2	1,000–3,500	0.50	1,200–4,200
0.5	400–1,400	1.0	600–2,100
1.0	200–700	2.0	300–1,050
2.0	100–350	5.0	120–420
5.0	40–140	10.0	60–210
10.0	20–70	20.0	30–105
20.0	10–35	50.0	12–42
50.0	4–14	100	6–21
100	0–7	300	0–7

Source: C.N. Sawyer and P.L. McCarty, *Chemistry for Sanitary Engineers,* 2nd ed., McGraw-Hill, New York, 1967.

8.3–3 CONVERSION FACTORS FOR 5 DAY BOD

Table A

Time, days	Temperature, °C 17	18	19	20	21	22	23	24	25	26
1	4.03	3.75	3.57	3.35	3.11	2.98	2.77	2.64	2.48	2.33
2	2.20	2.09	1.95	1.85	1.72	1.64	1.57	1.50	1.41	1.34
3	1.61	1.55	1.44	1.37	1.30	1.24	1.18	1.09	1.07	1.02
4	1.32	1.25	1.19	1.14	1.08	1.03	0.99	0.94	0.90	0.87
5	1.15	1.09	1.03	1.00	0.96	0.91	0.88	0.85	0.81	0.78
6	1.04	0.99	0.96	0.92	0.88	0.84	0.81	0.78	0.75	0.73
7	0.96	0.92	0.88	0.86	0.82	0.79	0.77	0.74	0.72	0.69
8	0.91	0.88	0.84	0.81	0.79	0.76	0.74	0.71	0.69	0.67
9	0.87	0.84	0.81	0.78	0.76	0.73	0.71	0.69	0.67	0.65
10	0.84	0.81	0.78	0.76	0.74	0.71	0.71	0.68	0.66	0.64
11	0.82	0.79	0.77	0.74	0.72	0.71	0.68	0.67	0.65	0.63
12	0.79	0.77	0.75	0.73	0.71	0.69	0.67	0.66	0.64	0.63
13	0.79	0.77	0.74	0.72	0.70	0.68	0.67	0.65	0.64	0.62
14	0.77	0.75	0.73	0.71	0.69	0.68	0.66	0.65	0.63	0.62
15	0.77	0.74	0.73	0.71	0.69	0.67	0.66	0.65		0.61
16	0.76	0.74	0.72	0.70	0.68	0.67	0.65	0.64		
17	0.75	0.73	0.72	0.70	0.68	0.67				
18	0.75	0.73	0.71	0.70	0.68	0.66				
19	0.75	0.73	0.71	0.69	0.68					
20	0.74	0.72	0.70	0.68	0.67					

Table B

Time, days	Temperature, °C 27	28	29	30	31	32	33	34	35	36	37.5
1	2.22	2.10	1.97	1.87	1.77	1.67	1.61	1.50	1.43	1.35	1.25
2	1.27	1.21	1.15	1.10	1.05	0.99	0.97	0.91	0.88	0.84	0.79
3	0.98	0.94	0.89	0.86	0.82	0.79	0.79	0.73	0.70	0.68	0.64
4	0.83	0.81	0.77	0.74	0.72	0.70	0.68	0.65	0.63	0.61	0.58
5	0.75	0.73	0.70	0.68	0.66	0.64	0.63	0.60	0.59	0.57	0.55
6	0.71	0.68	0.66	0.64	0.63	0.61	0.60	0.58	0.56	0.55	0.53
7	0.67	0.65	0.64	0.62	0.60	0.59	0.58	0.56	0.55	0.54	0.52
8	0.65	0.63	0.62	0.60	0.59	0.58	0.57	0.55	0.54	0.53	0.52
9	0.63	0.62	0.60	0.59	0.58	0.57	0.56	0.55	0.54	0.53	0.51
10	0.63	0.61	0.60	0.59	0.57	0.56	0.56	0.54	0.53	0.52	
11	0.62	0.61	0.59	0.58	0.57	0.56	0.56				
12	0.62	0.60	0.59	0.58	0.57	0.56	0.55				
13	0.61	0.60	0.58	0.58	0.56	0.55					
14	0.61	0.60		0.57							
15	0.61	0.59									
16	0.60										
17											
18											
19											
20											

Note: To obtain 5 day 20°C. BOD Multiply by factor.

Source: Manual for Sewage Plant Operators, Texas Water and Sewage Works Association, 1955.

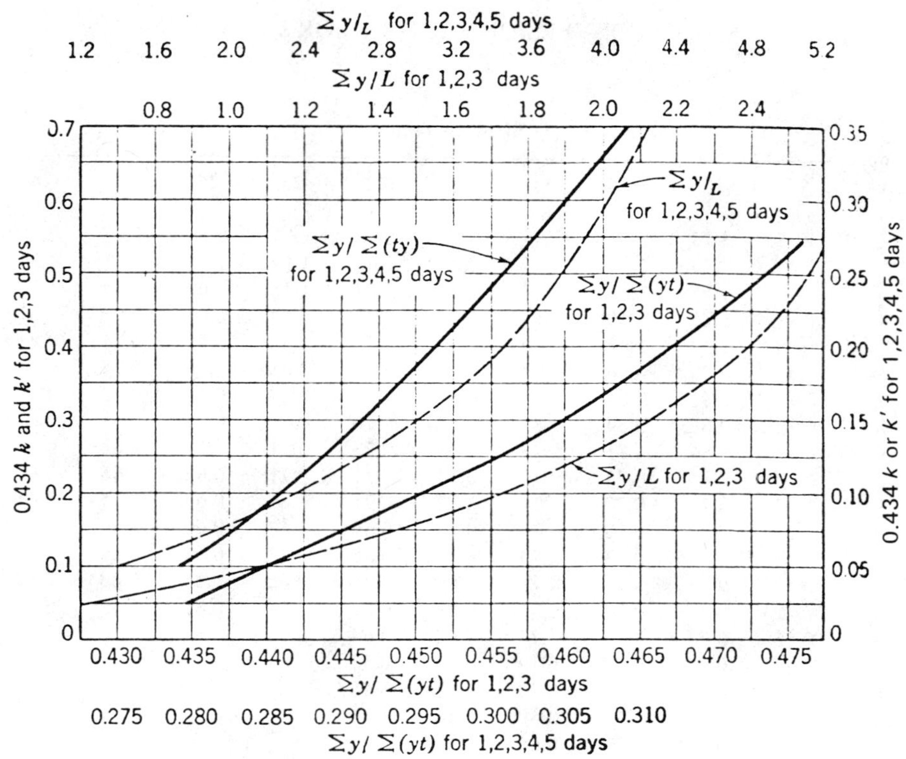

8.3–4 Estimation of BOD Parameters from Early Data.

Note: Values of $k' = 0.434k$ for calculated magnitudes of $\Sigma y/\Sigma(yt)$ and values of $\Sigma y/L$ for these values of k' for three-day and five-day sequences.

To find the magnitudes of the rate of reaction k and k' and the first-stage BOD L, the necessary calculations are $\sum_{i=0}^{n} y_i$ and $\sum_{i=0}^{n} (t_i y_i)$ from $\sum_{i=0}^{n} y_i / \sum_{i=0}^{n} (t_i y_i)$ values on the figure. Read 0.434 k or k' then $\sum_{i=1}^{n} yi/L$. L can be calculated.

Source: F.W. Moore, H.A. Thomas, and W.B. Snow, *Sewage Ind. Wastes, 22*:1343, 1950.

8.3–5 BOD REACTION VELOCITY CONSTANTS

Test	Dilution water	Reaction-velocity constant/day			Lag period, days		
		0.5°C	5°C	20°C	0.5°C	5°C	20°C
Settled domestic sewage	Bicarbonate	0.036	0.059	0.18	1.5	3.7	−0.2
Settled domestic sewage	Phosphate	0.037	0.051	0.12	2.6	1.0	−0.7
Aberjona Brook Water		0.012	0.017	0.026	−8.2	−8.1	−0.4
Cotton filtered domestic sewage	Bicarbonate	–	0.066	0.17	–	3.3	0.0
Cotton filtered domestic sewage	Bicarbonate	–	0.048	0.19	–	1.1	0.1

Source: H.A. Thomas, Jr., *Sewage Ind. Wastes, 28*(1):1, 1951.

8.3–6 PARTICLE SIZE EFFECT ON BOD RATE

	Particle size, μm	Percent of 5-day raw sewage BOD	k_1
Raw sewage solids	–	–	0.22
Settleable	>100	15	0.08
Supracolloidal	100–1	25	0.09
Colloidal	<1	15	0.22
Soluble	–	40	0.39

Source: H. Heukelekian and E. Weisberg, *Sewage Ind. Wastes, 28*(4):558, 1956.

8.3–7 SYNTHETIC SUBSTRATES FOR BOD AND WASTE WATER STUDIES

Component	Concentrations, mg/l									
	1[a]	2[b]	2[c]	3[e]	4[h]	5[a,i]	6[j,k]	7[l]	7[m]	7[n]
D-glucose	300	5,000	10–1,000	660	5,000	360	3,600			
Sucrose	300									
Peptone	300									
Urea	50									
Meat extract	200									
$NH_4H_2PO_4$	1,000									
NaCl	15									
Kcl	7									
$MgSO_4 \cdot 7H_2O$	5	200	100					22.5	0.01	513
$CaCl_2$	7	15	7.5	5.0			1.0	27.5		50
Distilled water, ml	1,000	d	d	d	d	d	d	d	d	d
1 *M* potassium phosphate buffer, pH 7.0, ml/l		30	10							
$NaNO_3$		800	400					182		1,000
$FeCl_3 \cdot 6H_2O$		1	0.5	0.667	2.5					
$MnSO_4 \cdot H_2O$		20	10	6.67	50					
Tap water		100	100		100	1,000				
$(NH_4)_2SO_4$				333.35	2,500		1.0		2.50	
$MgSO_4 \cdot H_2O$				66.65	500					
Yeast extract				25.0			90			
Buffer										
KH_2PO_4				73.60[f] 368.0[g]		9	2.50	8.5	0.085	
K_2HPO_4				144.66[f] 723.32[g]				21.75		250

[a] Synthetic sewage.
[b] Synthetic waste used for growing acclimated seed.
[c] Synthetic waste used in Warburg flasks.
[d] Make up to 1 liter.
[e] Feed substrate.
[f] Low pH unit buffers.
[g] Normal pH unit buffers.
[h] Synthetic waste.
[i] COD = 1200 mg/l; used for anaerobic studies.
[j] Dilute substrate for use in aeration studies.
[k] Values in millimoles/liter except for glucose and yeast which are in mg/l.
[l] Standard dilution water.
[m] Lea-Nichols dilution water.
[n] Used for algae dilution medium.

8.3–7 SYNTHETIC SUBSTRATES FOR BOD AND WASTE WATER STUDIES *(Continued)*

Component	Concentrations, mg/l									
	1[a]	2[b]	2[c]	3[e]	4[h]	5[a,i]	6[j,k]	7[l]	7[m]	7[n]
$CaCl_2 \cdot 2H_2O$					37.5					
1 *M* phosphate buffer, pH 6–8					60					
Tryptose broth						420				
Castile soap						146				
NH_4HCO_3						136				
$NaH_2PO_4 \cdot H_2O$						9	1.50	33.4		
Fe						1				
Mn						1				
Co						1				
Cu						1				
Mo						1				
Zn						1				
$C_6H_{12}O_6$							20.00			
Na_3 citrate							0.50			
NH_4Cl							23.00	1.7		50
$MgCl_2 \cdot 6H_2O$							1.00			
$FeCl_3$							0.10	0.15		3
$MnCl_2 \cdot 4H_2O$							0.02			
$ZnCl_2$							0.02			
$CuCl_2 \cdot 2H_2O$							0.01			
$CoCl_2 \cdot 6H_2O$							0.01			

[a] Synthetic sewage.
[b] Synthetic waste used for growing acclimated seed.
[c] Synthetic waste used in Warburg flasks.
[d] Make up to 1 liter.
[e] Feed substrate.
[f] Low pH unit buffers.
[g] Normal pH unit buffers.
[h] Synthetic waste.
[i] COD = 1200 mg/l; used for anaerobic studies.
[j] Dilute substrate for use in aeration studies.
[k] Values in millimoles/liter except for glucose and yeast which are in mg/l.
[l] Standard dilution water.
[m] Lea-Nichols dilution water.
[n] Used for algae dilution medium.

8.3–7 SYNTHETIC SUBSTRATES FOR BOD AND WASTE WATER STUDIES *(Continued)*

	Concentrations, mg/l									
Component	1[a]	2[b]	2[c]	3[e]	4[h]	5[a,i]	6[j,k]	7[l]	7[m]	7[n]
$(NH_4)_6 MO_7O_{24} \cdot 4H_2O$									0.002	
$Na_2B_4O_7 \cdot 10H_2O$									0.003	
Cu_3PO_4										0.01
$NaHCO_3$										300

[a] Synthetic sewage.
[b] Synthetic waste used for growing acclimated seed.
[c] Synthetic waste used in Warburg flasks.
[e] Feed substrate.
[h] Synthetic waste.
[i] COD = 1200 mg/l; used for anaerobic studies.
[j] Dilute substrate for use in aeration studies.
[k] Values in millimoles/liter except for glucose and yeast which are in mg/l.
[l] Standard dilution water.
[m] Lea-Nichols dilution water.
[n] Used for algae dilution medium.

1. T.O. Rogers and H.A. Wilson, *J. Water Pollut. Control Fed., 38*(6):990, 1966.
2. M.N. Bhatla *et al., J. Water Pollut. Control Fed., 38*(9):1443, 1966.
3. G. Bower and L. Gaddis, *J. Water Pollut. Control Fed., 41*(2):R61, 1969.
4. B.S. Rao *et al., J. Water Pollut. Control Fed., 38*(5):794, 1966.
5. R.R. Dague *et al., J. Water Pollut. Control Fed., 38*(2):220, 1966.
6. E.P. Willimon and J.F. Andrews, *J. Water Pollut. Control Fed., 41*(1):99, 1969.
7. M.M. Allen *et al., J. Water Pollut. Control Fed., 36*(8):1053, 1964.

8.3–8 BOD OF ORGANIC ACIDS AND THEIR SALTS

Compound	Concentration, ppm	BOD, g/g	Incubation time, days	Method	Seed	Ref.
Formic acid	–	0.19	5	St. dil.	Sewage	1
Formic acid	–	0.27	5	St. dil.	Sewage	2
Formic acid	–	0.152	5	St. dil.	Sewage	3
Formic acid	–	0.02	5	St. dil.	Sewage	4
Sodium formate	–	0.096	5	St. dil.	Sewage	5
Sodium formate	–	0.04	5	St. dil.	Sewage	1
Calcium formate	–	0.20	10	St. dil.	Sewage	6
Acetic acid	–	0.88	5	St. dil.	Sewage	1
Acetic acid	–	0.85	5	St. dil.	Sewage	2
Acetic acid	–	0.77	5	St. dil.	Sewage	7
Acetic acid	–	0.53	5	St. dil.	Sewage	8
Acetic acid	–	0.586	5	St. dil.	Sewage	3
Acetic acid	–	0.45	5	St. dil.	–	9
Acetic acid	–	0.34	5	St. dil.	Sewage	4
Acetic acid	–	0.52	5	St. dil.	Sewage	10
Acetic acid	–	0.62	5	St. dil.	Sewage	11
Sodium acetate	2.5–10	0.27	5	St. dil.	Sewage	2
Sodium acetate	–	0.67	5	St. dil.	Sewage	7
Sodium acetate	2.5–20	0.33	5	St. dil.	Sewage	12
Sodium acetate	865	0.39	5	Sierp	Sewage[a]	12
Sodium acetate	250	0.38	5	Warburg	Sewage[a]	12
Sodium acetate	–	0.11, 0.17 0.35, 0.43	5	St. dil.	sp. cult.	5
Sodium acetate	–	0.32	5	St. dil.	Sewage	4
Calcium acetate	–	0.43	5	St. dil.	Sewage	1
Dichloracetic acid	–	0	5	St. dil.	Sewage	8
Propionic acid	2	0.36	5	St. dil.	sp. cult.	13
Propionic acid	2	0.69	10	St. dil.	sp. cult.	13
Propionic acid	–	0.837	5	St. dil.	Sewage	3
Sodium propionate	–	0.52	5	St. dil.	Sewage	14
Sodium propionate	–	0.36	5	St. dil.	sp. cult.	5
Sodium propionate	1,000	0.61	5	Manom.	Sewage	14
Butyric acid	2	0.34	5	St. dil.	sp. cult.	13
Butyric acid	2	0.74	10	St. dil.	sp. cult.	13
Butyric acid	–	0.90	5	St. dil.	–	9
Butyric acid	–	0.89	5	St. dil.	Sewage	3
Sodium butyrate	–	0.41	5	St. dil.	sp. cult.	5
Sodium isobutyrate	–	0.43	5	St. dil.	sp. cult.	5
Valeric acid	–	1.059	5	St. dil.	Sewage	3
Sodium valerate	–	0.48	5	St. dil.	sp. cult.	5
Sodium isovalerate	–	0.57	5	St. dil.	sp. cult.	5
Sodium caproate	–	0.69	5	St. dil.	sp. cult.	5
Glucono delta lactone	–	0.52	5	St. dil.	Sewage	4
Sodium heptylate	–	0.33, 0.82, 0.44	5	St. dil.	sp. cult.	5

[a] 10%.

8.3–8 BOD OF ORGANIC ACIDS AND THEIR SALTS *(Continued)*

Compound	Concentration, ppm	BOD, g/g	Incubation time, days	Method	Seed	Ref.
Sodium caprylate	–	0.76	5	St. dil.	sp. cult.	5
Nonylic acid	–	0.59	5	St. dil.	Sewage	3
Palmitic acid	–	1.07	5	St. dil.	Sewage	3
Sodium palmitate	–	1.02, 0.45, 0.40	5	St. dil.	sp. cult.	5
Stearic acid	–	0.786	5	St. dil.	Sewage	3
Sodium stearate	–	0.45, 0.51	5	St. dil.	Sewage	5
Sodium stearate	2.5–20	1.20	5	St. dil.	Sewage	12
Sodium stearate	865	1.40	5	Sierp	Sewage[a]	12
Sodium stearate	250	1.72	5	Warburg	Sewage[a]	12
Glycolic acid	–	0.175	5	St. dil.	Sewage	3
Oxalic acid	–	0.14	5	St. dil.	Sewage	15
Oxalic acid	–	0.10	5	St. dil.	Sewage	3
Sodium oxalate	2.5–20	0.08	5	St. dil.	Sewage	12
Sodium oxalate	865	0.16	5	Sierp	Sewage[a]	12
Sodium oxalate	250	0.085	5	Warburg	Sewage[a]	12
Lactic acid	–	0.63	5	Warburg	Sewage	3
Lactic acid	2	0.64	5	St. dil.	sp. cult.	13
Lactic acid	2	0.88	10	St. dil.	sp. cult.	13
Succinic acid	–	0.64	5	Warburg	Sewage	3
Succinic acid	2	0.21	5	St. dil.	sp. cult.	13
Succinic acid	2	0.31	10	St. dil.	sp. cult.	13
Fumaric acid	–	0.60, 0.70	5	Warburg	Sewage	16
Fumaric acid	3.75–7.5	0.57	5	Warburg	Sewage	2
Malic acid	–	0.57	5	Warburg	Sewage	2
Malic acid	2.5–10	0.34	5	Warburg	Sewage	2
Maleic acid	–	0.38	5	Warburg	Sewage	2
Maleic anhydride	3.75–7.5	0.60	5	Warburg	Sewage	2
Maleic anhydride	–	0.40	5	Warburg	Sewage	16
Tartaric acid	–	0.30	5	Warburg	Sewage	3
Mucic acid	–	0.245	5	Warburg	Sewage	3
Citric acid	–	0.40	5	Warburg	Sewage	3
Sodium citrate	2.5–20	0.32	5	Warburg	Sewage	12
Sodium citrate	865	0.39	5	Sierp	Sewage[a]	12
Sodium citrate	250	0.39	5	Warburg	Sewage[a]	12
Undecylenic acid	–	0.537	5	St. dil.	Sewage	3
Sodium oleate	–	1.29	5	St. dil.	Sewage	14
Sodium oleate	–	0.47, 0.56	5	St. dil.	sp. cult.	5
Sodium oleate	1,000	1.43	5	Manom.	Sewage	14
Benzoic acid	–	1.40	10	St. dil.	Sewage	6
Benzoic acid	–	1.34	5	St. dil.	Sewage	8
Benzoic acid	–	1.40	5	St. dil.	Sewage	3

[a] 10%.

8.3–8 BOD OF ORGANIC ACIDS AND THEIR SALTS *(Continued)*

Compound	Concentration, ppm	BOD, g/g	Incubation time, days	Method	Seed	Ref.
Sodium benzoate	–	1.13	5	St. dil.	Sewage	12
Sodium benzoate	440	1.36	5	St. dil.	Sewage	12
Benzoyl benzoic acid	–	.001	5	St. dil.	Sewage	16
Salicylic acid, ortho	–	0.95	5	St. dil.	Sewage	3
Mandelic acid	–	0.31	5	St. dil.	Sewage	3
Phthalic acid	–	0.85	5	St. dil.	Sewage	2
Phthalic acid	–	0.89	5	St. dil.	Sewage	2
Phthalic acid	–	1.44	5	St. dil.	Sewage	3
Phthalic anhydride	–	1.20	5	St. dil.	Sewage	2
Phthalic anhydride	1–4	1.26	5	St. dil.	Sewage	2
Phthalic anhydride	–	0.72	5	St. dil.	Sewage	1
Anthranilic acid	–	1.32	5	St. dil.	Sewage	8
Sulfanilic acid	–	1.11	5	St. dil.	Sewage	3
Toluene sulfonic acid	–	0	5	St. dil.	Sewage	8
Anthraquinone sulfonic acid	–	0	5	St. dil.	Sewage	8
Gallic acid	–	0.08	5	St. dil.	Sewage	3
Tannic acid	–	0.31	5	St. dil.	Sewage	3
Tannic acid	–	0.46	5	St. dil.	–	9
Methacrylic acid	–	0.89	5	St. dil.	Sewage	1
Methyl methacrylic acid	–	0.14	5	St. dil.	Sewage	1
Ethylene diamine tetraacetic acid	–	0.01	5	St. dil.	Sewage	11

[a] 10%.

Source: H. Heukelekian and M.C. Rand, *Sewage Ind. Wastes, 27*(9):1040, 1955.

REFERENCES

1. W. Staniar, *Plant Engineering Handbook,* McGraw-Hill, New York, 1950.
2. R.W. Hess, Private Communication.
3. B. Meissner, *Wasserwirtschaft-Wassertechnik, 4*:166, 1954.
4. W. Stafford and H.J. Northrup, Proc. Am. Assn. Text. Chem. Colorests, May 23, 1955.
5. G.E. Symons and A.M. Buswell, *Ind. Eng. Chem., 24*:460, 1932.
6. E.J. Mills, Jr. and V.T. Stack, Jr., Proc. 8th Purdue Ind. Waste Conf., Lafayette, Ind., 1953.
7. Southwest Research Institute, *Eng. News-Rec., 146*(6):*44*, 1951.
8. I. Gellman, "Studies on Biochemical Oxidation of Sewage, Industrial Wastes and Organic Compounds," Ph.D. Thesis, Rutgers University, 1952.
9. R.S. Weston, "Activated Carbon in Sewage and Industrial Waste Treatment," Master's Thesis, New York University, 1939.
10. J.W. Masselli and M.G. Burford, *J. Water Pollut. Control Fed.,* 1109, 1954.
11. J.W. Masselli and M.G. Burford, "Pollution Sources in Wool Scouring and Finishing Mills and Their Reduction through Process and Process Chemical Changes," Rpt. New England Inter. Water Pollut. Control Comm., 1954.
12. I. Gellman, "The Direct Oxygen Utilization Method and Its Application to Study of Oxidation Characteristics of Wastes," Master's Thesis, Rutgers University, 1950.
13. C.E. Zobell, *Biol. Bull., 78*:388, 1940.
14. I. Gellman and H. Heukelekian, *J. Water Pollut. Control Fed., 23*:1267, 1951.
15. M.G. Burford *et al.,* "Industrial Waste Surveys of Two New England Cotton Finishing Mills," Rpt. New England Inter. Water Pollut. Control Comm., 1953.
16. H.G. Swope and M. Kenna, *Sewage Ind. Wastes Eng., 21*:467, 1950.

8.3–9 BOD OF ALCOHOLS

Compound	Concentration, ppm	BOD, g/g	Incubation time, days	Method	Seed	Ref.
Methyl alcohol	–	0.80	5	St. dil.	Sewage	1
Methyl alcohol	1–5	1.12	5	St. dil.	Sewage	2
Methyl alcohol	–	0.90	5	St. dil.	Sewage	3
Methyl alcohol	1–10	0.76	5	St. dil.	Sewage	4
Methyl alcohol	2.5–7.5	0.77	5	St. dil.	Sewage	5
Methyl alcohol	–	0.80	5	St. dil.	Sewage	6
Methyl alcohol	–	0.90	5	St. dil.	Sewage	7
Methyl alcohol	–	0.93	5	St. dil.	Sewage	7
Methyl alcohol	–	0.90	5	St. dil.	Sewage	8
Methyl alcohol	500–1,500	0.13	5	Sierp	Sewage[a]	5
Methyl alcohol	50–250	0.80	5	Warburg	Sewage[a]	5
Methyl alcohol	1,000	0.60	5	Warburg	Sewage[a]	7
Methyl alcohol	–	0.77	5	St. dil.	–	9
Methyl alcohol	–	0.82	5	St. dil.	–	10
Ethyl alcohol	–	1.67	5	St. dil.	Sewage	2
Ethyl alcohol	–	1.44	5	St. dil.	Sewage	3
Ethyl alcohol	1–10	1.07	5	St. dil.	Sewage	4
Ethyl alcohol	–	1.25	5	St. dil.	Sewage	7
Ethyl alcohol	2.5–7.5	1.14	5	St. dil.	Sewage	5
Ethyl alcohol	–	1.40	5	St. dil.	Sewage	6
Ethyl alcohol	–	1.26	5	St. dil.	Sewage	6
Ethyl alcohol	–	1.29	5	St. dil.	Sewage	8
Ethyl alcohol	1,000	1.50	5	Warburg	Sewage[a]	7
Ethyl alcohol	500–1,500	1.37	5	Sierp	Sewage[a]	5
Ethyl alcohol	50–250	1.52	5	Warburg	Sewage[a]	5
Ethyl alcohol	2	1.03	5	St. dil.	sp. cult.	11
Ethyl alcohol	2	1.56	10	St. dil.	sp. cult.	11
Ethyl alcohol	–	1.00	5	St. dil.	Sewage	10
Ethyl alcohol	–	0.93	5	St. dil.	Sewage	1
n-Propyl alcohol	–	0.47	5	St. dil.	Sewage	10
n-Propyl alcohol	–	1.5	5	St. dil.	Sewage	3
iso-Propyl alcohol	–	1.29, 1.59	5	St. dil.	Sewage	7
iso-Propyl alcohol	–	1.29	5	St. dil.	Sewage	6
iso-Propyl alcohol	–	1.42	5	St. dil.	Sewage	8
iso-Propyl alcohol	1,000	0.075	5	Warburg	Sewage[a]	6
Allyl alcohol	–	1.60	10	St. dil.	Sewage	12
Glycerol	–	0.65	5	St. dil.	Sewage	5
Glycerol	–	0.80	5	St. dil.	Sewage	6
Glycerol	–	0.617	5	St. dil.	Sewage	8
Glycerol	150–500	0.78	5	Sierp	Sewage[a]	5
Glycerol	50–250	0.83	5	Warburg	Sewage[a]	5
Glycerol	2	0.81	5	St. dil.	sp. cult.	11
Glycerol	2	0.98	10	St. dil.	sp. cult.	11
Glycerol	–	0.64	5	St. dil.	Sewage	13
n-Butyl alcohol	–	1.45	5	St. dil.	Sewage	3
n-Butyl alcohol	1–5	1.10	5	St. dil.	Sewage	4
n-Butyl alcohol	–	1.5, 1.65	5	St. dil.	Sewage	7
n-Butyl alcohol	1–10	1.45	5	St. dil.	Sewage	4
n-Butyl alcohol	2.5–7.5	1.30	5	St. dil.	Sewage	5
n-Butyl alcohol	–	1.50	5	St. dil.	Sewage[a]	6
n-Butyl alcohol	1,000	1.79	5	Warburg	Sewage[a]	7

[a] 10%.

8.3–9 BOD OF ALCOHOLS (*Continued*)

Compound	Concentration, ppm	BOD, g/g	Incubation time, days	Method	Seed	Ref.
n-Butyl alcohol	1–5	1.92	5	St. dil.	Sewage	14
n-Butyl alcohol	500–1,500	1.67	5	Sierp	Sewage[a]	5
n-Butyl alcohol	50–250	2.04	5	Warburg	Sewage[a]	5
iso-Butyl alcohol	–	1.66	5	St. dil.	Sewage	8
iso-Butyl alcohol	–	0.07	5	St. dil.	Sewage	10
n-Amyl alcohol	–	1.50	5	St. dil.	Sewage	15
n-Amyl alcohol	–	1.61	5	St. dil.	Sewage	3
iso-Amyl alcohol	–	1.50	5	St. dil.	Sewage	3
iso-Amyl alcohol	–	1.62	5	St. dil.	Sewage	8
Amyl alcohol	–	1.23	5	St. dil.	Sewage	16
Amyl alcohol	10–20	1.10	5	St. dil.	Sewage	4
Amyl alcohol	–	1.26	5	St. dil.	Sewage	8
Amyl alcohol	1,000	1.54	5	Sierp	Sewage[a]	16
Amyl alcohol	1,000	1.59	5	Warburg	Sewage[a]	16
Amyl alcohol	500–1,500	1.59	6	Sierp	Sewage[a]	5
Allyl alcohol	–	0.20	5	St. dil.	Sewage	1
Cyclohexanol	–	0.08	5	St. dil.	Sewage	8
Mannitol	–	0.68	5	St. dil.	Sewage	8
Octyl alcohol	–	1.09	5	St. dil.	Sewage	8
Ethylene glycol	–	0.16	5	St. dil.	Sewage	1
Ethylene glycol	12	0.68	5	St. dil.	Sewage	17
Ethylene glycol	12	0.93	10	St. dil.	Sewage	17
Ethylene glycol	–	1.10	10	St. dil.	Sewage	12
Ethylene glycol	–	0.49	5	St. dil.	Sewage	8
Ethylene chlorohydrin	–	0.50	10	St. dil.	Sewage	12
Diethylene glycol	–	nil	10	St. dil.	Sewage	12
Diethylene glycol	–	0.02	5	St. dil.	Sewage	1
Diethylene glycol	–	0.06	5	St. dil.	Sewage	13
Triethylene glycol	–	0.50	10	St. dil.	Sewage	12
Tetraethylene glycol	–	0.50	10	St. dil.	Sewage	12
Polyethylene glycol	–	0.08, 0.30	10	St. dil.	Sewage	12
Ethylene glycol monoethyl ether	–	1.1	10	St. dil.	Sewage	12
1,5 Pentanediol	–	2.0	10	St. dil.	Sewage	12
Methyl phenyl carbinol	–	1.0	10	St. dil.	Sewage	12
Methyl phenyl carbinol redistilled	–	2.1	10	St. dil.	Sewage	12
Benzyl alcohol	–	1.55	5	St. dil.	Sewage	8

[a] 10%.

Source: H. Heukelekian and M.C. Rand, *Sewage Ind. Wastes,* *27*(9):1040, 1955.

8.3–9 BOD OF ALCOHOLS (*Continued*)

REFERENCES

1. C.B. Lamb and G.F. Jenkins, Proc. 7th Purdue Ind. Waste Conf., Lafayette, Ind., 1952.
2. E.J. Mills, Jr. and V.T. Stack, Jr., Proc. 8th Purdue Ind. Waste Conf., Lafayette, Ind., 1953.
3. M.B. Ettinger, R.J. Lishka, and R.C. Kroner, *Ind. Eng. Chem., 46*:791, 1954.
4. H.G. Swope and M. Kenna, *Sewage Ind. Wastes Eng., 21*:467, 1950.
5. R.W. Hess, Private Communication.
6. Southwest Research Institute, *Eng. News-Rec., 146*(6):44, 1951.
7. C.N. Sawyer *et al., S. Water Pollut. Control Fed., 22*:26, 1950.
8. W.L. Lea and M.S. Nichols, *Sewage Works J., 8*:435, 1936.
9. E.W. Lee and W.J. Oswald, *J. Water Pollut. Control Fed., 26*:1097, 1954.
10. A.W. Moore and C.C. Ruchhoft, *J. Water Pollut. Control Fed., 23*:705, 1951.
11. G.L. Symons and A.M. Buswell, *Ind. Eng. Chem., 24*:460, 1932.
12. H. Heukelekian and I. Gellman, *J. Water Pollut. Control Fed., 23*:1546, 1951.
13. I. Gellman and H. Heukelekian, *J. Water Pollut. Control Fed., 23*:1267, 1951.
14. W. Rudolfs, R. Manganelli, and I. Gellman, *Sewage Works J., 21*:605, 1949.
15. Unpublished Data from Department of Sanitation Files, Rutgers University, New Brunswick, N.J.
16. H. Heukelekian, *Sewage Works J. 19*:612, 1947.
17. I. Gellman and H. Heukelekian, *J. Water Pollut. Control Fed., 22*:1321, 1950.
18. I. Gellman and H. Heukelekian, *J. Water Pollut. Control Fed.,* in manuscript.
19. I. Gellman, "The Direct Oxygen Utilization Method and Its Application to Study of Oxidation Characteristics of Wastes," Master's Thesis, Rutgers University, 1950.
20. I. Gellman, "Studies on Biochemical Oxidation of Sewage, Industrial Wastes and Organic Compounds," Ph.D. Thesis, Rutgers University, 1952.
21. G.E. Symons and A.M. Buswell, *Ind. Eng. Chem., 24*:460, 1932.
22. B. Meissner, *Wasserwirtschaft-Wassertechnik, 4*:166, 1954.
23. C.E. Zobell, *Biol. Bull., 78*:388, 1940.
24. R.S. Weston, "Activated Carbon in Sewage and Industrial Waste Treatment," Master's Thesis, New York University, 1939.
25. W. Staniar, *Plant Engineering Handbook,* McGraw-Hill, New York, 1950.
26. S. Kaufman, Lederle Lab. Div. American Cyanamid Co., Private Communication.
27. W. Stafford and H.J. Northrup, Proc. Am. Assn. Text. Chem. Colorests, May 23, 1955.
28. J.W. Masselli and M.G. Burford, *J. Water Pollut. Control Fed., 26*: 1109, 1954.
29. J.W. Masselli and M.G. Burford, "Pollution Sources in Wool Scouring and Finishing Mills and Their Reduction through Process and Process Chemical Changes," Rpt. New England Inter. Water Pollut. Control Comm., 1954.
30. M.G. Burford, *et al.,* "Industrial Waste Surveys of Two New England Cotton Finishing Mills," Rpt. New England Inter. Water Pollut. Control Comm., 1953.

8.3–10 BOD OF ALDEHYDES, KETONES AND QUINONES

Compound	Concentration, ppm	BOD, g/g	Incubation time, days	Method	Seed	Ref.
Formaldehyde	–	1.06	5	St. dil.	Sewage	1
Formaldehyde	–	0.64	5	St. dil.	Sewage	2
Formaldehyde	2.5–10	0.33	5	St. dil.	Sewage	3
Formaldehyde	1.7–20	0.45	5	St. dil.	Sewage	4
Formaldehyde	260	1.10	5	Manom.	Sewage[a]	5
Formaldehyde	260	0.57	5	Manom.	Sewage[b]	5
Formaldehyde	440	0	5	Sierp	Sewage[c]	6
Formaldehyde	130	1.00	7	Warburg	Sewage[a]	7
Formaldehyde	260	1.10	5	Warburg[d]	Sewage[e]	7
Formalin, 35 percent	–	0.37	5	St. dil.	Sewage	8
Acetaldehyde	–	1.27	5	St. dil.	Sewage	1
Butyraldehyde	–	1.06	5	St. dil.	Sewage	9
n-Butyraldehyde	–	1.62	5	St. dil.	Sewage	1
iso-Butyraldehyde	–	1.62	5	St. dil.	Sewage	1
Acetaldol	–	0.90	10	St. dil.	Sewage	10
Crotonaldehyde	–	1.30	10	St. dil.	Sewage	10
Furfural	1.7–20	0.77	5	St. dil.	Sewage	6
Furfural	440	0.28	5	Sierp	Sewage[c]	6
Benzaldehyde	–	1.50	10	St. dil.	Sewage	10
Acetone	–	1.20	10	St. dil.	Sewage	10
Acetone	–	1.39	5	St. dil.	Sewage	1
Acetone	10	0.31	5	St. dil.	Sewage	3
Acetone	1.7–20	0.05–0.72	5	St. dil.	Sewage	6
Acetone	12	1.43	5	St. dil.	Sewage	11
Acetone	12	1.63	10	St. dil.	Sewage	11
Acetone	–	0.544	5	St. dil.	Sewage	2
Acetone	440	0	5	Sierp	Sewage	6
Acetone	–	0.69	5	St. dil.	Sewage	12
Methyl ethyl ketone	–	2.14	5	St. dil.	Sewage	1
Diethyl ketone	–	1.00	10	St. dil.	Sewage	10
Methyl isobutyl ketone	–	0.12	5	St. dil.	Sewage	9
Methyl isobutyl ketone	–	0.6	10	St. dil.	Sewage	10
Methyl isobutyl ketone	–	2.14	5	St. dil.	Sewage	1
Methyl *n*-amyl ketone	–	0.50	10	St. dil.	Sewage	10
Methyl phenyl ketone	–	1.40	10	St. dil.	Sewage	10
Methyl phenyl ketone	–	0.518	5	St. dil.	Sewage	10
1:4 Naphtho quinone	–	0.81	5	St. dil.	Sewage	13

[a] 50%.
[b] 5%.
[c] 10%.
[d] Adapted.
[e] 25–50%.

Source: H. Heukelekian and M.C. Rand, *Sewage Ind. Wastes, 27*(9):1040, 1955.

8.3–10 BOD OF ALDEHYDES, KETONES AND QUINONES *(Continued)*

REFERENCES

1. Southwest Research Institute, *Eng. News-Rec., 146*(6):44, 1951.
2. B. Meissner, *Wasserwirtschaft-Wassertechnik, 4*:166, 1954.
3. H. Heukelekian, *Sewage Works J., 19*:612, 1947.
4. E.W. Lee and W.J. Oswald, *J. Water Pollut. Control Fed., 26*:1097, 1954.
5. I. Gellman and H. Heukelekian, *J. Water Pollut. Control Fed., 22*:1321, 1950.
6. I. Gellman, "The Direct Oxygen Utilization Method and Its Application to Study of Oxidation Characteristics of Wastes," Master's Thesis, Rutgers University, 1950.
7. I. Gellman, "Studies on Biochemical Oxidation of Sewage, Industrial Wastes and Organic Compounds," Ph.D. Thesis, Rutgers University, 1952.
8. W. Stafford and H.J. Northrup, Proc. Am. Assn. Text. Chem. Colorests, May 23, 1955.
9. C.B. Lamb and G.F. Jenkins, Proc. 7th Purdue Ind. Waste Conf., Lafayette, Ind., 1952.
10. E.J. Mills, Jr. and V.T. Stack, Jr., Proc. 8th Purdue Ind. Waste Conf., Lafayette, Ind., 1953.
11. Unpublished Data from Department of Sanitation Files, Rutgers University, New Brunswick, N.J.
12. W. Staniar, *Plant Engineering Handbook,* McGraw-Hill, New York, 1950.
13. R.W. Hess, Private Communication.

8.3–11 BOD OF AMINES AND AMIDES

Compound	Concentration, ppm	BOD, g/g	Incubation time, days	Method	Seed	Ref.
Ethanolamine	–	0.78	5	St. dil.	Sewage	1
Monoethanolamine	2.5	0.80, 1.10	5	St. dil.	Sewage	2
Diethanolamine	2.5	0.10	5	St. dil.	Sewage	2
Triethanolamine	–	nil	10	St. dil.	Sewage	3
Triethanolamine	–	0.01	5	St. dil.	Sewage	4
Triethanolamine	–	0	5	St. dil.	Sewage	2
Aniline	–	1.49	5	St. dil.	Sewage	5
Aniline	–	2.26	5	St. dil.	Sewage	6
Aniline	1.5–3	1.47	5	St. dil.	Sewage	6
Aniline	–	1.50	5	St. dil.	Sewage	4
Diethylaniline	–	0	5	St. dil.	Sewage	1
Melamine	–	0, 0.006	5	St. dil.	Sewage	7
Butylamine hydrochloride	–	1.25	5	St. dil.	Sewage	7
Hexamethylene tetramine	–	0.015, 0.026	5	St. dil.	Sewage	7
p-Toluidine	2.5	1.63	5	St. dil.	Sewage	6
p-Toluidine	–	1.44	5	St. dil.	Sewage	5
o-Toluidine	1.25–2.5	1.43	5	St. dil.	Sewage	6
o-Toluidine	–	0.24	5	St. dil.	Sewage	5
Morpholine	–	nil	10	St. dil.	Sewage	3
Morpholine	–	.02	5	St. dil.	Sewage	2
Diamino pyridine	–	0	5	St. dil.	Sewage	5
Monoethyl aniline	–	0.048	5	St. dil.	Sewage	5
Dimethyl aniline	–	0.25	5	St. dil.	Sewage	5
Diethyl aniline	–	0	5	St. dil.	Sewage	5
Xylidine	–	0	5	St. dil.	Sewage	5
Naphthylamine	–	0.89	5	St. dil.	Sewage	5
Acetanilide	–	1.20	10	St. dil.	Sewage	3
n-Acetyl ethanolamine	–	1.10	10	St. dil.	Sewage	3
n-Acetyl morpholine	–	nil	10	St. dil.	Sewage	3
Acrylonitrile	–	0.70	10	St. dil.	Sewage	3
Formamide	–	0	5	St. dil.	Sewage	8

Source: H. Heukelekian and M.C. Rand, *Sewage Ind. Wastes, 27*(9):1040, 1955.

REFERENCES

1. I. Gellman, "Studies on Biochemical Oxidation of Sewage, Industrial Wastes and Organic Compounds," Ph.D. Thesis, Rutgers University, 1952.
2. C.B. Lamb and G.F. Jenkins, Proc. 7th Purdue Ind. Waste Conf., Lafayette, Ind., 1952.
3. E.J. Mills, Jr. and V.T. Stack, Jr., Proc. 8th Purdue Ind. Waste Conf., Lafayette, Ind., 1953.
4. W. Stafford and H.J. Northroup, Proc. Am. Assn. Text. Chem. Colorests, May 23, 1955.
5. B. Meissner, *Wasserwirtschaft-Wassertechnik, 4*:166, 1954.
6. R.W. Hess, Private Communication.
7. H.G. Swope and M. Kenna, *Sewage Ind. Wastes Eng., 21*:467, 1950.
8. W. Staniar, *Plant Engineering Handbook,* McGraw-Hill, New York, 1950.

8.3–12 BOD OF CARBOHYDRATES

Compound	Concentration, ppm	BOD, g/g	Incubation time, days	Method	Seed	Ref.
Glucose	–	0.53	5	St. dil.	Sewage	1
Glucose	–	0.62	5	St. dil.	Sewage	2
Glucose	–	0.63	5	St. dil.	Sewage	3
Glucose	–	0.53	5	St. dil.	sp. cult.	4
Glucose	–	0.64	5	St. dil.	Sewage	5
Glucose	2.5–20	0.64	5	St. dil.	Sewage	5
Glucose	–	.50	5	St. dil.	Sewage	6
Glucose	1,000	0.74	5	Manom.	Sewage	5
Glucose	865	0.77	5	Sierp	Sewage	7
Glucose	250	0.72	5	Warburg	Sewage	7
Glucose	2	0.78	5	St. dil.	sp. cult.	8
Glucose	2	1.01	10	St. dil.	sp. cult.	8
Glucose	5	0.74	5	St. dil.	Sewage	9
Sucrose	1.25–7	0.688	5	St. dil.	Sewage	10
Sucrose	2.5–20	0.70	5	St. dil.	Sewage	7
Sucrose	865	0.67	5	Sierp	Sewage	7
Sucrose	250	0.70	5	Warburg	Sewage	7
Sucrose	–	0.49	5	St. dil.	Sewage	11
Sucrose	5	0.76	5	St. dil.	Sewage	9
Lactose	–	0.45	5	St. dil.	–	11
Lactose	2.5–20	0.52	5	St. dil.	Sewage	7
Lactose	–	0.55	5	St. dil.	Sewage	6
Lactose	865	0.72	5	Sierp	Sewage	7
Lactose	250	0.65	5	Warburg	Sewage	7
Starch	2	0.22	5	St. dil.	sp. cult.	8
Starch	2	0.59	10	St. dil.	sp. cult.	8
Starch	2.5–20	0.51	5	St. dil.	Sewage	7
Starch	865	0.71	5	Sierp	Sewage	7
Starch	250	0.67	5	Warburg	Sewage	7
Starch	1,000	0.69	5	Manom.	Sewage	5
Starch	–	0.51	5	St. dil.	Sewage	5
Starch	–	0.63	5	St. dil.	Sewage	12
Starch	–	0.68	5	St. dil.	Sewage	6
Starch	20	0.39	5	St. dil.	Sewage	12
Starch	20	0.55	10	St. dil.	Sewage	12
Starch modified	20	0.15	5	St. dil.	Sewage	12
Starch modified	20	0.43	10	St. dil.	Sewage	12
Starch (corn)	–	1.25	5	St. dil.	Sewage	13
Starch (corn)	–	0.60	5	St. dil.	Sewage	9
Starch (corn)	–	0.50	5	St. dil.	Sewage	14
Starch (wheat)	–	0.55	5	St. dil.	Sewage	14
Dextrin	–	.40	5	St. dil.	Sewage	6
Dextrin	–	.50	5	St. dil.	Sewage	1
Cellulose	2	0.08	5	St. dil.	sp. cult.	23
Cellulose	2	0.21	10	St. dil.	sp. cult.	23
Cellulose, alpha	10,000	0.15	5	Sierp	Sewage	15
Cellulose, alpha	10,000	0.40	10	Sierp	Sewage	15
Cellulose, alpha	10,000	0.89	20	Sierp	Sewage	15
Cellulose, alpha	10,000	1.1	80	Sierp	Sewage	15
Cellulose, alpha	10,000	0.70	5	Sierp	Adapted	15

8.3–12 BOD OF CARBOHYDRATES *(Continued)*

Compound	Concentration, ppm	BOD, g/g	Incubation time, days	Method	Seed	Ref.
Cellulose, alpha	10,000	1.03	10	Sierp	Adapted	12
Cellulose, alpha	10,000	1.1	20	Sierp	Adapted	12
Methylcellulose	165	0.008	5	St. dil.	Sewage	12
Carboxymethylcellulose	–	0.3	5	St. dil.	Sewage	14
Lignin	2	0	5	St. dil.	sp. cult.	8
Lignin	2	0	5	St. dil.	sp. cult.	8

Source: H. Heukelekian and M.C. Rand, *Sewage Ind. Wastes, 27*(9):1040, 1955.

REFERENCES

1. W. Stafford and H.J. Northrup, Proc. Am. Assn. Text. Chem. Colorests, May 23, 1955.
2. Southwest Research Institute, *Eng. News-Rec., 146*(6):44, 1951.
3. C.N. Sawyer *et al., J. Water Pollut. Control Fed., 22*:26, 1950.
4. W.L. Lea and M.S. Nichols, *Sewage Works J., 8*:435, 1936.
5. I. Gellman and H. Heukelekian, *Sewage Ind. Wastes, 23*, 1267, 1951.
6. B. Meissner, *Wasserwirtschaft-Wassertechnik, 4*:166, 1954.
7. I. Gellman, "The Direct Oxygen Utilization Method and Its Application to Study of Oxidation Characteristics of Wastes," Master's Thesis, Rutgers University, 1950.
8. C.E. Zobell, *Biol. Bull., 78*:388, 1940.
9. S. Kaufman, Lederle Lab. Div. American Cyanamid Co., Private Communication.
10. R.W. Hess, Private Communication.
11. R.S. Weston, "Activated Carbon in Sewage and Industrial Waste Treatment," Master's Thesis, New York University, 1939.
12. Unpublished Data from Department of Sanitation Files, Rutgers University, New Brunswick, N.J.
13. W. Staniar, *Plant Engineering Handbook,* McGraw-Hill, New York, 1950.
14. M.G. Burford *et al.,* "Industrial Waste Surveys of Two New England Cotton Finishing Mills," Rpt. New England Inter. Water Pollut. Control Comm., 1953.

8.2–13 BOD OF DYES

Compound	Concentration, ppm	BOD, g/g	Incubation time, days	Method	Seed	Ref.
Methyl orange	–	0	5	St. dil.	Sewage	1
Aniline yellow	–	0	5	St. dil.	Sewage	1
Tropeolin 0 0	–	0	5	St. dil.	Sewage	1
Congo red	–	0	5	St. dil.	Sewage	1
Malachite green	–	0	5	St. dil.	Sewage	1
Methyl violet	–	0	5	St. dil.	Sewage	1
Phloxin	–	0	5	St. dil.	Sewage	1
Bromthymol blue	–	0	5	St. dil.	Sewage	1
Phenol red	–	0	5	St. dil.	Sewage	1

Source: H. Heukelekian and M.C. Rand, *Sewage Ind. Wastes, 27*(9):1040, 1955.

REFERENCE

1. B. Meissner, *Wasserwirtschaft-Wassertechnik, 4*:166, 1954.

8.3–14 BOD OF ESTERS

Compound	Concentration, ppm	BOD, g/g	Incubation time, days	Method	Seed	Ref.
Ethyl formate	–	0.5	10	St. dil.	Sewage	1
Ethyl acetate	–	1.13	5	St. dil.	Sewage	2
Ethyl acetate	–	0.29	5	St. dil.	Sewage	3
Ethyl acetate	1,000	1.14	4	Sierp	Sewage	4
Ethyl acetate	1,000	0.76	4	Sierp	Sewage	5
Ethyl acetate	1,000	0.81	4	Sierp	Sewage	5
Ethyl acetate	1,000	0.80	4	Sierp	Sewage	5
Ethyl acetate	1,000	0.83	4	Sierp	Adapted	5
Ethyl acetate	1,000	1.14	4	Sierp	Sewage[a]	5
Ethyl acetate	1.7–20	0.57	5	St. dil.	Sewage	5
Ethyl acetate	440	0.92	5	Sierp	Sewage[b]	5
Ethyl acetate	–	0.66	5	St. dil.	Sewage	6
Ethyl acetate	–	0.86	5	St. dil.	Sewage	7
Butyl acetate	–	0.52	5	St. dil.	Sewage	6
Butyl acetate	–	0.15	5	St. dil.	Sewage	3
Amyl acetate	1.7–20	0.31	5	St. dil.	Sewage	5
Amyl acetate	440	0.88	5	Sierp	Sewage[b]	5
Dibutyl phthalate	–	0.43	5	St. dil.	Sewage	7
Isopropyl acetate	–	0.26	5	St. dil.	Sewage	6

[a] Buffered.
[b] 10%.

Source: H. Heukelekian and M.C. Rand, *Sewage Ind. Wastes, 27*(9):1040, 1955.

REFERENCES

1. E.J. Mills, Jr. and V.T. Stack, Jr., Proc. 8th Purdue Ind. Waste Conf., Lafayette, Ind., 1953.
2. R.W. Hess, Private Communication. Southwest Research Institute, *Eng. News-Rec., 146*(6):44, 1951.
3. B. Meissner, *Wasserwirtschaft-Wassertechnik, 4*:166, 1954.
4. H. Heukelekian and I. Gellman, *J. Water Pollut. Control Fed., 23*:1546, 1951.
5. I. Gellman, "The Direct Oxygen Utilization Method and Its Application to Study of Oxidation Characteristics of Wastes," Master's Thesis, Rutgers University, 1950.
6. C.B. Lamb and G.F. Jenkins, Proc. 7th Purdue Ind. Waste Conf., Lafayette, Ind., 1952.
7. W. Staniar, *Plant Engineering Handbook,* McGraw-Hill, New York, 1950.

8.3–15 BOD OF ETHERS AND POLYMERS

Compound	Concentration, ppm	BOD, g/g	Incubation time, days	Method	Seed	Ref.
Ethyl ether	–	0.03	5	St. dil.	Sewage	1
Diethyl ether	–	nil	10	St. dil.	Sewage	2
Ethylene glycol monomethyl ether (methyl cellosolve)	–	1.10	10	St. dil.	Sewage	2
Ethylene glycol diethyl ether	–	0.10	10	St. dil.	Sewage	2
Ethylene glycol monoethyl ether	5–10	1.58	5	St. dil.	Sewage	3
Diethylene glycol monomethyl ether acetate (methyl carbitol acetate)	–	1.10	10	St. dil.	Sewage	2
Diethylene glycol diethyl ether	–	0.10	10	St. dil.	Sewage	2
Polyethylene glycol	–	0.08, 0.03	10	St. dil.	Sewage	2
Polyethylene glycol diacetate	–	0.20	10	St. dil.	Sewage	2
Dioxane	–	nil	10	St. dil.	Sewage	2
Iso-amyl ether	–	0	5	St. dil.	Sewage	4

Source: H. Heukelekian and M.C. Rand, *Sewage Ind. Wastes, 27*(9):1040, 1955.

REFERENCES

1. H. Heukelekian, *Sewage Works J., 19*:612, 1947.
2. E.J. Mills, Jr. and V.T. Stack, Jr., Proc. 8th Purdue Ind. Waste Conf., Lafayette, Ind., 1953.
3. S. Kaufman, Lederle Lab. Div. American Cyanamid Co., Private Communication.
4. B. Meissner, *Wasserwirtschaft-Wassertechnik, 4*:166, 1954.

8.3–16 BOD OF HETEROCYCLIC COMPOUNDS

Compound	Concentration, ppm	BOD, g/g	Incubation time, days	Method	Seed	Ref.
Pyridine	–	1.47	5	St. dil.	Sewage	1
Pyridine	1.7–20	0	5	St. dil.	Sewage	2
Pyridine	440	0	5	Sierp	Sewage[a]	2
Pyridine	–	0	5	St. dil.	Sewage	3
Pyridine	–	1.15	5	St. dil.	Sewage	4
Quinoline	–	1.71	5	St. dil.	Sewage	4

[a] 10%.

Source: H. Heukelekian and M.C. Rand, *Sewage Ind. Wastes, 27*(9):1040, 1955.

REFERENCES

1. M.B. Ettinger, R.J. Lishka, and R.C. Kroner, *Ind. Eng. Chem., 46*:791, 1954.
2. I. Gellman, "The Direct Oxygen Utilization Method and Its Application to Study of Oxidation Characteristics of Wastes," Master's Thesis, Rutgers University, 1950.
3. I. Gellman, "Studies on Biochemical Oxidation of Sewage, Industrial Wastes and Organic Compounds," Ph.D. Thesis, Rutgers University, 1952.
4. B. Meissner, *Wasserwirtschaft-Wassertechnik, 4*:166, 1954.

8.3–17 BOD OF HYDROCARBONS

Compound	Concentration, ppm	BOD, g/g	Incubation time, days	Method	Seed	Ref.
Gasoline	2.5–5	0.078	5	St. dil.	Sewage	1
Kerosene	1–15	0.53	5	St. dil.	Sewage	1
Benzene	–	1.20	10	St. dil.	Sewage	2
Benzene	–	0	5	St. dil.	Sewage	3
Toluene	–	0	5	St. dil.	Sewage	3
Xylene	–	0	5	St. dil.	Sewage	3
Naphthalene	–	0	5	St. dil.	Sewage	3
α-Methylnaphthalene		0	5	St. dil.	Sewage	3
Tetrahydronaphthalene (tetralin)		0	5	St. dil.	Sewage	3
Halogenated hydrocarbons						
Ethylene dichloride	–	0.05	10	St. dil.	Sewage	3
Ethylene dichloride	–	0.002	5	St. dil.	Sewage	1
Chloroform	–	0.008	5	St. dil.	Sewage	1
Chloroform	–	0	5	St. dil.	Sewage	2
Carbon tetrachloride	–	0	5	St. dil.	Sewage	3
Monochlorobenzene	–	0.03	5	St. dil.	Sewage	4
Nitrated Hydrocarbons						
Nitrobenzene	1.7–20	0	5	St. dil.	Sewage	5
Nitrobenzene	440	0	5	Sierp	Sewage[a]	5

[a] 10%.

Source: H. Heukelekian and M.C. Rand, *Sewage Ind. Wastes, 27*(9):1040, 1955.

REFERENCES

1. H. Heukelekian, *Sewage Works J., 19*:612, 1947.
2. E.J. Mills, Jr. and V.T. Stack, Jr., Proc. 8th Purdue Ind. Waste Conf., Lafayette, Ind., 1953.
3. B. Meissner, *Wasserwirtschaft-Wassertechnik, 4*:166, 1954.
4. J.W. Masselli and M.G. Burford, "Pollution Sources in Wool Scouring and Finishing Mills and Their Reduction through Process and Process Chemical Changes," Rpt. New England Inter. Water Pollut. Control Comm., 1954.
5. I. Gellman, "The Direct Oxygen Utilization Method and Its Application to Study of Oxidation Characteristics of Wastes," Master's Thesis, Rutgers University, 1950.

8.3–18 BOD OF PHENOLIC COMPOUNDS

Compound	Concentration, ppm	BOD, g/g	Incubation time, days	Method	Seed	Ref.
Phenol	–	2.1	10	St. dil.	Sewage	1
Phenol	–	1.71	5	St. dil.	Sewage	2
Phenol	1.25–2.5	1.81	5	St. dil.	Sewage	2
Phenol	–	1.78	5	St. dil.	Sewage	3
Phenol	–	1.40	5	St. dil.	Sewage	4
Phenol	1,000	2.0	5	Manom.	Adapted	5
Phenol	1,000	0.05	5	Manom.	Sewage	5
Phenol	500	0.54	4	Sierp	Sewage[a]	6
Phenol	500	1.84	4	Sierp	Adapted[a]	6
Phenol	1,000	1.88	5	Sierp	Adapted[a]	6
o-Cresol	–	1.64	5	St. dil.	Sewage	3
m-Cresol	–	1.70	5	St. dil.	Sewage	3
p-Cresol	1.7–20	1.40	5	St. dil.	Sewage	6
p-Cresol	440	1.48	5	Sierp	Sewage[b]	6
Xylenol 1:3:5	–	0.82	5	St. dil.	Sewage	3
Xylenol 1:3:4	–	1.50	5	St. dil.	Sewage	3
Pyrocatechol	–	0.69	5	St. dil.	Sewage	3
Resorcinol	–	1.15	5	St. dil.	Sewage	3
Hydroquinone	–	0.48	5	St. dil.	Sewage	3
Pyrogallol	–	0.016	5	St. dil.	Sewage	3
Phloroglucinol	–	0.47	5	St. dil.	Sewage	3
Guiacol	–	1.40	5	St. dil.	Sewage	3
Naphthol, alpha	–	1.69	5	St. dil.	Sewage	3
Naphthol, alpha	–	1.75	5	St. dil.	Sewage	3
Naphthol, beta	0.75–3	1.79	5	St. dil.	Sewage	2
Oxyquinoline	–	0	5	St. dil.	Sewage	3

[a] 50%.
[b] 10%.

Source: H. Heukelekian and M.C. Rand, *Sewage Ind. Wastes, 27*(9):1040, 1955.

REFERENCES

1. E.J. Mills, Jr. and V.T. Stack, Jr., Proc. 8th Purdue Ind. Waste Conf., Lafayette, Ind., 1953.
2. R.W. Hess, Private Communication.
3. B. Meissner, *Wasserwirtschaft-Wassertechnik, 4*:166, 1954.
4. R.S. Weston, "Activated Carbon in Sewage and Industrial Waste Treatment," Master's Thesis, New York University, 1939.
5. H. Heukelekian and I. Gellman, *J. Water Pollut. Control Fed., 23*:1546, 1951.
6. I. Gellman, "The Direct Oxygen Utilization Method and Its Application to Study of Oxidation Characteristics of Wastes," Master's Thesis, Rutgers University, 1950.

8.3–19 BOD OF PROTEINS, AMINO ACIDS, AND THEIR DEGRADATION PRODUCTS

Compound	Concentration, ppm	BOD, g/g	Incubation time, days	Method	Seed	Ref.
Casein	–	0.25	5	St. dil.	Sewage	1
Casein	–	0.77	5	St. dil.	Sewage	2
Casein	1,000	0.95	5	Manom.	Sewage	2
Casein	2.5–20	0.77	5	St. dil.	Sewage	2
Casein	865	1.01	5	Sierp	Sewage	3
Casein	250	0.89	5	Warburg	Sewage	3
Casein	–	0.57	5	St. dil.	Sewage	4
Casein	–	1.17	5	St. dil.	Sewage	5
Gelatin	–	0.88	5	St. dil.	Sewage	5
Gelatin	2.5–20	0.55	5	St. dil.	Sewage	3
Gelatin	865	0.66	5	Sierp	Sewage	3
Gelatin	250	0.69	5	Warburg	Sewage	3
Gelatin	10	0.92	5	St. dil.	Sewage	6
Peptone	–	0.15	5	St. dil.	Sewage	4
Peptone	5	0.88	5	St. dil.	Sewage	6
Uric acid	–	0.30	5	St. dil.	Sewage	4
Indol	–	2.07	5	St. dil.	Sewage	7
Skatol	–	1.51	5	St. dil.	Sewage	7
Glycine	–	0.55	5	St. dil.	Sewage	4
Glycine	5	0.52	5	St. dil.	Sewage	6
Alanine	–	0.94	5	St. dil.	Sewage	4
Aspartic acid	–	0.58	5	St. dil.	Sewage	4
Glutamic acid	–	0.64	5	St. dil.	Sewage	8
Glutamic acid	–	0.45	ult.	St. dil.	Sewage[a]	9
Glutamic acid	–	0.42	ult.	St. dil.	Sewage[a]	9

[a] Frozen.

Source: H. Heukelekian and M.C. Rand, *Sewage Ind. Wastes, 27*(9):1040, 1955.

REFERENCES

1. W. Stafford and H.J. Northrup, Proc. Am. Assn. Text. Chem. Colorests, May 23, 1955.
2. I. Gellman and H. Heukelekian, *J. Water Pollut. Control Fed., 23*:1267, 1951.
3. I. Gellman, "The Direct Oxygen Utilization Method and Its Application to Study of Oxidation Characteristics of Wastes," Master's Thesis, Rutgers University, 1950.
4. B. Meissner, *Wasserwirtschaft-Wassertechnik, 4*:166, 1954.
5. W. Staniar, *Plant Engineering Handbook,* McGraw-Hill, New York, 1950.
6. S. Kaufman, Lederle Lab. Div. American Cyanamid Co., Private Communication.
7. E.J. Mills, Jr. and V.T. Stack, Jr., Proc. 8th Purdue Ind. Waste Conf., Lafayette, Ind., 1953.
8. C.N. Sawyer *et al., J. Water Pollut. Control Fed., 22*:26, 1950.
9. E.W. Lee and W.J. Oswald, *J. Water Pollut. Control Fed., 26*:1097, 1954.

8.3–20 BOD OF SOAPS AND DETERGENTS (COMMERCIAL)

Compound	Concentration, ppm	BOD, g/g	Incubation time, days	Method	Seed	Ref.
Ivory Snow	–	1.41	5	St. dil.	Sewage	1
Lux Flakes	–	1.47	5	St. dil.	Sewage	1
Super Suds	–	0.97	5	St. dil.	Sewage	1
Dreft	–	0.49	5	St. dil.	Sewage	1
Nacconol N.R. (Swerl)	–	0	5	St. dil.	Sewage	1
Nacconol N.R.	–	.02	5	St. dil.	Sewage	2
Tween 80	–	0.35	5	St. dil.	Sewage	3
Sodium alginate	4	0.36	5	St. dil.	Sewage	3

Source: H. Heukelekian and M.C. Rand, *Sewage Ind. Wastes, 27*(9):1040, 1955.

REFERENCES

1. W. Rudolfs, R. Manganelli, and I. Gellman, *Sewage Works J., 21*:605, 1949.
2. M.G. Burford *et al.,* "Industrial Waste Surveys of Two New England Cotton Finishing Mills," Rpt. New England Inter. Water Pollut. Control Comm., 1953.
3. Unpublished Data from Department of Sanitation Files, Rutgers University, New Brunswick, N.J.

8.3–21 BOD AND COD VALUES ON PARAFFINIC HYDROCARBONS USING APHA STANDARD METHODS

	5 day BOD, ppm	COD, ppm	TOD,[a] ppm
n-Hexane	0	18	2,328
n-Heptane	0	132	2,402
n-Decane	28	59	2,544
n-Dodecane	37	22	2,662
n-Hexadecane	53	40	2,683
n-Heptadecane	60	0	2,692

[a] TOD–Theoretical oxygen demand.

Note: Test concentrations: 1 ml/l, diluted 1:300.

Source: Reprinted, with publisher's permission, from *J. Am. Water Works Assn., 62*(12):784, © 1970 by the American Water Works Association, Inc., 2 Park Ave., New York, N.Y. 10016.

8.3–22 OXYGEN DEMANDS OF INDUSTRIAL TOXIC ORGANIC COMPOUNDS

Chemical and seed	5 day BOD, mg/l	COD, mg/l	Ratio COD/BOD
Acrylonitrile with acrylonitrile seed	721,000	1,612,500	2.23
Acrylamide monomer with acrylamide monomer seed	972,000	1,300,000	1.34
Acrylamide monomer with acrylonitrile seed	217,000	–	
Acrylamide monomer with sewage seed	168,500	–	
Oxydipropionitrile	Undetermined	–	

Source: A.B. Cherry *et al., Sewage Ind. Wastes, 28*(9):1145, 1956.

8.3–23 OXYGEN UPTAKE WITH PHENOL-ADAPTED CULTURE

Compound	Test time, min	Oxygen consumed, μl: Endogenous	Oxygen consumed, μl: Cells + test compound	Oxygen consumed, μl: Cells + phenol[b] 90 min	Test compound, mg/l	Ratio endogenous to:[a] Test compound[c]	Ratio endogenous to:[a] Phenol[d]
Phenols							
Phloroglucinol	210	69	79	320	60	1.1	11.9
Pyrogallol	210	80	0	359	100	<End.	9.2
Catechol	180	125	380	368	↓	3.0	4.7
Resorcinol	↓	125	377	368	↓	3.0	4.7
Quinol	↓	125	274	368	↓	2.2	4.7
m-Chlorophenol	↓	47	113	334	↓	2.4	12.8
p-Chlorophenol	180	47	127	334	100	2.7	12.8
2,4-Dichlorophenol	210	57	106	345	60	1.9	12.8
2,6-Dichlorophenol	170	65	104	386	100	1.6	10.7
2,4,6-Trichlorophenol	180	47	103	334	↓	2.2	12.8
2,4-Dibromophenol	150	70	92	356	↓	1.3	8.5
o-Aminophenol	210	57	138	345	↓	2.4	12.8
m-Aminophenol	180	125	249	368	↓	2.0	4.7
p-Aminophenol	↓	125	184	368	↓	1.5	4.7
2,4-Dimethylphenol	↓	47	173	334	↓	3.7	12.8
2,6-Dimethylphenol	↓	47	87	334	↓	1.9	12.8
3,4-Dimethylphenol	↓	47	236	334	↓	5.0	12.8
3,5-Dimethylphenol	180	47	117	334	↓	2.5	12.8
o-Cresol	120[e]	36	443	345	↓	12.3	12.8
m-Cresol	120[e]	94	506	368	↓	5.4	4.7
p-Cresol	90[e]	78	382	368	↓	4.9	4.7
Dimethyldihydroresorcinol	150	70	69	356	↓	<End.	8.5
Orcinol	150	70	142	356	100	2.0	8.5
6-Chloro-*m*-cresol	170	65	146	386	80	2.2	10.7
4,6-Dichloro-*m*-cresol	170	65	89	386	80	1.4	10.7
6-Chloro-2-methylphenol	170	65	131	386	80	2.0	10.7
4-Chloro-2-methylphenol	180	67	160	331	80	2.4	9.5
4-Chloro-3-methylphenol	210	69	195	320	60	2.8	11.9
Resorcinol monoacetate	190	55	261	336	100	4.7	10.8
o-Nitrophenol	210	58	112	336	100	1.9	11.2
m-Nitrophenol	210	58	130	336	100	2.2	11.2

[a] Endogenous considered as 1.0.
[b] Phenol test reached endogenous level in 90 min with 100 mg/l phenol.
[c] Endogenous O_2 uptake at time shown in column 1 used.
[d] 90 min endogenous result used in calculation.
[e] Ratio calculated at time when activity was at or near the endogenous level.

8.3–23 OXYGEN UPTAKE WITH PHENOL-ADAPTED CULTURE *(Continued)*

Compound	Test time, min	Oxygen consumed, μl: Endogenous	Oxygen consumed, μl: Cells + test compound	Oxygen consumed, μl: Cells + phenol[b] 90 min	Test compound, mg/l	Ratio endogenous to:[a] Test compound[c]	Ratio endogenous to:[a] Phenol[d]
p-Nitrophenol	180	125	179	368	100	1.4	4.7
2,4-Dinitrophenol	210	58	131	336	60	2.3	11.2
2,6-Dinitrophenol	↓	69	127	320	60	1.8	11.9
2,4,6-Trinitrophenol	↓	58	82	336	100	1.4	1.2
2,4,6-Trinitroresorcinol	210	58	66	336	60	1.1	1.2
2-Chloro-4-nitrophenol	230	165	249	384	60	1.5	4.9
4-Chloro-2-nitrophenol	180	67	190	331	100	2.8	9.5
2,6-Dichloro-4-nitrophenol	210	69	104	320	100	1.5	11.9
4,6-Dinitro-*o*-cresol	↓	58	92	336	100	1.6	11.2
2,4,6-Trinitro-*m*-cresol	↓	58	76	336	60	1.3	11.2
2,6-Dinitrothymol	210	58	85	336	60	1.5	11.2
Thymol	180	125	173	368	100	1.4	4.7
p-Tertiarybutylphenol	190	55	153	336	100	2.8	10.8
2,4-Ditertiaryamylphenol	190	55	143	336	80	2.6	10.8
o-Phenylphenol	200	52	119	275	100	2.3	9.6
a-Phenyl-*p*-cresol	190	55	136	336	80	2.5	10.8
m-Phenoxyphenol	190	55	64	336	100	1.2	10.8
Benzyl Alcohols							
Benzyl alcohol	210	69	98	320	100	1.4	11.9
DL-*a*-Methylbenzylalcohol	210	69	75	320	100	1.1	11.9
Heterocyclic Compounds							
Quinoline	180	67	140	331	100	2.1	9.5
Cyclic Compounds							
Cyclohexanol	210	69	71	320	100	1.0	11.9
Shikimic acid	180	55	220	346	100	4.0	10.6
Benzoic and Other Acids							
Benzoic acid	90[e]	56	296	355	100	5.3	6.3
p-Hydroxybenzoic acid	120[e]	71	281	355	↓	4.0	6.3
2,5-Dihydroxybenzoic acid	230	62	190	344	↓	3.1	13.3
3,4-Dihydroxybenzoic acid	90[e]	26	137	344	↓	5.3	13.3

[a] Endogenous considered as 1.0.
[b] Phenol test reached endogenous level in 90 min with 100 mg/l phenol.
[c] Endogenous O_2 uptake at time shown in column 1 used.
[d] 90 min endogenous result used in calculation.
[e] Ratio calculated at time when activity was at or near the endogenous level.

8.3–23 OXYGEN UPTAKE WITH PHENOL-ADAPTED CULTURE *(Continued)*

Compound	Test time, min	Oxygen consumed, μl: Endogenous	Oxygen consumed, μl: Cells + test compound	Oxygen consumed, μl: Cells + phenol[b] 90 min	Test compound, mg/l	Ratio endogenous to:[a] Test compound[c]	Ratio endogenous to:[a] Phenol[d]
Salicylic acid	180	97	302	355	↓	3.1	6.3
Gallic acid	210	80	191	359	↓	2.4	9.2
o-Nitrobenzoic acid	210	83	84	342	↓	1.0	8.5
m-Nitrobenzoic acid	210	83	96	342	↓	1.2	8.5
p-Nitrobenzoic acid	180	55	64	346	↓	1.2	10.6
2,5-Dinitrobenzoic acid	210	83	90	342	↓	1.1	8.5
3,4-Dinitrobenzoic acid	210	83	95	↓	100	1.1	8.5
3,5-Dinitrobenzoic acid	210	83	108	↓	60	1.3	8.5
2,4,6-Trinitrobenzoic acid	180	53	61	↓	100	1.2	11.0
3,5-Dinitrosalicylic acid	180	83	104	342	↓	1.3	8.5
o-Chlorobenzoic acid	180	125	119	368	↓	<End.	4.7
m-Chlorobenzoic acid	190	55	85	336	↓	1.5	10.8
o-Toluic acid	230	62	53	344	↓	<End.	13.3
m-Toluic acid	180[e]	50	265	↓	↓	5.3	↓
	210	62	61	↓	↓	<End.	↓
Benzenesulfonic acid	230	62	58	↓	↓	<End.	↓
p-Toluenesulfonic acid	230	62	63	344	↓	1.0	13.3
Vanillic acid	180	55	86	346	↓	1.6	10.6
Syringic acid	180	55	58	316	100	1.1	10.6
DL-Mandelic acid	210	57	54	345	60	<End.	12.8
Tannic acid	180	55	76	346	100	1.3	10.6
Phthalic acid	210	80	94	359	↓	1.2	9.2
Anthranilic acid	210	53	68	230	↓	1.3	11.5
DL-Phenylalanine	210	53	197	230	↓	3.7	11.5
Tyrosine	90[e]	27	174	230	100	6.4	11.5
Benzaldehydes and Amides							
Benzaldehyde	90[e]	57	253	376	100	4.4	6.6
p-Hydroxybenzaldehyde	90	57	271	376	↓	4.8	6.6
m-Nitrobenzaldehyde	210	53	91	230	↓	1.7	11.5
p-Nitrobenzaldehyde	210	53	76	230	↓	1.4	11.5

[a] Endogenous considered as 1.0.
[b] Phenol test reached endogenous level in 90 min with 100 mg/l phenol.
[c] Endogenous O_2 uptake at time shown in column 1 used.
[d] 90 min endogenous result used in calculation.
[e] Ratio calculated at time when activity was at or near the endogenous level.

8.3–23 OXYGEN UPTAKE WITH PHENOL-ADAPTED CULTURE *(Continued)*

Compound	Test time, min	Oxygen consumed, μl			Test compound, mg/l	Ratio endogenous to:[a]	
		Endogenous	Cells + test compound	Cells + phenol[b] 90 min		Test compound[c]	Phenol[d]
Vanillin	180	55	161	346	100	2.9	10.6
Benzamide	210	53	67	230	100	1.3	11.5
Benzenes							
Benzene	170	53	87	353	100	1.6	11.8
Chlorobenzene	170	53	71	353	80	1.3	11.8
o-Dichlorobenzene	210	80	87	359	60	1.1	9.2
1,3,5-Trichlorobenzene	180	125	146	368	100	1.2	4.7
Nitrobenzene	170	53	32	353	↓	<End.	11.8
m-Dinitrobenzene	210	57	102	345	↓	1.8	12.8
p-Dinitrobenzene	180	65	97	395	↓	1.5	11.0
1,3,5-Trinitrobenzene	↓	65	127	↓	↓	2.0	11.0
2,4-Dinitrotoluene	↓	65	168	↓	↓	2.6	11.0
2,4,6-Trinitrotoluene	↓	65	160	395	100	2.5	11.0
Toluene	180	67	133	331	80	2.0	9.5
Aniline	210	53	120	230	100	2.3	11.5
p-Phenylenediamine	210	57	93	345	60	1.6	12.8
N,N-Dimethyl-*p*-phenylenediamine	190	55	83	336	100	1.5	10.8
m-Nitroaniline	210	57	136	345	100	2.3	12.8
p-Nitroaniline	210	53	80	230	100	1.5	11.5

[a] Endogenous considered as 1.0.
[b] Phenol test reached endogenous level in 90 min with 100 mg/l phenol.
[c] Endogenous O_2 uptake at time shown in column 1 used.
[d] 90 min endogenous result used in calculation.
[e] Ratio calculated at time when activity was at or near the endogenous level.

Source: C.W. Chambers *et al., J. Water Pollut. Control Fed., 35*(12):1517, 1963.

8.3–24 PERCENT THEORETICAL OXIDATION OF 500 PPM CARBOHYDRATES, ORGANIC ACIDS, AMINO ACIDS AND AROMATICS BY PHENOL ACCLIMATED ACTIVATED SLUDGE AFTER 12 HR OF AERATION

	Theoretical oxidation, %
Carbohydrates	
Glucose	21
Fructose	16
Sucrose	16
Arabinose	18
Xylose	19
Organic Acids	
Formic	66
Acetic	50
Propionic	40
Butyric	30
Heptanoic	27
Octanoic	20
Pyruvic	30
Citric	46
Succinic	57
Fumaric	41
Malic	46
Glycolic	32
Amino Acids	
Glycine	58[a]
Alanine	40[a]
Valine	36[a]
Leucine	24[a]
Norleucine	20[a]
Phenylalanine	24[a]
Threonine	34[a]
Glutamic Acid	38[a]
Arginine	40[a]
Histidine	13[a]
Tryptophan	21[a]
Hydroxyproline	36[a]
Methionine	16
Aromatics	
Phenol	33
o-Cresol	31
m-Cresol	30
p-Cresol	26
2,4-dimethylphenol	16
2,6-dimethylphenol	6
3,4-dimethylphenol	27
3,5-dimethylphenol	0
p-Amino benzoic acid	2[a]
Benzene	33[a]
Toluene	47[a]
Ethyl benzene	27[a]
n-Propylbenzene	10[a]
n-Butylbenzene	4[a]
s-Butylbenzene	1
t-Butylbenzene	1
s-Amylbenzene	1
t-Amylbenzene	1

[a] Adaptive metabolism.

Source: R.E. McKinney *et al., Sewage Ind. Wastes, 28*(4):547, 1956.

8.3–25 COMPARISON OF MEASURED BOD TO THEORETICAL OXYGEN DEMAND (TOD)

Group	Compound	BOD_5, mg/mg	BOD_{20}, mg/mg	TOD, mg/mg	BOD_5:TOD, %	BOD_{20}:TOD, %
Aliphatics	Methanol	1.12	1.18	1.50	75	79
	Ethanol	1.58	1.78	2.08	76	86
	Ethylene glycol	0.36	0.91	1.26	29	72
	Isopropanol	0.16	1.68	2.39	7	70
	Maleic acid	0.64	0.76	0.83	77	92
	Acetone	0.81	1.78	2.20	37	81
	Methyl ethyl ketone	1.81	1.92	2.44	74	79
	Ethyl acetate	1.24	1.43	1.82	68	79
	Oxalic acid	0.16	0.16	0.18	89	89
	Group average	–	–	–	55.8	79.8
Aromatics	Toluene	0.86	1.02	3.13	28	32
	Benzaldehyde	1.62	1.78	2.42	67	74
	Benzoic acid	1.45	1.52	1.96	74	78
	Hydroquinone	1.00	1.15	1.89	53	61
	o-Cresol	1.76	2.16	2.52	70	86
	Group average	–	–	–	58.4	66.2
Nitrogenous	Monoethanolamine	0.83	1.00	2.49	34	40
	Acrylonitrile	nil	1.21[a]	3.17	0	38
	Aniline	1.42	2.02	3.18	44	64
	Group average	–	–	–	26	47.3
Resistant	*tert*-Butanol	0	0	2.59	0	0
	Diethylene glycol	0.15	0.32	1.51	10	21
	Pyridine	0.06	2.02	3.13	2	65
	Group average	–	–	–	4	28.7

[a] BOD_{30}.

Source: R.H.F. Young *et al., J. Water Pollut. Control Fed., 40*(8)(Part 2):R354, 1968.

8.3–26 OXYGEN UPTAKE, LAG, AND TIMES OF BACTERIAL GROWTH AND CARBON UTILIZATION

O_2 uptake classification	Compound	Lag, days	Bacterial peak, days	Time of maximum respirative synthesized carbon ratio, days
Readily degraded	Acetone	2	3	4–5
	Benzaldehyde	0	4	4
	Benzoic acid	0	2	3
	o-Cresol	1	3	4
	Ethanol	0	2	7
	Ethyl acetate	1	2	7
	Hydroquinone	1	3	9
	Maleic acid	1	2	9
	Methanol	1	2	9
	Methyl ethyl ketone	1	3	3
	Oxalic acid	0	2	9
Acclimation	Acrylonitrile	12	13	>9
	Aniline	3	5	>9
	Ethylene glycol	4	7	>9
	Isopropanol	5	7	>9
	Monoethanolamine	2	3	7
	Pyridine	9	2	>9
Slightly oxidized	tert-Butanol	–	3	9
	Diethylene glycol	0	3	9
	Toluene	1	2	9
Control	Seed blank	0	2	4
	Glucose-glutamic acid	0	2	5

Source: R.H.F. Young *et al., J. Water Pollut. Control Fed., 40*(8)(Part 2):R354, 1968.

Section 9

Monitoring

9. Introduction
Pages 697–704

9.1 Public Water Supplies, Wells, and Ground Water
Pages 705–737

9.2 Irrigation Waters, Pesticides, and Salinity Build-up
Pages 738–761

9.3 Heavy Metals
Pages 762–781

9. MONITORING

This section is concerned with monitoring activities related to water quality including water pollution problems. Typical data on the quality of rivers, lakes, and well waters are given in Section 1. Readers are referred to the U.S. Environmental Protection Agency STORET system for data on specific waters at particular times.

Requirements for monitoring systems are examined here with a special emphasis on problems of current importance.

9–1 PARAMETER COVERAGE AND SAMPLING FREQUENCY

U.S. Water Quality Monitoring Stations

Parameter	No. of Stations Measuring Given Parameter at Indicated Frequency							
	Continuous	Daily	Weekly	Monthly	Quarterly	Annually	Other	Total
Temperature	871	822	529	2,270	832	1,064	1,276	7,664
Specific conductance	241	432	97	1,636	491	1,287	1,045	5,229
Turbidity	31	285	332	1,053	615	57	623	2,996
Color	14	87	205	939	611	1,122	747	3,725
Odor	9	32	14	389	406	13	248	1,111
pH (field)	77	55	156	745	372	122	721	2,251
pH (lab)	16	451	268	1,914	749	1,398	1,111	5,907
EH	18	0	0	7	0	0	1	26
Suspended solids	0	2	48	328	213	2	45	594
Other physical analyses	12	19	73	146	179	41	193	663
Dissolved solids	42	188	232	1,495	624	1,187	1,063	4,831
Chloride	10	121	118	1,196	354	552	931	3,282
Nutrients (nitrogen)	14	32	243	1,266	507	896	775	3,733
Nutrients (phosphorus)	23	58	135	853	609	889	918	3,485
Common ions	20	412	119	1,775	555	1,357	1,288	5,526
Hardness	6	362	225	1,490	674	1,310	907	4,974
Radiochemical	21	3	37	339	448	31	265	1,144
Dissolved oxygen	85	45	345	1,379	500	913	681	3,948
Other gases	10	86	13	34	0	5	18	166
Minor elements	0	12	3	164	45	149	175	548
Pesticides	1	1	7	69	11	24	80	193
Detergents	0	1	85	363	463	33	114	974
Biological oxygen demand	0	12	239	866	460	30	176	1,783
Carbon (total, dissolved, etc)	0	4	4	22	9	31	4	65
Coliforms	13	240	423	1,261	637	160	345	3,079
Other microorganisms	2	13	185	252	128	4	113	697
Biologic	6	3	102	127	60	13	76	387
Sediment concentration (suspended)	25	291	96	269	41	9	328	1,059
Particle size (suspended)	5	56	7	134	12	51	315	580
Particle size (bed load material)	0	0	0	10	2	6	33	42
Other sediment	1	8	5	97	2	6	71	190

Source: Federal Water Quality Administration, based on data provided by U.S. Geological Survey, 1968.

9–2 PARAMETERS MONITORED IN SURFACE WATERS

Uniform 100%	**Frequent, 99–50%**	**Infrequent, 49–20%**	**Least frequent, 19–0%**
Dissolved oxygen	Radioactivity	Arsenic	Bottom deposits
pH	Parameters listed in Public Health Service *Drinking Water Standards*	Barium	Chromium (trivalent)
Coliform	Total dissolved solids	Cadmium	Electrical conductance
Temperature		Chromium (hexavalent)	Ammonia
Floating solids		Cyanide	Acidity
Settleable solids		Fluoride	Alkalinity
Turbidity color		Lead	Carbon chloroform extract (CCE)
Taste-odor		Selenium	Hydrogen sulfide
Toxic substances		Silver	Pesticides
		Suspended solids	Sodium
		Turbidity	Iron
		Chloride	Plankton
		Copper	Foaming substances
		Nitrate	Boron
		Phenols	Manganese
		Phosphate	Hardness
		Sulfate	Biochemical oxygen demand (BOD)
		Color	Methylene blue active substances
			Zinc

Source: W.T. Sayers, *Environ. Sci. Technol.*, *5*(2):114, 1971.

9–3 SUBSTANCES ANALYZED FOR WATER USED FOR SPECIFIC PURPOSES

Component	Public water supply	Industrial water	Fish and wildlife	Livestock	Irrigation
Alkalinity	x	x	x		
Acidity		x			
Hardness	x	x	x		
pH	x	x	x		x
Total dissolved solids	x	x			
Total suspended solids		x			
Temperature	x	x			
Bicarbonate		x			
Chloride	x	x			x
Fluoride	x	x		x	
Nitrate	x	x			
Phosphate	x	x			
Sulfate	x	x			
Organics (CCl_4 extractable)		x			
COD		x			
Hydrogen sulfide		x			
Dissolved oxygen	x		x		
Carbon dioxide			x		
Total dissolved materials			x	x	x
Salinity			x		
Turbidity	x		x		
Color	x		x		
Cyanide	x		x		
Pesticides	x		x		
Toxic substances			x		
Oil	x		x		
ABS			x		
Odor					
Nitrite	x				
Herbicides	x				
Phenols	x				
Grease	x				
MBAS (surfactants)	x				
Carbon chloroform extracts (CCE)	x				
Ammonia	x				
Radionuclides	x		x	x	x
Gross beta	x			x	
Radium 226	x			x	
Strontium 90	x			x	
Arsenic	x			x	x
Barium	x				x
Boron	x				
Cadmium	x			x	
Copper	x				
Iron	x				
Lead	x			x	
Manganese	x				
Selenium	x			x	
Silver	x				
Uranium	x				
Dissolved organic compounds				x	
Sodium adsorption ratio					x
Metals (aluminum to zinc)	x		x		x
Zinc					
Chromium	x			x	

Source: D.G. Ballinger, *Water Sewage Works,* Ref. No. 1968, p. R38.

9–4 EXAMPLES OF WASTE PARAMETERS SAMPLED FOR INDUSTRIAL WASTES

Paper and Allied Products

Color	Chromium
Specific conductance	Copper
Bromide	Lead
Chloride	Mercury
Sulfate	Nickel
Sulfite	Sodium
Phenols	Zinc
Total organic carbon	Total coliform

Soap, Detergents, and Cleaning Preparations, Perfumes, Cosmetics, and Other Toilet Preparations

Color	Cyanide
Specific conductance	Sulfate
Turbidity	Sulfide
Oil and grease	Phenols
Chloride	Surfactants
	Total coliform

Fertilizers

Color	Cadmium
Turbidity	Chromium
Hardness	Copper
Oil and grease	Iron
Chloride	Lead
Fluoride	Manganese
Nitrite	Mercury
Pesticides	Nickel
Phosphorus-ortho	Zinc
Sulfate	Radioactivity
Aluminum	Total coliform
Arsenic	Fecal coliform

Iron and Steel (partial list)

Iron	Phenols
Oil and grease	Turbidity
pH	Sulfate
Chloride	

Plating (partial list)

Cyanides	Cadmium
Chromium	Copper
Nickel	Silver
pH	

Source: D.G. Ballinger, *Environ. Sci. Technol.,* 6(2):130, 1972.

REFERENCE

1. D.G. Ballinger, *Water Sewage Works,* R38, 1968.

9–5 SUMMARY OF PARAMETERS AND TESTING METHODS

Parameter	Method	Optimum range, mg/l	Accuracy, % bias	Analyst level	Typical charge/ sample, $	Laboratory instrument	On-line instru- ment
Solids	Gravimetric			Technician	10–30		
Color	Visual comparison						
Turbidity	90° scatter photometer					Turbidimeter	Yes
Radioactivity	Gross α, β, ν, tritium			Expert	>50	Counters, spectrometers	No
Specific conductance	Wheatstone bridge	0.5–2,000 μmho	– 5	Technician	3–5	Wheatstone bridge	Yes
Biochemical oxygen demand	5 day, 20°C	2–500	[a]	Technician	30		
Chemical oxygen demand	Dichromate reflux	40–500	– 4	Technician	15–30		
Ammonia	Distillation-nesslerization or auto-analyzer	0.05–2.0	– 5	Skilled	5–15	Auto analyzer	Yes[b]
Kjeldahl nitrogen	Digestion-distillation or auto analyzer						
Nitrate	Brucine sulfate or auto analyzer	0.1–2.0	+ 4	Skilled			
Nitrite	Diazotization or auto analyzer						
Total phosphorus	Persulfate digestion or auto analyzer	0.01–0.5	+ 3	Skilled		Auto analyzer	No
Phosphorus-ortho	Single reagent, stannous chloride, or auto analyzer	0.01–0.5	+ 3	Skilled		Auto analyzer	No
Phenols	Colorimetric	0.5–5.0	[a]	Skilled	15–30	Visible spectrophotometer	Yes[b]
Total organic carbon	Combustion-infrared			Technician	30	Combustion-Infrared	Yes[b]
Pesticides	Gas chromatography	0.01–0.2 μg/l	–20	Skilled	>50	Gas chromatograph	No
Alkalinity	Electrometric titration or auto analyzer			Technician	3–12		
Arsenic	Atomic absorption	0.01–0.5	0	Skilled	9–25	AA spectrophotometer	Yes[b]
Bromide	Colorimetric						
Chloride	Mercuric nitrate titration or auto analyzer	10–400	– 0.5	Technician	5–7	Auto analyzer or electrode	Yes
Copper	Atomic absorption	0.1–10	–10	Skilled	9–25	AA spectrophotometer	Yes[b]
Cyanide	Silver nitrate titration or pyridine pyrazalone			Technician	30	Visible spectrophotometer	No
Fluoride	SPADNS with distillation or probe						
Hardness	EDTA titration, auto analyzer, or atomic absorption	10–400	– 1				
Lead	Atomic absorption	0.1–10	–12	Skilled	9–25	AA spectrophotometer	Yes[b]
Sulfate	Turbidimetric or auto analyzer	10–500	– 3				
Mercury	Flameless atomic absorption	0.2–10 μg/l	–13	Skilled	20–30	Flameless atomic absorption or mercury meter	Yes[b]

[a] True accuracy cannot be determined for these tests.
[b] Reliability has not been established.

Source: D.G. Ballinger, *Environ. Sci. Technol.*, *6*(2):131, 1972.

9–6 PERFORMANCE CHARACTERISTICS OF SENSORS FOR CONTINUOUS MONITOR SYSTEMS

Parameter	Sensor	Range		Response	Accuracy	Depth	Automatic temperature compensation	Required stability, weeks
		From	To					
Temperature	Thermistor or thermocouple	−10°C–+50°C	0°C–100°C	40% in 10 sec–2 min	±0.01°–±0.1°C	submerged–10,000 ft	–	4
Dissolved oxygen	Polarographic	0.5–15 ppm	0–50 ppm	90% in 6 sec–90% in 20 sec	±1%–±5% of scale	submerged–500 ft	yes	2
Conductivity	Platinized electrode/Wheatstone bridge	0–1,400 mho/cm	2.5 mho/cm	–	±0.05%–±1% of scale	submerged–1,000 ft	yes	–
pH	Potentiometric glass electrode – Ag/AgCl	0	14	98% in 2 sec–98% in 20 sec	±0.005–±0.1 pH	submerged–200 ft	yes	4
Solar radiation	–	0.–2.8 cal/cm² min	0–5 cal/cm² min	98% in 20 sec	±1% of scale	0–20 ft		–
Oxidation reduction potential	Platinum electrode–Ag/AgCl potentiometric	0–±1,400 mv	–	–	±1% of scale	–	–	–
Chloride	Specific electrode–Ag/AgCl potentiometric	1–10,000	–	90% in 20 sec	–	–	yes	4
Turbidity	Optical-light scattering (automatic cleaning)	0–1,000 JCU	0–10,000 JCU	30 sec	±5%–±2% of scale	submerged–200 ft	yes	2

REFERENCES

1. H.W. Porterfield, *Oceanogr. Int.*, *5*(10):22, 1970.
2. D.G. Ballinger, *Environ. Sci. Technol.*, *2*(8):606, 1968.

9–7 DATA LOST BY FWPCA AUTOMATIC MONITORS IN 629 DAYS

Parameter	Data lost due to sensor, %	Data lost due to recorder, %	Data lost due to pump, %	Data lost due to power, %	Total data lost, %
Dissolved oxygen	5	7	8	1	21
pH	21	7	8	1	37
Turbidity	0	7	8	1	16
Conductivity	0	7	8	1	16
Temperature	0	7	8	1	16
Solar radiation intensity	2	7	0	1	10

Source: D.G. Ballinger, *Environ. Sci. Technol.*, *2*(8):609, 1968.

9–8 SAMPLE PRESERVATION

Parameter	Preservative	Maximum holding period
Acidity-alkalinity	Refrigeration at 4°C	24 hr
BOD	Refrigeration at 4°C[a]	6 hr
Calcium	None required	
COD	2 ml/liter H_2SO_4	7 days
Chloride	None required	
Color	Refrigeration at 4°C	24 hr
Cyanide	NaOH to pH 10	24 hr
Dissolved oxygen	Determine on site[b]	No holding
Fluoride	None required	
Hardness	None required	
Metals, total	5 ml/liter HNO_3	6 months
Metals, dissolved	Filtrate: 3 ml/liter 1:1 HNO_3	6 months
Nitrogen, ammonia	40 mg/liter $HgCl_2$ –4°C	7 days
Nitrogen, Kjeldahl	40 mg/liter $HgCl_2$ –4°C	Unstable
Nitrogen, nitrate-nitrite	40 mg/liter $HgCl_2$ –4°C	7 days
Oil and grease	2 ml/liter H_2SO_4 –4°C	24 hr
Organic carbon	2 ml/liter H_2SO_4 (pH 2)	7 days
pH	None available	
Phenolics	1.0 g $CuSO_4$ + H_3PO_4 to pH 4.0–4°C	24 hr
Phosphorus	40 mg/liter $HgCl_2$ –4°C	7 days
Solids	None available	
Specific conductance	None required	
Sulfate	Refrigeration at 4°C	7 days
Sulfide	2 ml/liter Zn acetate	7 days
Threshold odor	Refrigeration at 4°C	24 hr
Turbidity	None available	

[a] Slow freezing techniques (to –25°C) can be used for preserving samples to be analyzed for organic content.
[b] For some methods of determination, 4–8 hr preservation can be accomplished with 0.7 ml concentration H_2SO_4 and 20 mg NaN_3. Refer to *Standard Methods for the Examination of Water and Waste Water,* Am. Publ. Health Assn., 1971.

Source: *Methods for Chemical Analysis of Water and Wastes,* U.S. Department of the Interior, Federal Water Pollution Control Administration, 1969.

9–9 SUGGESTED SAMPLING OR COMPOSITING SCHEDULE

Characteristic	High variability	Low variability
BOD[a]	4 hr	12 hr
COD or TOC[a]	2 hr	8 hr
Suspended solids	8 hr	24 hr
Alkalinity or acidity	1 hr grab	8 hr grab
pH	Continuous	4 hr grab
Nitrogen and phosphorus[b]	24 hr	24 hr
Heavy metals	4 hr	24 hr
Temperature	2 hr	8 hr

[a] The compositing schedule where continuous samplers are not used depends on variability, i.e., 15 min for high variability to 1 hr for low variability.
[b] Does not apply to nitrogen or phosphorus wastes (e.g., fertilizer).

Source: W.W. Eckenfelder, *Water Quality Engineering for Practicing Engineers,* Barnes and Noble, New York, 1970.

9–10 ANALYSIS SCHEDULE FOR ACTIVATED SLUDGE STUDY

Analysis	Frequency	Raw waste	Mixed liquor	Effluent
BOD, mg/l (filtered and unfiltered)	3/week	x		x
COD, mg/l (filtered and unfiltered)	3/week	x		x
TOC (if desired), mg/l (filtered and unfiltered)	3/week	x		x
Suspended solids, mg/l	Daily		x	x
Volatile suspended solids, mg/l	3/week		x	x
pH	Daily		x	
Color, turbidity	3/week			x
Nitrogen and phosphorus	1/week	x		x
Microscopic examination	1/week		x	
Specific compounds or ions	3/week	x		x
Settling characteristics	1/week		x	

Source: Adapted from W.W. Eckenfelder and D.L. Ford, *Water Pollution Control,* Jenkins Publishing, Austin, Texas, 1970 in Metcalf & Eddy, *Wastewater Engineering,* McGraw-Hill, New York, 1972.

9–11 ANALYSIS SCHEDULE FOR ANAEROBIC TREATMENT STUDY

Analysis	Frequency	Raw waste	Mixed liquor	Centrifuged effluent
COD	3/week	x		x
BOD	3/week	x		x
Volatile acids	3/week			x
pH	Daily		x	
Alkalinity	Daily		x	
Ammonia nitrogen	1/week			x
Gas analysis (% CH_4, % CO_2)	3/week			

Source: Metcalf & Eddy, Inc., *Wastewater Engineering,* McGraw-Hill, New York, 1972.

9.1 PUBLIC WATER SUPPLIES, WELLS, AND GROUND WATER

9.1–1 REPRESENTATIVE EXAMPLES OF WATER QUALITY OF SOME MAJOR U.S. CITIES

City and state	Population,[a] 1,000's	Source[b]	Treatment[c]	Range of hardness, ppm	Total dissolved solids, ppm
Birmingham, Ala.	441	S	C, Cl	61–120	101–250
Los Angeles, Cal.	2,458	M	S, Cl	61–120	251–500
San Diego, Cal.	600	S	C, S, Cl	> 180	> 500
San Francisco, Cal.	1,600	S	Cl, F	< 61	< 100
Washington, D.C.	1,100	S	C, Cl, F	61–120	101–250
Jacksonville, Fla.	247	G	Cl	> 180	251–500
Miami, Fla.	550	G	C, S, Cl, F	61–120	101–250
Atlanta, Ga.	600	S	C, Cl	< 61	< 100
Chicago, Ill.	4,425	S	C, Cl, F	121–180	101–250
Boston, Mass.	2,000	S	Cl	< 61	< 100
Detroit, Mich.	3,078	S	C, Cl	61–120	101–250
New York, N.Y.	8,350	M	C, Cl	< 61	< 100

[a] 1960 census.
[b] S, surface; G, ground; M, mixed.
[c] C, clarification; Cl, chlorination; S, softening; F, fluoridation.

Source: L.E. Ciaccio, *Water and Water Pollution Handbook,* Marcel Dekker, New York, 1971.

REFERENCE

1. C.N. Durfor and E.J. Becker, *J. Am. Water Works Assn., 56*:237, 1964.

9.1–2 CHARACTERISTICS OF SPECIFIC WATER SUPPLIES

Table A

Location	Sample	Total dissolved solids	Silica, SiO_2	Iron, Fe	Calcium, Ca^{++}	Magnesium, Mg^{++}	Sodium and potassium, as Na^+	Bicarbonate, HCO_3^-	Sulfate, SO_4^{-2}
Gainesville, Fla.	Tap water	108	15	trace	15	8.8	29	61	18
Dallas, Texas	Raw water (River)	280(191–305)	3.3(2.0–5.0)[a]		52(40–58)	4.7(3.6–5.5)		140(112–154)	52(27–67)
Van Camp, Ind.	Raw well water	471	8	0.1	190	134		384	50
Tuscon Ariz.	Tap water (ground)	400			54	9		183	77
El Cerrito, Cal.	Tap-surface	50			8	1			2
Berkeley, Cal.	Tap-surface	82			16	4			12
Somis, Cal.	Tap-well	388			55	9			71
Soledad, Cal.	Tap-well	1,140			158	70			400
El Rio, Cal.	Tap-well	1,194			155	59			502
Oxnard, Cal.	Tap-well	1,401			192	67			659
Ann Arbor, Mich.	Tap water	146		0.1	25	5			
Prado, Cal.[b]	Santa Ana R.	566–809							
Rotterdam, Netherlands	Raw Rhine R., 1964	780(480–1,100)[a]							
	Rhine R., 1967	580(350–770)							
Fargo, Ga.	Suwannee R.			0.52	0.8	0.8			1
Gettysburg, Pa.	Well supply	2,100	12	0.16	12	1.7		721	324
	New well	1,900	3.0	0.25	11	3.6		765	436
Kansas City, Mo.	Surface–Missouri R.	400		0.0	60.6	13.8			112.8
Denison Dam, Texas	Red River[c]	829	12	0.02	94	22	150	135	188
Comite, La.	Comite R.[d]	75	16	0.58	3.7	1.3	6.8	16	3.7

[a] Values in parentheses represent range of values.
[b] Range of values obtained over 17 year period.
[c] Specific conductance 1,300 μmho at 25°C.
[d] Specific conductance 59 μmho at 25°C.

9.1–2 CHARACTERISTICS OF SPECIFIC WATER SUPPLIES *(Continued)*

Table B

Location	Chloride, Cl^-	Fluoride, F^-	Total hardness, as $CaCO_3$	pH	Carbonate $CO_3^=$	Hydromide $OH^=$	Noncarbonate hardness	Turbidity	Ref.
Gainesville, Fla.	20	0.63	74	9.3					1
Dallas, Texas	42(6 –70)		152(115–168)	8.1 (8.0 –8.2)	0	0	38(23–50)	79(11–921)	2
Van Camp, Ind.	16			7.3				3	3
Tuscon, Ariz.	24		171		0				4
El Cerito, Cal.	2								5
Berkeley, Cal.	11								5
Somis, Cal.	22								5
Soledad, Cal.	142								5
El Rio, Cal.	64								5
Oxnard, Cal.	82								5
Ann Arbor, Mich.	65	0.9	80	10.3					6
Predo, Cal.[b]	87–152		324–396						7
Rotterdam, Netherlands	224(92–364)		283(216–370)	7.25(6.95–7.50)					8
	146(51–248)		243(177–288)	7.55(7.3 –7.95)					
Fargo, Ga.	6		4	4.46				18	9
Gettysburg, Pa.	515		37	8.4	0.62				10
	440	0.12	43	7.9	0				
Kansas City, Mo.	22.3		208	8.1					11
Denison Dam, Texas	242	0.04	325		0				12
Comite, La.	6.2	0.7	15		0				12

[b] Range of values obtained over 17 year period.

Note: Concentrations reported in mg/l except pH and as marked.

REFERENCES

1. A.P. Black and C-L. Chen, *J. Am. Water Works Assn.*, *57*(3):354, 1965.
2. E.C. Rees *et al.*, *J. Am. Water Works Assn.*, *56*(3):301, 1964.
3. B.F. Willey *et al.*, *J. Am. Water Works Assn.*, *56*(4):475, 1964.
4. R.M. Jordan, *J. Am. Water Works Assn.*, *55*(6):771, 1963.
5. W.H. Bruvold *et al.*, *J. Am. Water Works Assn.*, *59*(5):547, 1967.
6. G.W. Foess and J.A. Borchardt, *J. Am. Water Works Assn.*, *61*(7):353, 1969.
7. D.C. Baier, *J. Am. Water Works Assn.*, *62*(2):106, 1970.
8. J.J. Rook and G. Oskam, *J. Am. Water Works Assn.*, *62*(4):249, 1970.
9. J.J. Maulding and R.H. Harris, *J. Am. Water Works Assn.*, *60*(4):460, 1968.
10. H.C. Preul, *J. Am. Water Works Assn.*, *54*(11):1389, 1962.
11. J.R. Popalisky and F. Pogge, *J. Am. Water Works Assn.*, *59*(9):1121, 1967.
12. W.W. Hastings, *Southwest Water Works J.*, *38*(7):34, 1956.

9.1–3 ANALYTICAL METHODS FOR WATER SUPPLIES

Compound	PHS limits, mg/l	Method	Minimum amount, μg	Detectable concentration, mg/l
Arsenic	Recommended 0.01	Silver diethyldithiocarbamate	1	0.03
	Rejection >0.05	Mercuric bromide strain	1	0.04
Barium	>1.0			
Boron	Recommended 1.0	Curcumin	0.2	0.2
	Rejection >5.0	Carmine	2.0	1.0
		Potentiometric	–	0.1
Cadmium	>0.01	Dithizone	0.5	0.02
Carbon chloroform extract (CCE)	>0.2		–	<.001
Chloride	250		–	~0.5
Chromium (hexavalent)	>0.05		–	0.003–0.005
Copper	1.0		–	0.02
Cyanide	Recommended 0.01 Rejection >0.20	New analytical method needed	–	–
Iron	0.30	Phenanthroline,	3	0.06
		tripyridine,	3	0.06
		extraction method	–	0.1
Lead	>0.05	Dithizone	2	0.08
Manganese	0.05	Persulfate,	5	0.059
		periodate method	5	0.05
Methylene blue active substances	0.5		10	0.025
Nitrogen (NO_3 or NO_2 form)	Nitrogen (nitrate or nitrite), 10	Nitrate: Phenoldisulfonic acid,	1	0.01
		Brucine,	–	1.0
		U.V. spectrophotometry,	–	0.02
		Polarography	–	{0.2 NO_3; 1 NO_2}
Phenols	0.001		1	0.002
Selenium	>0.01	Diaminobenzidine,	1	0.001
		distillation-diaminobenzidine,	1	0.001
		distillation-colored sol 10 μg or 0.001 mg/l	10	0.001
Silver	>0.05	Spectrographic method,	20	0.01
		dithizone method 0.2 μg or 0.002 mg/l	0.2	0.002
Sulfate	250		–	⩾1
Total dissolved solids	500 mg/l		–	a
Uranyl ion (UO_2^{++})	>5.0		–	0.00001
Zinc	5.0	Dithizone method,	1	0.1
		zincon,	1	0.02
		polarography	–	0.05

[a] No minimum detectable levels given dependent on sample size, etc.

Source: Reprinted, with publisher's permission, from *J. Am. Water Works Assn., 62*(3):172, © 1970, by the American Water Works Association, Inc., 2 Park Ave., New York, N.Y. 10016.

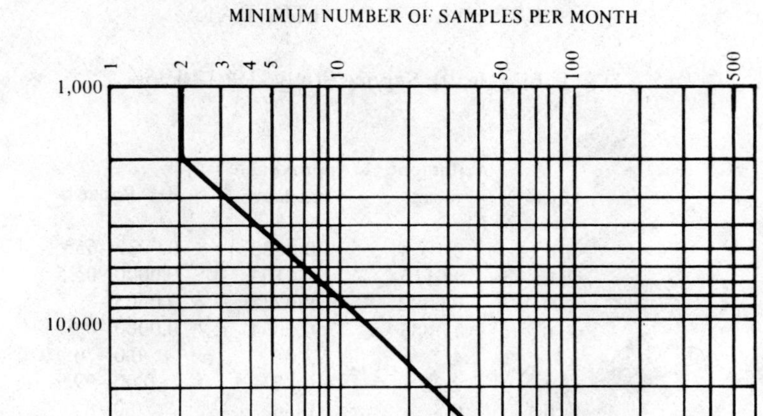

9.1–4 Recommended Minimum Monthly Samples for Bacteria Counts for Water Supply Systems.

Source: U.S. Public Health Service.

9.1–5 WATER QUALITY DATA

U.S. Public Health Service Study, 1961–1966

Component	Goal	Arithmetic mean	Approximate median	Range	No. samples	Less than values
Aluminum, mg/l	<0.05	0.1054	0.0583	0.0033–65[a]	91	
Arsenic, mg/l	<0.05	<0.0117	0.0104	0.0000–0.45	565	515
Barium, mg/l	<1	0.0403	0.0228	0.0005–3	558	
Cadmium, mg/l	<0.01	<0.0085	0.0064	0.0000–0.09	559	536
Color units	<3	4.2	2.3	0.0–70	218	
Turbidity, J.T.U.	<0.1	2.25	0.88	0.00–99.9	227	
Odor, units	0	0.5	0.56	0.0–8	112	
Carbon alcohol extract, mg/l	<0.1	<0.13	0.124	0.1–0.9	478	306
Chloride, mg/l	<250	40.7	21.3	0.7–605	273	
Carbon chloroform extract, mg/l	<0.04	<0.067	0.056	0.005–0.423	540	367
Chromium, mg/l	<0.05	<0.0041	0–003	0.0000–0.045	557	468
Copper, mg/l	<0.2	<0.0293	0.00549	0.00000–0.98	560	122
Fluoride, mg/l	0.6–1.2	0.61	0.52	0.00–10.1	445	
Cyanide, mg/l	<0.2	0.00375	0.00375			
Iron, mg/l	<0.05	0.0806	0.0279	0.0000–1.92	470	
Lead, mg/l	<0.05	<0.0171	0.0125	0.0001–0.4	559	468
Methylene blue active substances, mg/l	<0.2	<0.046	0.033	0.000–0.64	556	151
Manganese, mg/l	<0.01	<0.0369	0.01	0.0000–2	492	244
Nitrate, NO_3, mg/l	<45	2.3	0.69	0.0–99.8	260	
Selenium, mg/l	<0.01	<0.007	0.01	0.0000–0.01	556	503
Sulfate (SO_4), mg/l	<250	57.2	39.3	0.0–400.9	258	
Silver, mg/l	<0.08	0.00083	0.00025		189	
Total dissolved solids, mg/l	<200	271	205	4–1,739	271	
Zinc, mg/l	<1	<0.62	0.39	0.00–7	559	
Hardness, mg/l	80	136	144	0–920	341	
Beta activity (displaced), pCi/l	<100	<7.1	4.4	0.1–300	367	84
Beta activity (suspended), pCi/l	<100	<2.5	–	0.0–35	343	299
90 Si., pCi/l	<10	<1.7	1	0.0–10	97	50

[a] 1962 data only.

Source: Reprinted, with publisher's permission, from *J. Am. Water Works Assn., 60*(7):764, © 1968, by the American Water Works Association, Inc., 2 Park Ave., New York, N.Y. 10016.

REFERENCE

1 C.N. Durfor and E. Becker, *J. Am. Water Works Assn., 56*:231, 1964.

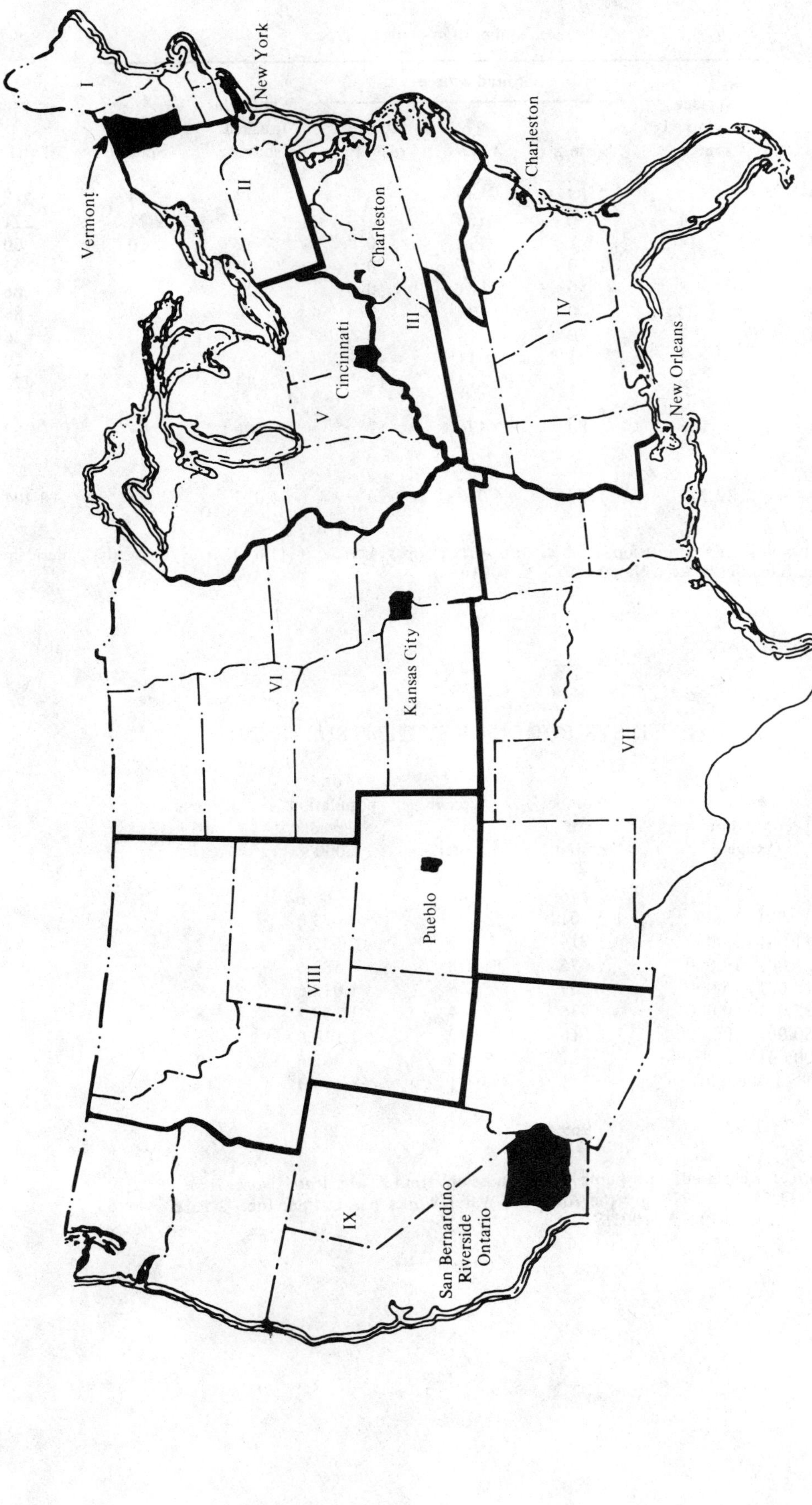

9.1–6 Community Water Supply Study, 1969: Size and Location of Regional Boundaries.

Source: Reprinted, with publisher's permission, from *J. Am. Water Works Assn.*, *62*(11):670, © 1970, by the American Water Works Association, Inc., 2 Park Ave., New York, N.Y. 10016.

9.1–7 SUMMARY OF WATER SUPPLY SYSTEM TYPES BY REGION

Region	Population, 1,000's	Number of systems by type: Surface and mixed source	Ground source: Springs	Ground source: Wells	Ground source: Mixture of both	Wholesale finished source	Special	Total
I	307.2	73	67	40	23	1	14	218
II	12,356.3	14	0	135	1	30	41	221
III	229.3	14	1	12	0	3	0	30
IV	251.1	1	0	16	0	3	2	22
V	1,366.0	8	0	31	0	19	8	66
VI	1,383.5	23	0	24	0	34	7	88
VII	1,085.4	7	0	14	0	2	3	26
VIII	111.5	4	1	11	1	0	3	20
IX	1,113.4	22	11	213	12	14	6	278
Total	18,203.8	166	80	496	37	106	84	969
Study population served, 1,000's		12,777	149	4,060	29	1,093	94	18,204

Source: Reprinted, with publisher's permission, from *J. Am. Water Works Assn.*, *62*(11):670, © 1970, by the American Water Works Association, Inc., 2 Park Ave., New York, N.Y. 10016.

9.1–8 SIZE DISTRIBUTION OF SYSTEMS SURVEYED

Population range	Number of systems	Percent of total	Total population served, 1,000's	Percent of total
⩽ 500	446	46	87.6	< 1
501–1,000	101	10	73.0	< 1
1,001–5,000	214	22	532.3	3
5,001–10,000	75	8	539.6	3
10,001–25,000	59	6	1,018.6	6
25,001–50,000	36	4	1,380.2	8
50,001–100,000	16	2	1,110.0	6
100,001–1,000,000	21	2	5,362.5	29
> 1,000,000	1	<1	8,100.0	44
Total	969		18,203.8	

Source: Reprinted, with publisher's permission, from *J. Am. Water Works Assn.*, *62*(11):670, © 1970, by the American Water Works Association, Inc., 2 Park Ave., New York, N.Y. 10016.

9.1–9 WATER TREATMENT PRACTICES IN SYSTEMS SURVEYED

Treatment practice	Type of system, number of systems of each: Surface water and mixed source	Ground water source: Springs	Ground water source: Wells	Ground water source: Mixture of both	Wholesale finished water source	Special system	Population served, 1,000's (number of systems of each): ⩽0.5	0.5–5	5–10	10–25	25–50	50–100	>100	Overall system totals: Number	Percent
Do not disinfect, clarify, or buy, chlorinated water	13	65	330	25	10	53	318	131	17	20	8	2	0	496	51
Disinfection only or buy chlorinated water	64	15	108	12	93	19	98	117	38	24	19	6	9	311	32
Clarification[a] and disinfection	88	0	52	0	3	12	29	63	19	15	8	8	13	155	16
Clarification[a] without disinfection	1	0	6	0	0	0	1	4	1	0	1	0	0	7	1
System totals, number	166	80	496	37	106	84	446	315	75	59	36	16	22	969	100

[a] Clarification is the removal of suspended material by coagulation, sedimentation and filtration.

Source: Reprinted, with publisher's permission, from *J. Am. Water Works Assn.*, *62*(11):670, © 1970, by the American Water Works Association, Inc., 2 Park Ave., New York, N.Y. 10016.

9.1–10 WATER QUALITY EVALUATION BY SOURCE AND COMMUNITY SIZE

	Overall system totals		Type of system, % of system totals by type						Population served, 1,000's % of size totals[a]						
			Surface water and mixed source	Ground water			Whole-sale finished water source	Special systems							
	Number	Percent		Springs	Wells	Mixture of both			≤0.5	0.5–5	5–10	10–25	25–50	50–100	>100
Met drinking water standards	572	59	57	45	57	54	76	65	50	67	64	71	67	62	73
Exceeded constituent limits	397	41	43	55	43	46	24	35	50	33	36	29	33	38	27
Exceeded recommended but not mandatory limits	238	25	25	14	29	11	19	23	26	21	25	20	30	51	27
Exceeded mandatory limits	159	16	18	41	14	35	5	12	24	12	11	9	3	7	0
System totals, number	969	–	166	80	496	37	106	84	446	315	75	59	36	16	22

[a] Percent of 18.2 million survey population in the four above groups was:

Met drinking water standards	86
Exceeded constitutent limits	14
Exceeded recommended limits	12
Exceeded mandatory limits	2

Source: Reprinted, with publisher's permission, from *J. Am. Water Works Assn., 62*(11):670, © 1970, by the American Water Works Association, Inc., 2 Park Ave., New York, N.Y. 10016.

9.1–11 FACILITIES EVALUATION BY SOURCE AND COMMUNITY SIZE

	Overall system totals		Type of system, % of system totals by type						Population served, 1,000's % of system totals by type[b]						
			Surface water and mixed source	Ground water			Whole-sale finished water source	Special systems							
	Number	Percent		Springs	Wells	Mixture of both			⩽0.5	0.5–5	5–10	10–25	25–50	50–100	>100
Essentially free of major deficiencies	422	44	15	28	51	30	60	58	39	44	49	49	64	50	64
Major deficiencies	547	56	85	72	49	70	40	42	61	56	51	51	36	50	36
Inadequate source protection, in the absence of disinfection or buying chlorinated water	184	19	8	41	18	38	22	13	27	17	7	5	6	0	0
Inadequate disinfection, if practiced	155	16	35	11	14	17	4	13	12	20	28	24	6	13	9
Inadequate control of disinfection, if practiced or if buying chlorinated water	65	7	11	4	6	5	5	10	8	7	8	5	3	0	0
Inadequate clarification, if practiced	43	4	16	NA[a]	3	NA[a]	0	2	1	7	8	8	3	12	14
Inadequate control of clarification, if practiced	60	6	33	NA[a]	<1	NA[a]	0	5	3	11	4	5	3	6	10
Low (<20 psi) pressure in some or all areas of the dist. system	249	26	36	51	22	32	15	15	32	17	24	22	25	31	32
System totals	969	100	166	80	496	37	106	84	446	315	75	59	36	16	22

[a] Not applicable.
[b] 31% of the survey population was served drinking water from systems that were essentially free of major deficiencies.

Source: Reprinted, with publisher's permission, from *J. Am. Water Works Assn.*, *62*(11):670, © 1970, by the American Water Works Association, Inc., 2 Park Ave., New York, N.Y. 10016.

9.1–12 BACTERIOLOGICAL SURVEILLANCE EVALUATION BY SOURCE AND COMMUNITY SIZE

	Overall system totals		Type of system, % of system totals by type						Population served, 1,000's, % of system totals[a]						
			Surface water and mixed source	Ground water			Whole-sale finished water source	Special systems							
	Number	Percent		Springs	Wells	Mixture of both			⩽0.5	0.5–5	5–10	10–25	25–50	50–100	>100
Met bacteriological surveillance criteria	100	10	21	2	9	2	8	7	4	18	12	10	3	12	36
Did not meet bacteriological surveillance criteria	869	90	79	98	91	98	92	93	96	82	88	90	97	88	64
Did not collect samples at a rate in the bacteriological surveillance criteria	827	85	68	98	85	95	92	94	94	74	83	85	92	88	64
System totals, number	969	100	166	80	496	37	106	84	446	315	75	59	36	16	22

[a] 60% of the survey population were served drinking water by systems over which a bacteriological surveillance program met the criteria.

Source: Reprinted, with publisher's permission, from *J. Am. Water Works Assn., 62*(11):670, © 1970, by the American Water Works Association, Inc., 2 Park Ave., New York, N.Y. 10016.

9.1–13 SUMMARY OF RESPONSES ON ADEQUACY OF SURFACE WATER SOURCES

Item[a]	Systems responding	Adequate, %	Inadequate, %	
Quantity	124	87	13	
Bacteriological quality	112	80	20	
Chemical quality	113	83	17	
Physical quality	114	83	17	
Source protection	117	55	45	
		Improved, %	**Deteriorated, %**	**Same, %**
Changes in source quality during last 3 yr	123	15	26	59

[a] 124 water supply systems were classified as surface water source for this analysis.

Source: Reprinted, with publisher's permission, from *J. Am. Water Works Assn.*, *62*(11):670, © 1970, by the American Water Works Association, Inc., 2 Park Ave., New York, N.Y. 10016.

9.1–14 OPERATOR EXPERIENCE AND TRAINING

Item	Percent responses
Education level – high school or better	84
Water treatment training short course or better	39
Number of previous positions	
None	58
Training in sanitary microbiology	
None	63
On-the-job	14
Training in chemistry (systems with more than disinfection)	
At short school or higher level	54
None or on-the-job	46
Full time operators	47
Staff adequate in (operator's opinion)	
Number	83
Knowledge	89
Salary	
$7,500/yr or less	84
$2,000/yr or less	37
Operators with more than 20 yr experience	43
Operators with less than 2 yr experience	15
Operators with less than 2 yr experience who are also part time operators	95

Source: Reprinted, with publisher's permission, from *J. Am. Water Works Assn., 62*(11):670, © 1970, by the American Water Works Association, Inc., 2 Park Ave., New York, N.Y. 10016.

9.1–15 EFFECT OF SIZE ON OPERATION AND WATER QUALITY

Item	≤1,000	1,001–5,000	>50,000
Supplies having bacteriological laboratory	2	7	73
Supplies with operators having more than 5 yr experience	53	75	94
Supplies with certified operators	11	39	88
Supplies that met the bacteriological criteria 11 of the previous 12 months of record	7	12	24

Note: All data calculated as percent of responses to the individual questions.

Source: Reprinted, with publisher's permission, from *J. Am. Water Works Assn., 62*(11):670, © 1970, by the American Water Works Association, Inc., 2 Park Ave., New York, N.Y. 10016.

9.1–16 SUMMARY OF WATER QUALITY EVALUATION

	Population group served			
	500 or less	501–100,000	Greater than 100,000	All populations
No. systems	446	501	22	969
Evaluation of systems, %				
Met drinking water standards	50	67	73	59
Exceeded recommended limits	26	22	27	25
Exceeded mandatory limits	24	11	0	16
Survey population in each group in thousands	88	4,552	13,463	18,203

Source: Reprinted, with publisher's permission, from *J. Am. Water Works Assn., 62*(11):670, © 1970, by the American Water Works Association, Inc., 2 Park Ave., New York, N.Y. 10016.

9.1–17 PRINCIPAL CONSTITUENT LIMITS EXCEEDED

969 Systems

From recommended list	No. systems on average exceeding limit	From mandatory list	No. systems on average exceeding limit
Iron	96	Coliform organisms	120
Total dissolved solids	95		
Manganese	90		
Fluoride	52	Fluoride	24
Sulfate	25	Lead	14
Nitrate	19		

Source: Reprinted, with publisher's permission, from *J. Am. Water Works Assn.*, *62*(11):670, © 1970, by the American Water Works Association, Inc., 2 Park Ave., New York, N.Y. 10016.

9.1–18 DISTRIBUTION OF WATER SUPPLY SYSTEMS WHERE DELIVERED WATER EXCEEDED THE COLIFORM DENSITY LIMIT

Region[a]	No. systems	No. systems exceeding	Exceeding, %
I	218	63	29
II	221	12	5
III	30	2	7
IV	22	0	0
V	66	0	0
VI	88	8	9
VII	26	0	0
VIII	20	4	20
IX	278	31	11
Totals	969	120	12

[a] See 9.1–6 for regional locations.

Source: Reprinted, with publisher's permission, from *J. Am. Water Works Assn.*, *62*(11):670, © 1970, by the American Water Works Association, Inc., 2 Park Ave., New York, N.Y. 10016.

9.1–19 TYPE OF WATER SUPPLY SYSTEMS THAT EXCEEDED THE COLIFORM DENSITY LIMIT

Type of system	No. systems	No. systems exceeding	Exceeding, %
Surface and mixed source	166	26	16
Spring source	80	30	37
Well source	496	41	8
Mixed well and spring wholesale source	37	11	30
Finished source	106	5	5
Special systems	84	7	8
Totals	969	120	12

Source: Reprinted, with publisher's permission, from *J. Am. Water Works Assn.*, *62*(11):670, © 1970, by the American Water Works Association, Inc., 2 Park Ave., New York, N.Y. 10016.

9.1–20 WATER SUPPLY SYSTEMS IN VARIOUS POPULATION SIZES WHERE DELIVERED WATER EXCEEDED THE COLIFORM DENSITY LIMIT

Population range	No. systems	No. systems exceeding	Systems exceeding, %	Survey population exceeding, %
≤500	446	79	18	15
501–5,000	315	29	10	8
5,001–10,000	75	5	7	7
10,001–25,000	59	5	9	7
25,001–50,000	36	1	3	3
50,001–100,000	16	1	6	6
>100,000	22	0	0	0
Totals	969	120	12	2

Source: Reprinted, with publisher's permission, from *J. Am. Water Works Assn.*, *62*(11):670, © 1970, by the American Water Works Association, Inc., 2 Park Ave., New York, N.Y. 10016.

9.1–21 WATER SUPPLY AND DISTRIBUTION SAMPLES WHERE AVERAGE DELIVERED WATER EXCEEDED SPECIFIC CONSTITUENT LIMITS[a]

Constituent	Constituent limit	Maximum concentration found	No. samples exceed-ing limit	Exceeding limit, %	No. systems exceeding con-stituent limit	All water supplies in study, %	No. systems exceeding one or more ad-ditional constituent limit
			Recommended				
ABS	0.5 mg/l	0.41 mg/l	0	0	0	0	0
Arsenic (As)	0.01 mg/l	0.10 mg/l	10	0.4	2	<1	2
Boron (B)	1.0 mg/l	3.28 mg/l	20	0.8	9	1	9
Chloride (Cl^-)	250 mg/l	1,950 mg/l	31	1.2	9	1	9
Color	15 units	49 mg/l	17	0.7	7	1	5
Copper (Cu)	1.0 mg/l	8.35 mg/l	42	1.6	11	1	4
CCE	0.200 mg/l	0.56 mg/l	33	1.2[d]	12[e]	1	2
Cyanide (Cn^{-2})	0.01 mg/l	0.008 mg/l	0	0	0	0	0
Fluoride (F^-)	Varies	4.40 mg/l	118	4.5	52	5	35
Iron (Fe)	0.3 mg/l	26.0 mg/l	223	8.6	96	10	52
Manganese (Mn)	0.05 mg/l	1.32 mg/l	211	8.1	90	9	45
Iron and manganese					159	16	56
Nitrate (NO_3)	45 mg/l	127 mg/l	53	2.1	19	3	15
Radium-226	3 pc/l	135.9 pc/l	16	0.6	6	1	5
Strontium-90	10 pc/l	2 pc/l	0	0	0	0	0
Sulfate (SO_4)	250 mg/l	770 mg/l	47	1.8	25	3	25
TDS	500 mg/l	2,760 mg/l	219	8.5	95	10	62
Turbidity	5 units or 1 unit[b]	53 units	60	2.4	26	3	17
Zinc (Zn)	5 mg/l	13.0 mg/l	8	0.3	1	<1	1
			Mandatory				
Arsenic (As)	0.05 mg/l	0.10 mg/l	5	0.2	2	<1	1
Barium (Ba)	1.0 mg/l	1.55 mg/l	2	<0.1[d]	1	<1	0
Boron (B)	5.0 mg/l	3.28 mg/l	0	0	0	0	0
Cadmium (Cd)	0.01 mg/l	3.94 mg/l	4	0.2	3	<1	3
Chromium (hexavalent)[c]	0.05 mg/l	0.079 mg/l	5	0.2	4	<1	4
Coliform organisms	1/100 ml	2,000/100 ml	228	8.8	120	12	34
Cyanide (CN^{-2})	0.20 mg/l	0.008 mg/l	0	0	0	0	0
Fluoride (F^-)	Varies	4.40 mg/l	55	2.2	24	2	17
Gross beta activity	1,000 pc/l	154 pc/l	0	0	0	0	0
Lead (Pb)	0.05 mg/l	0.64 mg/l	37	1.4	14	1	6
Selenium (Se)	0.01 mg/l	0.07 mg/l	10	0.4	5	1	5
Silver (Ag)	0.05 mg/l	0.03 mg/l	0	0	0	0	0

[a] Constituent in 2,595 distribution samples. Constituent in 969 water supply systems.
[b] Treated by more than disinfection.
[c] During this study total chromium was measured and compared to the standard.
[d] These constituents were evaluated only on selected samples. The remainder were assumed not to exceed the limit.
[e] Measured in only 110 selected supplies, including 94 surface supplies.

Source: Adapted from Community Water Supply Study, 1969 analysis national survey findings. Reprinted, with publisher's permission, from *J. Am. Water Works Assn., 62*(11):670, © 1970, by the American Water Works Association, Inc., 2 Park Ave., New York, N.Y. 10016.

9.1–22 DEFICIENCES RELATED TO HEALTH HAZARDS

Source and Treatment

Source	Percent[a] of systems
Inadequate quantity (922)	9
Restricted water use (900)	19
Deteriorating surface raw water quality (124)	26
Inadequate source protection (496)	19
Treatment	
Plans and specs, not inspected by the state (655)	29
By-pass possible (375)	52
Common walls (613)	6
Inadequate clarification beyond disinfection (162)	24
Inadequate control of clarification beyond disinfection (162)	36
Inadequate disinfection (311)	50
Inadequate control of disinfection (311)	22
Inadequate records (872)	44
Inadequate maintenance (901)	20

Distribution Systems

Distribution	Percent[a] of systems
Pressure less than 20 lb/in.^{2} (969)	26
Inadequate distribution facilities (845)	9
Uncovered finished water reservoirs (806)	14
No chlorine residual in distant part of distribution system (457)	57

Source and Treatment

Distribution (cont.)	Percent[a] of systems
No disinfection of new or repaired mains (879)	51
No cross connections control ordinance (854)	54
No effective cross connection control program (837)	87
Cross connection control limited to new construction only (889)	43
No cross connection control program for continuous reinspection (879)	89
Inadequate progress in cross connection control (826)	89
Interconnected with other water systems of unknown quality without protection (742)	6

Surveillance and Personnel

Surveillance	Percent[a] of systems
Does not meet bacteriological surveillance standards (969)	90
No sanitary survey within the last 3 yr (969)	64
Personnel (chief operator)	
Less than short school level training (969)	61
Inadequate bacteriological training (969)	77
Inadequate chemical training (162)	46

[a] Percent with deficiencies calculated on basis of supplies where applicable–number in parentheses.

Source: Reprinted, with publisher's permission, from *J. Am. Water Works Assn.*, *62*(11):670, © 1970, by the American Water Works Association, Inc., 2 Park Ave., New York, N.Y. 10016.

9.1–23 FACILITY DEFICIENCIES

	Population group served			
	500 or less	501–100,000	Greater than 100,000	All populations
No. systems	**446**	**501**	**22**	**969**
No major deficiencies, %	39	47	64	44
Some major deficiencies, %	61	53	36	56

Source: Reprinted, with publisher's permission, from *J. Am. Water Works Assn.*, *62*(11):670, © 1970, by the American Water Works Association, Inc., 2 Park Ave., New York, N.Y. 10016.

9.1–24 CARBON CHLOROFORM EXTRACT DATA

Type of system	No. systems sampled of this type	Exceeding limit, %
Well source	12	8
Mixed ground and surface source	4	0
Surface source	94	11
All types	110	10

Source: Reprinted, with publisher's permission, from *J. Am. Water Works Assn.*, *62*(11):670, © 1970 by the American Water Works Association, Inc., 2 Park Ave., New York, N.Y. 10016.

9.1–25 NITROGEN CONCENTRATION IN SOUTHERN CALIFORNIA WATER SUPPLIES

County and supply	NO_3 Nitrogen				NO_2 Nitrogen			NH_3 Nitrogen				Organic Nitrogen			
	<1.0 ppm	1.0–9.0 ppm	10–20 ppm	>20 ppm	0.000–0.009 ppm	0.01–0.10 ppm	>0.10 ppm	0.00–0.24 ppm	0.25–0.49 ppm	0.50–1.0 ppm	>1.0 ppm	0.00–0.24 ppm	0.25–0.49 ppm	0.50–1.0 ppm	>1.0 ppm
San Luis Obispo															
Community	11	16	6	6	28	9	2	35	2	1	1	17	12	9	1
Rural	14	28	5	2	34	14	1	43	0	4	2	22	15	12	0
Santa Barbara															
Community	6	18	2	2	18	7	3	25	2	1	0	10	14	3	1
Rural	18	32	8	0	44	12	2	51	2	2	3	24	25	8	1
Ventura															
Community	16	27	2	2	29	13	5	35	9	3	0	30	7	5	5
Rural	20	20	2	0	29	12	1	36	4	1	1	28	10	3	1
Los Angeles															
Community	26	33	14	1	36	37	1	36	24	8	6	48	17	7	2
Rural	10	18	0	1	13	15	1	21	5	2	1	21	5	3	0
San Bernardino															
Community	25	36	5	3	55	12	0	44	12	10	2	46	18	2	1
Rural	11	28	3	1	34	5	4	33	8	2	0	14	22	5	2
Orange															
Community	31	32	0	1	23	40	1	44	12	7	1	49	10	3	2
Rural	21	14	0	0	6	29	0	12	7	12	4	24	6	4	1
Riverside															
Community	22	30	8	3	53	6	4	53	6	4	0	41	18	4	0
Rural	20	19	1	1	29	10	2	36	2	0	3	29	9	2	1
San Diego															
Community	15	16	2	1	21	13	0	31	3	0	0	25	6	3	0
Rural	29	29	8	0	23	39	4	56	7	2	1	43	20	1	2

Source: Reprinted, with publisher's permission, from *J. Am. Water Works Assn.*, *55*(5):615, © 1963 by the American Water Works Association, Inc., 2 Park Ave., New York, N.Y. 10016.

9.1–26 HIGHEST NITROGEN CONCENTRATIONS REPORTED

	Concentration, ppm			
County	NO_3 Nitrogen	NO_2 Nitrogen	NH_3 Nitrogen	Organic nitrogen
San Luis Obispo	48.0	0.32	8.8	1.50
Santa Barbara	68.0	0.17	2.4	1.88
Ventura	33.2	0.15	3.2	2.15
Los Angeles	25.4	0.11	5.0	2.80
San Bernardino	75.0	0.60	3.8	3.31
Orange	20.0	0.15	6.5	1.90
Riverside	27.2	1.20	1.4	1.66
San Diego	31.0	0.58	1.0	1.65

Source: Reprinted, with publisher's permission, from *J. Am. Water Works Assn.*, *55*(5):615, © 1963 by the American Water Works Association, Inc., 2 Park Ave., New York, N.Y. 10016.

9.1–27 SPECTROGRAPHIC ANALYSES OF THREE MISSOURI TREATED WATER SUPPLIES

	Concentration, μg/l		
		St. Louis	
Element	Kansas City[a]	Howard Bend Plant[b]	Chain of Rocks Plant[c]
Silver	<0.28	<0.22	<0.23
Aluminum	58	850	>2,000
Boron	82	37	43
Barium	85	59	77
Chromium	1.2	1.5	2.0
Copper	7.4	8.5	2.5
Iron	24	95	39
Lithium	25	15	15
Manganese	<2.8	4.3	3.4
Molybdenum	4.4	5.4	7.0
Nickel	3.6	4.1	4.8
Lead	5.2	2.8	3.0
Rubidium	no data	<2.2	<2.3
Strontium	270	150	250
Titanium	<2.8	2.4	15
Vanadium	8.2	8.9	11

[a] Sample taken July 26, 1961.
[b] Sample taken October 18, 1961; data represent 34% of St. Louis supply.
[c] Sample taken October 19, 1961; data represent 66% of St. Louis supply.

Source: Reprinted, with publisher's permission, from *J. Am. Water Works Assn.*, *55*(8):1019, © 1963 by the American Water Works Association, Inc., 2 Park Ave., New York, N.Y. 10016.

9.1–28 MINOR ELEMENT ANALYSES

Denver Water Supply Surface Waters

Element or chemical	Date of collection	Moffat Treatment Plant, μg/l[a]		Domestic taps, μg/l		North Marston Plant, μg/l[b]		South Marston Plant, μg/l[b]		Kassler Plant, μg/l[b]		Kassler Galleries, μg/l[c]		Domestic taps, μg/l	
		Raw	Finished	No. 1 Finished	No. 2 Finished	Raw	Finished	Raw	Finished	Raw	Finished	Raw	Finished	No. 3 Finished	No. 4 Finished
Aluminum	5/13/66	35	140	74	100	74	120	70	120	80	47	27	23	130	150
	6/10/66	30	170	130	220	95	160	50	95	75	47	53	44	250	80
	7/ 8/66	40	140	240	220	60	210	40	90	40	35	20	20	210	210
	8/ 5/66	36	190	170	130	71	220	50	95	520	40	29	36	220	90
	9/16/66	17	260	160	150	70	190	60	150	80	16	10	6	200	340
Barium	5/13/66	21	22	21	27	48	50	45	50	44	53	70	70	47	51
	6/10/66	18	17	21	25	50	50	48	53	52	54	54	62	48	53
	7/ 8/66	20	25	23	20	40	45	48	39	56	52	45	50	44	48
	8/ 5/66	17	17	18	17	37	35	39	43	50	42	60	64	39	43
	9/16/66	16	18	21	17	48	45	48	45	45	38	50	45	48	40
Boron	5/13/66	11	10	10	14	30	30	27	26	26	26	37	38	26	28
	6/10/66	14	11	12	12	22	22	24	28	30	28	33	34	27	26
	7/ 8/66	14	15	13	13	22	22	26	23	26	27	21	34	24	32
	8/ 5/66	14	13	13	12	25	22	26	26	25	25	31	34	18	26
	9/16/66	7	6	9	7	17	15	14	14	15	16	30	26	18	19
Chromium	5/13/66	ND[d]	ND	ND	ND	ND	ND	ND	ND	ND	ND	6	6	ND	ND
	6/10/66	ND	ND	ND	ND	ND	ND	ND	ND	ND	ND	ND	ND	ND	ND
	7/ 8/66	ND	ND	ND	ND	ND	ND	ND	ND	ND	ND	3	3	ND	ND
	8/ 5/66	ND	ND	ND	ND	ND	ND	ND	ND	ND	ND	6	6	ND	ND
	9/16/66	ND	ND	ND	ND	ND	ND	ND	ND	ND	ND	7	5	ND	ND
Copper	5/13/66	24	14	4	16	120	20	8	8	5	10	5	5	18	9
	6/10/66	26	8	3	13	85	12	6	4	4	8	3	4	22	7
	7/ 8/66	26	6	3	3	45	8	4	4	4	8	2	2	7	4
	8/ 5/66	34	7	3	5	50	7	4	3	5	8	3	4	7	5
	9/16/66	65	3	3	3	35	6	2	2	4	6	2	2	7	4

[a] Source: Upper Frazer and Williams Fork Rivers.
[b] Sources: watershed of South Platte River south of Denver, and the Dillon Reservoir, which collects water from the Snake River, Blue River, and Tenmile Creek.
[c] Source: underground seepage from the South Platte River and the Kassler filter beds.
[d] Not detected.

9.1–28 MINOR ELEMENT ANALYSES *(Continued)*

Element or chemical	Date of collection	Moffat Treatment Plant, μg/l[a] Raw	Moffat Finished	Domestic taps, μg/l No. 1 Finished	Domestic taps No. 2 Finished	North Marston Plant, μg/l[b] Raw	North Marston Finished	South Marston Plant, μg/l[b] Raw	South Marston Finished	Kassler Plant, μg/l[b] Raw	Kassler Plant Finished	Kassler Galleries, μg/l[c] Raw	Kassler Galleries Finished	Domestic taps, μg/l No. 3 Finished	Domestic taps No. 4 Finished
Iron	5/13/66	46	5	4	15	10	6	8	6	70	82	14	17	3	15
	6/10/66	5	5	16	52	5	4	4	7	70	100	22	32	60	100
	7/ 8/66	39	8	9	26	8	4	6	4	18	60	15	11	5	7
	8/ 5/66	40	6	15	10	10	6	10	4	530	80	6	20	12	30
	9/16/66	23	30	11	35	5	7	13	7	32	21	10	10	30	95
Lithium	5/13/66	0.9	1	1	1	11	10	8	9	6	9	7	6	7	11
	6/10/66	0.5	0.7	2	2	10	10	11	10	12	10	11	10	10	9
	7/ 8/66	0.4	0.7	0.8	0.8	8	8	8	7	8	7	9	9	7	8
	8/ 5/66	0.4	0.5	0.6	0.5	6	5	6	5	5	5	6	7	4	5
	9/16/66	–	–	–	–	–	–	–	–	–	–	–	–	–	–
Manganese	5/13/66	4	ND[d]	2	3	ND	6	9	ND	23	ND	ND	ND	4	ND
	6/10/66	3	1	3	4	ND	ND	34	ND	24	ND	ND	38	ND	ND
	7/ 8/66	3	5	4	3	4	2	56	8	12	4	ND	ND	4	4
	8/ 5/66	10	3	5	4	ND	ND	29	7	39	ND	ND	ND	ND	5
	9/16/66	3	3	4	4	ND	ND	11	ND	20	ND	ND	ND	ND	ND
Molybdenum	5/13/66	2	2	2	2	5	6	7	9	2	2	10	10	6	8
	6/10/66	2	2	2	3	6	6	8	7	14	4	11	12	6	6
	7/ 8/66	1	1	1	1	14	15	10	9	54	27	14	15	18	25
	8/ 5/66	2	2	2	2	60	51	36	38	61	95	41	39	39	65
	9/16/66	1	1	1	1	140	140	130	140	260	240	100	85	120	190
Rubidium	5/13/66	0.6	0.4	0.4	0.5	1	0.9	0.7	0.9	0.9	0.9	0.9	0.9	0.7	1
	6/10/66	0.9	0.7	0.8	0.8	2	2	2	0.9	4	1	ND	ND	1	1
	7/ 8/66	0.5	0.3	0.6	0.5	0.7	0.8	0.6	0.7	0.9	0.8	0.6	0.5	0.6	0.9
	8/ 5/66	0.5	0.5	0.5	0.4	1	1	1	1	2	1	0.9	0.9	1	1
	9/16/66	–	–	–	–	–	–	–	–	–	–	–	–	–	–

[a] Source: Upper Frazer and Williams Fork Rivers.
[b] Sources: watershed of South Platte River south of Denver, and the Dillon Reservoir, which collects water from the Snake River, Blue River, and Tenmile Creek.
[c] Source: underground seepage from the South Platte River and the Kassler filter beds.
[d] Not detected.

9.1–28 MINOR ELEMENT ANALYSES *(Continued)*

Element or chemical	Date of collection	Moffat Treatment Plant, μg/l[a]		Domestic taps, μg/l		North Marston Plant, μg/l[b]		South Marston Plant, μg/l[b]		Kassler Plant, μg/l[b]		Kassler Galleries, μg/l[c]		Domestic taps, μg/l	
		Raw	Finished	No. 1 Finished	No. 2 Finished	Raw	Finished	Raw	Finished	Raw	Finished	Raw	Finished	No. 3 Finished	No. 4 Finished
Strontium	5/13/66	43	51	47	64	210	200	210	220	190	220	280	300	230	260
	6/10/66	29	36	60	75	260	250	270	240	260	310	260	340	280	310
	7/ 8/66	45	50	55	40	260	260	280	240	330	320	280	300	340	320
	8/ 5/66	46	50	58	50	330	310	330	280	250	280	360	370	220	300
	9/16/66	35	40	45	35	260	250	240	240	190	190	260	250	260	210
Titanium	5/13/66	1	ND	ND	ND	ND	ND	ND	ND	ND	ND	ND	ND	ND	ND
	6/10/66	ND	ND	ND	ND	ND	ND	ND	ND	ND	ND	9	ND	ND	ND
	7/ 8/66	3	2	ND	ND	ND	ND	ND	ND	ND	ND	ND	ND	ND	ND
	8/ 5/66	2	0.7	2	1	ND	ND	ND	ND	26	ND	ND	ND	ND	ND
	9/16/66	1	3	ND	1	ND	ND	ND	ND	4	ND	ND	ND	ND	ND
Zinc[e]	5/13/66	210	180	80	10	10	10	20	10	10	50	20	80	10	<10
	6/10/66	20	20	30	20	10	20	10	10	10	40	20	10	20	10
	7/ 8/66	20	<10	30	<10	<10	<10	<10	10	<10	40	<10	<10	<10	10
	8/ 5/66	20	10	40	10	10	<10	10	<10	<10	40	<10	<10	<10	10
	9/16/66	20	10	50	10	<10	<10	10	<10	<10	40	<10	<10	<10	10

[a] Source: Upper Frazer and Williams Fork Rivers.
[b] Sources: watershed of South Platte River south of Denver, and the Dillon Reservoir, which collects water from the Snake River, Blue River, and Tenmile Creek.
[c] Source: underground seepage from the South Platte River and the Kassler filter beds.
[d] Not detected.
[e] Determined by atomic absorption.

Source: Reprinted, with publisher's permission, from *J. Am. Water Works Assn., 61*(2):61, © 1969, by the American Water Works Association, Inc., 2 Park Ave., New York, N.Y. 10016.

9.1–29 SUMMARY OF TRACE ELEMENTS IN 380 FINISHED WATERS

Element	Frequency of detection, %	Minimum, μg/l	Maximum, μg/l	Mean, μg/l
Zinc	77.0	3	2,010	79.2
Cadmium	0.2	12	12	12
Iron	83.4	2	1,920	68.9
Manganese	58.7	0.5	450	25.5
Copper	65.2	1	1,056	43.0
Silver	6.1	0.3	5	2.2
Lead	18.1	3	139	33.9
Chromium	15.2	1	29	7.5
Barium	99.7	1	172	28.6
Molybdenum	29.9	3	1,024	85.9
Aluminum	47.8	3	1,600	179.1
Beryllium	1.1	0.02	0.17	0.1
Nickel	4.6	1	490	34.2
Cobalt	0.5	22	29	26.0
Vanadium	3.4	14	222	46.1
Strontium	100	1	960	105.0
Boron	100	1	1,900	92.2

Source: J.F. Kopp in *Trace Substances in Environmental Health,* D.D. Hemphill, Ed., University of Missouri, Columbia, 1969.

9.1–30 SUMMARY OF VIOLATIONS OF THE WATER QUALITY CRITERIA

Metal	Limit, μg/l[a]	No. violations	Mean of those values which exceeded the limits, μg/l
Cadmium	10	6	39
Arsenic	50	41	91
Iron	300	25	>676
Manganese	50	74	>586
Lead	50	27	71
Chromium^{+6}	50	4	94
Zinc	5,000	0	–
Copper	1,000	0	–
Silver	50	0	–
Barium	1,000	0	–

[a] As set by the Water Quality Criteria for Public Water Supplies.

Source: J.F. Kopp in *Trace Substances in Environmental Health,* D.D. Hemphill, Ed., University of Missouri, Columbia, 1969.

9.1–31 THRESHOLD ODOR RESULTS ON SAMPLES OF ORGANIC EXTRACT

	Threshold odor, ppb			
	Raw water		Tap water	
Sampling point[a]	Total extract	Neutral fraction	Total extract	Neutral fraction
Yankton, S.D.			720	512
Omaha, Nebr.	430	324	304	204
St. Joseph, Mo.	304	430	256	362
Kansas City, Mo.	218	430	304	512
Lexington, Mo.	256	362	362	362
Jefferson City, Mo.	180	92	151	92
St. Louis City (Howard Bend), Mo.	362	180	430	256
St. Louis County (North Plant), Mo.	128	180	256	256

[a] Samples collected February 17–March 5, 1958.

Source: Reprinted, with publisher's permission, from *J. Am. Water Works Assn.*, *53*(11):1392, © 1961 by the American Water Works Association, Inc., 2 Park Ave., New York, N.Y. 10016.

9.1–32 AVERAGE CONCENTRATION AMOUNT OF CCE IN RAW AND TAP WATERS AT EACH SAMPLING POINT

1958

	CCE concentration, ppb		
Sampling point	Raw water	Tap water	Organics, lb/day
Yankton, S.D.	37	39	4,100
Omaha, Nebr.	36	47	4,900
St. Joseph, Mo.	33	38	3,900
Kansas City, Mo.	41	36	4,800
Lexington, Mo.	36	45	10,100
Jefferson City, Mo.	32	46	7,300
St. Louis City (Howard Bend), Mo.	41	48	17,200
St. Louis County (North Plant), Mo.	36	53	16,200

Source: Reprinted, with publisher's permission, from *J. Am. Water Works Assn.*, *53*(11);1392, © 1961, by the American Water Works Association, Inc., 2 Park Ave., New York, N.Y. 10016.

9.1–33 CONCENTRATION OF ORGANIC MATTER EXTRACTED FROM WELL WATERS

Town	Total volume filtered, gal	Solvent used	Total extract,[a] mg/l	COD, mg/g extract
Oakwood, Ill.	1,073	chloroform	2.17	–
		water	0.99	–
		ethanol	1.14	2.79
		ethanol + NH_3	1.13	–
		ethanol + HCl	–	–
		Total	5.43	–
Philo, Ill.	950	chloroform	0.13	–
		water	0.08	–
		ethanol	0.94	0.62
		ethanol + NH_3	0.35	0.22
		ethanol + HCl	–	
		Total	1.50	–
Clinton, Ill.	1,426	chloroform	0.34	–
		water	0.68[b]	–
		ethanol	2.00	–
		ethanol + NH_3	0.71	–
		ethanol + HCl		–
		Total	3.75	–
Atwood, Ill.	1,293	chloroform	0.34	–
		water	1.72	–
		ethanol	1.36	–
		ethanol + NH_3	1.15	–
		ethanol + HCl		–
		Total	4.57	–
Clinton, Ill.	1,153	chloroform	0.55	1.67
		water	1.35	1.04
		ethanol	3.31	2.31
		ethanol + NH_3	2.08	1.39
		ethanol + HCl		–
		Total	7.29	–

[a] Ethanol plus HCl extract not included.
[b] Samples lost, values used are from Clinton Experiment 5.

Source: Reprinted, with publisher's permission, from *J. Am. Water Works Assn.*, *59*(2):227, © 1967, by the American Water Works Association, Inc., 2 Park Ave., New York, N.Y. 10016.

9.1–34 PESTICIDES IN 1 GALLON SAMPLES OF WATER[a]

Location	Date	Lindane	BHC's	Aldrin	p,p′-DDE	p,p′-DDT	Dieldrin	Endrin	Heptachlor	HCE	Chlordane
Finished Drinking Water											
Algiers	03/31/65 08/25/65			0/7	1/12	1/12	10/12	4/12			
Carrolton	03/31/65 08/25/65			0/6	0/12	0/12	6/12	3/12			
Carville	03/29/65 12/14/67	1/1	1/1	1/19	1/19	1/19	9/19	1/19	1/10	1/10	
Vicksburg	03/29/65 05/25/67	4/6	5/6	3/45	23/45	23/45	21/45	7/45	1/24	6/37	0/6
Cape Girardeau	04/19/65 06/07/67	6/6	6/6	9/39	7/39	6/39	18/39	3/39	0/13	0/13	0/6
Omaha	05/04/65 05/23/67	3/5	4/5	2/19	1/19	1/19	4/19	1/19	0/19	0/19	0/6
St. Louis	12/11/66 05/25/67	6/7	6/7	2/7	1/7	0/7	5/7	1/7	0/7	0/7	1/6[b]
Burlington	03/24/65 06/06/67	5/6	5/6	2/11	4/11	3/11	3/11	0/11	0/6	1/6	0/5
Jefferson City	05/30/66 05/26/67	4/7	5/7	2/15	5/15	4/15	5/15	0/15	0/15	0/15	1/6
Kansas City	03/19/65 05/23/67	5/9	5/9	16/52	10/54	10/54	24/54	4/54	1/33	13/54	12/28[c]
Total		34/47	37/47	37/220	53/233	49/233	105/233	24/233	3/127	21/161	14/63
Raw Water											
Carville	12/14/66 01/05/67	2/2	2/2[b]	0/2	1/2	1/2	0/2	0/2	0/2	0/2	0/1
Vicksburg	02/27/67 05/25/67	4/4	4/4	1/4	2/4	2/4	2/4	1/4	1/4	0/4	1/4
Cape Girardeau	03/29/65 06/07/67	4/8	5/8	4/8	3/8	2/8	5/8	2/8	0/8	0/8	0/6

[a] Ratio of number of samples with detectable concentrations to total number of samples.
[b] One sample assayed at 0.5 ppb.
[c] Four samples assayed from 8–0.25 ppb.

9.1–34 PESTICIDES IN 1 GALLON SAMPLES OF WATER[a] *(Continued)*

Location	Date	Lindane	BHC's	Aldrin	p,p'-DDE	p,p'-DDT	Dieldrin	Endrin	Heptachlor	HCE	Chlordane
Raw Water (continued)											
Jefferson City	05/30/66 05/26/67	4/7	6/7	5/15	3/15	4/15	8/15	0/15	0/15	1/15	1/7
Kansas City	05/04/66 05/23/67	7/7	7/7	12/22	3/22	3/22[b]	10/22	1/22	1/22	4/22	9/22[d]
Omaha	04/20/66 05/23/67	4/6	5/6	4/16	3/16	3/16	4/16	1/16	0/16	0/16	0/6
Total		25/34	29/34	26/67	15/67	15/67	29/67	5/67	2/67	5/67	11/46

[a] Ratio of number of samples with detectable concentrations to total number of samples.
[b] One sample assayed at 0.5 ppb.
[c] Four samples assayed from 8–0.25 ppb.
[d] Two samples assayed from 0.5–0.3 ppb.

Source: M.L. Schafer *et al.*, *Environ. Sci. Technol.*, *3*(12):1261, 1969.

9.1–35 PESTICIDES IN 100 Ml SAMPLES OF WATER[a]

Location	Date	Aldrin	p,p′-DDE	p,p′-DDT	Dieldrin	Endrin
Finished Drinking Water						
Algiers	03/03/64 12/17/64	0/2	5/19[b]	5/19[b]	2/19	14/19[b]
Carrolton	03/03/64 12/17/64		3/20	3/20	7/20	14/20[c]
Carville	04/16/64 06/08/66	4/26[b]	13/39[c]	12/39[b]	14/39	24/39[d]
Vicksburg	03/31/64 06/30/66	4/41	29/41[e]	28/41[e]	23/41[d]	34/41[f]
Cape Girardeau	06/02/64 03/14/66	2/25	13/25[b]	14/25[b]	13/25[c]	11/25[d]
Burlington	06/30/64 03/22/66	0/10	2/10	2/10	2/10	4/10[b]
St. Louis	06/11/64 06/16/66	3/32	14/32[b]	14/32[b]	17/32[b]	10/32[b]
Jefferson City	06/04/64 07/16/64	1/7	3/7	3/7	7/7	6/7
Kansas City	06/11/64 05/23/66	3/10	1/10	1/10	0/10	5/12[d]
Omaha	06/01/64 03/07/66	4/20	7/20	6/20	6/20	10/20[d]
Total		21/173	90/223	88/223	91/223	132/225
Raw Water						
Carville	06/17/74 03/10/66	0/3	5/10	5/10	6/10	3/10[b]
Cape Girardeau	06/02/64 03/14/66	1/12	2/12	2/12	4/12	4/12[d]
Kansas City	04/27/66 05/23/66	0/2	0/2	0/2	0/2	0/2
Total		1/17	7/24	7/24	10/24	7/24

[a] Ratio of number of samples with detectable concentrations to total number of samples.
[b] One sample assayed >0.1 ppb.
[c] Two samples assayed >0.1 ppb.
[d] Three samples assayed >0.1 ppb.
[e] Four samples assayed >0.1 ppb.
[f] Six samples assayed >0.1 ppb.

Source: M.L. Schafer *et al., Environ. Sci. Technol., 3*(12): 1261, 1969.

9.1–36 ORGANIC COMPOUNDS FOUND IN MUNICIPAL WATER SUPPLY

Evansville, Indiana

Compound	Evidence[a]
Bromodichloromethane	A
Toluene	A, B, C
Chlorodibromomethane	A
Tetrachloroethylene	A, B, C
Ethylbenzene[b]	A, B, C
Xylene[b]	A, B, C
Styrene	A, B, C
Bromoform	A, B, C
Bis(2-chloroethyl)ether	A, B, C
Bis(2-chloroisopropyl)ether	A, B, C, D
Hexachloroethane	A, B
Hexachlorobenzene	A, B
Chlorohydroxybenzophenone[c]	A

[a] A – interpretation of mass spectrum, B – comparison with mass spectrum of authentic sample, C – retention time match on two different glc columns, and D – comparison with ir spectrum of authentic sample.
[b] Analysis on a 1/1/10 Bentone 34, Silicone DC 430, Chromosorb W 60–80 mesh column indicated a mixture of xylene isomers and ethylbenzene.
[c] Substitution is on the same ring. This component was found in the acid fraction.

Source: R.D. Kleopfer and B.J. Fairless, *Environ. Sci. Technol., 6*(12):1036, 1972.

9.1–37 METHANE EQUIVALENT FOR NATURAL AND MUNICIPAL WATERS

Source of water	Methane equivalent, mg/l
Natural:	
North Fork Matilija Creek	0.017
Santa Clara River	0.043
Lake Tahoe	0.016
Municipal:	
Alameda, Cal.	0.007
Los Angeles, Cal.	0.099
San Diego, Cal.	0.052
San Luis Obispo, Cal.	0.047
Santa Ana, Cal.	0.026
Thousand Oaks, Cal.	0.032
Flagstaff, Arik.	0.026
Albuquerque, N.M.	0.033
Reno, Nev.	0.060

Source: I. Lysyj *et al., J. Water Pollut. Control Fed., 40*(5):R185, 1968.

9.1–38 STUDIES OF ABS CONTAMINATION OF GROUND WATER IN HOUSING DEVELOPMENTS[a]

Location	Distance of well from sewage system, ft		No. wells sampled	No. wells containing ABS	Max ABS concentration, ppm	No. wells containing indicated concentration			Ref.
	Range	Mean				≥ 0.1 ppm	≥ 0.5 ppm	≥ 1.0 ppm	
Alabama									
Huntsville		50 ±	11	6	0.4	6	0	0	1
Connecticut	75–100		5	4	4.4[b]	4	3	2	2
Florida									
Dade County	50–75	60	21	6	0.2	6	0	0	1
Kansas[c]									3
Massachusetts[c]									4
Minnesota									
Coon Rapids									
Daily and Hurter Subdivision		50 ±	4	2	0.1	2	0	0	1
Thompson Park		50 ±	7	3	0.4	3	0	0	1
Other			96	22	1.2				5
Windward Heights		60 ±	4	4	0.1				1

[a] Each unit serviced by its own well and subsurface sewage disposal system.
[b] Based on 20% of household detergent concentration.
[c] Several instances of contamination reported.

9.1–38 STUDIES OF ABS CONTAMINATION OF GROUND WATER IN HOUSING DEVELOPMENTS[a] *(Continued)*

Location	Distance of well from sewage system, ft		No. wells sampled	No. wells containing ABS	Max ABS concentration, ppm	No. wells containing indicated concentration			Ref.
	Range	Mean				⩾ 0.1 ppm	⩾ 0.5 ppm	⩾ 1.0 ppm	
New Mexico									
Bernalillo County	20–100	50	20	9	0.8	9	1	0	1
New York									
Suffolk County									
Amityville Harbor			31	17					6
Babylon			20	19					6
Brentwood		100	27	2	0.2	2	0	0	7
Center Moriches			65	29					6
Copiague	35–80		186	60	3.3				8
Farmingdale	30–80	80	11	0	< 0.1	0	0	0	7
Fire Island[d]			24	2					6
Islip			47	8					6
Lindhurst			45	5					6
Middle Island	30–100	45	21	0	< 0.1	0	0	0	7
N. Lindhurst			54	45	4.5				6
Shirley	60–90	75	6	0	< 0.1	0	0	0	7
Smithtown	30–100	70	36	1	0.1	1	0	0	7
W. Islip			161	52					6
New Jersey					9.0[e]				9
Pennsylvania[f]									
Rhode Island									
Portsmouth	30–150	50	25	24	5.0	24	19	13	10
Virginia									
Norfolk		50	36	27	4.1	27	8	6	1
Wisconsin									
La Crosse			13	10	0.4	10	0	0	1
Other			140	63	10				11

[d] Summer colony; only small amounts of detergents used.
[e] 30% of well waters examined contained ABS; reference standard other than AASGP formulation.
[f] Many underground pools known to contain detergents in some quantity.

Source: Reprinted, with publisher's permission, from *J. Am. Water Works Assn., 52*(11):1354, © 1960 by the American Water Works Association, Inc., 2 Park Ave., New York, N.Y. 10016.

REFERENCES

1. R.T. Kiser, *A Reconnaissance Study of Anionic Surfactants in Ground Water,* Federal Housing Administration, 1959.
2. I. Newell and F. Almquist, *J. New Engl. Water Works Assn., 74*:61, 1960.
3. R.L. Culp, Private communication.
4. W. Taylor, Private communication.
5. F.L. Woodward, Private communication.
6. H. Davids and J. Flynn, *Report on the Status of Ground Water Quality and Related Factors in Suffolk County,* Suffolk County of Health, Riverhead, N.Y., 1960.
7. H.E. Lauman *et al., Study of Synthetic Detergents in Ground Water, Suffolk County Area,* Long Island Home Builders Institute, Inc., Bethpage, N.Y. (unpublished).
8. J. Flynn, A. Andreoli, and A. Guerrera, *J. Am. Water Works Assn., 50*(12):1551, 1958.
9. M. Alpers, Private communication.
10. J. Deluty, *Public Health Repts., 75*:75, 1960.
11. M.S. Nichols, Private communication.

9.1–39 REPORTED INSTANCES OF ABS CONTAMINATION OF GROUND WATER BY OTHER THAN UNDERGROUND SEWAGE DISPOSAL

Location	Date	Type of well	Probable source of contamination	Apparent ABS concentration, ppm	Approximate distance of travel, ft	Ref.
California	1959	Private	Sewage and industrial waste discharged to dry stream bed	0.6		
Colorado						
Derby	1957	Municipal	Tractor repair garage lagoon	0.6	1,000	1
Commercetown	1959	Industrial	Induced infiltration of sewage contaminated water	a	1,000	
Connecticut						
Plainville	1958	Municipal	Industrial waste, holding pond	0.4–4.0	300	2, 3
				0.3–2.0	500	
				0.1–0.9	700	
Other			Induced infiltration from contaminated stream	0.6	800	2
Florida						
Diana	1959	Industrial	Subsurface disposal of treated laundry wastes			4
Other	1958	Irrigation	Subsurface disposal of treated laundry wastes	10	60	4
Illinois						
Peoria	1959	Test well	Recharge of contaminated	0.7	200	5
		Industrial	river water	0.2	1,800	
Kansas						
El Dorado	1956	Standby municipal	Induced infiltration of con-	2.6		6
Augusta		wells	taminated river water			
Nebraska						
Kearny	1954–55	Private	Sewage oxidation pond	2.0	800	7
				1.5	1,000	
				0.3	4,000	
Washington						
Tieton	1958	Private	Sewage oxidation pond	0.2	300	8
Wisconsin						
Lac Du Flambeau	1959	Private	Holding pond for laundry wastes	a	100	9

[a] Foaming indicated presence of ABS.

Source: Reprinted, with publisher's permission, from *J. Am. Water Works Assn.*, *52*(11):1354, © 1960 by the American Water Works Association, Inc., 2 Park Ave., New York, N.Y. 10016.

REFERENCES

1. F.M. Middleton, *Report on Analysis of Organic Contaminats Recovered From Town and Country Mutual Water Co. Well at Commercetown, Colo.,* Robert A. Taft San. Eng. Center, Cincinnati, Ohio (unpublished).
2. I. Newell and F. Almquist, *J. New Engl. Water Works Assn.*, *74*:61, 1960.
3. F.O. Almquist, Private communication.
4. D. Lee, Private communication.
5. F.M. Middleton, *ABS in Sample From Illinois River and Wells in Peoria Area,* Unpublished data obtained in cooperative study with Illinois State Water Survey, Robert A. Taft San. Eng. Center, Cincinnati, Ohio, 1959.
6. D.F. Metzler, *J. Am. Water Works Assn.*, *48*(48):999, 1956.
7. J. Neel and G. Hopkins, *J. Water Pollut. Control Fed.*, *28*:1326, 1956.
8. R.H. Bogan, Private communication.
9. C.R. Bowman, Private communication.

9.1–40 EFFECT OF WASTE WATER INJECTION ON GROUND WATER QUALITY

Constituent, mg/l	Average injection water	Native ground water	Distance from injection well, ft 100	245	545
Lambda Aquifer[a]					
Calcium	105	40	150	165	180
Magnesium	30	7	31	27	29
Sodium	255	35	250	230	100
Potassium	30	3	22	12	6
Bicarbonate	225	185	355	330	260
Sulfate	450	40	435	400	305
Chloride	280	12	275	250	180
Organic nitrogen	1	0	0.6	1	0.6
Ammonia N	23	0	16	6	0
Nitrite N	0	0	0.005	0.008	0.016
Nitrate N	0.5	0	0.05	0.10	0.2
Phosphate	0.3	0	0.03	0.03	0.05
Chemical oxygen demand	54	0	34	23	13
Carbon chloroform extract	1.1	0.03	0.7	0.4	
Methylene blue active substances	2	0	1.1	1.0	0.7
Volatile solids	100		110	135	170
Iron	0.05	0.02	0.07	0.08	0.05
Manganese	0.05	0	1.45	0.73	0.03
Fluoride	0.7	0.8	0.5	0.3	0.4
Boron	1	0.1	1.1	1.0	0.5
Silica	16	18	18	18	14
Dissolved oxygen	6.5	1	0.2	0.2	0.3
Carbon dioxide	65	2	35	30	10
Total dissolved solids	1,300	235	1,360	1,255	930
pH, units	6.6	7.7	6.9	7.0	7.3
Color, units	15	0	13	8	0
Odor, threshold odor number	40	0	40	40	8
Beta Aquifer[b]					
Calcium	55	45	75	70	60
Magnesium	15	6	16	15	11
Sodium	160	35	155	90	50
Potassium	18	2	8	4	3
Bicarbonate	190	190	245	235	205
Sulfate	220	35	210	135	75
Chloride	150	12	135	80	40
Organic nitrogen	0.5	1.4	0.6	1.1	0
Ammonia N	13	1.0	5	1.1	0
Nitrite N	0.1	0.06	–	–	–
Nitrate N	0.3	0.1	0.1	0.2	0.3
Phosphate	0.15	0	0.02	0.08	–
Chemical oxygen demand	25	0	17	7	5
Iron	0.06	0	0.05	0.05	0.06
Manganese	0.15	0	0.43	0.03	0.03
Fluoride	0.4	1.0	0.6	0.4	0.6
Boron	0.7	0.1	0.7	0.1	0.3
Silica	14	18	18	16	15
Dissolved oxygen	4.0	1.3	0.3	0.3	0.2
Carbon dioxide	35	1.7	20	4	
Total dissolved solids	800	230	725	510	350
pH, units	6.7	7.6	7.0	7.4	–
Color, units	40	0	0	0	–
Odor, threshold odor number	20	0	8	4	3

[a] Transmissibility (coefficient of permeability aquifer thickness), 85,000 gal/day ft.
[b] Transmissibility (coefficient of permeability aquifer thickness), 150,000 gal/day ft.

Note: All data are from Orange County, Cal.

Source: Reprinted, with publisher's permission, from *J. Am. Water Works Assn.*, *62*(3):203, © 1970, by the American Water Works Association, Inc., 2 Park Ave., New York, N.Y. 10016.

9.1–41 DATA ON ACREAGE FOR ARTIFICIAL RECHARGE OF GROUND WATER

State	Recharge area, acres	Benefitted area used for recharge, %	Distance from pumping center, mi
Arkansas	5		0
California	30,000	1	0–40
Colorado	48.0		
Illinois	3.0		0.1
Indiana	2.0	40	0
Utah	8.0		10
Washington	40.0	2	0.7

Source: Reprinted, with publisher's permission, from *J. Am. Water Works Assn., 55*(6):705, © 1963, by the American Water Works Association, Inc., 2 Park Ave., New York, N.Y. 10016.

9.1–42 GROUND WATER RECHARGED BY IRRIGATION WATER

State	Amount, %
California	20
Colorado	50
Florida	25
Idaho	60–25
Mississippi	slight
Montana	10–40
Nevada	10
New Jersey	15
Oregon	5–50
South Dakota	> 5
Texas	slight
Wyoming	20–25

Source: Reprinted, with publisher's permission, from *J. Am. Water Works Assn., 55*(6):705, © 1963 by the American Water Works Association, Inc., 2 Park Ave., New York, N.Y. 10016.

9.1–43 INFILTRATION RATES FOR GROUND WATER RECHARGED WITH IRRIGATION DRAINAGE

State	Basins, ft/day	Injection wells, gal/min
Arizona		500
California	0.1–10	58–540
Utah	1.0	
Illinois	4–170	
New Jersey	0.08–0.52	200
Washington	19.8	660
Utah	0.19–13.7	

Source: Reprinted, with publisher's permission, from *J. Am. Water Works Assn., 55*(6):705, © 1963, by the American Water Works Association, Inc., 2 Park Ave., New York, N.Y. 10016.

9.2 *IRRIGATION WATER, PESTICIDES, AND SALINITY BUILD-UP*

9.2–1 COMPOSITION OF REPRESENTATIVE IRRIGATION WATERS

Water	EC,[a] μmho /cm at 25°C	Total concentration, mg/l	Total concentration, mEq/l	B, mg/l	Ca	Mg	Na	K	CO_3 + HCO_3	SO_4	Cl	NO_3	SAR	pH_c
					mEq/l									
							Individual rivers							
San Joaquin, Biola, Cal.	60	46	0.52	0.10	0.23	0.08	0.19	0.02	0.36	0.06	0.09	0.02	0.5	9.5
Rogue, Grants Pass, Ore.	82	74	0.82	0.03	0.39	0.23	0.20	–	0.79	0.04	0.06	0.01	0.4	8.9
Sacramento, Knights Landing, Cal.	148	99	1.42	0.10	0.65	0.41	0.34	0.02	1.10	0.14	0.17	0.01	0.5	8.6
Snake, Heise, Idaho	371	218	3.85	0.05	2.34	0.98	0.48	0.05	2.69	0.90	0.31	0.02	0.4	7.7
Missouri, Nebraska City, Neb.	694	456	7.26	0.13	2.89	1.49	2.74	0.14	3.05	3.75	0.51	0.01	1.8	7.6
Colorado, Lees Ferry, Ariz.	864	547	8.80	0.11	3.84	1.72	3.13	0.11	2.72	4.37	1.72	0.02	1.9	7.6
Arkansas, Ralston, Okla.	1,260	724	11.75	–	2.74	0.94	8.07	–	1.77	1.98	7.95	0.05	5.9	7.9
Salt, below Stewart Mt. Dam, Ariz.	1,520	832	14.04	0.27	2.94	1.15	9.79	0.16	2.80	1.23	10.01	0.01	6.8	7.7
Sevier, Lynndyl, Utah	1,950	1,190	20.37	0.31	3.76	5.80	10.66	0.15	5.44	6.23	8.46	0.14	4.9	7.4
Pecos, Shumla, Tex.	3,160	2,000	31.48	0.22	7.18	5.51	18.79	–	2.54	8.59	20.57	0.03	7.4	7.4
Gila, Gellespie Dam, Ariz.	8,160	5,620	87.02	2.60	14.87	10.94	60.90	0.31	3.18	27.90	56.96	0.34	17.0	7.2

[a] Electrical conductivity.

Source: Reprinted, with publisher's permission, from *J. Am. Water Works Assn.*, *57*(2):156, © 1965, by the American Water Works Association, Inc., 2 Park Ave., New York, N.Y. 10016.

REFERENCE

1. S.K. Love, U.S. Geological Survey Water Supply Papers, U.S. Department of the Interior, 1954–1966.

9.2–2 CHEMICAL COMPOSITION OF RIVER WATER BEFORE AND AFTER INTENSIVE IRRIGATION DEVELOPMENTS

Constituent, mEq/l	Colorado River 1893	Colorado River 1962	Missouri River 1906	Missouri River 1953	Rio Grande River 1893	Rio Grande River 1945
Ca	3.3	4.5	3.2	3.1	2.95	4.5
Mg	1.1	2.4	1.6	1.5	0.67	1.5
NA	8.5	5.8	2.1	2.6	2.85	6.3
$CO_3 + HCO_3$	2.8	2.7	3.3	3.1	2.26	3.6
Cl	5.2	3.5	0.3	0.4	1.52	3.4
SO_4	4.8	6.8	3.5	3.7	2.61	5.3
Total solute contents	774	814	514	523	380	850
Millions of irrigated acres and year:	0.9 (1902)	3.5 (1950)	2.9 (1902)	5.1 (1950)	0.5 (1902)	2.5 (1950)

Source: H.L. Parkinson and H.R. McDonald, Irrig. Drain. Spec. Conf., Billings, Montana, Am. Soc. Civ. Eng., 1965.

9.2–3 POUDRE RIVER QUALITY CHANGES DUE TO IRRIGATION

Item	1902 Data Distance from mouth, mi		1949 Data Distance from mouth, mi			
	55	9	50	50	14	14
Approximate flow, ft^3/sec	a	24[b]	32	32	45	45
Date sampled	November 3	August 11	June 15	June 20	August 26	September 13
Total hardness, ppm	18	920	25	17	816	918
Alkalinity as $CaCO_3$, ppm	20	190	28	22	258	286
Noncarbonate hardness, ppm	0	730	0	0	558	632
Sulfate as $SO_4^{=}$, ppm	3.4	940				
Chloride as Cl^-, ppm	1.5	40				
Total dissolved solids, ppm	37	1,571				

[a] Sample was taken at a time when there was no flood water and after a period of good weather . . . the water then flowing represented the normal water of the Poudre.
[b] Flow in July 27, 1902, at or near sampling station.[1]

Source: Reprinted, with publisher's permission, from *J. Am. Water Works Assn.*, *45*(11):1160, © 1953, by the American Water Works Association, Inc., 2 Park Ave., New York, N.Y. 10016.

REFERENCE

1. W.P. Headden, Colorado Agric. Expt. Sta. Bul. No. 82, 1903.

9.2–4 EFFECTS OF IRRIGATION ON WATER QUALITY

Salinity, Yakima River, 1951[a]

Sampling section	Miles traveled	Dissolved salts, mg/l
Easton	0	41
Cle Elum	17	49
Ellensburg	44	63
Rosa Dam	71	51
Wapato Dam	92	74
Tuppenish	108	326[b]
Granger	115	348[b]
Sunnyside	128	358[b]

[a] Data from Washington Pollution Control Commission.
[b] Major return flows enter above this location.

Source: E.F. Eldridge, *J. Water Pollut. Control Fed.*, *35*(5):614, 1963.

9.2–5 EFFECTS OF IRRIGATION

Water Quality Changes, Rio Grande River, 1945

Item	Otowi Bridge[a]	San Marcial[b]	Elephant Butte[c]	Caballo Dam	Leasburg Dam	El Paso	County Line[d]	Fort Quitman[e]
	Downstream							
Total discharge, acre ft	1,131,500	813,750	839,690	882,920	814,170	568,900	280,620	207,710
Electrical conductivity, (EC x 10^6 at 25°C)	280	567	662	740	778	1,210	2,880	3,200
Sodium, %[f]	24	41	42	43	44	51	60	62
Boron, ppm	0.03	0.10	0.11	0.11	0.12	0.18	0.32	0.33
Dissolved solids, ppm	190	380	430	480	510	802	1,870	2,060
Calcium (as Ca^{++}), ppm	33.5	51.4	58.1	62.8	64.8	89.4	173	175
Magnesium (as Mg^{++}), ppm	6.0	9.7	12	13.1	13.7	18.8	36.0	44.1
Total hardness (as $CaCO_3$), ppm	108	168	194	210	218	300	579	620
Alkalinity ($CO_3^{=}$ + HCO_3^-) (as $CaCO_3$), ppm	86	124	133	144	142	190	220	206
Carbonate hardness (as $CaCO_3$), ppm	86	124	133	144	142	190	220	206
Noncarbonate hardness (as $CaCO_3$), ppm	22	44	61	66	76	110	359	414
Sodium (as Na^+), ppm	16	52.9	63.5	73.6	78.9	146	407	461
Sulfate (as $SO_4^{=}$), ppm	47	112	149	156	169	254	457	497
Chloride (as Cl^-), ppm	5	32	35	50	54	122	554	644
Nitrate (as NO_3), ppm	Trace	1	Trace	0.6	0.6	0.6	1	0.6

[a] Near Santa Fe, N.M.
[b] Upper end of Elephant Butte Reservoir.
[c] Reservoir outlet.
[d] Hudspeth County, Texas.
[e] Approximately 80 miles below International Dam at El Paso.
[f] 100 Na/(Na + K + Ca + Mg), concentrations in equivalents per million.

Source: Reprinted, with publisher's permission, from *J. Am. Water Works Assn., 45*(11):1160, © 1953, by the American Water Works Association, Inc., 2 Park Ave., New York, N.Y. 10016.

9.2–6 EFFECT OF IRRIGATION ON TOTAL AND PERMANENT HARDNESS

Location	Total hardness, mg/l[a]		Permanent hardness, mg/l[a]
	Above irrigation	Below irrigation	
Rio Grande, Texas	111	631	500
Yakima River, Wash.	33	134	50
Sunnyside Project, Wash.	40	299	188
Arkansas River	212	890	650
Sutter Basin, Cal.	72	480	208

[a] As $CaCO_3$.

Source: E.F. Eldridge, *J. Water Pollut. Contr. Fed.*, *35*(5):614, 1963.

9.2–7 NITRATE CONTENT OF DRAINAGE WATER[a,b]

Location	Nitrate as NO_3, mg/l
Drain at Oxnard Road	28.7
Ditch near Port Hueneme	21.0
Ditch at Bridge 520	49.6
Revilon Slough	56.9
Ditch, Hueneme Road	68.1
Ditch west of West Road	64.5
Ditch south of Nauman Road	84.0
Ditch on Pleasant View Road	74.0

[a] Data from California Water Resources Board for Oxnard Plain Area.
[b] Water applied had a NO_3 content of 3 mg/l.

Source: E.F. Eldridge, *J. Water Pollut. Contr. Fed.*, *35*(5):614, 1963.

9.2–8 TURBIDITY AND COLOR OF RETURN FLOWS

Yakima Basin

Sampling point	Turbidity units	Color units
Yakima River above diversions	2	4
Teanaway Creek	113	207
Selah Drain	10	13
Selah-Moxee Canal	18	36
Moxee Drain	239	160
Drain 2, Wapato	93	30
Toppenish Creek	24	33
Corral Creek	41	48
Snipes Creek	178	70
Spring Creek Drain	119	66
Grandview Drain	45	38
Granger Drain	33	38
Sulfur Creek Drain	19	21
Sulfur Creek Drain and Wasteway	39	28
Yakima River below diversions	14	19

Source: E.F. Eldridge, *J. Water Pollut. Contr. Fed.*, *35*(5):614, 1963.

9.2–9 CHEMICAL COMPOSITION OF AGRICULTURAL WASTE WATERS

Irrigated Fields of the San Joaquin Valley, California

Constituent	Concentration, mg/l	
	Tile drains	Open drains
Ca	440	57
Mg	260	32
Na	2,000	120
K	–	4
HCO_3	200	220
SO_4	3,600	170
Cl	1,200	130
B	28	0.8
Total dissolved solids	6,400	650
Total N	29	6.5
Total P	0.3	0.5

Source: G. Walton, in *Agricultural Waste Waters*, L.D. Doneen, Ed., Water Resources Center, University of California, Los Angeles, 1966, p. 273.

9.2–10 COMPOSITIONS OF IRRIGATION AND TILE DRAINAGE WATERS

San Joaquin Valley, California

Water	EC,[a] mmho/ cm	B, mg/l	HCO_3	mEq/l						
				Cl	SO_4	NO_3	Na	Ca	Mg	K
Irrigation water	0.7	0.1	1.6	3.7	1.4	0.12	3.7	1.1	1.6	0.2
Tile effluent, field A	6.0	6.5	5.1	18.0	52.0	1.64	32.7	27.4	14.7	0.4
Tile effluent, field B	9.7	13.5	5.2	19.8	95.7	3.18	79.5	21.9	23.0	0.5

[a] Electrical conductivity.

Source: L.D. Doneen, Paper 4002, Water Science Engineering, University of California, Davis, 1966.

9.2–11 CHEMICAL COMPOSITION OF COLUSA BASIN DRAIN WATER AND SACRAMENTO RIVER WATER AT KNIGHTS LANDING, CALIFORNIA

Constituent	Range in concentration, mg/l	
	Colusa Basin Drain[1]	Sacramento River[2]
Total dissolved solids	246–792	95–131
SO_4	46–243	1.9–12.0
Cl	21–113	3.9–9.2
Total N	0.7–2.1	0–0.4
Total P	0.4–0.8	

Source: L.E. Ciaccio, *Water and Water Pollution Handbook,* Marcel Dekker, New York, 1971.

REFERENCES

1. W. Jopling, in *Agricultural Waste Waters,* L.D. Doneen, Ed., Water Resources Center, University of California, Los Angeles, 1966, p. 144.
2. S.K. Love, U.S. Geological Survey Water Supply Papers, U.S. Department of the Interior 1264, 1954; 1362, 1955; 1430, 1958; 1465, 1959; 1485, 1960; 1524, 1960; 1575, 1961; 1699, 1963; 1946, 1966.

9.2–12 AVERAGE COMPOSITIONS OF APPLIED WATER AND SURFACE AND SUBSURFACE DRAINAGE WATERS DURING THE IRRIGATION AND NONIRRIGATION SEASONS *

Yakima Valley, Washington, 1959–1960

Constituent or characteristic	Applied water	Subsurface drain water	Surface drain water
Temperature, C	16.0	13.3 (10.9)[a]	17.9 (8.1)[a]
Dissolved oxygen, mg/l	10.2	6.8 (7.5)	9.0 (11.2)
pH	8.1	7.7 (7.7)	8.2 (8.0)
HCO_3 alkalinity, mg/l as $CaCO_3$	46.0	218.0 (248)	138.0 (223)
CO_3 alkalinity, mg/l as $CaCO_3$	1.0	0 (0)	2.0 (3)
Hardness as $CaCO_3$	46.0	186.0 (226)	121.0 (205)
Turbidity units	37.0	12.0 (8)	130.0 (35)
Color units	22.0	12.0 (7)	38.0 (16)
EC,[b] μmho/cm	83.0	420.0 (441)	283.0 (406)
Cl, mg/l	1.0	12.0 (19)	8.0 (16)
NO_3 as N mg/l	0.25	2.5 (2.5)	0.8 (1.9)
Total Kjeldahl N mg/l	0.27	0.32 (0.3)	0.3 (0.3)
Chem. oxygen demand, mg/l	7.0	9.0 (4)	10.0 (5)
Soluble PO_4, mg/l	0.21	0.7 (0.7)	0.6 (0.6)
Total PO_4, mg/l	0.32	0.9 (0.8)	0.8 (0.8)
SO_4, mg/l	5.4	39.0 (50)	37.0 (67)
Ca, mg/l	10.0	44.0 (51)	31.0 (45)
Mg, mg/l	5.0	20.0 (24)	12.0 (22)
Na, mg/l	4.1	38.0 (46)	26.0 (46)
K, mg/l	1.4	4.7 (5.3)	5.3 (6.0)
Coliforms/100 ml	1,070.0	103.0 (90)	10,600.0 (1170)

* Adapted from Reference 1.
[a] All figures in parentheses refer to the nonirrigation season.
[b] Electrical conductivity.

Source: L.E. Ciaccio, *Water and Water Pollution Handbook,* Marcel Dekker, New York, 1971.

REFERENCE

1. R.O. Sylvester and R.W. Seabloom, *A Study on the Character and Significance of Irrigation Return Flows,* University of Washington, Department of Civil Engineering, 1962.

9.2–13 INCREMENTAL SALT CONCENTRATION ATTRIBUTABLE TO SPECIFIC SOURCES

Colorado River at Hoover Dam

Factor	TDS increment, mg/l
Natural sources	
Diffuse sources	274
Point sources (mineral springs, wells, etc)	69
Irrigation	
Consumption	88
Leaching	165
Municipal and industrial sources	10
Water exports	22
Evaporation and phreatophytes	97
Total	725

Note: 1942–1961 period of record adjusted to 1960 condition.

Source: U.S. Geological Survey Professional Paper 441, *Water Resources of the Upper Colorado River Basin,* 1965; U.S. Department of the Interior, Progress Report No. 3, *Quality of Water, Colorado River Basin,* January 1967; Federal Water Pollution Control Administration records on open file.

9.2–14 COMPARISON OF THE SALINITY IN IRRIGATION AND DRAINAGE WATERS FROM SELECTED IRRIGATION DISTRICTS

	Salinity (mineralization as electrical conductivity)	
Irrigation district	Irrigation water	Drainage water
Middle Rio Grande Valley		
Albuquerque Division[1]	481	1,070
Lower Rio Grande		
Rincon, Division[1]	834	1,480
Mesilla Division[1]	869	1,410
El Paso Division[1]	1,200	3,680
Pecos River, Carlsbad Area[2]	4,360	7,800
Colorado River Water, Meloland		
Imperial Valley, California[3]	1,120	11,770
Wapato Project, Washington		
(Yakima River water)[3]	137	296

Source: *The Cost of Clean Water,* U.S. Government Printing Office.

REFERENCES

1. U.S. Geological Survey Water Supply Paper 839, 1938.
2. U.S. National Resources Planning Board, Pecos River Joint Investigation, Reports of Participating Agencies, 1942.
3. Unpublished data, U.S. Salinity Laboratory, Riverside, California.

9.2–15 INCREASE IN SALTS AND IONS AS A RESULT OF IRRIGATION

Location[a]	Number of times greater than in irrigation water						
	Salinity	Ca	Mg	Na	HCO_3	SO_4	Cl
1	10.8	5.1	7.5	21.4	2.4	10.5	129.0
2	7.0	–	–	–	4.3	4.7	3.0
3	5.0	4.4	4.2	8.1	1.8	1.9	
4	5.9	3.7	5.5	14.7	1.8	8.6	11.5
5	7.1	4.6	6.9	12.7	2.9	4.6	40.0
6	5.6	4.4	14.0	3.5	1.6	15.0	Decrease
7	7.8	5.3	6.0	17.2	5.2	18.0	8.0

[a] The numbers in this column are defined as follows:
1 Rio Grande between Otowi Bridge and Fort Quitman.
2 Yakima River between Easton and Prosser.
3 Average of 1943-44 input and output on an irrigation district.
4 Average of 1940-41 data on Arkansas River between Pueblo and Holly.
5 Average of five years of data on input and output of an irrigated area.
6 Data collected on a short stretch of the Colorado River between Cameo and Grand Junction.
7 Input and output data of an irrigation project in Boise River Basin.

Source: E.F. Eldridge, *Return Irrigation Water – Characteristics and Effects,* U.S. Public Health Service, Region IX, May 1, 1960.

9.2–16 SALT BALANCE IN SOME IRRIGATED VALLEYS

Item	Mesilla Valley, N.M.–Texas	El Paso Valley, Texas	Yuma Valley, Ariz.
Irrigated area, acres	80,000	60,000	40,000
Irrigation water, acre ft	744,380	496,113	259,917
Drainage water, acre ft	496,113	173,219	57,095
Percent of irrigation water	66.6	34.9	22.0
Salts in irrigation water			
ton/acre ft	0.805	1.226	0.956
ppm[a]	592	900	704
Salts in drainage water			
ton/acre ft	1.226	2.666	1.993
ppm[a]	900	1,970	1,470
Output of salts, % input	101.5	75.9	45.8
Output salt concentration, % input	152	218	208

[a] Based on a table in C.S. Scafield, *J. Agric. Res., 61*:17, 1940.
[b] Calculations by P.D. Haney and T.W. Bendixen, *J. Am. Water Works Assn., 45*(11):1160, 1953.

Source: Reprinted, with publisher's permission, from *J. Am. Water Works Assn., 45*(11):1160, © 1953, by the American Water Works Association, Inc., 2 Park Ave., New York, N.Y. 10016.

9.2–17 PESTICIDES IDENTIFIED IN ANALYSES OF DRAINS

San Joaquin Valley, California

Pesticide	No. times detected	Concentration, ng/l		
		Max	Min	Avg
Panoche Drain[a]				
Chlorinated hydrocarbons				
DDE	12	150	10	64
DDD and/or DDT	54	5,700	60	606
Dieldrin	1	120	120	120
Heptachlor epoxide	6	100	10	24
Lindane	7	220	10	67
Toxaphene	60	7,900	100	2,009
Thiodan	1	210	210	210
Methoxychlor	1	450	450	450
Summation of identified chlorinated hydrocarbons	61	11,400	100	2,548
Thiophosphates				
Baytex	4	160	30	87
Ethion	11	1,200	15	265
Malathion	7	320	55	119
Methyl parathion	3	6,400	300	2,530
Parathion	19	3,600	20	520
Thimet	1	30	30	30
Summation of identified thiophosphates	33	6,400	15	654
Tile Drainage Effluents[b]				
BHC	2	1,500	300	900
DDE	16	130	10	25
DDD and/or DDT	62	600	10	78
Dieldrin	13	130	10	35
Heptachlor epoxide	25	40	10	14
Lindane	22	340	10	35
Toxaphene	13	950	130	528
TCBC	1	50	50	50

[a] 61 analyses.
[b] 66 analyses.

Source: Adapted from W.A. Johnston *et al., J. Water Resources Res., 3*:525, 1967 in L.E. Ciaccio, *Water and Water Pollution Handbook,* Marcel Dekker, New York, 1971.

9.2–18 PERSISTENCE OF PESTICIDES IN RIVER WATER

Compound	Original compound found,[a] %				
	0-time	1 wk	2 wk	4 wk	8 wk
Organochlorine Compounds					
BHC	100	100	100	100	100
Heptachlor	100	25	0	0	0
Aldrin	100	100	80	40	20
Heptachlor epoxide	100	100	100	100	100
Telodrin	100	25	10	0	0
Endosulfan	100	30	5	0	0
Dieldrin	100	100	100	100	100
DDE	100	100	100	100	100
DDT	100	100	100	100	100
DDD	100	100	100	100	100
Chlordane (technical)	100	90	85	85	85
Endrin	100	100	100	100	100
Organophosphorus Compounds					
Parathion	100	50	30	< 5	0
Methyl parathion	80	25	10	0	0
Malathion	100	25	10	0	0
Ethion	100	90	75	50	50
Trithion	90	25	10	0	0
Fenthion	100	50	10	0	0
Dimethoate	100	100	85	75	50
Merphos	0	0	0	0	0
Merphos recovery as Def	100	50	30	10	< 5
Azodrin	100	100	100	100	100
Carbamate Compounds					
Sevin	90	5	0	0	0
Zectran	100	15	0	0	0
Matacil	100	60	10	0	0
Mesurol	90	0	0	0	0
Baygon	100	50	30	10	5
Monuron	80	40	30	20	0
Fenuron	80	60	20	0	0

[a] Pesticide concentrations were 10 μg/l. Recoveries were rounded off to the nearest 5%.

Source: J.W. Eichelberger and J.J. Lichtenberg, *Environ. Sci. Technol.*, *5*(6):543, 1971.

9.2–19 NUMBER OF OCCURRENCES OF INSECTICIDES

Pesticides	Missouri R. at Nebraska City, Nebr.	Arkansas R. below John Martin Reservoir, Colo.	Arkansas R. at Van Buren, Ark.	Brazos R. at Richmond, Texas	Colorado R. at Wharton, Texas	Rio Grande below Anzalduas Dam, Texas
Aldrin	1	–	1	–	–	–
DDD	–	1	2	2	–	5
DDE	–	3	2	2	2	4
DDT	2	1	2	3	2	1
Dieldrin	5	3	5	3	3	5
Endrin	1	–	–	–	1	3
Heptachlor	1	3	2	1	1	4
Heptachlor epoxide	2	3	1	2	2	2
Lindane	4	8	5	3	3	8
Totals	16	22	20	16	14	32

Pesticides	Colorado R. at Yuma, Ariz.	Sacramento R. at Verona, Calif.	Yakima R. at Kiona, Wash.	Snake R. at King Hill, Idaho	Columbia R. near The Dalles, Ore.	Totals
Aldrin	–	–	–	1	1	4
DDD	–	1	2	–	–	13
DDE	2	–	3	–	–	18
DDT	1	–	1	1	–	14
Dieldrin	1	2	–	1	1	29
Endrin	1	–	–	1	–	7
Heptachlor	–	–	1	1	–	14
Heptchlor epoxide	2	2	2	1	1	20
Lindane	3	3	4	1	4	46
Totals	10	8	13	7	7	165

Source: E. Brown and Y.A. Nishioka, *J. Pest. Monit.*, *2*:38, 1967.

9.2–20 INCIDENTS OF WATER POLLUTION BY ORGANIC PESTICIDES

Location	Year	Pesticide	Source	Remarks	Ref.
Montebello, Cal.	1945	2,4-D	IW[a]	Well waters contaminated for several years with the phenol	1, 2
Saskatchewan River, Canada	1949–51	DDT	DA[b]	Blackfly larvae control	3
Tennessee River, Ala.	1950	toxaphene	R[c]	Fish kill	4
Detroit River, Mich., and Lake St. Clair, Ont.	1953	DDT	R	Recovered from raw and treated drinking water	5, 6
St. Lucie County, Fla.	1955	dieldrin	DA	Sandfly larvae control	7, 8
Snake, Missouri, and Mississippi rivers	1957	aldrin DDT	R	Isolated by carbon filter technique	5, 6
Great Lakes	1957	TFM	DA	Sea lamprey control	9
Lake Hopatcong, N.J.	1957–58	2,4-D 2,4,5-T	DA	Aquatic-plant control	10
Dickinson Reservoir, N.D.	1958	toxaphene, rotenone	DA	Trash fish control	11
Dutchess County, N.Y.	1958–60	2,4,5-T	DA	Aquatic-plant control	12
Denver, Colo.	1959	2,4-D	IW	Ground water contaminated	13, 14, 15
Clayton Lake, N.M.	1959	toxaphene	DA	Distribution study	16
Various lakes in Illinois	1960	endothal	DA	Aquatic-plant control	17
Farm pond, S.C.	1960	parathion	R	Insect control in peach orchard	18

[a] Industrial waste water.
[b] Direct application.
[c] Runoff.

Source: Reprinted, with publisher's permission, from *J. Am. Water Works Assn., 56*(3):267, © 1964, by the American Water Works Association, Inc., 2 Park Ave., New York, N.Y. 10016.

REFERENCES

1. H.A. Swenson, *Proc. Soc. Water Treat., 11*:84, 1962.
2. A.N. Sayre and V.T. Stringfield, *J. Am. Water Works Assn., 40*(11):1152, 1948.
3. F.J.H. Freeden *et al., Can. J. Agric. Sci., 33*:379, 1953.
4. L.A. Young and H.P. Nicholson, *Progr. Fish Cult., 13*:193, 1951.
5. A.A. Rosen and F.M. Middleton, *Anal. Chem., 31*:1729, 1959.
6. F.M. Middleton and J.J. Lichtenberg, *Ind. Eng. Chem., 52*:99A, 1960.
7. H.P. Nicholson, *Am. Water Works Assn., 51*:981, 1959.
7. H.P. Nicholson, *J. Am. Water Works Assn., 51*:981, 1959.
8. R.W. Harrington and W.L. Bidlingmayer, *J. Wildlife Manage., 22*:176, 1958.
9. M.A. Smith, *Anal. Chem., 32*:167, 1960.
10. R.R. Younger, *Proc. Northeast Weed Control Conf., 13*:322, 1959.
11. J.M. Cohen *et al., J. Am. Water Works Assn., 53*:233, 1961.
12. M. Pierce, *Proc. Northeast Weed Control Conf., 15*:545, 1961.
13. T.R. Walker, Bull. 72, Geological Society of America, 1961.
14. G. Walton, *Public Health Aspects of the Contamination of Ground Water in South Platte River Basin in Vicinity of Henderson, Colo.*, U.S. Public Health Service, 1959.
15. G. Walton, *Public Health Aspects of the Contamination of Ground Water in the Vicinity of Derby, Colo.*, Robert A. Taft San. Eng. Center, Cincinnati, 1961.
16. B.J. Kallman, *Trans. Am. Fisheries Soc., 91*:14, 1962.
17. R.C. Hiltibran, *Weeds, 10*:17, 1962.
18. H.P. Nicholson *et al., Trans. Am. Fisheries Soc., 91*:213, 1962.

9.2–21 RESIDUES OF PESTICIDES IN AIR AND RAINWATER

Residues (pp 10^{-12})[a]

Location	Ref.	No. sites	Medium	DDT and related compounds		γ BHC		Aldrin		Dieldrin		Chlordane		Heptachlor		Toxaphene	
				Max	Mean	Max	Mean	Max	Mean	Max	Mean	Max	Mean	Max	Mean	Max	Mean
Great Britain	1	1	Air	13.0	13.0	–	11.0	–	–	–	21.0	–	–	–	–	–	–
Great Britain	2	7 (4 dates)	Rainwater	79.3	79.3	230.0	60.3	–	–	35.0	7.6	–	–	–	–	–	–
Great Britain	3	1 (3 dates)	Rainwater	–	–	120.0	97.0	–	–	36.0	28.0	–	–	–	–	–	–
				1.0	1.0	164.0	100.0	–	–	25.0	20.0	–	–	–	–	–	–
				3.0	3.0	52.0	29.0			16.0	9.0						
U.S.	4	6	Air (in agricultural communities)	0.003	0.004	–	–	0.004	0.0006	–	–	0.006	0.002	–	–	0.015	0.002
		1	Air (nonurban community)	4.2	4.2	–	–	–	–	–	–	0.03	0.016	–	–	–	–
		1	Air (eastern city)	56.1	37.3	–	–	–	–	–	–	–	–	–	–	–	–
U.S.	5, 6	(3 dates)	Rainwater	102.0	210.0	70.0	23.0	–	–	–	–	–	–	–	–	–	–
			Dust in water	800,000.0	600,000	–	–	–	–	–	3,000	–	500,000	–	40,000	–	–
Barbados	7	1 (15 dates)	Dust in air	45,000.0	38,280	–	–	–	-	8,100	2,186	–	–	–	–	–	–

[a] ng/l for water
μg/m³ for air

Source: C.E. Edwards, *Persistent Pesticides in the Environment*, CRC Press, Cleveland, 1971.

REFERENCES

1. D.C. Abbott *et al., Nature, 211:*259, 1966.
2. K.R. Tarrant and J.O'G. Tatton, *Nature, 219:*(5155):725, 1968.
3. G.A. Wheatley and J.A. Hardman, *Nature, 207:*486, 1965.
4. E.C. Tabor, *Trans. N.Y. Acad. Sci. Ser.,* 2(28(5)):569, 1966.
5. S.R. Weibel *et al., J. Am. Water Works Assn., 58:*1075, 1966.
6. J.M. Cohen and C. Pinkerton, *Adv.Chem. Ser., 60:*163, 1966.
7. R.W. Riseborough *et al., Science (NY), 159:*1233, 1968.

9.2–22 PESTICIDE RESIDUES IN WATER SURVEYS

(pp 10^{-12}) ng/l

Location	Ref.	No. sites	Type of water	DDT and related compounds		γ BHC		Aldrin		Dieldrin		Chlordane		Heptachlor and H. epoxide		Endrin	
				Max	Mean	Max	Mean	Max	Mean	Max	Mean	Max	Mean	Max	Mean	Max	Mean
U.S.A.	1	99	Major river basins	149.0	8.2	4.0	t	–	–	68.0	6.9	–	–	155.0	6.3	116.0	2.41
	2	10	Mississippi river delta	720.0	112.0	120.0	28.0	30.0	5.0	60.0	10.0	–	–	10.0	2.0	4,230.0	541.0
	3	109	Major rivers	127.0	8.3	56.0	2.2	–	–	167.0	5.9	75.0	0.1	19.0	0.1	69.0	3.6
	4	48	Water areas (Utah)	32.4	9.7	–	–	–	–	–	–	–	–	–	–	–	–
	5	82	Water areas (California)	22.0	0.62	0.15	0.01	–	–	–	–	–	–	t	t	–	–
	6	97	Major river basins	102.0	9.3	–	–	85.0	0.9	118.0	7.5	–	–	–	–	94.0	5.5
	7	11	Western streams	145.0	10.3	20.0	2.8	5.0	0.2	15.0	2.3	–	–	105.0	2.6	40.0	1.4
Great Britian	8	–	Sewage effluents	130.0	36.0	–	–	–	–	300.0	200.0	–	–	–	–	–	–
	9	21	Sewage effluents	800.0	130.9	390.0	92.5	–	–	1,900.0	145.0	–	–	–	–	–	–
	9	9	Yorkshire rivers	908.0	64.6	180.0	38.6	–	–	630.0	114.0	–	–	–	–	–	–
	9	7	British rivers	15.0	1.6	98.0	18.7	–	–	40.0	3.3	–	–	–	–	–	–

Source: C.E. Edwards, *Persistent Pesticides in the Environment,* CRC Press, Cleveland, 1971.

REFERENCES

1. W.W. Breidenbach *et al., Publ. Health Rept., 82:*139, 1967.
2. *Monitoring Agricultural Pesticide Residues,* U.S. Department of Agriculture, 1966, 53.
3. R.S. Green, G.C. Gunnerson, and J.J. Lichtenberg, in *Agriculture and the Quality of our Environment,* Am. Assn., Adv. Sci., 1966, p. 137.
4. S.L. Warnick, R.F. Gaufin, and A.R. Gaufin, *J. Am. Water Works Assn., 58:*601, 1966.
5. J.O. Keith and E.G. Hunt, *Trans. 31st N. Amer. Wildlife Res. Conf,,* 150, 1966.
6. L. Weaver *et al., Publ. Health Rept., 80:*481, 1965.
7. E. Brown and Y.A. Nishioka, *Pest. Mon. J., 1*(2):38, 1967.
8. A.V. Holden and K. Marsden, *J. Proc. Sew, Purif., 4:*295, 1966.
9. G.F. Lowden, C.L. Saunders, and R.W. Edwards, *P. Treat: Exam.,* (in press).

9.2–23 CONCENTRATION OF PESTICIDE RESIDUES EXTRACTED FROM ALGAE AFTER 7 DAYS

Algae	Pesticide in medium before test, μg/ml	Pesticide recovered in algae, μg/ml			
		Dieldrin	p,p′-DDT	Aldrin	Endrin
Microcystis aeruginosa	1	175	230	225	200
Anabaena cylindrica	1	180	268	120	222
Scenedesmus quadricauda	1	120	134	143	156
Oedogonium species	1	127	270	131	140

Source: Reprinted, with publisher's permission, from *J. Am. Water Works Assn., 61*(7):360, © 1969, by the American Water Works Association Inc., 2 Park Ave., New York, N.Y. 10016.

9.2–24 CONCENTRATION OF DIELDRIN IN THE ORGANISMS OF VARIOUS TROPHIC LEVELS OF A MARINE ECOSYSTEM

Trophic level	Organism	Dieldrin concentration, ppm	
1	Seaweed	0.001	
2	Microzooplankton and crustacea	0.02	
3	Crustacea and fish	0.03	
4	Fish	0.02	
	Birds	eggs	liver
	Shags	1.2	1.6
	Kittiwakes	0.1	–
	Terns	0.2	–
5	Cormorants eggs	1.2	

Source: J. Robinson, *Chem. Brit., 4*(4):158, 1968.

9.2–25 MOVEMENT OF INSECTICIDES FROM SOIL OR WATER INTO PLANTS

Insecticide	Crop	Source of insecticide	Residues: Amount in source μg/l	Residues: Amount in plant mg/l	Residues: Concentration or dilution factor[a]	Ref.
Toxaphene	Aquatic plants	Water	0.41	0.21	512.0	1
DDT	Aquatic vegetation	Water	200.0	75.0	375.0	2
DDT	Aquatic vegetation	Water	20.0	31.0	1,550.0	3
Organo-chlorines	Aquatic plants	Water	0.45	1.0	2,220.0	4
Organo-chlorines	Aquatic plants	Water	0.35	1.1	3,171.0	4
Organo-chlorines	Aquatic plants	Water	0.23	0.8	3,478.0	4
Organo-chlorines	Aquatic plants	Water	0.30	30.3	100,000.0	4
DDT	Algae and moss	Water	0.33	0.01	33.0	5
DDT	Algae	Water	5.8	0.002	0.34	6
Chlordane	Algae	Water	6.6	0.013	1.97	6
Endrin	Algae	Water	10.5	0.007	0.66	6
DDT	Vascular plants	Water	5.8	0.003	0.52	6
Chlordane	Vascular plants	Water	6.6	0.003	0.45	6
Endrin	Vascular plants	Water	10.5	0.006	0.57	6
			mg/kg	mg/kg		
Aldrin	Carrot roots	Muck soil	8.36	0.01 (root)	0.0013	7
Aldrin	Carrot roots	Clay soil	0.48	0.01 (root)	0.021	7
Dieldrin	Carrot roots	Muck soil	3.9	0.02	0.0051	7
Dieldrin	Carrot roots	Clay soil	0.48	0.11	0.23	7
Heptachlor	Rutabagas roots	Soil	0.320	0.024	0.075	8
Heptachlor	Wheat foliage	Seed treated	543.0	0.015	0.036	9
Aldrin & dieldrin	Wheat foliage	Soil	1.8	0.014	0.0077	10
Aldrin & dieldrin	Cucumber fruit	Soil	3.7	0.113	0.031	11
Heptachlor	Cucumber fruit	Soil	3.8	0.091	0.024	11
Dieldrin	Cucumber fruit	Soil	1.4	0.043	0.031	11
Aldrin	Alfalfa	Soil	0.84	0.009	0.011	11
Heptachlor	Alfalfa	Soil	0.78	0.028	0.036	11
Aldrin	Carrot roots	Soil	0.94	0.32	0.34	12
Aldrin	Potato tuber	Soil	0.94	0.07	0.074	12
Heptachlor	Carrot roots	Soil	0.49	0.36	0.73	12
Heptachlor	Potato tuber	Soil	0.49	0.05	0.10	12
DDT	Alfalfa foliage	Soil	1.39	0.113	0.08	13

[a] Concentration factor = $\frac{\text{concentration in animal}}{\text{concentration in water}}$.

Source: C.E. Edwards, *Persistent Pesticides in the Environment,* CRC Press, Cleveland, 1971.

REFERENCES

1. L.C. Terriere, *J. Agr. Food Chem., 14*:66, 1965.
2. R.A. Crocker and A.J. Wilson, *Trans. Am. Fish Soc., 94*(2):152, 1965.
3. W.R. Bridges, B.J. Kallman, and A.K. Andrews, *Trans. Am. Fish Soc., 92*:421, 1963.
4. J.O. Keith, *J. Appl. Ecol. (Suppl. 3),* 71, 1966.
5. G.L. Mack *et al., N.Y. Fish Game J., 11*(2):148, 1964.
6. P.J. Godsil and W.C. Johnson, *Pest. Mon. J., 1*(4):21, 1968.
7. H. Hurtig and C.R. Harris, *Conf. Pollut. Environ.,* Montreal, 1966, 16.
8. J.G. Saha and W.W.A. Stewart, *Can. J. Plant Sci., 47*:79, 1967.
9. R.H. Burrage and J.G. Saha, *Can. J. Plant Sci., 47*:114, 1967.
10. J.G. Saha and M. McDonald, *J. Agr. Food Chem., 15*:205, 1967.
11. E.P. Lichtenstein and K.R. Schulz, *J. Agr. Food Chem., 13*(1):57, 1965.
12. E.P. Lichtenstein, T.W. Fuhremann, and K.R. Schulz, *J. Agr. Food Chem., 16*(2):348, 1968.
13. G.W. Ware, B.J. Estesen, and W.P. Cahill, *Pest. Mon. J., 2*(3):129, 1968.

9.2–26 CONCENTRATION OF RESIDUES FROM WATER INTO AQUATIC INVERTEBRATES

Organism	Insecticide	Amount: In water (pp 10^{-9})[a]	Amount: In animals (pp 10^{-6})[b]	Concentration factor[c]	Ref.
Aquatic invertebrates	Toxaphene	0.63	1.43	2,270	1
Shrimp	DDT	0.5	0.14	2,800	2
Crayfish	DDT	20.0	1.47	73	3
Crayfish	DDT	20.0	2.0	100	4
Crayfish	DDT	24.0	2.32	97	5
Crayfish	DDT	20.0	0.33	16.5	6
Plankton	DDT	20.0	5.0	250	7
Plankton	DDT	0.3	5.0	16,666	8
Eastern oyster	DDT	10.0	151.0	15,100	9
Eastern oyster	DDT	1.0	30.0	30,000	
Eastern oyster	DDT	0.1	7.0	70,000	
Pacific oyster	DDT	1.0	20.0	20,000	
Hand clam	DDT	1.0	3.0-9.0	3,000 – 9,000	
Eastern oyster	Dieldrin	1.0	3.5	3,500	
Commercial shrimp	DDT	0.5	0.14	280	
Sea squirt	DDT	100.0	20.0	200	
		10.0	10.0	1,000	
Sea squirt	DDT	0.1	20.0	200,000	
Sea squirt	DDT	0.01	10.0	1,000,000	
Sea hare	DDT	0.01	1.78	178,000	
Crabs	DDT	50.0	7.2	144	
Snails	DDT	50.0	74.0	1,480	
Oysters	BHC	0.33	0.006	18.2	10
	DDT	0.55	0.033	60.0	
	Dieldrin	0.44	0.006	13.6	
	Heptachlor	0.55	0.002	3.6	
Clams	DDT	5.8	0.008	1.4	11
	Chlordane	6.6	0.006	0.9	
	Endrin	10.5	0.013	1.2	
Common clam	DDT	0.1	7.0	70,000	12
Oyster	Dieldrin	1.0	3.5	3,500	13
Hooked mussel	DDT	1.0	24.0	24,000	
Eastern oyster	DDT	1.0	26.0	26,000	
Pacific oyster	DDT	1.0	20.0	20,000	
European oyster	DDT	1.0	15.0	15,000	
Crested oyster	DDT	1.0	23.0	23,000	
Northern quahogs	DDT	1.0	6.0	6,000	

[a] μg/l.
[b] mg/kg.
[c] Concentration factor = $\frac{\text{concentration in animal}}{\text{concentration in water}}$.

Source: C.E. Edwards, *Persistent Pesticides in the Environment*, CRC Press, Cleveland, 1971.

REFERENCES

1. L.C. Terriere *et al., J. Agr. Food Chem., 14*:66, 1965.
2. P.A. Butler, in *Effects of Pesticides on Fish and Wildlife,* U.S. Department of the Interior, *Fish Wildlife Circ., 226*:65, 1965.
3. O.B. Cope, *J. Appl. Ecol. (Suppl.), 3*:33, 1966.
4. W.R. Bridges, B.J. Kallman, and A.K. Andrews, *Trans. Am. Fish Soc., 92*:421, 1963.
5. H. Cole *et al., Bull. Environ. Contam. Toxicol., 2*:127, 1967.
6. *A Review of Fish and Wildlife Service Investigations During the Calendar Year,* U.S. Department of the Interior, *Fish Wildlife Serv. Circ., 167*:1963, 109.

9.2–26 CONCENTRATION OF RESIDUES FROM WATER INTO AQUATIC INVERTEBRATES *(Continued)*

7. E.G. Hunt and A.I. Bischoff, *Calif. Fish Game, 46*:91, 1960.
8. J.O. Keith, *U.S. Fish Wildlife Serv., Denver Wildlife Res. Center Ann. Progr. Rept.*, 1964, 14.
9. *A Review of Fish and Wildlife Service Investigations During the Calendar Year,* U.S. Department of the Interior, *Fish Wildlife Serv. Circ., 199*:1964, 129.
10. G.E. Burdick *et al., Trans. Am. Fish. Soc., 93*:127, 1964.
11. P.J. Godsil and W.C. Johnson, *Pest. Mon. J., 1*(4):21, 1968.
12. V.L. Loosanoff, in *Research in Pesticides,* Academic Press, New York, 1965, 135.
13. P.A. Butler, *J. Appl. Ecol. (Suppl.), 3*:253, 1966.

9.2–27 CONCENTRATION OF RESIDUES FROM WATER TO FISH

		Amount of residue			
Organism	**Insecticide**	**In water (pp 10^{-9})[a]**	**In animal (pp 10^{-6})[b]**	**Concentration factor[c]**	**Ref.**
Rainbow trout	DDT	20.0	4.15	207	1
Black bullhead	DDT	20.0	3.11	155	1
Bluegill	Heptachlor	50.0	15.7	314	1
Catfish	Aldrin and dieldrin	0.044	0.07	1,590	2
Catfish	Aldrin and dieldrin	0.009	0.04	4,444	2
Catfish	Aldrin and dieldrin	0.021	0.02	952	2
Catfish	Aldrin and dieldrin	0.007	0.01	1,428	2
Buffabfish	Aldrin and dieldrin	0.023	0.09	3,913	2
Buffabfish	Aldrin and dieldrin	0.007	0.21	30,000	2
Scaled sardine	DDT	0.1	0.11	1,100	3
Rainbow trout	Toxaphene	0.41	7.72	18,829	4
Fish	DDT	0.30	1.0 – 6.4	3,333 – 21,333	5
Fish	Endrin	0.10	7.0	70,000	6
Fish	Toxaphene	1.0 – 4.0	0.8 – 2.5	2,000–2,500	7
Fish (5 spp.)	DDT	30–40	4–58	130–1,450	8
Bullhead trout	DDT	20	2–4	100–200	9
Fathead minnows	Endrin	0.015	0.15	10,000	10
Croakers	DDT	0.1	2.0	20,000	11
Pinfish	DDT	1.0	12.0	12,000	11
Pinfish	DDT	0.1	4.0	40,000	11
Fish	Dieldrin and DDT	10.0	0.1–1.0	10–100	12
Trout	Dieldrin	2.3	7.7	3,300	13
Chubs	DDT	5.8	0.029	5	14
Chubs	Chlordane	6.6	0.008	1.2	14
Chubs	Endrin	10.5	0.050	4.7	14
Trout	DDT	20.0	4.0	200.0	15
Bluegills	Heptachlor	50.0	56.8	1,130.0	16
Fish	DDT	0.015	12.44	829,300	17

[a] μg/l.
[b] mg/kg
[c] $\text{Concentration factor} = \frac{\text{Concentration in animal}}{\text{Concentration in water}}$.

9.2–27 CONCENTRATION OF RESIDUES FROM WATER TO FISH *(Continued)*

Organism	Insecticide	Amount of residue: In water (pp 10^{-9})[a]	Amount of residue: In animal (pp 10^{-6})[b]	Concentration factor[c]	Ref.
Fish	DDT	0.11	3.85	35,000	17
White catfish	DDD	14.0	30.4–129.0	2,172–9,214	18
Largemouth bass	DDD	14.0	19.7–25.0	1,407–1,785	18
Brown bullhead	DDD	14.0	15.5–24.8	1,107–1,771	18
Black crappie	DDD	14.0	5.4–115.0	386–8,214	18
Bluegill	DDD	14.0	6.6–10.0	471– 714	18
Sacramento blackfish	DDD	14.0	10.9–17.6	778–1,257	18
Brook trout	DDT	24.0	17.3	710.0	19

[a] μg/l.
[b] mg/kg
[c] Concentration factor = $\frac{\text{Concentration in animal}}{\text{Concentration in water}}$.

Source: C.E. Edwards, *Persistent Pesticides in the Environment,* CRC Press, Cleveland, 1971.

REFERENCES

1. O.B. Cope, *J. Appl. Ecol. (Suppl.), 3*:33, 1966.
2. B.I. Sparr, *et al., Adv. Chem. Ser., 60*:146, 1966.
3. P.A. Butler, in *Effects of Pesticides on Fish and Wildlife,* U.S. Department of the Interior, *Fish Wildlife Circ., 226*:65, 1965.
4. L.C. Terriere, *et al., J. Agr. Food Chem., 14*:66, 1965.
5. J.O. Keith, *U.S. Fish Wildlife Serv., Denver Wildlife Res. Center Ann. Progr. Rept.,* 1964, 14.
6. E. Langer, *Science, 144*:35, 1964.
7. B.J. Kallman, O.B. Cope, and R.J. Navarre, *Trans. Am. Fish Soc., 91*:14, 1962.
8. R.A. Crocker and A.J. Wilson, *Trans. Am. Fish Soc., 94*(2):152, 1965.
9. W.R. Bridges, B.J. Kallman, and A.K. Andrews, *Trans. Am. Fish Soc., 92*:421, 1963.
10. D.I. Mount, and G.J. Putnicki, *Trans. 31st N. Amer. Wildlife Nat. Res. Conf.,* 177, 1966.
11. D.J. Hansen, *Ann. Rept. Bur. Comm. Fish Circ., 247*:10, 1966.
12. A.V. Holden and K. Marsden, *J. Proc. Sew. Purif., 4*:295, 1966.
13. A.V. Holden, *J. Appl. Ecol. (Suppl.), 3*:45, 1966.
14. P.J. Godsil and W.C. Johnson, *Pest. Mon. J., 1*(4):21, 1968.
15. *A Review of Fish and Wildlife Service Investigations During the Calendar Year,* U.S. Department of the Interior, *Fish Wildlife Serv. Circ., 167,* 1963, 109.
16. *A Review of Fish and Wildlife Service Investigations During the Calendar Year,* U.S. Department of the Interior, *Fish Wildlife Serv. Circ., 199,* 1964, 129.
17. G.L. Mack *et al., N.Y. Fish Game J., 11*(2):148, 1964.
18. E.G. Hunt and A.I. Bischoff, *Calif. Fish Game, 46*:91, 1960.
19. H. Cole *et al., Bull. Environ. Contam. Toxicol., 2*:127, 1967.

9.2–28 EFFECT OF STORM SIZE ON ATRAZINE LOSSES IN WASHOFF

				Atrazine in washoff					
				Applied 1 hr before rain			Applied 96 hr before rain		
Storm size, in.[a]	Rainfall duration, min	Soil loss, ton/acre[b]	Water loss, in.[b]	Total lost, lb/acre	% lost	Avg concentration, ppm	Total lost, lb/acre	% lost	Avg concentration, ppm
0.5	12	0.26	0.07	0.13	4.3	7.94	0.06	2.0	3.66
1.25	30	1.84	0.61	0.36	12.0	2.54	0.16	5.3	1.13
2.50	60	5.30	1.54	0.50	17.0	1.39	0.22	7.3	0.62

[a] Storm sizes of 0.5, 1.25, and 2.50 inches represent a relatively common, a 1 year frequency, and a 10 year frequency storm, respectively, for Watkinsville, Georgia.
[b] Average of losses on all plots.

Source: A.W. White *et al., Environ. Sci. Technol., 1*(9):743, 1967.

9.2–29 ATRAZINE CONTENT IN WASHOFF AS RELATED TO TEMPERATURE CONDITIONS AND REPLICATIONS

Replication no.	Air temperature, °F.	Soil temperature, °F.	Atrazine in washoff ppm	Atrazine in washoff lb/acre
Atrazine Applied 1 Hour Before Rain				
1	88[a]	105[b]	1.10	0.41
2	74[a]	80[b]	1.67	0.60
Atrazine Applied 96 Hours Before Rain				
1	86[c]	107[d]	0.54	0.20
2	78[c]	92[d]	0.70	0.24

[a] Maximum air temperature recorded at nearby weather station on day of run.
[b] Soil temperature recorded at 1 in. depth just prior to run.
[c] Mean maximum air temperature recorded at nearby weather station for 4 days prior to run.
[d] Mean maximum soil temperature recorded at 1 in. depth for 4 days prior to run.

Source: A.W. White *et al., Environ. Sci. Technol., 1*(9):743, 1967.

9.2–30 ADSORPTION OF 2,4-D COMPOUNDS BY VARIOUS CLAYS

Clay	Sodium salt of 2,4-D[a]	Isopropyl[a]	Butyl ester[a]	Isooctyl[a]	2,4-DCP[b]
	Adsorption/g clay, μg				
Kaolinite	38	27	23	22	10
Bentonite	25	145	114	97	61
Illite	29	85	75	71	52

[a] Concentration expressed as the acid equivalent of 2,4-D.
[b] Concentration expressed as itself.

Source: Reprinted, with publisher's permission, from *J. Am. Water Works Assn., 61*(4):483, © 1969, by the American Water Works Association, Inc., 2 Park Ave., New York, N.Y. 10016.

9.2–31 SUMMARY OF DISTRIBUTION DATA FOR PARATHION

Sorbent	Sorbent concentration, ppm	Approximate parathion distribution[a]
Sandy soil	10,000	1×10^0
High clay soil	10,000	2×10^0
Algae	900	1.5×10^1
Coal	10,000	5×10^2
Activated carbon	200	1.5×10^4

[a] Sorbed parathion in ppm at an equilibrium concentration of parathion in water of 100 ppb.

Source: Reprinted, with publisher's permission, from *J. Am. Water Works Assn., 61*(4):483, © 1969, by the American Water Works Association, Inc., 2 Park Ave., New York, N.Y. 10016.

9.2–32 AMOUNTS OF DDT AND TOXAPHENE APPLIED TO WATERSHEDS

Watershed	Yr of insecticide application	Insecticide applied, lb		Cropping sequence, 1964, 1965, 1966
		DDT/acre	Toxaphene/acre	
Y-10	1957	0.5	1.0	Sorghum,
	1960	1.5	3.0	oats-clover,
	1963	3.0	6.0	cotton
Total		5.0	10.0	
Y-6	1958	3.0	6.0	Cotton,
	1961	4.0	8.0	sorghum,
	1964	2.0	4.0	oats-clover
Total		9.0	18.0	
Y-8	1959	5.0	10.0	Oats-clover,
	1962	3.0	6.0	cotton,
	1965	3.0	6.0	sorghum
Total		11.0	22.0	

Source: A.R. Swoboda *et al., Environ. Sci. Technol., 5*(2):141, 1971.

9.2–33 DDT (*p,p*-DDT, DDE, AND DDD) AND TOXAPHENE FOUND IN HOUSTON BLACK SOIL

Field	Application, lb/acre	Last application	Quantity found, lb/acre		Recovery, %		Found in top 12 in., %
			0–12 in.	0–60 in.	0–12 in.	0–60 in.	
			DDT				
Y-10	5	1963	0.446	0.754	8.9	15.1	59.2
Y-6	9	1964	0.840	1.388	9.3	15.4	60.5
Y-8	11	1965	0.939	1.267	8.5	11.5	74.1
			Toxaphene				
Y-10	10	1963	1.90	2.14	19.0	21.4	88.8
Y-6	18	1964	2.60	2.73	14.4	15.2	95.2
Y-8	22	1965	2.52	2.71	11.5	12.3	93.0

Source: A.R. Swoboda *et al., Environ. Sci. Technol., 5*(2):141, 1971.

9.2–34 ABS ADSORPTION AND OTHER PROPERTIES OF VARIOUS SOILS

Soil	Retention at sieve no.[8]	Mean grain diameter, μ	Specific weight, g/cm³	Bulk density, g/cm³	Porosity, %	ABS adsorption, μg/g	Water ABS front velocity ratio
Siliceous materials							
Ottawa sand	30	705	2.65	1.77	33	35	38
Pennsylvanian	70	250	2.66	1.59	40	260	208
	100	177	2.66	1.57	41	235	181
	200	105	2.66	1.46	45	385	251
Mississippian	200	105	2.66	1.51	43	770	536
Glauconitic	200	105	2.76	1.55	44	720	507
Calcareous materials							
High purity	100	177	2.71	1.52	44	375	260
Oolitic	100	177	2.71	1.46	47	590	367
Silty clays							
Peoria clay	200[a]	14	2.79	1.41	50	2,600	1,465
Illite	200[a]	10	2.78	1.36	51	3,000	1,600
Bentonite	200[a]	4.7	2.76	1.09	60	8,000	2,900

[a] Passing sieve number shown.

Source: Reprinted, with publisher's permission, from *J. Am. Water Works Assn., 56*(1):89, © 1964, by the American Water Works Association, Inc., 2 Park Ave., New York, N.Y. 10016.

9.2–35 ABS ADSORPTION ON PARTICULATE SURFACES

Adsorbent	Adsorbate ABS mg/l	ABS adsorption capacity, mg/g	Ref.
Natural stream silt	2.5	26.3	Renn and Barada[1]
Talc	5	2.75	Renn and Barada[1]
Precipitated $CaCO_3$	5	1.4	Renn and Barada[1]
China clay	5	0.87	Renn and Barada[1]
Diatomite	5	0.71	Renn and Barada[1]
Silica gel	5	0.65	Renn and Barada[2]
Alkaline garden soil	5	0.003	Renn and Barada[1]
Sands	1	0.001	Jenkins and Klein[2]
Sandy loam	1	0.022	Jenkins and Klein[2]
Ottawa sand	10	0.004	Ewing, Lefke, and Banerji[3]
Newton sand	10	0.01–0.02	Robeck[4]
Ottawa sand	100	0.144	Wayman[5]
Kaolinite	10	0.25	Wayman[6]
Illite	10	0.2	Wayman[6]
Montmorillonite	10	0.1	Wayman[6]
Newton sand	10	0.01	Robeck, Cohen, Sayers, and Woodward[7]

Source: V.V. Volk and M.L. Jackson, *J. Water Pollut. Control Fed., 40*(2) (Part I):205, 1968.

REFERENCES

1. C.E. Renn and M.F. Barada, *Sew. and Ind. Wastes, 31*(7):850, 1959.
2. D. Jenkins and S.A. Klein, *Calif. Univ. News Quarterly, San. Eng. Res. Lab., 11*(9):9, 1961.
3. B.B. Ewing *et al.*, Synthetic detergents in groundwater and soils, Progress Report, Illinois University Dept. Civil Engineering, 1960.
4. G.G. Robeck, Proc. 1961 Symp. on Ground Water Contamination, Robert A. Taft Gen. Eng. Center Tech. Report W61-5, 193, 1961.
5. C.H. Wayman, *U.S. Geol. Surv. Prof. Pap.*, 450-C, C137, 1962.
6. C.H. Wayman, *Proc. Int. Clay Conf., 1*:329, 1963.
7. G.G. Robeck *et al., J. Water Pollut. Control Fed., 35*(10):1225, 1963.

9.2–36 TOXAPHENE CONCENTRATION IN CORES TAKEN AT MAXIMUM WATER DEPTH, 5m

Ottman Lake and Silver Lake, Wisc.

Date cored	Days after first treatment	Toxaphene concentration in section, μg/g 0–5 cm	5–10 cm	10–15 cm	15–20 cm
Ottman Lake					
10/22/1967	26	1.5	0.8	a	a
1/25/1968	121	7.9	2.6	0.7	a
4/ 5/1968	191	15.8	2.0	0.8	a
6/28/1968 I	275	8.8	1.1	0.7	a
6/28/1968 II	275	9.5	0.9	a	a
6/28/1968 III	275	9.3	1.1	0.5	a
Silver Lake					
10/22/1967	25	2.1	0.9	a	a
1/25/1968	120	4.1	1.2	1.0	a
6/28/1968	275	4.8	0.8	0.6	a

[a] Below the detectable limit of 0.5 μg/g dry weight.

Source: G.D. Vieth and G.F. Lee, *Environ. Sci. Technol.*, 5, 232, 1971.

9.2–37 ACID DISSOCIATION CONSTANT VALUES FOR HERBICIDES[a]

Thermodynamic convention	pK 2,4-D	2,4,5-T	2-CPA	4-CPA	MCPA	MCPB	Fenac	Dalapon	Ref.
c,d	2.73[c]	2.88[d]	2.92[c]	2.95[c]	3.05[d]	4.80[c]	3.70[c]	1.84[c]	1[b]
?	2.81	2.57	–	–	–	–	–	–	2[b]
Concentration	2.96	–	–	–	–	–	–	–	3[b]
Inf. Dil.	–	–	2.99	3.02	–	–	–	–	4[e]
?	2.64	–	–	–	3.28	–	–	–	5[e]
Inf. Dil.	2.90	2.83	–	3.01	3.11	–	–	–	6[e]
Inf. Dil.	2.92	–	–	–	–	–	–	–	7[e]
Inf. Dil.	–	–	3.05	3.10	–	–	–	–	8[f]
?	3.28	–	–	–	–	–	–	–	9
–	3.31	3.14	–	2.36	–	–	3.87	1.53	10[g]
–	2.63	3.14	–	3.56	2.90	4.86 (60°)	–	–	11[g]
	2.80	3.46 (60°)			3.40	6.21			
	3.22 (60°)								
	3.31								

[a] Temperature = 25°C unless otherwise stated.
[b] Potentiometric measurements.
[c] I = 0.05.
[d] I = 0.03.
[e] Conductometric measurements.
[f] E.M.F. measurements.
[g] Literature review.

Source: N.H. Nelson and S.D. Faust, *Environ. Sci. Technol.*, *3*(11):1187, 1969.

9.2–37 ACID DISSOCIATION CONSTANT VALUES FOR HERBICIDES *(Continued)*

REFERENCES

1. N.H. Nelson and S.D. Faust, *Environ. Sci. Technol.*, *3*(11):1187, 1969.
2. J. van Overbeek, R. Blondeau, and V. Horne, *Plant Physiol.*, *26*:687, 1951.
3. R.T. Wedding, L.C. Erickson, and B.L. Brannaman, *Plant Physiol.*, *29*:64, 1954.
4. O. Behaghel, *J. Prakt. Chem.*, *114*:287, 1926.
5. J.A.A. Ketelaar, H.R. Gersmann, and M. Beck, *Rec. Trav. Chim.*, *71*:497, 1952.
6. M. Matell and S. Lindenfors, *Acta Chem. Scand.*, *11*:324, 1957.
7. R.L. Wershaw, M.C. Goldberg, and D.J. Pinckney, *Water Resources Res.*, *3*:511, 1967.
8. N.V. Hayes and G.E.K. Branch, *J. Am. Chem. Soc.*, *65*:1555, 1943.
9. L.J. Audus, *New Phytologist*, *48*:97, 1949.
10. V.H. Freed, *Chemistry of Herbicides*, Tech. Paper 2165, Oregon Agr. Expt. Station, Corvallis, Ore., 1964.
11. G.W. Bailey and J.L. White, *Residue Rev.*, *10*:14, 1965.

9.2–38 FREUNDLICH ISOTHERM PARAMETERS FOR ADSORPTION OF PESTICIDES ON CLAY AND SANDS

System	pH	Temperature, °C	Slope	Intercept, log K
Silica gel-isocil	5.2	0	1.0	0.58
		25	1.0	0.31
Silica gel-isocil	3.5	0	0.8	0.97
		25	1.0	0.31
Silica gel-bromacil	4.6	0	0.8	1.03
		25	0.8	0.68
Illite-isocil	7.4	0	0.8	0.79
		25	0.9	0.3
Illite-bromacil	7.0	0	0.9	0.48
		25	0.9	0.42
Montmorillonite-isocil	6–7	0	1.1	0.49
		25	1.2	0.1
Montmorillonite-bromacil	6–7	0	1.3	-0.03
		25	1.3	-0.14
Kaolinite-isocil	6–7	0	1.1	-0.36
		25	1.2	-1.07
Humic acid-isocil	6–7	0	0.7	2.19
		25	0.7	2.05
Humic acid-bromacil	6–7	0	0.7	2.1
		25	0.7	1
Aquifer sands-dieldrin[b]	–	5	1.0[a]	-2.3[a]
		40	0.82	-2.45
Aquifer sands-lindane[b]	–	5	1.05[a]	-3.5[a]
		40	0.94	-3.7
Silica sand-dieldrin[b]	–	5	0.95	-2.67
Silica sand-lindane[b]	–	5	1.14	-4.24
Ottawa sand-lindane[b]	–	5	0.97	-4.5

[a] Average of 3 separate determinations.
[b] Data are based on 24 hr uptake values.

Source: F.R. Baucher and G.F. Lee, *Environ. Sci. Technol.*, *6*(6):538, 1972.

9.3 *HEAVY METALS*

9.3–1 CONCENTRATION OF TRACE ELEMENTS IN U.S. RIVERS

1958–1962

Element	Colorado River[a]		Columbia River[b]		Mississippi River[c]		Missouri River[d]		Ohio River[e]	
	Range μg/l	Frequency of detection percent	Range μg/l	Frequency of detection percent	Range μg/l	Frequency of detection percent	Range μg/l	Frequency of detection percent	Range μg/l	Frequency of detection percent
Ba	20–200	79	4–60	94	20–100	100	10–200	96	20–200	94
Cr	10–30	12	1–10	87	3–20	23	8–10	10	4–10	20
Cu	2–50	61	1–50	100	1–100	96	1–50	85	1–100	91
Fe	10–300	82	2–90	100	6–100	95	4–200	92	5–100	100
Mn	–	0	3–5	13	9–50	11	–	6	6–30	17
Mo	8–40	82	1–9	90	3–10	68	7–20	65	3–10	37
Ni	10–30	9	2–10	48	3–10	23	6–9	8	3–30	31
Pb	–	6	4–50	29	10–50	11	–	3	10–30	11
V	–	9	–	3	–	0	–	2	–	0
Ag	–	0	–	0	–	12	0.2–20	57	0.6–1	60

[a] Cd (30) and S (100) occurred once each; Be, Sn, Bi, Zn, and Co were never found in positive amounts.
[b] Other positive values were; Cd (3), Sn (2), and Zn (900, 2300); Be, Sb, Bi, and Co were never found in positive amounts.
[c] Other positive values were; Sn (3, 8, 10) and Cd (7); Be, Sb, Bi, Co. and Zn were never found in positive amounts.
[d] Co (8) occurred once; Cd, Be, Sn, Sb, Bi, and Zn were never found in positive amounts were; Be, Bi, Cd, Co, Sb, Sn, and Zn.

Source: Reprinted, with publisher's permission, from *J. Am. Water Works Assn.*, *57*(2):156, © 1965, by the American Water Works Association, Inc., 2 Park Ave., New York, N.Y. 10016.

9.3–2 CONCENTRATIONS IN TRACE METALS

Great Lakes, 1958–62

Element[a]	No. positive values[b]	Range, µg/l	Unusual values, µg/l	Frequency of detection, %
Ba	42	2–50	200	95
Cr	14	1–9	40	32
Cu	43	1–50	70,200	98
Fe	43	3–90	–	98
Mn	2	5–5	–	4
Mo	22	1–8	–	50
Ni	17	1–10	60	39
Pb	9	3–20	90[c]	20
V	0	–	–	0
Ag	1	1	–	14

[a] Other positive values were Sn (2) and Co (2); Cd, Be, Sb, Bi, and Zn were never found in positive amounts.
[b] Data for Ag based on seven samples, for all other elements 44 samples.
[c] Exceeds USPHS limit.

Source: Reprinted, with publisher's permission, from *J. Am. Water Works Assn.*, *57*(2):150, © 1965, by the American Water Works Association, Inc., 2 Park Ave., New York, N.Y. 10016.

9.3–3 SPECTROGRAPHIC ANALYSIS OF TRACE METALS FROM LAKE TAHOE AND ITS TRIBUTARIES

Trace metal	Tahoe City	Upper Truckee River	Trout Creek	Logan Creek	Incline Creek
	Concentration, µg/l				
Aluminum	16	5.3	11	17	26
Berylium	< 0.3	< 0.29	< 0.36	< 0.29	< 0.29
Bismuth	< 0.3	< 0.14	< 0.18	< 0.14	< 0.14
Cadmium	< 0.7	< 0.71	≦ 0.91	< 0.71	< 0.71
Cobalt	< 0.6	< 0.71	< 0.91	< 0.71	< 0.71
Chromium	< 0.07	< 0.71	< 0.91	< 0.71	< 0.71
Copper	trace	≦ 0.71	8.7	< 0.71	< 0.71
Iron	5.3	29.	51.	≦ 143	143
Gallium	< 2.8	< 2.9	< 3.6	< 2.9	< 2.9
Germanium	< 0.3	< 0.14	< 0.18	< 0.14	< 0.14
Manganese	2.6	≦ 0.71	15.	≦ 0.71	≦ 0.71
Molybdenum	0.51	< 0.14	0.95	< 0.14	< 0.14
Nickel	< 0.3	0.53	0.27	< 0.14	0.86
Lead	< 0.06	< 0.71	< 0.91	< 0.71	< 0.71
Titanium	< 0.6	< 0.29	1.3	< 0.29	< 0.29
Vanadium	15	1.4	1.8	1.7	1.7
Zinc	< 14.	< 2.9	< 3.6	< 2.9	< 2.9

Note: Where levels were below detection the analytical limit is indicated in the left column.

Source: C.R. Goldman *et al., J. Water Pollut. Control Fed., 37*(7):1055, 1965.

9.3–4 TRACE METAL CONCENTRATIONS IN VARIOUS NATURAL AND WASTE WATERS DETERMINED BY ANODIC STRIPPING VOLTAMMETRY

		Concentration, μg/l				
			Free		Acid exchangeable	
Sample	**Location**	**Cd**	**Pb**	**Cu**	**Pb**	**Cu**
1	Rouge River 0.5 mile (0.8 km) above Ford Motor Co. turning basin	<0.1	1.1	2	15	27
2	Rouge River turning basin	7.4	2.2	14	37	108
3	Rouge River mouth	1.5	0.4	8	11	19
4	Detroit River, Trenton Channel	0.8	0.4	11	6	29
5	Detroit River, Livingston Channel	1.9	0.4	11	6	29
6	Lake Erie, central basin, surface water	–	–	–	1.8	6.8
7	Lake Erie, central basin, bottom water	0.3	0.07	0.5	1.6	1.8
8	Lake Michigan, Waukegan, Ill.	1.0	0.2	0.8	3.3	19
9	Lake Michigan, Ludington, Mich.	–	–	–	0.6	0.4

Source: K.H. Mancy *et al., Instrumental Analysis for Water Pollution Control,* Ann Arbor Science, Ann Arbor, 1971.

9.3–5 TRACE METAL DATA FROM NATURAL WATERS

High Altitude Lakes and Rivers

	Zn		Cd		Pb		Cu	
	Concentration, μg/l	E_p, V	Concentration, μg/l	E_p, V	Concentration, μg/l	E_p, V	Concentration, μg/l	E_p, V
Trinity River, Shasta-Trinity National Forest, Cal.[a]	0.3	−0.955	0.2	−0.555	2.5	−0.370	2.2	+0.07
Sulphur Springs, Kittitas County, Wash.	3.2	−0.933	added	−0.525	1.8	−0.340	b	b
Park Lake, Kittitas County, Wash.[c]	4.3	−0.945	1.0	−0.525	1.0	−0.325	0.2	+0.070
Rachel Lake, Kittitas County, Wash.	0.4	−0.950	0.2	−0.550	<0.2	−0.350	0.6	+0.070
Roosevelt Lake, Okanogan County, Wash.	1.6	−0.935	0.3	−0.530	1.6	−0.350	1.3	+0.090

[a] Unidentified peak at −0.170.
[b] Not possible to determine because of a −0.25-V shift in the mercury dissolution.
[c] Unidentified peak at −0.165.

Source: A. Zirano and M.L. Healy, *Environ. Sci. Technol.*, 6(3):243, 1972.

9.3–6 ANALYSIS OF SELECTED NATURAL WATER SAMPLES

3 Electrode Polarograph

No.[a]	Source of sample[b]	pH[c]	Pb(II), ppm	Cd(II), ppm	Zn(II), ppm	Fe(III), ppm
1(a)	Sport-fishing stream	6.0	N[d]	N	0.02	1.22
1(b)			N	N	N	1.48
2(a)	Artificial impoundment	7.5	0.01	N	0.02	0.04
2(b)			N	N	N	0.06
3(a)	Surface runoff	6.5	0.01	0.02	N	0.30
3(b)			N	0.02	N	0.06
4(a)	Polluted creek	5.5	0.02	N	0.15	1.35
4(b)			N	N	0.09	1.70
5(a)	Great lake	7.5	N	N	0.24	0.08
5(b)			N	N	0.32	0.11
6(a)	Large river	6.5	0.03	N	0.11	0.62
6(b)			0.02	N	0.15	0.70
7(a)	Farm well	7.0	N	0.02	0.28	0.18
7(b)			N	0.01	0.20	0.26
8(a)	Strip mine pit	3.5	0.05	0.01	0.75	2.44
8(b)			0.07	0.01	0.60	2.86
9(a)	Industrial effluent	6.5	0.78	0.42	0.45	0.56
9(b)			0.80	0.38	0.54	0.70
10(a)	Mine drainage slough	5.0	1.40	0.32	3.42	0.25
10(b)			1.44	0.30	3.65	0.20

[a] Samples designated (a) analyzed after 200–10 preconcentration; those designated (b) analyzed without preconcentration.
[b] Samples collected at average depth of 3 ft under surface of water with Esmarch APHA-type Water Sampling Apparatus.
[c] pH of original sample, estimated to nearest 0.5 pH with EM-Reagents Non-Bleeding Indicator Sticks (0–6 pH).
[d] N indicates that the metal was not measurable with the apparatus and technique used.

Source: D.M. Coleman *et al., Environ. Sci. Technol., 6*(5):452, 1972.

9.3–7 EXCHANGEABLE Zn, Cu, AND Ni CONTENTS OF SOME BOTTOM SEDIMENTS[a]

Sediment source	Sample	Zn, mE/100 g	Cu, mE, 100 g	Ni, mE/100 g	Organic matter, %
Delaware River	A	0.025	0.010	0.003	1.56
	B	0.021	0.010	0.001	4.63
	C	0.083	0.010	0.001	10.56
	D	0.090	0.010	0.007	7.02
	E	0.112	0.020	0.010	11.26
	F	0.381	0.030	0.230	11.28
Barnegat Bay	A	0.001	0.012	0.006	12.80
	B	0.000	0.018	0.018	22.00
	C	0.000	0.009	0.012	23.75
Freshwater impoundments, Pennsylvania	A	0.012	0.012	ND[b]	7.38
	B	0.012	0.015	ND	6.36
	C	0.008	0.010	ND	1.65
	D	0.013	0.011	ND	6.36
	E	0.012	0.013	ND	4.41
	F	0.009	0.010	ND	2.25

[a] Calculated as divalent ions.
[b] Not determined.

Source: S.J. Toth and A.N. Ott, *Environ. Sci. Technol.*, *4*:939, 1970.

9.3–8 TRACE CONTAMINANTS IN PRECIPITATION AND IN UNTREATED WATER SUPPLIES

Element	Avg concentration in precipitation,[a] ppm	Avg concentration in untreated surface water supplies,[b] ppm	Precipitation USPH limit (1962), ppm	$\frac{A}{C}$ x 100, %
Pb	0.034	0.017	0.05	68
Mn	0.012	0.070	0.05	24
Cu	0.021	0.021	1.0	2
Zn	0.107	–	5.0	2
Ni	0.0043	0.0065		

[a] Average of samples collected from September 1966 through January 1967.
[b] Average computed.[1]

Source: A.L. Lazrus, E. Lorange, and J.P. Lodge, Jr., *Environ. Sci. Technol.*, *4*(1):55, 1970.

REFERENCE

1. C. Durfor and E. Becker, U.S. Geological Survey Water Supply Paper No. 1812, 1964.

9.3–9 VERTICAL DISTRIBUTION OF ARSENATE AND PHOSPHATE AT THREE STATIONS ON R/V TRIDENT CRUISE TR-068, 1969

Depth, m		Arsenate, μmol/l		Phosphate, μmol/l, mean
		Replicates	Mean	
Station 13, TR-068 6/17/69 Caribbean 20°27′N 83°36′W[a]	495	0.05, 0.08, 0.04, 0.06	0.06 ± 0.02	1.13 ± 0.01
	991	0.10, 0.11, 0.11, 0.11	0.11 ± 0.01	1.83 ± 0.01
	1,488	0.15,[b] 0.10, 0.10, 0.10	0.10 ± 0.00	1.49 ± 0.01
	1,985	0.03, 0.07, 0.07, 0.06	0.06 ± 0.02	1.36 ± 0.01
	2,481	0.00,[b] 0.05, 0.04, 0.05	0.05± 0.01	1.34 ± 0.01
	2,974	0.02, 0.01, 0.05, 0.05	0.03 ± 0.02	1.32 ± 0.01
	3,473	0.04, 0.05, 0.05, 0.02	0.04 ± 0.02	1.32 ± 0.01
	3,971	0.04, 0.05, 0.06, 0.05	0.05 ± 0.01	1.31 ± 0.01
Station 15, TR-068 4/19/69 Gulf of Mexico 23°41′N 89°33′W[a]				
	25	0.08, 0.06, 0.05, 0.02	0.05 ± 0.03	0.02 ± 0.01
	50	0.04, 0.05	0.05 ± 0.01	0.00 ± 0.00
	75	0.13,[b] 0.06, 0.06, 0.07	0.06 ± 0.01	0.04 ± 0.01
	100	0.13,[b] 0.01, 0.02, 0.01	0.01 ± 0.01	0.25 ± 0.01
	125	0.04, 0.04, 0.03, 0.02	0.03 ± 0.01	0.45 ± 0.01
	151	0.05, 0.03, 0.04, 0.04	0.04 ± 0.01	0.45 ± 0.01
Station 48, TR-068 4/28/69 Western Atlantic 31°08′N 71°05′W[a]				
	2	0.02, 0.04, 0.03, 0.03	0.03 ± 0.01	0.01 ± 0.01
	974	0.05, 0.06, 0.04, 0.05	0.05 ± 0.01	1.54 ± 0.01
	1,962	0.06, 0.06, 0.04, 0.03	0.05 ± 0.01	1.23 ± 0.01
	2,956	0.04, 0.00, 0.03, 0.02	0.02 ± 0.02	1.26 ± 0.01
	3,944	0.05, 0.01, 0.04, 0.03	0.03 ± 0.02	1.27 ± 0.01
	4,938	0.01, 0.04, 0.04, 0.02	0.03 ± 0.01	1.36 ± 0.01
	4,943	0.01, 0.08, 0.04, 0.05	0.05 ± 0.04	1.36 ± 0.01
Freshwater sources				
Rowe, Mass.,[c] 8/12/68	4		0.22 ± 0.03	1.81 ± 0.10
Providence River, 12/23/69	3		0.43 ± 0.10	14.1 ± 0.10

[a] Stations on *R/V Trident* Cruise TR-068, 1969.
[b] Q, values rejected on basis of "Q" test.[1]
[c] Stream in Davis Mine, an abandoned iron pyrite mine.

Source: D.L. Johnson, *Environ. Sci. Technol.*, *5*(5):411, 1971.

REFERENCE

1. R.B. Dean and W.J. Dixon, *Anal. Chem.*, *23*:636, 1951.

9.3–10 MERCURY IN INDUSTRIAL EFFLUENTS

CEC and ECS Values of Bottom Sediments (mE/100 g)

			Exchangeable cations, mE/100 g									Unsaturation, mE/100 g		
Sediment source	Sample	Total capacity, mE/100 g	Fe[a]	Mn[a]	Na	K	Ca	Mg	H[b]	H[c]	H[b]	H[c]	pH
Hudson River	A	24.7	1.0	1.3	2.4	0.4	3.9	2.4	15.6	13.3	63.1	53.8	ND[d]
	B	34.0	0.3	1.7	3.7	0.6	9.1	5.9	14.7	12.7	43.2	37.3	ND
	C	34.0	0.6	1.0	5.3	0.6	6.4	5.0	17.3	12.7	50.6	37.3	ND
	D	25.8	0.4	1.2	3.0	0.5	6.2	4.1	11.9	10.7	46.1	41.4	ND
	E	25.8	0.3	1.6	2.1	0.3	7.5	3.5	12.4	10.5	48.0	40.6	ND
	F	31.8	0.4	1.9	2.0	0.3	6.4	2.7	20.4	18.1	64.1	56.9	ND
Delaware River	A	31.3	1.4	0.6	1.5	0.7	5.0	6.3	17.8	15.8	56.8	50.4	7.1
	B	24.7	9.4	0.3	0.5	5.4	5.4	3.5	15.0	5.3	60.5	21.4	6.3
	C	43.4	2.9	0.2	0.9	0.6	4.7	4.7	32.5	29.4	74.9	70.0	6.5
	D	20.8	8.1	0.5	0.3	0.3	4.5	2.1	13.6	0.0	65.3	0.0	6.4
	E	33.8	4.2	0.2	0.2	0.4	3.8	2.0	26.9	22.5	79.5	66.5	6.4
	F	24.5	8.0	0.5	0.2	0.5	5.1	2.7	15.8	7.3	64.5	29.8	6.5
Freshwater impoundments, New Jersey	A	30.4	0.7	0.3	0.4	0.5	4.4	6.6	18.5	12.9	60.8	42.4	6.0
	B	36.0	1.2	1.0	0.4	0.5	3.4	6.5	25.2	23.0	70.0	63.9	6.5
	C	7.1	0.4	0.2	0.3	0.2	1.0	0.8	4.8	4.2	67.6	60.0	5.6
	D	8.1	1.7	0.7	0.2	0.1	1.6	1.1	5.1	2.7	63.0	33.3	5.5
	E	6.8	0.6	0.5	0.2	0.1	3.9	0.5	2.1	1.0	30.8	14.7	6.9
Freshwater impoundments, Pennsylvania	A	21.5	12.2	9.7	0.1	0.3	4.1	1.1	15.9	0.0	73.8	0.0	5.2
	B	43.0	5.0	2.7	0.4	0.3	7.3	1.0	34.0	26.3	80.0	61.1	5.4
	C	46.2	5.5	5.1	0.4	0.3	4.0	1.5	40.0	29.4	86.5	63.6	5.8
	D	28.4	4.0	7.1	0.6	0.4	4.3	1.3	21.8	10.7	76.7	37.6	5.8
	E	33.0	15.8	2.5	0.2	0.3	5.3	2.8	25.4	6.1	73.9	18.4	4.3
Chesapeake Bay	A	42.0	ND	0.7	32.9	0.9	1.3	5.5	1.4	–	3.3	–	7.4
	B	35.9	ND	0.5	26.0	0.8	1.2	5.0	2.9	–	8.3	–	7.2
	C	40.6	ND	0.9	26.4	0.7	1.7	5.0	6.8	–	17.7	–	7.5
	D	42.9	ND	0.6	32.9	0.8	1.3	5.0	2.9	–	6.9	–	7.3
	E	33.5	ND	0.8	21.0	0.8	1.8	5.2	3.7	–	26.5	–	7.4
Barnegat Bay	A	100.0	12.4	0.2	7.6	2.7	5.6	15.3	69.8	56.2	69.8	56.2	6.0
	B	85.5	11.6	0.1	10.8	2.7	10.7	25.0	36.3	24.6	40.0	30.0	6.4
	C	97.1	3.0	0.1	13.0	2.7	10.7	19.1	51.6	48.5	53.0	50.3	6.0

[a] West sample values; drying sediments reduces exchangeable Fe and Mn contents to almost trace levels.
[b] CEC = (Na+K+Ca+Mg).
[c] CEC = (Fe+Mn+Na+K+Ca+Mg).
[d] Not determined.

Source: S.J. Toth and A.N. Ott, *Environ. Sci. Technol., 4*(11):935, 1970.

9.3–11 SOLUBILIZATION OF LEAD BY NTA

	Lead concentration, ppm[a]		
	Mamaroneck Reservoir bottom	Crystal Lake bottom	Rt. 80 shoulder
Total Pb	36	120	160
After 1 hr			
0 ppm NTA[b]	<0.02	0.03	<0.02
2	0.02	0.05	0.05
20	0.07	0.61	0.40
After 24 hr			
0 ppm NTA	0.02	<0.02	0.02
2	0.08	0.05	0.04
20	0.03	0.43	0.15

[a] Lead reported as mg/kg of sediment or as mg/kg of solution.
[b] NTA as trisodium salt.

Source: C.D. Gregor, *Environ. Sci. Technol.*, *6*(3):278, 1972.

9.3–12 MERCURY CONTENT OF NATURAL WATERS

Description	Range, ppb	Mean, ppb
Rainwater	0.05–0.48	0.2
Coal mine waters (Donets Basin, U.S.S.R.)	1–10	
Normal stream, river, and lake waters	0.01–0.1 (17)[a]	0.03
Stream and river waters near mercury deposits	0.5–100	
Oceans and seas	0.005–5.0	0.1
Hot springs and certain mineral waters	<0.01–2.5 (50)[b]	0.10
Normal ground waters	0.01–0.10	0.05
Ground waters and mine waters near polymetallic sulfide deposits	1–1,000	
Oil field and other saline waters	0.1–230	

[a] (17) Upper limit; values above 0.1 ppb probably due to natural or industrial contamination.
[b] (50) Probable value in hot springs presently precipitating mercury minerals.

Source: I.R. Jonasson and R.W. Boyle, Geochemistry of Mercury, Proc. Royal Society of Canada, *Symp. Mercury in Man's Environment*,. Ottawa, 1971.

9.3–13 LAHAVE RIVER AND ESTUARY ANALYSES

Station	Depth, meters	Salinity, ppt	Dissolved Hg, ppb	Total particulate matter, g/l	Hg in suspended particulate matter, ppm, dry weight	Particulate Hg, ppb
3 (mean)	0	0.0	0.036	0.001	8.05	0.008
2, 4	0	1.9	0.090	0.001	10.9	0.01
2, 4	1.5	16.2	0.380	0.003	3.93	0.01
2, 4	6.5	26.6	0.134	0.007	7.12	0.05
2, 4 (mean)		14.9	0.201	0.004	7.32	0.03
2, 4	Bottom sediments, 0.27 ppm					
1, 5	0	5.4	0.175	0.002	3.59	0.01
1, 5	1.5	8.8	0.057	0.002	6.39	0.01
1, 5	7.5	28.4	0.091	0.017	7.58	0.13
1, 5 (mean)		14.2	0.108	0.007	5.85	0.04
1, 5	Bottom sediments, 1.06 ppm					
6	0	24.9	0.057	0.003	6.32	0.02
6	2.0	29.0	0.056	0.003	20.2	0.06
6	6.0	28.4	0.058	0.003	7.05	0.02
6 (mean)		27.4	0.057	0.003	11.2	0.03
6	Bottom sediments, 0.12 ppm					
7, 8	0	26.5	0.054	0.002	14.0	0.03
7, 8	2.0	28.7	0.076	0.002	30.2	0.06
7, 8	10.5	29.6	0.046	0.002	34.4	0.07
7, 8 (mean)		28.3	0.059	0.002	26.2	0.05
7, 8	Bottom sediments, 0.09 ppm					
9, 10	2.0	30.0	0.071	0.002	20.2	0.04
9, 10	20	30.3	0.088	0.002	5.12	0.01
9, 10	240	34.0	0.058	0.003	2.04	0.01
9, 10	Bottom sediments, 0.14 ppm					

Source: R.E. Cranston and D.E. Buckley, *Environ. Sci. Technol.*, *6*(3):274, 1972.

9.3–14 MERCURY IN A SIZE-FRACTIONATED BOTTOM SEDIMENT[a]

LaHave River

Diameter, μ	Mean specific surface area,[b] cm^2/g	Size fraction in sample, weight %	Hg, ppm	Total concentration, %
> 60	400	2	0.51	1
20–60	600	20	0.44	12
4–20	2,000	49	0.68	44
< 4	12,000	29	1.12	43

[a] Size fractionation by sedimentation based on settling times of equivalent spheres in deionized water (Stoke's Law).
[b] Specific surface area calculated on basis of spheres having density of 2.5 g/cm^3.

Source: R.E. Cranston and D.E. Buckley, *Environ. Sci. Technol.*, *6*(3):274, 1972.

9.3–15 MERCURY IN INDUSTRIAL EFFLUENTS

Sample	Dissolved Hg, ppb	Hg in suspended particulate matter, ppm, dry weight	Hg in bottom sediments, ppm, dry weight
Paper mill (settling pond)	0.080	10.0	–
Paper mill (effluent)	2.0–3.4	5.6	1.16
Fertilizer plant	2.6–4.0	32.0	0.56
Smelting plant	2.0–4.0	–	0.71
Chlor-alkali plant	80–2,000	14.0	2.02

Source: R.E.Cranston and D.E. Buckley, *Environ. Sci. Technol.*, *6*(3):274, 1972.

9.3–16 MERCURY CONCENTRATIONS IN BOTTOM SEDIMENTS AND ORGANISMS

Collected Off La Jolla and Palos Verdes

	Station[a]					
Sample	La Jolla	Palos Verdes (I)	Palos Verdes (II)	Palos Verdes (III)	Palos Verdes (IV)	Palos Verdes (V)
Sediment	0.02	0.3	0.2	1.0	0.3	0.2
Crab, *Cancer*	0.8	1.0	1.2	0.5	2.3	
anthonyi		1.1			0.9	
Whelk, *Kelletia*	1.5	0.9	1.8	2.3	1.6	3.2
kelletia	0.9			1.6	2.5	
	2.1					
Rock scallop,	0.7	1.0	0.4			0.5
Hinnites	1.6	0.6				
multirugosus						

[a] ppm, dry weight basis.

Source: D.H. Klein and E.D. Goldberg, *Environ. Sci. Technol.*, *4*(9):765, 1970.

9.3–17 BIOLOGICAL MAGNIFICATION OF MERCURY IN THE AQUATIC FOOD CHAIN

	No. samples	Range of values	Arithmetic mean	More numerous organisms
Algae eaters	39	0.01–0.18	0.05	Zooplankton; snails; mayfly nymphs
Zooplankton eaters	9	0.01–0.07	0.04	Insect larvae; minnows
Omnivores	9	0.14–0.16	0.45	Insect larvae and adults; scuds
Detritus eaters	12	0.13–0.89	0.54	Worms; clams; insect larvae
Predators	25	0.01–5.82	0.73	Insect larvae and adults; frogs

Source: A. Hamilton, "Mercury Levels in Canadian Fish," in a paper by E.G. Bligh presented at Royal Society of Canada, *Int. Symp. Mercury in Man's Environment,* Ottawa, Canada, February 15–16, 1971, p. 87.

9.3–18 MERCURY ABSORBED BY INDIVIDUAL GOLDFISH

Mercury in water, ppm	Mercury absorbed by individual fish, ppm
0.001	0.25, 0.25, 0.25
0.01	0.48, 0.54, 0.58
0.1	4.6, 3.2, 3.4

Note: Fish were exposed for 81 hr to 1 liter mercuric chloride solution at the concentrations indicated.

Source: C.E. McKone, *Environ. Sci. Technol., 5*(11):1139, 1971.

9.3–19 METHYLMERCURY AND TOTAL MERCURY LEVELS IN FISH

Food product	Year	No. samples	Average level of methyl-mercury, ppb	Percent of average level of total mercury, ppb
Muscle tissue of perch	1966	1	200	91
Muscle tissue of perch	1966	1	2,990	92
Muscle tissue of pike	1966	1	3,110	93
Muscle tissue of pike	1966	1	550	98
Muscle tissue of cod	1966	1	55	86
Muscle tissue of cod	1966	1	22	85

Source: F.M. D'Itri, *The Environmental Mercury Problem,* CRC Press, Cleveland, 1972.

REFERENCE

1. G. Westöo, in *Chemical Fallout,* M.W. Miller and G.G. Berg, Eds., Charles C Thomas, Springfield, Ill., 1969.

9.3–20 Hg CONTENT IN VARIOUS FISH ORGANS AND TISSUE

Sample source	Hg found, ppm
Edible tissue	0.48
Brain	0.23
Stomach	0.17
Kidney	0.36
Liver	0.87
Heart	0.44

Source: J.M. Rottschafer, *Environ. Sci. Technol.*, *5*(4):337, 1971.

9.3–21 Hg CONTENT IN FISH FROM VARIOUS GEOGRAPHICAL LOCATIONS

Type of fish	Source	Hg found, ppm
Large mouth bass	Bonham State Park Lake, Texas	0.14
White fish	Northern Lake Huron	0.05–0.15[a]
Bluegill	Sandy Bottom Lake, Mich.	0.40
Rock bass	Sandy Bottom Lake, Mich.	0.42
Coho salmon	S.W. Lake Erie	0.48
White bass	Western Lake Erie	0.60
Yellow perch	Western Lake Erie	0.17
Yellow perch	Eastern Lake Erie	0.34
Walleyed pike	Lake St. Clair, Mich.	1.56

[a] Several different fish.

Source: J.M. Rottschafer *et al.*, *Environ. Sci. Technol.*, *5*(4):337, 1971.

9.3–22 MERCURY LEVELS IN FISH NORMALLY EATEN IN NORWAY AND SWEDEN

Species	No. determinations	Mercury levels, ppb Range	Mean	Ref.
Sweden				
Bream	1	–	37	1
Burbot	9	10–74	42	1
Carp	1	–	11	1
Cod	9	24–110	50	1
Haddock	4	16–60	31	1
Halibut	1	–	89	1
Perch	25	130–3,950	1,300	1
Pike	28	150–5,200	1,090	1
Pike-perch	6	170–2,550	1,000	1
Plaice	4	37–64	50	1
Roach	1	–	200	1
Salmon trout	6	22–130	70	1
Whitefish	12	35–460	290	1
Whiting	1	–	48	1
Norway				
Brown trout	40	90–7,380	1,450	2
Char	6	140–550	327	2
Perch	28	130–6,080	1,130	2
Cod	24	120–3,180	1,070	2

Source: F.M. D'Itri, *The Environmental Mercury Problem*, CRC Press, Cleveland, 1972.

REFERENCES

1. G. Westöo, *Var. Foeda, 19*:1, 1967.
2. B. Underdal and T. Hastein, *OiKos, 22*:101, 1971.

9.3–23 MERCURY LEVELS IN CANADIAN ATLANTIC COAST SEAFOODS

Species	Average level of mercury, ppb	Ref.
Cod	27–80	1
Cod	20–230	2
Clam	20–110	2
Crab	160	1
Crab	60–150	2
Flounder	70–170	2
Flounder	62–170	1
Haddock	70–100	2
Haddock	55	1
Herring	20–90	2
Herring meal	20–140	2
Lobster	80–200	2
Oyster	20–140	2
Oyster	68	1
Swordfish	820–1,000	2
Tuna	33–860	2
Mackerel	170	1
Sole	43	1
Perch	180	1
Shad	60	1
Scallops	9–68	1

Source: F.M. D'Itri, *The Environmental Mercury Problem,* CRC Press, Cleveland, 1972.

REFERENCES

1. R.E. Jervis *et al., Mercury Residues in Canadian Foods, Fish, Wildlife,* University of Toronto, Canada, 1970.
2. E.G. Bligh, Proc. Royal Soc. Canada, *Int. Symp. on Mercury in Man's Environment,* Ottawa, 1971.

9.3–24 MERCURY LEVELS IN FISH TAKEN IN LAKE ERIE IN THE VINICITY OF THE RAISIN RIVER

Species	No. determinations	Mercury levels, ppb	
		Range	Mean
Perch	31	200–860	490
Goldfish	18	110–490	280
Carp	8	80–330	190
Sheepshead	1	–	1,700
Carp-goldfish	2	450–900	670
Pumpkinseed	1	–	400
Carpsucker	1	–	180

Source: C.F. Annett, *et al., Sources of Mercury Pollution and an Estimation of the Total Mercury Content in Some Lake Erie Fish and Sediments in the Vicinity of the Raisin River.* Michigan Academy of Science, in press.

9.3–25 MERCURY LEVELS IN FISH

Lake Lansing, Michigan

Species	No. determinations	Mercury levels, ppb	
		Range	Mean
Bluegills	21	100–960	360
Tiger musky	2	270–860	560
Northern pike	11	120–780	400
Yellow bullhead	8	120–880	480
Brown bullhead	3	50–130	100
Black bullhead	1	–	70
Black crappie	4	120–860	470
Perch	1	–	120
Mud pickerell	1	–	1,010
Lake chubsucker	1	–	150
Warmouth	1	–	510
Redhorse sucker	1	–	100

Source: C.F. Annett *et al., Sources of Mercury Pollution and an Estimation of the Total Mercury Content in Some Lake Erie Fishes and Sediments in the Vicinity of the Rasin River,* Michigan Academy of Science, in press.

9.3–26 VARIATION OF SELENIUM CONCENTRATION WITH LOCATION

Lake Mendota, Dane County, Wisconsin

Station no.[a]	Depth, m	Se found, ppm
3	15	1.5
2	19	1.2
23	20	1.3
32	24	1.5
12	10	0.5
24	21	1.7
32	24	1.8

[a] Locations of stations reported by J.J. Delfino *et al.*, *Environ. Sci. Technol.*, *3*:1189, 1969.

Source: J.H. Wiersma and G.F. Lee, *Environ. Sci. Technol.*, *5*(12):1203, 1971.

9.3–27 SELENIUM CONCENTRATION IN MENDOTA CORE

Depth, cm	Se, ppm
5–10	1.8
10–15	2.3
15–20	2.0
20–25	1.9
25–30	2.2
30–35	2.2
35–40	2.4
40–45	2.3
45–50	1.9
50–55	2.2
55–60	2.2
60–65	1.1
65–70	1.0
70–75	1.1
75–80	1.4
80–85	0.8
85–90	1.3
90–95	1.2

Source: J.H. Wiersma and G.F. Lee, *Environ. Sci. Technol.*, *5*(12):1203, 1971.

9.3–28 STREAMFLOW AND VANADIUM CONCENTRATION

Tributaries to Lake Powell

Tributary	Average streamflow during period of study, ft^3/sec	Contribution to inflow to Lake Powell, %	Average vanadium concentration, μg/l
Green River	3,726	39	0.9
San Juan River	4,300	46	7.5
Colorado River	1,396	15	4.6

Source: Reprinted, with publisher's permission, from *J. Am. Water Works Assn.*, *61*(2):85, © 1969, by the American Water Works Association, Inc., 2 Park Ave., New York, N.Y. 10016.

9.3–29 VANADIUM TRACES ALONG THE COLORADO RIVER

Sources	1966									1967					
	April	May	June	July	Aug	Sept	Oct	Nov	Dec	Jan	Feb	March	April	May	June
Tributaries															
Flaming Gorge Green River	a	1.0	1.8	0.6	0.4	1.3	0.7	0.9	0.9	1.2	0.7	0.9	0.7	0.6	0.6
Cedar Hill Animas River	0.4	0.3	0.2	0.2	–	–	0.2	0.4	0.3	0.4	–	–	–	–	–
Shiprock San Juan River	2.1	0.6	–	–	11.3	–	3.9	–	–	7.3	1.3	–	9.0	7.3	25.1
Loma Colorado River	3.6	4.3	3.9	5.9	4.8	6.2	3.5	–	–	–	–	8.9	4.8	1.9	3.3
The Colorado River															
Page	3.4	3.9	4.1	–	–	–	–	–	3.5	4.6	2.8	3.8	3.9	2.7	2
Hoover Dam	2.8	3.2	3.2	2.9	3.1	2.3	2.4	2.5	3.7	3.3	3.9	3.1	2.8	3.0	2
Parker Dam	3.1	3.7	1.7	1.9	–	2.5	2.7	2.7	3.2	3.8	2.0	2.6	2.7	3.6	3
Yuma	3.0	2.1	–	2.2	1.9	–	1.7	1.8	1.5	1.7	1.7	–	2.1	1.9	3

[a] Samples lost or not collected.

Source: Reprinted, with publisher's permission, from *J. Am. Water Works Assn., 61*(2):85, © 1969, by the American Water Works Association, Inc., 2 Park Ave., New York, N.Y. 10016.

9.3–30 ANALYSIS OF RIVER WATERS FOR ZINC

Chesapeake Bay Region

Location	Date of sampling	Zinc (Zn^{++}) concentration, ppb	pH	Total alkalinity as $CaCO_3$, ppm	Total hardness as $CaCO_3$, ppm	Calcium hardness as $CaCO_3$, ppm	Suspended solids concentration of unfiltered sample, ppm
Region North of Baltimore							
Big Gunpowder Falls River	3/9/60[a]	116	7.8	44	72	40	
Perry Hall, Md.		115					
Loch Raven Reservoir	7/8/61	63	9.0				
Prettyboy Reservoir	7/8/61	34	7.2				
Jones Falls	7/8/61	15	7.7				68
Butler, Md.							
Above Rockland	7/8/61	24	7.4				41
Bleach and Dye Co.							
Baltimore, Md.							
Susquehanna River	3/9/60	84	7.7	30	84	52	
Columbia, Pa.							
Millersburg, Pa.	9/27/60[a]	44	7.8				
		34					
Eastern Channel,	9/27/60	45	7.1				
2.5 mi below							
Danville, Pa.							
East Shore	9/27/60	155	5.6				
Danville, Pa.							
West Channel	9/27/60	120	5.9				
Danville, Pa.							
Region of Maryland's Eastern Shore							
Choptank River	4/3/60	73	6.7	12	28	20	
Denton, Md.							
Nanticoke River	4/3/60[a]	87	6.9	10	20	16	
Seaford, Del.		90					
Vienna, Md.	4/3/60[a]	89	6.7	7	32	20	
		86					
Pocomoke River	4/3/60[a]	60	6.5	6	24	12	
Pocomoke City, Md.		55					
Tuckahoe River	4/3/60	95	7.2				
Hillsboro, Md.							
Wicomico River	4/3/60	61	6.9	14	24	16	
Salisbury, Md.							
Region of Potomac River Basin							
Monocacy River	3/13/60[a]	85	7.7	56	88	64	
Frederick, Md.		81					
	5/23/61	280	8.1				1,860
		13					
	5/23/61	13	8.0				4
		20[b]					

[a] Two samples.
[b] Filtered sample.

9.3–30 ANALYSIS OF RIVER WATERS FOR ZINC (Continued)

Location	Date of sampling	Zinc (Zn^{++}) concentration, ppb	pH	Total alkalinity as $CaCO_3$, ppm	Total hardness as $CaCO_3$, ppm	Calcium hardness as $CaCO_3$, ppm	Suspended solids concentration of unfiltered sample, ppm
		Region of Potomac River Basin (*continued*)					
Potomac River Harper's Ferry, W. Va.	3/13/60	51	7.9	80	120	80	
	5/23/61	31	8.2				14
		32[b]					
Point of Rocks, Md.	3/13/60	65	7.8	40	64	44	
Shenandoah River Harper's Ferry, W. Va.	3/13/60[a]	61	8.1	114	140	88	
		61					
	5/23/61	27	8.1				9
		18[b]					
	5/23/61	580	8.3				2.570
		12[b]					
North Branch Front Royal, Va.	9/15/60	25	8.1	133	168	84	
	9/28/60	30	8.2				
	5/23/61	38	8.1				1,450
		32[b]					
	5/23/61	31	8.3				11
		29[b]					
South Branch, above Front Royal, Va.	4/11/60	92	7.8	71	96	68	
South Branch Front Royal, Va.	9/15/60[a]	20	7.9	139	164	96	
	9/28/60	19	8.3				
South Branch, above Front Royal, Va.	5/23/61	38	8.2				349
		12[b]					
	5/23/61	23	8.1				8
		14[b]					
South Branch Elkton, Va.	9/15/60	27	8.0	137	160	96	
South Branch, 2.5 mi south of Elkton Va.	5/23/61	17	8.2				28
		20[b]					
	5/23/61	21	8.0				19
		18[b]					
		Region South of Baltimore					
Appomattox River Macon, Va.	4/16/60[c]	94	7.2	26	32	20	
		98					
		95					
James River Goochland, Va.	4/17/60	55	7.6	37	52	36	
Mattaponi River Aylett, Va.	4/17/60	44	6.6	7	16	8	
Pamunkey River Old Church, Va.	4/17/60[c]	95	7.3	14	20	12	
		87					
		87					
Patuxent River Bowie, Md.	4/17/60[a]	60	7.0	14	28	20	
		56					
Upper Marlboro, Md.	6/28/61	20	7.7				76

[a] Two samples.
[b] Filtered sample.
[c] Three samples.

9.3–30 ANALYSIS OF RIVER WATERS FOR ZINC (*Continued*)

Location	Date of sampling	Zinc (Zn^{++}) concentration, ppb	pH	Total alkalinity as $CaCO_3$, ppm	Total hardness as $CaCO_3$, ppm	Calcium hardness as $CaCO_3$, ppm	Suspended solids concentration of unfiltered sample, ppm
Region South of Baltimore (*Continued*)							
Liberty Dam (face), Md.	7/9/61	29	7.6				13
Liberty Dam (back-water)	7/9/61	15	7.3				41
Rappahannock River Fredericksburg, Va.	4/17/60	61	7.2	16	49	16	
	5/10/61	24 21[b]	7.6				6
	5/10/61	26 15[b]	7.6				8
	5/10/61	24 21[b]	7.6				7
Port Royal, Va.	4/17/60	87	6.6	5	20	16	
Rivanna River Palmyra, Va.	4/17/60	76	7.0	15	24	12	

[a] Two samples.
[b] Filtered sample.

Source: Reprinted, with publisher's permission, from *J. Am. Water Works Assn., 56*(3):280, © 1964, by the American Water Works Association, Inc., 2 Park Ave., New York, N.Y. 10016.

9.3–31 ZINC ON BOTTOM SOLIDS

Streams of the Chesapeake Bay Region

Location	Date of sampling	Zinc content per gram of solids, μg
Big Gunpowder Falls		
Perry Hall, Md.	3/ 9/60	398
Potamac River		
Harper's Ferry, W. Va.	3/13/60	400
Point of Rocks, Md.	3/13/60	152
Shenandoah River		
Harper's Ferry, W. Va.	3/13/60	514
South Branch Front Royal, Va.	4/11/60	103
Potamac River		
Dalecarlia Treatment Plant Washington, D.C.	6/ 4/61	100

Source: Reprinted, with publisher's permission, from *J. Am. Water Works Assn., 56*(3):280, © 1964, by the American Water Works Association, Inc., 2 Park Ave., New York, N.Y. 10016.

9.3–32 ZINC CONCENTRATION AND pH OF RIVER WATERS

Southeastern U.S.

Location	Date of sampling	Zinc (Zn^{++}) concentration, ppb	pH
Nassau River, at bridge on U.S. Rte. 17 in Florida	4/7/61	64	7.4
St. Mary's River, on U.S. Rte. 17 at Georgia-Florida border	4/7/61	66	6.9
Santee River at bridge on U.S. Rte. 17 in South Carolina	4/8/61	36	7.3
Roanoke River near U.S. Rte. 301 at Weldon Hatchery, Weldon, N.C.	4/9/61	37	7.7

Source: Reprinted, with publisher's permission, from *J. Am. Water Works Assn.*, *56*(3):280, © 1964, by the American Water Works Association, Inc., 2 Park Ave., New York, N.Y. 10016.

Appendix

A–1 ATOMIC, MOLECULAR AND EQUIVALENT WEIGHTS AND CONVERSION FACTORS TO $CaCO_3$ OF SUBSTANCES FREQUENTLY APPEARING IN THE CHEMISTRY OF WATER

Substance	Formula	Molecular weight	Equivalent weight	Multiplying factor considering molecular weight of $CaCO_3$ as 100. Substance to $CaCO_3$ equivalent	$CaCO_3$ equivalent to substance
Aluminum	Al	27.0	9.0	5.56	0.18
Aluminum chloride	$AlCl_3$	133.	44.4	1.13	0.89
Aluminum chloride	$AlCl_3.6H_2O$	241.	80.5	0.62	1.61
Aluminum sulfate	$Al_2(SO_4)_3 \cdot 18H_2O$	66.4	111.1	0.45	2.22
Aluminum sulfate	$Al_2(SO_4)_3$ (anhydrous)	342.1	57.0	0.88	1.14
Aluminum hydrate	$Al(OH)_3$	78.0	26.0	1.92	0.52
Alumina	Al_2O_3	101.9	17.0	2.94	0.34
Sodium aluminate	$Na_2Al_2O_4$	163.9	27.3	1.83	0.55
Ammonium alum	$Al_2(SO_4)_3(NH_4)_2SO_4 \cdot 24H_3O$	906.6	151.1	0.33	3.02
Potassium alum	$Al_2(SO_4)_3K_2SO_4 \cdot 24H_2O$	948.8	156.1	0.32	3.12
Ammonia	NH_3	17.0	17.0	2.94	0.34
Ammonium (ion)	NH_4	18.0	18.0	2.78	0.36
Ammonium chloride	NH_4Cl	53.5	53.5	0.94	1.07
Ammonium hydroxide	NH_4OH	35.1	35.1	1.43	0.70
Ammonium sulfate	$(NH_4)_2SO_4$	132.	66.1	0.76	1.32
Barium	Ba	137.4	68.7	0.73	1.37
Barium carbonate	$BaCO_3$	197.4	98.7	0.51	1.97
Barium chloride	$BaCl_2 \cdot 2H_2O$	244.3	122.2	0.41	2.44
Barium hydroxide	$Ba(OH)_2$	171.	85.7	0.59	1.71
Barium oxide	BaO	153.	76.7	0.65	1.53
Barium sulfate	$BaSO_4$	233.4	116.7	0.43	2.33
Calcium	Ca	40.1	20.0	2.50	0.40
Calcium bicarbonate	$Ca(HCO_3)_2$	162.1	81.1	0.62	1.62
Calcium carbonate	$CaCO_3$	100.08	50.1	1.00	1.00
Calcium chloride	$CaCl_2$	111.0	55.5	0.90	1.11
Calcium hydrate	$Ca(OH)_2$	74.1	37.1	1.35	0.74
Calcium hypochlorite	$Ca(ClO)_2$	143.1	35.8	0.70	1.43
Calcium oxide	CaO	56.1	28.0	1.79	0.56
Calcium sulfate	$CaSO_4$ (anhydrous)	136.1	68.1	0.74	1.36
Calcium sulfate	$CaSO_4 \cdot 2H_2O$ (gypsum)	172.2	86.1	0.58	1.72
Calcium nitrate	$Ca(NO_3)_2$	164.1	82.1	0.61	1.64
Calcium phosphate	$Ca_3(PO_4)_2$	310.3	51.7	0.97	1.03
Carbon	C	12.0	3.00	16.67	0.06
Chlorine (ion)	Cl	35.5	35.5	1.41	0.71
Copper (cupric)	Cu	63.6	31.8	1.57	0.64
Copper sulfate (cupric)	$CuSO_4$	160.	80.0	0.63	1.60
Copper sulfate (cupric)	$CuSO_4 \cdot 5H_2O$	250.	125.	0.40	2.50
Iron (ferrous)	Fe″	55.8	27.9	1.79	0.56
Iron (ferric)	Fe‴	55.8	18.6	2.69	0.37
Ferrous carbonate	$FeCO_3$	116.	57.9	0.86	1.16
Ferrous hydroxide	$Fe(OH)_2$	89.9	44.9	1.11	0.90

A–1 ATOMIC, MOLECULAR AND EQUIVALENT WEIGHTS AND CONVERSION FACTORS TO $CaCO_3$ OF SUBSTANCES FREQUENTLY APPEARING IN THE CHEMISTRY OF WATER (*Continued*)

				Multiplying factor considering molecular weight of $CaCO_3$ as 100.	
Substance	**Formula**	**Molecular weight**	**Equivalent weight**	**Substance to $CaCO_3$ equivalent**	**$CaCO_3$ equivalent to substance**
Ferrous oxide	FeO	71.8	35.9	1.39	0.72
Ferrous sulfate	$FeSO_4$ (anhydrous)	151.9	76.0	0.66	1.52
Ferrous sulfate	$FeSO_4 \cdot 7H_2O$	278.0	139.0	0.36	2.78
Ferrous sulfate	$FeSO_4$ (anhydrous)	151.9	151.9		
Ferric chloride	$FeCl_3$	162.	54.1	0.93	1.08
Ferric chloride	$FeCl_3 \cdot 6H_2O$	270.	90.1	0.56	1.80
Ferric hydroxide	Fe $(OH)_3$	107.	35.6	1.41	0.71
Ferric oxide	Fe_2O_3	160.	26.6	1.88	0.53
Ferric sulfate (ferrisul)	$Fe_2\ (SO_4)_3$	399.9	66.7	0.75	1.33
Ferrous or ferric	Fe or Fe	55.8	55.8		
Ferrous sulfate	$FeSO_4$	151.9	151.9		
Fluorine	F	19.0	19.0	2.66	0.38
Hydrogen (ion)	H	1.01	1.01	50.0	0.02
Iodine	I	127.	127.	0.40	2.54
Lead	Pb	207.	104.	0.48	2.08
Magnesium	Mg	24.3	12.2	4.10	0.24
Magnesium oxide	MgO	40.3	20.2	2.48	0.40
Magnesium bicarbonate	Mg $(HCO_3)_3$	146.3	73.2	0.68	1.46
Magnesium carbonate	$MgCO_3$	84.3	42.2	1.19	0.84
Magnesium chloride	$MgCl_2$	95.2	47.6	1.05	0.95
Magnesium hydrate	$Mg(OH)_2$	58.3	29.2	1.71	0.58
Magnesium nitrate	$Mg(NO_3)_2$	148.3	74.2	0.67	1.48
Magnesium phosphate	$Mg_3(PO_4)_2$	262.9	43.8	1.14	0.88
Magnesium sulfate	$MgSO_4$	120.4	60.2	0.83	1.20
Manganese (manganous)	Mn''	54.9	27.5	1.82	0.55
Manganese (manganic)	Mn'''	54.9	18.3	2.73	0.37
Manganese chloride	$MnCl_2$	125.8	62.9	0.80	1.26
Manganese dioxide	MnO_2	86.9	21.7	2.30	0.43
Manganese hydrate	$Mn(OH)_2$	89.0	44.4	1.13	0.89
Manganic oxide	Mn_2O_3	158.	26.3	1.90	0.53
Manganous oxide	MnO	70.9	35.5	1.41	0.71
Nitrate (ion)	NO_3	62.0	62.0	0.81	1.24
Nitric acid	HNO_3	63.0	63.0	0.79	1.26
Nitrogen (valence 3)	N'''	14.0	4.67	10.8	0.093
Nitrogen (valence 5)	N'''''	14.0	2.80	17.9	0.056
Oxygen	O	16.0	8.00	6.25	0.16
Phosphorus (valence 3)	P'''	31.0	10.3	4.76	0.21
Phosphorus (valence 5)	P'''''	31.0	6.20	8.33	0.12

A–1 ATOMIC, MOLECULAR AND EQUIVALENT WEIGHTS AND CONVERSION FACTORS TO $CaCO_3$ OF SUBSTANCES FREQUENTLY APPEARING IN THE CHEMISTRY OF WATER (*Continued*)

				Multiplying factor considering molecular weight of $CaCO_3$ as 100.	
Substance	**Formula**	**Molecular weight**	**Equivalent weight**	**Substance to $CaCO_3$ equivalent**	**$CaCO_3$ equivalent to substance**
Potassium	K	39.1	39.1	1.28	0.78
Potassium carbonate	K_2CO_3	138.	69.1	0.72	1.38
Potassium chloride	KCl	74.6	74.6	0.67	1.49
Potassium hydroxide	KOH	56.1	56.1	0.88	1.12
Silver chloride	AgCl	143.3	143.3	0.35	2.87
Silver nitrate	$AgNO_3$	169.9	169.9	0.29	3.40
Silica	SiO_2	60.1	30.0	1.67	0.60
Silicon	Si	28.1	7.03	7.14	0.14
Sodium	Na	23.0	23.0	2.18	0.46
Sodium bicarbonate	$NaHCO_3$	84.0	84.0	0.60	1.68
Sodium bisulfate	$NaHSO_4$	120.			
Sodium bisulfite	$NaHSO_3$	104.			
Sodium carbonate	Na_2CO_3	106.	53.0	0.94	1.06
Sodium carbonate	$Na_2CO_3 \cdot 10H_2O$	286.	143.	0.35	2.86
Sodium chloride	NaCl	58.5	58.5	0.85	1.17
Sodium hypochlorite	NaClO	74.5	37.3	0.67	1.49
Sodium hydrate	NaOH	40.0	40.0	1.25	0.80
Sodium nitrate	$NaNO_3$	85.0	85.0	0.59	1.70
Sodium nitrite	$NaNO_2$	69.0	34.5	0.73	1.38
Sodium oxide	Na_2O	62.0	31.0	1.61	0.62
Tri-sodium phosphate	$Na_3PO_4 \cdot 12H_2O$ (18.7% P_2O_5)	380.2	126.7	0.40	2.53
Tri-sodium phosphate (anhydrous)	Na_3PO_4 (43.2% · P_2O_5)	164.0	54.7	0.91	1.09
Di-sodium phosphate	$Na_2HPO_4 \cdot 12H_2O$ (19.8% P_2O_5)	358.2	119.4	0.42	2.39
Di-sodium phosphate (anhydrous)	Na_2HPO_4 (50% P_2O_5)	142.0	47.3	1.06	0.95
Mono-sodium phosphate	$NaH_2PO_4 \cdot H_2O$ (51.4% P_2O_5)	138.1	46.0	1.09	0.92
Mono-sodium phosphate (anhydrous)	NaH_2PO_4 (59.1% P_2O_5)	120.0	40.0	1.25	0.80
Meta-phosphate (hagan)	$NaPO_3$ (69% P_2O_5)	102.0	34.0	1.47	0.68
Sodium sulfate	$Na_2SO_4 \cdot 10H_2O$	322.1	161.1	0.31	3.22
Sodium sulfate	Na_2SO_4	142.1	71.0	0.70	1.42
Sodium thiosulfate	$Na_2S_2O_3$	158.1	158.1	0.63	1.59
Sodium tetrathionate	$Na_2S_4O_6$	270.2	135.1	0.37	2.71
Sodium sulfite	Na_2SO_3	126.1	83.0	0.79	1.27
Sulfur (valence 2)	S''	32.1	16.0	3.13	0.32
Sulfur (valence 4)	S''''	32.1	8.02	6.25	0.16
Sulfur (valence 6)	S''''''	32.1	5.34	9.10	0.11
Sulfur dioxide	SO_2	64.1	32.0		
Tin	Sn	119.			
Water	H_2O	18.0	9.00	5.56	0.18
Zinc	Zn	65.4	32.7	1.54	0.65

A–1 ATOMIC, MOLECULAR AND EQUIVALENT WEIGHTS AND CONVERSION FACTORS TO $CaCO_3$ OF SUBSTANCES FREQUENTLY APPEARING IN THE CHEMISTRY OF WATER (*Continued*)

				Multiplying factor considering molecular weight of $CaCO_3$ as 100.	
Substance	**Formula**	**Molecular weight**	**Equivalent weight**	**Substance to $CaCO_3$ equivalent**	**$CaCO_3$ equivalent to substance**
		Acid Radicals			
Bicarbonate	HCO_3	61.0	61.0	0.82	1.22
Carbonate	CO_3	60.0	30.0	1.67	.60
Carbon dioxide	CO_2	44.0	22.0	2.27	.44
Chloride	Cl	35.5	35.5	1.41	.71
Iodide	I	126.9	126.9	0.40	2.54
Nitrate	NO_3	62.0	62.0	0.81	1.24
Hydrate	OH	17.0	17.0	2.94	0.34
Phosphate	PO_4	95.0	31.7	1.58	0.63
Phosphorus oxide	P_2O_5	142.0	23.7	2.11	0.47
Sulfide	S	32.1	16.0	3.11	0.32
Sulfate	SO_4	96.1	48.0	1.04	0.96
Sulfur trioxide	SO_3	80.1	40.0	1.25	0.80
		Acids			
Hydrogen	H	1.0	1.0	50.00	0.02
Acetic acid	$HC_2H_3O_2$	60.1	60.1	0.83	1.20
Carbonic acid	H_2CO_3	62.0	31.0	1.61	0.62
Hydrochloric acid	HCl	36.5	36.5	1.37	0.73
Phosphoric acid	H_3PO_4	98.0	32.7	1.53	0.65
Sulfurous acid	H_2SO_3	82.1	41.1	1.22	0.82
Sulfuric acid	H_2SO_4	98.1	49.0	1.02	0.98
Hydrogen sulfide	H_2S	–	–	–	–
Manganous acid	H_2MnO_3	104.9	52.5	0.95	1.05

Source: *The Water and Waste Treatment Data Book,* © The Permutit Co., division Sybron Corp., New York, 1958, p. 34.

A–2 SOLUBILITY PRODUCT

The solubility product (or ion product constant) is the product of the concentrations of the ions in the saturated solution of a difficultly soluble salt. The concentrations are expressed as moles per liter of solution. The number of cations (or anions) resulting from the dissociation of one molecule of the salt, appears in the formula for calculations of the solubility product as the exponent of the concentration of the cation (or anion).

If two solutions, each containing one of the ions of a difficultly soluble salt, are mixed, no precipitation takes place unless the product of the ion concentrations in the mixture is greater than the solubility product.

In a solution containing two salts which yield a common ion the ratio of solubilities of the two salts is the ratio of the solubility products.

Substance	Solubility product at temperature noted
Aluminum hydroxide	4×10^{-13} (15°)
Aluminum hydroxide	1.1×10^{-15} (18°)
Aluminum hydroxide	3.7×10^{-15} (25°)
Barium carbonate	7×10^{-9} (16°)
Barium carbonate	8.1×10^{-9} (25°)
Barium chromate	1.6×10^{-10} (18°)
Barium chromate	2.4×10^{-10} (28°)
Barium fluoride	1.6×10^{-6} (9.5°)
Barium fluoride	1.7×10^{-6} (18°)
Barium fluoride	1.73×10^{-6} (25.8°)
Barium iodate, $Ba(IO_3)..2H_2O$	8.4×10^{-11} (10°)
Barium iodate, $Ba(IO_3)..2H_2O$	6.5×10^{-10} (25°)
Barium oxalate, $BaC_2O_4.3\frac{1}{2}H_2O$	1.62×10^{-7} (18°)
Barium oxalate, $BaC_2O_4.2H_2O$	1.2×10^{-7} (18°)
Barium oxalate, $BaC_2O_4.\frac{1}{2}H_2O$	2.18×10^{-7} (18°)
Barium sulfate	0.87×10^{-10} (18°)
Barium sulfate	1.08×10^{-10} (25°)
Barium sulfate	1.98×10^{-10} (50°)
Cadmium oxalate $CdC_2O_4.3H_2O$	1.53×10^{-8} (18°)
Cadmium sulfide	3.6×10^{-29} (18°)
Calcium carbonate (calcite)	0.99×10^{-8} (15°)
Calcium carbonate (calcite)	0.87×10^{-8} (25°)
Calcium fluoride	3.4×10^{-11} (18°)
Calcium fluoride	3.95×10^{-11} (26°)
Calcium iodate, $Ca(IO_3)_2.\ 6H_2O$	22.2×10^{-8} (10°)
Calcium iodate, $Ca(IO_3)_2.6H_2O$	64.4×10^{-8} (18°)
Calcium oxalate, $CaC_2O_4.H_2O$	1.78×10^{-9} (18°)
Calcium oxalate, $CaC_2O_4.H_2O$	2.57×10^{-9} (25°)
Calcium sulfate	2.45×10^{-5}(25°C)
Calcium tartrate, $CaC_4H_4O_6.2H_2O$	0.77×10^{-6} (18°)
Cobalt sulfide	3×10^{-26} (18°)
Cupric iodate	1.4×10^{-7} (25°)
Cupric oxalate	2.87×10^{-8} (25°)
Cupric sulfide	8.5×10^{-45} (18°)
Cuprous bromide	4.15×10^{-8} (18–20°)
Cuprous chloride	1.02×10^{-6} (18–20°)
Cuprous iodide	5.06×10^{-12} (18–20°)
Cuprous sulfide	2×10^{-47} (16–18°)
Cuprous thiocyanate	1.6×10^{-11} (18°)
Ferric hydroxide	1.1×10^{-36} (18°)
Ferrous hydroxide	1.64×10^{-14} (18°)
Ferrous oxalate	2.1×10^{-7} (25°)
Ferrous sulfide	3.7×10^{-19} (18°)
Lead carbonate	3.3×10^{-14} (18°)
Lead chromate	1.77×10^{-14} (18°)
Lead fluoride	2.7×10^{-8} (9°)
Lead fluoride	3.2×10^{-8} (18°)
Lead fluoride	3.7×10^{-8} (26.6°)
Lead iodate	5.3×10^{-14} (9.2°)
Lead iodate	1.2×10^{-13} (18°)
Lead iodate	2.6×10^{-13} (25.8°)

Substance	Solubility product at temperature noted
Lead iodide	7.47×10^{-9} (15°)
Lead iodide	1.39×10^{-8} (25°)
Lead oxalate	2.74×10^{-11} (18°)
Lead sulfate	1.06×10^{-8} (18°)
Lead sulfide	3.4×10^{-28} (18°)
Lithium carbonate	1.7×10^{-3} (25°)
Magnesium ammonium phosphate	2.5×10^{-13} (25°)
Magnesium carbonate	2.6×10^{-5} (12°)
Magnesium fluoride	7.1×10^{-9} (18°)
Magnesium fluoride	6.4×10^{-9} (27°)
Magnesium hydroxide	1.2×10^{-11} (18°)
Magnesium oxalate	8.57×10^{-5} (18°)
Manganese hydroxide	4×10^{-14} (18°)
Manganese sulfide	1.4×10^{-15} (18°)
Mercuric sulfide	4×10^{-53} to 2×10^{-49} (18°)
Mercurous bromide	1.3×10^{-21} (25°)
Mercurous chloride	2×10^{-18} (25°)
Mercurous iodide	1.2×10^{-28} (25°)
Nickel sulfide	1.4×10^{-24} (18°)
Potassium acid tartrate $[K^+][HC_4H_4O_6^-]$	3.8×10^{-4} (18°)
Silver bromate	3.97×10^{-5} (20°)
Silver bromate	5.77×10^{-5} (25°)
Silver bromide	4.1×10^{-13} (18°)
Silver bromide	7.7×10^{-13} (25°)
Silver carbonate	6.15×10^{-12} (25°)
Silver chloride	0.21×10^{-10} (4.7°)
Silver chloride	0.37×10^{-10} (9.7°)
Silver chloride	1.56×10^{-10} (25°)
Silver chloride	13.2×10^{-10} (50°)
Silver chloride	215×10^{-10} (100°)
Silver chromate	1.2×10^{-12} (14.8°)
Silver chromate	9×10^{-12} (25°)
Silver cyanide $[Ag^+][Ag(CN)^-_2]$	2.2×10^{-12} (20°)
Silver dichromate	2×10^{-7} (25°)
Silver hydroxide	1.52×10^{-8} (20°)
Silver iodate	0.92×10^{-8} (9.4°)
Silver iodide	0.32×10^{-16} (13°)
Silver iodide	1.5×10^{-16} (25°)
Silver sulfide	1.6×10^{-49} (18°)
Silver thiocyanate	0.49×10^{-12} (18°)
Silver thiocyanate	1.16×10^{-12} (25°)
Strontium carbonate	1.6×10^{-9} (25°)
Strontium fluoride	2.8×10^{-9} (18°)
Strontium oxalate	5.61×10^{-8} (18°)
Strontium sulfate	2.77×10^{-7} (2.9°)
Strontium sulfate	3.81×10^{-7} (17.4°)
Zinc hydroxide	1.8×10^{-14} (18–20°)
Zinc oxalate, $ZnC_2O_4.2H_2O$	1.35×10^{-9} (18°)
Zinc sulfide	1.2×10^{-23} (18°)

Source: *Handbook of Chemistry and Physics,* 53rd ed., CRC Press, Cleveland, 1972, p. B232.

A–3 ELECTROCHEMICAL SERIES

Table A

Reaction	Potential, volts
$Ag^{+} + e^{-} \rightarrow Ag$	0.7996
$Ag^{+2} + e^{-} \rightarrow Ag^{+1}$(4f $HClO_4$)	1.987
$AgAc + e^{-} \rightarrow Ag + Ac^{-}$	0.64
$AgBr + e^{-} \rightarrow Ag + Br^{-}$	0.0713
$AgBrO_3 + e^{-} \rightarrow Ag + BrO_3^{-}$	0.680
$AgC_2O_4 + 2e^{-} \rightarrow Ag + C_2O_4^{-2}$	0.4776
$AgCl + e^{-} \rightarrow Ag + Cl^{-}$	0.2223
$AgCN + e^{-} \rightarrow Ag + CN^{-}$	−0.02
$Ag_2CO_3 + 2e^{-} \rightarrow 2Ag + CO_3^{-2}$	0.4769
$Ag_2CrO_4 + 2e^{-} \rightarrow 2Ag + CrO_4^{-2}$	0.4463
$Ag_4Fe(CN)_6 + 4e^{-} \rightarrow 4Ag + Fe(CN)_6^{-4}$	0.1943
$AgI + e^{-} \rightarrow Ag + I^{-}$	−0.1519
$AgIO_3 + e^{-} \rightarrow Ag + IO_3^{-}$	0.3551
$Ag_2MoO_4 + 2e^{-} \rightarrow 2Ag + MoO_4^{-2}$	0.49
$AgNO_2 + e^{-} \rightarrow Ag + NO_2^{-}$	0.59
$Ag_2O + H_2O + 2e \rightarrow 2Ag + 2OH^{-}$	0.342
$Ag_2O_3 + H_2O + 2e^{-} \rightarrow 2AgO + 2OH^{-}$	0.74
$2AgO + H_2O + 2e^{-} \rightarrow Ag_2O + 2OH^{-}$	0.599
$AgOCN + e^{-} \rightarrow Ag + OCN^{-}$	0.41
$Ag_2S + 2e^{-} \rightarrow 2Ag + S^{-2}$	−0.7051
$Ag_2S + 2H^{+} + 2e^{-} \rightarrow 2Ag + H_2S$	−0.0366
$AgSCN + e^{-} \rightarrow Ag + SCN^{-}$	0.0895
$Ag_2SeO_3 + 2e \rightarrow 2Ag + SeO_3^{-2}$	0.3629
$Ag_2SO_4 + 2e^{-} \rightarrow 2Ag + SO_4^{-2}$	0.653
$Ag_2(WO_4) + 2e^{-} \rightarrow 2Ag + WO_4^{-2}$	0.466
$Al^{+3} + 3e^{-} \rightarrow Al$ (0.1f NaOH)	−1.706
$H_2AlO_3^{-} + H_2O + 3e^{-} \rightarrow Al + 4OH^{-}$	−2.35
$As + 3H^{+} + 3e^{-} \rightarrow AsH_3$	−0.54
$As_2O_3 + 6H^{+} + 6e^{-} \rightarrow 2As + 3H_2O$	0.234
$HAsO_2 + 3H^{+} + 3e^{-} \rightarrow As + 2H_2O$	0.2475
$AsO_2^{-} + 2H_2O + 3e^{-} \rightarrow As + 4OH^{-}$	−0.68
$H_3AsO_4 + 2H^{+} + 2e^{-} \rightarrow HAsO_2 + 2H_2O$(1f HCl)	0.58
$AsO_4^{-3} + 2H_2O + 2e^{-} \rightarrow AsO_2^{-} + 4OH^{-}$	−0.71
$AsO_4^{-3} + 2H_2O + 2e^{-} \rightarrow AsO_2^{-} + 4OH^{-}$ (1f NaOH)	−0.08
$Au^{+} + e^{-} \rightarrow Au$	1.68
$Au^{+3} + 2e^{-} \rightarrow Au^{+1}$	1.29
$Au^{+3} + 3e^{-} \rightarrow Au$	1.42
$AuBr_2^{-} + e^{-} \rightarrow Au + 2Br^{-}$	0.963
$AuBr_4^{-} + 3e^{-} \rightarrow Au + 4Br^{-}$	0.858
$AuCl_4^{-} + 3e^{-} \rightarrow Au + 4Cl^{-}$	0.994
$Au(OH)_3 + 3H^{+} + 3e^{-} \rightarrow Au + 3H_2O$	1.45
$H_2BO_3^{-} + 5H_2O + 8e^{-} \rightarrow BH_4^{-} + 8OH^{-}$	−1.24
$H_2BO_3^{-} + H_2O + 3e^{-} \rightarrow B + 4OH^{-}$	−2.5
$H_3BO_3 + 3H^{+} + 3e^{-} \rightarrow B + 3H_2O$	−0.73
$Ba^{+2} + 2e^{-} \rightarrow Ba$	−2.90
$Ba^{+2} + 2e^{-} \rightarrow Ba(Hg)$	−1.570
$Ba(OH)_2 \cdot 8H_2O + 2e^{-} \rightarrow Ba + 2OH^{-} + 8H_2O$	−2.97
$Be^{+2} + 2e^{-} \rightarrow Be$	−1.70
	(−1.85)
$Be_2O_3^{-2} + 3H_2O + 4e^{-} \rightarrow 2Be + 6OH^{-}$	−2.28
$Bi(Cl)_4^{-} + 3e^{-} \rightarrow Bi + 4Cl^{-}$	0.168
$Bi_2O_3 + 3H_2O + 6e^{-} \rightarrow 2Bi + 6OH^{-}$	−0.46
$Bi_2O_4 + 4H^{+} + 2e^{-} \rightarrow 2BiO^{+} + 2H_2O$	1.59
$BiO^{+} + 2H^{+} + 3e^{-} \rightarrow Bi + H_2O$	0.32
$BiOCl + 2H^{+} + 3e \rightarrow Bi + Cl^{-} + H_2O$	0.1583
$BiOOH + H_2O + 3e^{-} \rightarrow Bi + 3OH^{-}$	−0.46
$Br_2(aq) + 2e^{-} \rightarrow 2Br^{-}$	1.087
$Br_2(l) + 2e^{-} \rightarrow 2Br^{-}$	1.065
$HBrO + H^{+} + e^{-} \rightarrow 1/2Br_2 + H_2O$	1.59
$HBrO + H^{+} + 2e^{-} \rightarrow Br^{-} + H_2O$	1.33
$2HBrO + 2H^{+} + 2e^{-} \rightarrow Br_2(l) + 2H_2O$	1.6
$BrO^{-} + H_2O + 2e^{-} \rightarrow Br^{-} + 2OH^{-}$(1f NaOH)	0.70
$BrO_3^{-} + 6H^{+} + 5e^{-} \rightarrow 1/2Br_2 + 3H_2O$	1.52
$BrO_3^{-} + 6H^{+} + 6e^{-} \rightarrow Br^{-} + 3H_2O$	1.44
$BrO_3^{-} + 3H_2O + 6e^{-} \rightarrow Br^{-} + 6OH^{-}$	0.61
$C_6H_4O_2 + 2H^{+} + 2e \rightarrow C_6H_4(OH)_2$	0.6992
$Ca^{+} + e^{-} \rightarrow Ca$	−3.02
$Ca^{+2} + 2e^{-} \rightarrow Ca$	−2.76
Calomel Electrode; Molal KCl	0.2800
*Calomel Electrode, N KCl	0.2807
*Calomel Electrode 0.1 N KCl	0.3337
*Calomel Electrode, Sat'd. KCl	0.2415
*Calomel Electrode, Sat'd NaCl	0.2360
$Ca(OH)_2 + 2e^{-} \rightarrow Ca + 2OH^{-}$	−3.02
$Cb_2O_5 + 10H^{+} + 10e^{-} \rightarrow 2Cb + 5H_2O$	−0.62
$Cd^{+2} + 2e^{-} \rightarrow Cd$	−0.4026
$Cd^{+2} + 2e^{-} \rightarrow Cd(Hg)$	−0.3521
$Cd(OH)_2 + 2e^{-} \rightarrow Cd(Hg) + 2OH^{-}$	−0.761
	(−0.81)
$CdSO_4 \cdot 8/3H_2O + 2e^{-} \rightarrow Cd(Hg) + CdSO_4$(sat'd aq)	−0.4346
$Ce^{+3} + 3e^{-} \rightarrow Ce$	−2.335
$Ce^{+3} + 3e^{-} \rightarrow Ce(Hg)$	−1.4373
$Ce^{+4} + e^{-} \rightarrow Ce^{+3}$	1.4430
	(1.61)
$Ce^{+4} + e^{-} \rightarrow Ce^{+3}$(0.5f H_2SO_4)	1.4587
$CeOH^{+3} + H^{+} + e^{-} \rightarrow Ce^{+3} + H_2O$	1.7134
$Cl_2(g) + 2e^{-} \rightarrow 2Cl^{-}$	1.3583
$HClO + H^{+} + e^{-} \rightarrow 1/2Cl_2 + H_2O$	1.63
$HClO + H^{+} + 2e^{-} \rightarrow Cl^{-} + H_2O$	1.49
$ClO^{-} + H_2O + 2e^{-} \rightarrow Cl^{-} + 2OH^{-}$	0.90
$ClO_2 + e^{-} \rightarrow ClO_2^{-}$	1.15
$ClO_2 + H^{+} + e^{-} \rightarrow HClO_2$	1.27
$HClO_2 + 2H^{+} + 2e^{-} \rightarrow HClO + H_2O$	1.64
$HClO_2 + 3H^{+} + 3e^{-} \rightarrow 1/2Cl_2 + 2H_2O$	1.63
$HClO_2 + 3H^{+} + 4e^{-} \rightarrow Cl^{-} + 2H_2O$	1.56
$ClO_2^{-} + H_2O + 2e^{-} \rightarrow ClO^{-} + 2OH^{-}$	0.59
$ClO_2^{-} + 2H_2O + 4e^{-} \rightarrow Cl^{-} + 4OH^{-}$	0.76
$ClO_2(aq) + e^{-} \rightarrow ClO_2^{-}$	0.954
$ClO_3^{-} + 2H^{+} + e^{-} \rightarrow ClO_2 + H_2O$	1.15
$ClO_3^{-} + 3H^{+} + 2e \rightarrow HClO_2 + H_2O$	1.21
	(1.23)
$ClO_3^{-} + 6H^{+} + 5e^{-} \rightarrow \frac{1}{2}Cl_2 + 3H_2O$	1.47
$ClO_3^{-} + 6H^{+} + 6e^{-} \rightarrow Cl^{-} + 3H_2O$	1.45
$ClO_3^{-} + H_2O + 2e^{-} \rightarrow ClO_2^{-} + 2OH^{-}$	0.35
$ClO_3^{-} + 3H_2O + 6e^{-} \rightarrow Cl^{-} + 6OH^{-}$	0.62
$ClO_4^{-} + 2H^{+} + 2e^{-} \rightarrow ClO_3^{-} + H_2O$	1.19
$ClO_4^{-} + 8H^{+} + 7e^{-} \rightarrow \frac{1}{2}Cl_2 + 4H_2O$	1.34
$ClO_4^{-} + 8H^{+} + 8e^{-} \rightarrow Cl^{-} + 4H_2O$	1.37
$ClO_4^{-} + H_2O + 2e^{-} \rightarrow ClO_3^{-} + 2OH^{-}$	0.17
$(CN)_2 + 2H^{+} + 2e^{-} \rightarrow 2HCN$	0.37
$2HCNO + 2H^{+} + 2e \rightarrow (CN)_2 + 2H_2O$	0.33
$(CNS)_2 + 2e^{-} \rightarrow 2CNS^{-}$	0.77
$Co^{+2} + 2e^{-} \rightarrow Co$	−0.28
$Co^{+3} + e^{-} \rightarrow Co^{+2}$(3f HNO_3)	1.842
$CO_2 + 2H^{+} + 2e^{-} \rightarrow HCOOH$	−0.2
$2CO_2 + 2H^{+} + 2e^{-} \rightarrow H_2C_2O_4$	−0.49
$Co(NH_3)_6^{+3} + e^{-} \rightarrow Co(NH_3)_6^{+2}$	0.1
$Co(OH)_2 + 2e^{-} \rightarrow Co + 2OH^{-}$	−0.73
$Co(OH)_3 + e^{-} \rightarrow Co(OH)_2 + OH^{-}$	0.2 (0.17)
$Cr^{+2} + 2e \rightarrow Cr$	−0.557
$Cr^{+3} + e^{-} \rightarrow Cr^{+2}$	−0.41
$Cr^{+3} + 3e^{-} \rightarrow Cr$	−0.74
$Cr^{+6} + 3e^{-} \rightarrow Cr^{+3}$(2f H_2SO_4)	1.10
$Cr^{+6} + 3e^{-} \rightarrow Cr^{+3}$(1f NaOH)	−0.12
$Cr_2O_7^{-2} + 14H^{+} + 6e^{-} \rightarrow 2Cr^{+3} + 7H_2O$	1.33
$CrO_2^{-} + 2H_2O + 3e^{-} \rightarrow Cr + 4OH^{-}$	−1.2
$HCrO_4^{-} + 7H^{+} + 3e^{-} \rightarrow Cr^{+3} + 4H_2O$	1.195
$CrO_4^{-2} + 4H_2O + 3e^{-} \rightarrow Cr(OH)_3 + 5OH^{-}$	−0.12
$Cr(OH)_3 + 3e^{-} \rightarrow Cr + 3OH^{-}$	−1.3
$Cs^{+} + e^{-} \rightarrow Cs$	−2.923
$Cu^{+} + e^{-} \rightarrow Cu$	0.522
$Cu^{+2} + 2CN^{-} + e^{-} \rightarrow Cu(CN)_2^{-}$	1.12
$Cu^{+2} + e \rightarrow Cu^{+}$	0.158
	(0.167)
$Cu^{+2} + 2e^{-} \rightarrow Cu$	0.3402
$Cu^{+2} + 2e \rightarrow Cu(Hg)$	0.345
$CuI_2 + e^{-} \rightarrow Cu + 2I^{-}$	0.00
$Cu_2O + H_2O + 2e^{-} \rightarrow 2Cu + 2OH^{-}$	−0.361
$Cu(OH)_2 + 2e^{-} \rightarrow Cu + 2OH^{-}$	−0.224
$2Cu(OH)_2 + 2e^{-} \rightarrow Cu_2O + 2OH^{-} + H_2O$	−0.09
$D^{+} + e^{-} \rightarrow 1/2D_2$	−0.0034
$2D^{+} + 2e^{-} \rightarrow D_2$	−0.044
$Eu^{+3} + e^{-} \rightarrow Eu^{+2}$	−0.43
$1/2F_2 + e^{-} \rightarrow F^{-}$	2.85
$1/2F_2 + H^{+} + e^{-} \rightarrow HF$	3.03
$F_2 + 2e^{-} \rightarrow 2F^{-}$	2.87
$F_2O + 2H^{+} + 4e^{-} \rightarrow H_2O + 2F^{-}$	2.1
$Fe^{+2} + 2e^{-} \rightarrow Fe$	−0.409
$Fe^{+3} + 3e^{-} \rightarrow Fe$	−0.036
$Fe^{+3} + e^{-} \rightarrow Fe^{+2}$	0.770
$Fe^{+3} + e^{-} \rightarrow Fe^{+2}$(1f HCl)	0.770
$Fe^{+3} + e^{-} \rightarrow Fe^{+2}$(1f $HClO_4$)	0.747
$Fe^{+3} + e^{-} \rightarrow Fe^{+2}$(1f H_3PO_4)	0.438
$Fe^{+3} + e^{-} \rightarrow Fe^{+2}$(0.5f H_2SO_4)	0.679

A–3 ELECTROCHEMICAL SERIES (*Continued*)

Table A

Reaction	Potential, volts
$Fe(CN)_6^{-3} + e^- \rightarrow Fe(CN)_6^{-4}$ (0.01f NaOH)	0.46
$Fe(CN)_6^{-3} + e^- \rightarrow Fe(CN)_6^{-4}$ (1f H_2SO_4)	0.69
$FeO_4^{-2} + 8H^+ + 3e^- \rightarrow Fe^{+3} + 4H_2O$	1.9
$Fe(OH)_3 + e^- \rightarrow Fe(OH)_2 + OH^-$	−0.56
Fe (phenanthroline)$_3^{+3} + e^- \rightarrow$ Fe(ph)$_3^{+2}$	1.14
Fe (phenanthroline)$_3^{+3} + e^- \rightarrow$ Fe(ph)$_3^{+2}$(2f H_2SO_4)	1.056
$Ga^{+3} + 3e^- \rightarrow Ga$	−0.560
$H_2GaO_3^- + H_2O + 3e^- \rightarrow Ga + 4OH^-$	−1.22
$GeO_2 + 2H^+ + 2e^- \rightarrow GeO + H_2O$	−0.12
$H_2GeO_3 + 4H^+ + 4e^- \rightarrow Ge + 3H_2O$	−0.13
$2H^+ + 2e^- \rightarrow H_2$	0.0000
$1/2H_2 + e^- \rightarrow H^-$	−2.23
$2H_2O + 2e^- \rightarrow H_2 + 2OH^-$	−0.8277
$H_2O_2 + 2H^+ + 2e^- \rightarrow 2H_2O$	1.776
$HfO^{+2} + 2H^+ + 4e^- \rightarrow Hf + H_2O$	−1.68
$HfO_2 + 4H^+ + 4e^- \rightarrow Hf + 2H_2O$	−1.57
$HfO(OH)_2 + H_2O + 4e^- \rightarrow Hf + 4OH^-$	−2.60
$Hg^{+2} + 2e^- \rightarrow Hg$	0.851
$2Hg^{+2} + 2e^- \rightarrow Hg_2^{+2}$	0.905
$1/2Hg_2^{+2} + e^- \rightarrow Hg$	0.7986
$Hg_2^{+2} + 2e^- \rightarrow 2Hg$	0.7961
$Hg_2(AcO)_2 + 2e^- \rightarrow 2Hg + 2AcO^-$	0.5113
$Hg_2Br_2 + 2e^- \rightarrow 2Hg + 2Br^-$	0.1396
$Hg_2Cl_2 + 2e^- \rightarrow 2Hg + 2Cl^-$	0.2682
$Hg_2Cl_2 + 2e^- \rightarrow 2Hg + 2Cl^-$ (0.1f NaOH)	0.3419
	(0.268)
$Hg_2HPO_4 + H^+ + 2e^- \rightarrow 2Hg + H_2PO_4^-$	0.639
$Hg_2I_2 + 2e^- \rightarrow 2Hg + 2I^-$	−0.0405
$Hg_2O + H_2O + 2e^- \rightarrow 2Hg + 2OH^-$	0.123
$HgO + H_2O + 2e^- \rightarrow Hg + 2OH^-$	0.0984
$Hg_2SO_4 + 2e^- \rightarrow 2Hg + SO_4^{-2}$	0.6158
$HO_2 + H^+ + e^- \rightarrow H_2O_2$	1.5
$I_2 + 2e^- \rightarrow 2I^-$	0.535
$I_3^- + 2e^- \rightarrow 3I^-$	0.5338
$In^{+2} + e^- \rightarrow In^{+1}$	−0.40
$In^{+3} + e^- \rightarrow In^{+2}$	−0.49
$In^{+3} + 2e^- \rightarrow In^{+1}$	−0.40
$In^{+3} + 3e^- \rightarrow In$	−0.338
$H_3IO_6^{-2} + 2e^- \rightarrow IO_3^- + 3OH^-$	~0.70
$H_5IO_6 + H^+ + 2e^- \rightarrow IO_3^- + 3H_2O$	~1.7
$HIO + H^+ + e^- \rightarrow 1/2I_2 + H_2O$	1.45
$HIO + H^+ + 2e^- \rightarrow I^- + H_2O$	0.99
$IO^- + H_2O + 2e^- \rightarrow I^- + 2OH^-$	0.49
$IO_3^- + 6H^+ + 5e^- \rightarrow 1/2I_2 + 3H_2O$	1.195
$IO_3^- + 6H^+ + 6e^- \rightarrow I^- + 3H_2O$	1.085
$2IO_3^- + 12H^+ + 10e^- \rightarrow I_2 + 6H_2O$	1.19
$IO_3^- + 2H_2O + 4e^- \rightarrow IO^- + 4OH^-$	0.56
$IO_3^- + 3H_2O + 6e^- \rightarrow I^- + 6OH^-$	0.26
$IrCl_6^{-2} + e^- \rightarrow IrCl_6^{-3}$	1.02
$IrCl_6^{-3} + 3e^- \rightarrow Ir + 6Cl^-$	0.77
$Ir_2O_3 + 3H_2O + 6e^- \rightarrow 2Ir + 6OH^-$	0.1
$K^+ + e^- \rightarrow K$	−2.924
	(−2.923)
$La^{+3} + 3e^- \rightarrow La$	−2.37
$La(OH)_3 + 3e^- \rightarrow La + 3OH^-$	−2.76
$Li^+ + e^- \rightarrow Li$	−3.045
	(−3.02)
$Mg^{++} + 2e^- \rightarrow Mg$	−2.375
$Mg(OH)_2 + 2e^- \rightarrow Mg + 2OH^-$	−2.67
$Mn^{+2} + 2e^- \rightarrow Mn$	−1.029
$Mn^{+3} + e^- \rightarrow Mn^{+2}$	1.51
$MnO_2 + 4H^+ + 2e^- \rightarrow Mn^{+2} + 2H_2O$	1.208
$MnO_4^- + e^- \rightarrow MnO_4^{-2}$	0.564
$MnO_4^- + 4H^+ + 3e^- \rightarrow MnO_2 + 2H_2O$	+1.679
$MnO_4^- + 8H^+ + 5e^- \rightarrow Mn^{+2} + 4H_2O$	1.491
$MnO_4^- + 2H_2O + 3e^- \rightarrow MnO_2 + 4OH^-$	0.588
$MnO_4^{-1} + 2H_2O + 3e^- \rightarrow MnO_2 + 4OH^-$	0.58
$Mn(OH)_2 \rightarrow Mn + 2OH^-$	−1.47
$Mn(OH)_3 + e^- \rightarrow Mn(OH)_2 + OH^-$	−0.40
$H_2MoO_4 + 6H^+ + 6e^- \rightarrow Mo + 4H_2O$	0.0
$N_2 + 2H_2O + 4H^+ + 2e^- \rightarrow 2NH_3OH^+$	−1.87
$3N_2 + 2H^+ + 2e^- \rightarrow 2HN_3$	−3.1
$N_2H_5^+ + 3H^+ + 2e^- \rightarrow 2NH_4^+$	1.27
$N_2O + 2H^+ + 2e^- \rightarrow N_2 + H_2O$	1.77
$H_2N_2O_2 + 2H^+ + 2e^- \rightarrow N_2 + 2H_2O$	2.65
$N_2O_4 + 2e^- \rightarrow 2NO_2^-$	0.88
$N_2O_4 + 2H^+ + 2e^- \rightarrow 2HNO_2$	1.07
$N_2O_4 + 4H^+ + 4e^- \rightarrow 2NO + 2H_2O$	1.03
$Na^+ + e^- \rightarrow Na$	−2.7109
	(−2.712)

Reaction	Potential, volts
$Nb^{+5} + 2e^- \rightarrow Nb^{+3}$(2fHCl)	0.344
$Nd^{+3} + 3e^- \rightarrow Nd$	−2.246
$2NH_3OH^+ + H^+ + 2e^- \rightarrow N_2H_5^+ + 2H_2O$	1.42
$Ni^{+2} + 2e^- \rightarrow Ni$	−0.23
$Ni(OH)_2 + 2e^- \rightarrow Ni + 2OH^-$	−0.66
$NiO_2 + 4H^+ + 2e^- \rightarrow Ni^{+2} + 2H_2O$	1.93
$NiO_2 + 2H_2O + 2e^- \rightarrow Ni(OH)_2 + 2OH^-$	0.49
$2NO + 2e^- \rightarrow N_2O_2^{-2}$	0.10
$2NO + 2H^+ + 2e^- \rightarrow N_2O + H_2O$	1.59
$2NO + H_2O + 2e^- \rightarrow N_2O + 2OH^-$	0.76
$HNO_2 + H^+ + e^- \rightarrow NO + H_2O$	0.99
$2HNO_2 + 4H^+ + 4e^- \rightarrow H_2N_2O_2 + 2H_2O$	0.80
$2HNO_2 + 4H^+ + 4e^- \rightarrow N_2O + 3H_2O$	1.27
	(1.29)
$NO_2^- + H_2O + e^- \rightarrow NO + 2OH^-$	−0.46
$2NO_2^- + 2H_2O + 4e^- \rightarrow N_2O_2^{-2} + 4OH^-$	−0.18
$2NO_2^- + 3H_2O + 4e^- \rightarrow N_2O + 6OH^-$	0.15
$NO_3^- + 3H^+ + 2e^- \rightarrow HNO_2 + H_2O$	0.94
$NO_3^- + 4H^+ + 3e^- \rightarrow NO + 2H_2O$	0.96
$2NO_3^- + 4H^+ + 2e^- \rightarrow N_2O_4 + 2H_2O$	0.81
$NO_3^- + H_2O + 2e^- \rightarrow NO_2^- + 2OH^-$	0.01
$2NO_3^- + 2H_2O + 2e^- \rightarrow N_2O_4 + 4OH^-$	−0.85
$Np^{+3} + 3e^- \rightarrow Np$	−1.9
$Np^{+4} + e^- \rightarrow Np^{+3}$(1fHClO$_4$)	0.155
$Np^{+5} + e^- \rightarrow Np^{+4}$(1fHClO$_4$)	0.739
$Np^{+6} + e^- \rightarrow Np^{+5}$(1fHClO$_4$)	1.137
$1/2O_2 + 2H^+(10^{-7}M) + 2e^- \rightarrow H_2O$	0.815
$O_2 + 2H^+ + 2e^- \rightarrow H_2O_2$	0.682
$O_2 + 4H^+ + 4e^- \rightarrow 2H_2O$	1.229
$O_2 + H_2O + 2e^- \rightarrow HO_2^- + OH^-$	−0.076
$O_2 + 2H_2O + 2e^- \rightarrow H_2O_2 + 2OH^-$	−0.146
$O_2 + 2H_2O + 4e^- \rightarrow 4OH^-$	0.401
$O_3 + 2H^+ + 2e^- \rightarrow O_2 + H_2O$	2.07
$O_3 + H_2O + 2e^- \rightarrow O_2 + 2OH^-$	1.24
$O_{(g)} + 2H^+ + 2e^- \rightarrow H_2O$	2.42
$OH + e^- \rightarrow OH^-$	1.4
$HO_2^- + H_2O + 2e^- \rightarrow 3OH^-$	0.87
$OsO_4 + 8H^+ + 8e^- \rightarrow Os + 4H_2O$	0.85
$P + 3H^+ + 3e^- \rightarrow PH_3(g)$	−0.04
$P + 3H_2O + 3e^- \rightarrow PH_3(g) + 3OH^-$	−0.87
$Pb^{+2} + 2e^- \rightarrow Pb$	−0.1263
	(−0.126)
$Pb^{+2} + 2e^- \rightarrow Pb(Hg)$	−0.1205
$PbBr_2 + 2e^- \rightarrow Pb(Hg) + 2Br^-$	−0.275
$PbCl_2 + 2e^- \rightarrow Pb(Hg) + 2Cl^-$	−0.262
$PbF_2 + 2e^- \rightarrow Pb(Hg) + 2F^-$	−0.3444
$PbHPO_4 + H^+ + 2e^- \rightarrow Pb(Hg) + HPO_4^=$	−0.2448
$PbI_2 + 2e^- \rightarrow Pb(Hg) + 2I^-$	−0.358
$PbO + H_2O + 2e^- \rightarrow Pb + 2OH^-$	−0.576
$PbO_2 + 4H^+ + 2e^- \rightarrow Pb^{+2} + 2H_2O$	1.46
$HPbO_2^- + H_2O + 2e^- \rightarrow Pb + 3OH^-$	−0.54
$PbO_2 + H_2O + 2e^- \rightarrow PbO + 2OH^-$	0.28
$PbO_2 + SO_4^{-2} + 4H^+ + 2e^- \rightarrow PbSO_4 + 2H_2O$	1.685
$PbSO_4 + 2e^- \rightarrow Pb + SO_4^{-2}$	−0.356
$PbSO_4 + 2e^- \rightarrow Pb(Hg) + SO_4^{-2}$	−0.3505
$Pd^{+2} + 2e^- \rightarrow Pd$	0.83
$Pd^{+2} + 2e^- \rightarrow Pd$(1fHCl)	0.623
$Pd^{+2} + 2e^- \rightarrow Pd$(4f HClO$_4$)	0.987
$PdCl_4^{-2} + 2e^- \rightarrow Pd + 4Cl^-$	0.623
$PdCl_6^{-2} + 2e^- \rightarrow PdCl_4^{-2} + 2Cl^-$	1.29
$Pd(OH)_2 + 2e^- \rightarrow Pd + 2OH^-$	0.1
$H_2PO_2^- + e^- \rightarrow P + 2OH^-$	−1.82
$H_3PO_2 + H^+ + e^- \rightarrow P + 2H_2O$	−0.51
$H_3PO_3 + 2H^+ + 2e^- \rightarrow H_3PO_2 + H_2O$	−0.50
	(−0.59)
$H_3PO_3 + 3H^+ + 3e^- \rightarrow P + 3H_2O$	−0.49
$HPO_3^{-2} + 2H_2O + 2e^- \rightarrow H_2PO_2^- + 3OH^-$	−1.65
$HPO_3^{-2} + 2H_2O + 3e^- \rightarrow P + 5OH^-$	−1.71
$H_3PO_4 + 2H^+ + 2e^- \rightarrow H_3PO_3 + H_2O$	−0.276
	(−0.2)
$PO_4^{-3} + 2H_2O + 2e^- \rightarrow HPO_3^{-2} + 3OH^-$	−1.05
$Pt^{+2} + 2e^- \rightarrow Pt$	~1.2
$PtCl_4^{-2} + 2e^- \rightarrow Pt + 4Cl^-$	0.73
$PtCl_6^{-2} + 2e^- \rightarrow PtCl_4^{-2} + 2Cl^-$	0.74
$Pt(OH)_2 + 2e^- \rightarrow Pt + 2OH^-$	0.16
$Pu^{+4} + e^- \rightarrow Pu^{+3}$(1f HClO$_4$)	0.982
$Pu^{+5} + e^- \rightarrow Pu^{+4}$(0.5f HCl)	1.099
$Pu^{+6} + e^- \rightarrow Pu^{+5}$(1f HClO$_4$)	0.9184
$Pu^{+6} + 2e^- \rightarrow Pu^{+4}$(1f HCl)	1.052
Quinhydrone Elec. H^+, a = 1*	0.6995

A–3 ELECTROCHEMICAL SERIES (*Continued*)

Table A

Reaction	Potential, volts
$Rb^{+} + e^{-} \rightarrow Rb$	−2.925
	(−2.99)
$Re^{+3} + 3e^{-} \rightarrow Re$	0.3
$ReO_4^{-} + 4H^{+} + 3e^{-} \rightarrow ReO_2 + 2H_2O$	0.51
$ReO_2 + 4H^{+} + 4e^{-} \rightarrow Re + 2H_2O$	0.26
$ReO_4^{-} + 2H^{+} + e^{-} \rightarrow ReO_{3(c)} + 2H_2O$	0.768
$ReO_4^{-} + 4H_2O + 7e^{-} \rightarrow Re + 8OH^{-}$	−0.81
$ReO_4^{-} + 8H^{+} + 7e^{-} \rightarrow Re + 4H_2O$	0.367
$Rh^{+4} + e^{-} \rightarrow Rh^{+3}$	1.43
$Rh(Cl)_6^{-3} + 3e^{-} \rightarrow Rh + 6Cl^{-}$	0.44
$Ru^{+3} + e^{-} \rightarrow Ru^{+2}(0.1f\ HClO_4)$	−0.11
$Ru^{+3} + e^{-} \rightarrow Ru^{+2}(1.6f\ HCl)$	−0.084
$Ru^{+4} + e^{-} \rightarrow Ru^{+3}(0.1f\ HClO_4)$	0.49
$Ru^{+4} + e^{-} \rightarrow Ru^{+3}(2f\ HCl)$	0.858
$RuO_2 + 4H^{+} + 4e^{-} \rightarrow Ru + 2H_2O$	−0.8
$RuO_4^{-} + e^{-} \rightarrow RuO_4^{-2}$	0.59
$RuO_{4(c)} + e^{-} \rightarrow RuO_4^{-}$	1.00
$S + 2e^{-} \rightarrow S^{-2}$	−0.508
$S + 2H^{+} + 2e^{-} \rightarrow H_2S_{(aq)}$	0.141
$S + H_2O + 2e^{-} \rightarrow HS^{-} + OH^{-}$	−0.478
$S_2O_6^{-2} + 4H^{+} + 2e^{-} \rightarrow 2H_2SO_3$	0.6
$S_2O_8^{-2} + 2e^{-} \rightarrow 2SO_4^{-2}$	2.0
	(2.05)
$S_4O_6^{=} + 2e^{-} \rightarrow 2S_2O_3^{-2}$	0.09
	(0.10)
$Sb + 3H^{+} + 3e^{-} \rightarrow H_3Sb$	−0.51
$Sb^{+5} + 2e^{-} \rightarrow Sb^{+3}(3.5f\ HCl)$	0.75
$Sb_2O_3 + 6H^{+} + 6e^{-} \rightarrow 2Sb + 3H_2O$	0.1445
	(0.152)
$Sb_2O_5 + 4H^{+} + 4e^{-} \rightarrow Sb_2O_3 + 2H_2O$	0.69
$Sb_2O_{5(s)} + 6H^{+} + 4e^{-} \rightarrow 2SbO^{+} + 3H_2O$	0.64
$SbO^{+} + 2H^{+} + 3e^{-} \rightarrow Sb + 2H_2O$	0.212
$SbO_2^{-} + 2H_2O + 3e^{-} \rightarrow Sb + 4OH^{-}$	−0.66
$SbO_3^{-} + H_2O + 2e^{-} \rightarrow SbO_2^{-} + 2OH^{-}$	−0.59
$Sc^{+3} + 3e^{-} \rightarrow Sc$	−2.08
$Se + 2e^{-} \rightarrow Se^{-2}$	−0.78
$Se + 2H^{+} + 2e^{-} \rightarrow H_2Se(aq)$	−0.36
$H_2SeO_3 + 4H^{+} + 4e^{-} \rightarrow Se + 3H_2O$	0.74
$SeO_3^{-2} + 3H_2O + 4e^{-} \rightarrow Se + 6OH^{-}$	−0.35
$SeO_4^{-2} + 4H^{+} + 2e^{-} \rightarrow H_2SeO_3 + H_2O$	1.15
$SeO_4^{-2} + H_2O + 2e^{-} \rightarrow SeO_3^{-2} + 2OH^{-}$	0.03
$SiF_6^{-2} + 4e^{-} \rightarrow Si + 6F^{-}$	−1.2
$SiO_2 + 4H^{+} + 4e^{-} \rightarrow Si + 2H_2O$	−0.84
$SiO_3^{-2} + 3H_2O + 4e^{-} \rightarrow Si + 6OH^{-}$	−1.73
$Sn^{+2} + 2e^{-} \rightarrow Sn$	−0.1364
$Sn^{+4} + 2e^{-} \rightarrow Sn^{+2}$	0.15
$Sn^{+4} + 2e^{-} \rightarrow Sn^{+2}(0.1f\ HCl)$	0.070
$Sn^{+4} + 2e^{-} \rightarrow Sn^{+2}(1f\ HCl)$	0.139
$HSnO_2^{-} + H_2O + 2e^{-} \rightarrow Sn + 3OH^{-}$	−0.79
$Sn(OH)_6^{-2} + 2e^{-} \rightarrow HSnO_2^{-} + 3OH^{-} + H_2O$	−0.96
$2H_2SO_3 + H^{+} + 2e^{-} \rightarrow HS_2O_4^{-} + 2H_2O$	−0.08
$H_2SO_3 + 4H^{+} + 4e^{-} \rightarrow S + 3H_2O$	0.45
$2SO_3^{-2} + 2H_2O + 2e^{-} \rightarrow S_2O_4^{-2} + 4OH^{-}$	−1.12
$2SO_3^{-2} + 3H_2O + 4e^{-} \rightarrow S_2O_3^{-2} + 6OH^{-}$	−0.58
$SO_4^{-2} + 4H^{+} + 2e^{-} \rightarrow H_2SO_3 + H_2O$	0.20
$2SO_4^{-2} + 4H^{+} + 2e^{-} \rightarrow S_2O_6^{-2} + H_2O$	−0.2
$SO_4^{-2} + H_2O + 2e^{-} \rightarrow SO_3^{-2} + 2OH^{-}$	−0.92
$Sr^{+2} + 2e^{-} \rightarrow Sr$	−2.89

Reaction	Potential, volts
$Sr^{+2} + 2e^{-} \rightarrow Sr(Hg)$	−1.793
$Sr(OH)_2 \cdot 8H_2O + 2e^{-} \rightarrow Sr + 2OH^{-} + 8H_2O$	−2.99
$Ta_2O_5 + 10H^{+} + 10e^{-} \rightarrow 2Ta + 5H_2O$	−0.71
$TcO_4^{-} + 4H^{+} + 3e^{-} \rightarrow TcO_{2(c)} + 2H_2O$	0.738
$Te + 2e^{-} \rightarrow Te^{-2}$	−0.92
$Te + 2H^{+} + 2e^{-} \rightarrow H_2Te(Ag)$	−0.69
	(−0.72)
$Te^{+4} + 4e^{-} \rightarrow Te(2.5f\ HCl)$	0.63
$TeO_2 + 4H_2 + 4e^{-} \rightarrow Te + 2H_2O$	0.593
$TeO_3^{-2} + 3H_2O + 4e^{-} \rightarrow Te + 6OH^{-}$	−0.02
$TeO_4^{-} + 8H^{+} + 7e^{-} \rightarrow Te + 4H_2O$	0.472
$H_6TeO_{6(s)} + 2H^{+} + 2e^{-} \rightarrow TeO_{2(s)} + 4H_2O$	1.02
$Th^{+4} + 4e^{-} \rightarrow Th$	−1.90
$ThO_2 + 4H^{+} + 4e^{-} \rightarrow Th + 2H_2O$	−1.80
$ThO_2 + 2H_2O + 4e^{-} \rightarrow Th + 4OH^{-}$	−2.64
$Ti^{+2} + 2e^{-} \rightarrow Ti$	−1.63
$Ti^{+3} + e^{-} \rightarrow Ti^{+2}$	−2.0
$TiO_2 + 4H^{+} + 4e^{-} \rightarrow Ti + 2H_2O$	−0.86
$Ti(OH)^{+3} + H^{+} + e^{-} \rightarrow Ti^{+3} + H_2O$	0.06
$Tl^{+} + e^{-} \rightarrow Tl$	−0.3363
$Tl^{+} + e^{-} \rightarrow Tl(Hg)$	−0.3338
$Tl^{+3} + e^{-} \rightarrow Tl^{+2}$	−0.37
$Tl^{+3} + 2e^{-} \rightarrow Tl^{+1}$	1.247
$Tl^{+3} + 2e^{-} \rightarrow Tl^{+1}(1f\ HCl)$	0.783
$TlBr + e^{-} \rightarrow Tl(Hg) + Br^{-}$	−0.606
$TlCl + e^{-} \rightarrow Tl(Hg) + Cl^{-}$	−0.555
$TlI + e^{-} \rightarrow Tl(Hg) + I^{-}$	−0.769
$Tl_2O_3 + 3H_2O + 4e^{-} \rightarrow 2Tl^{+} + 6OH^{-}$	0.02
$TlOH + e^{-} \rightarrow Tl + OH^{-}$	−0.3445
$Tl(OH)_3 + 2e^{-} \rightarrow TlOH + 2OH^{-}$	−0.05
$Tl_2SO_4 + 2e^{-} \rightarrow Tl(Hg) + SO_4^{-2}$	−0.4360
$U^{+3} + 3e^{-} \rightarrow U$	−1.8
$U^{+4} + e^{-} \rightarrow U^{+3}$	−0.61
$U^{+4} + e^{-} \rightarrow U^{+3}(1f\ HClO_4)$	−0.631
$U^{+5} + e^{-} \rightarrow U^{+4}(1f\ HCl)$	1.02
$U^{+6} + e^{-} \rightarrow U^{+5}(1f\ HClO_4)$	0.063
$UO_2^{+} + 4H^{+} + e^{-} \rightarrow U^{+4} + 2H_2O$	0.62
$UO_2^{+2} + e^{-} \rightarrow UO_2^{+}$	0.062
$UO_2^{+2} + 4H^{+} + 2e^{-} \rightarrow U^{+4} + 2H_2O$	0.334
$UO_2^{+2} + 4H^{+} + 6e^{-} \rightarrow U + 2H_2O$	−0.82
$V^{+2} + 2e^{-} \rightarrow V$	−1.2
$V^{+3} + e^{-} \rightarrow V^{+2}$	−0.255
$V^{+5} + e^{-} \rightarrow V^{+4}(1f\ NaOH)$	−0.74
$VO^{+2} + 2H^{+} + e^{-} \rightarrow V^{+3} + H_2O$	0.337
$VO_2^{+} + 2H^{+} + e^{-} \rightarrow VO^{+2} + H_2O$	1.00
$V(OH)_4^{+} + 2H^{+} + e^{-} \rightarrow VO^{+2} + 3H_2O$	1.00
$V(OH)_4^{+} + 4H^{+} + 5e^{-} \rightarrow V + 4H_2O$	−0.25
$W_2O_5 + 2H^{+} + 2e^{-} \rightarrow 2WO_2 + H_2O$	−0.04
$WO_2 + 4H^{+} + 4e^{-} \rightarrow W + 2H_2O$	−0.12
$WO_3 + 6H^{+} + 6e^{-} \rightarrow W + 3H_2O$	−0.09
$2WO_3 + 2H^{+} + 2e^{-} \rightarrow W_2O_5 + H_2O$	−0.03
$Y^{+3} + 3e^{-} \rightarrow Y$	−2.37
$Zn^{+2} + 2e^{-} \rightarrow Zn$	−0.7628
$Zn^{+2} + 2e^{-} \rightarrow Zn(Hg)$	−0.7628
$ZnO_2^{-2} + 2H_2O + 2e^{-} \rightarrow Zn + 4OH^{-}$	−1.216
$ZnSO \cdot 7H_2O + 2e^{-} \rightarrow Zn(Hg) + SO_4^{-2}(\text{Sat'd } ZnSO_4)$	−0.7993
$ZrO_2 + 4H^{+} + 4e \rightarrow Zr + 2H_2O$	−1.43
$ZrO(OH)_2 + H_2O + 4e^{-} \rightarrow Zr + 4OH^{-}$	−2.32

Note: Values listed are Standard Reduction Potentials.

Source: *Handbook of Chemistry and Physics,* 53rd ed., CRC Press, Cleveland, 1972, p. D-111. Compiled by J.F. Hunsberger.

A–3 ELECTROCHEMICAL SERIES

Table B

This table is divided into two parts. The first part lists reduction reactions which are positive with respect to the potential of the Standard Hydrogen Electrode. In this first part of Table II the reduction reactions are written in increasing order of the positive potential, beginning with 0.00 and ending with +3.03 volts. The second part of this Table II lists reduction reactions having reduction potentials more negative than that of the Standard Hydrogen Electrode. This second part of the table lists reactions in their order of increasing negative potential, beginning with 0.00 and ending with −3.1 volts.

PART I

Reaction	Potential, volts
$2H^{+} + 2e^{-} \rightarrow H_2$	0.0000
$CuI_2^{-} + e^{-} \rightarrow Cu + 2I^{-}$	0.00
$H_2MoO_4 + 6H^{+} + 6e^{-} \rightarrow Mo + 4H_2O$	0.0
$NO_3^{-} + H_2O + 2e^{-} \rightarrow NO_2^{-} + 2OH^{-}$	0.01
$Tl_2O_3 + 3H_2O + 4e^{-} \rightarrow 2Tl^{+} + 6OH^{-}$	0.02
$SeO_4^{-2} + H_2O + 2e^{-} \rightarrow SeO_3^{-2} + 2OH^{-}$	0.03
$Ti(OH)^{+3} + H^{+} + e^{-} \rightarrow Ti^{+3} + H_2O$	0.06
$UO_2^{+2} + e^{-} \rightarrow UO_2^{+}$	0.062
$U^{+6} + e^{-} \rightarrow U^{+5}$(1f $HClO_4$)	0.063
$Sn^{+4} + 2e^{-} \rightarrow Sn^{+2}$(0.1f HCl)	0.070
$AgBr + e^{-} \rightarrow Ag + Br^{-}$	0.0713
$AgSCN + e^{-} \rightarrow Ag + SCN^{-}$	0.0895
$S_4O_6^{=} + 2e^{-} \rightarrow 2S_2O_3^{-2}$	0.09
	(0.10)
$HgO + H_2O + 2e^{-} \rightarrow Hg + 2OH^{-}$	0.0984
$2NO + 2e^{-} \rightarrow N_2O_2^{-2}$	0.10
$Ir_2O_3 + 3H_2O + 6e^{-} \rightarrow 2Ir + 6OH^{-}$	0.1
$Pd(OH)_2 + 2e^{-} \rightarrow Pd + 2OH^{-}$	0.1
$Co(NH_3)_6^{+3} + e^{-} \rightarrow Co(NH_3)_6^{+2}$	0.1
$Hg_2O + H_2O + 2e^{-} \rightarrow 2Hg + 2OH^{-}$	0.123
$Sn^{+4} + 2e^{-} \rightarrow Sn^{+2}$(1f HCl)	0.139
$Hg_2Br_2 + 2e^{-} \rightarrow 2Hg + 2Br^{-}$	0.1396
$S + 2H^{+} + 2e^{-} \rightarrow H_2S_{(aq)}$	0.141
$Sb_2O_3 + 6H^{+} + 6e^{-} \rightarrow 2Sb + 3H_2O$	0.1445
	(0.152)
$2NO_2^{-} + 3H_2O + 4e^{-} \rightarrow N_2O + 6OH^{-}$	0.15
$ReO_4^{-} + 8H^{+} + 7e^{-} \rightarrow Re + 4H_2O$	0.15
$Sn^{+4} + 2e^{-} \rightarrow Sn^{+2}$	0.15
$Np^{+4} + e^{-} \rightarrow Np^{+3}$(1f $HClO_4$)	0.155
$Cu^{+2} + e \rightarrow Cu^{+}$	0.158
	(0.167)
$BiOCl + 2H^{+} + 3e \rightarrow Bi + Cl^{-} + H_2O$	0.1583
$Pt(OH)_2 + 2e^{-} \rightarrow Pt + 2OH^{-}$	0.16
$Bi(Cl)_4^{-} + 3e^{-} \rightarrow Bi + 4Cl^{-}$	0.168
$ClO_4^{-} + H_2O + 2e^{-} \rightarrow ClO_3^{-} + 2OH^{-}$	0.17
$Ag_4Fe(CN)_6 + 4e^{-} \rightarrow 4Ag + Fe(CN)_6^{-4}$	0.1943
$SO_4^{-2} + 4H^{+} + 2e^{-} \rightarrow H_2SO_3 + H_2O$	0.20
$Co(OH)_3 + e^{-} \rightarrow Co(OH)_2 + OH^{-}$	0.2
	(0.17)
$SbO^{+} + 2H^{+} + 3e^{-} \rightarrow Sb + 2H_2O$	0.212
$AgCl + e^{-} \rightarrow Ag + Cl^{-}$	0.2223
*Calomel Electrode, Sat'd NaCl	0.2360
$As_2O_3 + 6H^{+} + 6e^{-} \rightarrow 2As + 3H_2O$	0.234
*Calomel Electrode, Sat'd. KCl	0.2415
$HAsO_2 + 3H^{+} + 3e^{-} \rightarrow As + 2H_2O$	0.2475
$IO_3^{-} + 3H_2O + 6e^{-} \rightarrow I^{-} + 6OH^{-}$	0.26
$ReO_2 + 4H^{+} + 4e^{-} \rightarrow Re + 2H_2O$	0.26
$Hg_2Cl_2 + 2e^{-} \rightarrow 2Hg + 2Cl^{-}$	0.2682
Calomel Electrode, Molal KCl	0.2800
$PbO_2 + H_2O + 2e^{-} \rightarrow PbO + 2OH^{-}$	0.28
*Calomel Electrode, N KCl	0.2807
$Re^{+3} + 3e^{-} \rightarrow Re$	0.3 ~
$BiO^{+} + 2H^{+} + 3e^{-} \rightarrow Bi + H_2O$	0.32
$2HCNO + 2H^{+} + 2e \rightarrow (CN)_2 + 2H_2O$	0.33
*Calomel Electrode 0.1 N KCl	0.3337
$UO_2^{+2} + 4H^{+} + 2e^{-} \rightarrow U^{+4} + 2H_2O$	0.334
$VO^{+2} + 2H^{+} + e^{-} \rightarrow V^{+3} + H_2O$	0.337
$Cu^{+2} + 2e^{-} \rightarrow Cu$	0.3402
$Hg_2Cl_2 + 2e^{-} \rightarrow 2Hg + 2Cl^{-}$(0.1fNaOH)	0.3419
	(0.268)
$Ag_2O + H_2O + 2e^{-} \rightarrow 2Ag + 2OH^{-}$	0.342
$Nb^{+5} + 2e \rightarrow Nb^{+3}$(2f HCl)	0.344
$Cu^{+2} + 2e \rightarrow Cu(Hg)$	0.345
$ClO_3^{-} + H_2O + 2e^{-} \rightarrow ClO_2^{-} + 2OH^{-}$	0.35
$AgIO_3 + e^{-} \rightarrow Ag + IO_3$	0.3551
$Ag_2SeO_3 + 2e \rightarrow 2Ag + SeO_3^{-2}$	0.3629
$ReO_4^{-} + 8H^{+} + 7e^{-} \rightarrow Re + 4H_2O$	0.367
$(CN)_2 + 2H^{+} + 2e^{-} \rightarrow 2HCN$	0.37
$O_2 + 2H_2O + 4e^{-} \rightarrow 4OH^{-}$	0.401
$AgOCN + e^{-} \rightarrow Ag + OCN^{-}$	0.41
$Fe^{+3} + e^{-} \rightarrow Fe^{+2}$(1f H_3PO_4)	0.438
$Rh(Cl)_6^{-3} + 3e^{-} \rightarrow Rh + 6Cl^{-}$	0.44
$Ag_2CrO_4 + 2e^{-} \rightarrow 2Ag + CrO_4^{-2}$	0.4463
$H_2SO_3 + 4H^{+} + 4e^{-} \rightarrow S + 3H_2O$	0.45
$Fe(CN)_6^{-3} + e^{-} \rightarrow Fe(CN)_6^{-4}$(0.01f NaOH)	0.46
$Ag_2(WO_4) + 2e^{-} \rightarrow 2Ag + WO_4^{-2}$	0.466
$TeO_4^{-} + 8H^{+} + 7e^{-} \rightarrow Te + 4H_2O$	0.472
$Ag_2CO_3 + 2e^{-} \rightarrow 2Ag + CO_3^{-2}$	0.4769
$AgC_2O_4 + 2e^{-} \rightarrow Ag + C_2O_4^{-2}$	0.4776
$Ag_2MoO_4 + 2e^{-} \rightarrow 2Ag + MoO_4^{-2}$	0.49
$IO^{-} + H_2O + 2e^{-} \rightarrow I^{-} + 2OH^{-}$	0.49
$NiO_2 + 2H_2O + 2e^{-} \rightarrow Ni(OH)_2 + 2OH^{-}$	0.49
$Ru^{+4} + e^{-} \rightarrow Ru^{+3}$(0.1f $HClO_4$)	0.49
$ReO_4^{-} + 4H^{+} + 3e^{-} \rightarrow ReO_2 + 2H_2O$	0.51
$Hg_2(AcO)_2 + 2e^{-} \rightarrow 2Hg + 2AcO^{-}$	0.5113
$Cu^{+} + e^{-} \rightarrow Cu$	0.522
$I_3^{-} + 2e^{-} \rightarrow 3I^{-}$	0.5338
$I_2 + 2e^{-} \rightarrow 2I^{-}$	0.535
$IO_3^{-} + 2H_2O + 4e^{-} \rightarrow IO^{-} + 4OH^{-}$	0.56
$MnO_4^{-} + e^{-} \rightarrow MnO_4^{-2}$	0.564
$MnO_4^{-1} + 2H_2O + 3e^{-} \rightarrow MnO_2 + 4OH^{-}$	0.58
$H_3AsO_4 + 2H^{+} + 2e^{-} \rightarrow HAsO_2 + 2H_2O$(1f HCl)	0.58
$MnO_4^{-} + 2H_2O + 3e^{-} \rightarrow MnO_2 + 4OH^{-}$	0.588
$AgNO_2 + e^{-} \rightarrow Ag + NO_2^{-}$	0.59
$ClO_2^{-} + H_2O + 2e^{-} \rightarrow ClO^{-} + 2OH^{-}$	0.59
$RuO_4^{-} + e^{-} \rightarrow RuO_4^{-2}$	0.59
$TeO_2 + 4H_2 + 4e^{-} \rightarrow Te + 2H_2O$	0.593
$2AgO + H_2O + 2e^{-} \rightarrow Ag_2O + 2OH^{-}$	0.599
$S_2O_6^{-2} + 4H^{+} + 2e^{-} \rightarrow 2H_2SO_3$	0.6
$BrO_3^{-} + 3H_2O + 6e^{-} \rightarrow Br^{-} + 6OH^{-}$	0.61
$Hg_2SO_4 + 2e^{-} \rightarrow 2Hg + SO_4^{-2}$	0.6158
$ClO_3^{-} + 3H_2O + 6e^{-} \rightarrow Cl^{-} + 6OH^{-}$	0.62
$UO_2^{+} + 4H^{+} + e^{-} \rightarrow U^{+4} + 2H_2O$	0.62
$Pd^{+2} + 2e^{-} \rightarrow Pd$(1f HCl)	0.623
$PdCl_4^{-2} + 2e^{-} \rightarrow Pd + 4Cl^{-}$	0.623
$Te^{+4} + 4e^{-} \rightarrow Te$(2.5f HCl)	0.63
$Hg_2HPO_4 + H^{+} + 2e^{-} \rightarrow 2Hg + H_2PO_4^{-}$	0.639
$AgAc + e^{-} \rightarrow Ag + Ac^{-}$	0.64
$Sb_2O_{5(s)} + 6H^{+} + 4e^{-} \rightarrow 2SbO^{+} + 3H_2O$	0.64
$Ag_2SO_4 + 2e^{-} \rightarrow 2Ag + SO_4^{-2}$	0.653
$Fe^{+3} + e^{-} \rightarrow Fe^{+2}$(0.5f H_2SO_4)	0.679
$AgBrO_3 + e^{-} \rightarrow Ag + BrO_3^{-}$	0.680
$O_2 + 2H^{+} + 2e^{-} \rightarrow H_2O_2$	0.682
$Fe(CN)_6^{-3} + e^{-} \rightarrow Fe(CN)_6^{-4}$(1f H_2SO_4)	0.69
$Sb_2O_5 + 4H^{+} + 4e^{-} \rightarrow Sb_2O_3 + 2H_2O$	0.69
$C_6H_4O_2 + 2H^{+} + 2e \rightarrow C_6H_4(OH)_2$	0.6992
Quinhydrone Elec. H^{+}, a = 1	0.6995
$BrO^{-} + H_2O + 2e^{-} \rightarrow Br^{-} + 2OH^{-}$(1f NaOH)	0.70
$H_3IO_6^{-2} + 2e^{-} \rightarrow IO_3^{-} + 3OH^{-}$	~0.70
$PtCl_4^{-2} + 2e^{-} \rightarrow Pt + 4Cl^{-}$	0.73
$TcO_4^{-} + 4H^{+} + 3e^{-} \rightarrow TcO_{2(c)} + 2H_2O$	0.738
$Np^{+5} + e^{-} \rightarrow Np^{+4}$(1f $HClO_4$)	0.739
$Ag_2O_3 + H_2O + 2e^{-} \rightarrow 2AgO + 2OH^{-}$	0.74
$H_2SeO_3 + 4H^{+} + 4e^{-} \rightarrow Se + 3H_2O$	0.74
$PtCl_6^{-2} + 2e^{-} \rightarrow PtCl_4^{-2} + 2Cl$	0.74
$Fe^{+3} + e^{-} \rightarrow Fe^{+2}$(1f $HClO_4$)	0.747
$Sb^{+5} + 2e^{-} \rightarrow Sb^{+3}$(3.5f HCl)	0.75
$ClO_2^{-} + 2H_2O + 4e^{-} \rightarrow Cl^{-} + 4OH$	0.76
$NiO_2 + 2H_2O + 2e^{-} \rightarrow Ni(OH)_2 + 2OH^{-}$	0.76
$2NO + H_2O + 2e^{-} \rightarrow N_2O + 2OH^{-}$	0.76
$ReO_4^{-} + 2H^{+} + e^{-} \rightarrow ReO_{3(c)} + 2H_2O$	0.768
$(CNS)_2 + 2e^{-} \rightarrow 2CNS^{-}$	0.77
$Fe^{+3} + e^{-} \rightarrow Fe^{+2}$	0.770
$Fe^{+3} + e^{-} \rightarrow Fe^{+2}$(1f HCl)	0.770
$IrCl_6^{-3} + 3e^{-} \rightarrow Ir + 6Cl^{-}$	0.77
$Tl^{+3} + 2e^{-} \rightarrow Tl^{+1}$(1f HCl)	0.783
$Hg_2^{+2} + 2e^{-} \rightarrow 2Hg$	0.7961
$1/2Hg_2^{+2} + e^{-} \rightarrow Hg$	0.7986
$Ag^{+} + e^{-} \rightarrow Ag$	0.7996
$2HNO_2 + 4H^{+} + 4e^{-} \rightarrow H_2N_2O_2 + 2H_2O$	0.80
$2NO_3^{-} + 4H^{+} + 2e^{-} \rightarrow N_2O_4 + 2H_2O$	0.81
$1/2O_2 + 2H^{+}(10^{-7}M) + 2e^{-} \rightarrow H_2O$	0.815

A–3 ELECTROCHEMICAL SERIES (*Continued*)

Table B

Reaction	Potential, volts
$Pd^{+2} + 2e^- \rightarrow Pd$	0.83
$OsO_4 + 8H^+ + 8e^- \rightarrow Os + 4H_2O$	0.85
$Hg^{+2} + 2e^- \rightarrow Hg$	0.851
$AuBr_4 + 3e^- \rightarrow Au + 4Br$	0.858
$Ru^{+4} + e^- \rightarrow Ru^{+3}$(2f HCl)	0.858
$TiO_2 + 4H^+ + 4e^- \rightarrow Ti + 2H_2O$	0.86
$HO_2^- + H_2O + 2e^- \rightarrow 3OH^-$	0.87
$N_2O_4 + 2e^- \rightarrow 2NO_2^-$	0.88
$ClO^- + H_2O + 2e^- \rightarrow Cl^- + 2OH$	0.90
$2Hg^{+2} + 2e^- \rightarrow Hg_2^{+2}$	0.905
$Pu^{+6} + e^- \rightarrow Pu^{+5}$(1f $HClO_4$)	0.9184
$NO_3^- + 3H^+ + 2e^- \rightarrow HNO_2 + H_2O$	0.94
$ClO_2(aq) + e^- \rightarrow ClO_2^-$	0.954
$NO_3^- + 4H^+ + 3e^- \rightarrow NO + 2H_2O$	0.96
$AuBr_2^- + e^- \rightarrow Au + 2Br^-$	0.963
$Pu^{+4} + e^- \rightarrow Pu^{+3}$(1f $HClO_4$)	0.982
$Pd^{+2} + 2e^- \rightarrow Pd$(4f $HClO_4$)	0.987
$HIO + H^+ + 2e^- \rightarrow I^- + H_2O$	0.99
$HNO_2 + H^+ + e^- \rightarrow NO + H_2O$	0.99
$AuCl_4^- + 3e^- \rightarrow Au + 4Cl^-$	0.994
$RuO_{4(c)} + e^- \rightarrow RuO_4^-$	1.00
$VO_2^+ + 2H^+ + e^- \rightarrow VO^{+2} + H_2O$	1.00
$V(OH)_4^+ + 2H^+ + e^- \rightarrow VO^{+2} + 3H_2O$	1.00
$H_6TeO_{6(s)} + 2H^+ + 2e^- \rightarrow TeO_{2(s)} + 4H_2O$	1.02
$IrCl_6^{-2} + e^- \rightarrow IrCl_6^{-3}$	1.02
$U^{+5} + e^- \rightarrow U^{+4}$(1f HCl)	1.02
$N_2O_4 + 4H^+ + 4e^- \rightarrow 2NO + 2H_2O$	1.03
$Pu^{+6} + 2e^- \rightarrow Pu^{+4}$(1f HCl)	1.052
$Fe(phenanthroline)_3^{+3} + e^- \rightarrow Fe(ph)_3^{+2}$(2f H_2SO_4)	1.056
$Br_{2(l)} + 2e^- \rightarrow 2Br^-$	1.065
$N_2O_4 + 2H^+ + 2e^- \rightarrow 2HNO_2$	1.07
$IO_3^- + 6H^+ + 6e^- \rightarrow I^- + 3H_2O$	1.085
$Br_{2(aq)} + 2e^- \rightarrow 2Br^-$	1.087
$Pu^{+5} + e^- \rightarrow Pu^{+4}$(0.5f HCl)	1.099
$Cr^{+6} + 3e^- \rightarrow Cr^{+3}$(2f H_2SO_4)	1.10
$Cu^{+2} + 2CN^- + e^- \rightarrow Cu(CN)_2^-$	1.12
$Np^{+6} + e^- \rightarrow Np^{+5}$(1f $HClO_4$)	1.137
$Fe(phenanthroline)_3^{+3} + e^- \rightarrow Fe(ph)_3^{+2}$	1.14
$SeO_4^{-2} + 4H^+ + 2e^- \rightarrow H_2SeO_3 + H_2O$	1.15
$ClO_2 + e^- \rightarrow ClO_2^-$	1.15
$ClO_3^- + 2H^+ + e^- \rightarrow ClO_2 + H_2O$	1.15
$ClO_4^- + 2H^+ + 2e^- \rightarrow ClO_3^- + H_2O$	1.19
$2IO_3^- + 12H^+ + 10e^- \rightarrow I_2 + 6H_2O$	1.19
$HCrO_4^- + 7H^+ + 3e^- \rightarrow Cr^{+3} + 4H_2O$	1.195
$IO_3^- + 6H^+ + 5e^- \rightarrow 1/2I_2 + 3H_2O$	1.195
$Pt^{+2} + 2e^- \rightarrow Pt$	~1.2
$MnO_2 + 4H^+ + 2e^- \rightarrow Mn^{+2} + 2H_2O$	1.208
$ClO_3^- + 3H^+ + 2e \rightarrow HClO_2 + H_2O$	1.21
	(1.23)
$O_2 + 4H^+ + 4e^- \rightarrow 2H_2O$	1.229
$O_3 + H_2O + 2e^- \rightarrow O_2 + 2OH^-$	1.24
$Tl^{+3} + 2e^- \rightarrow Tl^{+1}$	1.247
$ClO_2 + H^+ + e^- \rightarrow HClO_2$	1.27
$2HNO_2 + 4H^+ + 4e^- \rightarrow N_2O + 3H_2O$	1.27
	(1.29)
$N_2H_5^+ + 3H^+ + 2e^- \rightarrow 2NH_4^+$	1.27
$Au^{+3} + 2e^- \rightarrow Au^{+1}$	~1.29
$PdCl_6^{-2} + 2e^- \rightarrow PdCl_4^{-2} + 2Cl^-$	1.29
$HBrO + H^+ + 2e^- \rightarrow Br^- + H_2O$	1.33
$Cr_2O_7^{-2} + 14H^+ + 6e^- \rightarrow 2Cr^{+3} + 7H_2O$	1.33
$ClO_4^- + 8H^+ + 7e^- \rightarrow 1/2Cl_2 + 4H_2O$	1.34
$Cl_{2(g)} + 2e^- \rightarrow 2Cl^-$	1.3583
$ClO_4^- + 8H^+ + 8e^- \rightarrow Cl^- + 4H_2O$	1.37
$OH + e^- \rightarrow OH^-$	1.4
$Au^{+3} + 3e^- \rightarrow Au$	1.42
$2NH_3OH^+ + H^+ + 2e^- \rightarrow N_2H_5^+ + 2H_2O$	1.42
$Rh^{+4} + e^- \rightarrow Rh^{+3}$	1.43
$BrO_3^- + 6H^+ + 6e^- \rightarrow Br^- + 3H_2O$	1.44
$Ce^{+4} + e^- \rightarrow Ce^{+3}$	1.4430
	(1.61)
$Au(OH)_3 + 3H^+ + 3e^- \rightarrow Au + 3H_2O$	1.45
$ClO_3^- + 6H^+ + 6e^- \rightarrow Cl^- + 3H_2O$	1.45
$HIO + H^+ + e^- \rightarrow 1/2I_2 + H_2O$	1.45
$Ce^{+4} + e^- \rightarrow Ce^{+3}$(0.5f H_2SO_4)	1.4587
$PbO_2 + 4H^+ + 2e^- \rightarrow Pb^{+2} + 2H_2O$	1.46
$ClO_3^- + 6H^+ + 5e^- \rightarrow 1/2Cl_2 + 3H_2O$	1.47
$HClO + H^+ + 2e^- \rightarrow Cl^- + H_2O$	1.49
$MnO_4^- + 8H^+ + 5e^- \rightarrow Mn^{+2} + 4H_2O$	1.491
$HO_2 + H^+ + e^- \rightarrow H_2O_2$	1.5
$Mn^{+3} + e^- \rightarrow Mn^{+2}$	1.51
$BrO_3^- + 6H^+ + 5e^- \rightarrow 1/2Br_2 + 3H_2O$	1.52
$HClO_2 + 3H^+ + 4e^- \rightarrow Cl^- + 2H_2O$	1.56
$Bi_2O_4 + 4H^+ + 2e^- \rightarrow 2BiO^+ + 2H_2O$	1.59
$HBrO + H^+ + e^- \rightarrow 1/2Br_2 + H_2O$	1.59
$2NO + 2H^+ + 2e^- \rightarrow N_2O + H_2O$	1.59
$2HBrO + 2H^+ + 2e^- \rightarrow Br_{2(l)} + 2H_2O$	1.6
$HClO_2 + 3H^+ + 3e^- \rightarrow 1/2Cl_2 + 2H_2O$	1.63
$HClO + H^+ + e^- \rightarrow 1/2Cl_2 + H_2O$	1.63
$HClO_2 + 2H^+ + 2e^- \rightarrow HClO + H_2O$	1.64
$MnO_4^- + 4H^+ + 3e^- \rightarrow MnO_2 + 2H_2O$	1.679
$Au^+ + e^- \rightarrow Au$	1.68
$PbO_2 + SO_4^{-2} + 4H^+ + 2e^- \rightarrow PbSO_4 + 2H_2O$	1.685
$H_5IO_6 + H^+ + 2e^- \rightarrow IO_3^- + 3H_2O$	~1.7
$CeOH^{+3} + H^+ + e^- \rightarrow Ce^{+3} + H_2O$	1.7134
$N_2O + 2H^+ + 2e^- \rightarrow N_2 + H_2O$	1.77
$H_2O_2 + 2H^+ + 2e^- \rightarrow 2H_2O$	1.776
$Co^{+3} + e^- \rightarrow Co^{+2}$(3f HNO_3)	1.842
$FeO_4^{-2} + 8H^+ + 3e^- \rightarrow Fe^{+3} + 4H_2O$	1.9
$NiO_2 + 4H^+ + 2e^- \rightarrow Ni^{+2} + 2H_2O$	1.93
$Ag^{+2} + e^- \rightarrow Ag^{+1}$(4f $HClO_4$)	1.987
$S_2O_8^{-2} + 2e^- \rightarrow 2SO_4^{-2}$	2.0
	(2.05)
$O_3 + 2H^+ + 2e^- \rightarrow O_2 + H_2O$	2.07
$F_2O + 2H^+ + 4e^- \rightarrow H_2O + 2F^-$	2.1
$O_{(g)} + 2H^+ + 2e^- \rightarrow H_2O$	2.42
$H_2N_2O_2 + 2H^+ + 2e^- \rightarrow N_2 + 2H_2O$	2.65
$1/2F_2 + e^- \rightarrow F^-$	2.85
$F_2 + 2e^- \rightarrow 2F^-$	2.87
$1/2F_2 + H^+ + e^- \rightarrow HF$	3.03

PART 2

Reaction	Potential, volts
$2H^+ + 2e^- \rightarrow H_2$	0.0000
$D^+ + e^- \rightarrow 1/2D_2$	−0.0034
$AgCN + e^- \rightarrow Ag + CN^-$	−0.02
$TeO_3^{-2} + 3H_2O + 4e^- \rightarrow Te + 6OH^-$	−0.02
$2WO_3 + 2H^+ + 2e^- \rightarrow W_2O_5 + H_2O$	−0.03
$Fe^{+3} + 3e^- \rightarrow Fe$	−0.036
$Ag_2S + 2H^+ + 2e^- \rightarrow 2Ag + H_2S$	−0.0366
$P + 3H^+ + 3e^- \rightarrow PH_3(g)$	−0.04
$W_2O_5 + 2H^+ + 2e^- \rightarrow 2WO_2 + H_2O$	−0.04
$Hg_2I_2 + 2e^- \rightarrow 2Hg + 2I^-$	−0.0405
$2D^+ + 2e^- \rightarrow D_2$	−0.044
$Tl(OH)_3 + 2e^- \rightarrow TlOH + 2OH^-$	−0.05
$O_2 + H_2O + 2e^- \rightarrow HO_2^- + OH^-$	−0.076
$AsO_4^{-3} + 2H_2O + 2e^- \rightarrow AsO_2^- + 4OH^-$(1f NaOH)	−0.08
$2H_2SO_3 + H^+ + 2e^- \rightarrow HS_2O_4^- + 2H_2O$	−0.08
$Ru^{+3} + e^- \rightarrow Ru^{+2}$(1–6f HCL)	−0.084
$2Cu(OH)_2 + 2e^- \rightarrow Cu_2O + 2OH^- + H_2O$	−0.09
$WO_3 + 6H^+ + 6e^- \rightarrow W + 3H_2O$	−0.09
$Ru^{+3} + e^- \rightarrow Ru^{+2}$(0.1f $HClO_4$)	−0.11
$Cr^{+6} + 3e^- \rightarrow Cr^{+3}$(1f NaOH)	−0.12
$CrO_4^{-2} + 4H_2O + 3e^- \rightarrow Cr(OH)_3 + 5OH^-$	−0.12
$GeO_2 + 2H^+ + 2e^- \rightarrow GeO + H_2O$	−0.12
$WO_2 + 4H^+ + 4e^- \rightarrow W + 2H_2O$	−0.12
$Pb^{+2} + 2e^- \rightarrow Pb(Hg)$	−0.1205
$Pb^{+2} + 2e^- \rightarrow Pb$	−0.1263
	(0.126)
$H_2GeO_3 + 4H^+ + 4e^- \rightarrow Ge + 3H_2O$	−0.13
$Sn^{+2} + 2e^- \rightarrow Sn$	−0.1364
$O_2 + 2H_2O + 2e^- \rightarrow H_2O_2 \rightarrow 2OH^-$	−0.146
$AgI + e^- \rightarrow Ag + I$	−0.1519
$2NO_2^- + 2H_2O + 4e^- \rightarrow N_2O_2^{-2} + 4OH^-$	−0.18
$CO_2 + 2H^+ + 2e^- \rightarrow HCOOH$	−0.2
$2SO_4^{-2} + 4H^+ + 2e^- \rightarrow S_2O_6^{-2} + 2H_2O$	−0.2
$Cu(OH)_2 + 2e^- \rightarrow Cu + 2OH^-$	−0.224

A–3 ELECTROCHEMICAL SERIES (*Continued*)

Table B

Reaction	Potential, volts
$Ni^{+2} + 2e^- \rightarrow Ni$	−0.23
$PbHPO_4 + H^+ + 2e^- \rightarrow Pb(Hg) + HPO_4$	−0.2448
$V(OH)_4^+ + 4H^+ + 5e^- \rightarrow V + 4H_2O$	−0.25
$V^{+3} + e^- \rightarrow V^{+2}$	−0.255
$PbCl_2 + 2e^- \rightarrow Pb(Hg) + 2Cl^-$	−0.262
$PbBr_2 + 2e^- \rightarrow Pb(Hg) + Br^-$	−0.275
$H_3PO_4 + 2H^+ + 2e^- \rightarrow H_3PO_3 + H_2O$	−0.276
	(−0.2)
$Co^{+2} + 2e^- \rightarrow Co$	−0.28
$Tl^+ + e^- \rightarrow Tl(Hg)$	−0.3338
$Tl^+ + e^- \rightarrow Tl$	−0.3363
$In^{+3} + 3e^- \rightarrow In$	−0.338
$PbF_2 + 2e^- \rightarrow Pb(Hg) + 2F^-$	−0.3444
$TlOH + e^- \rightarrow Tl + OH^-$	−0.3445
$SeO_3^{-2} + 3H_2O + 4e^- \rightarrow Se + 6OH^-$	−0.35
$PbSO_4 + 2e^- \rightarrow Pb(Hg) + SO_4^{-2}$	−0.3505
$Cd^{+2} + 2e^- \rightarrow Cd(Hg)$	−0.3521
$PbSO_4 + 2e^- \rightarrow Pb + SO_4^{-2}$	−0.356
$PbI_2 + 2e^- \rightarrow Pb(Hg) + 2I^-$	−0.358
$Se + 2H^+ + 2e^- \rightarrow H_2Se(aq)$	−0.36
$Cu_2O + H_2O + 2e^- \rightarrow 2Cu + 2OH^-$	−0.361
$Tl^{+3} + e^- \rightarrow Tl^{+2}$	−0.37
$In^{+2} + e^- \rightarrow In^{+1}$	−0.40
$In^{+3} + 2e^- \rightarrow In^{+1}$	−0.40
$Mn(OH)_3 + e^- \rightarrow Mn(OH)_2 + OH^-$	−0.40
$Cd^{+2} + 2e^- \rightarrow Cd$	−0.4026
$Fe^{+2} + 2e^- \rightarrow Fe$	−0.409
$Cr^{+3} + e^- \rightarrow Cr^{+2}$	−0.41
$Eu^{+3} + e^- \rightarrow Eu^{+2}$	−0.43
$CdSO_4 \cdot 8/3H_2O + 2e^- \rightarrow Cd(Hg) + CdSO_4$ (sat'd aq)	−0.4346
$Tl_2SO_4 + 2e^- \rightarrow Tl(Hg) + SO_4^{-2}$	−0.4360
$Bi_2O_3 + 3H_2O + 6e^- \rightarrow 2Bi + 6OH^-$	−0.46
$BiOOH + H_2O + 3e^- \rightarrow Bi + 3OH^-$	−0.46
$NO_2^- + H_2O + e^- \rightarrow NO + 2OH^-$	−0.46
$S + H_2O + 2e^- \rightarrow HS^- + OH^-$	−0.478
$H_3PO_3 + 3H^+ + 3e^- \rightarrow P + 3H_2O$	−0.49
$In^{+3} + e^- \rightarrow In^{+2}$	−0.49
$2CO_2 + 2H^+ + 2e^- \rightarrow H_2C_2O_4$	−0.49
$H_3PO_3 + 2H^+ + 2e^- \rightarrow H_3PO_2 + H_2O$	−0.50
	(−0.59)
$S + 2e^- \rightarrow S^{-2}$	−0.508
$H_3PO_2 + H^+ + e^- \rightarrow P + 2H_2O$	−0.51
$Sb + 3H^+ + 3e^- \rightarrow H_3Sb$	−0.51
$As + 3H^+ + 3e^- \rightarrow AsH_3$	−0.54
$HPbO_2^- + H_2O + 2e^- \rightarrow Pb + 3OH^-$	−0.54
$TlCl + e^- \rightarrow Tl(Hg) + Cl^-$	−0.555
$Cr^{+2} + 2e^- \rightarrow Cr$	−0.557
$Ga^{+3} + 3e^- \rightarrow Ga$	−0.560
$Fe(OH)_3 + e^- \rightarrow Fe(OH)_2 + OH^-$	−0.56
$PbO + H_2O + 2e^- \rightarrow Pb + 2OH^-$	−0.576
$2SO_3^{-2} + 3H_2O + 4e^- \rightarrow S_2O_3^{-2} + 6OH^-$	−0.58
$SbO_3^- + H_2O + 2e^- \rightarrow SbO_2^- + 2OH^-$	−0.59
$TlBr + e^- \rightarrow Tl(Hg) + Br^-$	−0.606
$U^{+4} + e^- \rightarrow U^{+3}$	−0.61
$Cb_2O_5 + 10H^+ + 10e^- \rightarrow 2Cb + 5H_2O$	−0.62
$U^{+4} + e^- \rightarrow U^{+3}$(1 f$HCLO_4$)	−0.631
$Ni(OH)_2 + 2e^- \rightarrow Ni + 2OH^-$	−0.66
$SbO_2^- + 2H_2O + 3e^- \rightarrow Sb + 4OH^-$	−0.66
$AsO_2^- + 2H_2O + 3e^- \rightarrow As + 4OH^-$	−0.68
*$Te + 2H^+ + 2e^- \rightarrow H_2Te(Ag)$	−0.69
	(−0.72)
$Ag_2S + 2e^- \rightarrow 2Ag + S^{-2}$	−0.7051
$Ta_2O_5 + 10H^+ + 10e^- \rightarrow 2Ta + 5H_2O$	−0.71
$AsO_4^{-3} + 2H_2O + 2e^- \rightarrow AsO_2^- + 4OH^-$	−0.71
$Co(OH)_2 + 2e^- \rightarrow Co + 2OH^-$	−0.73
$H_3BO_3 + 3H^+ + 3e^- \rightarrow B + 3H_2O$	−0.73
$Cr^{+3} + 3e^- \rightarrow Cr$	−0.74
$V^{+5} + e^- \rightarrow V^{+4}$(1f NaOH)	−0.74
$Cd(OH)_2 + 2e^- \rightarrow Cd(Hg) + 2OH^-$	−0.761
	(−0.81)
$Zn^{+2} + 2e^- \rightarrow Zn$	−0.7628
$Zn^{+2} + 2e^- \rightarrow Zn(Hg)$	−0.7628
$TlI + e^- \rightarrow Tl(Hg) + I^-$	−0.769
$Se + 2e^- \rightarrow Se^{-2}$	−0.78
$HSnO_2^- + H_2O + 2e^- \rightarrow Sn + 3OH^-$	−0.79
$ZnSO_4 \cdot 7H_2O + 2e^- \rightarrow Zn(Hg) + SO_4^{-2}$ (Sat'd $ZnSO_4$)	−0.7993
$RuO_2 + 4H^+ + 4e^- \rightarrow Ru + 2H_2O$	−0.8
$ReO_4^- + 4H_2O + 7e^- \rightarrow Re + 8OH^-$	−0.81
$N_2 + 2H_2O + 4H^+ + 2e^- \rightarrow 2NH_3OH^+$	−1.87
$Th^{+4} + 4e^- \rightarrow Th$	−1.90
$Np^{+3} + 3e^- \rightarrow Np$	−1.9
$Ti^{+3} + e^- \rightarrow Ti^{+2}$	−2.0
$Sc^{+3} + 3e^- \rightarrow Sc$	−2.08
$1/2H_2 + e^- \rightarrow H^-$	−2.23
$Nd^{+3} + 3e^- \rightarrow Nd$	−2.246
$Be_2O_3^{-2} + 3H_2O + 4e^- \rightarrow 2Be + 6OH^-$	−2.28
$ZrO(OH)_2 + H_2O + 4e^- \rightarrow Zr + 4OH^-$	−2.32
$Ce^{+3} + 3e^- \rightarrow Ce$	−2.335
$H_2AlO_3^- + H_2O + 3e^- \rightarrow Al + 4OH^-$	−2.35
$Y^{+3} + 3e^- \rightarrow Y$	−2.37
$La^{+3} + 3e^- \rightarrow La$	−2.37
$Mg^{++} + 2e^- \rightarrow Mg$	−2.375
$H_2BO_3^- + H_2O + 3e^- \rightarrow B + 4OH^-$	−2.5
$HfO(OH)_2 + H_2O + 4e^- \rightarrow Hf + 4OH^-$	−2.60
$ThO_2 + 2H_2O + 4e^- \rightarrow Th + 4OH^-$	−2.64
$Mg(OH)_2 + 2e^- \rightarrow Mg + 2OH^-$	−2.67
$Na^+ + e^- \rightarrow Na$	−2.7109
	(−2.712)
$Ca^{+2} + 2e^- \rightarrow Ca$	−2.76
$La(OH)_3 + 3e^- \rightarrow La + 3OH^-$	−2.76
$Sr^{+2} + 2e^- \rightarrow Sr$	−2.89
$Ba^{+2} + 2e^- \rightarrow Ba$	−2.90
$Cs^+ + e^- \rightarrow Cs$	−2.923
$K^+ + e^- \rightarrow K$	−2.924
	(−2.923)
$Rb^+ + e^- \rightarrow Rb$	−2.925
	(−2.99)
$Ba(OH)_2 \cdot 8H_2O + 2e^- \rightarrow Ba + 2OH^- + 8H_2O$	−2.97
$Sr(OH)_2 \cdot 8H_2O + 2e^- \rightarrow Sr + 2OH^- + 8H_2O$	−2.99
$Ca(OH)_2 + 2e^- \rightarrow Ca + 2OH^-$	−3.02
$Ca^+ + e^- \rightarrow Ca$	−3.02
$Li^+ + e^- \rightarrow Li$	−3.045
	(−3.02)
$3N_2 + 2H^+ + 2e^- \rightarrow 2HN_3$	−3.1
$UO_2^{+2} + 4H^+ + 6e^- \rightarrow U + 2H_2O$	−0.82
$2H_2O + 2e^- \rightarrow H_2 + 2OH^-$	−0.8277
$SiO_2 + 4H^+ + 4e^- \rightarrow Si + 2H_2O$	−0.84
$2NO_3^- + 2H_2O + 2e^- \rightarrow N_2O_4 + 4OH^-$	−0.85
$TiO_2 + 4H^+ + 4e^- \rightarrow Ti + 2H_2O$	−0.86
$P + 3H_2O + 3e^- \rightarrow PH_3(g) + 3OH^-$	−0.87
$SO_4^{-2} + H_2O + 2e^- \rightarrow SO_3^{-2} + 2OH^-$	−0.92
$Te + 2e^- \rightarrow Te^{-2}$	−0.92
$Sn(OH)_6^{-2} + 2e^- \rightarrow HSnO_2^- + 3OH^- + H_2O$	−0.96
$Mn^{+2} + 2e^- \rightarrow Mn$	−1.029
$PO_4^{-3} + 2H_2O + 2e^- \rightarrow HPO_3^{-2} + 3OH^-$	−1.05
$2SO_3^{-2} + 2H_2O + 2e^- \rightarrow S_2O_4^{-2} + 4OH^-$	−1.12
$CrO_2^- + 2H_2O + 3e^- \rightarrow Cr + 4OH^-$	−1.2
$SiF_6^{-2} + 4e^- \rightarrow Si + 6F^-$	−1.2
$V^{+2} + 2e^- \rightarrow V$	−1.2
$ZnO_2^- + 2H_2O + 2e^- \rightarrow Zn + 4OH^-$	−1.216
$H_2GaO_3^- + H_2O + 3e^- \rightarrow Ga + 4OH^-$	−1.22
$H_2BO_3^- + 5H_2O + 8e^- \rightarrow BH_4^- + 8OH^-$	−1.24
$Cr(OH)_3 + 3e^- \rightarrow Cr + 3OH^-$	−1.3
$ZrO_2 + 4H^+ + 4e^- \rightarrow Zr + 2H_2O$	−1.43
$Ce^{+3} + 3e^- \rightarrow Ce(Hg)$	−1.4373
$Mn(OH)_2 \rightarrow Mn + 2OH^-$	−1.47
$Ba^{+2} + 2e^- \rightarrow Ba(Hg)$	−1.570
$HfO_2 + 4H^+ + 4e^- \rightarrow Hf + 2H_2O$	−1.57
$Ti^{+2} + 2e^- \rightarrow Ti$	−1.63
$HPO_3^{-2} + 2H_2O + 2e^- \rightarrow H_2PO_2^- + 3OH^-$	−1.65
$HfO^{+2} + 2H^+ + 4e^- \rightarrow Hf + H_2O$	−1.68
$Be^{+2} + 2e^- \rightarrow Be$	−1.70
	(−1.85)
$Al^{+3} + 3e^- \rightarrow Al$(0.1f NaOH)	−1.706
$HPO_3^{-2} + 2H_2O + 3e^- \rightarrow P + 5OH^-$	−1.71
$SiO_3^{-2} + 3H_2O + 4e^- \rightarrow Si + 6OH^-$	−1.73
$Sr^{+2} + 2e^- \rightarrow Sr(Hg)$	−1.793
$ThO_2 + 4H^+ + 4e^- \rightarrow Th + 2H_2O$	−1.80
$U^{+3} + 3e^- \rightarrow U$	−1.8
$H_2PO_2^- + e^- \rightarrow P + 2OH^-$	−1.82

Note: Values listed are Standard Reduction Potentials.

Source: *Handbook of Chemistry and Physics*, 53rd ed., CRC Press, Cleveland, 1972, p. D-113. Compiled by J.F. Hunsberger.

A–4 DISSOCIATION CONSTANTS OF ORGANIC BASES IN AQUEOUS SOLUTION

No.	Compound	Temp. °C	Step	pK_a	K_a
a1	Acetamide	25		0.63	2.34×10^{-1}
a2	Acridine	20		5.58	2.63×10^{-6}
a3	α-Alanine	25		2.345	4.52×10^{-3}
a4	Alanine, glycyl-	25		3.153	7.03×10^{-4}
a5	Alanine, methoxy-(DL)	25		2.037	9.18×10^{-3}
a6	Alanine, phenyl	25	2	9.18	6.61×10^{-10}
a7	Allothreonine	25	1	2.108	7.80×10^{-3}
	Allothreonine	25	2	9.096	8.02×10^{-10}
a8	*n*-Amylamine	25		10.63	2.34×10^{-11}
a9	Aniline	25		4.63	2.34×10^{-5}
a10	Aniline, *n*-allyl	25		4.17	6.76×10^{-5}
a11	Aniline, 4-(p-aminobenzoyl)	25	1	2.932	1.17×10^{-3}
a12	Aniline, 4-benzyl	25		2.17	6.76×10^{-3}
a13	Aniline, 2-bromo	25		2.53	2.95×10^{-3}
a14	Aniline, 3-bromo	25		3.58	2.63×10^{-4}
a15	Aniline, 4-bromo	25		3.86	1.38×10^{-4}
a16	Aniline, 4-bromo-N,N-dimethyl	25		4.232	5.86×10^{-5}
a17	Aniline, o-chloro	25		2.65	2.24×10^{-3}
a18	Aniline, m-chloro	25		3.46	3.47×10^{-4}
a19	Aniline, p-chloro	25		4.15	7.08×10^{-5}
a20	Aniline, 3-chloro-N,N-dimethyl	20		3.837	1.46×10^{-4}
a21	Aniline, 4-chloro-N,N-dimethyl	20		4.395	4.03×10^{-5}
a22	Aniline, 3,5-dibromo	25		2.34	4.57×10^{-3}
a23	Aniline, 2,4-dichloro	22		2.05	8.91×10^{-3}
a24	Aniline, N,N-diethyl	22		6.61	2.46×10^{-7}
a25	Aniline, N,N-dimethyl	25		5.15	7.08×10^{-6}
a26	Aniline, N,N-dimethyl-3-nitro	25		2.626	2.37×10^{-3}
a27	Aniline, N-ethyl	24		5.12	7.59×10^{-6}
a28	Aniline, 2-fluoro	25		3.20	6.31×10^{-4}
a29	Aniline, 3-fluoro	25		3.50	3.16×10^{-4}
a30	Aniline, 4-fluoro	25		4.65	2.24×10^{-5}
a31	Aniline, 2-iodo	25		2.60	2.51×10^{-3}
a32	Aniline, N-methyl	25		4.848	1.41×10^{-5}
a33	Aniline, 4-methylthio	25		4.35	4.46×10^{-5}
a34	Aniline, 3-nitro	25		2.466	3.42×10^{-3}
a35	Aniline, 4-nitro	25		1.0	1.00×10^{-1}
a36	Aniline, 2-sulfonic acid	25	2	2.459	3.47×10^{-3}
a37	Aniline, 3-sulfonic acid	25	2	3.738	1.82×10^{-4}
a38	Aniline, 4-sulfonic acid	25	2	3.227	5.92×10^{-4}
a39	*o*-Anisidine	25		4.52	3.02×10^{-5}
a40	*m*-Anisidine	25		4.23	5.89×10^{-5}
a41	*p*-Anisidine	25		5.34	4.57×10^{-6}
a42	Arginine	25	1	1.8217	1.51×10^{-2}
		25	2	8.9936	1.01×10^{-9}
a43	Asparagine	20	1	2.213	6.12×10^{-3}
		20	2	8.85	1.41×10^{-9}
a44	Asparagine, glycyl	25	1	2.942	1.14×10^{-3}
		18	2	8.44	3.63×10^{-9}
a45	DL-Aspartic acid	1	1	2.122	7.55×10^{-3}
		1	2	4.006	1.00×10^{-4}
a46	Azetidine (Trimethylimidine)	25		11.29	5.12×10^{-12}
a47	Aziridine	25		8.01	9.77×10^{-9}
b1	Benzene, 4-aminoazo	25		2.82	1.51×10^{-3}
b2	Benzene, 2-aminoethyl (β-Phenylamine)	25		9.84	1.45×10^{-10}
b3	Benzene, 4-dimethylaminoazo	25		3.226	5.94×10^{-4}
b4	Benzidine	30	1	4.66	2.19×10^{-5}
		30	2	3.57	2.69×10^{-4}
b5	Benzimidazole	25		5.532	2.94×10^{-6}
b6	Benzimidazole, 2-ethyl	25		6.18	6.61×10^{-7}
b7	Benzimidazole, 2-methyl	25		6.19	6.46×10^{-7}
b8	Benzimidazole, 2-phenyl	25	1	5.23	5.89×10^{-6}
		25	2	11.91	1.23×10^{-12}
b9	Benzoic acid, 2-amino (Anthranilic acid)	25	1	2.108	7.80×10^{-3}
		25	2	4.946	1.13×10^{-5}
b10	Benzoic acid, 4-amino	25	1	2.501	3.15×10^{-3}
		25	2	4.874	1.33×10^{-5}
b11	Benzylamine	25		9.33	4.67×10^{-10}
b12	Betaine	0		1.83	1.48×10^{-2}
b13	Biphenyl, 2-amino	22		3.82	1.51×10^{-4}
b14	Bornylamine(trans-)	25		10.17	6.76×10^{-11}
b15	Brucine	25	1	8.28	5.24×10^{-9}
b16	Butane, 1-amino-3-methyl	25		10.60	2.51×10^{-11}
b17	Butane, 2-amino-2-methyl	19		10.85	1.41×10^{-11}
b18	Butane, 1,4-diamino (Putrescine)	10	1	11.15	7.08×10^{-12}
		10	2	9.71	1.95×10^{-10}
b19	*n*-Butylamine	20		10.77	1.69×10^{-11}
b20	*t*-Butylamine	18		10.83	1.48×10^{-11}
b21	Butyric acid, 4-amino	25	1	4.0312	9.31×10^{-5}
		25	2	10.5557	2.78×10^{-11}
b22	*n*-Butyric acid, glycyl-2-amino	25	1	3.1546	7.01×10^{-4}
c1	Cacodylic acid	25	1	1.57	2.69×10^{-2}
			2	6.27	5.37×10^{-7}
c2	β-Chlortriethyl-ammonium	25		8.80	1.59×10^{-9}
c3	Cinnoline	20		2.37	4.27×10^{-3}
c4	Codeine	25		8.21	6.15×10^{-9}
c5	Cyclohexaneamine, *n*-butyl	25		11.23	5.89×10^{-12}
c6	Cyclohexylamine	24		10.66	2.19×10^{-11}
c7	Cystine	30	1	1.90	1.25×10^{-2}
		30	2	8.24	5.76×10^{-9}
d1	*n*-Decylamine	25		10.64	2.29×10^{-11}
d2	Diethylamine	40		10.489	3.24×10^{-11}
d3	Diisobutylamine	21		10.91	1.23×10^{-11}
d4	Diisopropylamine	28.5		10.96	1.09×10^{-11}
d5	Dimethylamine	25		10.732	1.85×10^{-11}
d6	*n*-Diphenylamine	25		0.79	1.62×10^{-1}
d7	*n*-Dodecaneamine (Laurylamine)	25		10.63	2.35×10^{-11}
e1	*d*-Ephedrine	10		10.139	7.26×10^{-11}
e2	*l*-Ephedrine	10		9.958	1.10×10^{-10}
e3	Ethane, 1-amino-3-methoxy	10		9.89	1.29×10^{-10}
e4	Ethane, 1,2-bismethylamino	25	1	10.40	3.98×10^{-11}
		25	2	8.26	5.50×10^{-9}
e5	Ethanol, 2-amino	25		9.50	3.16×10^{-10}
e6	Ethylamine	20		10.807	1.56×10^{-11}
e7	Ethylenediamine	0	1	10.712	1.94×10^{-11}
		0	2	7.564	2.73×10^{-8}
g1	*l*-Glutamic acid	25	1	2.13	7.41×10^{-3}
		25	2	4.31	4.90×10^{-5}
e2	Glutamic acid, α-monoethyl	25	1	3.846	1.42×10^{-4}
		25	2	7.838	1.45×10^{-8}
e3	*l*-Glutamine	—		9.28	5.25×10^{-10}
e4	*l*-Glutathione	25	2	3.59	2.57×10^{-4}
e5	Glycine	25	1	2.3503	4.46×10^{-3}
		25	2	9.7796	1.68×10^{-10}
e6	Glycine, *n*-acetyl	25		3.6698	2.14×10^{-4}
e7	Glycine, dimethyl	5		10.3371	4.60×10^{-11}
e8	Glycine, glycyl	25		3.1397	7.25×10^{-4}
e9	Glycine, glycylglycyl	25	1	3.225	5.96×10^{-4}
		25	2	8.090	8.13×10^{-9}
e10	Glycine, leucyl	25	1	3.25	5.62×10^{-4}
		25	2	8.28	5.25×10^{-9}
e11	Glycine, methyl (Sarcosine)	25	1	2.21	6.16×10^{-3}
		25	2	10.12	7.58×10^{-11}
g12	Glycine, phenyl	25	1	1.83	1.48×10^{-2}
		25	2	4.39	4.07×10^{-5}
g13	Glycine, N,*n*-propyl	25	1	2.35	4.46×10^{-3}
		25	2	10.19	6.46×10^{-11}
g14	Glycine, tetraglycyl	20	1	3.10	7.94×10^{-4}
		20	2	8.02	9.55×10^{-9}
g15	Glycylserine	25	1	2.9808	1.04×10^{-3}
		25	2	8.38	4.17×10^{-9}
h1	Hexadecaneamine	25		10.63	2.35×10^{-11}
h2	Heptane, 1-amino	25		10.66	2.19×10^{-11}
h3	Heptane, 2-amino	19		10.88	1.58×10^{-11}
h4	Heptane, 2-methylamino	17		10.99	1.02×10^{-11}
h5	Hexadecaneamine	25		10.61	2.46×10^{-11}

A–4 DISSOCIATION CONSTANTS OF ORGANIC BASES IN AQUEOUS SOLUTION (*Continued*)

No.	Compound	Temp. °C	Step	pK_a	K_a
h6	Hexamethylenediamine	0	1	111.857	1.39×10^{-12}
		0	2	0.762	1.73×10^{-11}
h7	Hexanoic acid, 6-amino.	25	1	4.373	4.23×10^{-5}
		25	2	10.804	1.57×10^{-11}
h8	*n*-Hexylamine	25		10.56	2.75×10^{-11}
h9	*dl*-Histidine	25	1	1.80	1.58×10^{-2}
		25	2	6.04	9.12×10^{-7}
		25	3	9.33	4.67×10^{-10}
h10	Histidine, β-alanyl (Carnosine)	20	1	2.73	1.86×10^{-3}
		20	2	6.87	1.35×10^{-7}
		20	3	9.73	1.48×10^{-10}
i1	Imidazol	25		6.953	1.11×10^{-7}
i2	Imidazol, 2,4-dimethyl	25		8.359	5.50×10^{-9}
i3	Imidazol, 1-methyl (Oxalmethyline)	25		6.95	1.12×10^{-7}
i4	Indane, 1-amino (*d*-1-Hydrindamine	22.5		9.21	6.17×10^{-10}
i5	Isobutyric acid, 2-amino	25	1	2.357	4.30×10^{-3}
		25	2	10.205	6.23×10^{-11}
i6	Isoleucine	25	1	2.318	4.81×10^{-3}
		25	2	9.758	1.74×10^{-10}
i7	Isoquinoline (Leucoline)	20		5.42	3.80×10^{-6}
i8	Isoquinoline, 1-amino	20		7.59	2.57×10^{-8}
i9	Isoquinoline, 7-hydroxy	20	1	5.68	2.09×10^{-6}
		20	2	8.90	1.26×10^{-9}
l1	*L*-Leucine	25	1	2.328	4.70×10^{-3}
		25	2	9.744	1.80×10^{-10}
l2	Leucine, glycyl	25		3.18	6.61×10^{-4}
m1	Methionine	25	1	2.22	6.02×10^{-3}
		25	2	9.27	5.37×10^{-10}
m2	Methylamine	25		10.657	2.70×10^{-11}
m3	Morphine	25		8.21	6.16×10^{-9}
m4	Morpholine	25		8.33	4.67×10^{-9}
n1	Naphthalene, 1-amino-6-hydroxy	25		3.97	1.07×10^{-4}
n2	Naphthalene, dimethylamino	25		4.566	2.72×10^{-5}
n3	α-Naphthylamine	25		3.92	1.20×10^{-4}
n4	β-Naphthylamine	25		4.16	6.92×10^{-5}
n5	α-Naphthylamine, *n*-methyl	27		3.67	2.13×10^{-4}
n6	Neobornylamine(cis-)	25		10.01	9.77×10^{-11}
n7	Nicotine	25	1	8.02	9.55×10^{-9}
		25	2	3.12	7.59×10^{-4}
n8	*n*-Nonylamine	25		10.64	2.29×10^{-11}
n9	Norleucine	25		2.335	4.62×10^{-3}
o1	Octadecaneamine	25		10.60	2.51×10^{-11}
o2	Octylamine	25		10.65	2.24×10^{-11}
o3	Ornithine	25	1	1.705	1.97×10^{-2}
		25	2	8.690	2.04×10^{-9}
p1	Papaverine	25		6.40	3.98×10^{-7}
p2	Pentane, 3-amino	17		10.59	2.57×10^{-11}
p3	Pentane, 3-amino-3-methyl	16		11.01	9.77×10^{-12}
p4	*n*-Pentadecylamine	25		10.61	2.46×10^{-11}
p5	Pentanoic acid, 5-amino(Valeric acid)	25	1	4.270	5.37×10^{-5}
		25	2	10.766	1.71×10^{-11}
p6	Perimidine	20		6.35	4.47×10^{-7}
p7	Phenanthridine	20		5.58	2.63×10^{-6}
p8	1,10-Phenanthroline	25		4.84	1.44×10^{-5}
p9	*o*-Phenetidine (2-Ethoxyaniline)	28		4.43	3.72×10^{-5}
p10	*m*-Phenetidine (3-Ethoxyaniline)	25		4.18	6.60×10^{-5}
p11	*p*-Phenetidine (4-Ethoxyaniline)	28		5.20	6.31×10^{-6}
p12	α-Picoline	20		5.97	1.07×10^{-6}
p13	β-Picoline	20		5.68	2.09×10^{-6}
p14	γ-Picoline	20		6.02	9.55×10^{-7}
p15	Pilocarpine	30		6.87	1.35×10^{-7}
p16	Piperazine	23.5	1	9.83	1.48×10^{-10}
		23.5	2	5.56	2.76×10^{-6}
p17	Piperazine, 2,5-dimethyl(trans-)	25	1	9.66	2.19×10^{-10}
		25	2	5.20	6.31×10^{-6}
p18	Piperidine	25		11.123	7.53×10^{-12}
p19	Piperidine, 3-acetyl	25		3.18	6.61×10^{-4}
p20	Piperidine, 1-*n*-butyl	23		10.47	3.39×10^{-11}
p21	Piperidine, 1,2-dimethyl	25		10.22	6.03×10^{-11}
p22	Piperidine, 1-ethyl	23		10.45	3.55×10^{-11}
p23	Piperidine, 1-methyl	25		10.08	8.32×10^{-11}
p24	Piperidine, 2,2,6,6-tetramethyl	25		11.07	8.51×10^{-12}
p25	Piperidine, 2,2,4-trimethyl	30		11.04	9.12×10^{-12}
p26	Proline	25	1	1.952	1.11×10^{-2}
		25	2	10.640	2.29×10^{-11}
p27	Proline, hydroxy	25	1	1.818	1.52×10^{-2}
		25	2	9.662	2.18×10^{-10}
p28	Propane, 1-amino-2,2-dimethyl	25		10.15	7.08×10^{-11}
p29	Propane, 1,2-diamino	25	1	9.82	1.52×10^{-10}
		25	2	6.61	2.46×10^{-7}
p30	Propane, 1,3-diamino	10	1	10.94	1.15×10^{-11}
		10	2	9.03	9.33×10^{-10}
p31	Propane, 1,2,3-triamino	20	1	9.59	2.57×10^{-10}
		20	2	7.95	1.12×10^{-8}
p32	Propanoic acid, 3-amino (β-Alanine)	25	1	3.551	2.81×10^{-4}
		25	2	10.238	5.78×10^{-11}
p33	Propylamine	20		10.708	1.96×10^{-11}
p34	Pteridine	20		4.05	8.91×10^{-5}
p35	Pteridine, 2-amino-4, 6-dihydroxy	20	2	6.59	2.57×10^{-7}
		20	3	9.31	4.90×10^{-10}
p36	Pteridine, 2-amino-4-hydroxy	20	1	2.27	5.37×10^{-3}
		20	2	7.96	1.10×10^{-8}
p37	Pteridine, 6-chloro	20		3.68	2.09×10^{-4}
p38	Pteridine, 6-hydroxy-4-methyl	20	1	4.08	8.32×10^{-5}
			2	6.41	3.89×10^{-7}
p39	Purine	20	1	2.30	5.01×10^{-3}
		20	2	8.96	1.10×10^{-9}
p40	Purine, 6-amino (Adenine)	25	1	4.12	7.59×10^{-5}
		25	2	9.83	1.48×10^{-10}
p41	Purine, 2-dimethylamino	20	1	4.00	1.00×10^{-4}
		20	2	10.24	5.75×10^{-11}
p42	Purine, 8-hydroxy	20	1	2.56	2.75×10^{-3}
		20	2	8.26	9.49×10^{-9}
p43	Pyrazine	27		0.65	2.24×10^{-1}
p44	Pyrazine, 2-methyl	27		1.45	3.54×10^{-2}
p45	Pyrazine, methylamino.	25		3.39	4.07×10^{-4}
p46	Pyridazine	20		2.24	5.76×10^{-3}
p47	Pyrimidine, 2-amino	20		3.45	3.54×10^{-4}
p48	Pyrimidine, 2-amino-4, 6-dimethyl	20		4.82	1.51×10^{-5}
p49	Pyrimidine, 2-amino-5-nitro	20		0.35	4.46×10^{-1}
p50	Pyridine	25		5.25	5.62×10^{-6}
p51	Pyridine, 2-aldoxime	20	1	3.59	2.57×10^{-4}
		20	2	10.18	6.61×10^{-11}
p52	Pyridine, 2-amino	20		6.82	1.51×10^{-7}
p53	Pyridine, 4-amino	25		9.1141	7.69×10^{-10}
p54	Pyridine, 2-benzyl	25		5.13	7.41×10^{-6}
p55	Pyridine, 3-bromo	25		2.84	1.45×10^{-3}
p56	Pyridine, 3-chloro	25		2.84	1.45×10^{-3}
p57	Pyridine, 2,5-diamino	20		6.48	3.31×10^{-7}
p58	Pyridine, 2,3-dimethyl (2,3-Lutidine)	25		6.57	2.69×10^{-7}
p59	Pyridine, 2,4-dimethyl (2,4-Lutidine)	25		6.99	1.02×10^{-7}
p60	Pyridine, 3,5-dimethyl (3,5-Lutidine)	25		6.15	7.08×10^{-7}
p61	Pyridine, 2-ethyl	25		5.89	1.28×10^{-6}
p62	Pyridine, 2-formyl	20		3.80	1.59×10^{-4}
p63	Pyridine, 2-hydroxy (2-Pyridol)	20	1	0.75	9.82×10^{-1}
		20	2	11.65	2.24×10^{-12}

A–4 DISSOCIATION CONSTANTS OF ORGANIC BASES IN AQUEOUS SOLUTION (*Continued*)

No.	Compound	Temp. °C	Step	pK_a	K_a
p64	Pyridine, 4-hydroxy	20	1	3.20	6.31×10^{-4}
		20	2	11.12	7.59×10^{-12}
p65	Pyridine, methoxy	25		6.47	3.30×10^{-7}
p66	Pyridine, 4-methylamino	20		9.65	2.24×10^{-10}
p67	Pyridine, 2,4,6-trimethyl	25		7.43	3.72×10^{-8}
p68	Pyrrolidine	25		11.27	5.37×10^{-12}
p69	Pyrrolidine, 1,2-dimethyl	26		10.20	6.31×10^{-11}
p70	Pyrrolidine, *n*-methyl	25		10.32	4.79×10^{-11}
q1	Quinazoline	20		3.43	3.72×10^{-4}
q2	Quinazoline, 5-hydroxy	20	1	3.62	2.40×10^{-4}
		20	2	7.41	3.89×10^{-8}
q3	Quinine	25	1	8.52	3.02×10^{-9}
		25	2	4.13	7.41×10^{-5}
q4	Quinoline	20		4.90	1.25×10^{-5}
q5	Quinoline, 3-amino	20		4.91	1.23×10^{-5}
q6	Quinoline, 3-bromo	25		2.69	2.04×10^{-3}
q7	Quinoline, 8-carboxy	25		1.82	1.51×10^{-2}
q8	Quinoline, 3-hydroxy (3-Quinolinol)	20	1	4.28	5.25×10^{-5}
		20	2	8.08	8.32×10^{-9}
q9	Quinoline, 8-hydroxy (8-Quinolinol)	20	1	5.017	1.21×10^{-6}
		25	2	9.812	1.54×10^{-10}
q10	Quinoline, 8-hydroxy-5-sulfo	25	1	4.112	7.73×10^{-5}
		25	2	8.757	1.75×10^{-9}
q11	Quinoline, 6-methoxy	20		5.03	9.33×10^{-6}
q12	Quinoline, 2-methyl (Quinaldine)	20		5.83	1.48×10^{-6}
q13	Quinoline, 4-methyl (Lepidine)	20		5.67	2.14×10^{-6}
q14	Quinoline, 5-methyl	20		5.20	6.31×10^{-6}
q15	Quinoxaline (Quinazine)	20		0.56	3.63×10^{-1}
s1	Serine (2-amino-3-hydroxypropanoic acid)	25	1	2.186	5.49×10^{-3}
		25	2	9.208	6.19×10^{-10}
s2	Strychnine	25		8.26	5.49×10^{-9}
t1	Taurine (2-Aminoethane sulfonic acid)	25	2	9.0614	8.69×10^{-10}
t2	Tetradecaneamine (Myristilamine)	25		10.62	2.40×10^{-11}
t3	Thiazole	20		2.44	3.63×10^{-3}
t4	Thiazole, 2-amino	20		5.36	4.36×10^{-5}
t5	Threonine	25	1	2.088	8.16×10^{-3}
		25	2	9.10	7.94×10^{-10}
t6	*o*-Toluidine	25		4.44	3.63×10^{-5}
t7	*m*-Toluidine	25		4.73	1.86×10^{-5}
t8	*p*-Toluidine	25		5.08	8.32×10^{-6}
t9	1,3,5-Triazine, 2,4,6-triamino	25		5.00	1.00×10^{-5}
t10	Tridecaneamine	25		10.63	2.35×10^{-11}
t11	Triethylamine	18		11.01	9.77×10^{-12}
t12	Trimethylamine	25		9.81	1.55×10^{-10}
t13	Tryptophan	25	1	2.43	3.72×10^{-3}
		25	2	9.44	3.63×10^{-10}
t14	Tyrosine	25	2	9.11	7.76×10^{-10}
		25	3	10.13	7.41×10^{-11}
t15	Tyrosineamide	25		7.33	4.68×10^{-8}
u1	Urea	21		0.10	7.94×10^{-1}
v1	Valine	25	1	2.286	5.17×10^{-3}
		25	2	9.719	1.91×10^{-10}

Source: *Handbook of Chemistry and Physics,* 53rd ed., CRC Press, Cleveland, 1972, p. D-117.

A–5 DISSOCIATION CONSTANTS OF INORGANIC BASES IN AQUEOUS SOLUTIONS

Approximately 0.1–0.01 N

Compound	T°C	Step	K_b	pK_b
Ammonium hydroxide	25		1.79×10^{-5}	4.75
Arsenous oxide	25		1.1×10^{-4}	3.96
Beryllium hydroxide	25	2	5×10^{-11}	10.30
Calcium hydroxide	25	1	3.74×10^{-3}	2.43
Calcium hydroxide	30	2	4.0×10^{-2}	1.40
Deuteroammonium hydroxide	25		1.1×10^{-5}	4.96
Hydrazine	20		1.7×10^{-6}	5.77
Hydroxylamine	20		1.07×10^{-8}	7.97
Lead Hydroxide	25		9.6×10^{-4}	3.02
Silver Hydroxide	25		1.1×10^{-4}	3.96
Zinc Hydroxide	25		9.6×10^{-4}	3.02

Source: *Handbook of Chemistry and Physics,* 53rd ed., CRC Press, Cleveland, 1972, p. D-119.

A–6 DISSOCIATION CONSTANTS OF ORGANIC ACIDS IN AQUEOUS SOLUTIONS

Compound	T°C	Step	K	pK
Acetic	25		1.76×10^{-5}	4.75
Acetoacetic	18		2.62×10^{-4}	3.58
Acrylic	25		5.6×10^{-5}	4.25
Adipamic	25		2.35×10^{-5}	4.63
Adipic	25	1	3.71×10^{-5}	4.43
Adipic	25	2	3.87×10^{-5}	4.41
d-Alanine	25		1.35×10^{-10}	9.87
Allantoin	25		1.10×10^{-9}	8.96
Alloxanic	25		2.3×10^{-7}	6.64
α-Aminoacetic (glycine)	25		1.67×10^{-10}	9.78
o-Aminobenzoic	25		1.07×10^{-7}	6.97
m-Aminobenzoic	25		1.67×10^{-5}	4.78
p-Aminobenzoic	25		1.2×10^{-5}	4.92
o-Aminobenzosulfonic	25		3.3×10^{-3}	2.48
m-Aminobenzosulfonic	25		1.85×10^{-4}	3.73
p-Aminobenzosulfonic	25		5.81×10^{-4}	3.24
Anisic	25		3.38×10^{-5}	4.47
o-*β*-Anisylpropionic	25		1.59×10^{-5}	4.80
m-*β*-Anisylpropionic	25		2.24×10^{-5}	4.65
p-*β*-Anisylpropionic	25		2.04×10^{-5}	4.69
Ascorbic	24	1	7.94×10^{-5}	4.10
Ascorbic	16	2	1.62×10^{-12}	11.79
DL-Aspartic	25	1	1.38×10^{-4}	3.86
DL-Aspartic	25	2	1.51×10^{-10}	9.82
Barbituric	25		9.8×10^{-5}	4.01
Benzoic	25		6.46×10^{-5}	4.19
Benzosulfonic	25		2×10^{-1}	0.70
Bromoacetic	25		2.05×10^{-3}	2.69
o-Bromobenzoic	25		1.45×10^{-3}	2.84
m-Bromobenzoic	25		1.37×10^{-4}	3.86
n-Butyric	20		1.54×10^{-5}	4.81
iso-Butyric	18		1.44×10^{-5}	4.84
Cacodylic	25		6.4×10^{-7}	6.19
n-Caproic	18		1.43×10^{-5}	4.83
iso-Caproic	18		1.46×10^{-5}	4.84
Chloroacetic	25		1.40×10^{-3}	2.85
o-Chlorobenzoic	25		1.20×10^{-3}	2.92
m-Chlorobenzoic	25		1.51×10^{-4}	3.82
p-Chlorobenzoic	25		1.04×10^{-4}	3.98
α-Chlorobutyric	R.T.		1.39×10^{-3}	2.86
β-Chlorobutyric	R.T.		8.9×10^{-5}	4.05
γ-Chlorobutyric	R.T.		3.0×10^{-5}	4.52
o-Chlorocinnamic	25		5.89×10^{-5}	4.23
m-Chlorocinnamic	25		5.13×10^{-5}	4.29
p-Chlorocinnamic	25		3.89×10^{-5}	4.41
o-Chlorophenoxyacetic	25		8.91×10^{-4}	3.05
m-Chlorophenoxyacetic	25		8.51×10^{-4}	3.07
p-Chlorophenoxyacetic	25		7.94×10^{-4}	3.10
o-Chlorophenylacetic	25		1.18×10^{-5}	4.07
m-Chlorophenylacetic	25		7.25×10^{-5}	4.14
p-Chlorophenylacetic	25		6.46×10^{-5}	4.19
β-(*o*-Chlorophenyl) propionic	25		2.63×10^{-4}	4.58
β-(*m*-Chlorophenyl) propionic	25		2.57×10^{-5}	4.59
β-(*p*-Chlorophenyl) propionic	25		2.46×10^{-5}	4.61
α-Chloropropionic	25		1.47×10^{-3}	2.83
β-Chloropropionic	25		1.04×10^{-4}	3.98
cis-Cinnamic	25		1.3×10^{-4}	3.89
trans-Cinnamic	25		3.65×10^{-5}	4.44
Citric	18	1	7.10×10^{-4}	3.14
Citric	18	2	1.68×10^{-5}	4.77
Citric	18	3	8.4×10^{-6}	6.39
o-Cresol	25		6.3×10^{-11}	10.20
m-Cresol	25		9.8×10^{-11}	10.01
p-Cresol	25		6.7×10^{-11}	10.17
Crotonic (*trans*-)	25		2.03×10^{-5}	4.69
Cyanoacetic	25		3.65×10^{-3}	2.45
γ-Cyanobutyric	25		3.80×10^{-3}	2.42
o-Cyanophenoxyacetic	25		1.05×10^{-3}	2.98
m-Cyanophenoxyacetic	25		9.33×10^{-4}	3.03
p-Cyanophenoxyacetic	25		1.18×10^{-3}	2.93
Cyanopropionic	25		3.6×10^{-3}	2.44
Cyclohexane-1 : 1-dicarboxylic	25	1	3.55×10^{-4}	3.45
Cyclohexane-1 : 1-dicarboxylic	25	2	7.76×10^{-7}	6.11
Cyclopropane-1 : 1-dicarboxylic	25	1	1.51×10^{-2}	1.82
Cyclopropane-1 : 1-dicarboxylic	25	2	3.72×10^{-8}	7.43
DL-Cysteine	30	1	7.25×10^{-9}	8.14
DL-Cysteine	30	2	4.6×10^{-11}	10.34
L-Cystine	25	1	1.4×10^{-8}	7.85
L-Cystine	25	2	1.4×10^{-10}	9.85
Deuteroacetic (in D_2O)	25		5.5×10^{-6}	5.25
Dichloroacetic	25		3.32×10^{-2}	1.48
Dichloroacetylacetic	?		7.8×10^{-3}	2.11
Dichlorophenol (2,3-)	25		3.6×10^{-8}	7.44
Dihydroxybenzoic (2,2-)	25		1.14×10^{-3}	2.94
Dihydroxybenzoic (2,5-)	25		1.08×10^{-3}	2.97
Dihydroxybenzoic (3,4-)	25		3.3×10^{-5}	4.48
Dihydroxybenzoic (3,5-)	25		9.1×10^{-5}	4.04
Dihydroxymalic	25		1.2×10^{-2}	1.92
Dihydroxytartaric	25		1.2×10^{-2}	1.92
Dimethylglycine	25		1.3×10^{-10}	9.89
Dimethylmalic	25	1	6.83×10^{-4}	3.17
Dimethylmalic	25	2	8.72×10^{-7}	6.06
Dimethylmalonic	25		7.08×10^{-4}	3.15
Dinicotinic	25		1.6×10^{-3}	2.80
Dinitrophenol (2,4-)	15		1.1×10^{-4}	3.96
Dinitrophenol (3,6-)	15		7.1×10^{-6}	5.15
Diphenylacetic	25		1.15×10^{-4}	3.94
Ethylbenzoic	25		4.47×10^{-5}	4.35
Ethylphenylacetic	25		4.27×10^{-5}	4.37
Fluorobenzoic	17		1.25×10^{-3}	2.90
Formic	20		1.77×10^{-4}	3.75
Fumaric (*trans*-)	18	1	9.30×10^{-4}	3.03
Fumaric (*trans*-)	18	2	3.62×10^{-5}	4.44
Furancarboxylic	25		7.1×10^{-4}	3.15
Furoic	25		6.76×10^{-4}	3.17
Gallic	25		3.9×10^{-5}	4.41
Glutaramic	25		3.98×10^{-5}	4.60
Glutaric	25	1	4.58×10^{-5}	4.31
Glutaric	25	2	3.89×10^{-6}	5.41

A–6 DISSOCIATION CONSTANTS OF ORGANIC ACIDS IN AQUEOUS SOLUTIONS (*Continued*)

Compound	T°C	Step	K	pK
Glycerol	25		7×10^{-15}	14.15
Glycine	25		1.67×10^{-10}	9.87
Glycol	25		6×10^{-15}	14.22
Glycolic	25		1.48×10^{-4}	3.83
Heptanoic	25		1.28×10^{-5}	4.89
Hexahydrobenzoic	25		1.26×10^{-5}	4.90
Hexanoic	25		1.31×10^{-5}	4.88
Hippuric	25		1.57×10^{-4}	3.80
Histidine	25		6.7×10^{-10}	9.17
Hydroquinone	20		4.5×10^{-11}	10.35
o-Hydroxybenzoic	19	1	1.07×10^{-3}	2.97
o-Hydroxybenzoic	18	2	4×10^{-14}	13.40
m-Hydroxybenzoic	19	1	8.7×10^{-5}	4.06
m-Hydroxybenzoic	19	2	1.2×10^{-10}	9.92
p-Hydroxybenzoic	19	1	3.3×10^{-5}	4.48
p-Hydroxybenzoic	19	2	4.8×10^{-10}	9.32
β-Hydroxybutyric	25		2×10^{-5}	4.70
γ-Hydroxybutyric	25		1.9×10^{-5}	4.72
β-Hydroxypropionic	25		3.1×10^{-5}	4.51
γ-Hydroxyquinoline	20		3.1×10^{-10}	9.51
Iodoacetic	25		7.5×10^{-4}	3.12
o-Iodobenzoic	25		1.4×10^{-3}	2.85
m-Iodobenzoic	25		1.6×10^{-4}	3.80
Itaconic	25	1	1.40×10^{-4}	3.85
Itaconic	25	2	3.56×10^{-6}	5.45
Lactic	100		8.4×10^{-4}	3.08
Lutidinic	25		7.0×10^{-3}	2.15
Lysine	25		2.95×10^{-11}	10.53
Maleic	25	1	1.42×10^{-2}	1.83
Maleic	25	2	8.57×10^{-7}	6.07
Malic	25	1	3.9×10^{-4}	3.40
Malic	25	2	7.8×10^{-6}	5.11
Malonic	25	1	1.49×10^{-3}	2.83
Malonic	25	2	2.03×10^{-6}	5.69
DL-Mandelic	25		1.4×10^{-4}	3.85
Mesaconic	25	1	8.22×10^{-4}	3.09
Mesaconic	25	2	1.78×10^{-5}	4.75
Mesitylenic	25		4.8×10^{-5}	4.32
Methyl-*o*-aminobenzoic	25		4.6×10^{-6}	5.34
Methyl-*m*-aminobenzoic	25		8×10^{-6}	5.10
Methyl-*p*-aminobenzoic	25		9.2×10^{-6}	5.04
o-Methylcinnamic	25		3.16×10^{-5}	4.50
m-Methylcinnamic	25		3.63×10^{-5}	4.44
p-Methylcinnamic	25		2.76×10^{-5}	4.56
β-Methylglutaric	25		5.75×10^{-5}	4.24
n-Methylglycine	18		1.2×10^{-10}	9.92
Methylmalonic	25		1.17×10^{-4}	3.07
Methylsuccinic	25	1	7.4×10^{-5}	4.13
Methylsuccinic	25	2	2.3×10^{-6}	5.64
o-Monochlorophenol	25		3.2×10^{-9}	8.49
m-Monochlorophenol	25		1.4×10^{-9}	8.85
p-Monochlorophenol	25		6.6×10^{-10}	9.18
Naphthalenesulfonic	25		2.7×10^{-1}	0.57
α-Naphthoic	25		2×10^{-4}	3.70
β-Naphthoic	25		6.8×10^{-5}	4.17
α-Naphthol	25		4.6×10^{-10}	9.34
β-Naphthol	25		3.1×10^{-10}	9.51
Nitrobenzene	0		1.05×10^{-4}	3.98
o-Nitrobenzoic	18		6.95×10^{-3}	2.16
m-Nitrobenzoic	25		3.4×10^{-4}	3.47
p-Nitrobenzoic	25		3.93×10^{-4}	3.41
o-Nitrophenol	25		6.8×10^{-8}	7.17
m-Nitrophenol	25		5.3×10^{-9}	8.28
p-Nitrophenol	25		7×10^{-8}	7.15
o-Nitrophenylacetic	25		1.00×10^{-5}	4.00
m-Nitrophenylacetic	25		1.07×10^{-4}	3.97
p-Nitrophenylacetic	25		1.41×10^{-4}	3.85
o-*β*-Nitrophenylpropionic	25		3.16×10^{-5}	4.50
p-*β*-Nitrophenylpropionic	25		3.39×10^{-5}	4.47
Nonanic	25		1.09×10^{-5}	4.96
Octanoic	25		1.28×10^{-5}	4.89
Oxalic	25	1	5.90×10^{-2}	1.23
Oxalic	25	2	6.40×10^{-5}	4.19
Phenol	20		1.28×10^{-10}	9.89
Phenylacetic	18		5.2×10^{-5}	4.28
o-Phenylbenzoic	25		3.47×10^{-4}	3.46
γ-Phenylbutyric	25		1.74×10^{-5}	4.76
α-Phenylpropionic	25		2.27×10^{-5}	4.64
β-Phenylpropionic	25		4.25×10^{-5}	4.37
o-Phthalic	25	1	1.3×10^{-3}	2.89
o-Phthalic	25	2	3.9×10^{-6}	5.51
m-Phthalic	25	1	2.9×10^{-4}	3.54
m-Phthalic	18	2	2.5×10^{-5}	4.60
p-Phthalic	25	1	3.1×10^{-4}	3.51
p-Phthalic	16	2	1.5×10^{-5}	4.82
Picric	25		4.2×10^{-1}	0.38
Pimelic	25		3.09×10^{-5}	4.71
Propionic	25		1.34×10^{-5}	4.87
iso-Propylbenzoic	25		3.98×10^{-5}	4.40
2-Pyridinecarboxylic	25		3×10^{-6}	5.52
3-Pyridinecarboxylic	25		1.4×10^{-5}	4.85
4-Pyridinecarboxylic	25		1.1×10^{-5}	4.96
Pyrocatechol	20		1.4×10^{-10}	9.85
Quinolinic	25		3×10^{-3}	2.52
Resorcinol	25		1.55×10^{-10}	9.81
Saccharin	18		2.1×10^{-12}	11.68
Suberic	25		2.99×10^{-5}	4.52
Succinic	25	1	6.89×10^{-5}	4.16
Succinic	25	2	2.47×10^{-6}	5.61
Sulfanilic	25		5.9×10^{-4}	3.23
α-Tartaric	25	1	1.04×10^{-3}	2.98
α-Tartaric	25	2	4.55×10^{-5}	4.34
meso-Tartaric	25	1	6×10^{-4}	3.22
meso-Tartaric	25	2	1.53×10^{-5}	4.82
Theobromine	18		1.3×10^{-8}	7.89
Terephthalic	25		3.1×10^{-4}	3.51
Thioacetic	25		4.7×10^{-4}	3.33
Thiophenecarboxylic	25		3.3×10^{-4}	3.48
o-Toluic	25		1.22×10^{-4}	3.91
m-Toluic	25		5.32×10^{-5}	4.27
p-Toluic	25		4.33×10^{-5}	4.36
Trichloroacetic	25		2×10^{-1}	0.70
Trichlorophenol	25		1×10^{-6}	6.00
Trihydroxybenzoic (2,4,6-)	25		2.1×10^{-2}	1.68
Trimethylacetic	18		9.4×10^{-6}	5.03
Trinitrophenol (2,4,6-)	25		4.2×10^{-1}	0.38
Tryptophan	25		4.2×10^{-10}	9.38
Tyrosine	17		3.98×10^{-9}	8.40
Uric	12		1.3×10^{-4}	3.89
n-Valeric	18		1.51×10^{-5}	4.82
iso-Valeric	25		1.7×10^{-5}	4.77
Veronal	25		3.7×10^{-8}	7.43
Vinylacetic	25		4.57×10^{-5}	4.34
Xanthine	40		1.24×10^{-10}	9.91

Source: *Handbook of Chemistry and Physics*, 53rd ed., CRC Press, Cleveland, 1972, p. D-120.

A–7 DISSOCIATION CONSTANTS OF INORGANIC ACIDS IN AQUEOUS SOLUTIONS

Approximately 0.1–0.01 N

Compound	T°C	Step	K	pK
Arsenic	18	1	5.62×10^{-3}	2.25
Arsenic	18	2	1.70×10^{-7}	6.77
Arsenic	18	3	3.95×10^{-12}	11.60
Arsenious	25		6×10^{-10}	9.23
o-Boric	20	1	7.3×10^{-10}	9.14
o-Boric	20	2	1.8×10^{-13}	12.74
o-Boric	20	3	1.6×10^{-14}	13.80
Carbonic	25	1	4.30×10^{-7}	6.37
Carbonic	25	2	5.61×10^{-11}	10.25
Chromic	25	1	1.8×10^{-1}	0.74
Chromic	25	2	3.20×10^{-7}	6.49
Germanic	25	1	2.6×10^{-9}	8.59
Germanic	25	2	1.9×10^{-13}	12.72
Hydrocyanic	25		4.93×10^{-10}	9.31
Hydrofluoric	25		3.53×10^{-4}	3.45
Hydrogen sulfide	18	1	9.1×10^{-8}	7.04
Hydrogen sulfide	18	2	1.1×10^{-12}	11.96
Hydrogen peroxide	25		2.4×10^{-12}	11.62
Hypobromous	25		2.06×10^{-9}	8.69
Hypochlorous	18		2.95×10^{-8}	7.53
Hypoiodous	25		2.3×10^{-11}	10.64
Iodic	25		1.69×10^{-1}	0.77
Nitrous	12.5		4.6×10^{-4}	3.37

Compound	T°C	Step	K	pK
Periodic	25		2.3×10^{-2}	1.64
o-Phosphoric	25	1	7.52×10^{-3}	2.12
o-Phosphoric	25	2	6.23×10^{-8}	7.21
o-Phosphoric	18	3	2.2×10^{-13}	12.67
Phosphorous	18	1	1.0×10^{-2}	2.00
Phosphorous	18	2	2.6×10^{-7}	6.59
Pyrophosphoric	18	1	1.4×10^{-1}	0.85
Pyrophosphoric	18	2	3.2×10^{-2}	1.49
Pyrophosphoric	18	3	1.7×10^{-6}	5.77
Pyrophosphoric	18	4	6×10^{-9}	8.22
Selenic	25	2	1.2×10^{-2}	1.92
Selenious	25	1	3.5×10^{-3}	2.46
Selenious	25	2	5×10^{-8}	7.31
m-Silicic	R.T.	1	2×10^{-10}	9.70
m-Silicic	R.T.	2	1×10^{-12}	12.00
o-Silicic	30	1	2.2×10^{-10}	9.66
o-Silicic	30	2	2×10^{-12}	11.70
o-Silicic	30	3	1×10^{-12}	12.00
o-Silicic	30	4	1×10^{-12}	12.00
Sulfuric	25	2	1.20×10^{-2}	1.92
Sulfurous	18	1	1.54×10^{-2}	1.81
Sulfurous	18	2	1.02×10^{-7}	6.91
Telluric	18	1	2.09×10^{-8}	7.68
Telluric	18	2	6.46×10^{-12}	11.29
Tellurous	25	1	3×10^{-3}	2.48
Tellurous	25	2	2×10^{-8}	7.70
Tetraboric	25	1	$\sim 10^{-4}$	~4.00
Tetraboric	25	2	$\sim 10^{-9}$	~9.00

Source: *Handbook of Chemistry and Physics,* 53rd ed., CRC Press, Cleveland, 1972, p. D-121.

A–8 DISSOCIATION CONSTANTS (K_b) OF AQUEOUS AMMONIA

Temperature °C.	pK_b	K_b
0	4.862	1.374×10^{-5}
5	4.830	1.479×10^{-5}
10	4.804	1.570×10^{-5}
15	4.782	1.652×10^{-5}
20	4.767	1.710×10^{-5}
25	4.751	1.774×10^{-5}
30	4.740	1.820×10^{-5}
35	4.733	1.849×10^{-5}
40	4.730	1.862×10^{-5}
45	4.726	1.879×10^{-5}
50	4.723	1.892×10^{-5}

Note: Values of K_b accurate to ±0.005; determined by e.m.f. method by: R.G. Bates and G.D. Pinching, *J. Am. Chem. Soc.*, 72, 1393, 1950.

Source: *Handbook of Chemistry and Physics,* 53rd ed., CRC Press, Cleveland, 1972, p. D-121.

A–9 IONIZATION CONSTANT FOR WATER (K_ω)

$-\log_{10} K_\omega$	Temperature °C.	$-\log_{10} K_\omega$	Temperature °C.
14.9435	0	13.8330	30
14.7338	5	13.6801	35
14.5346	10	13.5348	40
14.3463	15	13.3960	45
14.1669	20	13.2617	50
14.0000	24	13.1369	55
13.9965	25	13.0171	60

Source: *Handbook of Chemistry and Physics,* 53rd ed., CRC Press, Cleveland, 1972, p. D-122.

A–10 IONIZATION CONSTANTS OF ACIDS IN WATER AT VARIOUS TEMPERATURES

Acids		0°	5°	10°	15°	20°	25°	30°	35°	40°	45°	50°
Formic	$K_A \cdot 10^4$	1.638	1.691	1.728	1.749	1.765	1.772	1.768	1.747	1.716	1.685	1.650
Acetic	$K_A \cdot 10^5$	1.657	1.700	1.729	1.745	1.753	1.754	1.750	1.728	1.703	1.670	1.633
Propionic	$K_A \cdot 10^5$	1.274	1.305	1.326	1.336	1.338	1.336	1.326	1.310	1.280	1.257	1.229
n-Butyric	$K_A \cdot 10^5$	1.563	1.574	1.576	1.569	1.542	1.515	1.484	1.439	1.395	1.347	1.302
Chloracetic	$K_A \cdot 10^3$	1.528	....	1.488	....	....	1.379	....	....	1.230	....	....
Lactic	$K_A \cdot 10^4$	1.287	....	....	....	....	1.374	....	....	....	....	1.270
Glycollic	$K_A \cdot 10^4$	1.334	....	....	....	....	1.475	....	....	....	....	1.415
Oxalic	$K_{2A} \cdot 10^5$	5.91	5.82	5.70	5.55	5.40	5.18	4.92	4.67	4.41	4.09	3.83
Malonic	$K_{2A} \cdot 10^6$	2.140	2.165	2.152	2.124	2.076	2.014	1.948	1.863	1,768	1.670	1.575
Phosphoric	$K_A \cdot 10^3$	8.968	....	....	....	....	7.516	....	....	....	....	5.495
Phosphoric	$K_{2A} \cdot 10^8$	4.85	5.24	5.57	5.89	6.12	6.34	6.46	6.53	6.58	6.59	6.55
Boric	$K_A \cdot 10^{10}$	....	3.63	4.17	4.72	5.26	5.79	6.34	6.86	7.38	....	8.32
Carbonic	$K_{1A} \cdot 10^7$	2.64	3.04	3.44	3.81	4.16	4.45	4.71	4.90	5.04	5.13	5.19
Phenol-sulfonic	$K_{2A} \cdot 10^{10}$	4.45	5.20	6.03	6.92	7.85	8.85	9.89	10.94	12.00	13.09	14.16
Glycine	$K_{1A} \cdot 10^7$	....	3.82	3.99	4.17	4.32	4.46	4.57	4.66	4.73	4.77	4.79
Citric	$K_{1A} \cdot 10^4$	6.03	6.31	6.69	6.92	7.21	7.45	7.66	7.78	7.96	7.99	8.04
	$K_{2A} \cdot 10^5$	1.45	1.54	1.60	1.65	1.70	1.73	1.76	1.77	1.78	1.76	1.75
	$K_{3A} \cdot 10^7$	4.05	4.11	4.14	4.13	4.09	4.02	3.99	3.78	3.69	3.45	3.28

Note: Reproducibility between various workers is about ± (0.01–0.02). 10^5. All values are on the m-scale.

Source: *Handbook of Chemistry and Physics,* 53rd ed., CRC Press, Cleveland, 1972, p. D-122.

A–11 EQUIVALENT CONDUCTANCES OF SOME ELECTROLYTES IN AQUEOUS SOLUTIONS AT 25°C

Compound	Concentration in gram equivalents per 1000 cm³							
	Infinite dilution	0.0005	0.001	0.005	0.01	0.02	0.05	0.1
$AgNO_3$	133.36	131.36	130.51	127.20	124.76	121.41	115.24	109.14
$BaCl_2$	139.98	135.96	134.34	128.02	123.94	119.09	111.48	105.19
$CaCl_2$	135.84	131.93	130.36	124.25	120.36	115.65	108.47	102.4
$Ca(OH)_2$	257.9			232.9	225.9	213.9		
$CuSO_4$	133.6	121.6	115.26	94.07	83.12	72.20	59.05	50.58
HCl	426.16	422.74	421.36	415.80	412.00	407.24	399.09	391.32
KBr	151.9			146.09	143.43	140.48	135.68	131.39
KCl	149.86	147.81	146.95	143.35	141.27	138.34	133.37	128.96
$KClO_4$	140.04	138.76	137.87	134.16	131.46	127.92	121.62	115.20
$K_3Fe(CN)_6$	174.5	166.4	163.1	150.7				
$K_4Fe(CN)_6$	184.5		167.24	146.09	134.83	122.82	107.70	97.87
$KHCO_3$	118.0	116.10	115.34	112.24	110.08	107.22		
KI	150.38			144.37	142.18	139.45	134.97	131.11
KIO_4	127.92	125.80	124.94	121.24	118.51	114.14	106.72	98.12
KNO_3	144.96	142.77	141.84	138.48	132.82	132.41	126.31	120.40
$KReO_4$	128.20	126.03	125.12	121.31	118.49	114.49	106.40	97.40

Compound	Concentration in gram equivalents per 1000 cm³							
	Infinite dilution	0.0005	0.001	0.005	0.01	0.02	0.05	0.1
$LaCl_3$	145.8	139.6	137.0	127.5	121.8	115.3	106.2	99.1
$LiCl$	115.03	113.15	112.40	109.40	107.40	104.65	100.11	95.86
$LiClO_4$	105.98	104.18	103.44	100.57	98.61	96.18	92.20	88.56
$MgCl_2$	129.40	125.61	124.11	118.31	114.55	110.04	103.08	97.10
NH_4Cl	149.7		146.8	143.5	141.28	138.33	133.29	128.75
$NaCl$	126.45	124.50	123.74	120.65	118.51	115.51	111.06	106.74
$NaClO_4$	117.48	115.64	114.87	111.75	109.59	106.96	102.40	98.43
NaI	126.94	125.36	124.25	121.25	119.24	116.70	112.79	108.78
$NaOOCCH_3$	91.0	89.2	88.5	85.72	83.76	81.24	76.92	72.80
$NaOOCC_2H_5$	85.9		83.5	80.9	79.1	76.6		
$NaOOCC_3H_7$	82.70	81.04	80.31	77.58	75.76	73.39	69.32	65.27
$NaOH$	247.8	245.6	244.7	240.8	238.0			
Na_2SO_4	129.9	125.74	124.15	117.15	112.44	106.78	97.75	89.98
$SrCl_2$	135.80	131.90	130.33	124.24	120.24	115.54	108.25	102.19
$ZnSO_4$	132.8	121.4	114.53	95.49	84.91	74.24	61.20	52.64

Source: *Handbook of Chemistry and Physics,* 53rd ed., CRC Press, Cleveland, 1972, p. D-123.

A–12 THE EQUIVALENT CONDUCTANCE OF THE SEPARATE IONS

Ion.	0°	18°	25°	50°	75°	100°	128°	156°
K	40.4	64.6	74.5	115	159	206	263	317
Na	26.	43.5	50.9	82	116	155	203	249
NH_4	40.2	64.5	74.5	115	159	207	264	319
Ag	32.9	54.3	63.5	101	143	188	245	299
½Ba	33.	55.	65.	104	149	200	262	322
½Ca	30.	51.	60.	98	142	191	252	312
⅓La	35.	61.	72.	119	173	235	312	388
Cl	41.1	65.5	75.5	116	160	207	264	318
NO_3	40.4	61.7	70.6	104	140	178	222	263
$C_2H_3O_2$	20.3	34.6	40.8	67	96	130	171	211

Ion.	0°	18°	25°	50°	75°	100°	128°	156°
½SO_4	41.	68.	79.	125	177	234	303	370
½C_2O_4	39.	63.	73.	115	163	213	27 !	336
⅓$C_6H_5O_7$	36.	60.	70.	113	161	214		
¼$Fe(CN)_6$	58.	95.	111.	173	244	321		
H	240.	314.	350.	465	565	644	722	777
OH	105.	172.	192.	284	360	439	525	592

Note: From Smithsonian Physical Tables.

Source: *Handbook of Chemistry and Physics,* 53rd ed., CRC Press, Cleveland, 1972, p. D-123.

A–13 ACTIVITY COEFFICIENTS OF ACIDS, BASES AND SALTS

Name	0.1	0.2	0.3	0.4	0.5	0.6	0.7	0.8	0.9	1.0
$AgNO_3$	0.734	0.657	0.606	0.567	0.536	0.509	0.485	0.464	0.446	0.429
$AlCl_3$	(0.337)	0.305	0.302	0.313	0.331	0.356	0.388	0.429	0.479	0.539
$Al_2(SO_4)_3$	(0.0350)	0.0225	0.0176	0.0153	0.0143	0.0140	0.0142	0.0149	0.0159	0.0175
$CdSO_4$	(0.150)	0.102	0.082	0.069	0.061	0.055	0.050	0.046	0.043	0.041
$CrCl_3$	(0.331)	0.298	0.294	0.300	0.314	0.335	0.362	0.397	0.436	0.481
$Cr(NO_3)_3$	(0.319)	0.285	0.279	0.281	0.291	0.304	0.322	0.344	0.371	0.401
$Cr_2(SO_4)_3$	(0.0458)	0.0300	0.0238	0.0207	0.0190	0.0182	0.0181	0.0185	0.0194	0.0208
CsBr	0.754	0.694	0.654	0.626	0.603	0.586	0.571	0.558	0.547	0.538
CsCl	0.756	0.694	0.656	0.628	0.606	0.589	0.575	0.563	0.553	0.544
CsI	0.754	0.692	0.651	0.621	0.599	0.581	0.567	0.554	0.543	0.533
$CsNO_3$	0.733	0.655	0.602	0.561	0.528	0.501	0.478	0.458	0.439	0.422
CsOAc	0.799	0.771	0.761	0.759	0.762	0.768	0.776	0.783	0.792	0.802
$CuSO_4$	(0.150)	0.104	0.083	0.071	0.062	0.056	0.052	0.048	0.045	0.043
HBr	0.805	0.782	0.777	0.781	0.789	0.801	0.815	0.832	0.850	0.871
HCl	0.796	0.767	0.756	0.755	0.757	0.763	0.772	0.783	0.795	0.809
$HClO_4$	0.803	0.778	0.768	0.766	0.769	0.776	0.785	0.795	0.808	0.823
HI	0.818	0.807	0.811	0.823	0.839	0.860	0.883	0.908	0.935	0.963
HNO_3	0.791	0.754	0.735	0.725	0.720	0.717	0.717	0.718	0.721	0.724
KBr	0.772	0.722	0.693	0.673	0.657	0.646	0.636	0.629	0.622	0.617
KCl	0.770	0.718	0.688	0.666	0.649	0.637	0.626	0.618	0.610	0.604
KCNS	0.769	0.716	0.685	0.663	0.646	0.633	0.623	0.614	0.606	0.599
KF	0.775	0.727	0.700	0.682	0.670	0.661	0.654	0.650	0.646	0.645
KI	0.778	0.733	0.707	0.689	0.676	0.667	0.660	0.654	0.649	0.645
KNO_3	0.739	0.663	0.614	0.576	0.545	0.519	0.496	0.476	0.459	0.443
KOAc	0.796	0.766	0.754	0.750	0.751	0.754	0.759	0.766	0.774	0.783

Name	0.1	0.2	0.3	0.4	0.5	0.6	0.7	0.8	0.9	1.0
KOH	0.798	0.760	0.742	0.734	0.732	0.733	0.736	0.742	0.749	0.756
LiBr	0.796	0.766	0.756	0.752	0.753	0.758	0.767	0.777	0.789	0.803
LiCl	0.790	0.757	0.744	0.740	0.739	0.743	0.748	0.755	0.764	0.774
$LiClO_4$	0.812	0.794	0.792	0.798	0.808	0.820	0.834	0.852	0.869	0.887
LiI	0.815	0.802	0.804	0.813	0.824	0.838	0.852	0.870	0.888	0.910
$LiNO_3$	0.788	0.752	0.736	0.728	0.726	0.727	0.729	0.733	0.737	0.743
LiOAc	0.784	0.742	0.721	0.709	0.700	0.691	0.689	0.688	0.688	0.689
$MgSO_4$	(0.150)	0.108	0.088	0.076	0.068	0.062	0.057	0.054	0.051	0.049
$MnSO_4$	(0.150)	0.106	0.085	0.073	0.064	0.058	0.053	0.049	0.046	0.044
NaBr	0.782	0.741	0.719	0.704	0.697	0.692	0.689	0.687	0.687	0.687
NaCl	0.778	0.735	0.710	0.693	0.681	0.673	0.667	0.662	0.659	0.657
$NaClO_4$	0.775	0.729	0.701	0.683	0.668	0.656	0.648	0.641	0.635	0.629
NaCNS	0.787	0.750	0.731	0.720	0.715	0.712	0.710	0.710	0.711	0.712
NaF	0.765	0.710	0.676	0.651	0.632	0.616	0.603	0.592	0.582	0.573
NaH_2PO_4	0.744	0.675	0.629	0.593	0.563	0.539	0.517	0.499	0.483	0.468
NaI	0.787	0.751	0.735	0.727	0.723	0.723	0.724	0.727	0.731	0.736
$NaNO_3$	0.762	0.703	0.666	0.638	0.617	0.599	0.583	0.570	0.558	0.548
NaOAc	0.791	0.757	0.744	0.737	0.735	0.736	0.740	0.745	0.752	0.757
NaOH	0.766	0.727	0.708	0.697	0.690	0.685	0.681	0.679	0.678	0.678
$NiSO_4$	(0.150)	0.105	0.084	0.071	0.063	0.056	0.052	0.047	0.044	0.042
RbBr	0.763	0.706	0.673	0.650	0.632	0.617	0.605	0.595	0.586	0.578
RbCl	0.764	0.709	0.675	0.652	0.634	0.620	0.608	0.599	0.590	0.583
RbI	0.762	0.705	0.671	0.647	0.629	0.614	0.602	0.591	0.583	0.575
$RbNO_3$	0.734	0.658	0.606	0.565	0.534	0.508	0.485	0.465	0.446	0.430
RbOAc	0.796	0.767	0.756	0.753	0.755	0.759	0.766	0.773	0.782	0.792
$TlNO_3$	0.702	0.606	0.545	0.500	....	....	....	....	....	....
$ZnSO_4$	(0.150)	0.104	0.083	0.071	0.063	0.057	0.052	0.048	0.046	0.043

Note: The above coefficients are valid only at 25°C. The concentrations are expressed as molalities.

Source: *Handbook of Chemistry and Physics,* 53rd ed., CRC Press, Cleveland, 1972, p. D-123.

A–14 SOLUBILITY OF GASES IN WATER

α, the absorption coefficient, is the volume of gas when reduced to 0° and 760 mm., absorbed by one volume of water when the pressure of the gas itself, without the aqueous tension, amounts to 760 mm.

q is the weight of gas in grams dissolved in 100 grams of water when the total pressure (i.e., the sum of the partial pressure of the gas plus the aqueous tension at the given temperature) is 760 mm.

l is the volume of gas in c.c. dissolved by one volume of water when the total pressure (i.e., the sum of the partial pressure of the gas plus the aqueous tension at the given temperature) is 760 mm.

Temp.° C	Nitrogen*		Oxygen		Hydrogen		Carbon Dioxide	
	α	q	α	q	α	q	α	q
0	0.02354	0.002942	0.04889	0.006945	0.02148	0.0001922	1.713	0.3346
1	0.02297	0.002869	0.04758	0.006756	0.02126	0.0001901	1.646	0.3213
2	0.02241	0.002798	0.04633	0.006574	0.02105	0.0001881	1.584	0.3091
3	0.02187	0.002730	0.04512	0.006400	0.02084	0.0001862	1.527	0.2978
4	0.02135	0.002663	0.04397	0.006232	0.02064	0.0001843	1.473	0.2871
5	0.02086	0.002600	0.04287	0.006072	0.02044	0.0001824	1.424	0.2774
6	0.02037	0.002537	0.04180	0.005918	0.02025	0.0001806	1.377	0.2681
7	0.01990	0.002477	0.04080	0.005773	0.02007	0.0001789	1.331	0.2589
8	0.01945	0.002419	0.03983	0.005632	0.01989	0.0001772	1.282	0.2492
9	0.01902	0.002365	0.03891	0.005498	0.01972	0.0001756	1.237	0.2403
10	0.01861	0.002312	0.03802	0.005368	0.01955	0.0001740	1.194	0.2318
11	0.01823	0.002263	0.03718	0.005246	0.01940	0.0001725	1.154	0.2239
12	0.01786	0.002216	0.03637	0.005128	0.01925	0.0001710	1.117	0.2165
13	0.01750	0.002170	0.03559	0.005014	0.01911	0.0001696	1.083	0.2098
14	0.01717	0.002126	0.03486	0.004906	0.01897	0.0001682	1.050	0.2032
15	0.01685	0.002085	0.03415	0.004802	0.01883	0.0001668	1.019	0.1970
16	0.01654	0.002045	0.03348	0.004703	0.01869	0.0001654	0.985	0.1903
17	0.01625	0.002006	0.03283	0.004606	0.01856	0.0001641	0.956	0.1845
18	0.01597	0.001970	0.03220	0.004514	0.01844	0.0001628	0.928	0.1789
19	0.01570	0.001935	0.03161	0.004426	0.01831	0.0001616	0.902	0.1737
20	0.01545	0.001901	0.03102	0.004339	0.01819	0.0001603	0.878	0.1688
21	0.01522	0.001869	0.03044	0.004252	0.01805	0.0001588	0.854	0.1640
22	0.01498	0.001838	0.02988	0.004169	0.01792	0.0001575	0.829	0.1590
23	0.01475	0.001809	0.02934	0.004087	0.01779	0.0001561	0.804	0.1540
24	0.01454	0.001780	0.02881	0.004007	0.01766	0.0001548	0.781	0.1493
25	0.01434	0.001751	0.02831	0.003931	0.01754	0.0001535	0.759	0.1449
26	0.01413	0.001724	0.02783	0.003857	0.01742	0.0001522	0.738	0.1406
27	0.01394	0.001698	0.02736	0.003787	0.01731	0.0001509	0.718	0.1366
28	0.01376	0.001672	0.02691	0.003718	0.01720	0.0001496	0.699	0.1327
29	0.01358	0.001647	0.02649	0.003651	0.01709	0.0001484	0.682	0.1292
30	0.01342	0.001624	0.02608	0.003588	0.01699	0.0001474	0.665	0.1257
35	0.01256	0.001501	0.02440	0.003315	0.01666	0.0001425	0.592	0.1105
40	0.01184	0.001391	0.02306	0.003082	0.01644	0.0001384	0.530	0.0973
45	0.01130	0.001300	0.02187	0.002858	0.01624	0.0001341	0.479	0.0860
50	0.01088	0.001216	0.02090	0.002657	0.01608	0.0001287	0.436	0.0761
60	0.01023	0.001052	0.01946	0.002274	0.01600	0.0001178	0.359	0.0576
70	0.00977	0.000851	0.01833	0.001856	0.0160	0.000102	—	—
80	0.00958	0.000660	0.01761	0.001381	0.0160	0.000079	—	—
90	0.0095	0.00038	0.0172	0.00079	0.0160	0.000046	—	—
100	0.0095	0.00000	0.0170	0.00000	0.0160	0.000000	—	—

* Atmospheric Nitrogen 98.815% Vol. N_2 + 1.185% Vol. A.

Temp.° C	Carbon Monoxide		Hydrogen Sulfide		Sulfur Dioxide		Nitric Oxide		Air**	
	α	q	α	q	l	q	α	q	cc/1000cc	% O_2 in dissolved air
0	0.03537	0.004397	4.670	0.7066	79.789	22.83	0.07381	0.009833	29.18	34.91
1	0.03455	0.004293	4.522	0.6839	77.210	22.09	0.07184	0.009564	28.42	34.87
2	0.03375	0.004191	4.379	0.6619	74.691	21.37	0.06993	0.009305	27.69	34.83
3	0.03297	0.004092	4.241	0.6407	72.230	20.66	0.06809	0.009057	26.99	34.78
4	0.03222	0.003996	4.107	0.6201	69.828	19.98	0.06632	0.008816	26.32	34.74
5	0.03149	0.003903	3.977	0.6001	67.485	19.31	0.06461	0.008584	25.68	34.69
6	0.03078	0.003813	3.852	0.5809	65.200	18.65	0.06298	0.008361	25.06	34.65
7	0.03009	0.003725	3.732	0.5624	62.973	18.02	0.06140	0.008147	24.47	34.60
8	0.02942	0.003640	3.616	0.5446	60.805	17.40	0.05990	0.007943	23.90	34.56
9	0.02878	0.003559	3.505	0.5276	58.697	16.80	0.05846	0.007747	23.36	34.52
10	0.02816	0.003479	3.399	0.5112	56.647	16.21	0.05709	0.007560	22.84	34.47
11	0.02757	0.003405	3.300	0.4960	54.655	15.64	0.05587	0.007393	22.34	34.43
12	0.02701	0.003332	3.206	0.4814	52.723	15.09	0.05470	0.007233	21.87	34.38
13	0.02646	0.003261	3.115	0.4674	50.849	14.56	0.05357	0.007078	21.41	34.34
14	0.02593	0.003194	3.028	0.4540	49.033	14.04	0.05250	0.006930	20.97	34.30
15	0.02543	0.003130	2.945	0.4411	47.276	13.54	0.05147	0.006788	20.55	34.25
16	0.02494	0.003066	2.865	0.4287	45.578	13.05	0.05049	0.006652	20.14	34.21
17	0.02448	0.003007	2.789	0.4169	43.939	12.59	0.04956	0.006524	19.75	34.17
18	0.02402	0.002947	2.717	0.4056	42.360	12.14	0.04868	0.006400	19.38	34.13
19	0.02360	0.002891	2.647	0.3948	40.838	11.70	0.04785	0.006283	19.02	34.08
20	0.02319	0.002838	2.582	0.3846	39.374	11.28	0.04706	0.006173	18.68	34.03
21	0.02281	0.002789	2.517	0.3745	37.970	10.88	0.04625	0.006059	18.34	33.99
22	0.02244	0.002739	2.456	0.3648	36.617	10.50	0.04545	0.005947	18.01	33.95
23	0.02208	0.002691	2.396	0.3554	35.302	10.12	0.04469	0.005838	17.69	33.90
24	0.02174	0.002646	2.338	0.3463	34.026	9.76	0.04395	0.005733	17.38	33.86
25	0.02142	0.002603	2.282	0.3375	32.786	9.41	0.04323	0.005630	17.08	33.82
26	0.02110	0.002560	2.229	0.3290	31.584	9.06	0.04254	0.005530	16.79	33.77
27	0.02080	0.002519	2.177	0.3208	30.422	8.73	0.04188	0.005435	16.50	33.73
28	0.02051	0.002479	2.128	0.3130	29.314	8.42	0.04124	0.005342	16.21	33.68
29	0.02024	0.002442	2.081	0.3055	28.210	8.10	0.04063	0.005252	15.92	33.64
30	0.01998	0.002405	2.037	0.2983	27.161	7.80	0.04004	0.005165	15.64	33.60
35	0.01877	0.002231	1.831	0.2648	22.489	6.47	0.03734	0.004757		
40	0.01775	0.002075	1.660	0.2361	18.766	5.41	0.03507	0.004394		
45	0.01690	0.001933	1.516	0.2110	—	—	0.03311	0.004059		
50	0.01615	0.001797	1.392	0.1883	—	—	0.03152	0.003758		
60	0.01488	0.001522	1.190	0.1480	—	—	0.02954	0.003237		
70	0.01440	0.001276	1.022	0.1101	—	—	0.02810	0.002668		
80	0.01430	0.000980	0.917	0.0765	—	—	0.02700	0.001984		
90	0.0142	0.00057	0.84	0.041	—	—	0.0265	0.00113		
100	0.0141	0.00000	0.81	0.000	—	—	0.0263	0.00000		

** Cubic centimeters of air (free from CO_2 and NH_3) dissd. in 1000 c.c. H_2O with barometer at 760 mm. (total pressure).

Source: *Handbook of Chemistry and Physics,* 44th ed., CRC Press, Cleveland, 1962, p. 1706.

A–15 SOLUBILITY OF AMMONIA IN WATER

Press. NH_3,mm	0°C		20°C		40°C	
	g/g	cm^3/cm^3	g/g	cm^3/cm^3	g/g	cm^3/cm^3
700	.....	....	0.497	652.9	.....	......
800	.....	....	0.544	714.6	0.329	429.6
900	0.997	1312	0.588	772.4	.....	......
1000	1.094	1440	0.629	826.2	0.386	504.0
1100	1.192	1569	0.669	878.8	.....	......
1200	1.288	1695	0.707	928.8	0.433	565.4
1300	1.388	1827	0.745	978.7	.....	......
1400	1.488	1958	0.781	1025.9	0.472	616.3
1500	1.588	2090	0.815	1070.6	.....	......
1600	1.688	2221	0.847	1112.6	0.508	663.3
1700	1.778	2340	0.877	1152.1	.....	......
1800	1.847	2431	0.906	1190.1	0.543	709.0
1900	.....	....	0.934	1226.9	.....	......
2000	.....	....	0.959	1259.7	0.577	753.4
2100	.....	....	0.984	1292.6	.....	......
2200	.....	....	1.007	1322.8	0.611	797.8
2300	.....	....	1.029	1351.7	.....	......
2400	.....	....	1.052	1381.8	0.644	840.9
2500	.....	....	1.074	1410.8	.....	......
2600	.....	....	1.096	1439.6	0.676	882.7
2700	.....	....	1.117	1467.3	.....	......
2800	.....	....	1.140	1497.4	0.706	921.7
2900	.....	....	1.162	1526.4	.....	......
3000	.....	....	1.185	1556.6	0.732	955.8
3100	.....	....	1.207	1585.5	.....	......
3200	.....	....	1.230	1615.7	0.758	989.8
3300	.....	....	.....	......	.....	......
3400	.....	....	.....	......	0.784	1023.7

Source: *Handbook of Chemistry and Physics,* 44th ed., CRC Press, Cleveland, 1962, p. 1706.

A–16 SOLUBILITIES OF VARIOUS GASES IN WATER

Henry's Law Constant *K*

$K \times 10^{-7}$ $K = P/X$ P = partial pressure mm. of Hg; X = mole fraction

Gas / References	t = 0°	10°	20°	30°	38°	40°	50°	60°	70°	80°
Argon.......... 1, 8, 14, 15, 17	1.65	2.18	2.58	3.02	3.41	3.49	3.76	3.92	4.12	4.25
Carbon dioxide.. 6, 7, 9, 12	.0555	.0788	.108	.139	.168	.173	.217	.258		
Helium......... 1, 3, 8, 12, 14, 15, 18, 21	10.0	10.5	10.9	11.1	11.0	10.9	10.5	10.3	9.88	
Hydrogen....... 6, 12	4.42	4.82	5.20	5.51	5.72	5.78	5.82	5.80	5.77	5.73
Krypton........ 2, 14, 15	0.853	1.20	1.52	1.85	2.13	2.18	2.43	2.66	2.83	2.94
Neon........... (2), 8	7.68	8.49	9.14	9.45	9.76	9.80	10.0			
Nitrogen........ 12, 16, 20, 22, 23, 24	4.09	4.87	5.75	6.68	7.51	7.60	8.20	8.70	9.20	
Oxygen......... 6, 10, 12, 13	1.91	2.48	2.95	3.52	4.04	4.14	4.50	4.84	5.13	5.28
Radon.......... 2, 14, 15	.186	.286	.391	.529	.651	.683	.839	.976	1.07	
Xenon.......... 2, 14, 15	.392	.555	.742	.945	1.12	1.16	1.31	1.46	1.59	1.66
Nitrous oxide... 6, 9, 10, 12	.074	.108	.155	.210	.242	.246	.279			
Acetylene....... 5, 25	.0555	.0716	.0900	.112	.131	.133				
Ethylene........ 5, 10	.370	.552	.753	1.00	1.21	1.23				

Source: *Handbook of Chemistry and Physics,* 44th ed., CRC Press, Cleveland 1962, p. 1708. Compiled by Hardin B. Jones.

A–16 SOLUBILITIES OF VARIOUS GASES IN WATER *(Continued)*

REFERENCES

1. Gösta Akerlof. The Solubilities of Noble Gases in Aqueous Salt Solutions at 25°C., Jour. Am. Chem. Soc. 57: 1196–1201, (1935).
2. Andreas von Antropoff. Experimentelle Untersuchung über die L·slichkeit der Edelgase in Flüssigkeiten. Z. f. Electrochemie 25: (1919).
3. H. P. Cady, H. M. Elsey and E. V. Berger. The Solubility of Helium in Water. Jour. Am. Chem. Soc. 44: 1456 (1922).
4. T. Estreicher. Z. Physik. Chem. 31: 176 (1899).
5. A. Grollman. The Solubility of Gases in Blood and Blood Fluids. Jour. Biol. Chem. 82: 317 (1929).
6. G. Giffcken. Beitrage zur Kenntnis der Loslichkeitbeeinflussung. Z. f. Phys. Chemie. 49: 257–302 (1904).
7. H. S. Harned and R. J. Davis. The Ionization Constant of Carbonic Acid in Water and the Solubility of Carbon Dioxide in Water and Aqueous Salt Solutions from 0° to 50°C. Jour. Am. Chem. Soc. 65: 2030–37 (1943).
8. A. Lannung. The Solubilities of Helium, Neon, and Argon in Water and Some Organic Solvents. Jour. Am. Chem. Soc. 52: 68–80 (1930).
9. A. A. Markham and K. A. Kobe. Solubility of Carbon Dioxide and Nitrous Oxide in Aqueous Salt Solutions. Jour. Am. Chem. Soc. 63: 449–54 (1941).
10. M. Nicloux. Dissolution Aqueuse des Gaz. Compt. Rend. Soc. de Biol. 97: 1553–1555 (1927).
11. N. W. Rakestraw and V. M. Emmel. The Solubility of Nitrogen and Argon in Sea Water. Jour. of Phys. Chem. 42: 1211–15 (1938).
12. M. Randall and C. F. Failey. The Activity Coefficients of Gases in Aqueous Salt Solutions. Chem. Rev. 4: 271–284 (1927).
13. J. Sendray, R. T. Dillon, and D. D. Van Slyke. Studies of Gas and Electrolyte Equilibria in Blood. XIX. The Solubility and Physical State of Uncombined Oxygen in Blood.
14. S. Valentiner. Über die Loslichkeit der Edelgase in Wasser. Z. f. Physik. 42: 253–64 (1927).
15. S. Valentiner. Bemerkung zu meiner Arbeitl: Über die Loslichkeit der Edelgase in Wasser. Z. f. Physik 61: 363–65 (1930).
16. D. D. Van Slyke, R. T. Dillon and R. Margaria. Nitrogen Solubility in Blood. Jour. Biol. Chem. 105: 571–598 (1934):
17. L. W. Winkler. Z. f. Physik Chem. 55: 344 (1906).
18. R. Wiebe and V. L. Gaddy. The Solubility of Helium in Water at 0°, 25°, 50°, and 75° and at Pressures up to 1000 atmospheres.
19. J. A. Hawkins and C. W. Shilling. Helium Solubility in Blood at Increased Pressure. J. Biol. Chem. 113: 649–54 (1936).
20. Coste. Review of Nitrogen Solubilities. J. Phys. Chem. 31: 81–7 (1927).
21. A. R. Behnke and O. D. Yarbrough. Respiratory Resistance, Oil Water Solubility, and Mental Effects of Argon, Compared with Helium and Nitrogen. Am. J. Physiol. 126: 409–15 (1939).
22. Winkler. Berichte d. Deu. Ch. Gel. 24: 3606 (1891).
23. Bohr and Bock. Wied. Ann. 44: 318 (1891).
24. Braun. Z. Physik Chem. 33: 732 (1900).
25. Othmer & White. Industrial and Eng. Chemistry 34: 952 (1942).

A–17 OSMOTIC PARAMETERS AND ELECTRICAL CONDUCTIVITIES OF AQUEOUS SOLUTIONS

For any aqueous solutions within the range, the two right hand columns provide values for either condosity (T) or specific conductance (γ) where the other value is given; the three left hand columns provide values for freezing point depression (Δ), osmolality (O) or osmosity (S) where any one of these values is given. Values in all four columns are interconvertible *only* for solutions of sodium chloride, in which case, by definition, osmosity, molarity (M) and condosity are always equal. Other concentrative properties of specific solutions may be found in conversion tables on preceding pages.

Table columns are:

Δ = freezing point depression, °C
O = osmolality (=Δ/1.86, Os/kg
S = osmosity, molar concentration of NaCl solution having same freezing point or osmotic pressure as given solution, g-mol/l
M = molarity of NaCl solution, g-mol/l
T = condosity, molar concentration of NaCl solution having same specific conductance at 20°C as given solution, g-mol/l
γ = specific conductance (electrical) at 20°C, mmho/cm.

A–17 OSMOTIC PARAMETERS AND ELECTRICAL CONDUCTIVITIES OF AQUEOUS SOLUTIONS *(Continued)*

Δ °C	O Os/Kg	S,M,T, g-mol/l	γ mmho/cm	Δ °C	O Os/Kg	S,M,T, g-mol/l	γ mmho/cm
0.000	0.000	0.000	0.0	1.881	1.011	0.550	46.1
0.037	0.020	0.010	1.0	1.916	1.030	0.560	46.8
0.072	0.039	0.020	2.0	1.950	1.048	0.570	47.6
0.107	0.057	0.030	3.0	1.985	1.067	0.580	48.3
0.142	0.076	0.040	4.0	2.019	1.086	0.590	49.1
0.176	0.095	0.050	5.0	2.054	1.104	0.600	49.8
0.211	0.113	0.060	5.9	2.088	1.123	0.610	50.6
0.245	0.132	0.070	6.9	2.123	1.141	0.620	51.3
0.279	0.150	0.080	7.8	2.157	1.160	0.630	52.0
0.314	0.169	0.090	8.7	2.192	1.178	0.640	52.8
0.348	0.187	0.100	9.7	2.226	1.197	0.650	53.5
0.382	0.205	0.110	10.6	2.261	1.216	0.660	54.2
0.416	0.224	0.120	11.4	2.296	1.234	0.670	55.0
0.450	0.242	0.130	12.3	2.330	1.253	0.680	55.7
0.484	0.260	0.140	13.2	2.365	1.271	0.690	56.4
0.518	0.279	0.150	14.1	2.400	1.290	0.700	57.1
0.552	0.297	0.160	14.9	2.434	1.309	0.710	57.9
0.586	0.315	0.170	15.8	2.469	1.328	0.720	58.6
0.620	0.333	0.180	16.6	2.504	1.346	0.730	59.3
0.654	0.352	0.190	17.5	2.539	1.365	0.740	60.0
0.688	0.370	0.200	18.3	2.574	1.384	0.750	60.7
0.722	0.388	0.210	19.2	2.609	1.403	0.760	61.4
0.756	0.407	0.220	20.0	2.644	1.421	0.770	62.1
0.790	0.425	0.230	20.8	2.679	1.440	0.780	62.9
0.824	0.443	0.240	21.6	2.714	1.459	0.790	63.6
0.858	0.461	0.250	22.4	2.749	1.478	0.800	64.3
0.892	0.480	0.260	23.3	2.784	1.497	0.810	65.0
0.926	0.498	0.270	24.1	2.819	1.515	0.820	65.7
0.960	0.516	0.280	24.9	2.854	1.534	0.830	66.3
0.994	0.534	0.290	25.7	2.889	1.553	0.840	67.0
1.028	0.553	0.300	26.5	2.924	1.572	0.850	67.7
1.062	0.571	0.310	27.3	2.959	1.591	0.860	68.4
1.096	0.589	0.320	28.1	2.995	1.610	0.870	69.1
1.130	0.607	0.330	28.9	3.030	1.629	0.880	69.8
1.164	0.626	0.340	29.7	3.065	1.648	0.890	70.5
1.198	0.644	0.350	30.5	3.101	1.667	0.900	71.2
1.232	0.662	0.360	31.3	3.136	1.686	0.910	71.8
1.266	0.681	0.370	32.1	3.171	1.705	0.920	72.5
1.300	0.699	0.380	32.9	3.207	1.724	0.930	73.2
1.334	0.717	0.390	33.7	3.242	1.743	0.940	73.9
1.368	0.736	0.400	34.4	3.278	1.762	0.950	74.5
1.402	0.754	0.410	35.2	3.313	1.781	0.960	75.2
1.436	0.772	0.420	36.0	3.349	1.801	0.970	75.9
1.470	0.791	0.430	36.8	3.385	1.820	0.980	76.5
1.505	0.809	0.440	37.6	3.420	1.839	0.990	77.2
1.539	0.827	0.450	38.4	3.456	1.858	1.000	77.8
1.573	0.846	0.460	39.2	3.492	1.877	1.010	78.5
1.607	0.864	0.470	39.9	3.527	1.896	1.020	79.2
1.641	0.882	0.480	40.7	3.563	1.916	1.030	79.8
1.676	0.901	0.490	41.5	3.599	1.935	1.040	80.5
1.710	0.919	0.500	42.3	3.635	1.954	1.050	81.1
1.744	0.938	0.510	43.0	3.671	1.974	1.060	81.8
1.778	0.956	0.520	43.8	3.707	1.993	1.070	82.4
1.813	0.975	0.530	44.6	3.743	2.012	1.080	83.1
1.847	0.993	0.540	45.3	3.779	2.032	1.090	83.7

A–17 OSMOTIC PARAMETERS AND ELECTRICAL CONDUCTIVITIES OF AQUEOUS SOLUTIONS *(Continued)*

Δ °C	O Os/Kg	S,M,T, g-mol/l	γ mmho/cm	Δ °C	O Os/Kg	S,M,T, g-mol/l	γ mmho/cm
3.815	2.051	1.100	84.3	5.86	3.15	1.650	117
3.851	2.070	1.110	85.0	5.89	3.17	1.660	117
3.887	2.090	1.120	85.6	5.93	3.19	1.670	118
3.923	2.109	1.130	86.2	5.97	3.21	1.680	119
3.959	2.129	1.140	86.9	6.01	3.23	1.690	119
3.995	2.148	1.150	87.5	6.05	3.25	1.700	120
4.032	2.168	1.160	88.1	6.09	3.27	1.710	120
4.068	2.187	1.170	88.8	6.13	3.29	1.720	121
4.104	2.207	1.180	89.4	6.16	3.31	1.730	121
4.141	2.226	1.190	90.0	6.20	3.34	1.740	122
4.177	2.246	1.200	90.6	6.24	3.36	1.750	122
4.214	2.265	1.210	91.3	6.28	3.38	1.760	123
4.250	2.285	1.220	91.9	6.32	3.40	1.770	123
4.287	2.305	1.230	92.5	6.36	3.42	1.780	124
4.323	2.324	1.240	93.1	6.40	3.44	1.790	125
4.360	2.344	1.250	93.7	6.44	3.46	1.800	125
4.393	2.362	1.260	94.3	6.48	3.48	1.810	126
4.429	2.381	1.270	94.9	6.52	3.50	1.820	126
4.466	2.401	1.280	95.6	6.56	3.52	1.830	127
4.503	2.421	1.290	96.2	6.60	3.55	1.840	127
4.539	2.441	1.300	96.8	6.64	3.57	1.850	128
4.576	2.460	1.310	97.4	6.67	3.59	1.860	128
4.613	2.480	1.320	98.0	6.71	3.61	1.870	129
4.650	2.500	1.330	98.6	6.75	3.63	1.880	129
4.687	2.520	1.340	99.2	6.79	3.65	1.890	129
4.724	2.540	1.350	99.8	6.83	3.67	1.900	130
4.761	2.560	1.360	100	6.87	3.70	1.910	130
4.798	2.580	1.370	101	6.91	3.72	1.920	131
4.835	2.600	1.380	102	6.95	3.74	1.930	131
4.872	2.620	1.390	102	6.99	3.76	1.940	132
4.910	2.640	1.400	103	7.03	3.78	1.950	132
4.947	2.660	1.410	103	7.07	3.80	1.960	133
4.984	2.680	1.420	104	7.11	3.82	1.970	133
5.02	2.70	1.430	104	7.15	3.85	1.980	134
5.06	2.72	1.440	105	7.19	3.87	1.990	134
5.10	2.74	1.450	106	7.23	3.89	2.000	135
5.13	2.76	1.460	106	7.27	3.91	2.010	135
5.17	2.78	1.470	107	7.32	3.93	2.020	136
5.21	2.80	1.480	107	7.36	3.95	2.030	136
5.25	2.82	1.490	108	7.40	3.98	2.040	137
5.28	2.84	1.500	109	7.44	4.00	2.050	137
5.32	2.86	1.510	109	7.48	4.02	2.060	137
5.36	2.88	1.520	110	7.52	4.04	2.070	138
5.40	2.90	1.530	110	7.56	4.06	2.080	138
5.44	2.92	1.540	111	7.60	4.09	2.090	139
5.47	2.94	1.550	111	7.64	4.11	2.100	139
5.51	2.96	1.560	112	7.68	4.13	2.110	140
5.55	2.98	1.570	112	7.72	4.15	2.120	140
5.59	3.00	1.580	113	7.76	4.17	2.130	141
5.63	3.02	1.590	114	7.81	4.20	2.140	141
5.66	3.05	1.600	114	7.85	4.22	2.150	142
5.70	3.07	1.610	115	7.89	4.24	2.160	142
5.74	3.09	1.620	115	7.93	4.26	2.170	142
5.78	3.11	1.630	116	7.97	4.29	2.180	143
5.82	3.13	1.640	116	8.01	4.31	2.190	143

A–17 OSMOTIC PARAMETERS AND ELECTRICAL CONDUCTIVITIES OF AQUEOUS SOLUTIONS *(Continued)*

Δ °C	O Os/Kg	S,M,T, g-mol/l	γ mmho/cm	Δ °C	O Os/Kg	S,M,T, g-mol/l	γ mmho/cm
8.05	4.33	2.200	144	10.45	5.62	2.750	168
8.10	4.35	2.210	144	10.50	5.64	2.760	168
8.14	4.38	2.220	145	10.54	5.67	2.770	169
8.18	4.40	2.230	145	10.59	5.69	2.780	169
8.22	4.42	2.240	146	10.63	5.72	2.790	170
8.26	4.44	2.250	146	10.68	5.74	2.800	170
8.31	4.47	2.260	147	10.73	5.77	2.810	170
8.35	4.49	2.270	147	10.77	5.79	2.820	171
8.39	4.51	2.280	148	10.82	5.82	2.830	171
8.43	4.53	2.290	148	10.86	5.84	2.840	172
8.47	4.56	2.300	148	10.91	5.87	2.850	172
8.52	4.58	2.310	149	10.96	5.89	2.860	172
8.56	4.60	2.320	149	11.00	5.92	2.870	173
8.60	4.62	2.330	150	11.05	5.94	2.880	173
8.64	4.65	2.340	150	11.10	5.97	2.890	173
8.69	4.67	2.350	151	11.14	5.99	2.900	174
8.73	4.69	2.360	151	11.19	6.02	2.910	174
8.77	4.72	2.370	152	11.24	6.04	2.920	175
8.82	4.74	2.380	152	11.28	6.07	2.930	175
8.86	4.76	2.390	153	11.33	6.09	2.940	175
8.90	4.79	2.400	153	11.38	6.12	2.950	176
8.94	4.81	2.410	153	11.43	6.14	2.960	176
8.99	4.83	2.420	154	11.47	6.17	2.970	176
9.03	4.85	2.430	154	11.52	6.19	2.980	177
9.07	4.88	2.440	155	11.57	6.22	2.990	177
9.12	4.90	2.450	155	11.62	6.25	3.000	178
9.16	4.92	2.460	156	11.71	6.30	3.020	178
9.20	4.95	2.470	156	11.81	6.35	3.040	179
9.25	4.97	2.480	156	11.90	6.40	3.060	180
9.29	5.00	2.490	157	12.00	6.45	3.080	181
9.33	5.02	2.500	157	12.10	6.50	3.100	181
9.38	5.04	2.510	158	12.20	6.56	3.120	182
9.42	5.07	2.520	158	12.29	6.61	3.140	183
9.47	5.09	2.530	159	12.39	6.66	3.160	183
9.51	5.11	2.540	159	12.49	6.71	3.180	184
9.55	5.14	2.550	160	12.59	6.77	3.200	185
9.60	5.16	2.560	160	12.69	6.82	3.220	185
9.64	5.18	2.570	160	12.79	6.87	3.240	186
9.69	5.21	2.580	161	12.89	6.93	3.260	187
9.73	5.23	2.590	161	12.99	6.98	3.280	187
9.78	5.26	2.600	162	13.09	7.04	3.300	188
9.82	5.28	2.610	162	13.19	7.09	3.320	189
9.86	5.30	2.620	163	13.29	7.15	3.340	189
9.91	5.33	2.630	163	13.39	7.20	3.360	190
9.95	5.35	2.640	163	13.49	7.26	3.380	190
10.00	5.38	2.650	164	13.60	7.31	3.400	191
10.04	5.40	2.660	164	13.70	7.37	3.420	192
10.09	5.42	2.670	165	13.80	7.42	3.440	192
10.13	5.45	2.680	165	13.91	7.48	3.460	193
10.18	5.47	2.690	165	14.01	7.53	3.480	193
10.22	5.50	2.700	166	14.12	7.59	3.500	194
10.27	5.52	2.710	166	14.22	7.65	3.520	195
10.31	5.55	2.720	167	14.33	7.70	3.540	195
10.36	5.57	2.730	167	14.43	7.76	3.560	196
10.41	5.59	2.740	168	14.54	7.82	3.580	196

A–17 OSMOTIC PARAMETERS AND ELECTRICAL CONDUCTIVITIES OF AQUEOUS SOLUTIONS *(Continued)*

Δ °C	O Os/Kg	S,M,T, g-mol/l	γ mmho/cm
14.65	7.87	3.600	197
14.75	7.93	3.620	197
14.86	7.99	3.640	198
14.97	8.05	3.660	198
15.08	8.11	3.680	199
15.19	8.16	3.700	199
15.30	8.22	3.720	200
15.41	8.28	3.740	200
15.52	8.34	3.760	201
15.63	8.40	3.780	201
15.74	8.46	3.800	202
15.85	8.52	3.820	202
15.96	8.58	3.840	203
16.07	8.64	3.860	203
16.19	8.70	3.880	204
16.30	8.76	3.900	204
16.41	8.82	3.920	205
16.53	8.89	3.940	205
16.64	8.95	3.960	206
16.76	9.01	3.980	206
16.87	9.07	4.000	206
16.99	9.13	4.020	207
17.11	9.20	4.040	207
17.22	9.26	4.060	208
17.34	9.32	4.080	208
17.46	9.39	4.100	209
17.58	9.45	4.120	209
17.70	9.51	4.140	209
17.82	9.58	4.160	210
17.94	9.64	4.180	210
18.06	9.71	4.200	210
18.18	9.77	4.220	211
18.30	9.84	4.240	211
18.42	9.90	4.260	212
18.55	9.97	4.280	212
18.67	10.04	4.300	212
18.79	10.10	4.320	213
18.92	10.17	4.340	213
19.04	10.24	4.360	213
19.17	10.30	4.380	214
19.29	10.37	4.400	214
19.42	10.44	4.420	214
19.55	10.51	4.440	215
19.67	10.58	4.460	215
19.80	10.65	4.480	215
19.93	10.72	4.500	216
20.06	10.78	4.520	216
20.19	10.85	4.540	216
20.32	10.92	4.560	216
20.45	10.99	4.580	217
20.58	11.07	4.600	217
20.71	11.14	4.620	217
20.85	11.21	4.640	218
20.98	11.28	4.660	218
21.11	11.35	4.680	218
		4.700	218
		4.750	219
		4.800	220
		4.850	220
		4.900	221
		4.950	221
		5.000	222
		5.050	222
		5.100	223
		5.150	223
		5.200	224
		5.250	224
		5.300	225

Source: *Handbook of Chemistry and Physics,* 53rd ed., CRC Press, Cleveland, 1972, p. D-224.

A–18 LOWERING OF VAPOR PRESSURE BY SALTS IN AQUEOUS SOLUTIONS

The table gives the reduction of the vapor pressure in millimeters due to the presence of the number of grammolecules of salt per liter of water given at the head of the columns, at the temperature 100° C, at which temperature the vapor pressure of pure water is 760 millimeters.

Substance	0.5	1.0	2.0	3.0	4.0	5.0	6.0	8.0	10.0
$Al_2(SO_4)_3$	12.8	36.5							
$AlCl_3$	22.5	61.0	179.0	318.0					
BaS_2O_6	6.6	15.4	34.4						
$Ba(OH)_2$	12.3	22.5	39.0						
$Ba(NO_3)_2$	13.5	27.0							
$Ba(ClO_3)_2$	15.8	33.3	70.5	108.2					
$BaCl_2$	16.4	36.7	77.6						
$BaBr_2$	16.8	38.8	91.4	150.0	204.7				
CaS_2O_8	9.9	23.0	56.0	106.0					
$Ca(NO_3)_2$	16.4	34.8	74.6	139.3	161.7	205.4			
$CaCl_2$	17.0	39.8	95.3	166.6	241.5	319.5			
$CaBr_2$	17.7	44.2	105.8	191.0	283.3	368.5			
$CdSO_4$	4.1	8.9	18.1						
CdI_2	7.6	14.8	33.5	52.7					
$CdBr_2$	8.6	17.8	36.7	55.7	80.0				
$CdCl_2$	9.6	18.8	36.7	57.0	77.3	99.0			
$Cd(NO_3)_2$	15.9	36.1	78.0	122.2					
$Cd(ClO_3)_2$	17.5								
$CoSO_4$	5.5	10.7	22.9	45.5					
$CoCl_2$	15.0	34.8	83.0	136.0	186.4				
$Co(NO_3)_2$	17.3	39.2	89.0	152.0	218.7	282.0	332.0		
$FeSO_4$	5.8	10.7	24.0	42.4					
H_3BO_3	6.0	12.3	25.1	38.0	51.0				
H_3PO_4	6.6	14.0	28.6	45.2	62.0	81.5	103.0	146.9	189.5
H_3AsO_4	7.3	15.0	30.2	46.4	64.9				
H_2SO_4	12.9	26.5	62.8	104.0	148.0	198.4	247.0	343.2	
KH_2PO_4	10.2	19.5	33.3	47.8	60.5	73.1	85.2		
KNO_3	10.3	21.1	40.1	57.6	74.5	88.2	102.1	126.3	148.0
$KClO_3$	10.6	21.6	42.8	62.1	80.0				
$KBrO_3$	10.9	22.4	45.0						
$KHSO_4$	10.9	21.9	43.3	65.3	85.5	107.8	129.9	170.0	
KNO_2	11.1	22.8	44.8	67.0	90.0	110.5	130.7	167.0	198.8
$KClO_4$	11.5	22.3							
KCl	12.2	24.4	48.8	74.1	100.9	128.5	152.2		
$KHCO_3$	11.6	23.6	59.0	77.6	104.2	132.0	160.0	210.0	255.0
KI	12.5	25.3	52.2	82.6	112.2	141.5	171.8	225.5	278.5
$K_2C_2O_4$	13.9	28.3	59.8	94.2	131.0				
K_2WO_4	13.9	33.0	75.0	123.8	175.4	226.4			
K_2CO_3	14.4	31.0	68.3	105.5	152.0	209.0	258.5	350.0	
KOH	15.0	29.5	64.0	99.2	140.0	181.8	223.0	309.5	387.8
K_2CrO_4	16.2	29.5	60.0						
$LiNO_3$	12.2	25.9	55.7	88.9	122.2	155.1	188.0	253.4	309.2
$LiCl$	12.1	25.5	57.1	95.0	132.5	175.5	219.5	311.5	393.5
$LiBr$	12.2	26.2	60.0	97.0	140.0	186.3	241.5	341.5	438.0
Li_2SO_4	13.3	28.1	56.8	89.0					
$LiHSO_4$	12.8	27.0	57.0	93.0	130.0	168.0			
LiI	13.6	28.6	64.7	105.2	154.5	206.0	264.0	357.0	445.0
Li_2SiFl_6	15.4	34.0	70.0	106.0					
$LiOH$	15.9	37.4	78.1						
Li_2CrO_4	16.4	32.6	74.0	120.0	171.0				
$MgSO_4$	6.5	12.0	24.5	47.5					
$MgCl_2$	16.8	39.0	100.5	183.3	277.0	377.0			
$Mg(NO_3)_2$	17.6	42.0	101.0	174.8					
$MgBr_2$	17.9	44.0	115.8	205.3	298.5				
$MgH_2(SO_4)_2$	18.3	46.0	116.0						
$MnSO_4$	6.0	10.5	21.0						
$MnCl_2$	15.0	34.0	76.0	122.3	167.0	209.0			
NaH_2PO_4	10.5	20.0	36.5	51.7	66.8	82.0	96.5	126.7	157.1
$NaHSO_4$	10.9	22.1	47.3	75.0	100.2	126.1	148.5	189.7	231.4
$NaNO_3$	10.6	22.5	46.2	68.1	90.3	111.5	131.7	167.8	198.8
$NaClO_3$	10.5	23.0	48.4	73.5	98.5	123.3	147.5	196.5	223.5
$(NaPO_3)_6$	11.6								
$NaOH$	11.8	22.8	48.2	77.3	107.5	139.1	172.5	243.3	314.0
$NaNO_2$	11.6	24.4	50.0	75.0	98.2	122.5	146.5	189.0	226.2
Na_2HPO_4	12.1	23.5	43.0	60.0	78.7	99.8	122.1		
$NaHCO_3$	12.9	24.1	48.2	77.6	102.2	127.8	152.0	198.0	239.4
Na_2SO_4	12.6	25.0	48.9	74.2					
$NaCl$	12.3	25.2	52.1	80.0	111.0	143.0	176.5		
$NaBrO_3$	12.1	25.0	54.1	81.3	108.8	136.0			
$NaBr$	12.6	25.9	57.0	89.2	124.2	159.5	197.5	268.0	
NaI	12.1	25.6	60.2	99.5	136.7	177.5	221.0	301.5	370.0
$Na_4P_2O_7$	13.2	22.0							
Na_2CO_3	14.3	27.3	53.5	80.2	111.0				
$Na_2C_2O_4$	14.5	30.0	65.8	105.8	146.0				
Na_2WO_4	14.8	33.6	71.6	115.7	162.6				
Na_3PO_4	16.5	30.0	52.5						
$(NaPO_3)_3$	17.1	36.5							
NH_4NO_3	12.8	22.0	42.1	62.7	82.9	103.8	121.0	152.2	180.0
$(NH_4)_2SiFl_6$	11.5	25.0	44.5						
NH_4Cl	12.0	23.7	45.1	69.3	94.2	118.5	138.2	179.0	213.8
NH_4HSO_4	11.5	22.0	46.8	71.0	94.5	118.	139.0	181.2	218.0
$(NH_4)_2SO_4$	11.0	24.0	46.5	69.5	93.0	117.0	141.8		
NH_4Br	11.9	23.9	48.8	74.1	99.4	121.5	145.5	190.2	228.5
NH_4I	12.9	25.1	49.8	78.5	104.5	132.3	156.0	200.0	243.5
$NiSO_4$	5.0	10.2	21.5						
$NiCl_2$	16.1	37.0	86.7	147.0	212.8				
$Ni(NO_3)_2$	16.1	37.3	91.3	156.2	235.0				
$Pb(NO_3)_2$	12.3	23.5	45.0	63.0					
$Sr(SO_3)_2$	7.2	20.3	47.0						
$Sr(NO_3)_2$	15.8	31.0	64.0	97.4	131.4				
$SrCl_2$	16.8	38.8	91.4	156.8	223.3	281.5			
$SrBr_2$	17.8	42.0	101.1	179.0	267.0				
$ZnSO_4$	4.9	10.4	21.5	42.1	66.2				
$ZnCl_2$	9.2	18.7	46.2	75.0	107.0	153.0	195.0		
$Zn(NO_3)_2$	16.6	39.0	93.5	157.5	223.8				

Note: From Smithsonian Tables.

Source: *Handbook of Chemistry and Physics,* 53rd ed., CRC Press, Cleveland, 1972, p. E-1.

CONVERSION FACTORS

Acidity (see Concentration)

Activated carbon column loading

gpm/ft² = 6.7904×10^{-4} m³/m²-sec
m³/m²-sec = 1.4726×10^{3} gpm/ft² (gal/min-ft²)

Activated carbon contacting rate

gpm/ft³ = 2.2280×10^{-3} m³/m³-sec
m³/m³-sec = 4.4883×10^{-2} x gpm/ft³ (gal/min-ft³)

Air permeability

g/cm²-sec = 10.0 kg/m²-sec
kg/m²-sec = 10×10^{-1} g/cm²-sec

Airport water consumption

gal/passenger-day = 2.2824×10^{7} m³/passenger-sec
m³/passenger-sec = 4.3813×10^{-8} gal/passenger-day

Algae density on lake bottoms

kg/m² = 2.2046 lb/m²
kg/m² = 2.0482×10^{-1} lb/ft²
lb/ft² = 4.8824 kg/m²
lb/m² = 4.5359×10^{-1} kg/m²

Alkalinity (see Concentration)

Application rate

lb/acre = 1.1209×10^{-4} kg/m²
kg/m² = 8.9218×10^{3} lb/acre

Area

acre = 4.04686×10^{3} m²
cm² = 1.0×10^{-4} m²
ft² = 9.2903×10^{-2} m²
in.² = 6.4516×10^{-4} m²
km² = 1.0×10^{6} m²
m² = 2.4711×10^{-4} acres
m² = 1.0×10^{4} cm²
m² = 1.0764×10^{1} ft²
m² = 1.550×10^{3} in.²
m² = 1.0×10^{-6} km²
m² = 3.8610×10^{-7} mi²
m² = 1.0×10^{6} mm²
m² = 1.1960 yd²
mi² = 2.590×10^{6} m²
mm² = 1.0×10^{-6} m²
yd² = 8.3613×10^{-1} m²

Backwash required for filtration

kgal/ft² = 4.0746×10^{-2} m³/m²
m³/m² = 2.4542×10^{1} kgal/ft²

Bacteria concentration

ml⁻¹ = 1.000028×10^{6} m⁻³
m⁻³ = 9.99972×10^{-7} ml⁻¹
(area std. units per milliliter)

Bacteria density

MPN = 9.99972×10^{3} m⁻³

BOD equivalents of chemical compounds

g/g = 1.0 kg/kg
kg/kg = 1.0 g/g
mg/mg = 1.0 kg/kg
kg/kg = 1.0 mg/mg

Circulation (see Flow rate)

Clarifier loading (see Activated carbon column loading)

Coliform count (see Bacteria concentration)

Coliform discharge factor

E. coli/person-day = 8.640×10^{4} *E. coli*/person sec
E. coli/person-sec = 1.1574×10^{-5} *E. coli*/person day

Concentration

cc/liter = 9.99972×10^{-4} m³/m³
Ci/m³ = 1.000028×10^{10} pCi/liter
eq/m³ = 1.000028 meq/l
gal/mil gal = 1.0×10^{-6} m³/m³
gpg (grains/gal) = 1.7118×10^{1} mg/l
g/cm³ = 1.0×10^{3} kg/m³
g/l* = 1.0 ppt
g/liter = 9.99972×10^{-1} kg/m³
gram atoms/liter = 1.000028 kg atoms/m³
kg atom/m³ = 1.000028 g atom/liter
kg/kg = 1.0×10^{6} mg/kg
kg/kg = 1.0×10^{3} mg/g
kg/kg = 1.0×10^{6} μg/g
kg/m³ = 1.000028 g/liter
kg/m³ = 1.000028×10^{3} mg/liter
kg/m³ = 1.000028×10^{10} ng/liter
kg/m³ = 1.000028×10^{6} μg/l
kg/m³ = 1.000028×10^{3} μg/ml
kg-mole/m³ = 9.99972×10^{-1} mol/l
m³/m³ = 1.0×10^{3} ml/liter
m³/m³ = 9.99972×10^{4} cm³/liter
m³/m³ = 1.0×10^{6} gal/mil gal
m³/m³ = 1.0×10^{6} μl/l
me/l (meq/l) = 1.000028 eq/m³
me/l = 5.0×10^{1} mg/liter as $CaCO_3$
mg/g = 1.0×10^{-3} kg/kg
mg/kg = 1.0×10^{-6} kg/kg
mg/l* = 5.8416×10^{-2} grain/gal
mg/l* = 1.0 ppm

*This is approximately true for small concentrations of dissolved solids such that the density of the solution is not far from 1.0 g/cm³.

CONVERSION FACTORS (*Continued*)

mg/liter	=	9.99972×10^{-4} kg/m^3
mg/liter as $CaCO_3$	=	2.0×10^{-2} meq/liter
mg/l as $CaCO_3$	=	2.0×10^{-2} me/liter
ml/liter	=	1.0×10^{-3} m^3/m^3
mol/l	=	1.0×10^{6} μ mol/liter
mol/liter	=	1.0×10^{3} mmole/liter
mol/liter (m)	=	1.0 kg–mole/m^3
mmole/liter (mM)	=	1.0×10^{-3} mol/l
ng/l	=	9.99972×10^{-10} kg/m^3
p Ci/l	=	9.99972×10^{-10} Ci/m^3 (radioactive subst.)
ppb*	=	1.0 μg/l
ppm*	=	1.0 mg/l
ppm	=	1.0 g/m^3
ppt*	=	1.0 g/l
μg/g	=	1.0×10^{-6} kg/kg
μg/l*	=	1.0 ppb
μg/l	=	9.99972×10^{-7} kg/m^3
μg/ml	=	9.99972×10^{-4} kg/m^3
μl/l	=	1.0×10^{-6} m^3/m^3
μ mole/liter (m)	=	1.0×10^{-6} mol/liter

Costs

dollars/kg	=	9.0718×10^{2} dollars/ton
dollars/ton	=	1.1023×10^{-3} dollars/kg
¢/kgal	=	2.6417 ¢/m^3
¢/kgal	=	1.0×10^{1} dollars/mil gal
¢/m^3	=	3.7854 ¢/kgal
dollars/mil gal	=	1.0×10^{-1} ¢/kgal

Density

g/cm^3	=	1.0×10^{3} kg/m^3
kg/m^3	=	1.0×10^{-3} g/cm^3

Depth (see Length)

Design manpower

man hr/mil gal	=	2.6917×10^{-4} man hr/m^3
man hr/m^3	=	3.7854×10^{3} man hr/mil gal

Dissolved solids load

ton/mi^2 yr	=	3.5027×10^{-4} kg/m^2 yr
kg/m^2 yr	=	2.8550×10^{3} ton/mi^2 yr

Energy

Btu	=	2.5200×10^{2} gram cal
Btu	=	2.9330×10^{-4} kWh
gram cal	=	3.9663×10^{-3} Btu
gram cal	=	1.1639×10^{-6} kWh
kWh	=	3.4095×10^{3} Btu
kWh	=	8.5918×10^{5} gram cal

Energy consumption in treatment plants

g-cal/m^3	=	4.3995×10^{-6} kWh/kgal
kWh/kgal	=	2.2730×10^{5} g-cal/m^3

Flow rate

acre ft/yr	=	3.9113×10^{-5} m^3/sec
bil gal/day (bgd)	=	4.3813×10^{1} m^3/sec
bil gal/yr (bgy)	=	1.3817×10^{-1} m^3/sec
ft^3/sec	=	2.8317×10^{-2} m^3/sec
gal/day (gpd)	=	4.3813×10^{-8} m^3/sec
gal/min (gpm)	=	6.3090×10^{-5} m^3/sec
gal/yr (gpy)	=	1.3817×10^{-10} m^3/sec
kgal/day (kgd)	=	4.3813×10^{-5} m^3/sec
mil gal/day (mgd)	=	4.3813×10^{-2} m^3/sec
m^3/sec	=	2.5566×10^{4} acre ft/yr
m^3/sec	=	2.2824×10^{-2} bgd
m^3/sec	=	7.2376 gby
m^3/sec	=	3.5315×10^{1} ft^3/sec
m^3/sec	=	2.2824×10^{7} gpd
m^3/sec	=	1.5850×10^{4} gal/min
m^3/sec	=	7.2376×10^{8} gal/yr
m^3/sec	=	2.2824×10^{4} kgd
m^3/sec	=	2.2824×10^{1} mgd

Gas consumption at sewage treatment plants

kcf/mil gal	=	1.3368×10^{-4} m^3(gas)/m^3(water)
m^3/m^3	=	kft^3/mil gal$\cdot$(7.4805×10^{-6})

Heat supply

g-cal/cm^2 day	=	1.0×10^{1} kg-cal/m^2 day
kg-cal/m^2 day	=	1.0×10^{-1} g-cal/cm^2 day
Btu/ft^2 day	=	2.7125×10^{-1} g-cal/cm^2 day
g-cal/cm^2 day	=	3.6867 · Btu/ft^2 day

Hydrolysis rate constant (*k*)

sec^{-1}	=	6.0×10^{1} min^{-1}
min^{-1}	=	1.6667×10^{-2} sec^{-1}
hr^{-1}	=	2.778×10^{-4} sec^{-1}
sec^{-1}	=	3.600×10^{3} hr^{-1}
day^{-1}	=	1.1574×10^{5} sec^{-1}
sec^{-1}	=	8.6400×10^{4} day^{-1}

Ion exchange capactiy

meq/100 g	=	1.00×10^{-2} eq/kg
eq/kg	=	1.0×10^{-2} meq/100 g

Land area per unit population

(acres/100 population)	=	4.0469×10^{1} m^2/capita
m^2/capita	=	2.4711×10^{-2} acres/100 population

CONVERSION FACTORS (*Continued*)

Length

Angstrom (Å)	=	1.0×10^{-10} mm
cm	=	1.0×10^{-2} m
ft	=	3.048×10^{-1} m
in.	=	2.54×10^{-2} m
km	=	1.000×10^{3} m
mi	=	1.6093×10^{3} m
mil	=	2.54×10^{-5} m
mm	=	1.0×10^{-3} m
mμ	=	1.0×10^{-9} m
nm	=	1.0×10^{-9} m
μ (μm)	=	1.0×10^{-6} m
yd	=	9.1440×10^{-1} m
m	=	1.0×10^{10} Angstroms (Å)
m	=	1.0×10^{2} cm
m	=	3.2808 ft
m	=	3.9370×10^{1} in.
m	=	6.2137×10^{-4} mi
m	=	3.9370×10^{4} mil
m	=	1.0×10^{3} mm
m	=	1.0×10^{9} mμ
m	=	1.0×10^{9} nm
m	=	1.0×10^{6} μm
m	=	1.0936 yds
m	=	1.0×10^{-3} km

Mass

g	=	1.0×10^{-3} kg
gr	=	6.4799×10^{-5} kg
lb (avdp)	=	4.5359×10^{-1} kg
mg	=	1.0×10^{-6} kg
oz (avdp)	=	2.8350×10^{-2} kg
Metric ton	=	1.0×10^{3} kg
ng	=	1.0×10^{-12} kg
short ton	=	9.0718×10^{2} kg
μg	=	1.0×10^{-9} kg
kg	=	1.0×10^{3} g
kg	=	1.5432×10^{4} gr
kg	=	2.2046 lb (avdp)
kg	=	1.0×10^{6} mg
kg	=	3.5274×10^{1} (avdp)
kg	=	1.0×10^{-3} metric ton
kg	=	1.0×10^{12} ng
kg	=	1.1023×10^{-3} short ton
kg	=	1.0×10^{9} μg

Membrane permeability

g/cm sec	=	1.0×10^{-1} kg/m sec
kg/m sec	=	1.0×10^{1} g/m sec
cm^2/sec	=	1.0×10^{-4} m^2/sec
m^2/sec	=	1.0×10^{4} cm^2/sec

Membrane productivity

gal/ft^3/day	=	1.3368×10^{-1} m^3/m^3 day
m^3/m^3 day	=	7.4805×10^{-1} gal/ft^3 day

Membrane packing factor

ft^2 (membrane)/ft^3	=	3.2808 m^{-1}
m^{-1}	=	3.0480×10^{-1}

MPN (see Bacterial density)

Oil consumption at treatment plant

gal/mil gal	=	1.0×10^{-6} m^3/m^3
m^3/m^3	=	1.0×10^{6} gal/mil gal

Oxygenation capacity

D.O. (mg/l)/kWh	=	2.7777×10^{-10} kg/m^3 Joule
kg/m^3 Joule	=	3.6000×10^{9} (mg/l)/kWh

Permeability

gal/day ft^2	=	$4.0740 \times 10^{-2} \times 10^{1}$ m^3/day m^2
l/day cm^2	=	1.0×10^{1} m^3/day m^2
m^3/day m^2	=	2.4542×10^{1} gal/day ft^2
m^3/day m^2	=	1.0×10^{-1} l/day cm^2

Plankton concentration (see Bacterial density)

Population density

persons/acre	=	2.4711×10^{-4} persons/m^2
persons/m^2	=	4.0469×10^{3} persons/acre

Power

Btu/hr	=	2.92875×10^{-4} kW
kg-cal/hr	=	1.163895×10^{-3}
kW	=	3.4144×10^{3} Btu/hr
kW	=	8.5918×10^{2} kg-cal/hr

Precipitation (see also Flow rate units)

in.	=	2.54×10^{-2} m
m	=	3.9370×10^{1} in.
mm	=	1.0×10^{-3} m
m	=	1.0×10^{3} mm

Pressure

psi (lb/in^2)	=	7.031×10^{2} kg/m^2
psi	=	6.8046×10^{-2} atmosphere
kg/m^2	=	1.4223×10^{-3} psi
ft (H_2O)	=	3.0479×10^{1} kg/m^2
kg_f/m^2	=	3.2808×10^{-1} ft (H_2O)
in. (H_2O)	=	2.5399×10^{1} kg/m^2
kg/m^2	=	3.9370×10^{-2} in. (H_2O)
mm (Hg)	=	1.3150×10^{-3} atmosphere

Productivity of organisms

mg carbon/m^3 day	=	1.0×10^{-6} kg carbon/m^3 day
kg carbon/m^3 day	=	1.0×10^{6} mg carbon/m^3 day

CONVERSION FACTORS (*Continued*)

Radiation (solar)

g-cal/cm^2 day = 1.0 x 10^1 kg-cal/m^2 day
g-cal/cm^2 day = 1.1594 x 10^{-1} g-cal/m^2 sec
kg-cal/m^2 day = 1.0 x 10^{-1} g-cal/cm^2 day

Resistivity

ohm-cm = 1.0 x 10^{-2} ohm-meters
ohm-meters = 1.0 x 10^2 ohm-cm
ohm-cm = 6.0153 x 10^6 circ mil-ohms/ft
circ mil-ohm/ft = 1.662426122 x 10^{-7} ohm-cm

Runoff (see also Precipitation)

in./yr = 2.54 x 10^{-2} m/yr
m/yr = 3.9370 x 10^1 in./yr
in. = 2.54 x 10^{-2} m
m = 3.9370 x 10^1 in.
mm = 1.0 x 10^{-3} m
m = 1.0 x 10^3 mm

Screen size

meshes/in. = 3.9370 x 10^1 meshes/m
meshes/m = 2.54 x 10^{-2} meshes/in.

Sediment load

kg/m^2 = 1.0 x 10^3 metric tons/km^2
kg/m^2 = 8.921 x 10^3 lb/acre
lb/acre = 1.1208 x 10^{-4} kg/m^2
metric tons/km^2 = 1.0 x 10^{-3} kg/m^2

Snow evaporation and melting

g/cm^2 day = 1.0 x 10^1 kg/m^2 day
kg/m^2 day = 1.0 x 10^1 g/cm^2 day

Solubility (see Concentration)

Specific conductance (electrical)

µmho/cm = 1.0 x 10^{-4} mho/m
mho/m = 1.0 x 10^4 µmho/cm

Specific surface area

cm^2/g = 1.0 x 10^{-1} m^2/kg
m^2/g = 1.0 x 10^3 m^2/kg
m^2/kg = 1.0 x 10^{-3} m^2/g
m^2/kg = 1.0 x 10^1 cm^2/g

Storage space required for treatment chemicals

ft^2/ton = 1.186735 x 10^{-2} m^2/kg
m^2/kg = 8.42648 x 10^1 ft^2/ton

*Conversion factor applies only to SiO_2 turbidity.

Temperature

°F = 1.80 °C + 32
°C = (5/9) (°F−32)

Time

day = 8.6400 x 10^4 sec
hr = 3.6000 x 10^3 sec
month = 2.5920 x 10^6 sec
sec = 1.1574 x 10^{-5} day
sec = 2.7778 x 10^{-4} hr
sec = 3.8580 x 10^{-7} month
sec = 1.1653 x 10^{-6} week
sec = 3.2150 x 10^{-8} yr
week = 6.0480 x 10^5 sec
yr = 3.1104 x 10^7 sec

Toxic dosages (see also Concentration)

g/animal = 1.0 x 10^{-3} kg/animal

Treatment rate (herbicides, fertilizers, etc.)

gal/ft^3 sec = 1.3368 x 10^{-1} m^3/m^3 sec
m^3/m^3 sec = 7.4805 gal/ft^3 sec
lb/ft^3 sec = 1.6018 x 10^1 kg/m^3 sec
kg/m^3 sec = 6.2428 x 10^{-2} lb/ft^3 sec
lb/acre = 1.1208 x 10^{-4} kg/m^2
kg/m^2 = 8.9217 x 10^3 lb/acre
kg/ha = 1.0 x 10^{-4} kg/m^2
kg/m^2 = 1.0 x 10^4 kg/ha

Turbidity

Turbidity units = 1.0 JTU
Jackson candle units = 1.0 JTU
JTU = 1.0 ppm SiO_2 *
ppm (turbidity) = 1.0 JTU*

Utilization

hr/day = 3.6000 x 10^{-4} sec/day
sec/day = 2.778 x 10^3 hr/day
sec/sec = 8.6400 x 10^{-5} sec/day
sec/day = 1.1574 x 10^4 sec/sec

Velocity

ft/day = 3.527777 x 10^{-6} m/sec
ft/min = 5.0800 x 10^{-3} m/sec
in/day = 2.9398 x 10^{-7} m/sec
m/sec = 2.2369 x 10^{-7} mi/hr
m/sec = 3.9016 x 10^6 in./day
mi/hr = 4.4704 x 10^{-1} m/sec
m/sec = 1.9685 x 10^{-2} ft/min
m/sec = 2.83464567 x 10^5 ft/day

CONVERSION FACTORS (*Continued*)

Virus input

PFU/ml = 9.99972×10^5 PFU/m^3
PFU/m^3 = 1.000028×10^6 PFU/ml
PFU/l = 9.99972×10^2 PFU/m^3
PFU/m^3 = 1.000028×10^{-3} PFU/l

Voltage

mV = 1.0×10^{-3} V
V = 1.0×10^3 mV

Volume

acre-ft = 1.2335×10^3 m^3
bbl = 1.1924×10^1 m^3
bushel = 3.5239×10^{-2} m^3
ft^3 = 2.8317×10^{-2} m^3
gal = 3.7854×10^{-3} m^3
km^3 = 1.0×10^9 m^3
liter = 1.000028×10^{-3} m^3
m^3 = 8.1071×10^{-4} acre ft
m^3 = 8.3864 bbl
m^3 = 2.8378×10^1 bushels
m^3 = 3.5315×10^1 ft^3
m^3 = 2.6417×10^2 gal
m^3 = 1.0×10^{-9} km^3
m^3 = 9.99972×10^2 liter
m^3 = 2.6417×10^{-4} mil gal
m^3 = 9.99972×10^5 ml
m^3 = 9.99972×10^8 μl
mi^3 = 2.4259×10^{-10} m^3
mil gal = 3.7854×10^3 m^3
ml = 1.000028×10^{-6} m^3
μl = 1.00028×10^{-9} m^3

Wash off (see Sediment load)

Water demand

gal/cap-day = 3.7854×10^{-3} m^3/cap-day
m^3/cap-day = 2.6417×10^2 gal/cap-day

Water flux

g/cm^2 sec atm = 1.0×10^1 kg/m^2-sec-atm
kg/m^2-sec-atm = 1.0×10^{-1} g/cm^2-sec-atm
gal/ft^2 day = 2.4542×10^4 m^3/m^2 day
m^3/m^2 day = 4.074×10^{-5} gal/ft^2 day

Water required per ton of product

gal/ton = m^3/kg 2.91200×10^5
m^3/kg = (gal/ton) (3.434067835)

Water usage (see Flow rate)

Weight loss in corrosion testing

oz/ft^2 = 3.0515 kg/m^2
kg/m^2 = 3.27706×10^{-1} oz/ft^2

ACRONYMS COMMONLY USED IN WATER SUPPLY AND WASTEWATER TREATMENT

AA – atomic absorption (spectrophotometry)
ABS – alkyl benzene sulfonate (surface active agent)
bgy – billion gallons/yr (gal/yr x 10^9)
BOD – biological oxygen demand (analysis, concentration)
CAE – carbon alcohol extract
CCE – carbon chloroform extract
cfs – cubic feet/sec (ft^3/sec)
COD – chemical oxygen demand (analysis, concentration)
DO – dissolved oxygen (concentration)
EC – electrical conductivity
ED – electrodialysis (treatment process)
FAC – free available chlorine
FMA – free mineral acidity
gpad – gallons per acre per day (application rate)
gpcd – gallons per capita per day
gpg – grains/gallon
JTU – Jackson Turbidity Unit
LAS – linear alkylate sulfate (surface active agent)
MBAS – methylene blue active substances (surface active agents)
MF – membrane filter (test for bacterial cell concentration)
mgd – million gal/day
mgpapd – million gal/acre-day (application rate)
MLSS – mixed liquor suspended solids (concentration, in activated-sludge treatment)
MLVSS – mixed liquor volatile suspended solids (concentration, in activated-sludge treatment)
MPN – most probable number (bacterial cell concentration)
NTA – nitrilo triacetic acid (surface active agent)
ÓA – oxygen absorption ($KMnO_4$ oxidation, analytical method)
ORP – oxidation-reduction potential (analysis)
ppb – part per billion (usually μg/l)
ppm – part per million (usually mg/l)
psi – lb (force)/$in.^2$ (pressure)
RO – reverse osmosis (treatment process)
RW – raw water
SS – suspended solids (concentration)
STPP – sodium tripolyphosphate (detergent ingredient)
TDS – total dissolved solids (concentration)
TKN – total Kjeldahl nitrogen (analysis, concentration)
TN – total nitrogen (concentration)
TOC – total organic carbon (analysis, concentration)
TOD – total oxygen demand, or theoretical oxygen demand (analysis)
UV – ultraviolet (spectrophotometry, absorbance)
VSS – volatile suspended solids (concentration)

Index

INDEX

A

ABS adsorption, 760
Acid dissociation constants, 761
Actinomycete, 466, 480
Activated carbon
  butanol adsorption, 467
  effect on pesticides, 468
  physical properties, 467
  removal of actinomycete odor, 466
Adsorption
  actinomycete odor, 466
  activated carbon, 404, 466–468
  butanol, 467
  pesticides removal, 468, 469
Aeration systems, 547, 548
Algae
  chemical analyses, 566, 567
  effect of copper, 346
  effect of surfactants, 361
  product quality, 571
  removal
    alum as flocculent, 568
    flocculation, 568, 570
    sand and anthrafilt filters, 571
    sedimentation, 569
  toxic response of freeze-dried cells, 567, 568
  types in lakes, 70
Alkyl-aryl sulfonate, effect on aquatic organisms, 359
Analytical methods
  brucine nitrate, 635, 636
  carbon columns, 650
  cocrystallization, 638
  electrometric, 652–656
  freeze concentration, 640
  luminol chemiluminescence, 649
  for manganese, 637
  MPN data, 642–648
  for nitrates, 635, 636
  oxygen demand, 664–693
  for phosphorus, 636
  sludges, 617–634
  solvent extraction, 639, 651
  spectrometric, 657–663, 724
  surface waters, 617–634
  thin layer chromatography, 651
  for trace elements, 638, 639, 649
  vanadomolybdophosphoric acid, 637
  waste waters, 617–634
Aquatic organisms
  effect of alkyl-aryl sulfonate, 359
  effect of copper, 346–348
  fish, 366–388
  insects, 362, 363
  shellfish, 364, 365
  TL_m values of various chemicals, 349–357
  toxicity of insecticides, 358
  water quality criteria, 345
Arsenic, toxicity to livestock, 344

B

Bacteria
  adsorption rate constants, 595
  coliforms
    distribution, 576
    Hawaiian watercress ponds, 573
    occurrence with *Salmonella*, 577
    in Ohio waters, 575
    sewage discharge factors, 572
  content in water, 572
  diameters, 557
  disappearance, 595
  effect of pH, 579
  fecal streptococcus distribution, 572–574
  luminol chemiluminescence systems, 649
  nucleic acid content, 578
  in recreational waters, 599, 600, 605–608
  sampling, 709
  sulfate-reducing, 576
  surveillance evaluation, 716
Bathing *see* Swimming
Biocide limits in drinking water, 291
Biological oxygen demand *see* Oxygen demand
BOD *see* Oxygen demand
Boron, in irrigation waters, 333
Brackish water
  municipal supplies, 106, 107
Bromine, germicidal activity, 482
Butanol adsorption, 467

C

California
  hydrographic boundaries, 147
  nitrogen in water supply, 723, 724
  San Joaquin Valley, 97, 742, 743, 747
  waste discharge requirements, 259
  water supply systems, 401
  water usage, 146
Carbon
  activated *see* Activated carbon
  CCE, 722, 729
  color body fractions, 463
  mucidone removal, 464, 465
Carbon chloroform extract
  limits exceeded, 722
  in raw and tap waters, 729
Cedar River
  animal-human populations, 609
  bacteriological data, 607, 608
  chemical analyses, 610, 611
Chemical oxygen demand *see* Oxygen demand
Chloramines, lethal activities, 480
Chlorination
  bacteriological quality of water, 476, 477
  disinfection vs. FAC residuals, 478, 479

increase in use, 474
postchlorination, 474, 476
Chlorine
effect of ammonia, 481
effect of pH on hydrolysis, 491
effect on actinomycete, 466, 480
effect on *E. coli*, 484
effect on mucidone and odor, 465
germicidal activity, 482
inactivation of virus, 482
lethal activities of chloramines, 480
limits in irrigation waters, 331
oxidation-reduction reactions, 491
Clackamas River
animal-human populations, 609
bacteriological data, 607, 608
chemical analyses, 610, 611
human use vs. coliform densities, 613
Cloud seeding, precipitation increases, 115
Coagulation
coagulant types, 434
density of mineral products, 438
electrolyte dosage, 452
GT values, 437, 438
sludge characteristics, 436
virus removal, 584–590
COD *see* Oxygen demand
Coefficient of permeability, 108
Coliforms
density limit exceeded, 719, 720
diameters, 557
drinking water standards, 278
in Hawaiian watercress ponds, 573
occurrence and distribution, 576
occurrence with *Salmonella*, 577
Ohio waters, 575
raw water standards, 279
removal by slow sand filtration, 489
in sewage discharge, 572
in wading pools, 574
Collector wells, 101, 102
Copper, effect on aquatic organisms, 346–348
Corrosion
cast iron, 541, 542
inhibitors, 500
and soil resistivity, 543
Cost data
chemicals, 399, 432
effect of salinity on plumbing, 537
industrial water usage, 198
recreational fishing, 603
water treatment
chemicals, 399, 432
plants, 415

D

DDT and DDE
content in meals, 294
storage in body fat, 292, 293
Defluoridation plants, 502
Desalting processes, 507, 508, 511
efficiencies, 549, 552
installations, 550
performance factors, 551

Detergents
BOD, 686
effect on shellfish, 364, 365
Diseases
methemoglobinemia, 289
poliovirus, 580, 581, 583, 585–587, 590–594
waterborne, 280
Disinfection
bromine, 482
chloramines, 480
chlorination, 474, 476–479
in communities of >25,000, 473
effect of chlorine residual, 477
vs. FAC residuals, 478, 479
globaline, 483, 484
phenol, 485
silver, 485
silver nitrate, 486–488
state practices, 475, 476
swimming pools, 602
Drainage waters
compositions, 743, 744
nitrate content, 742
pesticides concentrations, 747
salinity, 745
Drinking water
biocide limits, 291
chemical limits, 281–285
estimated use, 155
fluoride levels, 286, 287
pesticide limits, 291–294
physical characteristics, 275
quality goals, 273
radioactivity levels, 276
standards
chemicals, 281–285
coliforms, 278
trace organics, 290
trace element concentrations, 271

E

E. coli
effect of disinfectants, 484, 485
in Hawaiian watercress ponds, 573
radiation effects, 489, 490
in wading pools, 574
Economic data *see* Cost data
Electrical conductivity, 328
Electric power generation, 221–227
Electroanalytical methods
detection limits, 652
ion-selective electrodes, 653, 654
nitrogen analysis, 656
potentiometric titration, 653
stripping peak potential, 652

Enterobacteriaceae, 608
Escherichia coli see E. coli
Estuarine and marine waters, 109–114
Evaporation equations, 14

F

Fecal streptococcus
  in wading pools, 574
  in warm-blooded animal feces, 572
  in water sources, 573
Filtration
  backwash characteristics, 449, 450
  centrifuge data, 452
  diatomite, 439, 440
  downward flow, 449
  filtered water characteristics, 446, 458
  media performance, 447, 448
  mixed media, 447
  plant waste production, 450
  process wastes, 451
  sand, 447
  screening, 453
  sludge treatment, 452
  system characteristics, 446
  upflow, 449
Finished water
  chemical characteristics, 402
  trace element content, 728
Fire hydrant, flow and spacing, 162
Fish
  lethal concentration of chemicals, 369
  lethal oxygen levels, 368
  maximum water temperatures, 366
  mercury content, 663
  poisons, 564, 565
  survival in wood fibers, 367, 368
  tainting by phenolic compounds, 388
  TL_m values of petrochemicals, 377–382
  toxicity
    acids, 373
    acrylonitrile, 387
    CAE, 385
    CCE, 385, 387
    chemicals, 371–373
    cyanides, 373
    metals, 374, 375
    pesticides, 376
    trace organics 383, 384
  turbidities found to be fatal, 366
Fishing, recreational 603, 604
Fluoridation
  defluoridation plants, 502
  population using fluoridated water, 501
  worldwide status, 433
Fluoride, drinking water levels, 287
Foam separation, 506
Food industry, water quality criteria, 309–311
Fresh water
  classes, 252
  compared with other types, 83
  distribution, 5
  standards, 252
  steam electric power generation, 227

G

Glaciers
  ice properties, 119
  melt runoff, 116
  South Cascade, 116
  in the United States, 116
Green River
  animal-human populations, 609
  bacteriological data, 607, 608
  chemical analyses, 610, 611
Ground water
  ABS contamination, 734, 736
  bacteriological quality, 92
  chemical analyses, 91, 97–101, 103
  collector wells, 101, 102
  compared with other resources, 82
  composition, 86, 88
  contaminated and uncontaminated, 104
  different genetic types, 83
  effect of canals, 99
  effect of waste water injection, 737
  new well analysis, 100
  quality, 90
  recharging, 738
  salinity, 85
  shallow wells, 103
  sodium content, 107
  trace organics concentrations, 92–96
  types (diagram), 85

H

Herbicides
  acid dissociation constants, 761
  crop tolerance, 336
  in irrigation waters, 336
  see also Pesticides,
Hydroelectric power, water requirements, 221, 222

I

Ice, properties, 119
Impoundment
  ammonia oxidation, 75
  chemical analyses, 77–80
  effect on water quality, 73, 74
  nitrogen loss, 75
  see also Lakes and impoundments, reservoirs
Industrial water supply
  chemicals and allied products, 199
  cooling and steam generation, 324–327
  electric power generation, 221–227
  food manufacture, 209, 309–311
  manufacturing, 178–214, 312–323
  mining, 215–220

pulp and paper industry, 315–320
quality criteria, 295–308
requirements, 202
Inland waters, standards, 248
Insecticides
absorption by plants, 754
frequency of occurrence, 749
residues, 755, 756
see also Pesticides
Insects, aquatic, 362, 363
Interstate waters, standards, 248
Iron
corrosion inhibitors, 500
equilibrium constants, 460
hydrolysis reactions, 461
ion exchange treatment, 457
permanganate treatment, 456
pickup from plumbing, 540
pipe, bursting pressures, 541
raw, settled, and filtered waters, 458
relation of pH to precipitation, 456
removal, 456–462
sludge treatment, 459
solution forms, 460
stability diagram, 462
Irrigation
acreage, 240, 241
effects, 741
resultant increase in salts and ions, 746
river water before and after irrigation, 740
water usage, 234–241
Irrigation waters
boron tolerance, 333
chloride in soil solution, 331
composition, 739, 743
herbicides tolerance, 336
pesticides content, 335, 336
quality changes, 740
recharging ground water, 738
salinity, 745, 746
salt tolerances, 330, 332
trace element tolerances, 334

L

Lakes and impoundments
algae types, 70
chemical analysis, 60
chemical composition, 57, 58, 62–65
pesticide distribution, 71, 72
quality criteria, 274
strip mine, 66–69
water quality, 59, 60
Langelier index, 540
Lead, solubilization by NTA, 771

M

Manganese
analytic methods, 637
in drinking water, 289
equilibrium constants, 461
ion exchange treatment, 457
permanganate treatment, 456
relation of pH to precipitation, 456
removal, 456–462
sludge treatment, 459
stability field diagram, 461
taste threshold frequencies, 289
Marine pollution
categr
categories, 114
sources, 112
Mercury
biological magnification, 774
concentration
bottom sediment, 772, 773
fish, 663, 774, 775, 776
industrial effluents, 773
natural waters, 771
Metals
cation exchange properties, 770
effect on aquatic organisms, 362, 363
lead, 771
mercury
biological magnification, 774
in bottom sediment, 772, 773
in fish, 663, 774, 775, 776
in industrial effluents, 773
in natural waters, 771
in natural water, 766, 767
selenium, 777
spectrophotometric determination, 658–662
taste thresholds in water, 288, 289, 466
toxicity to fish, 374, 375
toxicity to insects, 363
trace elements, 763–768
vanadium, 777, 778
vertical distribution of arsenate and phosphate, 769
zinc, 768, 779, 781, 782
Zn, Cu, and Ni in bottom sediments, 768
Methane equivalent, 734
Methemoglobinemia, 289
Microorganisms
amebas, 556
bacteria, 572–579
chemicals for treatment, 558
Daphnia magna, 562
diameters, 557
effect of pH, 579
lethal thermal conditions, 565
nematodes, 556, 562, 563
resistance to sterilization, 489
surveillance evaluation, 716
swimming pool water, 600
tolerance to chemicals, 561
viruses, 580–595
in wading pools, 574
Microstraining, 503–505
Missouri River
chemical composition, 49
organics content, 54, 55
trace organics concentration, 92–94

Monitoring
analysis schedules, 704
compositing schedule, 704
data lost in 629 days, 703
metals
arsenate and phosphate, 769
ion exchange properties, 770
lead, 771
mercury, 771–776
selenium, 777
trace elements, 763–768
vanadium, 777, 778
zinc, 779, 781, 782
parameters and testing methods, 697–701
sample preservation, 703
sampling frequency, 697
sensor performance characteristics, 702
MPN index, 642–648

N

National Technical Advisory Subcommittee on Public Water Supplies, 260–267
Nematodes, in water supplies, 556, 562, 563
Nitrates
analytical methods, 635, 636
in drinking water, 289
Nitrogen
color body fractions, 463
concentration in water supply, 723, 724

O

Ocean water *see* Seawater
Odor
descriptions, 641
organic extract, 729
qualitv criteria, 261
removal, 404, 466
threshold odor numbers, 641
trace organics, 95
Ohio waters
coliform sampling, 575
organics content, 54
sulfate-reducing bacteria, 576
Organics
in Missouri waters, 92–95
in rivers, 54, 55
Osmosis, reverse, 509–517
Oxidation, 691
Oxygen demand
BOD
alcohols, 673
aldehydes, 676
amines and amides, 678
amino acids, 685
carbohydrates, 679
comparison to TOD, 692
dyes, 680
esters, 681
ethers and polymers, 682
heterocyclic compounds, 682
hydrocarbons, 683, 686
ketones, 676
measurable, 664
organic acids, 670
paraffinic hydrocarbons, 686
particle size effect, 666
phenolic compounds, 684
proteins, 685
quinones, 676
reaction velocity constants, 666
soaps and detergents, 686
synthetic substrates, 667
toxic organic compounds, 686
COD, 686
COD/TOC
raw water reservoirs, 79
relationship, 664
phenol-adapted culture, 687
toxic organic compounds, 686
trace organics in Missouri waters, 94
Oxygenation, 548

P

Particulates
behavior during microstraining, 503
characteristics, 503
coagulation product densities, 438
diameters, 557
size, 503
size effect on BOD rate, 666
Permeability coefficient, 108
Pesticides
2,4-D compound adsorption, 758
ABS adsorption, 760
absorption by plants, 754
acid dissociation constants, 761
analytical methods, 651
atrazine loss in washoff, 757, 758
BHC in body fat, 294
DDT and DDE
in body fat, 292, 293
in meals, 294
DDT and toxaphene, 759
dieldrin, 294, 753
in drainage waters, 747
in drinking water, 291–294
effect of activated carbon, 468
frequency of occurrence, 749
incidence in water pollution, 750
in irrigation waters, 335, 336
odor removal from water, 468
parathion distribution data, 759
quality criteria, 266
residues
in air and rainwater, 751
in algae, 753
in aquatic invertebrates, 755
in fish, 756
in lakes 71, 72

in water surveys, 752
in river water, 748
solvent extraction, 651
threshold odor concentrations, 294
TL_m values
fresh water organisms, 355
shrimp, 364
various animals, 349
toxaphene concentration, 761
toxicity to aquatic organisms, 358, 376
treatment of water containing, 563, 565,
in water samples, 731, 732
Petrochemicals, TL_m values, 377
pH
effect of biologically mediated reactions, 557
effect of permanganate oxidation, 472
effect on chlorine hydrolysis, 491
effect on iodine hydrolysis, 491
effect on microorganisms, 579
effect on mucidone removal, 465
virus inactivation, 583
Phosphorus, analytical methods, 636
Plumbing fixtures
air gaps, 536
cast iron, 541, 542
critical level settings, 532
cross connections, 532, 545
demand constant, 528
distance from contamination source to water supply, 537
distance from trap to vent, 536
effect of salinity, 537
flow rates, 527, 529
grease interceptor capacities, 534
hot and cold water demand, 525
hot water heating, 530
iron pickup, 540
Langelier index, 540
minimum number, 522
pipe sizes, 533
service life, 538
steel pipe, 543
supply demand, 525
trap sizes, 535
water supply protection, 534
water tanks, pipe sizes, 533
Plumbing hazards, 544, 545
Poliovirus
disappearance, 592–594
inactivation, 583
ozonation, 592
removal
coagulation, 585, 587
flocculation and sand filtration, 591
summary, 590
water softening precipitation processes, 580–583
Pollutants *see* Water pollutants
Potable water *see* Drinking water
Potassium permanganate
effect on mucidone and odor, 465
oxidation of pure compounds, 471, 472
Power generation, 221–227
Precipitation
increases by cloud seeding, 115
manganese and iron, 456
U.S. distribution, 6
Public water supply
airport requirements, 161, 162
bacteria counts, 709
California systems, 401
Caribbean, 167
CCE limits exceeded, 722
Central America, 167
characteristics, 22
chemical cost data, 399
coliform limit exceeded, 719, 720
commercial, 159, 160
constituent limits exceeded,719, 721
consumption vs. production, 152
deficiencies related to health hazards, 722
design data, 152
domestic usage, 155, 157
facility deficiencies, 722
fire hydrant flow and spacing, 162
general data, 392, 395, 397
hot water use, 156
institutional, 158
low-income housing use, 154
meters, 395
methane equivalent, 734
minor element analyses, 725
nitrogen concentration, 723
organic compounds content, 734
per capita demand, 151
population not served, 163, 165
quality criteria
alkalinity, 263
ammonia, 263
boron, 264
carbon chloroform extract, 266
coliforms and fecal coliforms, 262
color, 261
definitions, 260
dissolved oxygen, 264
filterable residue, 265
fluoride, 264
hardness, 264
methylene blue active substances, 266
nitrate plus nitrite, 265
odor, 261
oil and grease, 266
pesticides and herbicides, 266
pH, 265
phosphorus, 265
surface water, 268
temperature, 262
turbidity, 262
uranyl ion, 266
quality summary, 400
raw water quality, 399
regional usage, 148
residential demands, 153
school and camp use, 157

services, 395
sources, 391, 393
South America, 167
state usage, 149
system types, 712
trace elements and compounds, 270
treatment, 391, 393, 396, 713
see also Water supply, water needs, water usage

Q

Quality criteria
agricultural water supply, 328, 329
aquatic life, 345
boiler water, 326, 327
chemical and allied products industry, 314
cooling, 324, 325
farmstead water supplies, 339, 342
food industry, 309–311
hydraulic cement industry, 312
industrial water supply
color limits, 308
cooling and steam generation, 324–327
discussion, 295
food, 309–311
iron and steel, 313
leather tanning, 322
manufacturing, 312–323
pulp and paper industry, 315–320
standards, 298
surface waters, 296
tolerances, 300
irrigation waters, 330–338
lake water, 274
livestock watering, 343, 344
public water supplies
definitions, 260
National Technical Advisory Subcommittee criteria, 260–267
surface water, 268
steam generation, 324, 325
textile industry waters, 323

R

Radial collector wells, 101
Radiation
effects on *E. coli*, 489, 490
ultraviolet absorption coefficients, 490
Radioactivity levels in drinking water, 276
Raw water
characteristics, 398, 458
chemical characteristics, 402
coliform standards, 279
compared with mill water, 316
Reaeration, 548
Recreation
areas, 604
fishing, 603–613
hunting lands, 603
preferences, 604
swimming, 599–602
Recycling *see* Waste water
Reservoirs
aeration systems, 547
ammonia oxidation, 75
chemical analyses, 77–80
COD/TOC ratios, 79
effect on water quality, 73, 74
evaporation reduction, 551
nitrogen loss, 75
Reverse osmosis, 509–517
Rivers
composition, 37–45, 49, 743
irrigation effects, 740–742
laboratory analyses. 51
organics content, 54, 55
pesticide concentrations, 748
pollutant concentrations, 48–56
Rhine, 48, 74
Rio Grande, 741
sediment yield, 52
water quality, 46
Runoff, natural, 10

S

Salinity
ground water, 85
in irrigation water, 329
limits for livestock, 344
in soil solution, 328
San Joaquin Valley, 97, 742, 743, 745
Seawater
chemical composition, 109, 110
compared with other waters, 83
pollution, 112, 114
Sediment
mercury concentration, 772, 773
river, 52, 53
Zn, Cu, and Ni content, 768
Shellfish, 364, 365
Silver nitrate
germicidal action, 486–488
neutralization, 486
Sludge
analysis schedule, 704
analytical methods, 617–634
from coagulation, 436
iron and manganese, 459
treatment methods, 452, 459
Snow
melting, 118, 119
physical properties, 119
survey reports, 117
Soap *see* Detergents
South Cascade Glacier, melt runoff, 116
Spectroanalytical methods, 657–663
Standards
bacteriological, 277–294
coliform, 278, 279
drinking water

chemicals, 281–285
coliforms, 278
trace organics, 290
WHO, 126
fresh surface waters, 252
interstate waters, 248
physical, 275, 276
plumbing system, 532–545
Strip mine lakes, 66–69
Subsurface water *see* Ground water
Surface water
adequacy of sources, 717
analytical methods, 617–634
Canadian, 30
chemical analyses, 30–36
classes, 252
composition, 32
lakes and impoundments
algae types, 70
chemical composition, 57–65
pesticides, 71, 72
water quality, 59, 60
odor removal, 404
parameters, 698
quality criteria, 254–257, 268
rivers
composition, 37–45, 46
pollutant concentrations, 48–56
sediment yield, 52
standards, 252
trace elements, 36
United States, 32–36
Suspended solids
natural waters, 81
in rivers, 52, 53
Swimming pool water
bacteriological tests, 599, 602
chemical analysis, 601
disinfectants, 602
indicator organisms, 599, 600
iodine and iodate concentration, 601
pollution characteristics, 600

T

Taste threshold frequencies
manganous ion, 289
metals, 288
zinc ion, 288
Thermoelectric power, water usage, 223–226
Threshold odor concentrations, 564
TOC *see* Oxygen demand
Total organic carbon *see* Oxygen demand
Trace elements
analytical methods, 638, 639
concentrations
Great Lakes, 764
natural waters, 766
untreated water supplies, 768
U.S. rivers, 763
waste waters, 765
in drinking water, 271
in finished waters, 728
in irrigation waters, 334
in public water supplies, 270
Lake Tahoe and tributaries, 764
spectrographic analysis, 764
Trace organics
drinking water standards, 290
flow scheme, 649
Missouri waters, 92–95

U

Underground water *see* Ground water

V

Vanadium, concentration in rivers, 777
Viruses
disappearance, 592–594
inactivation by pH, 583
removal
coagulation, 584–590
lime-soda ash softening, 582
magnesium hydroxide precipitation, 581, 583
magnetic iron oxide, 590
preformed calcium carbonate, 580
preformed magnesium hydroxide, 582
straight lime softening, 580, 581, 583
water softening precipitation, 580–583

W

Waste water
agricultural, 742, 743
analytical methods, 617–634
California requirements, 259
characteristics, 120
coagulation sludge, 436
effect on ground water, 737
filtration process, 451
ionic increments, 123
mineral additions, 122
nitrate content, 742
nutrient contents, 125
parameters, 700
quality, 121, 126
recycling, 227–229
secondary effluent characteristics, 120
sludge treatment, 452
trace element concentration, 765
Water
beneficial use controls, 247, 257
brackish
chemical analysis, 105
manufacturing use, 196
chemical constituents, 15–29
consumption *see* Water usage
contaminated, 104
cooling, 208

data collection agencies, 11
distribution, 521–552
economic value, 247
estuarine and marine, 109–114
evaporation equations, 14
fresh *see* Fresh water
management problems, 8
marine, 109–114
see also Seawater
mineralized, 107
nutrient content, 125
quality criteria
alkalinity, 263
ammonia, 263
boron, 264
carbon chloroform extract, 266
coliforms and fecal coliforms, 262
color, 261
definitions, 260
dissolved oxygen, 264
filterable residue, 265
fluoride, 264
hardness, 264
methylene blue active substances, 266
nitrate plus nitrite, 265
odor, 261
oil and grease, 266
pesticides and herbicides, 266
pH, 265
phosphorus, 265
temperature, 262
turbidity, 262
uranyl ion, 266
reclaimed *see* Waste water
recreational
animal-human populations, 609
bacteriological data, 599, 600, 605–608
chemical analyses, 610, 611
fishing, 603–613
human use vs. coliform densities, 613
indicator organism data, 607
mean flow, 612
swimming, 599–602
recycled *see* Waste water
regions, U.S. (map), 7
reuse in industry, 228, 229
sources, 3–9
storage, 546–552
see also Reservoirs
surface *see* Surface water
trace element concentrations, 23
U.S. sources, 4–7
waste *see* Waste water
see also Specific type of water
Waterborne diseases, 280
Water needs
estimated requirements, 132
farm animals and poultry, 242, 243
general data, 131–147
industrial
electric power, 221–227
general data, 168–177
manufacturing, 178–214
mining, 215–220
irrigation and farm, 230–243
public supply, 148–167
see also Water usage
Water pollutants
ABS contamination, 734, 736
concentrations, 470
contaminated and uncontaminated waters, 104
detergents, 574
pesticides, 750
Rhodamine B dye, taste level in tap water, 292
rivers, 48–56
seawater, 112, 114
Water quality
agricultural
electrical conductivity, 328, 329
farmstead use, 339, 342
irrigation, 330–338
livestock use, 343, 344
boilers, 326, 327
changes from storage, 546
characteristics, 538
chemical industry, 314
city examples, 705
control for beneficial use, 247, 257
cooling, 324, 325
criteria, 254, 260–276, 728
data, 710
effect of impoundment, 73, 74
effect of soil types, 76
evaluation, by source and community, 714
evaluation summary, 718
farmstead use, 339, 342
hydraulic cement industry, 312
interstate water standards, 248
leather tanning industry, 322
livestock use, 343, 344
mill vs. raw water, 316
monitoring stations, 697
petroleum industry, 321
ponds and lakes, 59, 60
pulp and paper industry, 315, 319, 320
standards
industrial use, 298
interstate waters, 248
steam generation, 324, 325
surface water, 254–257
textile industry, 323
Washington State, 255
Water softeners, 492–498
Water sources
impoundments, 62–65
lakes, 57–65
regions, 7
rivers
composition, 37–45
pollutants, 48–56
United States
annual runoff, 10
distribution, 4, 5
management problems, 8

precipitation distribution, 6
use vs. supply, 9
world, 3, 5
Water storage, 546–552
Water supply
ameba concentration, 556
analytical methods, 708
bacteriologic criteria, 277
biological control problems, 555
brackish, 106, 107
characteristics, by city, 706
destratification installations, 550
ground water sources, 86, 88
industries
cooling and steam generation, 324–327
food, 309–311
manufacturing, 312–323
nematode concentrations, 556, 562, 563
predominant organisms, 556
regional boundaries (map), 711
systems size distribution, 712
world, 3
Water supply systems *see* Public water supply
Water treatment
adsorption-filtration, 470
chemical cost data, 399, 432
chemical dosage equivalents, 454
chemicals
coagulation, 417
phases, 418
properties, 419
storage and feeding, 424
chlorination, 473–482, 484
coagulation
chemicals, 417
density of mineral products, 438
destabilization, 435
GT values, 437, 438
sludge characteristics, 436
color reduction with ozone, 464
color removal, 462, 464
desalting, 507, 508, 511
design factors, 405
disinfection
chlorination, 473–482, 484
globaline, 483, 484
phenol, 485
silver, 485
silver nitrate, 486–488
effluent characteristics, 398
electrodialysis, 508, 509
equipment service lives, 416
facilities evaluation, 715
facilities with chemical laboratories, 395
filtration, 439–453
fluoridation, 433, 501, 502
foam separation, 506
ion exchange efficiency, 457
iron and manganese removal, 456–462
microstraining
microorganism removal, 504
particulate characteristics, 503
sludge treatment, 505
mucidone removal
carbon, 464, 465
chlorine, 465
potassium permanganate, 465
odor removal
carbon, 404, 464
chlorine, 465, 466
potassium permanganate, 465
permanganate oxidation, 471, 472
pesticide removal, 468, 469, 563
plants
description, 407
production factors, 410
unit consumptions, 412
unit costs, 415
process energies, 416, 417
public water supply, 391, 393, 713
radiation, 489, 490
reverse osmosis, 509–517
screening, 453
separation processes, 499
softening
analyses after treatment, 492
calcium chloride process, 496
cold lime process, 496
home water softeners, 493
ion exchange, 494, 498
lime hardness removal, 495
lime soda process, 496
magnesium reduction, 495
sodium concentration, 494
well water, 494
Water usage
1900–1980, 131
1950–1965, 138
1960, 139
1965, 133
1980 (projected), 135
2000 (projected), 136
2020 (projected), 137
airport, 161, 162
appliance, 155
Caribbean, 167
Central America, 167
commercial, 159, 160
demand constant, 528
demand rates, 151, 527
design data, 152
electric power, 221–227
excluding hydroelectric power
by regions (1965), 140
by regions (1970), 144
by states (1970), 141
farm animals and poultry, 242, 243
fire hydrant, 162
hot water, 156
household, 155, 157
industrial
annual intake, 197
economic data, 198
electric power generation, 221–227

manufacturing, 178–214
mining, 215–220
per product unit, 200
projected requirements, 197
pulp and paper, 211–214
recycling, 228, 229
by regions (1970), 168
by states, 170–177
steel, 210
institutional, 158
irrigation
by regions, 234, 239
by states, 236–241
low-income housing, 154
manufacturing
cost data, 198
by industry, 178–195, 200, 206–209
metals, 210
process water intake, 199
supply source, 196
mineral industries, 215–220
nonmetal mining, 220
pattern, 132
per capita demand, 151
public supply, 148–167
pulp and paper mills, 211–214
residential, 153
rural, 231–233
schools and camps, 157
service demand, 531
South America, 167
supply vs. demand, 146
at various establishments, 521
withdrawal rate change, 139
Well water
bacteriological quality, 92
chemical analyses, 97–101, 103
chemical characteristics, 91
collector wells, 101, 102
effect of canals, 99
effect of other wells, 99
organic matter content, 730
quality, 90
shallow wells, 103
see also Ground water
World Health Organization drinking water standards, 126

Z

Zinc
in drinking water, 288
in river water, 779, 781, 782
taste threshold frequencies, 288